GEOTHERMAL ENERGY: TURN ON THE POWER!

TRANSACTIONS, VOLUME 6

Geothermal Resources Council
1982 Annual Meeting
11 - 14 October 1982
San Diego, California

ISSN #0193-5933
ISBN #0-934412-56-1

Geothermal Resources Council • P.O. Box 1350 • Davis, CA 95617

NOTICE

The views and conclusions contained in this document are those of the specific authors of the papers in which they are contained and should not be interpreted as necessarily representing the official policies or recommendations of the Geothermal Resources Council.

GEOTHERMAL ENERGY: TURN ON THE POWER!

Geothermal Resources Council
1982 Annual Meeting

Robert G. Lacy
San Diego Gas and Electric Company
Chairman

PROGRAM COMMITTEE

Willard E. Johnson
WESTEC Services, Inc.
Program Chairman

Peter B. Sherwood
WESTEC Services, Inc.
Poster Session Chairman

Roger Bowers
Hunt Energy Corporation

Susan J. Brown
California Energy Commission

Eugene V. Ciancanelli
Cascadia Exploration Corp.

George Crane
Southern California Edison Company

Linda Fassbender
Battelle Northwest Laboratories

S. K. Garg
Systems Science and Software

Richard Higuera
Los Angeles Department of Water & Power

Mark E. Kehoe
Chevron Resources Company

Sharon Province
WESTEC Services, Inc.

John C. Rowley
Los Alamos National Laboratory

Subir K. Sanyal
GeothermEx, Inc.

Robert K. Swanson
Southwest Research Institute

Robert Verity
Republic Geothermal, Inc.

TABLE OF CONTENTS

RESERVOIR TECHNOLOGY

EXPLORATION-I: GEOLOGY

Geothermal Resources Council, TRANSACTIONS Vol. 6, October 1982

EVIDENCE OF FORMER HIGHER TEMPERATURES FROM ALTERATION MINERALS, BOSTIC 1-A WELL, MOUNTAIN HOME, IDAHO

Barbara Arney

Los Alamos National Laboratory
Earth and Space Sciences Division
Los Alamos, New Mexico 87545

ABSTRACT

Cuttings from the silicic volcanics in the Bostic 1-A well near Mountain Home, Idaho have been examined petrographically with the assistance of x-ray diffraction and electron microprobe analyses. Our results indicate that these rocks have been subjected to much higher temperatures than were observed in the well in 1974, when a static temperature log was run. It is not known to what extent the alteration may be due to greater depth of burial in the past, or whether it resulted from an early hydrothermal system of higher temperature than the one now observed.

INTRODUCTION

The Bostic 1-A well near Mountain Home, Idaho (S25, T4S, R8E) (Fig. 1) was originally drilled to 9676 ft as an oil and gas wildcat by Al Griffith in 1973. In 1974 Gulf Mineral Resources Co. took over the well as a geothermal prospect and ran additional logs. In 1978 the well was acquired by Union Oil Co., who attempted to clean it out. It was plugged and abandoned in October 1978.

The Bostic 1-A first came to our attention in 1977 when it was proposed as a site for the Los Alamos Geothermal Well Stimulation Program. At that time it was reported to have bottomed in

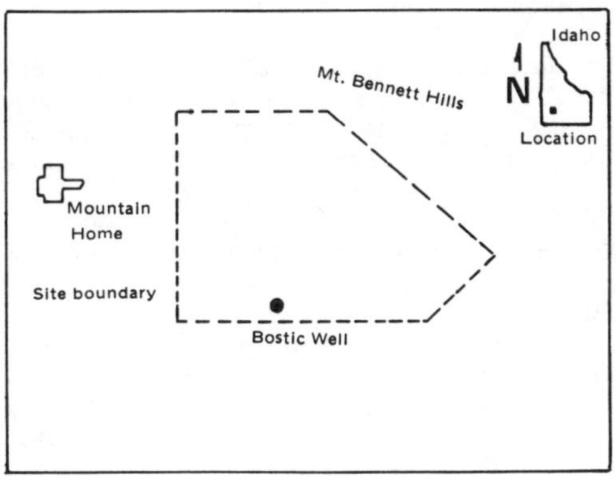

Figure 1. Map showing location of Bostic 1-A well in Idaho.

Mesozoic granite of the Idaho Batholith with a bottom hole temperature of 195°C and no fluid production.

When the area east of Mountain Home, Idaho was selected for further study as a potential hot dry rock (HDR) geothermal site, a study of the Bostic 1-A cuttings was begun. It was then discovered that the well in fact bottomed in silicic volcanics of the Idavada Formation instead of granite. Believing that Idaho Batholith lay beneath the silicic volcanics, as indeed it does just north of the HDR prospect (Malde et al., 1963), a more detailed study of the volcanic stratigraphy in the well and in the Mt. Bennett Hills was performed to allow an estimate of depth to granite (Gardner and Wood, in preparation; Arney et al., in preparation). Subsequent gravity modeling, however, suggests that no granite lies beneath the volcanics at the Bostic location. Rocks beneath the bottom of the Bostic 1-A well are basalts approximately 3 km thick, probably associated with the opening of the Snake River Plain (Arney et al., 1981).

IDAVADA VOLCANIC ROCKS IN THE BOSTIC 1-A

Silicic volcanics are first intercepted at a depth of 7150 ft (Fig. 2). Unlike Idavada Formation exposed in the Mt. Bennett Hills to the north (Malde et al., 1963; Gardner and Wood, in preparation), the siilicic volcanic sequence in the Bostic 1-A is interrupted by several basaltic intervals (Fig. 3). The thinner basalt intervals are very fresh, and are interpreted as dikes cutting the Idavada. The basalt units greater than 40 ft thick are more altered, contain abundant zeolites both as amygdale and fracture fillings, and are interpreted as interbedded flows. Some have oxidized tops and/or bottoms.

ALTERATION OF THE IDAVADA

The silicic volcanics are mostly very altered, with little of the original mineralogy remaining. Identification as Idavada is made on the basis of chlorite pseudomorphs after olivine and pyroxene, in which the original crystal morphology is well preserved. Idavada Volcanics as described by Malde et al. (1963) contain phenocrysts of andesine, clinopyroxene, hypersthene, and magnetite. Subsequent work by Gardner and

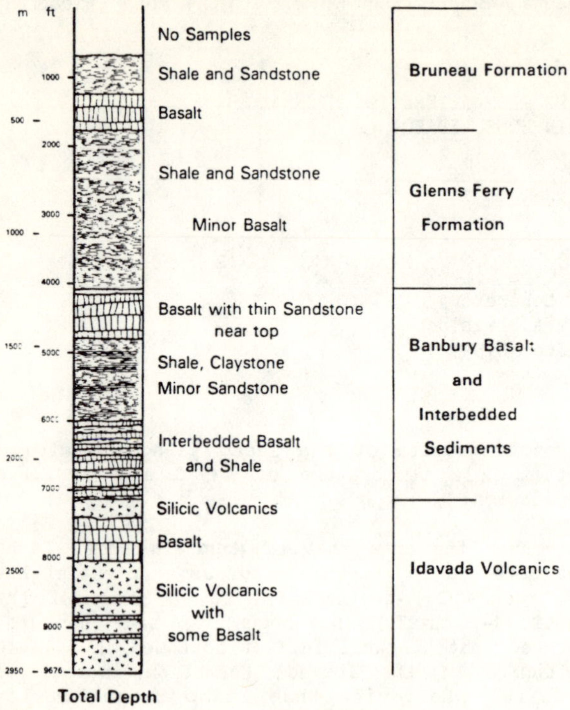

Figure 2. Bostic 1-A lithology log.

Wood (in preparation, 1982) has shown that the Idavada Volcanics in the Mt. Bennett Hills also contain olivine. The older Challis Volcanics are distinguished by the presence of biotite and/or hornblende (Malde et al., 1963). The cuttings from the Bostic 1-A contain no biotite or hornblende or psuedomorphs of these minerals.

Alteration of the silicic volcanics increases dramatically with depth. The uppermost silicic interval (from 7150-7420 ft)* contains saussuritized plagioclase, chlorite after pyroxene, sphene, and trace zircon. No potassium feldspar phenocrysts remain which can be analyzed by microprobe, but the orthoclase peaks are significant on the x-ray diffraction (XRD) patterns. Euhedral quartz phenocrysts are present in small amounts throughout the silicic interval. At this upper interval magnetite still survives and pyrite is not observed. Calcite occurs as vein fillings and partial replacement of plagioclase. Laumontite occurs primarily as fracture fillings. Analcime appears in the cuttings as euhedral crystals and in minor amounts on the XRD patterns.

The next silicic interval begins at 8170 ft, beneath a thick basalt and sedimentary sequence, and extends to 8440 ft. The plagioclase is almost completely recrystallized to albite plus calcite. Chlorite replaces pyroxene phenocrysts and occurs in the groundmass. Orthoclase is sufficiently preserved to be analyzed by microprobe. Small amounts of sphene are present. Magnetite still survives but is rounded. Pyrite is not observed. Druzy chartreuse crystals believed to be epidote

*Cuttings depths are not corrected for lag.

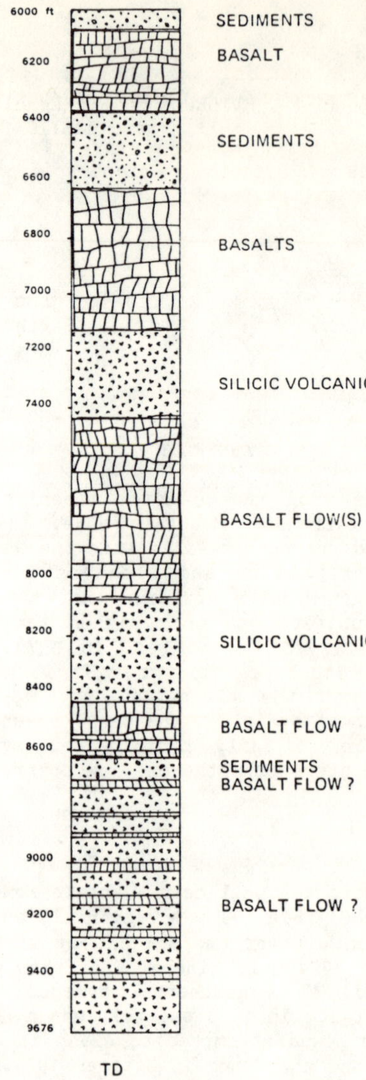

Figure 3. Detail of volcanic sequence below 6000 ft in Bostic 1-A.

first appear at 8410 ft. Laumontite peaks are much more prominent on the XRD pattern and analcime appears in trace amounts.

The next silicic interval occurs from 8673 to 8710 ft between two basalt and sediment sequences and is highly altered. Pyrite first appears at 8700 ft as euhedral crystals up to 0.25 mm diameter.

Beneath a thin basaltic and sedimentary sequence silicic volcanics occur again from 8750 to 9080 ft and are interrupted by three thin basaltic intervals interpreted as dikes, and possibly a thin sedimentary layer. This silicified sequence is characterized by complete replacement of plagioclase by albite plus calcite. Calcite and chlorite patches permeate the groundmass, and calcite fills fractures. Orthoclase decreases on the x-ray pattern and is not distinguishable in thin section. Anorthoclase appears at 8970 ft in the thin section along with two chlorite compositions reflecting the two original pyroxene compositions. Magnetite nearly disappears in this interval. Pyrite occurs as finely disseminated euhedral crystals increasing in abundance with depth. Epidote occurs as fine chartreuse coatings, especially toward the base of the interval.

There is a thin silicic interval from 9150 to 9220 ft which contains minor disseminated pyrite and abundant terminated quartz and epidote crystals coating surfaces and filling veins.

The final silicic sequence intercepted by the well extends from 9260 ft to TD at 9676 ft and is interrupted by a thin basalt interpreted to be a dike. The interval from 9400 to 9460 ft may be more intermediate in composition. In this final sequence chlorite is everywhere: after pyroxene, in cracks, and in the groundmass. There is no more calcite replacement of plagioclase or groundmass. Epidote appears as patches in the groundmass and in feldspar phenocrysts and as fracture fillings. Magnetite is completely altered. Pyrite occurs as finely disseminated crystals. Quartz occurs as fracture fillings, both in microfractures and as crystals up to 2 mm long.

DISCUSSION OF TEMPERATURE

A static temperature log run by Agnew and Sweet for Gulf in August, 1974 shows a nearly linear gradient and a bottom hole temperature at 8898 ft of 188.7°C (Fig. 4). Extrapolating the gradient to 9676 ft gives a temperature of 393°F or 200.5°C.

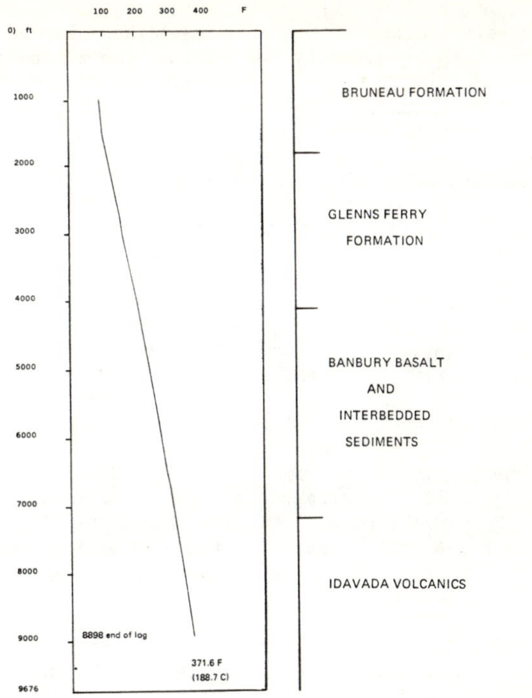

Figure 4. Static temperature survey of Bostic 1-A (August 15, 1974).

Secondary mineral assemblages in the cuttings suggest that the silicic volcanics in the well have seen significantly higher temperatures than those indicated by the static temperature survey. Alteration of the plagioclase to albite plus calcite is nearly complete by 8170 ft (176.6°C on survey). In silicic volcanics in the geothermal fields of New Zealand albite replaces andesine only at temperatures over 230°C (Browne and Ellis, 1970). Laumontite disappears below 8270 ft, but wairakite, which is the high temperature zeolite found in Wairakei, New Zealand, and Onikobe, Japan (Miyashiro, 1973) is not seen in any of the XRD patterns.

In the sedimentary rocks of Cerro Prieto, epidote forms at 230 to 250°C and is usually accompanied by wairakite (Elders et al., 1979). In silicic volcanics in New Zealand, epidote does not form below 250° at Wairakei and not below 260°C at Ohaki (Browne and Ellis, 1970). It first appears in the Bostic 1-A cuttings at 8410 ft (181°C on the temperature survey), though it does not form distinct peaks on the XRD until 9420 ft.

Though it is not possible to put an exact temperature on the alteration at the bottom of the well because of unknown pressure and fluid composition at the time, it seems reasonable to conclude that the temperature was once at least 250°C and probably 300°C. How much of the increased temperature was a result of greater depth of burial at the time, and how much can be attributed to a hotter hydrothermal system is unknown. Some idea of the age of the thermal event could possibly be obtained by dating the fresher basalts which are found in the Idavada sequence. It is not believed they were the direct cause of the alteration, as it does not seem more intense adjacent to the inferred dikes. They appear to have been emplaced after the main thermal event. A study of alteration in the basalts is planned.

SUMMARY

The top of the silicic volcanics in the Bostic 1-A well shows fairly advanced replacement of andesine by albite plus calcite, chlorite replacement of pyroxene, and minor amounts of laumontite and analcime. Laumontite increases then disappears below 8270 ft. Pyrite appears at 8700 ft. Saussuritization of plagioclase is complete by 8750 ft. Epidote first appears at 8410 ft and is fairly abundant by 9220 ft. The alteration assemblage at the bottom of the well is chlorite-epidote-quartz-albite, typical of greenschist metamorphism.

CONCLUSIONS

The albite-epidote-quartz-chlorite assemblage in the silicic volcanics at the bottom of the Bostic 1-A well suggest temperatures significantly higher than the 200°C measured in 1974. Both silicic volcanics and intercalated basalt flows show progressive alteration with depth but thinner basalt units interpreted as dikes appear much fresher. Presumably the highest thermal pulse had passed before the intrusion of these dikes. Alteration of the basalt intervals will be examined to verify this conclusion.

ACKNOWLEDGMENTS

I would like to thank Boise State University for loan of the Bostic 1-A cuttings. I would also like to thank David L. Bish for many interesting and helpful discussions and R. Ellen Semarge for help running the x-ray diffractometer. This work was supported by the U.S. Dept. of Energy, Division of Geothermal Energy.

REFERENCES

Arney, B. H., Belluomini, S. G., Gardner, J. N., 1982, Petrographic analysis and stratigraphic correlation of volcanic rocks in the Bostic 1-A well near Mountain Home, Idaho, Los Alamos National Laboratory report (in preparation).

Arney B. H., Goff, F. E., and Harding Lawson Associates, 1981, Evaluation of the hot dry rock geothermal potential of an area near Mountain Home, Idaho, Los Alamos National Laboratory Report LA-9365-HDR.

Browne, P. R. L. and Ellis, A. J., 1970, The Ohaki-Broadlands hydrothermal area, New Zealand: mineralogy and related geochemistry, Am. J. Sci. 269, p. 97-131.

Elders, W. A., Hoagland, J. R., McDowell, S. D., and Cobo R, J. M., 1979, Hydrothermal mineral zones in the geothermal reservoir of Cerro Prieto, _in_ Geology and Geothermics of the Salton Trough (W. A. Elders, ed.), Guidebook No. 5 for Geol. Soc. Am. 92nd Ann. Mtg., San Diego, CA, p. 36-43.

Gardner, J. N. and Wood, S. H., 1982, Geologic map of the Bennett Mountain area, Elmore County, Idaho, Los Alamos National Laboratory map, Scale 1:62,500 (in preparation).

Malde, H. E., Powers, H. A., and Marshall, C. H., 1963, Reconnaissance geologic map of west-central Snake River Plain, Idaho, U.S. Geol. Surv. Misc. Geol. Inv. Map I-373, Scale 1:125,000.

Miyashiro, A., 1973, _Metamorphism and metamorphic belts_, J. Wiley & Sons, N.Y., 156-162.

Geothermal Resources Council, TRANSACTIONS Vol. 6, October 1982

ALTERATION MINERALOGY OF THE AHUACHAPAN GEOTHERMAL FIELD

F. AUMENTO*, P.VIALE*, M.CHOUSSY**, A.SANTANA**

* ELC-Electroconsult, Milano, Italy
** CEL, San Salvador, El Salvador

ABSTRACT

Quantitative mineralogical studies of over 1,000 cores and cuttings from 31 wells drilled in a 2x3 km area of the Ahuachapan, El Salvador, Geothermal Field, together with structural and volcanological observations, have shown that geothermal fluids originating beneath the recent volcanics to the south of the field rise from the depth along open sections of the Southern Boundary Fault of the Central Valley. Their rise is hindered at a certain level by the volcanic products thermselves. The fluid flow is thus forced to alter course along transverse NNW-SSE faults which intercept the Boundary Faults beneath the volcanics and, to a lesser extent, along fractured horizons of the Ahuachapan Andesites. The fluids veer northwards along these new channelways, moving along two major NNW-SSE faults, with minor flows along other parallel faults and along the horizon of Ahuachapán Andesites. As the masses move north, a further hindrance is caused by an E-W fault acting both as a barrier to the northerly flow and as a channelway to any E-W an W-E flows. The result is that the water masses mix along this EW fault plane and its parallel equivalents. Reduced in mass and temperature, the fluid finds further conduits to the north along other open sections of the NNW-SSE faults. The flow is finally hindered by ever decreasing opening size along the faults and by an opposed influx of cold waters from the north.

INTRODUCTION

The Ahuachapan Geothermal Field has one of the longest histories of continuos operation of most commercial fields. Available for study are both systematic observations of its thermodynamic parameters from the very beginning of operation, and over 10,000 rock samples taken every two meters of perforation at each of the 31 wells drilled. Ten percent of available samples were studied optically and by X-ray diffraction to derive the lithology, the alteration history and the fluid motions.

SUB-SURFACE GEOLOGY

The down-hole lithology, deduced from the study of over 1,000 thin sections, includes:
The Balsamo Formation (Miocene). Over 700 m thick, considered to be the local basement.
The S.Salvador Formation (Quaternary). Readily subdivided as follows:
The Ahuachapan Andesites occurring throughout the area, from 470 m to 300 m thick, capped by 50-100 m of basaltic andesites. These Andesites are of importance to the field since, being competent rocks, when fractured tectonically they produce mm sized open fractures which tend to remain open, resulting in secondary permeability. The overlying Young Pyroclastics/Andesites form a continuous, variable layer from 150 to 520 m thick. This unit consists of interbedded pyroclastics and andesites. The latter are important to the field since they contain those production zones which do not occur within the Ahuachapan Andesites. The capping rocks, the Superficial Lavas/Tuffs produced by different eruptive centres, occur as a variety of rock types of more limited distribution. Towards the northeast, for example, wedges of andesite stem from the Holocene Cerro San Lazaro. To the south and east, instead, there are pyroclastics from the Laguna Verde volcanic centre.

STRUCTURE

Figure 1 shows that the field is cut by two sets of faults. The first produces a fan with its main axis approximately N-S; the second set includes a series of faults trending from W-E to NE-SW, associated with the formation of the central graben. Of geothermal interest are those faults which might still be open. Those associated with the graben system are older but of larger dimensions. The transverse set is smaller, possibly still active today. The latter may reactivate the older set of E-W faults, including parts of the marginal graben faults at least along selected segments. It is probable that the younger transverse faults are more open at the wide part of the fan, i.e. the

FIGURE 1

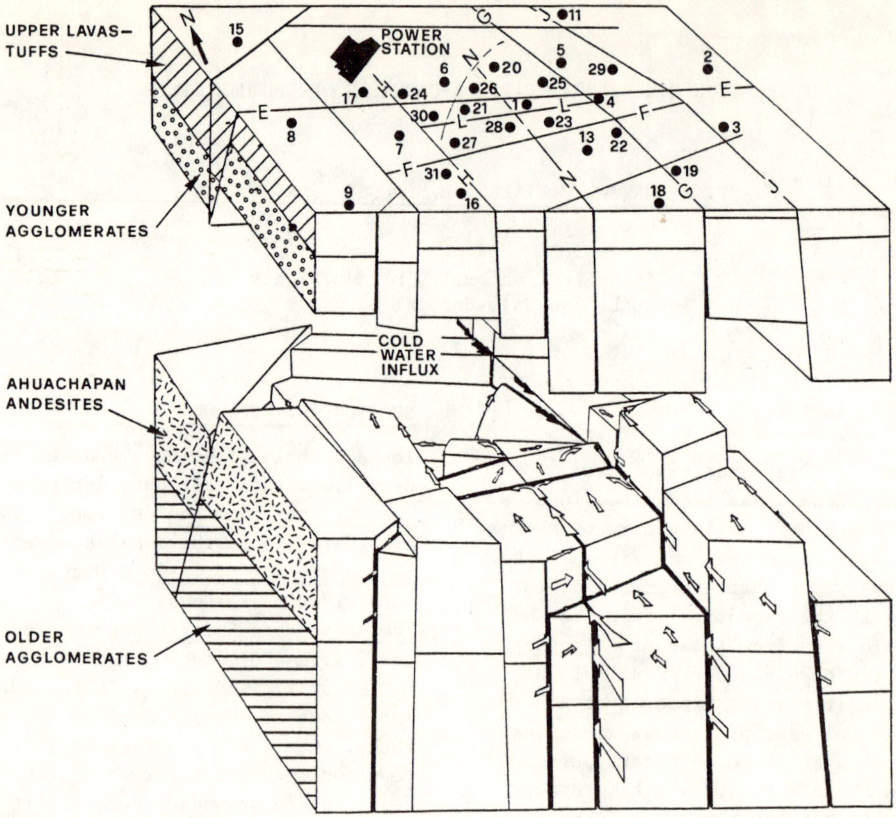

south, than at the narrow apex to the north.

ALTERATION MINERALOGY

The mineralogical data obtained optically and by X-ray diffraction was used in preparing the mineral concentration maps discussed below.

Total Whole Rock Alteration Figure 2 shows the sum of all the alteration mineral concentrations. Rock alteration is clearly at a maximum in the south-central part of the field, with rapid decreases to the E and W; to the north, fault E acts as a barrier, but alteration manages to pass this barrier along the northern extensions of faults H and J . The alteration maximum, which locally reaches 80-90%, generally lies beneath the contact between the Young Pyroclastics and the underlying Ahuachapan Andesites, although in the centre of the field the zone encroaches the former.

Clay Minerals Figure 3 shows a clay mineral maximum in the southwest. Northwards, the zone follows faults H and N up to transverse fault E, where it splits: one part swings westwards, the other follows NE along fault E, where it joins a second, separate clay mineral trend moving NW along fault J. Clays occur primarily in the top 500 m, the zones being especially thick in the north and south, thiming towards the centre of the field. At the extremity of the field the clay maxima occur within the Younger Agglomerates. In the centre, as these zones thin, they climb trough the agglomerates into the overlying Superficial Lavas and Tuffs, leaving the Ahuachapan Andesites clay-free.

Hematite The hematite zone (Figure 4) descends from the north along fault G. Deep zones of more than 20% hematite diminish in importance southwards, contemporaneously with a general elevation of the hematized levels. Most of the hematite occurs within the Younger Agglomerates, leaving the Superficial Lavas and Tuffs less oxidized and the Ahuachapan Andesites hematite-free.

Chlorites The chlorite distribution pattern shows a maximum bounded to the east and west by faults G and H respectively, and to the north by fault L. From this maximum, there emerge minor lobes along the northern sectors of faults H and M. The Superficial Lavas and Tuffs are not chloritized: chlorite first occurs within the Younger Agglomerates, with concentrations increasing to a maximum before decreasing rapidly, near the contact with the Ahuachapan Andesites.

Calcite Figure 5 shows a massive calcite distribution along fault G. The northward extension of maximum concentration is stopped by fault F, with small northerly lobes along J and N.

Quartz The quartz distribution map (Figure 6) is very similar to that of calcite, with a maximum along fault G. Fault control is stronger for these

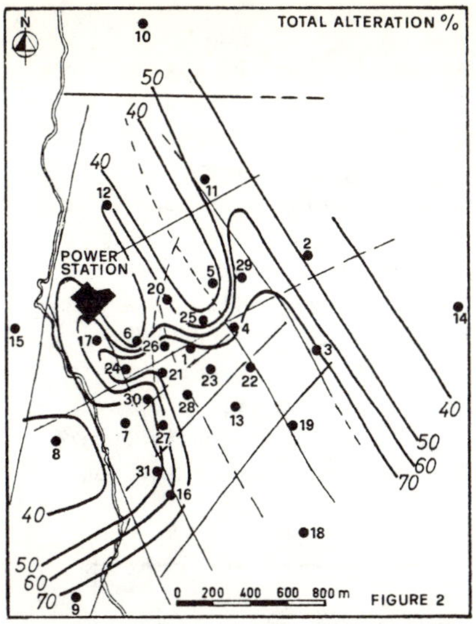

FIGURE 2

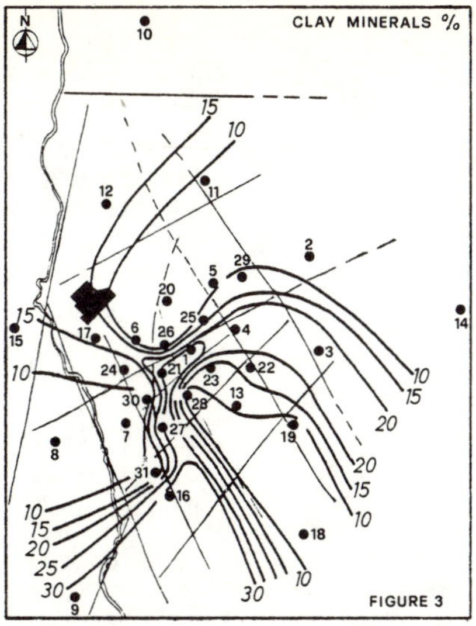

FIGURE 3

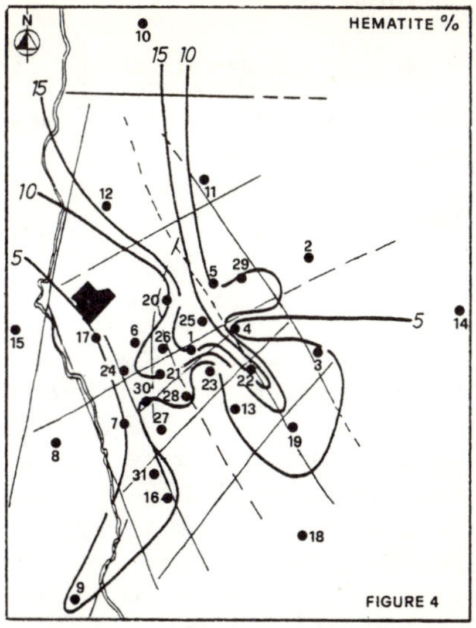

FIGURE 4

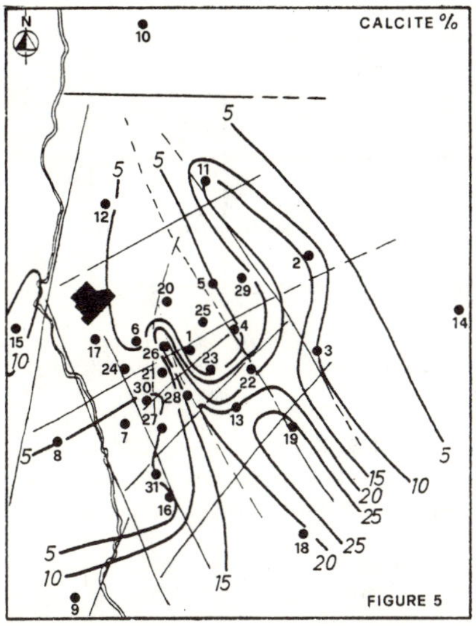

FIGURE 5

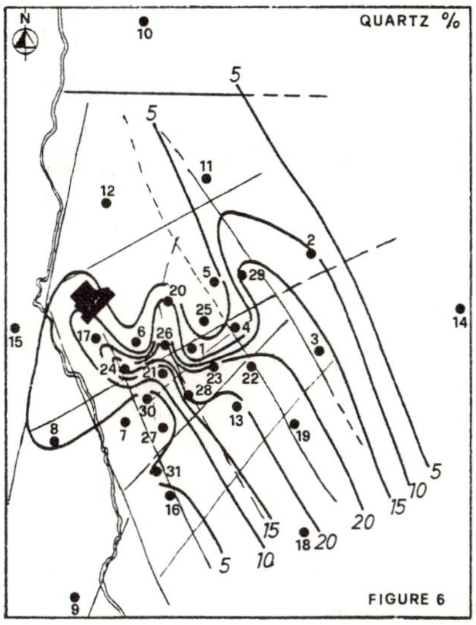

FIGURE 6

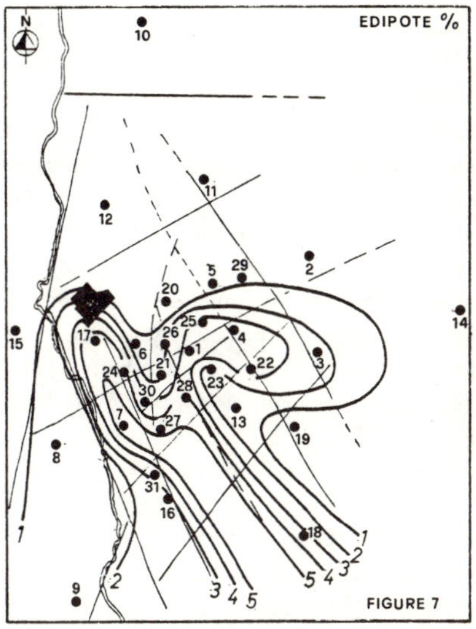

FIGURE 7

two minerals than for any other of the alteration minerals studied. In section, the restriction of quartz to the depths of the SE is clear; the mineral shows a steady north-westerly rise, commencing with high concentrations from within the Ahuachapan Andesites in the SE, to lesser amounts in the Younger Agglomerates in the centre of the field.

Epidote The epidote concentration (see Figure 7) resembles that of the chlorites. The maximum values, restricted to the south, are bounded by faults G and H until transverse fault F is reached when two lobes are formed. Therefore, both epidote and chlorite zones share the peculiarity of being bounded by fault planes rather than straddling them, as so clearly did calcite and quartz.

Zone Overlaps By observing simultaneously the different mineral zones in section, one notes the existence of a clear vertical mineral zonation. The clay minerals occupy the uppermost zone down to a few hundred meters above sea level. The next zone down, that of hematite, overlaps the lower part of the clays. Also overlapping the base of the clay mineral zone in places, but generally lying beneath it, is the chlorite zone. The latter cuts across the hematite zone in places, although massive chlorite zones are often hematite-free and vice versa. Vertical chlorite and calcite distributions are very similar. Quartz also follows calcite and chlorite, extending in upward swinging tongues beyond both these minerals; epidote occurs deeper than any of the other minerals.

Mineral Occurrence/Temperature Correlation Bearing in mind that the mineral distributions found do not necessarily reflect current temperatures but rather are the results of conditions throughout the life of the field, there are good correlations between mineral occurrence and measured temperatures.

MOVEMENT OF GEOTHERMAL FLUIDS

The detailed three-dimensional mineral distribution patterns permit us to map the flows of the geothermal fluids and to evaluate the different conditions which existed along the flows paths. Hematite, indicating cooler temperatures, enters the area from the north . It is interpreted as an entrance of cold water from the north, the water migrating primarily along fault G. Further south, the cold waters encounter an opposed migration of hot waters at depth from the south. Figure 1 shows the axis of movement of these cold northern waters. The remaining mineral zones, which include the range of temperature indicators from clays to epidote, oppose the hematite trend. They represent an entrance of hot waters from the depths south of holes AH-9 and 18. This hot fluid influx can be analyzed further by observing the behaviour of the different minerals. The clay minerals, marking the lowest temperature, span the whole entrance zone, exhibiting two maxima, one to the west and one to the east. The area in between is depleted in clay minerals. This depletion is accompanied by a parallel increase in the higher temperature minerals chlorite, epidote, quartz and calcite, the total alteration level remaining the same. Hence, the influx of hot fluids in the centre of the clay mineral zone must have either been greater or hotter, permitting the establishment of more elevated temperatures, which favoured the formation of higher temperature minerals. These influxes of fluids occurred along the maxima formed by the chlorite/epidote and quartz/calcite zones, along faults N and G respectively. Both faults must have remained active and open from the beginning of the geothermal system, and are still so active, with the difference that, whereas N was open only at depth, fault G periodically communicated with the surface as well, the latter causing sudden pressure drops to occur resulting in a temperature drop and the outgassing of dissolved gases from the geothermal waters. The drop in partial pressure of CO_2 precipitated calcite, whilst the associated temperature drop precipitated quartz, hence the quartz and calcite maxima along fault G. Figure 1 illustrates the flows entering the Ahuachapan field, their relative magnitude and their subsequent movement within the field. Some faults clearly act as passages, whilst others, or different sections of the same ones, as barriers. Localities well away from the faults will still be supplied by fluids due to the permeability of tectonically shattered Ahuachapan Andesites.

CONCLUSIONS

We have established the entrance of distinct bodies of water from the south of the field, along the parallel sets of faults C, H, N, G and J . The faults extend southwards below the Laguna Verde volcanic complex, intersecting at depth the major E-W graben boundary faults. Geothermal waters may be rising from depth along these regional E-W faults beneath the recently active volcanoes. Their upward migration will eventually be blocked by recent volcanic material which has covered the fault outcrops . However, they will also encounter the NNW-SSE transverse faults and the fractured Ahuachapan Andesite horizons. There, their flow will change direction from purely upwards to a primarily northerly and more gently upwards direction along the transverse faults and the andesite level.

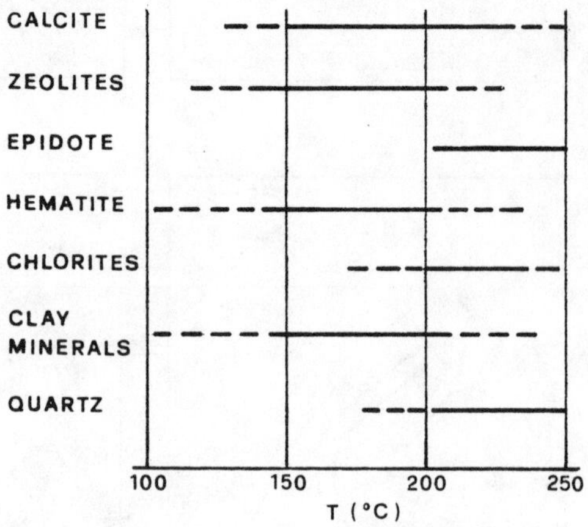

Figure 8. Mineral Occurrence/Temperature Relationship

SELECTED BIBLIOGRAPHY

AUMENTO, F., LIGUORI, P.E., CHOUSSY, M., SANTANA, A., CAMPOS, T., ESCOBAR, D., (1982) The Geothermal System of the Ahuachapán field. BHRA International Conference on Geothermal Energy, Vol 1.

BAILEY, D.G.F. (1969) Exploration for Geothermal Energy in El Salvador. Publicaciones Geológicas ICAITI, Vol 2, pp. 25-31

BODVARSSON, G. and BOLTON, R.S., (1971) A Study of Ahuachapán Geothermal Reservoir, U.N.D.P., Survey of Geothermal Resources in El Salvador pp. 1-20

C.E.L. (1981) El Campo Geotérmico de Ahuachapán durante los años 1979-1980 (CEL, Superintendencia de Recursos Geotérmicos). San Salvador, El Salvador, Abril 1981

C.E.L. (1981) Consideraciones Geoquímicas sobre la prueba de máxima extracción. Campo de Ahuachapán. (CEL, Subsección de Investigaciones Geotérmicas, Sal Salvador, El Salvador, Julio 1981)

ELC (1981) Informe Geológico-Vulcanológico. Campo Geotérmico de Ahuachapán (ELC-Electroconsult, Milan, Italia, Nov. 1981)

ELC (1981) Need for additional drilling. Planta Geotérmica de Ahuachapán (ELC-Electroconsult, Milan, Italia, Abril 1981)

JONSSON, J. (1970) Report on Geological Investigations in Ahuachapán area, Western El Salvador, Central America, U.N.D.P Report

WIESEMANN, G. (1975) Remarks on the Geologic Structure of the Republic of El Salvador, Central America; Mitt.Geol. Paläont, Hamburg

Geothermal Resources Council, TRANSACTIONS Vol. 6, October 1982

A REVIEW AND ANALYSIS OF GEOTHERMAL EXPLORATORY DRILLING RESULTS IN THE NORTHERN BASIN AND RANGE GEOLOGIC PROVINCE OF THE USA FROM 1974 THROUGH 1981

R. C. Edmiston

Anadarko Production Company

ABSTRACT

The northern Basin and Range Geologic Province of the western USA has been widely recognized as a highly prospective area for high temperature geothermal reservoirs. Yet only six apparent discoveries resulted from the drilling of 53 geothermal wildcat wells in this area from 1974 through 1981. This relatively low success rate can be partly attributed to the difficulty of developing accurate geological and geophysical models in this area prior to drilling. However, it may also indicate that large high temperature geothermal reservoirs may be less common in this area than thought previously.

INTRODUCTION

The northern Basin and Range Province of the western United States was recognized in the 1960's as having a high potential for geothermal energy development. This recognition was based on the presence of numerous hot springs, abnormally high regional geothermal gradients, geophysical indications of an anomalous thinning of the earth's crust and, in some localities, recent silicic volcanism. Prior to the leasing of federal lands for geothermal development, which commenced in 1974, drilling at several geothermal prospects had produced encouraging results including one confirmed discovery having a temperature in excess of 390°F. This early drilling activity has been reviewed by Koenig (1970). The U. S. Geological Survey as recently as 1978 estimated the geothermal potential of the northern Basin and Range Province exceeded that of any other geologic province in the U. S. as well as the combined proven and indicated potentials of both The Geysers and the Imperial Valley of California.

This paper reviews the results of the exploratory drilling which has taken place since 1974 for geothermal reservoirs with temperatures greater than 350°F, estimates the average exploration cost per discovery, and presents some conclusions concerning the relatively low exploration success rate compared to earlier expectations. The following types of wells have been excluded from these statistics: (1) wells intended for non-electric uses, (2) wells designed as re-injection wells, (3) all wells at the Raft River Low Temperature Demonstration Project. In this paper the year in which a well is reported is based on the date when drilling operations were terminated.

DEFINITIONS

In discussions of geothermal drilling, confusion usually arises concerning the distinctions between the terms "geothermal well", "geothermal observation well", and "temperature gradient hole". In common usage, the term "geothermal well" refers to a well which is designed to intersect and produce fluids from a known or suspected geothermal reservoir. The term "observation well" usually denotes either a dry hole which is being used for long term monitoring of subsurface temperatures and pressures or a slim hole designed to intersect and test, but not produce, a potential reservoir. Observation wells are normally drilled with oil field rotary rigs and include complete formation evaluation programs. The term "temperature gradient hole (or well)" denotes a relatively shallow slim hole with minimal casing which is intended to monitor temperatures above a suspected reservoir. Temperature gradient holes are usually drilled to depths of less than 3,000 ft by water well type rotary rigs at a fraction of the cost of an observation well. To avoid biasing the statistics pertaining to geothermal well drilling with the inclusion of hundreds of temperature gradient holes, all slim holes having depths of less than 3,000 ft are classified as temperature gradient holes and excluded from this paper.

Definitions are also needed in a discussion of geothermal drilling concerning the distinction between exploratory wells and development wells. In common oil field usage, an exploratory well is usually defined as a well drilled more than one mile from the closest well producing from either the formation to be tested or a deeper formation. The term "wildcat well" is often used interchangeably with the term "exploratory well". However, the term "wildcat well" usually carries the connotation of a particularly high risk exploratory well drilled in an unproven area. In order to highlight the statistics pertaining to the drilling of unproven thermal gradient anomalies, the term "geothermal wildcat well" is defined in this paper to include all production scale wells and observation wells drilled more than three miles from the closest successful geothermal well. The term "non-wildcat well" as used herein includes all development wells, confirmation wells, step-out wells and those exploratory wells drilled less than three miles from the closest successful well.

SUMMARY OF DRILLING ACTIVITY

A total of 82 geothermal wells satisfying the above criteria were drilled in the northern Basin and Range Province between the beginning of 1974 and the end of 1981. Of this total, 53 satisfy the definition of geothermal wildcat wells and 29 fall into the classification of non-wildcat wells. The locations of the wildcat wells drilled during this period and the resulting apparent discoveries are shown in Figure 1. Most of these wells were operated by major and independent oil companies which drilled 44 of the wildcat wells and 26 of the non-wildcat wells. The remaining wells were operated by independent geothermal companies, mining companies and a government laboratory.

The 53 wildcat wells resulted in six apparent discoveries for an apparent wildcat success ratio of 11.3%. Of the six apparent discoveries, only one, the Roosevelt geothermal field in southwestern Utah, is currently undergoing development for a commercial electric power plant. In addition, an announcement is expected soon concerning a pilot plant at the site of the only confirmed geothermal discovery made in the Basin and Range Province prior to 1974, the Beowawe field in north central Nevada. If one or more of the remaining five apparent discoveries should turn out to be unsuitable for development, the wildcat success ratio will decrease correspondingly. On the other hand, reworking of some of the existing dry wildcat wells could result in one or more additional discoveries and a proportionate increase in the apparent wildcat success ratio. Of the 29 wells classified as non-wildcats, 19 were apparently successful for an apparent non-wildcat success ratio of 65.5%.

The 53 wildcat wells were drilled at 35 separate prospects. Twenty-three of these prospects were tested with only one well, eight were tested with two wells, and four were tested with three or more wildcat wells. All of the discoveries during this period were made on either the first or second well. The apparent prospect success ratio, based on six apparent discoveries out of 35 drilled prospects during this period is 17.1%.

The average depth and number of wildcat and non-wildcat wells drilled annually from 1974 through 1981 are shown in Figure 2. With the exception of the year 1979, drilling activity was relatively constant from 1974 through 1981 with an average of nine wells per year consisting of six wildcats and three non-wildcats. 1979 was a record year for geothermal drilling in the northern Basin and Range Province with the drilling of 13 wildcat wells and six non-wildcats. This peak in activity was caused partly by the drilling of seven DOE supported wells during 1979. As this paper is being written, it appears 1982 will see the drilling of about two or three wildcats and six non-wildcats in this region, reversing the wildcat to non-wildcat ratio of previous years.

The average depth of both wildcat and non-wildcat wells drilled from 1974 through 1978 was relatively constant at about 5,600 ft. Although each year saw the drilling of one or more wells to depths of more than 7,000 ft a trend toward the drilling of deeper wells did not occur until after the discovery in late-1978 of a geothermal reservoir at a depth of more than 7,000 ft in the Dixie Valley area of northwestern Neveda. This trend towards deeper wildcat drilling was offset in 1979 by several relatively shallow wells. However in 1980 and 1981 the trend became more apparent as the average depth of wildcat wells increased to 7,642 ft. None of the relatively deep wildcat wells drilled following the Dixie Valley discovery are reported to have been successful. In fact, the two apparent discoveries made during this period both occurred at depths of less than 3,100 ft. The average depth of all successful geothermal wildcat wells drilled in the northern Basin and Range Province from 1974 through 1981 was a relatively shallow 3,382 ft.

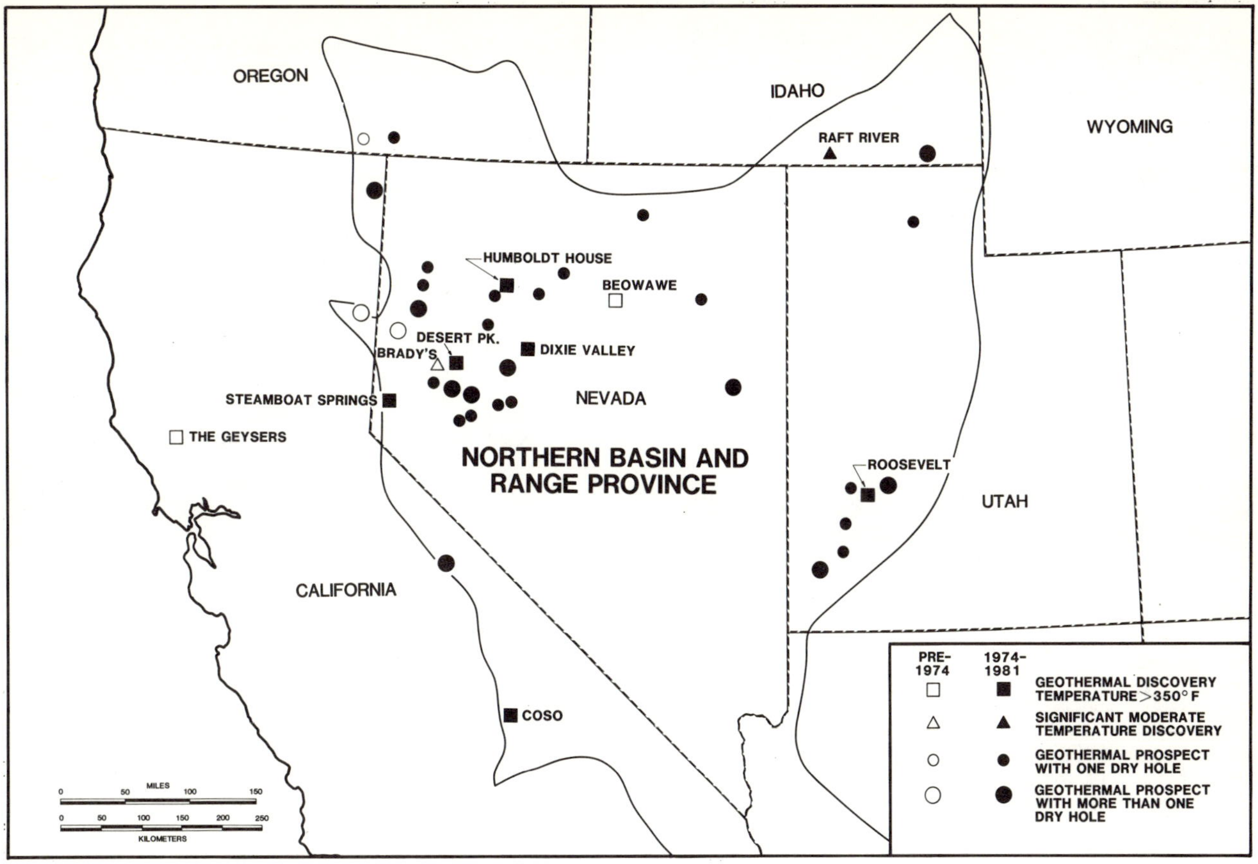

Figure 1. Map showing the location of geothermal discoveries and unsuccessful geothermal wildcat wells in the northern Basin and Range Province.

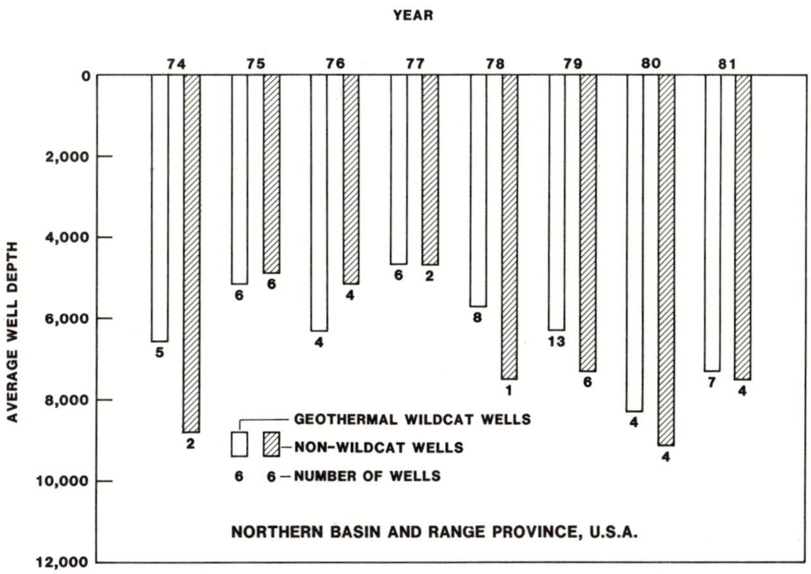

Figure 2. Average depth and number of geothermal wildcat and non-wildcat wells drilled in the northern Basin and Range Province from 1974 through 1981.

In 1977, the DOE initiated the DOE-Industry Coupled Program which eventually resulted in DOE bottom hole contributions to 15 new wells and the purchase of data from eight existing wells. These data are described by Glenn et al (1982). Of the 15 new DOE supported wells, ten fit the classification of wildcat wells. None of these wildcat wells and only one of the five non-wildcat wells, a step-out well at the Beowawe field, appear to have been successful. The DOE program may have encouraged the drilling of a few particularly high risk wells thereby lowering the success ratios in this region.

ANALYSIS AND CONCLUSIONS

The relatively low geothermal wildcat success ratio of 11.3%, while comparable to current oil and gas figures, is disappointing since some of the unsuccessful geothermal wildcats were drilled on highly rated prospects having hot springs for which the U. S. Geological Survey (1979) and private industry had estimated high subsurface temperatures and inferred large reservoir volumes. Combining this success ratio with the overall average wildcat depth from 1974 to 1981 of 6,212 ft and a current estimated dry hole drilling cost for this area of $200 per ft, the cumulative dry hole cost per discovery (at current prices) is estimated to be about $11 million.

In addition, for each prospect drilled, about $750,000 is typically spent on leasing, temperature gradient holes, and geophysics on that prospect and an additional $750,000 is spent on reconnaissance surveys and other prospects which never reach the drilling stage. Since the industry as a whole is currently making one discovery for every six prospects drilled, total non-drilling exploration costs amount to about $9 million per discovery bringing the total cost per discovery to about $20 million. This number can be slightly reduced by drilling observation wells in those areas where long casing strings are not required. It can be shown that an exploration cost per discovery of $20 million is not excessive in comparison to oil and gas exploration costs if the resulting discovery is promptly developed and is capable of producing at least 50 MW of electricity for 20 years.

The most commonly stated reason for the relatively low geothermal wildcat success rate in this region is the lack of a geophysical technique capable of locating geothermal reservoirs at depths of more than a few thousand feet with reasonable precision and reliability. The relatively shallow depth of most geothermal discoveries and the poor success rate for wildcat wells deeper than 5,000 ft support this argument. Conventional and high resolution seismic reflection surveys have been conducted at some prospects. These surveys have produced good coverage of the Tertiary basins but data quality deteriorates near the margins of the basins where most geothermal drilling takes place. Our limited understanding of the geology of geothermal reservoirs is also commonly cited as a reason for the low success rate in this particular province.

While there is truth in both of these arguments it should also be recognized that our starting hypothesis of numerous large geothermal reservoirs associated with the hot springs in the northern Basin and Range Province may be in error. Drilling results to date seem to indicate that, except in the vicinity of a few centers of recent silicic volcanism, geothermal reservoirs are generally smaller and lower in temperature than thought previously. It also appears that most hot springs are not points of leakage from large reservoirs. These observations do not mean that new discoveries will not be made. However, they do suggest that in the future companies may be more cautious in the selection and evaluation of geothermal prospects in this region.

REFERENCES

Glenn, W. E., Ross, H. P., and Atwood, J. W., 1982, Review of well logging in the Basin and Range Known Geothermal Resource Areas: Jour. of Pet. Tech., V. 34, p. 1104-1118.

Koenig, J. B., 1970, Geothermal exploration in the Western United States: U. N. Symp. on Dev. and Util. of Geothermal Resources, Pisa, Italy.

U. S. Geological Survey, 1979, Assessment of Geothermal Resources of the United States - 1978: USGS Circular 790

Geothermal Resources Council, TRANSACTIONS Vol. 6, October 1982

THERMAL AND TECTONIC HISTORY OF THE MINERAL MOUNTAINS INTRUSIVE COMPLEX

Stanley H. Evans, Jr. and Dennis L. Nielson

Earth Science Laboratory Division
University of Utah Research Institute
Salt Lake City, Utah 84108

ABSTRACT

Study of the Mineral Mountains intrusive complex was undertaken to decipher interrelationships of intrusion, uplift rate and structural control as related to the Roosevelt Hot Springs geothermal system. Results of fission track and K-Ar dating show that different units of the intrusive complex underwent similar cooling histories. Uplift rates calculated for the north (0.25 mm/yr) and central (0.5 mm/yr) portions differ substantially. This difference in uplift rate may have accounted for the development of faulting which has been important as structural controls on the geothermal system.

INTRODUCTION

Portions of the Mineral Mountains intrusive complex serve as host rocks for the Roosevelt Hot Springs geothermal system (Nielson et al., 1978). During our studies of the Roosevelt system it has become evident that thermal events which predate the present geothermal system have reset K-Ar dates of much of the intrusive complex as well as Precambrian rocks which also host portions of the geothermal reservoir. In addition, geologic evidence suggests that the intrusive complex has experienced very rapid uplift during its emplacement history. Faults developed during this rapid uplift are responsible for forming the structural controls of the geothermal reservoir.

This study was initiated to assess the interrelationships of intrusion, uplift rate, and structural development to better understand the genesis of the Roosevelt Hot Springs geothermal system.

The samples chosen for the present study represent the spectrum of rocks which are contained within the Mineral Mountains intrusive complex. These lithologies are shown in Figure 1. The northernmost unit sampled is also one of the oldest phases of the intrusive complex. Aleinikoff et al. (in press) have shown that this unit has a complex history but probably was emplaced 25 ± 4 m.y. ago. The second unit to be discussed is a quartz monzonite (Tqm) which forms a large pluton in the central portion of the Mineral Mountains. Most data indicate that the age of this unit is 20-22 m.y. The southernmost unit sampled for this study (Tbg) is a biotite granite from the southern portion of the intrusive complex. Its age is perhaps as young as 12 m.y.

ANALYTICAL TECHNIQUES

A number of samples from the Mineral Mountains were dated using both potassium-argon and fission track techniques. Mineral separates of hornblende, biotite, zircon and apatite were obtained using heavy liquid and magnetic separator techniques. Purity of separates was usually better than 98-99 percent.

Potassium-argon dating was accomplished using the standard ultra-high vacuum fusion method described in Dalrymple and Lanphere (1969). Argon isotopic composition was determined on a Reynolds-type mass spectrometer. Potassium analyses were done on a flame photometer using natural mineral standards. Uncertainties in the potassium-argon ages were calculated by the technique of Dalrymple and Lanphere (1969). Constants used in calculating ages were $\lambda_\varepsilon=0.581^{-10}$ yr^{-1}, $\lambda_\beta=4.962\times10^{-10}$ yr^{-1} and $^{40}K/K=1.167\times10^{-4}$ mole/mole (Steiger and Jäger, 1978). Results of potassium-argon dating are given in Table 1.

Table 1. Potassium-argon dates from the Mineral Mountains, Utah

Sample Number	Mineral	Unit	Weight(g)	%K	$^{40}Ar^*$, Mol/gm ($\times10^{11}$)	% $^{40}Ar_{at}$	T m.y.	±S m.y.
79-1	Biotite	Tbg	0.40065	7.21	13.082	62	10.4	0.5
79-1	Hornblende	Tbg	2.00465	0.77	1.588	69	11.8	0.6
79-153	Biotite	hd	0.50100	7.40	17.473	57	13.6	0.6
79-153	Hornblende	hd	2.50158	0.79	3.744	33	27.2	0.9
79-154	Biotite	Tqm	0.61965	7.00	14.777	49	12.1	0.5

Zircon and apatite separates were dated using techniques described by Naeser (1978). Samples were irradiated in the lazy susan of the TRIGA research reactor of the U.S. Geological Survey at

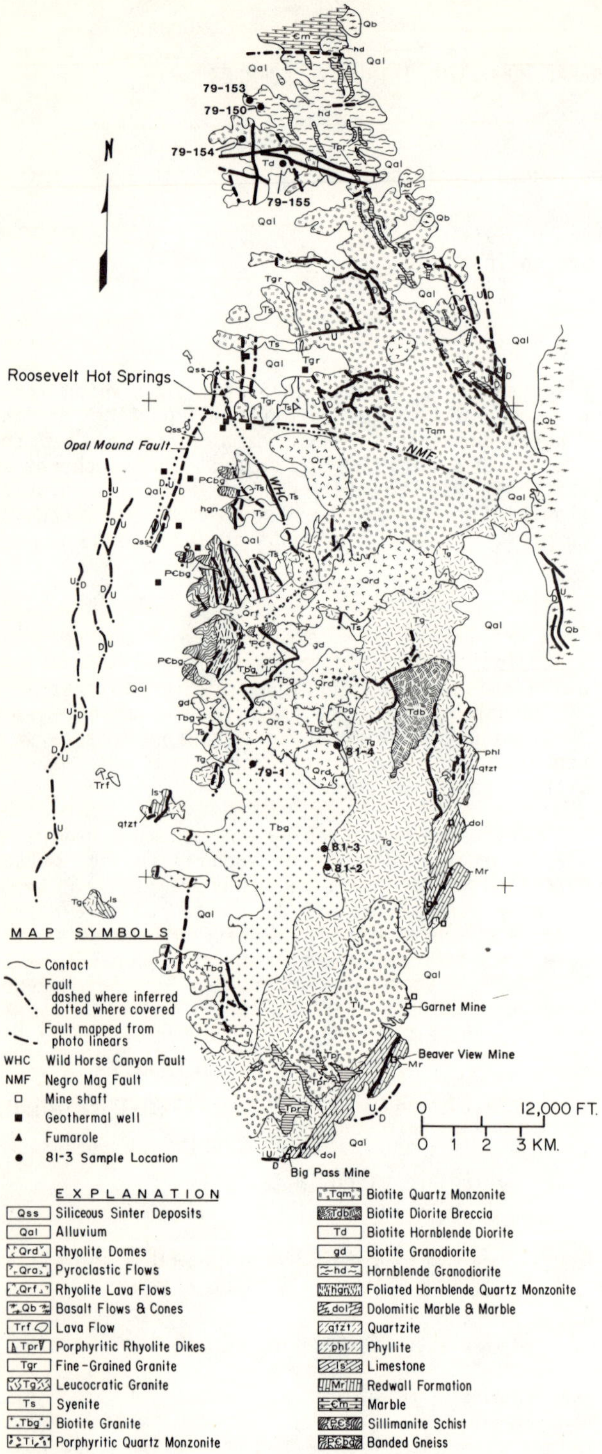

Figure 1. Geologic map of the Mineral Mountains intrusive complex, Utah. Sample locations for dating indicated by solid circles.

Denver, Colorado. Zircons were dated using the external detector method while apatites were dated by the population method. The decay constant used for spontaneous fission of ^{238}U was $\lambda_F = 7.03 \times 10^{-17}$ yr^{-1} (Naeser, 1979). The uncertainty for each fission track date was derived by the method given by Johnson et al. (1979). Results of fission track dating are given in Table 2.

RESULTS OF MINERAL DATING

Mineral dating of various units of the Mineral Mountains intrusive complex shows cooling histories which bear the effects of both simple cooling and tectonic uplift. The technique employed here is that of Harrison et al. (1979) and Harrison and McDougall (1980). Mineral ages are plotted against estimates of closure temperatures for the particular mineral in question. Estimates of closure temperatures are taken from Harrison and McDougall (1980). The closure temperatures used are: hornblende 530±40°C, biotite 280±40°C, zircon 175±25°C and apatite 105±10°C. It is important to note that these closure temperatures are dependent on the cooling history of a sample. Cooling rates on which the above estimates are based are: hornblende 100°C/m.y. and the other minerals 10°C/m.y. Justification for selection of these closure temperatures may be found in Harrison et al. (1979) and Harrison and McDougall (1980).

Cooling histories for three units of the Mineral Mountains intrusive complex are plotted in Figure 2 (sample locations are given on Figure 1). Sample 79-153, a hornblende granodiorite thought to be the northernmost unit of the intrusive complex, shows emplacement at 27 m.y. followed by slow cooling at a rate of approximately 18°C/m.y. until 9 m.y. ago. At this time cooling rapidly increased. After this interval of rapid cooling the rate slowed considerably up to the present.

Sample 79-1 shows a markedly different cooling history than sample 79-153. Here after initial emplacement at 12 m.y., rapid cooling took place followed by slower cooling until around 9 m.y. at which time cooling rate increased. After this interval of rapid cooling the rate decreased again as with sample 79-153.

The cooling history of a third sample is included here for comparison. Sample 79-154 is from the quartz monzonite that makes up a considerable portion of the intrusive complex. No emplacement date is available for this unit but cooling from the biotite closure temperature is similar to the two previously cited samples.

Apparent uplift rates for the northern and central parts of the intrusive complex may be obtained from samples collected along traverses designed for this purpose. The location of these traverses is given in Figure 1. Apparent uplift rates derived from both zircon and apatite ages are given in Table 3. It is clear that the

Table 2. Fission track dates from the Mineral Mountains, Utah

Sample Number	Mineral	ρ_S tracks/cm^2	ρ_S tracks	ρ_I tracks/cm^2	ρ_I tracks	Φ neutrons/cm^2	Φ tracks	T m.y.	±S m.y.	Number of Grains Counted	U,ppm
79-1	Zircon	$1.37x10^6$	92	$8.84x10^6$	297	$8.99x10^{14}$	1290	8.4	0.6	4	260
79-1	Apatite	$3.26x10^4$	75	$2.10x10^5$	483	$8.90x10^{14}$	1290	8.3	1.0	100/100	5
79-150	Apatite	$7.38x10^4$	85	$4.58x10^5$	366	$8.84x10^{14}$	1280	8.5	1.0	50/50	15
79-153	Zircon	$1.05x10^6$	64	$6.58x10^6$	200	$8.70x10^{14}$	1280	8.3	0.7	5	195
79-153	Apatite	$8.15x10^4$	94	$5.24x10^5$	604	$8.78x10^{14}$	1280	8.2	0.9	50/50	15
79-154	Zircon	$2.38x10^6$	178	$15.3x10^6$	573	$9.06x10^{14}$	1290	8.4	0.5	4	450
79-154	Apatite	$5.57x10^4$	128	$3.66x10^5$	841	$8.95x10^{14}$	1280	8.1	0.8	100/100	10
79-155	Zircon	$1.23x10^6$	172	$8.03x10^6$	560	$9.45x10^{14}$	1290	8.7	0.7	5	240
79-155	Apatite	$3.96x10^4$	91	$2.40x10^5$	553	$8.78x10^{14}$	1280	8.7	1.0	100/100	5
81-2	Zircon	$2.94x10^6$	329	$16.97x10^6$	950	$8.60x10^{14}$	1280	8.9	0.4	5	500
81-2	Apatite	$3.52x10^4$	18	$2.13x10^5$	109	$8.83x10^{14}$	1280	8.7	2.2	50/50	5
81-3	Zircon	$2.94x10^6$	165	$17.48x10^6$	492	$8.63x10^{14}$	1280	8.7	0.5	5	510
81-3	Apatite	$4.04x10^4$	93	$2.34x10^5$	538	$8.87x10^{14}$	1280	9.1	1.0	100/100	5
81-4	Zircon	$1.58x10^6$	189	$9.81x10^6$	590	$8.67x10^{14}$	1280	8.3	0.5	5	285
81-4	Apatite	$3.34x10^4$	77	$2.19x10^5$	504	$8.93x10^{14}$	1280	8.2	1.0	100/100	5

Neutron fluence, Φ, calculated using NBS glass standard 962 calibrated against Cu radiation.
For both spontaneous (ρ_S) and induced (ρ_I) track-densities both the density and actual number of tracks counted is given.
Uranium contents are rough estimates only.

apparent uplift rates derived from zircon data are twice those of apatites. This is attributable to the responsiveness of the two minerals to the combined effect of uplift and downward relaxation of isotherms. In order for an apparent uplift rate to equal the true uplift rate of a block of rock, isotherms must remain at a fixed level in the crust and denudation must equal uplift. Uplift data from apatite ages do match quite closely true uplift rates whereas uplift rates from zircon data are influenced strongly by relaxation of isotherms. This phenomenon has been demonstrated clearly by the incisive theoretical study of Parrish (1981). It is also apparent from Table 3 that the central portion of the Mineral Mountains has experienced more rapid apparent uplift than the northern portion. This is reflected in the more rugged topography of the central portion of the range.

Table 3. Apparent uplift rates for the north and central portions of the Mineral Mountains intrusive complex, Utah.

Sample	Apatite Age (m.y.)	Zircon Age (m.y.)	Elevation (Meters)
Northern Mineral Mountains			
79-150	8.5	--	1890
79-153	8.2	8.3	1830
79-154	8.1	8.4	1840
79-155	8.7	8.7	1990
Uplift Rate:	0.25 mm/yr	0.42 mm/yr	
Central Mineral Mountains			
79-1	8.3	8.4	2120
81-2	9.0	8.9	2760
81-3	9.1	8.7	2440
81-4	8.2	8.3	2130
Uplift Rate:	0.56mm/yr	1.08 mm/yr	

Uplift rates calculated from linear least squares regression of the data.

It is of interest that the apparent uplift rates are in reasonable agreement with sedimentation rates of 0.024 mm/yr to 0.132 mm/yr calculated by Hulen (1978) for the Roosevelt Hot Springs area.

DISCUSSION

This study has demonstrated that rapid but differential uplift rates have affected the Mineral Mountains intrusive complex. These data support field observations which suggest that the entire complex has ascended through the crust in a diapiric fashion throughout its evolution. The abrupt increase in rates at about 8-8.5 m.y. ago is thought to be the cause of low-angle denudation faulting which is responsible for much of the structural controls on the geothermal reservoir. The difference in uplift rates within the range may be responsible for east-west faulting, such as the Negro Mag Fault (Fig. 1), which is also an important reservoir control.

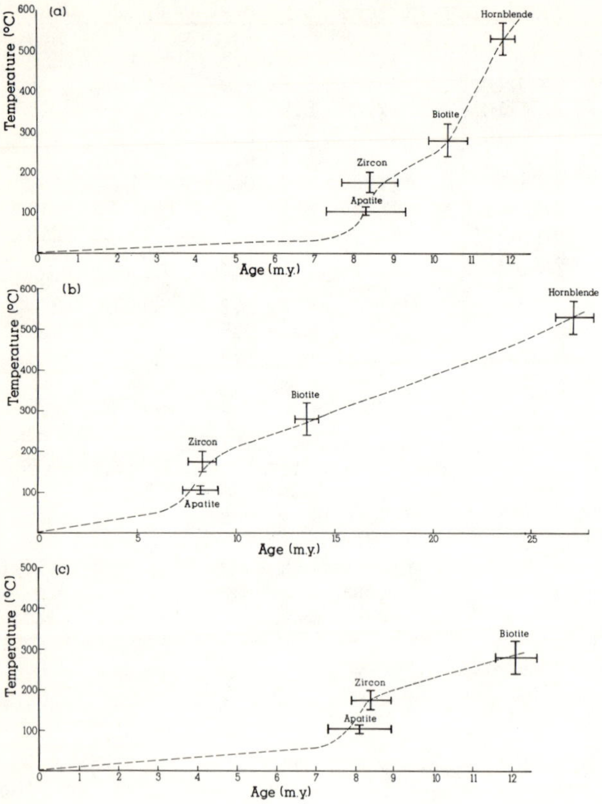

Figure 2. Cooling histories for three samples from the Mineral Mountains intrusive complex, Utah. K-Ar and fission track dates from Table 1 and Table 2 are plotted against closure temperature discussed in text. (a) sample 79-1 Biotite granite (Tbg); (b) sample 79-153 Hornblende granodiorite (hd); (c) 79-154 Quartz monzonite (Tqm). Error bars are those quoted in Tables 1, 2 and text. Dashed lines represent hypothetical cooling curve followed by each sample.

ACKNOWLEDGEMENTS

This work was performed under contract DE-AC07-80ID12079 from the Department of Energy to the Earth Science Laboratory/University of Utah Research Institute.

REFERENCES

Aleinikoff, J. N., Nielson, D. L., Hedge, C. E., and Evans, S. H., Jr., in press, Geochronology of Precambrian and Tertiary rocks in the Mineral Mountains, south-central Utah, in Shorter Contributions to Isotope Research: U.S. Geological Survey Professional Paper.

Dalrymple, G. B. and Lanphere, M. A., 1969, Potassium-Argon dating: W. H. Freeman and Co., San Francisco, 258 p.

Harrison, T. M., Armstrong, R. E., Naeser, C. W. and Harakal, J. E., 1979, Geochronology and thermal history of the Coast Plutonic Complex near Prince Rupert, British Columbia Can. J. Earth Sci., v. 16, p. 400-410.

Harrison, T. M., and McDougall, I., 1980, Investigations of an intrusive contact, northwest Nelson, New Zealand - I. Thermal, chronological and isotopic constraints: Geochim. Cosmochim. Acta, v. 44, p. 1985-2003.

Hulen, J. B., 1978, Stratigraphy and alteration, 15 shallow thermal gradient holes, Roosevelt Hot Springs and vicinity, Millard and Beaver Counties, Utah: Univ. of Utah Res. Inst., Earth Science Lab Rept., 15 p.

Johnson, N. M., McGee, V. E. and Naeser, C. W., 1979, A practical method of estimating standard error of age in the fission track dating method: Nuclear Tracks, v. 3, no. 3, p. 93-99.

Naeser, C. W., 1978, Fission-track dating: U. S. Geol. Survey Open-file Report 76-190 (revised 1978).

Naeser, C. W., 1979, Fission-track dating and geologic annealing of fission tracks, in Lectures in Isotope Geology, E. Jäger and J. C. Hunziker, eds.: Springer-Verlag, p. 154-169.

Nielson, D. L., Sibbett, B. S., McKinney, D. B., Hulen, J. B., Moore, J. N., and Samberg, S. M., 1978, Geology of Roosevelt Hot Springs KGRA, Beaver County, Utah: Univ. of Utah Res. Inst., Earth Science Lab Rept. No. 12, 121 p.

Parrish, R. R, 1981, Cenozoic thermal and tectonic history of the Coast Mountains of British Columbia as revealed by fission-track and geological data and quantitative thermal models: unpub. Ph.D. thesis, Univ. of British Columbia, 166 p.

Steiger, R. H., and Jager, E., 1978, Subcommission on Geochronology: Convention on the use of decay constants in geochronology and cosmochronology, in Contributions to the Geologic Time Scale: Amer. Assoc. of Petrol. Geol., Studies in Geology No. 6, 388 p.

Geothermal Resources Council, TRANSACTIONS Vol. 6, October 1982

PROSPECTING FOR NEAR VERTICAL AQUIFERS IN LOW TEMPERATURE GEOTHERMAL AREAS IN ICELAND

Olafur G. Flovenz and Ludvik S. Georgsson

National Energy Authority, Geothermal division
Grensasv.9, 108 Reykjavik, Iceland

ABSTRACT

This paper deals with the problem of locating permeable near vertical dykes, faults and fractures which serve as aquifers in low temperature areas in Iceland.

Four different methods are discussed, geological mapping, ground magnetic measurements, head-on resistivity profiling and shallow temperature gradient holes. Examples of successful use of the methods are given.

INTRODUCTION

The Mid-Atlantic ridge crosses Iceland from southwest to northeast. The neovolcanic zone in Iceland is the landward continuation of the ridge and is characterized by recent volcanism and rifting (fig.1). The Icelandic crust is mainly composed of basaltic lavas with some acid and intermediate rocks in volcanic centers. The lava pile dips generally towards the active volcanic zone. Regional heatflow is high in Iceland. The temperature gradient outside the volcanic zone varies between 50 and 150 °C/km, generally decreasing away from the volcanic zone (Palmason 1973).

The geothermal areas in Iceland are divided into two types: high temperature areas and low temperature areas. The former are characterized by temperatures exceeding 200°C at depths of few hundred meters. They are always associated with recent volcanism and are exclusively found within the neovolcanic zone. The heat source is usually connected with recent shallow intrusions. The low temperature areas have temperatures $\lesssim$ 150°C in the uppermost kilometer. Most of the major low temperature areas are found just outside the neovolcanic zone in rocks of quarternary age, but minor areas are found almost all over the country (fig.1).

The low temperature geothermal water is mainly used for space heating or greenhousing and today ≃ 75% of the population of Iceland are provided with geothermal space heating.

The geothermal water in the low temperature areas is meteoric and generally originates as precipitation in the central highlands. It percolates deep into the bedrock where it is heated by the high heat flow. It then flows laterally along permeable

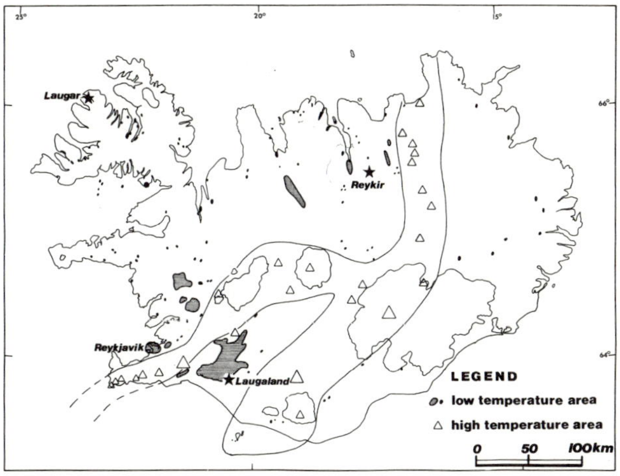

Figure 1. Geothermal areas in Iceland. Areas discussed in this paper are marked with an asterix. Modified after Saemundsson and Fridleifsson 1980.

dykes, faults, fractures and horizons driven by the hydrostatic gradient, before it appears on the surface in the lowlands (Einarsson 1942).

Intensive drilling in Iceland indicates that aquifers are mostly connected with some near vertical structures such as dykes, faults or fractures. It is common that hot springs are found at the intersection between two such structures where one acts as an aquifer and the other as an aquiclude. Thus it can be said that the main problem of low temperature geothermal exploration in Iceland is to locate dykes, faults and fractures, estimate their dip and find out which of these structures are permeable. Experience shows that only few of the numerous faults and dykes in Iceland are permeable. This paper deals with this problem.

GEOLOGICAL MAPPING

The first and the least expensive step in exploring a new geothermal area is detailed geological mapping. The main purpose is to map all visible structures and to find out the regional dip and strike.

The simplest case is when the hot water flows to the surface along an exposed dyke or fissure. It is usually assumed that dykes or fractures are

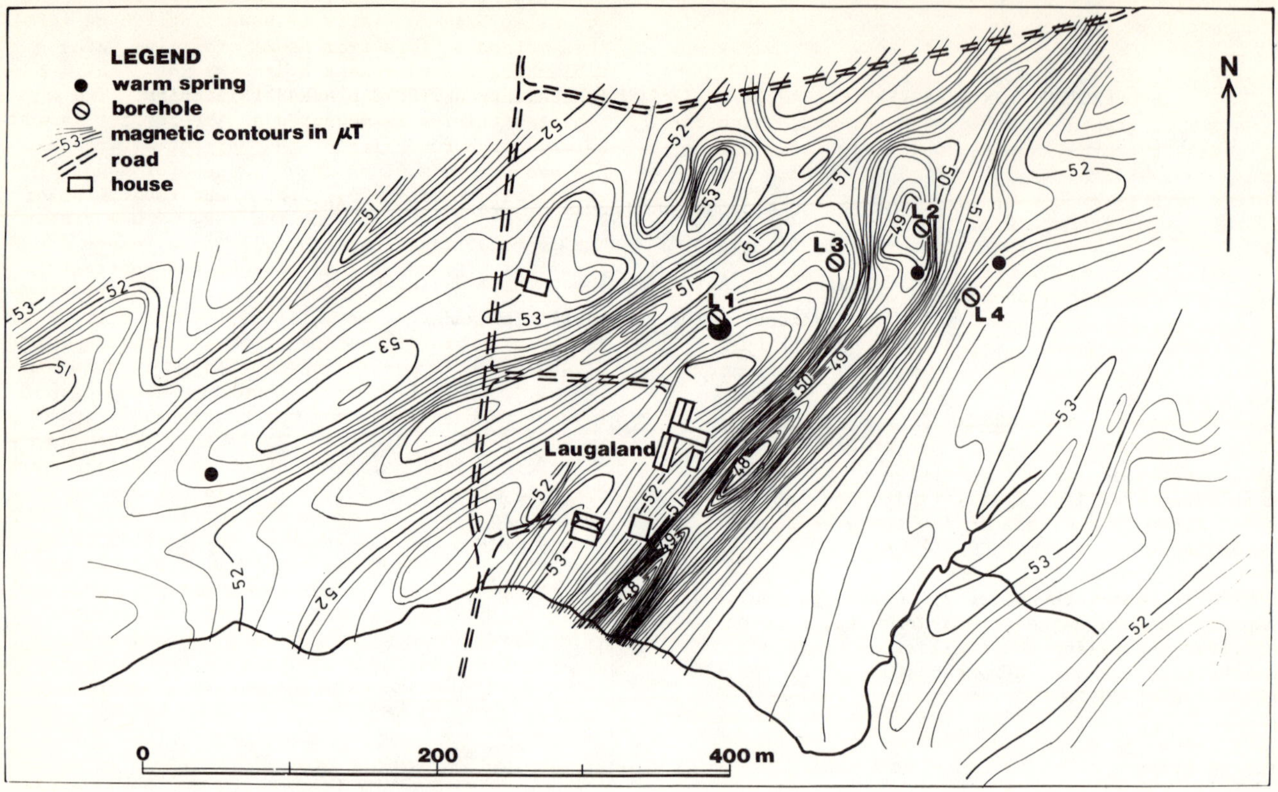

Figure 2. A magnetic contour map from Laugaland in southern Iceland

near perpendicular to the lava pile, and as the dip of the lava pile is usually better known, dips of near vertical structures can be inferred. The drillsite is then selected in order to cut a particular dyke or fault at a certain depth. This method often gives satisfactory results, but unfortunately it fails in many cases. In cases of failure or when the basement is covered with sediments or soil, geophysical exploration methods are necessary. It must however be stressed that knowledge of the regional geology is necessary to understand the nature of the geothermal systems and to interpret the results of the geophysical surveys.

GROUND MAGNETIC MEASUREMENTS

Ground magnetic measurements are widely used in geothermal prospecting in low-temperature areas in Iceland. The method is powerful to find dykes, faults and other linear structures when the basement is covered with relatively thin soil or sediments. The usual procedure is to measure the total magnetic field with a proton precession magnetometer along lines approximately perpendicular to the strike. Readings are normally made every 5 meters along the lines and a typical spacing between lines is 20-40 meters. The results are often easy to enterpret without any processing. In some cases, however, processing such as filtrering or removal of topographic effects is necessary.

A good example of the advantages of this method is shown in fig. 2. At Laugaland in S-Iceland four warm springs (T_{max} =43°C) are distributed along a 600 m straight line striking N75°E. Two old shallow wells (L-1 and L-2) yielded 4-5 l/s of nearly 50°C water. When geothermal exploration was restarted in 1977, it was concluded that this linear distribution of the warm springs was due to a fracture in the basement. Well no L-3 was sited to intersect this fracture. The result was, however, very poor, only 1 l/s from a 1308 m deep well. The temperature in the borehole was on the other hand encouraging since it indicated a geothermal system of approx. 90°C at 700-1000 m depth. Magnetic anomalies (fig. 2) indicate two reversely magnetized dykes crossing the line of the warm springs. One of the springs turned out to be at the intersection between the line of the warm springs and one of the dykes. A new well was sited to cut that dyke at 500-800 meters depth near this intersection. The well (844 m) was a success, yielding initially 20-25 l/s of 94°C hot water in free flow from aquifers at 750-837 m depth. Fig. 3 shows a model of

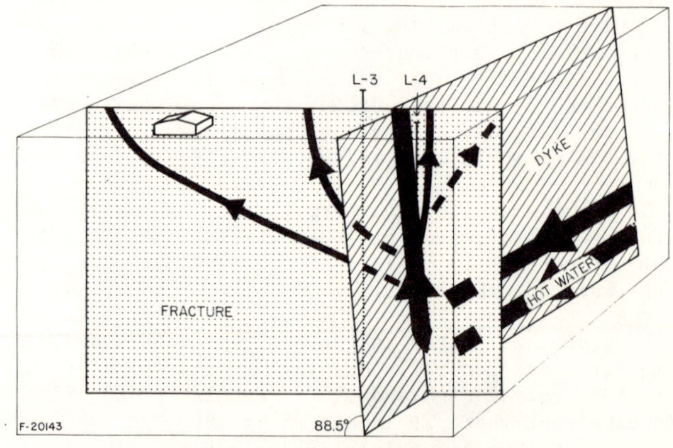

Fig. 3. Simplified model of the geothermal system at Laugaland in southern Iceland.

of the geothermal system based on surface exploration and well logging.

RESISTIVITY MEASUREMENTS

Opposite to ground magnetic measurements, which only give structural information, resistivity measurments can be directly related to geothermal parameters such as amount of interconnected pores and fissures, temperature and salinity of the geothermal fluid. In low temperature areas in Iceland, the salinity of the water is usually quite low. Hence the electrical conduction mostly takes place as surface conduction in waterfilled fractures and interconnected pores. This means that the bulk resisitivity, as measured for instance by Schlumberger soundings, is practically independent of the salinity and the temperature of the pore fluid as long as the conductivity of the fluid does not exceed certain limits (Flovenz 1980). In the basaltic crust of Iceland this limit seems to be of the order 1 siemens/m. Resistivity variations in the Icelandic crust do therefore primarily reflect variations in amount of interconnected pores and fractures in the rock.

Schlumberger soundings are quite useful in pointing out areas of low resistivity. They can be used to compare different areas and together with thermal gradient measurements they may be helpful in selecting promising fields for further investigation. The resolution of the Schlumberger soundings is, however, too limited to detect and map a single waterbearing dyke or fracture. The prospecting method which seems to be the most powerful one in locating permeable, near vertical structures is the so called head-on restivity profiling (Cheng 1980). This method differs from usual profiling with Schlumberger arrangement by the use of a third current electrode, (C), placed at infinity. The apparent resistivity is measured as the current is sent between each two pairs of the current electrodes, A,B and C. This gives three values for apparent resistivity called rho$_{AB}$, rho$_{AC}$ and rho$_{BC}$, two of which are independent. In the case of homogenous horizontally layered earth, all these values are identical. Near vertical resistivity contrasts, the curves for rho$_{AC}$ and rho$_{BC}$ versus location behave in a different way. A theoretical head-on resistivity profile across a low resistivity dyke covered with 25 m thick surface layer is shown in fig. 4. The plot of rho$_{AB}$ shows a minimum directly above the dyke but the curves for rho$_{AC-AB}$ (=rho$_{AC}$ - rho$_{AB}$) and rho$_{BC-AB}$ (=rho$_{BC}$ - rho$_{AB}$) have a different sign on each side of the dyke and cross directly above it. The picture is not always as obvious as this. Other values for the resistivity contrasts and the thickness of the surface layer may give a local maximum for rho$_{AB}$ above the low resistivity dyke, flanked by two minima. This could easily lead to misinterpretation of the rho$_{AB}$ data. The curves for rho$_{AC-AB}$ and rho$_{BC-AB}$ will on the other hand always intersect directly above the dyke, making interpretation safer and easier.

At the low temperature area in Laugar in Sugandafjordur in NW-Iceland this method was used to select a drillsite. The only surface manifestation of the geothermal field was a single warm spring. Drilling directly into the spring gave reasonably good results, but more water was needed. A ground magnetic survey showed that the warm spring was situated on a NW-striking fault and that this fault intersected a N-striking dyke 120 m east of the warm spring. A second well was drilled, intended to intersect the fault in the sama way as the former one, but 300 m westwards along the fault. A continuous core was taken. The fault turned out bo be inpermeable and no aquifers were observed in the 521 m deep well. At this stage head-on resistivity profiles were measured. Lines crossing the dyke showed a well defined minimum in rho$_{AB}$ just above the dyke and the curves for rho$_{AC-AB}$ and

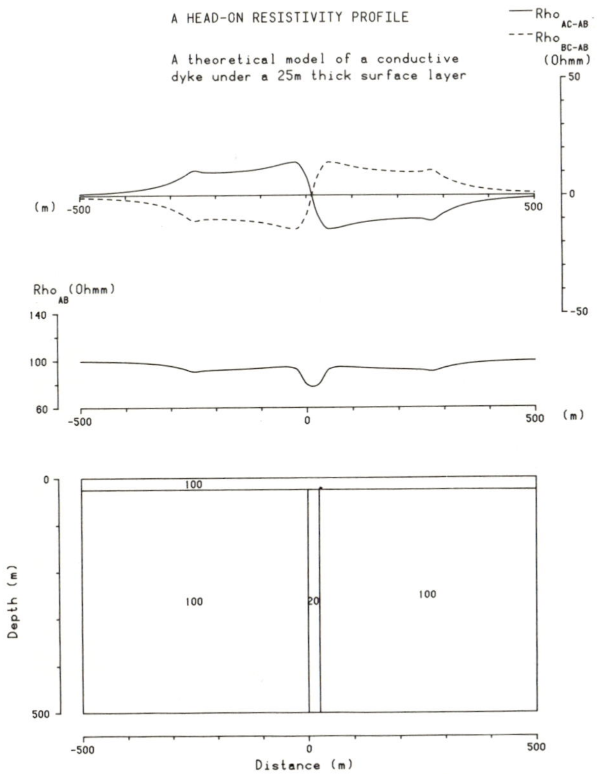

Fig. 4 Theoretical head-on resistivity profiles based on the finite difference algorithm by Dey (1976) with modifications by Halldor Halldorsson (Nat.En.Auth.)

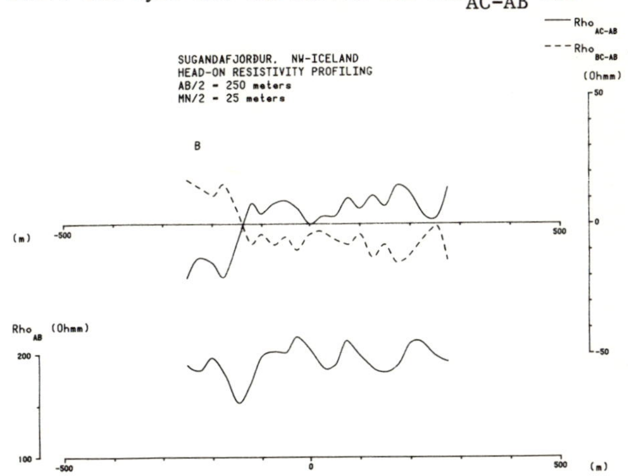

Fig. 5 Head-on resistivity profiles from Laugar in Sugandafjordur, NW-Iceland, measured across a permeable dyke.

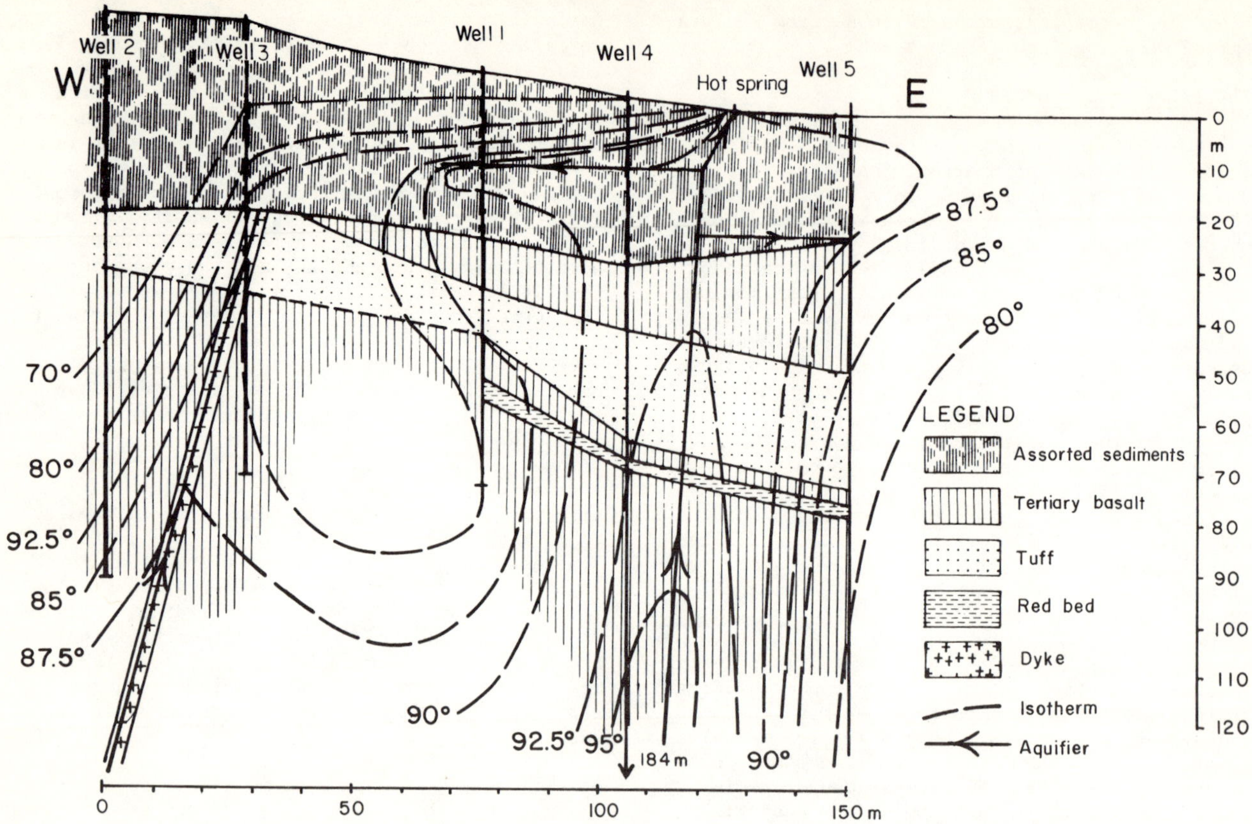

Fig. 6 A simplified geological cross-section through the wells at Reykir in Fnjoskadalur, N-Iceland (made by Asgrimur Gudmundsson, Nat.En.Auth.), and the location of the main aquifers as calculated from the temperature distribution.

rho_{BC-AB} did intersect at the same place. It was therefore concluded that the aquifers were associated with the dyke. This conclusion has been confirmed by drilling. A typical head-on profile across this dyke is shown in fig. 5. Profiles crossing the fault only showed low resistivity above the fault close to the dyke-fault intersection.

THERMAL METHODS

Measurements of thermal gradient in shallow boreholes can be quite useful in locating steeply dipping aquifers. This method is expensive since it demands drilling. An example of the use of such method may be taken from the low temperature field at Reykir in Fnjoskadalur in N-Iceland. About 7 l/s of 90°C hot water flow to the surface through 20-30 m thick sediments in a narrow walley. The Schlumberger soundings showed a north-south elongated body of low resistivity below the hot springs. The depth to the basement was too large for getting high resolution information from the ground magnetic survey, and the rugged topography made head-on resistivity profiling impossible. Therefore, five 80-197 m deep wells were drilled along an east-west striking line crossing the hot springs. The result is shown in fig. 6. Several small aquifers were cut in the wells but most of them were associated with horizontal layers. The temperatur distribution in the wells indicates strongly two steeply dipping aquifers. A computer program which computes the temperature distribution around aquifers with known temperature has been used to establish the temperature model shown in fig. 6. According to this model the hot springs are fed by a westward dipping aquifer. Consequently deepening of well 4 was recommended. This deepening will be done in late 1982.

REFERENCES

Cheng,Y.W. 1980, Location of nearsurface faults in geothermal prospects by the"combined, head-on resistivity profiling method". Proceedings of the New Zealand geothermal workshop 1980, 163-166.

Dey,A.1976, Resistivity modelling for arbitrarily shaped two dimensional structures. Part II: User's guide to the Fortran algorithm RESIS2D. Lawrence Berkeley Laboratory, report LBL-5283.

Einarsson,T. 1942, Über das Wesen der heissen Quellen Islands. Societas Scientiarum Islandica. v. 26, Reykjavik, 91 p.

Flovenz,O.G. 1980, Interpretation of resistivity soundings. Proceedings of a conference on Geothermal exploration held by the Geological society of Iceland in 1980, 51-53(in Icelandic).

Palmason,G. 1973, Kinematics and heat flow in a volcanic rift zone, with application to Iceland, Geophys. J.R. astr. Soc., 33, 451-481.

Saemundsson,K. and Fridleifsson,I.B. 1980, Application of geology in geothermal research in Iceland. Natturufraedingurinn, vol. 50, 157-188, (in Icelandic with an english summary).

GEOTHERMAL RESOURCE MAPS AND BOTTOM HOLE TEMPERATURE SURVEYS

William D. Gosnold, Jr.

Department of Geography and Geology
University of Nebraska at Omaha

ABSTRACT

Knowledge of subsurface temperatures obtained from heat flow and equilibrium temperature measurements were used as a basis for evaluation of the usefulness of the non-equilibrium temperatures of oil and gas wells for geothermal exploration. We find that the data in the Geothermal Gradient Map of North America do not conform to the thermal regime of Nebraska; and, that if the BHT data were the only data used, we would have underestimated our resource by about 80 %. Differences in geological sections and heat flow between Nebraska and the U.S. Gulf coast where the correction data for the USGS map were obtained are the reasons for the discrepency. Some significant inaccuracies in the BHT data may not be correctable.

INTRODUCTION

One of the major objectives of the U.S. Department of Energy State Coupled Resource Assessment Program has been to compile all existing data on subsurface temperatures. The ultimate goal of the various state-coupled projects is to produce a geothermal resource map based on these data. The most extensive data set for the states is the bottom hole temperature data for oil, gas, other mineral exploration and water wells. Consequently a primary task in many of the resource assessment projects has been to compile and analyze this data set. In five western-states-programs and three eastern-states-programs the geothermal resource map is to be largely derived from temperature gradients calculated from the BHT data.

During the course of the state-coupled program in Nebraska we have compared our BHT data to other data sets, i.e., equilibrium temperatures measured in deep wells and formation temperatures based on heat flow, thermal conductivity, and stratigraphic data. Consequently we have developed a basis for evaluation of the BHT data as a geothermal exploration tool. Some of the key questions in our evaluation are: Can the BHT data be corrected? How accurately do contour maps of the BHT derived gradients depict subsurface temperatures? Can these data be used to identify geothermal resources?

NATURE OF THE DATA

Most of the available BHT data are from the heading sheets of electric logs of oil and gas exploration wells. The major problems associated with using these data to produce temperature gradient maps, i.e., disequilibrium and inaccuracy, were discussed by Kehle et. al., (1970). The problem of disequilibrium may be corrected analytically if certain data, particularly the elapsed time since circulation and the surface temperature, are known. An important consideration of the correction is that it must be consistent with equilibrium temperature data. The correction for the U.S. Geological Survey Geothermal Gradient Map of North America (A.A.P.G., 1976) was based on data from the U.S. gulf coast and can be applied with accuracy only to locales with surface temperatures and geological sections corresponding to the original correction area. It is essential that each area for which BHT data are to be used for subsurface temperature studies have its own correction scheme.

The problem of inaccuracy is more difficult to correct. The usual approach is to apply statistical methods (Kehle, et. al., 1970; Harris et. al., 1981) on the assumption that, if enough data are compiled, statistical tests can identify and provide for the removal of bad data. Although many objections could be raised to various correction schemes, we consider only those that are pertinent to our study. In general, statistical filtering of data tends to eliminate outliers or anomalous data. In a tectonically stable area such as the Great Plains, high subsurface temperatures are not normally encountered and are not expected. If a localized geothermal area should occur in that area due to high radioactive heat generation in a buried pluton or upward convection in a confined aquifer, its existance might be rejected as bad data. Such may have been the case for high gradient areas in Nebraska (Figure 1) which do not appear on the A.A.P.G. map (Figure 2).

Another problem in correcting the data set that touches on both the analytical correction and the "bad data" problem is the uncertainty in mud temperatures. The correction scheme must be based

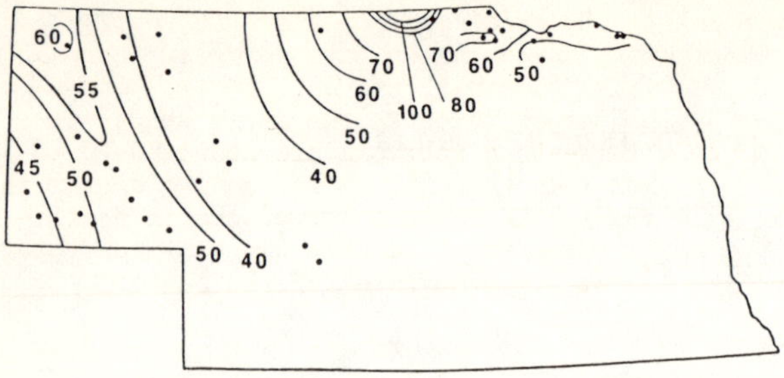

Figure 1. Contour map of the geothermal gradient existing above the Dakota Group (Cretaceous) in western and north central Nebraska. The units are °C/km and the contours were drawn on the basis of equilibrium temperatures measured at sites indicated by the dots.

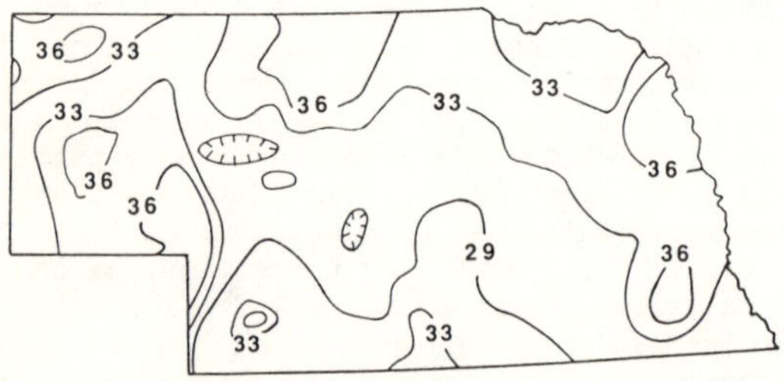

Figure 2. The Nebraska portion of the U.S.G.S. Geothermal Gradient Map of North America (A.A.P.G., 1976). The contours were converted from English units to S.I. units and rounded off to the nearest integer.

on the assumptions that the mud temperature was constant during circulation (Kehle <u>et. al.</u>, 1970) and that both the mud temperature and time elapsed since circulation were accurately recorded. The first assumption is generally valid unless an unusual occurrence such as a significant change in circulation rate happens. The two remaining assumptions are difficult to reconcile with our data. In a total of more than 14,000 well logs, we find few instances where the elapsed time since circulation was not recorded in one hour increments. The elapsed time is a critical factor in the correction and invalid numbers would make the correction worthless. The recorded mud temperatures are also suspect as indicated by a comparison of bottom hole temperatures to mud temperatures for a number of wells in the Denver-Julesburg basin (Figure 3). One phenomenon that apparently arises from the unscientific manner in which these data are recorded is the frequency of occurrence of certain temperatures. This phenomenon is observed for all areas of our study and raises questions about the validity of the data and the feasibility of correcting the data. It is especially disturbing that the "common" temperatures fall within the bulk of the data and would tend not to be rejected in statistical treatment of the data. Although we have not completely analyzed all of our data, at present we find that the question of how the data can be corrected verges on enigma.

The question of how accurately do contour maps of temperature gradients derived from the bottom hole temperatures depict subsurface temperatures is more easily answered. We have reasonably accurate data (Gosnold and Eversoll, 1981) on the temperatures on top of the Dakota Group (Cretaceous) in western Nebraska (Figure 4). Because most of the BHT data in western Nebraska are for wells drilled to the Dakota Group, a projection of temperature contours on top of the Dakota based on the BHT data (Figure 5) should compare reasonably well with Figure 4. Obviously the two figures do not compare well and the BHT derived temperatures underestimate the temperatures on the Dakota by a significant amount. The primary reason for the differences between Figures 4 & 5 is that the correction used for the BHT data, which are taken from the A.A.P.G. geothermal gradient map, does not apply to the subsurface of Nebraska.

The final question we consider is: Can the BHT data be used to identify geothermal resources? Because we have determined the occurrence of low-temperature resources by utilization of equilibrium temperatures, heat flow, and stratigraphic data (Gosnold and Eversoll, 1981), we have a basis for testing the usefulness of the BHT data. A limiting factor in this test is the definition of a low-temperature geothermal resource. We prefer a definition that considers the economics of utilizing a resource. However such a definition is difficult to construct on a fluctuating economic base.

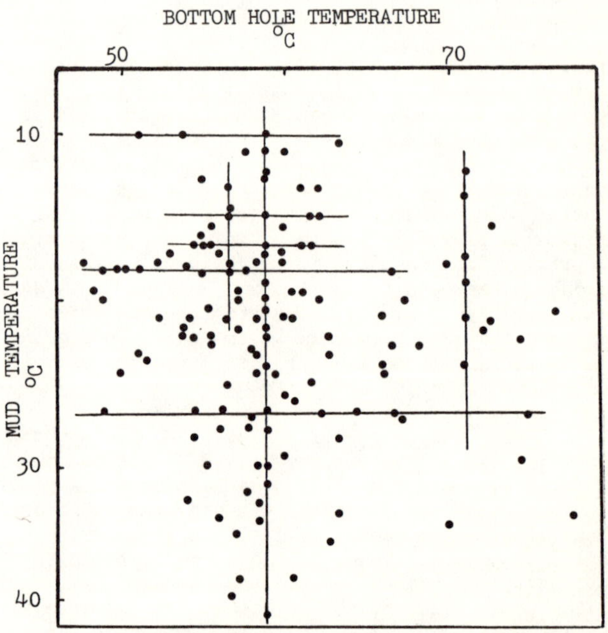

Figure 3. Bottom hole temperature vs. mud temperature for wells in the Denver-Julesburg basin. The straight lines emphasize commonly recorded temperatures that may not be accurate yet would not be removed from the data by statistical treatment.

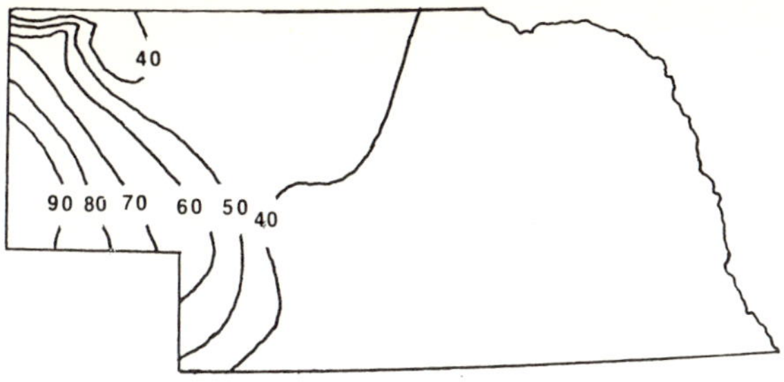

Figure 4. Temperatures on top of the Dakota Group based on equilibrium temperatures and heat flow data. Temperatures are in °C.

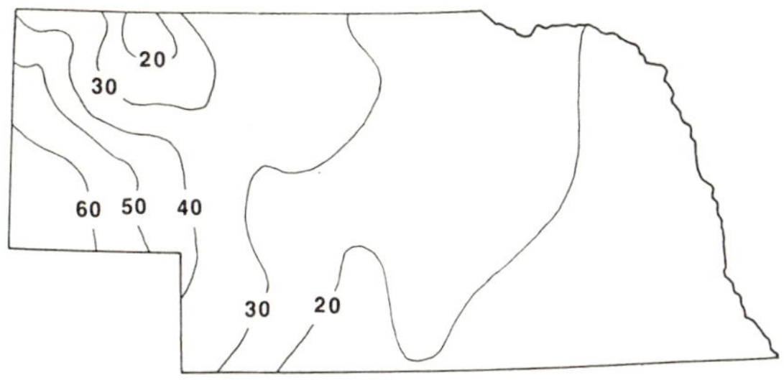

Figure 5. Temperatures on top of the Dakota Group based on the U.S.G.S. Geothermal Gradient Map of North America. Temperatures are in °C.

Consequently we have adopted the temperature-based definition proposed by the U.S. Geological Survey (Reed and Sorey, 1981), i.e., subsurface geothermal waters must have a temperature that is above a line having a slope of 25 °C/km and a surface intercept of 10° above the mean annual air temperature. On this basis we have identified a resource area (Figure 6) of about 107,000 km^2 in extent that contains on the order of 1000 x 10^{18} J of stored energy. Using the U.S.G.S. geothermal gradient map to perform the same evaluation identifies a resource area of about 20,000 km^2 that contains on the order of 170 x 10^{18} J of stored energy (Figure 7). The BHT-based evaluation underestimates our known resources by about 80 %.

DISCUSSION

The low estimates of subsurface temperatures derived from the U.S.G.S. geothermal gradient map are primarily caused by the inapplicability of the analytical correction scheme that was used to the subsurface geology of Nebraska. The factors that cause the correction scheme to be in error are the differences in geologic section and in the heat flow. In the gulf coast region for which the U.S.G.S. map correction was derived, the heat flow is about 30 mW/m^2 to 60 mW/m^2 (Lachenbruch and Sass, 1977). The geologic section in the gulf coast consists mostly of quartz-rich sands and should have a relatively high mean thermal conductivity for a sedimentary section. In the Denver-Julesburg basin in western Nebraska the heat flow ranges from 60 mW/m^2 to 100 mW/m^2 and the geologic section is predominately shales which have very low thermal conductivities (Gosnold, et. al., 1981). Consequently the temperatures in the depth ranges from 1 to 2 km in western Nebraska are about 20 °C to 30 °C higher than those at corresponding depths in the gulf coast. It is this difference in equilibrium temperatures that renders the gulf coast correction scheme inappropriate for the Denver-Julesburg basin.

The feasibility of developing a characteristic correction scheme for each sedimentary basin where bottom hole temperature studies might be used hinges on the acquisition of equilibrium temperature measurements. However, if one has the capability of making deep-hole temperature logs, the bottom hole temperature data are not necessary. If only a few equilibrium temperatures are known for a particular area and the bottom hole temperatures are to be used, there is one remaining caveat on the correction. Mud temperatures are largely a function of the bottom hole temperature which is, in turn, a function of the depth of the hole. Therefore it is necessary to have a correction scheme for limited depth ranges. Otherwise one could use a mud temperature of 40 °C to correct a bottom hole temperature in a 500 m hole in a normal geothermal gradient area.

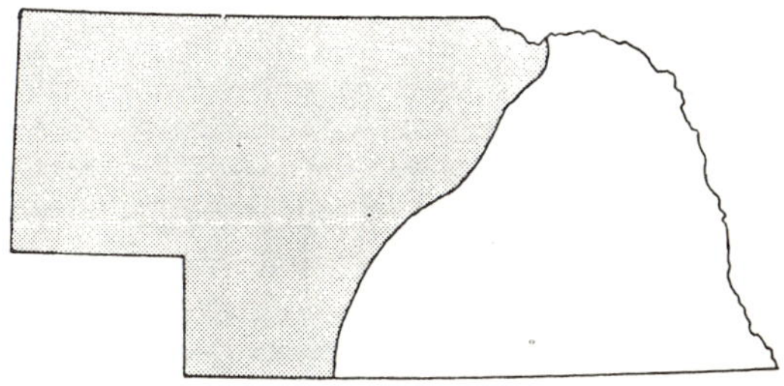

Figure 6. Occurrence of low-temperature geothermal resources in Nebraska (shaded area) based on heat flow and equilibrium temperature data. The resource area extends over 107,000 km^2 and contains on the order of 1000 x 10^{18} J of stored energy.

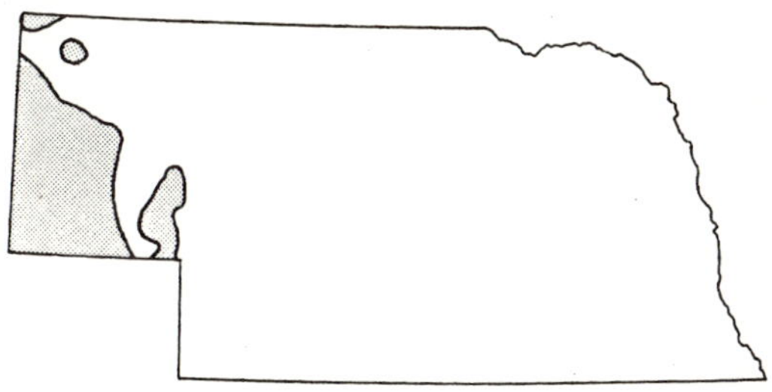

Figure 7. Occurrence of low-temperature geothermal resources in Nebraska (shaded area) based on the A.A.P.G. Geothermal Gradient Map of North America. The resource area extends over about 20,000 km^2 and contains on the order of 170 x 10^{18} J of stored energy.

ACKNOWLEDGEMENTS

Thanks are given to Duane Eversoll and Karen Messenger for compiling the bottom hole temperature data on Nebraska. Thanks are also given to Larry Carter and Jim McAuliffe for help in preparing the maps. This project was sponsored by the U.S. Department of Energy.

REFERENCES

AAPG-USGS, 1976, Geothermal Gradient Map of North America, U.S. Geological Survey, Arlington, Va.

Gosnold, W.D., Jr., and D.A. Eversoll, 1981, Usefulness of heat flow data in regional assessment of low-temperature geothermal resources with special reference to Nebraska, Geothermal Resources Council Transactions, v. 5, p.79-82.

Gosnold, W.D., Jr., D.J. Becker, and D.A. Eversoll, 1981, New heat flow data from Nebraska, EOS, v. 62, p. 1054.

Harris, K.L., F.L. Howell, L.M. Winczewski, B.L. Wartman, H.R. Umphrey, and S.B. Anderson, 1981 An evaluation of hydrothermal resources of North Dakota; Phase II, Final Technical Report: UND Engineering Experiment Station Bulletin No. 81-05-EES-02, 311 pp.

Kehle, R.O., R.J. Schoeppel, and R.D. Deford, 1970 The AAPG Geothermal Survey of North America, Geothermics, Special Issue 2, U.N. Symposium on the Development and Utilization of Geothermal Resources, V. 2, p. 358-367.

Lachenbruch, A.H., and J.A. Sass, 1977, Heat flow in the United States and the thermal regime of the crust, in The Earth's Crust, J.G. Heacock ed., American Geophysical Union Monograph 20, p. 626-675.

Reed, M.J., and M.A. Sorey, 1981, Low-temperature geothermal resource assessment of the United States: A progress report, Geothermal Resources Council Bulletin, v. 10, p. 11-14.

Geothermal Resources Council, TRANSACTIONS Vol. 6, October 1982

STRATIGRAPHIC PERMEABILITY IN THE BACA GEOTHERMAL SYSTEM, REDONDO CREEK AREA, VALLES CALDERA, NEW MEXICO

Jeffrey B. Hulen and Dennis L. Nielson

Earth Science Laboratory
University of Utah Research Institute
420 Chipeta Way; Salt Lake City, Utah 84108

ABSTRACT

Synthesis of surface and borehole data from the Baca geothermal system, as defined by drilling to date in the Redondo Creek area of the Valles Caldera, New Mexico, indicates that thermal fluid flow in the system is partially controlled by distinct stratigraphic aquifers. These aquifers are relatively thin, laterally restricted beds of non-welded tuff and tuffaceous sandstone, confined primarily to the Quaternary Bandelier Tuff. Recognition of the role of these rocks in reservoir definition at Baca should improve chances for future geothermal discoveries in the Valles Caldera and similar settings elsewhere.

* * *

INTRODUCTION

The Baca geothermal reservoir, as explored by Union Oil Company in the Redondo Creek area of the Valles Caldera, New Mexico (Fig. 1), is a liquid-dominated system with measured temperatures up to 330°C (Dondanville, 1978). Structural permeability in the Baca system has been well documented (Dondanville, 1978; Behrman and Knapp, 1980). Study of surface and borehole data and selected cuttings samples indicates that significant permeability at Baca is also developed along distinct stratigraphic aquifers. This paper will briefly describe these aquifers, document their common correlation with thermal fluid entries in Union wells, and consider implications for future exploration at Baca and other geothermal fields.

GEOLOGIC SETTING

For discussion of the regional geologic setting of the Valles Caldera, the reader is

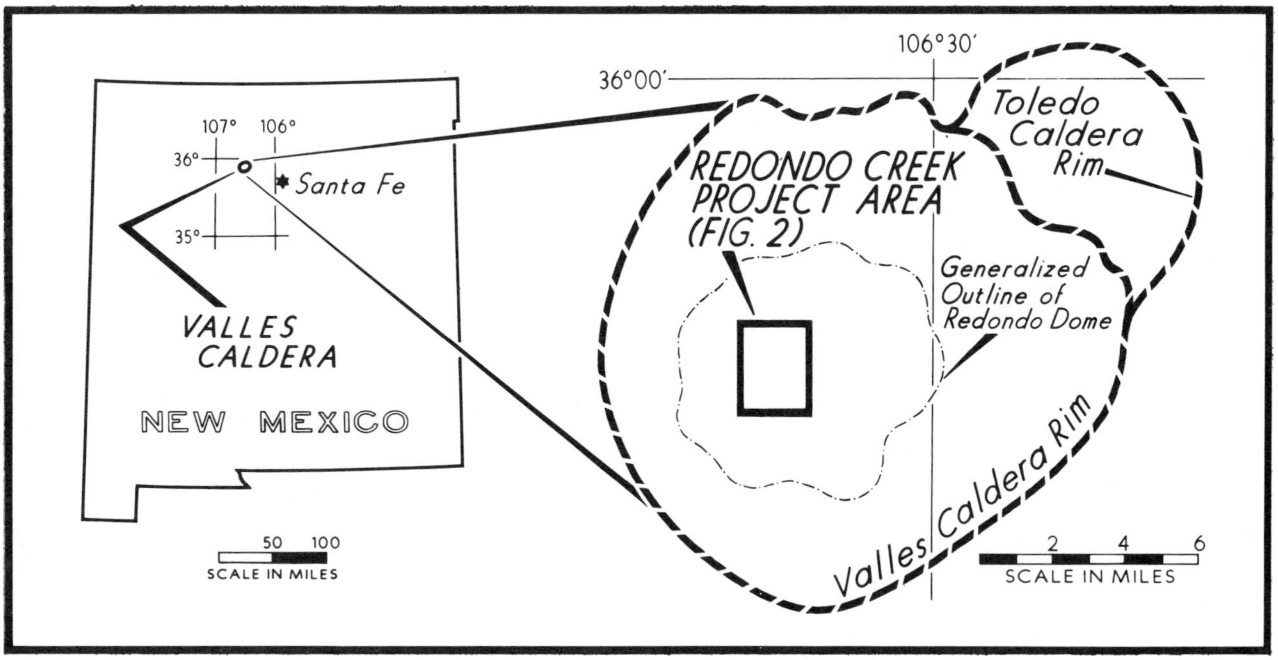

Figure 1. Location map showing position of the Baca geothermal system, Redondo Creek project area, in the Valles Caldera, New Mexico.

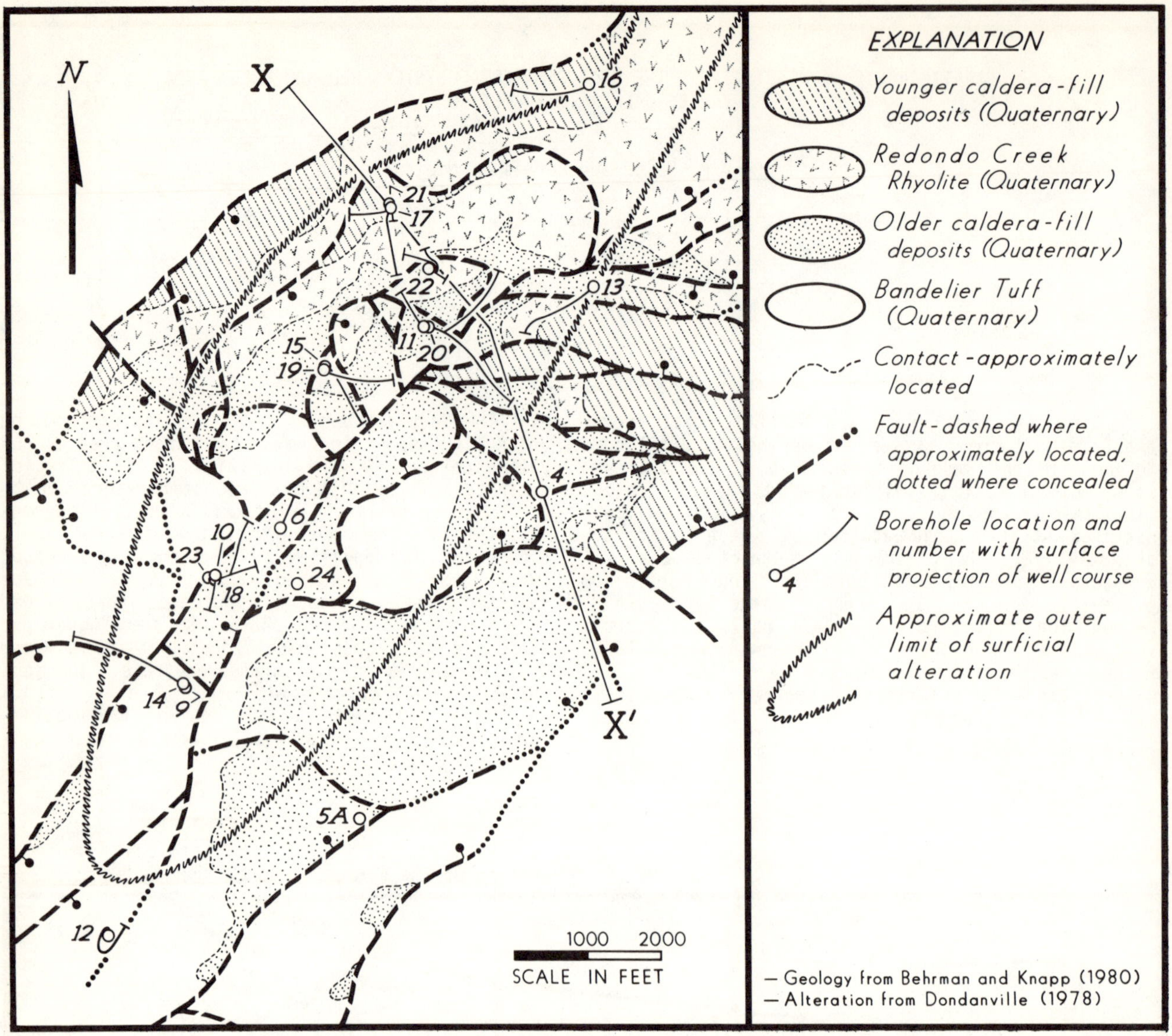

Figure 2. Geologic map of the Redondo Creek project area, Valles Caldera, New Mexico.

directed to Smith and Bailey (1968). The Redondo Creek project area is located along the apical graben of the caldera's resurgent dome, the Redondo dome (Fig. 1). The dome, graben and associated normal faults formed in response to the upward pressure of magma emplaced shortly after eruption of the Bandelier Tuff and resultant caldera collapse. To date, exploration wells at Redondo Creek have not intersected intrusive rocks attributable to resurgent dome formation.

Surface geology of the Redondo Creek project area has been mapped by Behrman and Knapp (1980; Fig. 2), who recognize four principal rock units, all of Quaternary age. The oldest of these units is the Bandelier Tuff, dominated by welded rhyolitic ash-flow tuff. The Bandelier is overlain by caldera-fill deposits (principally landslide debris) of two ages, locally separated by the Redondo Creek Rhyolite. All these rocks are surficially altered (Dondanville, 1978) and disrupted by northeast- and east-west-trending high-angle faults associated with the apical graben of the Redondo dome. Structure of the project area is further complicated by large slump blocks bound by low-angle faults.

SUBSURFACE STRATIGRAPHY, STRUCTURE AND PERMEABILITY IN THE REDONDO CREEK AREA

Figure 3 is a geologic section through the Redondo Creek project area, showing subsurface stratigraphy and interpreted structure as well as major thermal fluid entries in six Union wells. Beneath Quaternary caldera fill and rhyolite, this group of wells penetrated, in sequence: the Bandelier Tuff; the late Tertiary Paliza Canyon Formation, dominated by intermediate flow rocks;

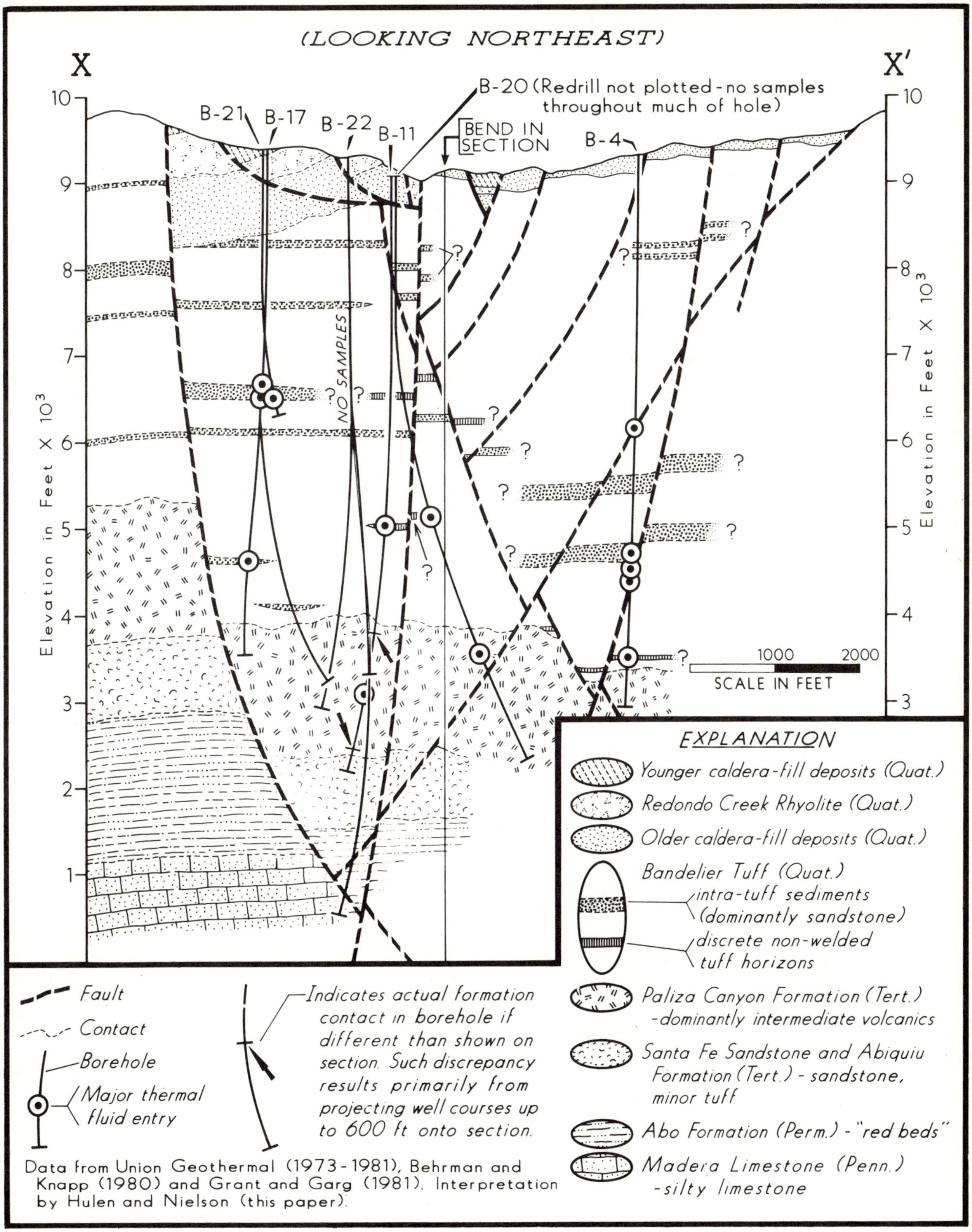

Figure 3. Geologic section through the Redondo Creek project area, Valles Caldera, New Mexico.

Middle to Late Tertiary sandstones and tuffs of the Santa Fe Sandstone and Abiquiu Formation; Permian Abo Formation "red-beds"; and the Pennsylvanian Madera Limestone. Well 12 (Fig. 2) drilled through the Madera into underlying Precambrian granite.

Thermal fluid entries identified in wells throughout the field during drilling (Union Oil Co., 1973-1981) or subsequent production testing (Grant and Garg, 1981; Garg and Riney, 1982) are confined primarily to the Bandelier Tuff, although a few occur in the underlying Paliza Canyon Formation. Many of these entries coincide with fault/fracture zones which can be correlated readily with mapped surface faults (Fig. 3). Other entries, however, appear to be stratigraphically controlled.

Scattered throughout the Bandelier Tuff in the Union wells are thin (generally <150 ft) tuffaceous sandstone and non-welded tuff beds (Fig. 3). These strata, accounting for less than 8% of the total formation thickness, nonetheless apparently provide a large share of the reservoir's total permeability. Of 23 major steam or hot water entries recognized in the Bandelier in 14 Baca wells during drilling (Union Oil Co., 1973-1981; Behrman and Knapp, 1980), seven occur within sandstone and five within non-welded tuff horizons. Garg and Riney (1982) have identified the 11 most productive of these entries in eight wells. Four of these entries are in non-welded tuff; a fifth coincides with tuffaceous sandstone.

DISCUSSION AND CONCLUSIONS

Permeable non-welded tuffs and tuffaceous sandstones of the Bandelier Tuff appear to be prominent fluid flow channels in the Baca geothermal system. Further investigation of these rocks as local stratigraphic aquifers should assist in selecting future geothermal exploration targets not only within the Valles Caldera but in comparable environments elsewhere.

Determination of the nature of permeability in Bandelier sandstones and non-welded tuffs must await detailed examination and analysis of cuttings samples. The original intergranular permeability of these units, for example, may be locally reduced or eliminated by interstitial deposition of secondary alteration minerals. Alternatively, original permeability may be increased by hydrothermal dissolution of unstable framework grains and cement, as documented for deep sandstone aquifers of the Cerro Prieto geothermal field (Lyons and van de Kamp, 1980). If sufficiently embrittled by alteration, portions of the sandstones and non-welded tuffs could be selectively ruptured, creating, as envisioned for the Bishop Tuff in the Long Valley geothermal system (Sorey et al., 1978), stratigraphically controlled fracture permeability.

ACKNOWLEDGEMENTS

This work was performed under contract DE-AC07-80ID12079 from the Department of Energy to the Earth Science Laboratory/University of Utah Research Institute. Illustrations were prepared by Connie Pixton. The manuscript was typed by Joan Pingree and Holly Baker.

REFERENCES

Behrman, P. G., and Knapp, R. B., 1980, Structure of the Redondo Creek area, Baca project, New Mexico--implications concerning the nature of permeability and production and recommendations for future drilling: Union Oil Co., Geoth. Div., Internal Rept., 43 p.

Dondanville, R. F., 1978, Geologic characteristics of the Valles Caldera geothermal system, New Mexico: Geoth. Resources Council, Trans., vol. 2, p. 157-160.

Garg, S., and Riney, T. D., 1982, Reservoir definition and conceptual model (for the Baca geothermal system) in Goldstein, N. E., ed., Final report of the Department of Energy review team for the Baca geothermal demonstration power plant project: Lawrence Berkeley Lab. Rept. LBL-14132 (in prep.), p. 32-39.

Grant, M. A., and Garg, S. K., 1981, Baca geothermal demonstration project--interpretation of downhole data from the Baca geothermal field: Systems, Science and Software, Topical Rept. No. DOE/ET/27163-12, 112 p.

Lyons, D. J., and van de Kamp, P. C., 1980, Subsurface geological and geophysical study of the Cerro Prieto geothermal field, Baja California, Mexico: Napa, California, GeoResource Associates, Rept. LBL-10540 for Lawrence Berkeley Laboratory, 95 p.

Smith, R. L., and Bailey, R. A., 1968, Resurgent cauldrons in Coats, R. R., Hay, R. L., and Anderson, C. A., eds., Studies in volcanology: Geol. Soc. America Mem. 116, p. 613-662.

Sorey, M. L., Lewis, R. E., and Olmstead, F. H., 1978, The hydrothermal system of Long Valley Caldera, California: U. S. Geol. Survey Prof. Paper 1044-A, 60 p.

Union Oil Company, Geothermal Division, 1973-1981, "Mud" logs (generalized lithology, drilling rate and temperature) for wells Baca 4-6 and 9-24.

GEOLOGY IN THE NIGORIKAWA GEOTHERMAL FIELD, MORI-MACHI, HOKKAIDO, JAPAN

Toshio Ide

Geothermal Development Division
Japan Metals and Chemicals Co., Ltd.

ABSTRACT

The Nigorikawa geothermal field is located in a krakatoan type caldera forming an approximate circular basin about 3 km in diameter. Geothermal manifestation concentrats in the northern half portion of the Nigorikawa Basin. The main rock units are composed of the pre-Tertiary Kamiiso Group, Miocene Ebiyagawa Formation and Pleistocene Nigorikawa caldera deposits. Geothermal fluids are stored in fractured zones of the caldera wall deposits and limestone and transverse faults of the Kamiiso Group. Based on porosity and permeability of the reservoir, the Nigorikawa geothermal system is divided into three zones of shallow warm water reservoir, high temperature water convective zone and solid (rock) conductive zone.

INTRODUCTION

A 50 MW geothermal power station will put into operation in November, 1982 in the Nigorikawa geothermal field, Mori-machi, Hokkaido, Japan (Fig.1). Since 1972, geothermal investigations as to surface geology, geochemistry, gravity, heat flow and others have been carried out by Japan Metals and Chemicals Co., Ltd. to make clear the configulation and extent of the reservoir and detect drilling targets. A total of 17 geothermal wells ranging in depth from 700 to 2,400 m have been drilled in the Nigorikawa geothermal field from 1977 to 1981. The temperature maximum attained in the geothermal wells is approximately 270°C, and sufficient geothermal steam for 50 MW power production has been obtained. This paper deals with geological structures affecting storage and movement of the geothermal fluids in the Nigorikawa geothermal field.

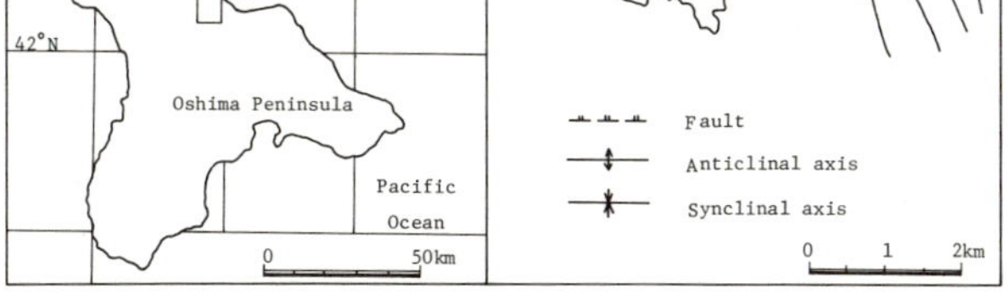

Fig.1 Index map of the Nigorikawa geothermal field

GEOLOGY

The rocks of the Nigorikawa geothermal field consist of the pre-Tertiary Kamiiso Group, Miocene Ebiyagawa Formation and Pleistocene Nigorikawa caldera deposits in ascending order.

The Kamiiso Group is composed of a thick (more than 2,000 m) alternating sequence of slate, chert, limestone and andesitic tuff. The Kamiiso Group, which is not exposed at the surface, is found from cores and cuttings obtained from geothermal wells at the depth of more than 450 m below the surface. Limestone of the Kamiiso Group yields bad-preserved fusulinids and conodonts, but the precise age can not be determined.

The Ebiyagawa Formation, the thickness of which is about 600 m, is composed of andesite lava, volcanic breccia, tuff, sandstone and mudstone. The Ebiyagawa Formation unconformably overlies the Kamiiso Group, and is assigned to be Middle Miocene in age from occurence of foraminifera. The Ebiyagawa Formation is characterized by the northnorthwest-southsoutheast-trending anticlinorium and same trending faults (Fig.1). At the northeastern end of the Nigorikawa Basin, the Ebiyagawa Formation was subjected to hydrothermal alteration, products of which include alunite, kaolinite, montmorillonite and amorphous silica.

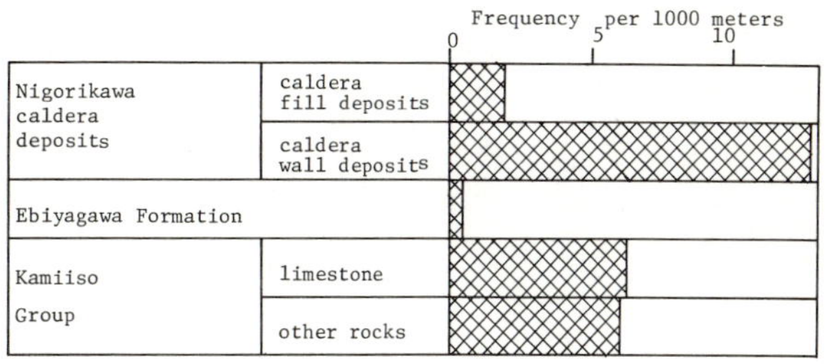

Fig.2 Frequency of circulation losses for each formation

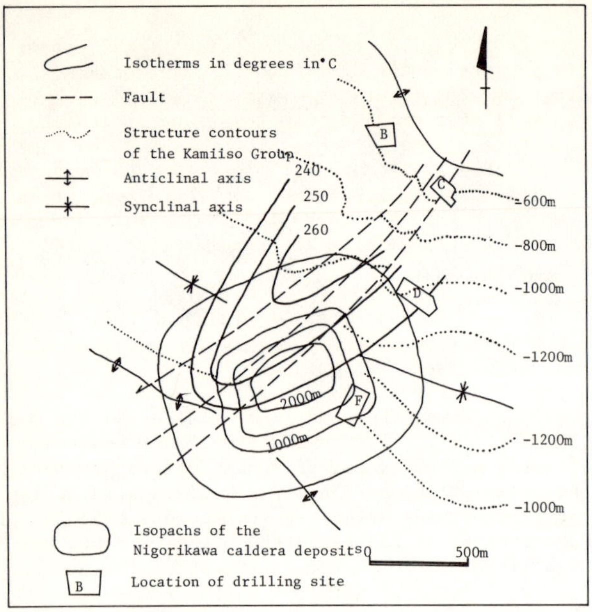

Fig.3 Temperature (°C) at 2000m depth

The Nigorikawa caldera deposits consist of the caldera fill deposits, caldera wall deposits, andesite dyke and pyroclastic flow deposits. The caldera fill deposits are divided into the following two parts. The upper part of them consists largely of unconsolidated mud, silt,and ash, whereas the lower part is made up of consolidated tuff breccia which is fall-back in origin. The results of geothermal loggings, gravimetric section seismic reflection and drilling data show that the caldera fill deposits have a funnel-shaped section with gentle dips in shallow part and steep dips in deeper part (Fig.3 and 4). The caldera wall deposits occupy the space between the caldera fill deposits and the Kamiiso Group or the Ebiyagawa Formation having several tens of meters in width. The caldera wall deposits are composed of volcanic breccia and tuff breccia containing many accidental large blocks derived from the country rocks (the Ebiyagawa Formation and the Kamiiso Group). Andesite dyke, which injected into the caldera fill deposits, is infered to be the root of the central dome lava of the Nigorikawa caldera. The pyroclastic flow deposits cover the surrounding areas of the Nigorikawa Basin, forming a pyroclastic plateau. They are mainly composed of hornblende-andesitic pumice and ash. The carbon-14 age determination indicates the pyroclastic flow deposits to be 12,900±270 B.P. (Sato, 1969).

FRACTURE SYSTEM

Circulation losses of mud water during drilling indicate the most adequate fracture location in the geothermal reservoir. Based on analysis on circulation losses, fracture system of the Nigorikawa geothermal field is composed of the caldera wall deposits and limestone and transverse faults of the Kamiiso Group.

The pressure measurement and tracer testing between wells through the caldera wall deposits indicate high permeability of about 7 darcys. Highly fractured rocks of the caldera wall deposits can be attributed to the violent eruption of magma.

Limestone of the Kamiiso Group has high frequency of circulation losses of mudwater caused by abundant fractures of networks of pores, cracks and caves (Fig.2). Some fractures in limestone are inferred to be solution cavity by hot water circulation. Limestone facies are prevailingly distributed in the geologic section of the Kamiiso Group. Limestone beds vary markedly in thickness as a result of original lenticular deposition having a maximum thickness of 200 m.

The transverse faults of the Kamiiso Group generally give excellent production of high temperature water and steam. They are northeast-southwest-trending transverse faults and dip approximately 80°NW, crossing the central part of the Nigorikawa Basin. The transverse faults are concealed but have been confirmed by circulation losses in connecting with fractured zone and abrupt change of rock facies. Moreover cores, taken at depth of 1,200 m from D-1 well, are brecciated and filled with calcite. Fig.3 shows that the isotherm contour of more than 240°C is consistent with the northeast-southwest-trending faults of the Kamiiso Group. This fact suggests that these faults could possibly provide deep conduit for geothermal fluids.

GEOTHERMAL SYSTEM IN THE NIGORIKAWA FIELD

Based on porosity and permeability of the reservoir, the Nigorikawa geothermal system is divided into three zones such as shallow warm water

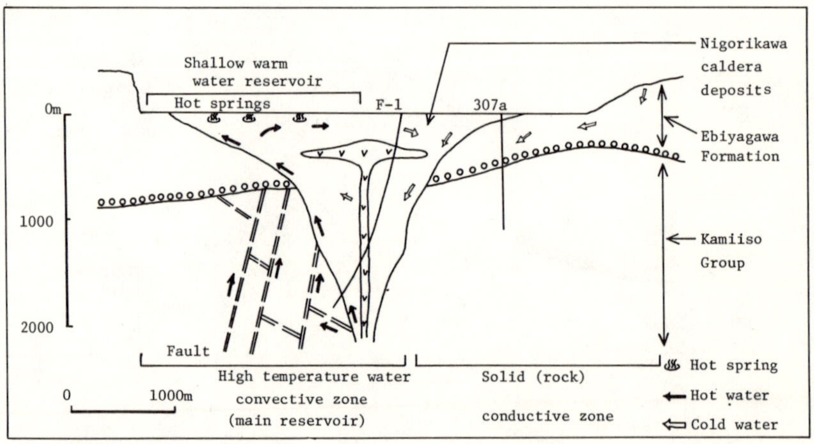

Fig.4 Conceptual model of the Nigorikawa geothermal system

reservoir, high temperature water convective zone, solid (rock) conductive zone (Fig.4).

The shallow warm water reservoir is filled with warm water at temperature of less than 100°C and acts as sources of surface hot spring discharge. Low temperature of the zone results in non-productive steam zone. Warm water which circulates in the upper part of the caldera fill deposits and the Ebiyagawa Formation is considered to be mixture of shallow meteoric water infiltrating from recharge area of southwestern part of the Nigorikawa Basin and deep high temperature water.

High temperature water convective zone closely connected with highly fractured zone in the transverse faults and limestone of the Kamiiso Group and the caldera wall deposits. As mentioned before, the transverse faults could possibly provide deep conduit for geothermal fluids and cause high temperture zone of more than 240°C at 2,000 m depth (Fig.3). The geothermal fluids flow in this zone is mainly controlled by nearly vertical transverse faults and the caldera wall deposits. Therefore, geothermal manifestation on the surface is accurately coincide with the high temperature water convective zone in the subsurface. Geothermal wells drilled into this zone produce a large amount of hot water and steam.

In the solid (rock) conductive zone, conductive heat transfer dominates through the Kamiiso Group lacking fractures. Temperature logging of the well 307a drilled into this zone indicates lenear temperature curve. The rocks of the Kamiiso Group take low permeability less than 10^{-4}m.darcys and low porosity less than 5%. Therefore, it is considered that the rocks lacking in fractures have low storage capability and transmissibility of fluids.

As a result, the Nigorikawa geothermal system is characterized by vertical-trending heat transportation along high dipping faults and the caldera wall deposits.

ACKNOWLEDGEMENT

The author is deeply indebted to Dr.Ko Sato for his kind guidance and critical reading of the manuscript. He is obliged to Dr. Hisayoshi Nakamura for guidance in the course of work, and also wishes to thank the engineers of the Geothermal Development Division of Japan Metals and Chemicals Co., Ltd. for their help with the investigations.

REFERENCE

Sato, H., 1969, Newly obtained ^{14}C data concerning volcanic activity in Hokkaido, Chishitsu News, Geol. Surv. Japan, No. 173, p. 30-35 (in Japanese).

GEOTHERMAL GRADIENT MAP OF THE UNITED STATES (Exclusive of Alaska and Hawaii)

Andrea Kron and John Stix

Earth and Space Sciences Division
Los Alamos National Laboratory
Los Alamos, New Mexico 87545

ABSTRACT

This paper describes an update of the Geothermal Gradient Map of the Conterminous United States (Kron and Heiken 1980) and compares the changes made since the first map. The second map presents a compilation of over 1700 wells that have been measured for temperature below 50 m and whose temperature/depth profiles are linear, or composed of linear segments which reflect changes in the thermal conductivity of the rocks rather than hydrology. The data are displayed at an enlarged scale of 1:2,500,000 and in a new format which shows the location, depth, and gradient of each well in a single color-coded symbol. This edition contains over two times the amount of data shown on the first map and is accompanied by a table, listing for each well its location, depth, gradient, heat flow (where available), thermal conductivity (where available), and a reference. Over 200 references have been consulted and are presented with the data.

BACKGROUND

Investigators in the late 1800's to early 1900's took advantage of deep oil wells (Hallock 1892) and mine shafts (Agassiz 1895; Spurr 1905) to observe the earth's temperatures and make estimates on the nature and thickness of the crust (Agassiz 1895). During the 1930's temperature measurements in the deep copper mines of Michigan were used to estimate the timing and extent of the last glacial epoch (Fisher et al. 1932; Hotchkiss and Ingersoll 1934). Attempts to correlate temperatures with geologic structures during the 1920's to 1930's resulted in numerous observations in deep wells throughout the oil fields of Texas, Oklahoma, Kansas, New Mexico, Wyoming, and California (Van Orstrand 1926; Van Orstrand 1934). Spicer (1964) continued to measure temperatures in deep oil wells and compiled measurements made from 1910 to 1945.

Interest in determining heat flow throughout the United States began in the 1940's with investigators often utilizing the deep well measurements of Van Orstrand (Benfield 1947; Birch 1947). Since the 1950's calculation of temperature gradients in wells, tunnels, and mine shafts has primarily been for determination of heat flow for understanding crustal dynamics (Roy et al. 1968; Sass and Munroe 1974).

The quest for geothermal energy during the 1970's and 1980's promoted extensive temperature surveys across the U.S. Measuring temperatures in wells has always been the preliminary exploration tool in searching for both hydrothermal and hot dry rock geothermal systems. In 1976 the American Association of Petroleum Geologists and the U.S. Geological Survey published a geothermal gradient map of north America using predominantly bottom-hole temperatures from oil and gas wells. The gradients were calculated using the average annual surface temperature as one point and a corrected temperature at the deepest depth as the second point. A more accurate method (and the one used for this map) for determining the geothermal gradient is to measure the temperature at regular depth intervals in a well that has attained thermal equilibrium, and calculate the slope of a least-squares line passing through a plot of the data points. A temperature log provides information that enables an interpretation of the conductive vs convective component of heat transfer in the underlying rocks.

Guffanti and Nathenson (1980) published a temperature gradient map of the U.S. containing data from over 200 wells that have been measured for temperature below 600 m. Their objective was to determine regional temperature gradients that can be expected to exist in a conductive thermal regime to a depth of two kilometers. This type of data is very useful for regional analyses but does not include information on local thermal anomalies which may also be of interest to the exploration community. The Kron and Stix (1982) map provides this additional information, however, each measurement represents conditions existing only within a particular well and cannot be considered a regional value or to extend beneath the depth of the well.

MAP FORMAT

The first map (Kron and Heiken 1980) was published at a scale of 1:5,000,000; each data point was labeled with a number representing its geothermal gradient. Data were contoured in five 15°C/km intervals and each area falling within a specific contour interval was color-coded. The shortcomings of this display are that data from wells of various depths and different geologic settings are weighted the same, and areas with no data are given values associated with their nearest neighbors. Due to the inherent variability of the data, and subsequent inability to correlate such data meaningfully, an improved method of displaying the information was developed.

The second map (Kron and Stix 1982) shows over 1700 data points, each representing a well or group of wells for which a least-squares gradient has been determined. The points are represented by a symbol, which reflects the depth of the deepest temperature measurement in the well, and by a color which represents a gradient interval. Each well is numbered and keyed to a table showing latitude, longitude, well depth, gradient, heat flow (where available), thermal conductivity (where available), and a reference. The data are not contoured and areas with no data are left blank. General trends can be noted by changes in the color of the symbols. Qualifying information such as well depth is readily distinguished.

CHOOSING, CALCULATING, AND AVERAGING GRADIENTS

The purpose of this map is to present gradients from wells that are in thermal equilibrium and appear to reflect a conductive regime. Only the linearity of the temperature/depth profile is used to determine if the gradient reflects a dominantly conductive regime. Other information concerning the hydrologic nature of the system is not considered. This method presents a consistent approach to evaluating data of highly variable quality, and data which can be very inconsistent over small distances due to local changes in hydrology and geology. Where temperature/depth profiles are not available, a quality rating by the author is considered. It is important to emphasize that gradients shown on this map reflect only the local conditions within a well and cannot be assumed to represent a regional value or to extend below the depth of actual measurement.

The following list specifies criteria used for selecting gradients for the map:

1. Wells must be deeper than 50 m.
2. Temperature/depth profiles should reflect minimum hydrologic disturbances, i.e., they should approximate a straight line, or consist of straight-line segments that reflect changes in the thermal conductivity of the rocks.
3. Gradients must be positive below the temperature inversion due to seasonal affects.
4. Only gradients given a high-quality rating by authors are used (generally, a least-squares line with a correlation coefficient of 10%).

Once a gradient has met the above criteria, the following methods were used to calculate a representative measurement for the well:

1. Gradients were calculated from measurements taken beneath the temperature inversion caused by changes in seasons.
2. A least-squares fit was calculated for straight-line segments of a temperature vs depth plot and a weighted average was determined from these segments.
3. In some cases, a straight line was visually drawn to approximate a least-squares fit, and the slope of the line (or straight-line segments) was calculated.
4. In Louisiana abundant deep, bottom-hole temperature (BHT) data are available from oil and gas wells. When BHT is plotted against depth for many wells in a region (see for e.g. Jam L. et al. 1969) a least-squares line through the data points approximates the average temperature gradient in that region. In Washington, Biggane (1981) found that a least-squares line through BHT's provided the best representation of the average gradient in the Yakima region. The location of such a measurement on the map is the approximate center point of the group of wells considered.

In areas where wells are clustered, an average gradient for the area was determined. If the gradients in an area are similar for similar depth intervals, the gradients from several boreholes were averaged. In areas where gradients differ greatly, an average was calculated from a representative sampling of the various anomalous gradients. The location listed in the table is a center point for all the boreholes averaged in a region; the depth represents the deepest well. If a well is significantly deeper than nearby wells, its gradient was chosen over the others.

ACKNOWLEDGMENTS

This work was funded by a grant from the Department of Energy, Division of Geothermal and Hydropower Technologies to the Hot Dry Rock Resource Evaluation program. G. Heiken provided invaluable assistance and guidance; D. T. Gambill (now of Union Geothermal) wrote the computer programs to compile and plot the data at Los Alamos. R. H. Smith, L. W. Miller, A. E. Theberge, Jr., and J. A. Ikelman of the National Geophysical and Solar-Terrestrial Data Center/National Oceanic and Atmospheric Administration, Boulder, Colorado, assisted in the cartographic preparation, printing, and publishing of the map. M. Guffanti of the U.S. Geological Survey, Menlo Park, California deserves special thanks for her tireless discussions and generous exchange of information.

REFERENCES

Agassiz, A., 1895, On Underground Temperatures at Great Depth, Amer. J. Sci., v. 50, 3rd series, pp. 503-505.

American Assoc. Petrol. Geol. and U.S. Geol. Survey, 1976, Geothermal Gradient Map of North America, scale 1:5,000,000.

Biggane, J. H., 1981, The Low Temperature Geothermal Resource of the Yakima Region - A Preliminary Report, State of Wash. Dept. Nat. Resour., Div. Geol. Earth Resour., Open File Rep. 81-7, 30 p.

Fisher, J., L. R. Ingersoll, and H. Vivian, 1932, Recent Geothermal Measurements in the Michigan Copper District, Amer. Inst. Min. Metall. Eng. Tech. Pub., v. 481, pp. 3-11.

Guffanti, M. and M. Nathenson, 1980, Preliminary Map of Temperature Gradients in the Conterminous United States, Geoth. Resour. Council Trans., v. 4, pp. 53-56.

Hallock, W., 1892, Preliminary Report of Observations at the Deep Well, Wheeling, West Virginia, Amer. J. Sci., v. 43, pp. 234-236.

Hotchkiss, W. O. and L. R. Ingersoll, 1934, Postglacial Time Calculations from Recent Geothermal Measurements in the Calumet Copper Mines, J. Geol., v. 42, pp. 113-122.

Jam L., P., P. A. Dickey, and E. Tryggvason, 1969, Subsurface Temperature in South Louisiana, Amer. Assoc. Petrol. Geol. Bull., v. 53, pp. 2141-2149.

Kron, A. and G. Heiken, 1980, Geothermal Gradient Map of the Conterminous United States, Los Alamos National Lab. Rep. LA-8476-MAP.

Kron, A. and J. Stix, 1982, Geothermal Gradient Map of the United States (Exclusive of Alaska and Hawaii), National Oceanic and Atmospheric Admin., in preparation.

Roy, R. F., E. R. Decker, D. D. Blackwell, and F. Birch, 1968, Heat Flow in the United States, J. Geophys. Res., v. 73, pp. 5207-5221.

Sass, J. H. and R. J. Munroe, 1974, Basic Heat-Flow Data from the United States, U. S. Geol. Survey Open File Rep. 74-9.

Spicer, H. C., 1964, A Compilation of Deep Earth Temperature Data: USA 1910-1945, U. S. Geol. Survey Open File Rep. 64-147.

Spurr, J. E., 1905, Geology of the Tonopah Mining District, U. S. Geol. Survey Prof. Paper 42, 295 p.

Van Orstrand, C. E., 1926, Some Evidence on the Variation of Temperature with Geologic Structure in California and Wyoming Oil Districts, Econ. Geol., v. 21, pp. 145-165.

Van Orstrand, C. E., 1934, Temperature Gradients, in Problems of Petroleum Geology, edited by W. E. Wrather and F. H. Lahee, Amer. Assoc. Petrol. Geol., pp. 989-1021.

AN ANALYSIS OF THE GEOTHERMAL POTENTIAL OF SOUTHWESTERN NEW MEXICO

Richard L. Lohse and Larry Icerman

New Mexico State University Energy Institute
Las Cruces, New Mexico 88003

ABSTRACT

Data sets for actual temperatures; bottom-hole temperature gradients; geotemperatures; tectonics; lineaments; and complete and residual Bouguer gravity were analyzed in order to assess the geothermal potential of southwestern New Mexico. Major NW-SE and N-S trends of structual weakness and control, suggested by these data, intersect in the general area of Dona Ana County and appear to be crossed by NE-SW trending secondary zones of weakness. Temperature data imply that there is a spatial relationship between intersections of these trends and potential geothermal anomalies, which suggests that the hydrothermal systems are predominately fault controlled.

INTRODUCTION

The location of the study area (see Fig. 1) includes the Gila Hot Springs, Kilbourne Hole, Lightning Dock, Lower Frisco Hot Springs, and Radium Springs KGRAs; the southern extent of the Socorro Peak KGRA, the well-known Faywood Hot Springs and Truth or Consequences areas, and the Las Cruces east mesa geothermal field. There are several users of the low-temperature geothermal energy resources within the study area, the largest of which is the New Mexico State University hot water and space heating project.

TEMPERATURE DATA

Measured Temperatures

Actual temperature data for springs and wells were compiled from USGS WATSTORE and GEOTHERM files; Chaturvedi (1980); Lohse (1980, 1982); Lohse et al. (1981a, 1981b); Morgan et al. (1980); and Swanberg (1980a, 1980b, 1980c). Data for approximately 50 springs and 845 wells with depths less than 150 meters, were screened for anomalous temperatures. Approximately 620 wells with depths greater than 10 meters were screened for anomalous bottom-hole temperature gradients. The recorded temperatures range

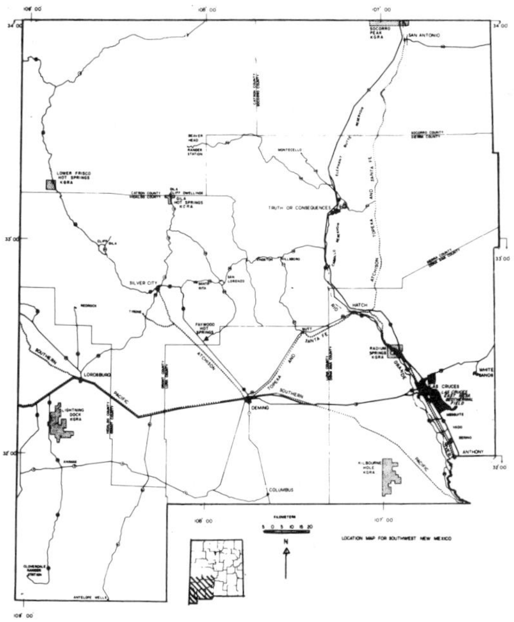

Figure 1. Location map of the study area.

from about 10° to 100°C for springs and for wells less than 150 m in depth. Approximately 50 wells have computed bottom-hole temperature gradients in excess of 100°C/km for an assumed surface temperature of 20°C, about 20 of these wells have gradients higher than 250°C/km. Normal temperature gradients for the study area are approximately 40°C/km.

Geothermometry Data

Data for the SiO_2, Na-K-Ca, and Na-K-Ca (Mg-corrected) geothermometers were compiled from Swanberg (1980a, 1980c, 1980d) and Jodry (1981) for approximately 605, 600, and 270 sites, respectively. About 35 locations have SiO_2 geotemperatures greater than 120°C, while Na-K-Ca and Na-K-Ca (Mg-corrected) geotemperatures above 200°C and 70°C were computed for 9 and 40 sites, respectively. Areas thought to be thermally anomalous are defined as those areas where both the SiO_2 and Na-K-Ca geothermometers are in agreement and indicate elevated base temperatures.

Geothermal Anomalies

The sites of the anomalous temperature data are shown in Figure 2. Anomalous geothermometry data are defined as sites that have SiO_2 geotemperatures ≧ 90°C and Na-K-Ca geotemperatures ≧ 120°C or the locations that have SiO_2 geotemperatures ≧ 90°C and Na-K-Ca (Mg-corrected) geotemperatures ≧ 70°C. Sites with Na-K-Ca geotemperatures ≧ 200°C were added to the latter group in order to increase the data base, since Mg corrections were not available for approximately one-half of the Na-K-Ca geothermometry data. Only these Na-K-Ca data were included to prevent the data from being biased by magnesium-rich geothermal fluids because, even after the application of the Mg correction, these waters would probably still indicate low to intermediate geothermal reservoir base temperatures.

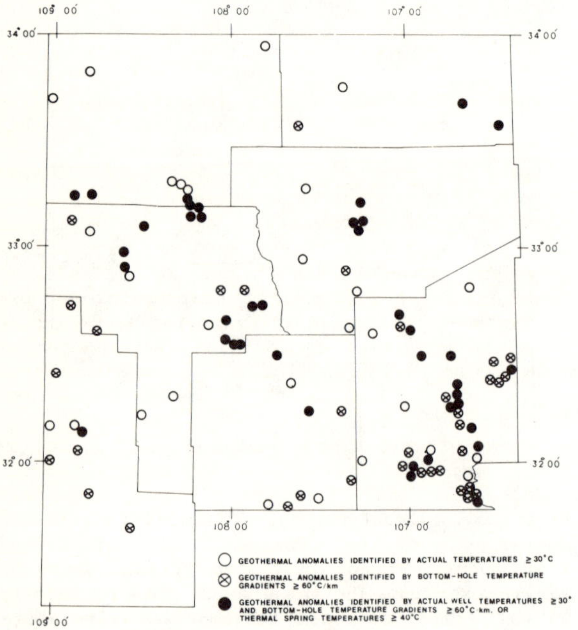

Figure 2. Sites characterized by anomalous actual well and spring temperatures (≧ 30°C), bottom-hole temperature gradients (≧ 60°C/km) of wells, or anomalous actual temperatures (≧ 40°C) of thermal springs.

Figure 3 shows a final reduction of the anomalous data and is a composite of Figure 2 and the geothermometry data. Figure 3 includes: (1) those sites from Figure 2 that have both actual well temperatures ≧ 30°C and bottom-hole temperature gradients ≧ 60°C/km, or thermal springs with temperatures ≧ 40°C; and (2) those sites which have SiO_2 geotemperatures ≧ 90°C and Na-K-Ca (Mg-corrected) geotemperatures ≧ 70°C, or have SiO_2 geotemperatures ≧ 90°C and Na-K-Ca geotemperatures ≧ 200°C. The locations in Figure 3 indicated by solid and open circles

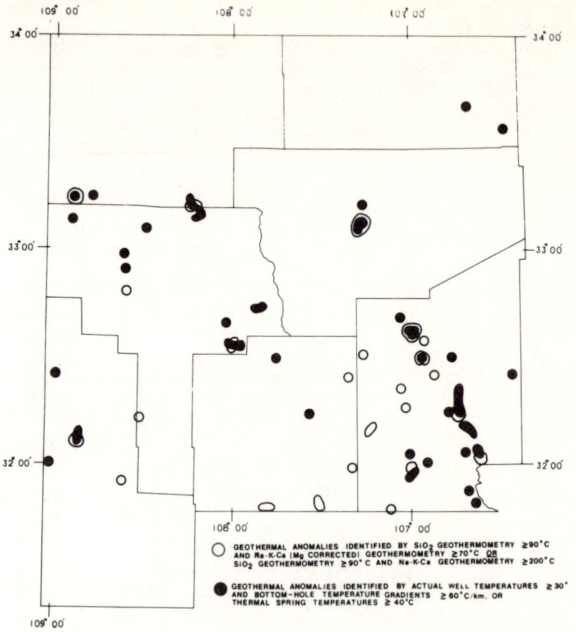

Figure 3. Sites characterized by anomalous geotemperatures, actual well temperatures, bottom-hole temperature gradients, and thermal spring temperatures.

simultaneously (i.e., conditions (1) and (2) above are both satisfied) are low to intermediate temperature geothermal resource areas.

GEOLOGICAL AND GEOPHYSICAL DATA

Tectonic Features

Figure 4 is a tectonic map illustrating two different periods of faulting during the Pliocene to Pleistocene epochs (~ 1 to 4 m.y.) and during the late Quaternary Period (< 1 m.y.). Northwest trending normal faults predominated during the Pliocene to Pleistocene epochs. The late Quaternary Period has been dominated by N-S trending normal faults that have persisted into the Holocene Epoch.

The majority of the Quaternary and late Pliocene volcanic rocks and their volcanic centers appear to be located in regions characterized by N-S trending normal faults. Most of the age dates associated with these volcanic rocks are approximately of the same age as the faults. The close agreement of age and location between faults and volcanic rocks suggests that the faulting and volcanism are contemporaneous.

Gravity Data

The residual Bouguer gravity map of Aiken et al. (1978), reproduced in Figure 5, clearly

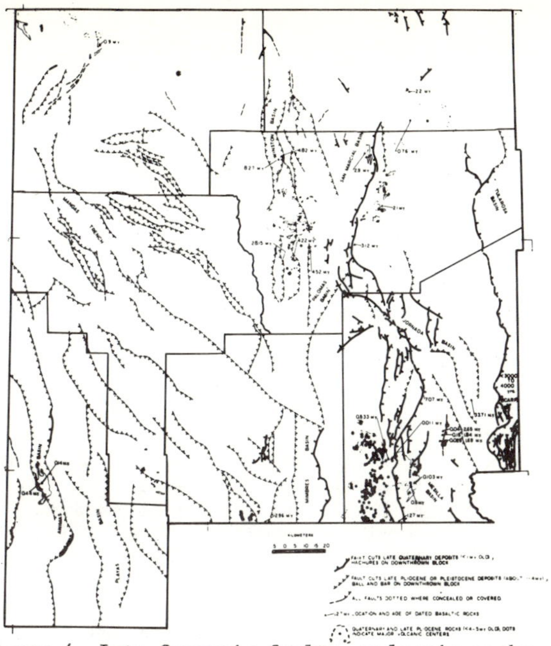

Figure 4. Late Cenozoic faults, volcanic rocks and vents, and locations and ages of dated basaltic rocks. The data are taken from Callender and Seager (1980), Seager et al. (1981), and Aldrich and Laughlin (1981).

shows strong NW-SE structural trends probably extending into the basement (Paleozoic and Precambrian strata ?), which pass through Grant, Luna, and Dona Ana counties. Also apparent in Figure 5 are N-S trending structural zones through Socorro, Sierra, and Dona Ana counties and the eastern one-third of Luna County. Dona Ana County is generally situated at the intersection of the two zones of structural weakness.

Lineament Data

A lineament study of New Mexico (Lepley 1982) identifies three major grains or zones of weakness through the study area: (1) a NW-SE grain, (2) a NE-SW grain, and (3) a N-S grain. A NW-SE grain crisscrossed with a NE-SW grain has also been observed by Ramberg et al. (1978).

DISCUSSION

Overlaying Figure 3 with Figure 4 shows a spatial relationship between fault and fracture zones and geothermal resource areas, which suggests that the geothermal reservoirs are fault-controlled hydrothermal systems (Lohse et al. 1981a). These data sets also suggest the existence of two major trends: (1) a NW-SE trend through Grant, Luna, and Dona Ana counties, and (2) a NNW-SSE trend across Dona Ana and Sierra counties. These two trends have been previously noted in Dona Ana County (Lohse et al. 1981a). The NW-SE trend is also present in the residual Bouguer gravity map (see Fig. 5), which suggests that the faulting and fracturing probably extend into the crystalline basement. The NNW-SSE trend does not appear to extend into the crystalline basement, unless this trend is actually a composite of N-S and NW-SE components, because no regional residual gravity gradient appears to coincide with this trend. The distribution of late Quaternary faulting also supports this conjecture of combined N-S and NW-SE components.

Figure 5. Residual Bouguer gravity map [from Aiken et al. (1978)].

The NE-SW trends identified by Lepley (1982) and Ramberg et al. (1978) are not readily apparent in Figure 5; however, these trends may be recognized easily in Figures 2 and 3. Comparison of the lineament data of Lepley (1982), the tectonic and gravity data shown in Figures 4 and 5, and the gravity data of Ramberg et al. (1978) with the thermal anomalies shown in Figure 2 and the identified potential geothermal areas shown in Figure 3 suggests that there is a strong spatial relationship between the intersections of the NW-SE, NE-SW, and N-S zones of weakness and geothermal anomalies.

CONCLUSIONS

Gravity, fault, lineament, and anomalous actual temperature, bottom-hole temperature-gradient, and geothermometry data suggest major NW-SE and N-S trends of structural weakness and control throughout southwestern New Mexico. These two trends intersect in the general area of Dona Ana County and appear to be crisscrossed and intersected by secondary zones of weakness that trend in a NE-SW direction. Anomalous temperature data indicate that there is a spatial relationship between points of intersections of these trends and potential geothermal anomalies, which suggests that the hydrothermal systems are fault controlled. The NW-SE trends do not appear to be characterized by nearly continuous hydrothermal activity and elevated near-surface temperatures as does the apparent NNW-SSE trend, which is believed to be composed of N-S and NW-SE components. Late Quaternary faulting, which appears to be associated with nearly continuous hydrothermal activity, is generally in a N-S direction; however, when pre-existing NW-SE fracture zones are intercepted, the Quaternary faulting has a tendency to propagate in a NW-SE direction. This observation, coupled with the observation that Quaternary volcanism appears to have a spatial and temporal relationship with Quaternary N-S faulting, suggests that the N-S trending features are probably of a tensional nature and that the volcanism and faulting are likely to have had a cause and effect relationship.

ACKNOWLEDGMENT

This work was supported by the U.S. Department of Energy, Contract DE-AS07-78ID01717.

REFERENCES

Aldrich, M.J., Jr., and Laughlin, A.W., 1981, Location, age, and rock type of volcanic rocks younger than 5 million years old in Arizona and New Mexico: Los Alamos National Laboratory, Report LA-8820-MS, 21 pp.

Aiken, C., Laughlin, A.W., and West, F.G., 1978, Residual Bouguer gravity anomaly map of New Mexico: Los Alamos National Laboratory, scale 1:500,000.

Callender, J.R., and Seager, W.R., 1980, Quaternary and late Pliocene geology, in State-coupled low temperature geothermal resource assessment program, fiscal year 1979: NTIS, p. 6/1-6/4.

Chaturvedi, L., 1980, Geothermal hydrology, in State-coupled low temperature geothermal resource assessment program, fiscal year 1979: NTIS, p. 8/1-8/22.

Jodry, D., 1981, personal communication.

Lepley, L.K., 1982, Lineament map of the state of New Mexico, in State-coupled low temperature geothermal resource assessment program, fiscal year 1981: New Mexico Energy Institute at New Mexico State University, p. 4/1-4/20.

Lohse, R.L., 1980, Thermal gradients, in State-coupled low temperature geothermal resource assessment program, fiscal year 1979: NTIS, p. 5/1-5/36.

Lohse, R.L., Morgan, P., and Swanberg, C.A., 1981a, A heat-flow study of Dona Ana County, southern Rio Grande Rift, New Mexico, in State-coupled low temperature geothermal resource assessment program, fiscal year 1980: New Mexico Energy Institute at New Mexico State University, p. 5/1-5/40.

Lohse, R.L., Morgan, P., and Swanberg, C.A., 1981b, Preliminary heat-flow assessment of southeast Luna County, New Mexico, in State-coupled low temperature geothermal resource assessment program, fiscal year 1980: New Mexico Energy Institute at New Mexico State University, p. 5/41-5/48.

Lohse, R.L., 1982, Collection of existing water well data in the southern Rio Grande Rift and southwestern New Mexico, in State-coupled low temperature geothermal resource assessment program fiscal year 1981: New Mexico Energy Institute at New Mexico State University, p. 5/1-5/20.

Morgan, P., Swanberg, C.A., and Lohse, R.L., 1980, Borehole temperature studies of the Las Alturas geothermal anomaly, New Mexico, in State-coupled low temperature geothermal resource assessment program, fiscal year 1979: NTIS, p. 9/1-9/14.

Ramberg, I.B., Cook, F.A., and Smithson, S.B., 1978, Structure of the Rio Grande Rift in southern New Mexico and west Texas based on gravity interpretation: Geological Society of America, Bulletin, v. 89, p. 107-123.

Seager, W.R., 1981, Geology of Organ Mountains and southern San Andres Mountains, New Mexico: New Mexico Bureau of Mines and Mineral Resources, Memoir 36, 97 pp.

Seager, W.R., Shafiquallah, M., and Clemons, R. E., 1981, Active fault analysis and radiometric dating of young basalts in southern New Mexico, in State-coupled low temperature geothermal resource assessment program, fiscal year 1980: New Mexico Energy Institute at New Mexico State University, p. 6/1-6/12.

Swanberg, C.A., 1980a, Estimated subsurface temperatures, in State-coupled low temperature geothermal resource assessment program, fiscal year 1979: NTIS, p. 1/1-1/17.

Swanberg, C.A., 1980b, Catalogue of thermal waters, in State-coupled low temperature geothermal resource assessment program, fiscal year 1979: NTIS, p. 2/1-2/4.

Swanberg, C.A., 1980c, Water quality of geothermal waters, in State-coupled low temperature geothermal resource assessment program, fiscal year 1979: NTIS, p. 2/5-2/38.

Swanberg, C.A., 1980d, personal communication.

Geothermal Resources Council, TRANSACTIONS Vol. 6, October 1982

A SENSITIVITY STUDY OF THE ECONOMIC PARAMETERS FOR THE GEYSERS GEOTHERMAL FIELD, CALIFORNIA

S.K. SANYAL AND M. CHE

GeothermEx, Inc.
Berkeley, California

ABSTRACT

This paper analyzes the sensitivity of the revenue to be derived by a developer at The Geysers to the variations in fifteen most important variables. The "cost" variables considered were drilling cost per well, redrilling cost per well, facilities cost per drilled well, facilities cost per redrilled well, gathering system cost per production well, disposal system cost per injection well and operating cost per well. The "rate" variables considered were deliverability decline rate, rate of drilling success, rate of redrilling success; rate of well loss per year due to mechanical causes, plant factor, steam price escalation rate, inflation rate and discount rate. Harmonic decline in well deliverability was assumed, based on the observed production decline and reported theoretical results. The results showed that cumulative revenue is most sensitive to the drilling cost per well among the cost variables, and to the steam price escalation rate among the rate variables.

INTRODUCTION

This study analyzes the sensitivity of the revenue to be derived by a developer at The Geysers to the various economic and resource variables. The revenue type (dependent variable) chosen for analysis is the cumulative present worth of the net revenue (R) over an assumed 30-year power plant life. The choice of independent variables (X_i; $i=1,2,\ldots,n$) for study and their typical values were based on the experience of the authors at The Geysers. The parameters chosen for the base case are listed in table 1, and discussed in the following section.

BASE CASE PARAMETERS

The base case assumes that a developer acquires a 400 acre lease for $5 million and spends $1 million in exploration and predevelopment activity. The first well proves productive after one re-drill with a total initial drilling cost of $3 million before the field is proven and then field development for 55 MWe is undertaken. The average well deliverability during initial development is 130,000 lbs. per hour.

The drilling cost considered includes costs for initial exploration well, production wells, redrills or workovers, replacement or make-up wells, and dry holes, including the conversion of dry holes to disposal wells. At The Geysers, it is a common practice to sidetrack and redrill dry holes one or more times before conceding that the hole is unproductive. Sidetracking often results in interception of sufficient fractures to change a dry hole into a producer. Drilling cost is assumed to be $1.9 million (all costs in this paper refer to 1982 dollars) per well. Redrilling a well is assumed to cost $800,000 per well. Drilling facilities costs are calculated at $120,000 per new well (new site) and $30,000 per redrill. These costs include new or revised pad and sump construction, road construction, well head completion for testing, well test equipment and fittings, etc. A 10% reserve production capacity was assumed to allow for well maintenance or production continuity in the event of an unforeseen well loss. Since typical geothermal wells have finite mechanical life, some 3 to 10% of the wells need to be worked over every year, except the first few years. A "well loss" rate of 5% per year after the first five years has been assumed. The "lost" wells are assumed to be brought back to production by workover, costing $600,000 each.

Initial Gathering System cost is estimated to be 10 million dollars. Wells are estimated to cost $60,000 each to bring on line. The disposal system (injection system) cost for each new injector to be brought into the injection system is $45,000. These gathering and injection system costs include costs of pipelines, valves, auxillary pumps, connectors, etc., required to carry steam and/or condensate from wellsite to generating plant or vice-versa. It is assumed that of the make-up wells drilled, on the average 66% are successful producers and are brought into the gathering system. The remaining 34% are redrilled. The chance of success in redrilling is assumed to be 50% on the average. Thus 50% of redrilled holes become successful producers. The remainder are unsuccessful and are included in the injection system, if required. A 10 to 1 ratio of producers to injectors has been assumed.

Operating costs are taken as $35,000 per

completed production well per year. The operational costs account for expenditures for staff and office operations, and routine maintenance. Base cost of steam is taken as $1 per 1,000 lbs. of steam (equivalent to 1.8¢/kw•hr). Steam price is escalated at 9% per year.

Other important assumptions are as follows:

1. 100% working interest.
2. 87.5% revenue interest (12.5% royalty rate).
3. Plant factor average of 85%.
4. All costs to escalate at 10% per year (estimated inflation rate) from 1981.
5. Taxes not considered.
6. Annual discount rate of 15%.

Productivity Decline Rate as a function of years of operation is the most difficult parameter to estimate. However, generalized trends are possible to establish. Budd (1973) presented three typical productivity decline curves corresponding to three well spacings (45 acres, 20 acres and 5 acres) at The Geysers based on a theoretical reservoir model. He presented the parameter q/q_i, where q is the production rate in a particular year of plant operation and q_i is the stabilized production rate at the beginning, as a function of the years of plant operation (figure 1). Budd (1973) presented the decline curves for the first 6 years of production; Figure 1 includes extrapolation of these curves to 36 years as explained below. Dykstra (1981) compiled the production data from 18 producing wells at The Geysers, the longest production history stretching over 8 years. He found that the average trend (figure 1) of production decline from these wells falls about midway between the 20 and 5 acre curves of Budd (1973). Dykstra (1981) did not identify all the wells used in his correlation nor the well spacings involved. However, he presents production data on 11 wells, which presumably are included in the 18 well-correlation mentioned in the report. An examination of the locations of these 11 wells shows that the well spacing around these wells ranges from as close as 2 acres to about 25 acres. Hence the fact that the empirical production decline trend reported by Dykstra (1981) falls between the 20 and 5 acre curves of Budd (1973) is not surprising. It should be noted here that before about 1970, much smaller well spacings were used at The Geysers than are used today, presumably because of the adverse effect of close well spacing on productivity decline experienced or estimated by the operators during the early development of The Geysers.

Dykstra (1981) also compared his empirical decline curve with the dimensionless decline type curves presented by Fetkovich (1973) and reported that the closest match obtained was with the harmonic type curve given by

$$\frac{q}{q_i} = \frac{1}{1 + D_i t} \qquad (1)$$

where t is the time (in years) and D_i is the decline rate. We have examined the productivity decline trends of Budd (1973) and concluded that (1) holds for all three spacings considered. Thus it was concluded that most probably a harmonic decline is a reasonable decline trend for wells at The Geysers, the parameter D_i in (1) being a function of the average well spacing. On this basis we have fitted Budd's (1973) and Dykstra's (1981) curves to (1) and extrapolated to 36 years (figure 1).

It should be realized that most probably Budd (1973) assumed a fixed well spacing throughout the plant life, that is, no assumption of reduced well spacing with make-up well drilling was made. But Dykstra (1981) used average data from wells covering a range of spacings (2 to 25 acres, approximately). The facts that all four curves show similar shape (harmonic decline) and that Dykstra's (1981) empirical data are apparently consistent with Budd's (1973) theoretical results give a reasonable basis for assuming decline curves given by (1) for The Geysers area. For the base case the decline rate corresponding to Budd's curve for 20-acre spacing (D_i = 0.068) was assumed. For each lease a value of D_i can be assumed from geological and reservoir engineering criteria.

COMPUTATIONAL APPROACH

Based on the steam requirement per MW•hr, plant capacity, initial average deliverability and a deliverability decline rate the required number of active wells for each year is calculated. From this, the numbers of standby wells, injectors and workovers required are calculated. Based on this requirement and a 30-year-average of the success rates in drilling and in redrilling, the numbers of wells drilled and redrilled and number of dry wells are calculated for each year. The total drilling cost (including redrilling and workover cost) is then calculated for each year. Facilities cost is calculated for drilled and redrilled holes for each year. Gathering system cost is calculated based on the number of successful producers each year. Disposal system cost each year is based on the number of new injectors drilled and the number of dry holes converted to injectors that year. Operations cost is based on the number of active wells (producers plus injectors) each year. The development cost (acquisition exploration, initial drilling, and gathering systems cost) is included in the 0th year, it being the last year before power goes on line. All unit costs are escalated at the annual inflation rate from the base year. The computer algorithm developed for this purpose checks at each step various logical criteria to preserve the most reasonable statistical correlation between the parameters involved.

Gross revenue is calculated based on plant size, steam requirement per MW, plant factor, unit steam price (escalated at a given rate from the base year price), royalty rate, any partnership formula, and an average 365.25 days per year. Net revenue is the difference between the gross revenue and total cost for the year. Present worth of the annual net revenue is calculated for semiannual compounding. The cumulative present worth of the net revenue (R) is calculated by adding annual net revenues for the 30 years of assumed plant life. For the base case, R is $60 million, with 1982 as the base year.

SENSITIVITY STUDY

Most of the input parameters used in calculating the economics of a project at The Geysers are subject to variation, either by design (such as a negotiated price escalation rate), unforeseen problems (such as plant factor), random nature of geothermal reservoirs (such as drilling success rate) or one's inability to estimate future trends (such as inflation rate). Hence an analysis of the sensitivity of a practical economic parameter such as R to the variable input parameters (x_i ; i = 1,2,...n) is illuminating. A sensitivity study of R to changes in 15 input parameters was done.

In this study a total of 100 cases were considered. The base case was that corresponding to the parameters listed in table 1. The various input variables can be grouped into "Cost" variables, i.e., those representing a cost item per well and "Rate" variables, i.e., those representing the rates and percentages but not dollar amounts. In order to illustrate the sensitivity of R to the various variables two "Spider Plots" were prepared. Figure 2 is a Spider Plot showing the sensitivity of R to rate variables. The central point represents the base case. The abscissa represents the input value of an independent variable in terms of its percentage difference from the base case value. Each curve on the Spider Plot depicts the change in R due to change in the value of a particular independent variable. The steeper the curve the more sensitive is R to that variable. Discount rate causes the most change in R, followed closely by steam price escalation rate and plant factor in order of decreasing effect. Deliverability decline rate and inflation rate have considerable impact on R. Rate of drilling success, rate of redrilling success and well loss rate (workover rate) have relatively modest impact on R. Figure 3 is a Spider Plot of sensitivity to Cost Variables. It is obvious from figure 3 that R is most sensitive to drilling cost per well followed by operating cost per well and redrilling cost per well. Disposal System Cost/well and facilities cost/well do not affect R significantly, while the rest of the cost variables have very small impact on R.

If $X_1, X_2, \ldots, X_n$ are the independent variables, then the revenue variable, R ($X_1, X_2, \ldots, X_n$), will show a total change dR due to changes in independent variables according to:

$$dR = \frac{\partial R}{\partial X_1} dX_1 + \frac{\partial R}{\partial X_2} dX_2 + \ldots + \frac{\partial R}{\partial X_n} dX_n \quad (2)$$

where $dX_1, dX_2, \ldots, dX_n$ are the changes in the independent variables. From (2),

$$dR = 100 \; \frac{S_1}{X_1} dX_1 + \frac{S_2}{X_2} dX_2 + \ldots + \frac{S_n}{X_n} dX_n \quad (3)$$

where dR is the difference in R from the base case caused by simultaneous changes in the independent variables $X_1, X_2, \ldots, X_n$ by amounts of $dX_1, dX_2, \ldots, dX_n$, respectively. $S_1, S_2, \ldots, S_n$ are the slopes, around the base case, of the "ribs" of the Spider Plots (figures 2 and 3) for the respective independent variables. Obviously, (3) shows that as the slope S_i increases the sensitivity of R to the ith variable increases.

Substituting values of S_i and X_i, where i = 1, 2,...,n, we get, from figures 2 and 3 and equation (3),

$$\begin{aligned} R = 37.817 &- 765.0\, dx_1 + 1500.0\, dx_2 \\ &+ 147.1\, dx_2 + 147.1\, dx_3 + 49.2\, dx_4 \\ &+ 20.0\, dx_5 - 180.0\, dx_6 - 610.0\, dx_7 \\ &- 1008.0\, dx_8 - 22.09\, dx_9 - 2.55\, dx_{10} \\ &- 7.73\, dx_{11} - 20.23\, dx_{12} - 23.58\, dx_{13} \\ &+ 7.74\, dx_{14} + 335.14\, dx_{15} \end{aligned} \quad (4)$$

where R is the cummulative present worth of the net revenue (in million dollars) for any scenario which differs from the base by changes in the independent parameter values of dx_1, dx_2, etc. Figures 2 and 3 show that apart from those for discount rate and deliverability decline rate, the "ribs" of the spider plots are relatively straight. Therefore, for these two parameters, particularly for discount rate, the changes (dx_i) should not be too large if (4) is to be used.

REFERENCES

Budd, C. F. Jr., 1973, Steam Production at The Geysers Geothermal Field, Geothermal Energy, Edited by Paul Kruger and Carl Otte, Stanford Univesity Press, California.

Dykstra, H., Reservoir Assessment of The Geysers Field--Reservoir Engineering, California Division of Oil and Gas, Sacramento, 1981.

Fetkovich, M. J., Decline Curve Analysis using Type Curves, SPE Paper 4629, Las Vegas, NV, 1973.

TABLE 1. INPUT PARAMETERS FOR THE BASE CASE

Plant Capacity (MW) = 55
Deliverability Decline Rate = .068
Initial Average Well Deliverability (LBS/HR) = 130,000
Steam Requirement (LBS/MW/HR) = 18,000
Productive Acreage (Acres) = 400
Number of Existing Producers = 1
Standby Well Fraction = .1
Success Rate in Drilling = .66
Success Rate in Redrilling = .5
Fraction of Wells Requiring Workover Per Year = .05
Ratio of Production to Injection Wells = 10
Cost of Drilling per Producing Well = \$1.9E+06
Cost of Redrilling Per Well = \$800,000
Cost of Drilling Per Injection Well = \$1.9E+06
Drilling Facilities Cost Per Well = \$120,000
Facilities Cost for Each Redrill = \$30,000
Cost of Gathering System Per Producer = \$60,000
Workover Cost Per Well = \$600,000
Acquisition Cost = \$4E+06
Initial Exploration Cost = \$500,000
Initial Drilling Cost = \$3E+06
Initial Gathering System Cost = \$1E+07
Initial Number of Injectors Available = 0
Disposal System Cost Per Well = \$45,000
Annual Inflation Rate = .1
Operating Cost Per Producer = \$35,000
Operating Cost Per Injector = \$35,000
Steam Price Per Pound = \$.001
Annual Rate of Steam Price Escalation = .09
Plant Factor = .85
Discount Rate for Cash Flow Calculation = .12
Total Development Cost = 1
Working Interest = 1
Revenue Interest = .875

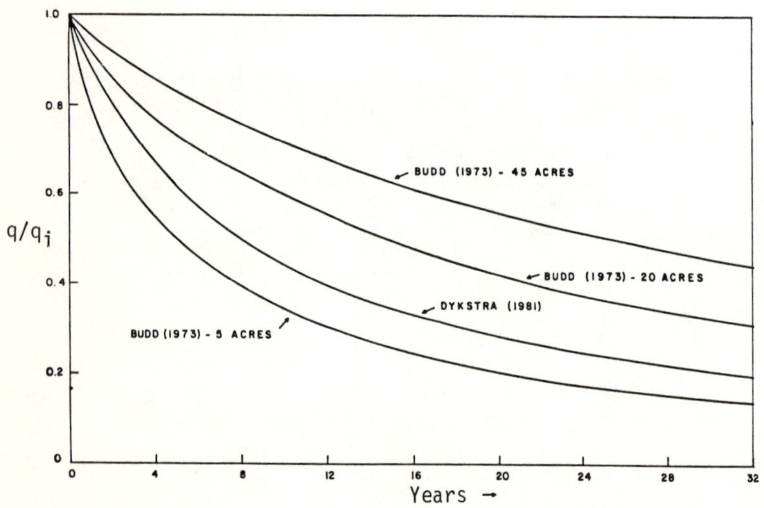

FIGURE 1. Deliverability Decline Curves for The Geysers Wells.

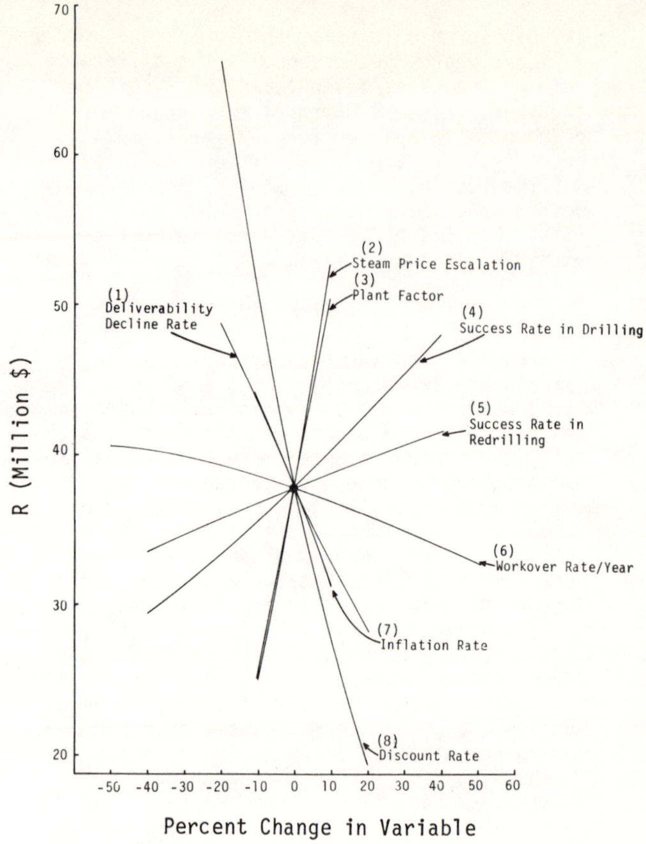

FIGURE 2. Sensitivity of R to Rate Variables

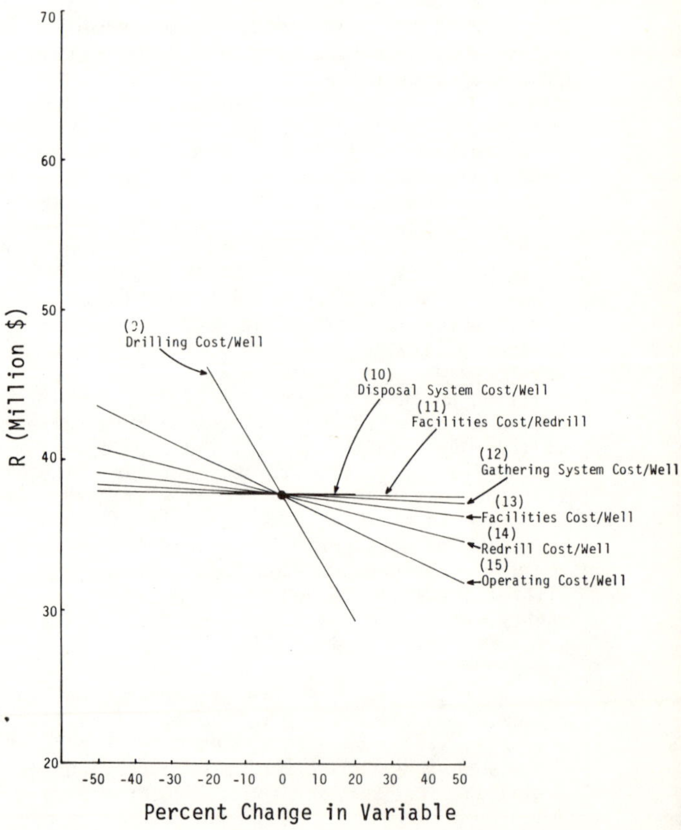

FIGURE 3. Sensitivity of R to Cost Variables.

Geothermal Resources Council, TRANSACTIONS Vol. 6, October 1982

GEOLOGY OF MACFARLANE'S SPRING THERMAL AREA, NEVADA

Bruce S. Sibbett*, Jon Zeisloft* and Roger L. Bowers**

*Earth Science Laboratory Div./University of Utah Research Institute, Salt Lake City, Utah
**Hunt Energy Corporation, Dallas, Texas

ABSTRACT

MacFarlane's Spring in northwestern Nevada is located near a recently active normal fault which has about 400 m of Quaternary offset. Tertiary basalt flows and tuffaceous sedimentary rocks near the spring have been faulted and tilted 22 to 40° to the east during the Tertiary. The 77°C water flows from a 4 m high and 180 m long travertine mound. Associated eroded travertine veins which cut Lake Lahontan deposits and wave-rounded travertine clasts indicate the hot spring has been active for thousands of years. Geothermometer calculations suggest a maximum reservoir temperature of less than 140°C.

INTRODUCTION

MacFarlane's Spring is located 94 km west of Winnemucca, Nevada, on the east edge of the Black Rock Desert (Fig. 1). The geology of the area around the hot springs and extending to the Jackson Mountains, 10 km to the east, was mapped to determine structures and lithologies which control the thermal system. Water samples were collected and analyzed to determine water chemistry and to calculate the reservoir temperature.

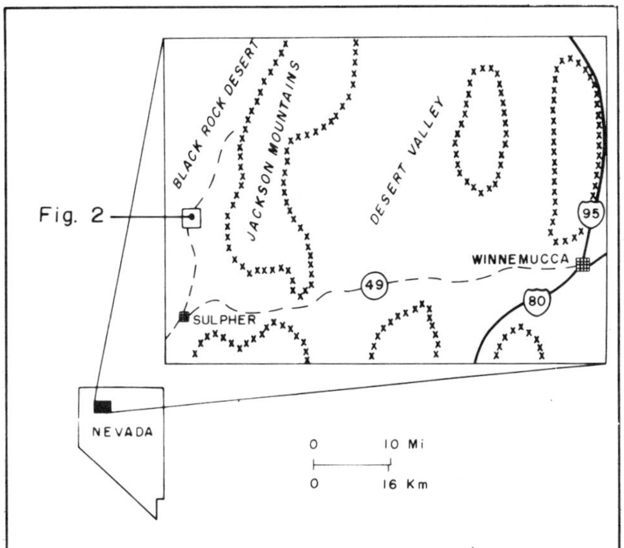

Figure 1. Location Map of MacFarlane's Spring

REGIONAL GEOLOGY

MacFarlane's Spring is located in the northwestern Basin and Range province, west of the Jackson Mountains which consist of Permian mafic volcanic and associated clastic rocks (the Happy Creek Volcanic Series), Mesozoic clastic and carbonate rocks, and Cretaceous to Tertiary intrusives (Willden, 1963). The Black Rock Desert is a large basin filled with Quaternary alluvium and Lake Lahontan sediments. About 3 km^2 of Tertiary basalt and tuffaceous sedimentary rocks are exposed 1 km southeast of the hot springs (Fig. 2). These Tertiary rocks are 5 km west of the Jackson Mountains.

LITHOLOGIES

Coarsely porphyritic basalt and andesite lava flows (Tba) in part vesicular to scoriaceous are exposed on the ridge 1 km south east of the hot spring. The flows maintain very uniform thickness of 3 to 6 meters. A stratigraphic thickness of at least 366 m is exposed on the ridge. These flows dip to the east 22-40 degrees and strike N10W to due N. If a 30 degree dip is assumed for the 550 m of volcanic rock penetrated in borehole 3452 (Table 1), this would indicate a stratigraphic thickness of 475 m. Lensoid voids in the basalt flows and flow breccias are filled with coarsely crystalline, yellow-brown calcite.

Table 1. Depth to the basalt flows (Tba) in thermal gradient holes.

Drill Hole	T.D.(m)	Depth to the Top of Basalt
3408	152	(not reached)
3409	610	454
3414	610	(not reached)
3415	122	(not reached)
3446	150	15
3447	150	85
3452	585	38

Source: Hunt Energy Corporation

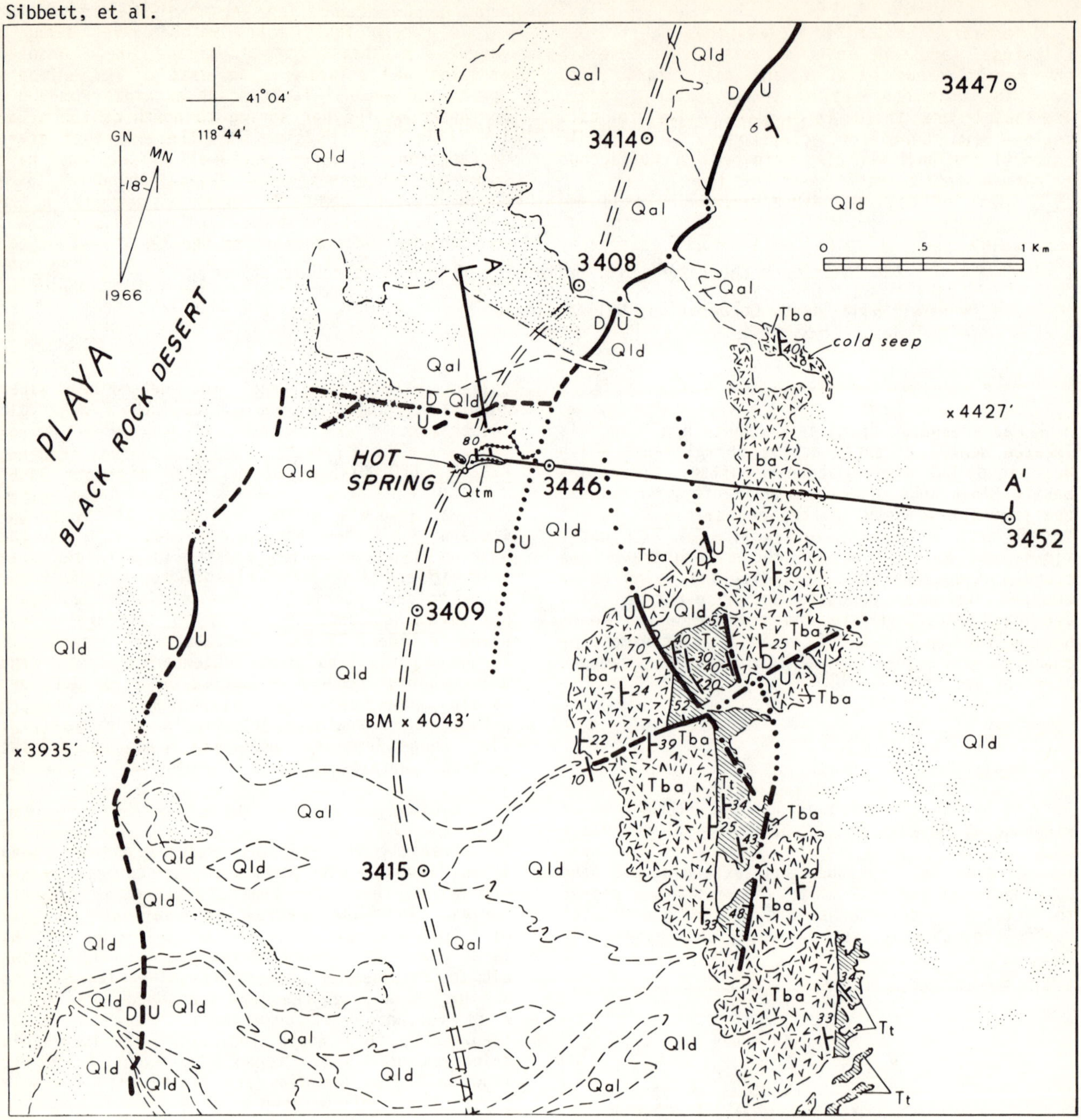

EXPLANATION

Qal	Alluvial fans & channels
Qtm	Travertine mound
Qld	Lake Lahontan deposits
Tt	Tuffaceous sedimentary rocks
Tba	Basalt and andesite flows

MAP SYMBOLS

- Contacts, dashed where approximate
- Faults, dashed—inferred, dotted—covered
- Photo linears of faults
- Travertine veins
- Strike and dip of beds (15)
- Sand dune fields
- 3409 Thermal gradient hole

FIGURE 2. GEOLOGIC MAP OF MACFARLANE'S SPRING AREA

Tertiary bentonitic claystone, tuffaceous siltstone, sandstone and conglomerate (Tt) overlie the basalt flows east of the hot spring (Fig. 2). The unit consists of three sediment types. The basal one third is a pale olive to pale yellow-brown bentonitic claystone. The middle third of the unit contains thin beds of tuffaceous siltstone and claystone with a few 1/2 to 2 m thick beds of coquina near the top. These beds seem to represent a shallow lacustrine environment. The upper third of the exposure consists of interbedded siltstone and sandstone exhibiting festoon crossbedding and conglomerate lenses. This portion of the deposit may have formed in a shallow near-shore lacustrine to fluvial environment.

This sequence of sedimentary rocks has been dated as late-middle Miocene to early Pliocene by diatoms collected from the south end of the Jackson Mountains (Willden, 1963). A coquina bed exposed in the study area contains mollusk shells similar to Goniobasis taylori (Dall, 1924), Pisidium leslieae and Goniobesis leslieae (Firby, 1967), all of which are late Miocene to early Pliocene. On the basis of comparable age and lithology the unit is correlated with the Truckee Formation (Morrison, 1964). The Tertiary rocks are overlain with angular unconformity by a Quaternary Lake Lahontan strand line and bar deposits of sand and gravel, and deep water sand, silt and clay.

STRUCTURE

Normal faults trending N10E to N20W form a small graben in the ridge east of the hot spring (Fig. 1). Truckee Formation beds have been warped into an asymmetrical syncline within the graben which has a stratigraphic throw of at least 305 m on the east side. The continuation of the east side fault to the south trends N10E and has offset the Tba-Tt contact 335 m down to the west. Tertiary tilting of the range to the east 22 to 40 degrees has overturned the west boundary fault of the graben (Fig. 3).

A recent fault scarp which curves along a general northeast trend cuts Lake Lahontan deposits and alluvium 1 km west of MacFarlane's Spring and another fault scarp extends from 500 m northeast of the hot spring to north of the study area (Fig. 2). Judging from the extent of scarp degradation, 1 m of vertical offset may have occurred as recently as several hundred to a thousand years ago. A dip of 56 degrees to the west with striations indicating dip slip movement was measured on a segment of the fault west of the hot spring. Greater offset occurring during late Pleistocene is indicated by over 6 m of offset of an algal tufa horizon in the lake beds 2.4 km northeast of the hot spring.

MacFarlane's Spring is located near a bend in the recently active fault where the fault trace turns from a N34°E trend to a westerly trend (Fig. 2). Most of the travertine veins parallel the westerly trending fault segment. Thermal gradient hole lithology logs (holes 3409, 3408 and 3446, Table 1) suggest a covered extension of the northeasterly trending portion of the fault continues to the south of the hot spring, offsetting the top of the basalt flows about 400 m down to the west (Fig. 3).

HOT SPRING DEPOSITS

MacFarlane's Spring thermal system has deposited a series of east-west trending travertine veins, which are dipping 80 degrees north to vertical. The present spring produces water of 77°C (170°F) at an estimated rate of several gallons per minute and issues from a central fissure atop the west end of a 4 m high, 180 m long travertine mound. The travertine veins cut lake sediments. The presence of wave-rounded travertine clasts indicate the hot springs were active during Lake Lahontan time. The orifice (1/3 m in diamter) of the hot spring is presently choked with loose calcite pisolites which appear to be forming as a result of agitation of the water by the rapid effervescence of the spring water. The gases escaping from the water have only a slight sulfurous odor and are thought to be

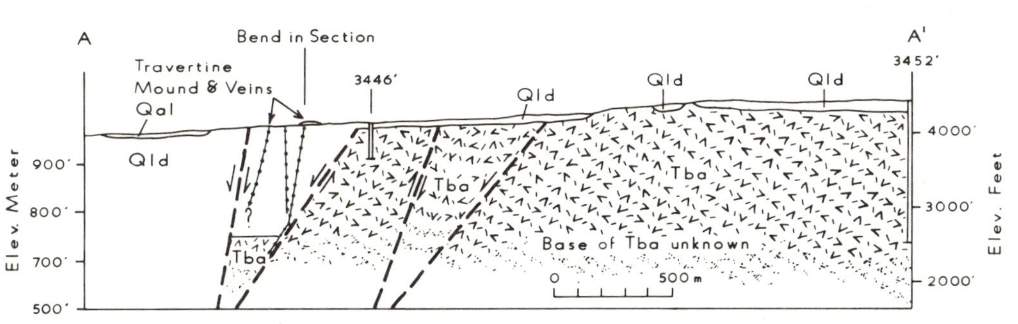

Figure 3. Geologic cross section. Depth of the Quaternary sediments west of borehole 3446 based on data from holes 3409, 3408 & 3414 (Table 1). For explanation of symbols see Figure 2.

largely carbon dioxide. On older, smaller veins, erosion has revealed several vertical funnel-shaped or cylindrical bodies of cemented pisolites, very similar in overall geometry to the present spring orifice. These pisolite occurrences provide evidence of genetic relationship between the older inactive travertine bodies and the currently active one. That, plus the presence of wave-rounded travertine clasts, indicates that the MacFarlane hot spring system has been active for a long period of time, well into the Lake Lahontan period (15-60 thousand years ago according to Morrison, 1964).

GEOCHEMISTRY

A water sample collected from the hot springs was analyzed to enable calculation of a reservoir temperature. The thermal waters are sodium-bicarbonate-chloride type and contain 3650 TDS. Results of Na-K-Ca and quartz mixing geothermometers are in close agreement yielding a maximum temperature of about 140°C. The Na/K, quartz and chalcedony geothermometers suggest a reservoir temperature in the range of 80 to 120°C.

CONCLUSION

The location of MacFarlane's Spring appears to be controlled by a flexure or split in a northeast-trending normal fault that has at least 400 m of Quaternary offset and a meter of offset in the last thousand years. The thermal anomaly is centered near borehole 3408 (Swanberg, this issue) which is near the projected intersection of northeast-trending Quaternary fault and the northwest-trending Tertiary graben.

The thermal system was active during Lake Lahontan time which was about 15-60 thousand years ago (Morrison, 1964). Geothermometer calculations indicate subsurface temperatures of 80 to 140°C. The structure and temperature data suggest deep circulation along Quaternary faults for the origin and control of the thermal waters. The Jackson Mountains to the east provide the most likely recharge area. Subsurface runoff from these mountains probably flows to great depth along the range front fault then rises as thermal waters to the west at MacFarlane's Spring.

ACKNOWLEDGEMENTS

The authors would like to thank Hunt Energy Corporation for the funding of this study and for permission to publish the results. We also thank David R. Cole for interpreting the water chemistry and calculating the geothermometers, and we thank John Atwood for identification of the fossils.

REFERENCES

Bowers, R. L., 1982, Exploration history of MacFarlane's Spring area: Geoth. Res. Council Transactions, v. 6 (in press).

Dall, W. H., 1924, Discovery of a Balkan fresh-water fauna in the Idaho Formation of Snake River Valley, Idaho: U. S. Geol. Surv. Prof. Paper 132, p. 109-114.

Firby, J. R., 1967, Pliocene non-marine mollusks from Contra Costa County, California: Proceedings of the California Academy of Sciences, v. 34, no. 14, p. 511-524.

Morrison, R. B., 1964, Lake Lahontan: Geology of Southern Carson Desert, Nev.: U.S. Geol. Surv., Prof. Paper. 401.

Swanberg, C. A., 1982, Downward continuation analysis of MacFarlane's Spring, Nevada: Geoth. Res. Council Transactions, v. 6 (in press).

Willden, R., 1963, General geology of the Jackson Mountains, Humboldt County, Nevada: U.S. Geol. Surv., Bull. 1141-D, 65 p.

Geothermal Resources Council, TRANSACTIONS Vol. 6, October 1982

MINERAL ZONATION AND STABLE ISOTOPE GEOCHEMISTRY OF A PRODUCTION WELL IN THE GEYSERS GEOTHERMAL FIELD, CALIFORNIA

Jeffrey N. Sternfeld* and Wilfred A. Elders

Institute of Geophysics and Planetary Physics
University of California, Riverside, CA 92521
*currently at Occidental Research Corp.
PO Box 19601 Irvine, CA 92713

ABSTRACT

Petrologic and stable isotope geochemical studies were conducted on well cuttings from a Sulphur Banks tract steam well in The Geysers. The observed hydrothermal mineral zonation is indicative of an extremely hot (>300°C) water-dominated geothermal system characterized by high fluid flow associated with fault contacts. Strata within the steam reservoir are not distinguished by diagnostic steam-derived secondary alteration products. However, a late stage generation of calcite was interpreted to have been deposited during the transition of the reservoir fluid to the vapor phase. The distribution of this material is restricted to a shallow zone (<600m) corresponding to the shallow steam anomaly reported in the Sulphur Banks area.

INTRODUCTION

The Geysers Steam Field is a vapor dominated field with reservoir temperatures of 240°C, located in the Mayacmas Mountains of Northern California approximately 110 km north-northwest of San Francisco. It has the largest installed electrical capacity of any geothermal field in the world, yet it is one of the least studied from a petrologic standpoint. Several authors have described surficial and near surface manifestations of hydrothermal activity. The only descriptions of deep subsurface samples, however, are those published by Moore (in Lockner et al. 1982) on isolated pieces of core taken from several wells. This paper is a synopsis of a comprehensive study of petrology and light stable isotope geochemistry of two wells (Sternfeld, 1981) that was conducted with the cooperation of Union Geothermal Co. The hydrothermal mineralization and the isotope behavior for ^{13}C and ^{18}O of a producing well are presented here.

GEOLOGIC SETTING

Geysers 1 (this designation is used at the request of Union Geothermal) is located in the Sulphur Banks tract of the Big Geysers production area (see Figure 1). The borehole was mud drilled to a depth of 900m and then air drilled to completion. Eight steam entries were encountered between 900 and 1390m (see Figure 2), with the major steam production originating below 1300m.

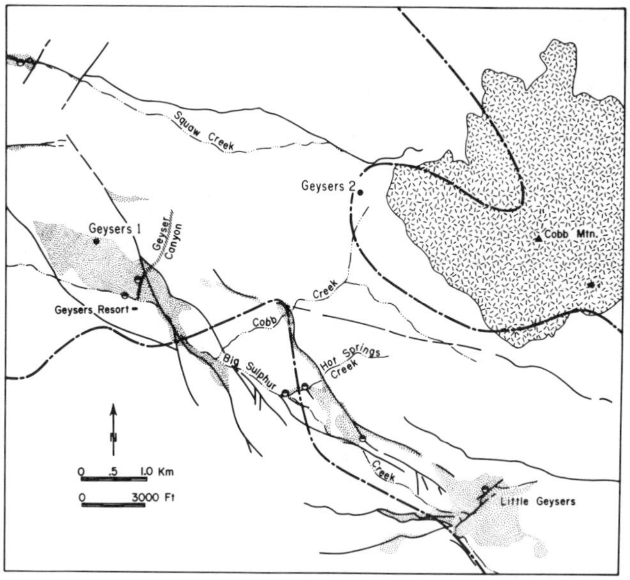

Figure 1. Map of Geysers Study Area Showing Well Site Locations
Stippled area is surface hydrothermal alteration; hatchered area represents Clear Lake Volcanics; dashed line marks extent of steam production (Ehni, 1981). Modified from McLaughlin (1978).

Geysers 1 penetrated a sequence of interbedded Franciscan graywackes and argillites that is capped by a block of cherty basaltic greenstone. The strata can be divided into five lithologic units based on the dual criteria of rock type and degree of apparent Franciscan metamorphism as suggested by Blake, et al. (1967) (see Figure 2). These unit contacts are probably thrust fault boundaries. Unit 1 consists of pillow basalts, chert and minor interbeds of volcaniclastic meta-sediments. Units 2 through 5 are composed of weakly metamorphosed lithic and quartz wackes with associated argillites that range from textural type 1.0 to 1.5.

HYDROTHERMAL MINERALIZATION

While the recognition of authigenic minerals was facilitated by the predominance of fracture-fill precipitation, the differentiation of hydrothermal phases of recent geothermal origin from Franciscan mineralization was more complicated. Fortunately, the graywacke units encountered in Geysers 1 were typical of low grade Franciscan metamorphism so the presence of calc-silicates, adularia and sulfides was not expected in these meta-sediments (Blake et al., 1967; Ernst, 1971). Metamorphic calcite and quartz veining could be distinguished from their geothermal counterparts by $\delta^{18}O$ values (Lambert, 1977) and also by the fortuitous presence of sulfide inclusions in the geothermal veins.

The distribution of geothermal minerals is distinctly zoned in Geysers 1 (see Figure 2). Mineralization progressively changes to higher temperature assemblages with increasing depth. This zonation is similar to the zonation observed in Broadlands, the Imperial Valley and Cerro Prieto (Browne, 1978) It represents the product of a water dominated system which must have predated the present vapor dominated system. Below 1000m, these mineral assemblages are characteristic of deposition at temperatures greater than 300°C. Fluid inclusion homogenization temperatures of 310-320°C measured from quartz vein material support this interpretation.

The geothermal minerals can be grouped into 3 assemblages (see Table 1) reflecting changing thermal and chemical regimes. The first assemblage is restricted to the upper 120m of the well bore and is indicative of near surface boiling related to self-sealing and surficial fumarolic activity. The second assemblage represents the initial hot water history of The Geysers. Its sulfide mineralization is similar to that in Broadlands - locally abundant disseminated idioblastic pyrite and sphalerite (characterized by chalcopyrite blebs) associated with vug quartz crystals (Browne, 1971). The deposition of the third assemblage, characterized by the appearance of pyrrhotite, is superimposed on the earlier formed minerals and represents chemical and temperature changes within the reservoir fluid during the transition to the vapor dominated stage. These late stage fluids are in disequilibrium with the pyrite as pyrite grains in close proximity to pyrrhotite-bearing veins exhibit evidence of dissolution and, rarely, replacement by pyrrhotite.

Table 1. Geothermal Mineral Assemblages.

I	Quartz + Calcite + Adularia + Pyrite + Marcasite
II	Quartz + Calcite + Adularia + Chlorite + Sphene + Epidote + Garnet + Diopside + Amphibole + Pyrite + Sphalerite + Chalcopyrite + Galena
III	Quartz + Calcite + Adularia + Pyrrhotite + Chalcopyrite + Sphalerite + Galena + Tetrahedrite-Tennantite + Gersdorffite

Major fracture zones occur intermittently throughout the borehole and are identified by significant concentrations of vein material. Most of these mineralized intervals are associated with the aforementioned lithologic unit boundaries. These are at 60 to 100m, 200 to 250m, between 550 and 650m, and between 1220 and 1350m. Two mineralized intervals, at 840m and 1585m, are not clearly of tectonic origin.

Figure 2. Mineral Zonation in Geysers 1.
Group I Sulfides: pyrite, marcasite, sphalerite, chalcopyrite, galena
Group II Sulfides: pyrrhotite, chalcopyrite, sphalerite, galena, marcasite, tetrahedrite-tennantite, gersdorffite

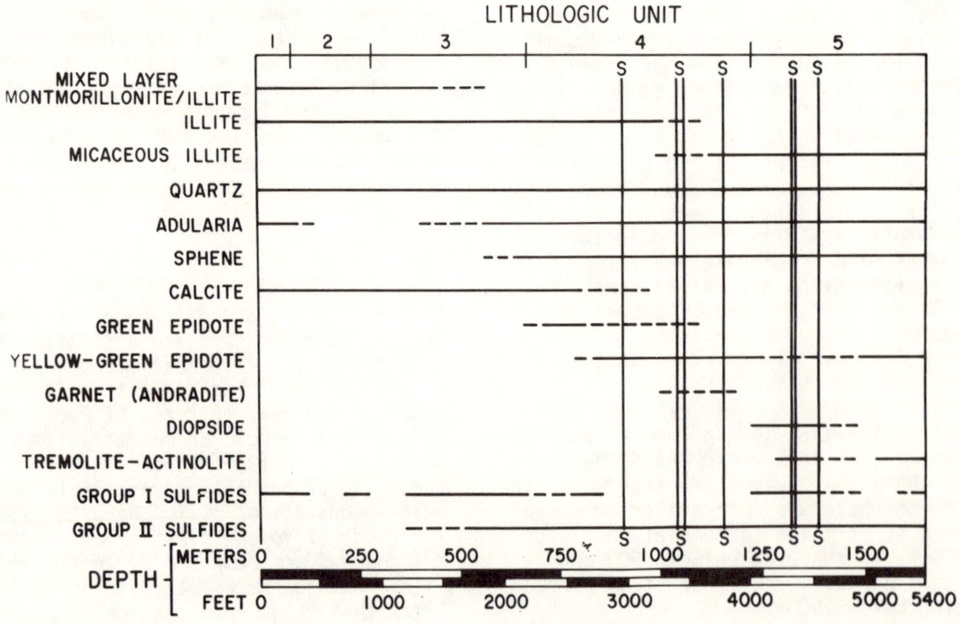

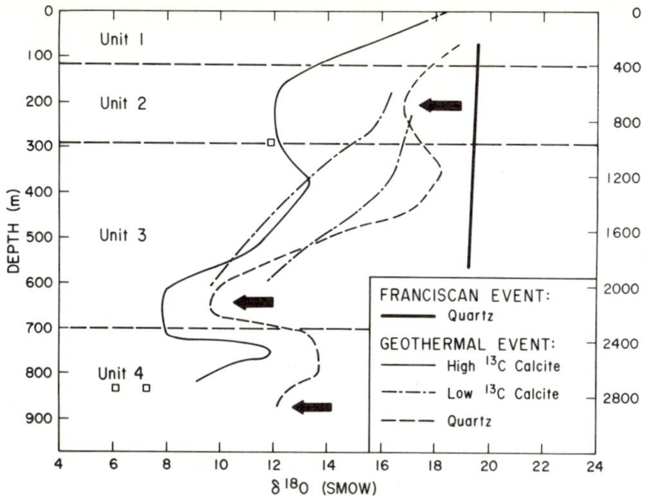

Figure 3. $\delta^{18}O$ Trends for Calcite and Quartz Vein Material from Geysers 1. Arrows indicate flow zones as indicated by substantial hydrothermal mineralization.

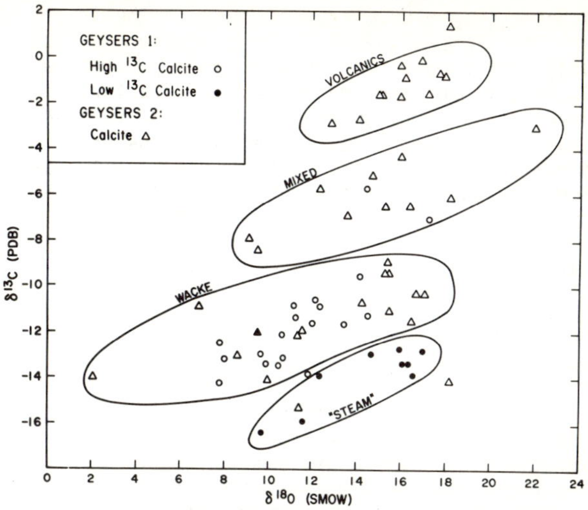

Figure 4. $\delta^{13}C$ Versus $\delta^{18}O$ for Calcite Vein Material from Geysers 1 and Geysers 2. Volcanics: Greenstone/Chert host rock; Wacke: Graywacke/Argillite host rock; Mixed: Interbedded Greenstone and Graywacke host rock; Steam: Late forming ^{13}C-depleted calcite

There is little correlation between the steam entries and the major mineralized zones. The primary steam production corresponds to the mineralized zone associated with the transition of Unit 5, but the minor entries occur within an interval of Unit 4 that is characterized by extremely sparse evidence for hydrothermal alteration. In addition, secondary mineral phases that are characteristic of steam alteration or steam condensate (i.e., cinnabar, sulfur, arsenic or calcite) were not observed in association with the steam entries.

STABLE ISOTOPE GEOCHEMISTRY

The $\delta^{18}O$ values for quartz and calcite samples from Geysers 1 plot along 5 curves (See Figure 3). The trends attributed to recent geothermal activity were differentiated from Franciscan (metamorphic) veining based on the pioneering study by Lambert (1977). Geothermal calcites can be further subdivided into two groups on the basis of their $\delta^{13}C$ values (see figure 4) and $\delta^{18}O$ behavior. There are no $\delta^{18}O$ values in calcite which are Franciscan in origin.

The geothermal quartz and high ^{13}C calcite trends are similar, suggesting co-genetic deposition. The two vein minerals, however, are not in continuous thermal equilibrium as determined by comparisons of Quartz-Calcite temperature calculations and fluid inclusion data. Both curves are characterized by sharp isotope shifts and reversals in association with the major fracture (mineralized) zones (see Figure 3). This pattern is indicative of localized high fluid flow and has been previously described from high flow zones in the Imperial Valley (Kendall, 1976).

An important relationship can be drawn from a comparison of the high ^{13}C calcite data and the calcite data of Geysers 2 (where no low ^{13}C calcites were identified). A graph of ^{13}C versus $\delta^{18}O$ values for these calcites can be divided into three fractionation curves (see Figure 4) which reflect the rock type in which they were found.

Calcites from greenstone host rocks ("volcanics") display δ^{13} values which are normal for marine carbonates (Fritz, 1976); calcites from metasediments ("wackes") have $\delta^{13}C$ values that are derived from marine organic material (Fritz, 1976) and calcite material recovered from greenstone-graywacke lithologic unit boundaries or greenstone rocks interbedded with sediments ("mixed") have intermediate $\delta^{13}C$ values. This heterogeneity is inherent to the deposition of the Franciscan Event carbonates (Lambert, 1977; Sternfeld, 1981) but is also preserved by the geothermal veining. Thus, it can be concluded that the carbon in the high ^{13}C calcite group was derived from the dissolution and remobilization of Franciscan calcite (resulting in the absence of Franciscan calcite in Geysers 1) and also that there was no uniform mixing of the reservoir fluids during the recent geothermal activity.

The $\delta^{13}C$ values of the low ^{13}C calcite group are more depleted than the main body of calcite data with similar $\delta^{18}O$'s and therefore plot along a unique fractionation line (see Figure 4). Their δO^{18} values are superimposed over the high ^{13}C calcite trend and plot along a continuous smooth curve that decreases with depth yet does not show the effects of high fluid flow zones (see Figure 3). This behavior indicates that the low ^{13}C calcites were deposited in an interconnected network of fractures in which temperature, not fluid flow, was the dominant variable (i.e., there was a low water to rock ratio) and that this carbonate generation superceded the high ^{13}C calcite group.

Another important clue pertaining to the origin of the low ^{13}C carbonates is their apparent relationship to the present steam reservoir. The $\delta^{13}C$ values of approximately -16% (PDB) derived from material originating at about 600m are similar to the $\delta^{13}C$ values of steam from producing wells in the Geysers (Huebner, 1981). Augmenting this data, homogenization temperatures for calcite fluid inclusions plot on a curve that is equivalent to the boiling point for pure water under vapor-static conditions in the depth interval corresponding to the low ^{13}C occurrences. These two lines of evidence indicate the presence of either a descending boiling water table under vapor-static pressures and/or the condensation of a vapor phase.

Both the vapor-static fluid inclusion temperatures and the steam-equilibrated $\delta^{13}C$ isotopes are restricted to a zone between 200 and 600m in Geysers 1. It is significant to note, however, that the Sulphur Banks area is underlain by a two layer steam reservoir. Steam is encountered in both a shallow steam anomaly and in a deeper regional fracture system (Lipman, et al., 1978). The "steam" zone above 600m in Geysers 1 coincides with the reported shallow steam anomaly.

DISCUSSION

The Truesdell and White (1973) model for The Geysers postulates a slowly descending boiling water table in equilibrium with the vapor phase. Evidence for a continuously descending boiling front is present only in the upper 600m of the well. Either Geysers 1 penetrated strata at depth unfavorable for the preservation of these features or, more probably, the shallow steam anomaly in The Geysers-Sulphur Bank area was the only steam cap formed by natural depletion. Consequently, a different mechanism must be called upon to produce steam in the deep fracture system. One possibility is vaporization induced by commercial exploitation (Weres, et al., 1977), another is decompression during major seismic events (Sternfeld, 1981).

CONCLUSION

There is ample evidence for the existence of a water-dominated system that preceded the present steam reservoir in The Geysers. Fluid temperatures exceeded 300°C and fluid flow preferentially infiltrated major fracture zones related primarily to thrust fault boundaries. The correlation between the mineralized high flow zones and steam entries is poor and there are no observable alteration products that indicate the presence of steam at depth. Above 600m in Geysers 1, an interval that coincides with the shallow steam anomaly observed in the Sulphur Banks area, late stage calcite veining displays characteristics that can be related to a descending boiling water table and/or to steam condensation.

ACKNOWLEDGEMENTS

Support of Union Geothermal and DOE Contract #DE-AT03-80SF11458 is gratefully acknowledged. Thanks to Linda Jankov for drafting the figures and Mary Keskinen for reading the manuscript.

REFERENCES

Blake, M., Irwin, W. and Coleman, R., 1967, Upside-down metamorphic zonation, blueschist facies, along a regional thrust in California and Oregon; U.S.G.S. Prof. Paper 575-C, pp. C1-C9.

Browne, P. R. L., 1971, Mineralization in the Broadlands Geothermal Field, Taupo Volcanic Zone, New Zealand; Soc. Mining Geol.; Japan, Spec. Issue 2, pp. 64-75.

Browne, P. R. L., 1978, Hydrothermal alteration in active geothermal Fields, Ann. Rev. Earth Planet. Sci, 6, pp. 229-50.

Ehni, W. J., 1981, Summary of 1980 geothermal drilling; western United States; Geothermal Energy 9 no. 9, pp. 4-15.

Ernst, W. G., 1971, Petrologic reconnaissance of Franciscan metagraywackes from the Diablo Range, Central California Coast Ranges; Jour Petrology, 12, pp. 413-437.

Fritz, P., 1976, Oxygen and carbon isotopes in ore deposits in sedimentary rocks; from Handbook of Strata-bound and Stratiform Ore Deposits, K. H. Wolf (ed); Elsevier-Scientific Publishing Co., Amsterdam, pp. 191-217.

Heubner, M., 1981, Carbon-13 isotope values for CO_2 and dissolved carbon species from selected springs and wells in the Clear Lake-Geysers region; from U.S.G.S. Prof. Paper 1141, pp. 211-213.

Kendall, C., 1976, Petrology and stable isotope geochemistry of three wells in the Buttes area of the Salton Sea geothermal field, Imperial Valley, California, USA. U.C. Riverside/ I.G.P.P. rept. 76/17.

Lambert, S. J., 1976, Stable isotope studies of some active hydrothermal systems; unpublished dissertation, Calif. Inst. of Technology.

Lipman, S. C., Strobel, C. J. and Gulati, M. S., 1978, Reservoir performance of The Geysers field; Geothermics, 7, pp. 209-219.

Lockner, D. A., Summers, R., Moore, D. and Byerlee, J. D., 1982, Laboratory measurements of reservoir rocks from the Geysers Geothermal Field, California; Int. J. of Rock Mechanics, v. 19, pp 65-80.

McLaughlin, R. J., 1978, Preliminary geologic map and structural sections of the central Mayacmas Mountains and The Geysers steam field, Sonoma, Lake and Mendocino Counties, California; U.S.G.S. Open File Rept., 78-389.

Sternfeld, J. N., 1981, The hydrothermal petrology and stable isotope geochemistry of two wells in The Geysers Geothermal Field, Sonoma County, California; U.C. Riverside/I.G.P.P. rept. 81/7.

Truesdell, A. H. and White, D. E., 1973, Production of superheated steam from vapor-dominated geothermal reservoirs; Geothermics, 2 no. 3-4, pp. 154-173.

Weres, O., Tsao, K. and Wood, B., 1977, Resource, technology and environment at The Geysers; Lawrence Berkeley Laboratory rept. LBL-5231.

Geothermal Resources Council, TRANSACTIONS Vol. 6, October 1982

GEOLOGY, RESISTIVITY, AND HYDROGEOCHEMISTRY OF THE OJO CALIENTE HOT SPRINGS AREA, NORTHERN NEW MEXICO

John Stix, Chris Pearson, Francois Vuataz, Fraser Goff, Jennifer East*, and Bernhard Hoffers

Earth and Space Sciences Division, Los Alamos National Laboratory
Los Alamos, New Mexico 87545
*Geology Department, University of Alaska, Fairbanks, Alaska 99701

ABSTRACT

Geothermal fluids of the Ojo Caliente area discharge from a northeast trending normal fault that juxtaposes Precambrian metarhyolite and Tertiary sediments. An electrical resistivity survey shows that the fluids emerge from the fault and flow as a plume of thermal water into cold aquifers east of the fault. Geochemistry of fluids indicates a maximum reservoir temperature at depth of 80°C with no suggestion of high temperature isotopic exchange between water and reservoir rocks. From this data, we believe the Ojo Caliente system is suitable only for small-scale direct use geothermal applications.

INTRODUCTION

The Ojo Caliente thermal area is located in the northwest part of the Española basin of north-central New Mexico. The basin is part of the Rio Grande rift, a major continental rift that stretches from southern Colorado to northern Mexico. The rift is characterized by active extensional tectonics, Tertiary to Quaternary volcanism, and high heat flow. The Ojo Caliente area lies on the eastern side of an exposed Precambrian horst and is in a region of high heat flow (>2.5 HFU; Reiter et al., 1975). The object of this investigation is to determine the relative size and reservoir temperature of the geothermal system at Ojo Caliente.

GEOLOGY AND STRUCTURE

Metarhyolite, amphibolite, schist, and subordinate gneiss of Precambrian age outcrop west of the village of Ojo Caliente. Paleozoic and Mesozoic sedimentary rocks do not outcrop, but unseen remnants may exist in the subsurface flanking the Precambrian horst. Tertiary basin-fill sediments are deposited nonconformably over the Precambrian basement. Volcanic pebble conglomerates of the Cordito Member of the Los Piños Formation, transported from the San Juan, Questa, and Taos volcanic centers (north and east of Ojo Caliente), underlie and interfinger with nonvolcanic sandstones of the Chama-El Rito Member, Tesuque Formation, Santa Fe Group. Westerly-derived, quartzose, aeolian sands of the Ojo Caliente Sandstone, Tesuque Formation, also of the Santa Fe Group, lie above the Los Piños and Chama-El Rito. Quaternary pediment and terrace gravels, loess, (spring) travertine, and alluvium unconformably overlie the Tertiary section. The hot springs emanate from alluvium and a single outcrop of sheared Precambrian metarhyolite overlain by Los Piños volcanic pebble conglomerates (Fig. 1).

We found little evidence of exposed Recent faulting in the Tertiary and Quaternary deposits to the west of Ojo Caliente. Faulting in the Los Pinos is infrequent; where it occurs, throw is not more than several meters. Several small, elevated deposits of (spring) travertine indicate more extensive movement of hot fluids, but no direct geologic evidence proves Quaternary faulting in this area. On the other hand, a major Pliocene fault, trending north-northeast and downthrown to the east, apparently runs beneath the alluvium in the Rio Ojo Caliente valley. This fault probably offsets Precambrian metarhyolite in the subsurface. Four lines of evidence point to the existence of such a fault. First, a relatively linear ridge of exposed metarhyolite that lies west of Ojo Caliente trends north-northeast and may be upfaulted against buried Los Piños hidden beneath the river alluvium. Second, the Ojo Caliente fault zone in the Tertiary section 5 km southwest of Ojo Caliente (May, 1979) strikes north-northeast and is aligned with the inferred fault in the Rio Ojo Caliente valley. Third, a fault 8 km northeast of Ojo Caliente, that down-faults the Tertiary to the east against Precambrian quartzite to the west, also follows this alignment. These structures are major faults with displacements of perhaps several hundred meters. Fourth, the location of the hot springs appears to be controlled by the intersection of pre-Tertiary northeast-trending Precambrian shear zones with the Pliocene valley fault (Fig. 1). From this evidence, we postulate that the Pliocene valley fault is a zone of normal faults downthrown to the east with cumulative offset of at least 100 m. Possibly, this zone is a reactivated Precambrian structure.

RESISTIVITY SOUNDINGS

In order to map the hot waters underlying the Ojo Caliente area, we conducted eight Schlumberger D.C. resistivity soundings in the Rio Ojo Caliente

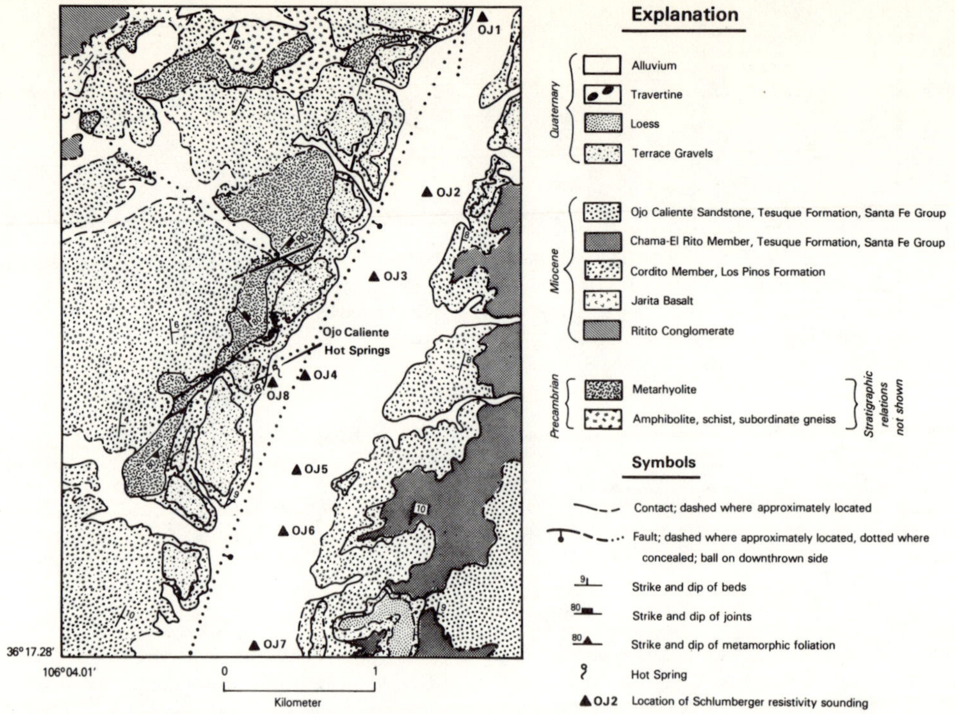

Figure 1. Simplified geologic map of the Ojo Caliente area, New Mexico showing locations of resistivity stations.

valley within a few kilometers of the hot springs. Figure 1 shows the location of the survey points. When interpreted, these soundings allow us to infer the earth's resistivity as a function of depth underneath the survey point. These surveys are effective in locating hot water because sediments saturated with geothermal fluids are much more conductive than sediments saturated with cold, fresh water. Methods of data acquisition and interpretation have been described in a previous paper of similar scope (Pearson and Goff, 1981).

We found an electrically conductive zone underneath the Rio Ojo Caliente valley with resistivities of less than 10 Ωm. This zone is centered at line OJ 4 near the hot springs where it comes within 10 m of the surface (Fig. 2). The depth to this conductive zone increases steadily to the north and south, disappearing 1 km to the north of the hot springs and 2 km to the south. The source of the geothermal waters apparently underlies the Ojo Caliente resort adjacent to the hot springs, since the low (>10 Ω-m) resistivities are closest to the surface here. Thus, we interpret the geothermal fluids to rise as a plume from a short segment of the Pliocene valley fault and then disperse eastward into the valley alluvium.

A zone of moderately conductive sediments (13-20 Ω-m) is found at intermediate depths (i.e., 10-40 m) south of the hot springs. This represents an outflow zone where geothermal waters in the valley mix with fresh cool water. This zone of mixing also may extend slightly to the north of Ojo Caliente. The resistive sediments that overlie the conducting zones represent sediments saturated with cool fresh ground water fed by the Rio Ojo Caliente. Line OJ 8 (Fig. 2), run at the hot springs several hundred meters west of OJ 4, also detected the geothermal plume. However, the slightly higher resistivities detected by OJ 8 may indicate calcite deposition in the underlying sediments. This reduces the porosity and thus increases the formation resistivity.

GEOCHEMISTRY OF THE THERMAL WATERS

On the site of the resort itself, five hot springs have been used for a very long time. Cold but mineralized water also flows from a seep located a few tens of meters west of the springs. Four shallow wells have been drilled during the past 40 years in the alluvium for water supply (Summers, 1976). All of them display evidence of mixture between thermal waters and cold ground water. One of these wells (the closest to the springs), which is not used and has no natural discharge, has a recorded bottom-hole temperature of 55°C at a depth of 26 m (J. Hunter, Los Alamos National Laboratory, unpublished data, 1982). The hot springs have a surface temperature ranging from 32° to 45°C and their total discharge varies from 200 to 1300 l/min, according to historical data (Summers, 1976).

All the emergences have the same type of chemistry and can be classified as sodium-bicarbonate waters with substantial amounts of silica, fluoride, lithium and boron (Table 1).

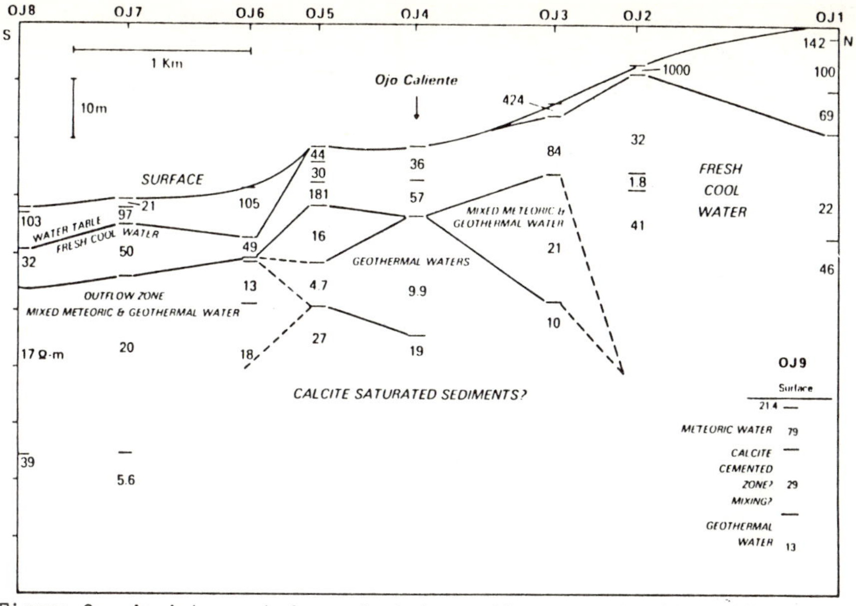

Figure 2. An interpreted geoelectric section near the Ojo Caliente area.

Table 1. Chemical analyses (mg/l), field pH, surface temperature (^{o}C) and chemical geothermometry (^{o}C) of selected samples from Ojo Caliente, New Mexico.

Sample[1]	Li	Na	K	Mg	Ca	HCO_3	SO_4	Cl	F	B	SiO_2
Iron Spring	2.2	953	29.0	6.60	21.0	2090	171	229	14.2	1.53	58.9
Lithia Spring	2.7	867	28.5	8.70	23.6	2070	173	227	13.5	1.47	58.4
Hot Well	5.5	927	27.8	2.90	7.3	1870	195	242	20.8	1.54	43.4
Well #1	-	496	16	16	40	907	155	149	7.5	-	-

Sample[1]	TDS[2]	pH	Temp.	T_{qtz}	T_{chal}	$T_{Na/Li}$	$T_{Na/K}$	T_{NaKCa}	$T_{NaKCaMg}$
Iron Spring	3576	6.40	43	110	80	127	133	149	73
Lithia Spring	3474	6.55	38	109	80	149	137	150	61
Hot Well	3343	6.85	54	95	65	206	132	156	92
Well #1	1787	7.80	32	-	-	-	136	137	42

[1]The first three samples were collected in December 1979, while Well #1 was obtained in August 1965 (Summers, 1976).

[2]TDS: total dissolved solids equals the sum of the constituents listed above.

Extensive travertine deposits are known in the area and the thermal water may not only dissolve silicates of Precambrian crystalline rocks, but also limestone from buried Paleozoic or Mesozoic sedimentary rocks during its ascent. Except discharge, the other physical and chemical parameters do not seem to go through severe seasonal variations and if one can deduce whether mixing and/or dilution between thermal and nonthermal waters occur, it should be quite constant. The first conclusion of the isotopic study shows that the warm waters of Ojo Caliente do not fit exactly on the general meteoric water line but on a kind of evaporation line with a slight enrichment of both oxygen-18 and deuterium (Fig. 3).

ESTIMATION OF THE RESERVOIR TEMPERATURE

The magnesium concentration of thermal waters often correlates in inverse ratio to the concentration of a non-reactive anion, like chloride, as shown by several other hot water systems (Vuataz, in prep.). A mixing model has been established for the springs and wells, using the data of the present study and all the historical data (Summers, 1976; Trainer and Lyford, 1979). The best-fit line of the magnesium versus chloride (Fig. 4A) may be extrapolated to a very small magnesium concentration, which could represent the unmixed thermal water at depth. The line intersects the chloride axis at a maximum value of 290 mg/l. A relation between temperature and chloride shows a good positive trend for the wells, while the springs display a conductive cooling of 8° to 23°C (Fig. 4B). Reporting the extrapolated chloride concentration on this diagram, a temperature of 68°C is obtained for the hot end-member. This temperature, around 13°C more than the maximum recorded should correspond to the thermal water at relatively shallow depth but before any mixing with the cold aquifer.

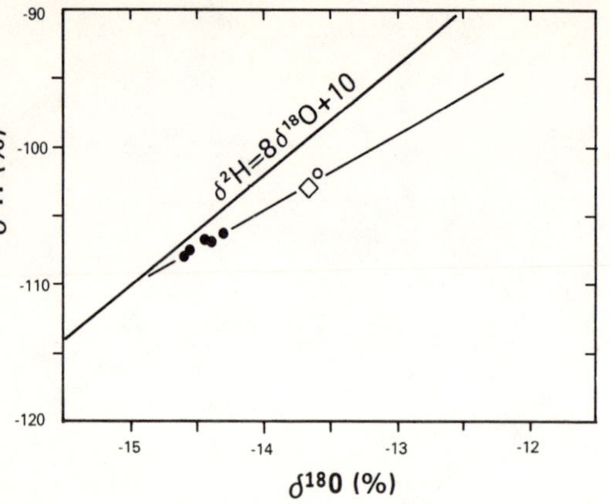

Figure 3. Oxygen-18 and deuterium in the thermal waters from Ojo Caliente. Symbols: diamond, unused hot well; solid circles, warm springs; open circles, shallow aquifer.

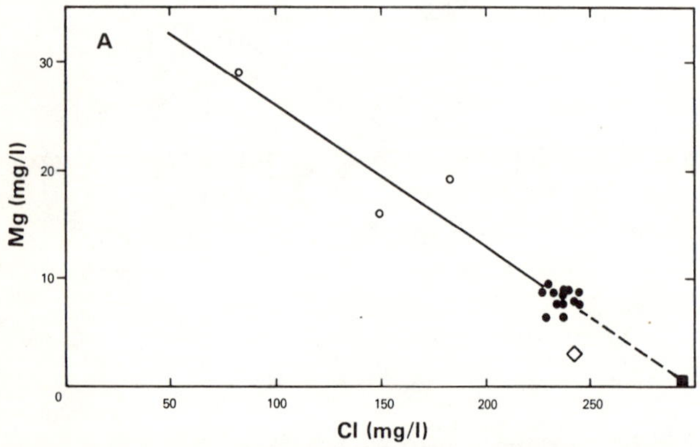

Figure 4A. Mixing model between magnesium and chloride. Symbols: box, extrapolated deep hot end-member; others same as Fig. 3.

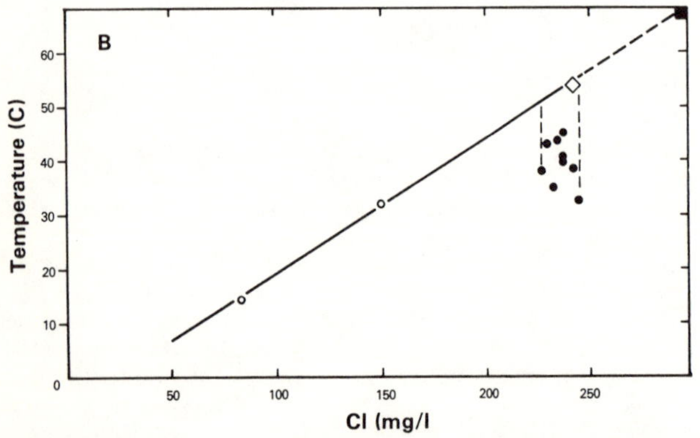

Figure 4B. Plot of temperature against chloride. Symbols same as Fig. 4A. The best-fit line is calculated only with the data of the wells, because the springs undergo conductive cooling.

The chemical geothermometry calculations do not give a clear picture of the reservoir temperature as yet (Table 1). At this point, the deep water reservoir is believed to reach a temperature of 80°C (±10°), by the chalcedony and the Na-K-Ca (Mg-corrected) geothermometers. This temperature is not far from the 68°C found with the mixing model. The hot water reservoir is probably fed by local ground waters infiltrating down through fractures and faults, and warmed by the regional heat-flow of the Rio Grande rift.

CONCLUSION

The geophysics and geology suggest that the thermal reservoir at Ojo Caliente is structurally controlled and of limited extent. The thermal waters emerge from a buried north-northeast trending Pliocene fault that is part of the Ojo Caliente fault zone. Thermal fluids are confined to deposits of alluvium in the Rio Ojo Caliente valley and mix with cool meteoric waters within the valley. Geochemistry of the fluids indicate a maximum reservoir temperature of 80°C, thus, it is our belief that the Ojo Caliente system is suitable only for small-scale direct use geothermal applications.

ACKNOWLEDGMENTS

George Mauro of Ojo Caliente Resort kindly helped and advised us throughout this project. This investigation was supported by U.S. Department of Energy, Office of Basic Energy Sciences and Division of Geothermal Energy.

REFERENCES

May, S. J., 1979, Neogene stratigraphy and structure of the Ojo Caliente-Rio Chama area, Espanola basin, New Mexico, New Mexico Geol. Soc. Guidebook, 30th Field Conf., p. 83-88.

Pearson, C. and Goff, F., 1981, A. Schlumberger resistivity study of the Jemez Springs region of northwestern New Mexico, Geotherm. Resources Council Trans., v. 5, p. 119-122.

Reiter, M., Edwards, C. L., Hartman, H., and Weidman, C., 1975, Terrestrial heat flow along the Rio Grande rift, New Mexico and southern Colorado, Geol. Soc. Am. Bull., v. 86, p. 811-818.

Summers, W. K., 1976, Catalog of thermal waters in New Mexico, New Mexico Bur. Mines and Miner. Resour. Hydrol. Report 4: 80 p.

Trainer, F. W. and Lyford, F. P., 1979, Geothermal hydrology in the Rio Grande rift, North-central New Mexico, New Mexico Geol. Soc. Guidebook, 30th Field Conf., Santa Fe Country, 1979: p. 229-306.

Vuataz, F. D., in prep., Hydrology, geochemistry and geothermal aspects of the thermal waters from Switzerland and adjacent alpine regions, Submitted to J. Volc. Geotherm. Res.

Geothermal Resources Council, TRANSACTIONS Vol. 6, October 1982

APPLICATION OF THE FLUID INCLUSION THERMOMETER TO SOME GEOTHERMAL FIELDS IN JAPAN

Sachihiro TAGUCHI and Masao HAYASHI

Research Institute of Industrial Science
Kyushu University 86
Kasuga City, Fukuoka 816, JAPAN

ABSTRACT

The homogenization temperature of fluid inclusions in hydrothermal quartz, anhydrite and calcite from some geothermal fields in Kyushu, Japan, has been measured. In the very active Hatchobaru geothermal field, it has revealed that the homogenizationtemperature concentrates on temperatures near the boiling point curve of water, and agrees well with the present reservoir temperatures. On the other hand, in the less active Kirishima geothermal field, inclusions have a wide range of homogenization temperature, and the minimum values of them are quite close to the present underground temperatures. This may indicate that the geothermal field has cooled down to some extent, and also that the fluid inclusion have been formed at various stages of the cooling process. Some considerations suggest that not only the primary inclusions but also the secondary ones can be useful, when they contained only liquid at the time of trapping.

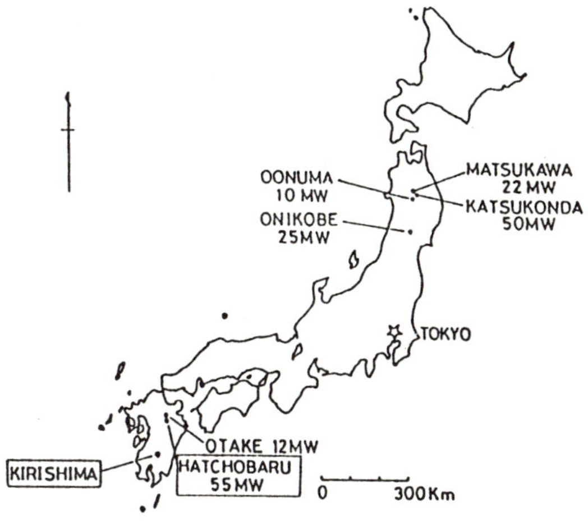

Fig. 1 Location map of investigated geothermal fields in Japan.

INTRODUCTION

The Hatchobaru geothermal field in central Kyushu , Japan, is one of the most active geothermal fields (Hayashi et al., 1981), and a 55 MW/h power station was constructed in 1977 (see Fig.1). In the Kirishima geothermal field in southern Kyushu, a large number of exploratory wells has been drilled with a view to the geothermal generation of electric power. The surveyed area, however, is not enough high temperature to generate electric power economically. In this paper, the under ground temperatures in the both of the active and less active fields, have been estimated using fluid inclusions from the fields, and the fluid inclusion method has been evaluated as a geothermal exploration tool.

EXPERIMENS

Hydrothermal minerals such as quartz (2-8mm long) , anhydrite and calcite (2x2-5x5 mm) were collected from cores and cuttings at various depths of wells, drilled by Kyushu Electric Power Co., Inc., in the Hatchobaru and the Kirishima geothermal fields. The minerals were polished on both faces.

The measurements of homogenization temperatures were carried out using a heating stage Mettler FP-52 at temperatures below 300°C, and a LINKAM TH600 above 300°C. The heating rate was 3°C/min, but more slowly near the homogenization temperature. The pressure correction for the homogenization temperature has not been applied based on our results (Taguchi et al., 1979).

RESULTS

As an example of active geothermal fields, the homogenization temperatures of Well H-4 in the Hatchobaru geothermal field are shown in Fig.2. The distribution of homogenization temperatures in the figure is shown according to Normal distribution with the following manners: the hight is proportional to the total numbers measured, and the width is in the range of three times of standard deviatios on one side. Accordingly, a tall distribution curve with a narrow range means a largr number of the homogenization temperatures of almost same values. As already reported (Taguchi et al., 1979), the homogenization temperature of minerals formed near the production zone of Well H-4, was 285°C on average, which is well in accordance with the chemically estimated temperatures (T_{sio2}=279°C, T_{na-k}=283°C and $T_{na-k-ca}$=285°C). The fluid inclu-

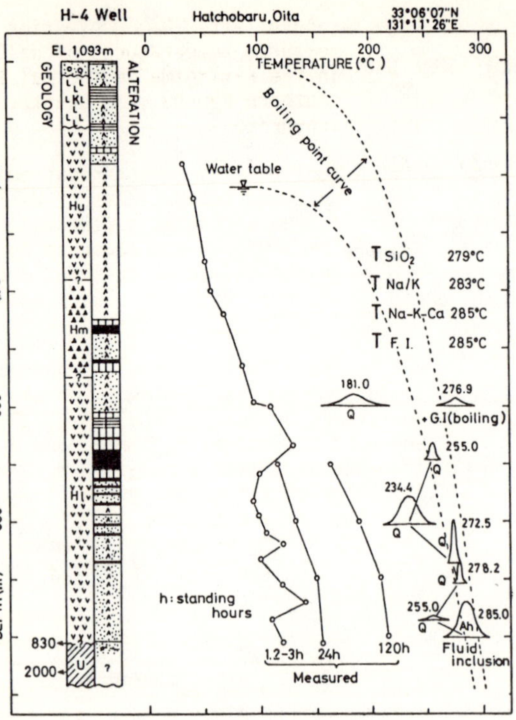

Fig. 2 Homogenization and measured temperatures of Production Well H-4 in the Hatchobaru geothermal field, central Kyushu. Q:quartz, Ah;anhydrite

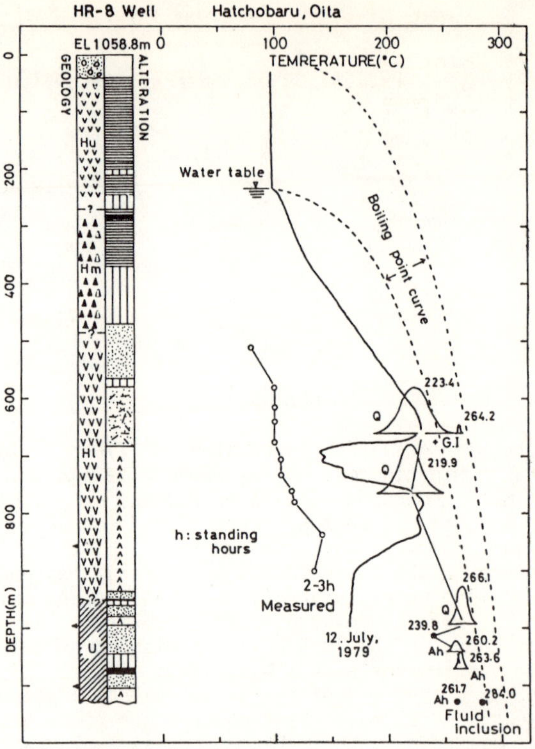

Fig. 5 Homogenization and measured temperatures of Reinjection Well HR-8 in the Hatchobaru geothermal field, central Kyusyu. The solid line (July, 1979) represents the temperature measured by a thermistor after 2.5 years injection, and the open circles tied by solid line by thermometer during drilling.

sion temperatures are likely to indicate directly the present temperature of the well. They are in a narrow range of about 20°C and at slightly inside of the boiling point curve.

The inclusion temperatures of Well KT-5 in the Kirishima geothermal field, are shown in Fig.3. One of the characteristics of the field is that the homogenization temperature has a wide range: usually from 60 to 100°C. Although the maximum homogenization temperature at a depth of 2100 m shows 320°C, the minimum temperature is about 220°C in well accordance with a measured temperature by thermometer. However, at depths from 900 to 1300 m, the measured temperatures are fairly lower than the minimum fluid inclusion temperatures, probably due to the inflow of cold ground water after drilling.

In Fig.4, the homogenization and measured temperatures of Well KT-7 in the Kirishima geothermal field are shown as an example of wells, which was strongly cooled down by the inflow of ground water. The minimum homogenization temperatures are quite close to the measured temperatures by a thermistor at the time, when the the well was drilled to a depth of 1100 m. However, after the well was drilled down deeper than 1100 m, the measured temperatures far decreased as much as 130°C at a depth of 1100 m. This suggests that cold ground water flowed into the well from shallower level.

Fig.5 shows the homogenization and measured temperatures of Well HR-8, one of injection well in the Hatchobaru geothermal field. HR-8 is located at 206 m NNE from production well H-4. The separated water had been injected into the well for two years and half until July, 1979. The fluid inclusion is about 260°C in homogenization temperature at ddepths greater than 900 m. The temperatures at same depths measured by a thermistor one weeek after stopping the injection in July, 1979, were 165 to 170°C, which are 90 to 95°C lower than the fluid inclusion temperatures. On the other hand, at shallower depths, the inclusion temperatures fairly agree well with the measured temperatures excepth depths of 700 m.

It is a common phenomenon through the above four wells that almost all of the maximum temperatures of fluid inclusions are slightly lower than the temperature of the boiling point curve of water, where water table is at the groung surface.

INTERPRETATION

There is a distinct difference in distribution of homogenization temperatures between the active and less active geothermal fields. That is, at

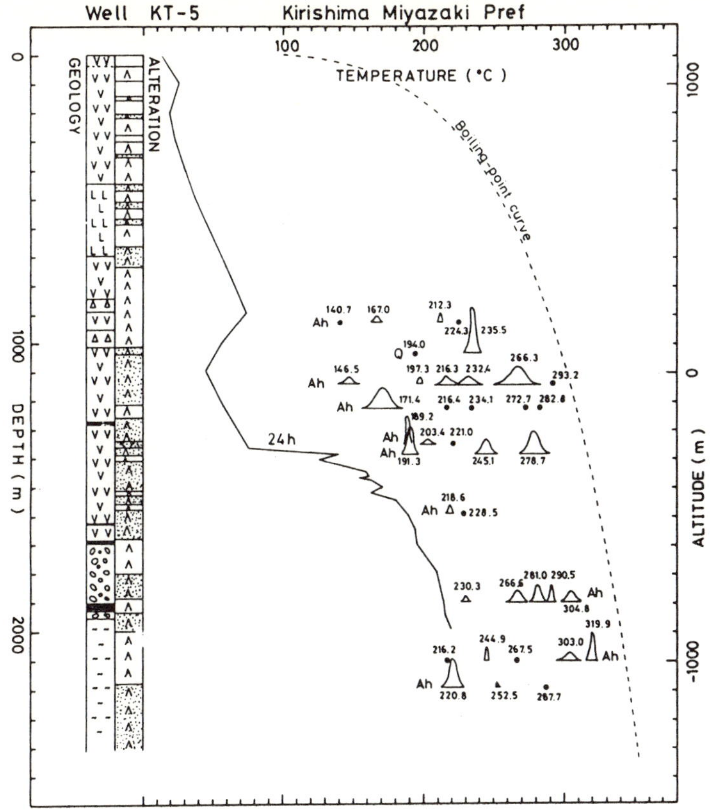

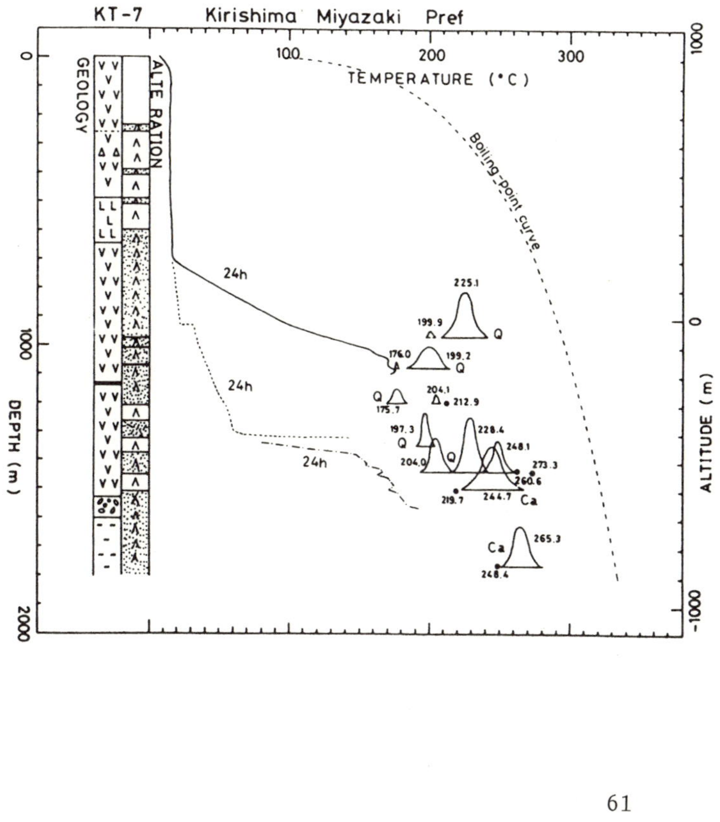

Fig. 3 (above left) Homogenization and measured temperatures of Well KT-5 in the Kirishima geothermal field, southern Kyushu. Q:quartz, Ah: anhydrite.

Fig. 4 (below left) Homogenization and measured temperatures of Well KT-7 in the Kirishima geothermal field. The temperatures measured by a in the course of drilling are shown. Q:quartz, Ca:calcite.

H-4 in the active Hatchobaru geothermal field, the homogenization temperature of the reservoir ranges from 277 to 300°C among 13 inclusions measured, and 9 of them are in a narrow range from 277 to 285°C, which values are quite close to the chemically estimated, 279 to 285°C. This may depend on young geothermal systems, which have not been scarcely cooled down so much.

On the other hand, in a less active field such as the Kirishima geothermal field, the homogenization temperature has wide ranges of distribution; for example in KT-5 it ranges over 80°C. The reasons may be due to that the field is on a cooling process, and that inclusions have been formed at various stages on the cooling process. The fact of cooling in the same field, was also confirmed by the consideration on the reduction of fission track length caused by geothermal heat. The maximum temperatures attained are at least 80 to 90°C greater than the present underground temperature (Taguchi et al., 1980).

The minimum homogenization temperatures of inclusions , often agree well with the temperature measured by a thermistor. These inclusions might be formed under the recent underground temperature, including not only the primary but also the secondary inclusions. Since the secondary inclusion is able to be formed under the present conditions after stopping of crystal growth, it is of particular importance for the estimation of the present under ground temperature. In such secondary inclusions that formed recently on a process of necking-down, the homogenizationtemperature should be close to the present temperature, when they trapped only liquid phase. Accordingly, in Well KT-5 , it may be estimated that the underground temperatures before drilling are close to the minimum homogenization temperatures rather than to the measured ones. In WEll KT-7, the similar phenomenon was observed. Thus,, when temperature of a borehole were disturbed by drilling , the minimum temperatures can

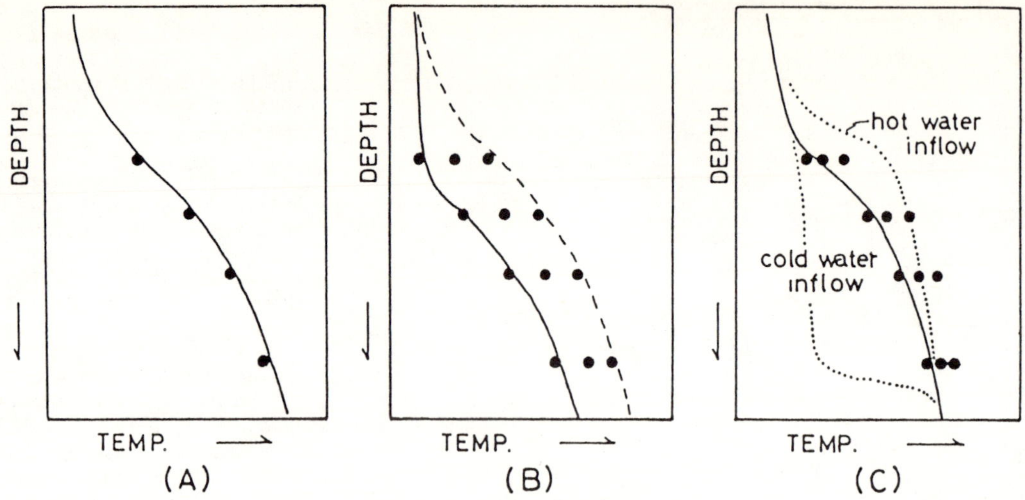

Fig. 6 Schematic relatinos among homogenization temperature, measured temperature and underground temperature. (A) shows a case of active geothermal field. The solid line is an underground temperature, and the solid circles indicate fluid inclusion temperatures. (B) is a case of less active field, where has been cooled down to some extent. The dashed linemeans the maximum underground temperature in the past. (C) is the case of a borehole, which was disturbed by the inflow of cold and hot water. The dotted lines are measured temperatures after the disterbance by cold and hot water.

be used to estimate the underground temperatures. As well as the cooling by drilling, the degree of cooling by injection is also able to evaluate using inclusions.

The relations among homogenization, underground and measured temperatures are schematically summarized as shown in Fig.5.

As mentioned above, not only the primary inclusions but also the secondary or pseudosecondary ones are useful to estimate the underground temperature of the present and also the past, even if the temperature conditions of a borehole are disturbed by drilling etc. The fluid inclusion method is also able to evalutae cooling degree by the distribution of homogenization temperatures.

CONCLUSION

The homogenization temperature of fluid inclusions from the Hatchobaru and the Kirishima geothermal fields has been measured. the following conclusion s have been obtained:

(1) In active geothermal fields such as Hatchobaru, fluid inclusions are in narrow ranges, and agree well with the present reservoir temperatures.

(2) In less active geothermal fields such as Kirishima , the homogenization temperatures show wide ranges of distribution. Of these fluid inclusion temperatures, the minimum value agrees well with the temperature measured by a thermistor. This suggest that the geothermal field has cooled down to some extent, and also that the fluid inclusions have been formed at various stages of the cooling process.

(3) On the estimation of underground temperature of geothermal fields, not only primary inclusions but also secondary or pseudosecondary ones are useful.

(4) The fluid inclusion method is applicable particularly when a borehole was disturbed by drilling, and also in the course of drilling.

REFERENCES

Taguchi, S., Fujino, T. and Hayashi, M. (1979) Homogenization temperature measurements of fluid inclusions in quartz and anhydrite from the Hachobaru geothermal field, Japan, and its applications for geothermal development, Geothermal Resources Council, Transactions, 3, 705-708.

Taguchi, S., Okaguchi, M. and Yamasaki, T (1980) Reduction of the length of fission tracks by geothermal heating and its application to thermal history, Report of Research Institute of Industrial Science, Kyushu Univ., (72), 21-26 (in Japanese with English abstract)

Hayashi, M., Taguchi, S. and Yamasaki, T. (1981) Activity index and thermal history of geothermal systems, Geothermal Resources Council, Transactions, 5, 177-180.

Geothermal Resources Council, TRANSACTIONS Vol. 6, October 1982

HAWAII GEOTHERMAL RESOURCE ASSESSMENT: 1982

Donald M. Thomas, Barry R. Lienert, Malcolm E. Cox,
James P. Kauahikaua*, and Mark Mattice

Hawaii Institute of Geophysics
2525 Correa Road, Honolulu, Hawaii 96822
*U.S. Geological Survey, Honolulu, Hawaii 96850

ABSTRACT

The Geothermal Resource Assessment Program of the Hawaii Institute of Geophysics has conducted a series of geochemical and geophysical surveys throughout the State of Hawaii since February 1978. The results compiled during this study have been used to prepare a map of potential geothermal resource areas throughout the state. Approximately thirteen separate locations on three islands have been studied in detail. Of these, four areas are known to have direct evidence of a geothermal anomaly (Kilauea East Rift Zone, Kilauea Southwest Rift Zone, Kawaihae, and Olowalu-Ukumehame) and three others are strongly suspected of having at least a low-temperature resource (Hualalai west flank, Haleakala Southwest Rift, and Lualualei Valley). In the remainder of the areas surveyed, the data obtained either were contradictory or gave no evidence of a geothermal resource.

INTRODUCTION

The Hawaii Geothermal Resource Assessment Program was initiated in February of 1978 with the objective of evaluating the potential of geothermal resources throughout the Hawaiian Islands chain. Field surveys have been conducted in about one dozen potential geothermal resource areas, and a geothermal resource map for Hawaii has been prepared. A summary of the data used for the resource map follows.

KILAUEA EAST RIFT ZONE

This area (Fig. 1) has been the site of frequent volcanic eruptive activity during historic times and has for several years been considered to have the greatest geothermal potential of any locale in the state. A substantial number of geophysical and geochemical studies have been conducted in the area (Druecker and Fan, 1976; Cox et al., 1980; Furumoto, 1975; Kauahikaua and Klein, 1978; Kroopnick et al., 1978), and most support the thesis that geothermal resources are present. There are currently at least three producible geothermal wells in the lower East Rift Zone; several more, in various locations along the rift, are planned for the near future. The Kilauea E.R.Z. is presently considered to be a known geothermal resource area with an estimated capacity sufficient to produce from 100 to 3,000 megawatt centuries of electrical power.

Hawaii Institute of Geophysics Contribution 1267

KILAUEA SOUTHWEST RIFT

This area has not been as volcanically active as the E.R.Z.; however, intrusive activity has been recorded along the Southwest Rift as recently as August 1981. Although relatively little geophysical or geochemical work has been done on the Southwest Rift, aeromagnetic (Godson et al., 1981) and resistivity (Hussong and Cox, 1967; and Kauahikaua and Mattice, 1981) anomalies are known to exist. In addition, steaming ground and a warm coastal spring have been mapped along the Southwest Rift. The area is considered to have a very high potential for a high-temperature resource within 3 km of the surface; however, no estimate of its capacity can be made with the data presently available.

SOUTHWEST RIFT ZONE OF MAUNA LOA

This area has experienced eruptive activity during recent geologic history, and a geothermal resource may be associated with this activity. However, a limited resistivity survey conducted recently across the lower Southwest Rift was unable to detect evidence of resistivity anomalies within 1 km of the surface (Kauahikaua and Mattice, 1981).

NORTH KONA

Mercury and radon surveys (Cox, 1980) have identified anomalies that may be associated with a thermal source on the Hualalai southwest flank, and infrared anomalies have been reported in the coastal waters near North Kona. Groundwater chemistry data (Cox and Thomas, 1979a) also suggest that some thermal alteration has occurred. Limited geophysical surveys, however, were unable to detect significant resistivity anomalies. If a geothermal resource is present in this location, it is highly probable that at best it will be of low-to-intermediate temperature.

HUALALAI WEST FLANK

Results of recent geologic mapping on the western flank of Hualalai (R. Moore, personal communication) suggest that this volcano has been

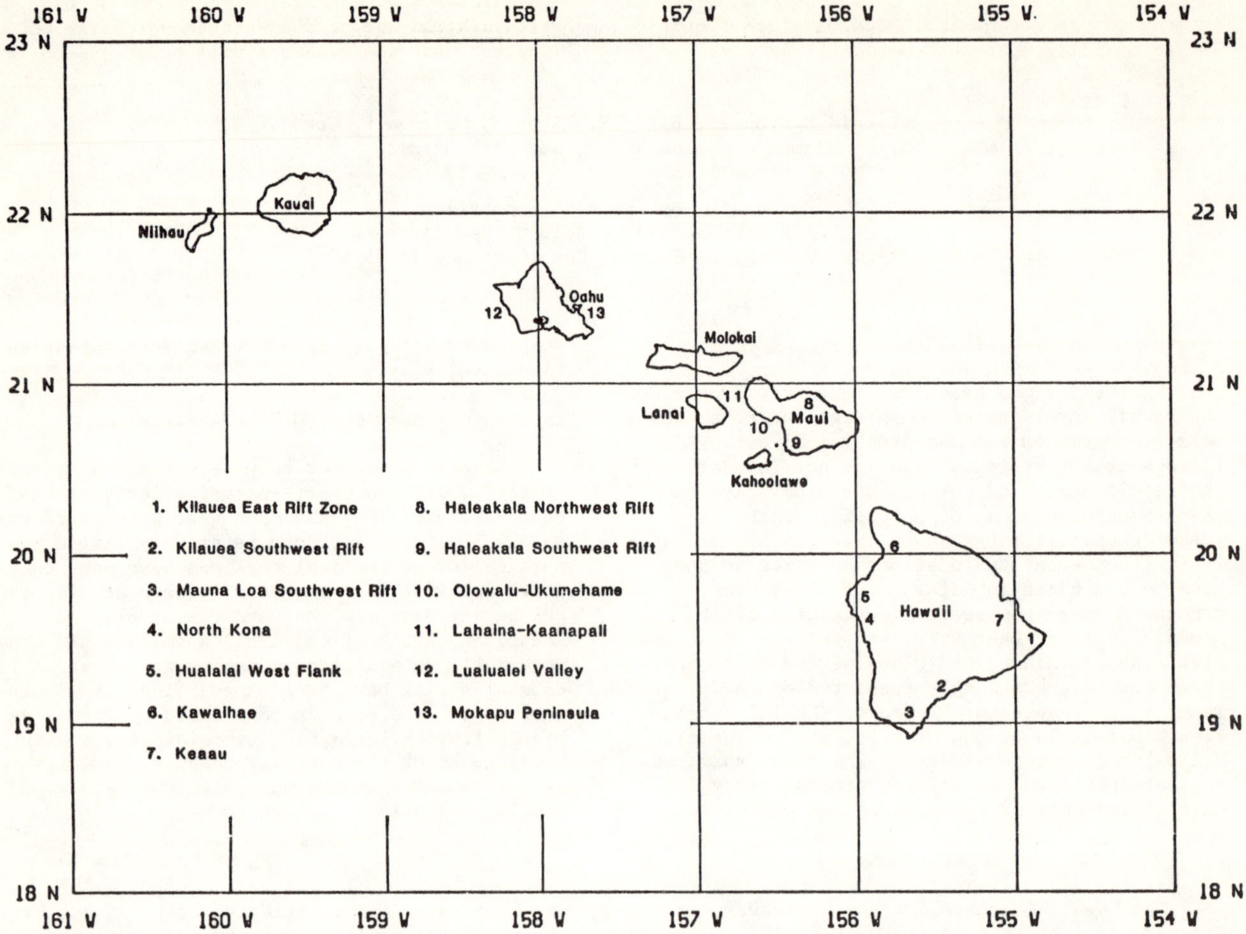

Fig. 1 Location map of field exploration areas evaluated under the Geothermal Assessment Program in Hawaii.

much more active during recent times than was originally thought. Mercury and radon surveys (Cox, 1980) conducted on the lower western flank have identified weak anomalies associated with the very diffuse rift zone, and resistivity soundings near the summit have detected lower than expected resistivities at 500- to 1,000-m depth. The probability that a moderate temperature resource is present within 3 km of the surface in this general area is considered to be about 20% to 30%.

KAWAIHAE

Mercury and radon surveys (Cox, 1980) have detected an elongate anomaly that extends to the east from Kawaihae Bay. Groundwater wells have slightly elevated water temperatures (Epp and Halunen, 1979) (approximately 5°-10°C above ambient) and a modified chloride/magnesium ratio (Cox and Thomas, 1979a, 1979b), suggesting that a thermal source is located in the near vicinity. Schlumberger soundings have identified an area having higher than expected resistivities to the northeast of Kawaihae (Kauahikaua and Mattice, 1981); it has been proposed that the resistive anomaly is the result of an intrusive body that may be associated with a nearly 80,000-yr-old eruptive vent. Although it is considered highly probable that a thermal resource is present near Kawaihae, it is unlikely that temperatures are more than 80° to 100°C.

KEAAU

Preliminary groundwater chemistry (Thomas et al., 1979) initially suggested that thermal fluids might be present; however, more recent time-domain electromagnetic soundings (Kauahikaua and Mattice, 1981) were unable to detect resistivity anomalies to depths of 5 km. The existence of a geothermal resource in this area is presently considered to be extremely unlikely.

HALEAKALA NORTHWEST RIFT

The results of mercury and radon surveys

(Cox, 1980) along the lower northwest rift initially suggested that a number of areas were anomalous. Although these data were substantiated by groundwater geochemical data, neither Schlumberger soundings nor self-potential surveys (Mattice and Lienert, 1980; Mattice, 1981) were able to detect any significant indications of a thermal source. An evaluation of all the data available suggests that the probability of a resource being present within 2 km of the surface on the Haleakala Northwest Rift is low.

HALEAKALA SOUTHWEST RIFT

The Haleakala Southwest Rift Zone was the site of the most recent, and only historical, eruption on Haleakala. Limited mercury and radon field surveys (Cox, 1980) across the upper and lower Haleakala southwest rift have detected significant concentration anomalies that are apparently associated with the rift zone. Data from passive seismic and electromagnetic studies similarly suggest that there may be a thermal resource along the southwest rift. There is a strong probability for a thermal resource along this rift zone.

OLOWALU-UKUMEHAME

A warm-water well (33°C) at the mouth of Ukumehame Canyon has been sampled and shows evidence of moderate thermal alteration of the saline component of the fluids present. Schlumberger resistivity soundings (Mattice and Lienert, 1980; Mattice, 1981) have mapped a warm-water lens within a few hundred meters of the surface, and electromagnetic sounding data suggest that the source of the thermal fluids may be located near the caldera boundary of West Maui volcano. The probability of a low-to-moderate temperature resource is considered to be very high; however, the age of the volcanism of this area indicates that the probability is substantially lower for a high-temperature resource.

LAHAINA KAANAPALI

Groundwater geochemical data and limited mercury and radon surveys (Cox, 1980) have identified moderate-to-low level anomalies in this area; Schlumberger soundings have not detected subsurface resistivities that are consistent with a thermal resource. The probability that a viable low-temperature resource is present is considered to be remote.

LUALUALEI VALLEY

Extensive geophysical and geochemical surveys have been conducted in Lualualei Valley, which encompasses the boundaries of the Waianae caldera (Cox et al., 1979). Groundwater chemistry data as well as mercury and radon survey results indicate that thermally altered ground waters are entering the near-surface environment along the inferred boundaries of the Waianae caldera. In addition, Schlumberger resistivity soundings have identified lower than expected resistivities at less than 1-km depth that may be due to the presence of thermal fluids. The probability that a geothermal resource is present is considered to be high, although the age of the Waianae volcanism would suggest that the resource will be of low-to-moderate temperature.

MOKAPU PENINSULA

Extensive geochemical and geophysical surveys (Cox et al., 1982) were conducted on the Mokapu Peninsula, located on the northern boundary of Koolau caldera. Although a few weak radon and mercury anomalies were identified, Schlumberger resistivity soundings were unable to detect resistivity structures consistent with the existence of a geothermal resource. An evaluation of the results obtained during this study suggests that it is unlikely that a geothermal resource is present in this part of Koolau Volcano.

Copies of the Geothermal Resources Map of Hawaii are available from:

Geothermal Assessment Program
Hawaii Institute of Geophysics
2525 Correa Rd.
Honolulu, HI 96822

ACKNOWLEDGMENTS

The assistance of Leslie Kajiwara, Dave Mills, Kevin Cuff, and Jan Danek in the preparation of the Geothermal Resources Map and the manuscript are gratefully acknowledged. This work was supported under Department of Energy Contracts DE-AC03-80SF 10819 and DE-AS-3-79ET27023 and State of Hawaii Contract 22-B-80-400-F-188-B-D02.

REFERENCES

Cox, M.E., 1980, Ground radon surveys for geothermal exploration in Hawaii: M.S. thesis, Univ. of Hawaii, 194 p.

Cox, M.E., Cuff, K.E., Lienert, B.R., Sinton, J.M. and Thomas, D.M., 1982, A preliminary geothermal evaluation of the Mokapu Peninsula on the island of Oahu, Hawaii: Hawaii Institute of Geophysics, Univ. of Hawaii Tech. Rept. HIG-82-2.

Cox, M.E., Cuff, K.E. and Thomas, D.M., 1980, Variations of ground radon concentrations with activity of Kilauea volcano, Hawaii: Nature, v. 288, no. 5786, p. 74-76.

Cox, M.E. and Thomas, D., 1979a, Cl/Mg ratios of shallow groundwaters as a regional geothermal indicator in Hawaii: Hawaii Institute of Geophysics, Univ. of Hawaii, Tech. Rept. HIG-79-9, 51 p.

----------, 1979b, Cl/Mg ratio of Hawaiian groundwaters as a regional geothermal indicator: GRC Transactions, v. 3, p. 145-148.

Cox, M.E., Sinton, J., Thomas, D., Mattic, M., Kauahikaua, J., Helstern, P. and Fan, P., 1979, Investigation of geothermal potential in the Waianae caldera area, Western Oahu, Hawaii: Hawaii Institute of Geophysics,

Univ. of Hawaii, Tech. Rept. HIG-79-8, 76 p.
Druecker, M. and Fan, P-f., 1976, Hydrology and chemistry of groundwater in Puna, Hawaii: Groundwater, v. 14, no. 5, p. 328-338.
Epp, D. and Halunen, A., Jr., 1979, Temperature profiles in wells on the island of Hawaii: Hawaii Institute of Geophysics, Univ. of Hawaii, Tech. Rept. HIG-79-7, 31 p.
Furumoto, A., 1975, A coordinated exploration program for geothermal sources on the island of Hawaii: Proc. Second U.N. Symp. Dev. Use Geothermal Resources, v. 2, p. 933-1003.
Godson, R.H., Zablocki, C.J., Pierce, H.A., Frayser, J.B., Mitchell, C.M. and Sneddon, R.A., 1981, Aeromagnetic map of the island of Hawaii (1:250 000) Map GP-946, U.S. Geological Survey.
Hussong, D.M. and Cox, D.C., 1967, Estimation of groundwater configuration near Pahala, Hawaii, using electrical resistivity techniques: Water Resources Research Center, Univ. of Hawaii, Tech. Rept. No. 17, 75 p.
Kauahikaua, J. and Klein, D., 1978, Results of electric surveys in the area of Hawaii geoghermal test well HGP-A: GRC Transactions, v. 2, p. 363-366.
Kauahikaua, J. and Mattice, M., 1981, Geophysical reconnaissance of prospective geothermal areas on the island of Hawaii using electrical methods: Hawaii Institute of Geophysics, Univ. of Hawaii, Tech. Rept. HIG-81-4, 50 p.
Kroopnick, P., Buddemeier, R., Thomas, D., Lau, L., and Bills, D., 1978, Hydrology and geochemistry of a Hawaiian geothermal system: HGP-A: Hawaii Institute of Geophysics, Univ. of Hawaii, Tech. Rept. HIG-78-6.
Mattice, M.D., 1981, Geothermal and groundwater exploration on Maui, Hawaii, by applying D.C. electrical soundings: M.Sc. thesis, Univ. of Hawaii, 96 p.
Mattice, M.D. and Lienert, B.R., 1980, Schlumberger survey of Maui Island, State of Hawaii, GRC Transactions, v. 4, p. 85-88.
Thomas, D., Cox, M., Erlandson, D. and Kajiwara, L., 1979, Potential geothermal resources in Hawaii: a preliminary regional survey: Hawaii Institute of Geophysics, Univ. of Hawaii, Tech. Rept. HIG-79-4.

Geothermal Resources Council, TRANSACTIONS Vol. 6, October 1982

GEOTHERMAL ENERGY RESOURCES OF THE LOWER SUSITNA BASIN, ALASKA

Donald L. Turner, Eugene M. Wescott, William Witte and Becky Petzinger

Geophysical Institute
University of Alaska
Fairbanks, AK 99701

ABSTRACT

Evidence from dry wildcat well temperatures and helium soil gas surveys suggests that a substantial warm water resource may be present in the Willow-Big Lake area of Alaska, and possibly even farther to the east. Discontinuous reservoirs may be present under an area of at least 40 square miles, probably fed by upwelling of thermal waters along Tertiary faults. A gravity survey indicates the presence of two ridges in the granitic basement. These ridges are believed to be overlain, in part, by geothermal reservoirs accessible to shallow (200-500 ft.) drilling.

Four dry wildcat wells (Figure 1) drilled in the lower Susitna basin have encountered anomalously high temperatures, suggesting that low temperature geothermal resources might be present which could be used for space heating and agriculture. Estimated temperature gradients range from 40 to 123°C/km. One well had an indicated maximum temperature of 76.7°C within 2050' depth. There are no known surface manifestations of a geothermal resource in the area but water wells have encountered warm saline water at a depth of 40 ft. in one locality. North of the Castle Mtn. Fault (Figure 2) the basin contains about 2,000 ft. of coal-bearing Tertiary nonmarine sediments overlain by glacial drift and underlain by a granodiorite basement. South of the fault the basin has been down-dropped and contains a much thicker sedimentary section (12-14,000 ft.).

A helium soil gas and water survey was conducted with an approximately one mile grid spacing in order to explore for hot water reservoirs at depth. Results are shown in Figure 1. The helium data corroborate the temperature anomalies in the three hot wells studied and suggest that discontinuous hot water reservoirs totalling at least 40 square miles may be present in the Willow-Big Lake area. Several areas in the proposed state capital site near Willow (Figure 1) contain substantial helium anomalies. The helium anomalies extend to within six miles of Wasilla, where our preliminary survey ended. It is possible that this anomaly trend may extend as far east as Wasilla, or even possibly farther to the east. There is a strong suggestion of elongate trends in the helium anomaly patterns, suggesting that these patterns could be controlled by Tertiary faults which are covered by the glacial drift that mantles the region.

A gravity survey of the area indicates that two ridges are present in the granodiorite basement (Figure 2). The regional gradient due to the Castle Mtn. Fault was removed from the data by use of a two dimensional fault model with a density contrast of 0.375 gm/cc. Depth control at a dry wildcat well allows calculation of depth to basement. The Tertiary nonmarine sedimentary section containing good aquifers is inferred to be thinner over the two basement ridges than in adjacent areas. Helium anomalies are present over both basement ridges. Geothermal aquifers are therefore likely to be encountered at shallower depths (200-500 ft.) over these ridges than in adjacent areas studied, and can probably be drilled cheaply with normal water well rigs.

We have proposed two possible models - radiogenic heating of aquifers by U and Th-rich basement pegmatites, and fault-controlled hydrothermal convection to account for the geothermal system in the area studied. Although both models are generally consistent with available field evidence, the very large extent and apparent orientation of discontinuous geothermal reservoirs appear to favor the fault model.

Our initial study suggests that a substantial warm water resource may be present in the Willow-Big Lake area (Turner and Wescott, 1982). However, large gaps exist in the survey data base we have used to delineate this resource, and the nature of the geothermal system supplying the reservoirs is not yet well understood. Reservoir depths and thicknesses are presently unknown. We have also been unable so far to determine the lateral extent of the suspected reservoir system, and, in particular, whether or not it extends to the rapidly growing population center of Wasilla. The potentially favorable economics resulting from the shallow predicted reservoir depths and the rapidly growing population of this area along the Parks Highway make a proposed follow-on exploration program very attractive. This is one of the few areas in Alaska where a sizeable future population could be located in the vicinity of a large geothermal resource that could be economically utilized.

We are attempting to obtain State of Alaska funding for a follow-on program of exploration geophysics and shallow drilling which will focus on providing a much better definition of the nature of the geothermal system and the lateral and vertical distribution of suspected geothermal reservoirs. We emphasize that the presence of large amounts of hot water has not yet been confirmed in the study area. Indirect evidence from our helium survey is very encouraging, but the actual confirmation of the suspected geothermal resource will require exploratory drilling and well testing. Drilling should be relatively inexpensive due to the shallow depths to suspected reservoirs inferred from the gravity survey.

Acknowledgments

Helicopter support was provided by the Alaska Division of Geological and Geophysical Surveys. Field camp facilities were kindly provided by the Alaska State Parks System. We are grateful to Steve Hackett and David Barnes for providing their unpublished gravity data. Financial support for this study was provided by the Division of Geothermal Energy of the U.S. Dept. of Energy under Cooperative Agreement DE-FC07-79-ET-27034.

References

Turner, Donald L. and Eugene M. Wescott, 1982, A preliminary investigation of the geothermal energy resources of the lower Susitna Basin, University of Alaska Geophys. Inst. Rept. UAG R-287, 50 p., 3 pl.

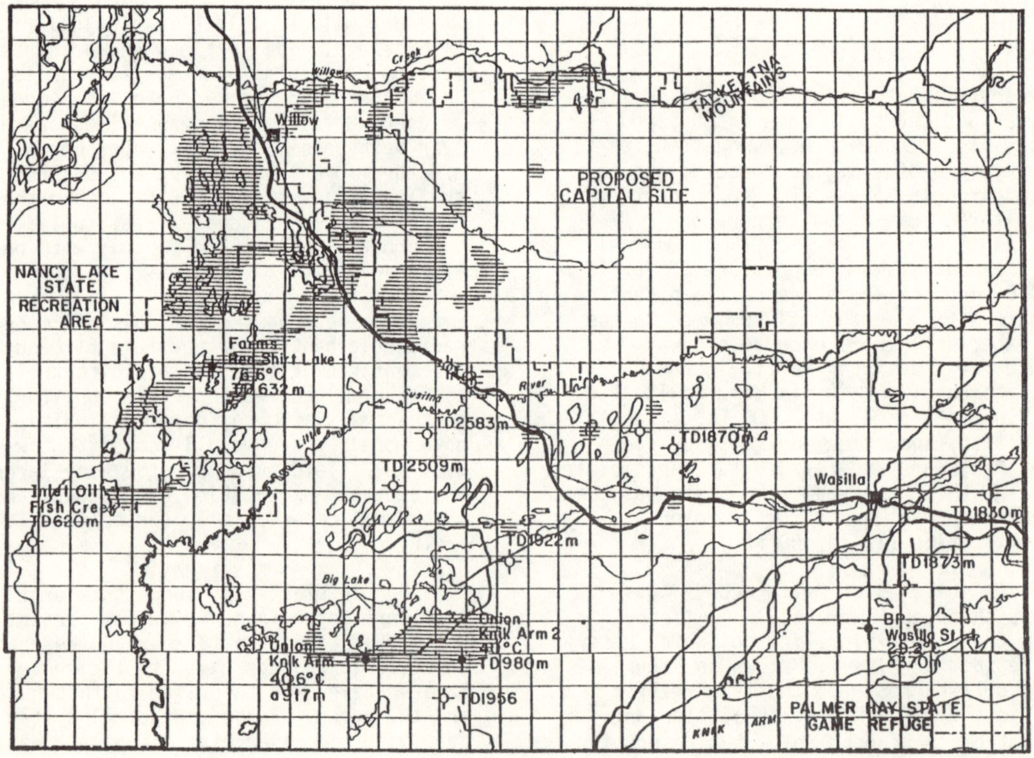

Figure 1. Map of the lower Susitna basin showing estimated areas of high helium concentration (shaded areas), and locations of wildcat wells.

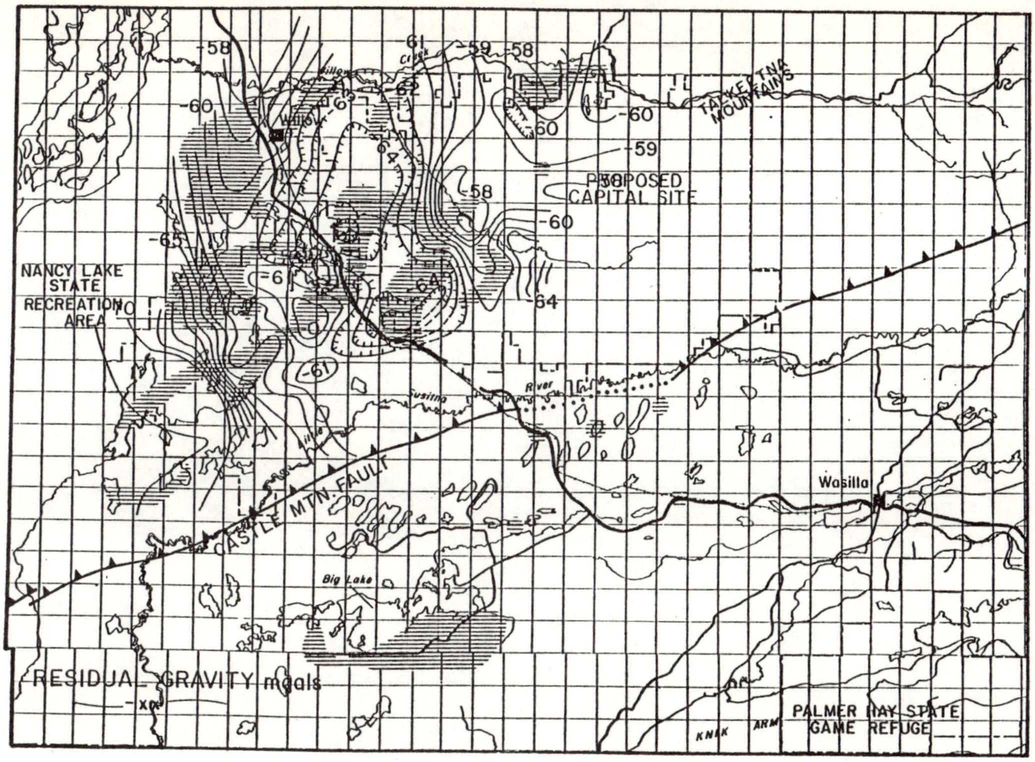

Figure 2. Residual gravity anomaly map of the lower Susitna basin Estimated areas of anomalously high helium concentrations are shaded.

EXPLORATION-II: GEOCHEMISTRY

Geothermal Resources Council, TRANSACTIONS Vol. 6, October 1982

EVALUATION OF THE GEOTHERMAL RESERVOIR ASSOCIATED WITH AUBURN AND JOHNSON HOT SPRINGS, UPPER STAR VALLEY, WYOMING

Michael C. Adams and Regina M. Capuano

Earth Science Laboratory
Univ. of Utah Research Institute, Salt Lake City, Utah

ABSTRACT

Auburn and Johnson Hot Springs are part of a low- to moderate- temperature geothermal resource in western Wyoming. The thermal water is sodium chloride in character, with a dissolved solids content ranging from 5250 to 5962 ppm, and nearly neutral pH. Comparison with other thermal waters from Alpine and Heise Hot Springs, which occur from 45 to 115 km north along the same structural trend, shows them to be chemically very similar and suggests that the mechanisms for generation of these thermal waters are similar. The predicted subsurface temperature and depth of circulation for the thermal waters are on the order of 92°C and 2600m, respectively. Other sodium chloride waters flow as cold springs from the halite-bearing Preuss sandstone just south of the area. These waters, which range in dissolved solids content from 672 to 270000 ppm, are chemically distinct from the thermal waters and do not appear to have had a thermal origin.

INTRODUCTION

Hot springs and travertine deposits are common along the margin of the north-trending valleys that separate the Caribou Range from the Salt River and Snake River Ranges in Idaho and Wyoming (Fig. 1). Auburn and Johnson Hot Springs, located in Upper Star Valley, mark the southern end of this belt of thermal activity. These springs are sodium chloride in character, and have low to moderate salinities and predicted reservoir temperatures of less than 100°C. Local ground waters are cold and relatively dilute with a chemical character ranging from calcium bicarbonate to calcium sulfate, distinctly different from the character of the thermal water. There are, however, several cold springs south of Auburn that have salinities up to two orders of magnitude higher than the hot springs and are also sodium chloride in character. In contrast to the thermal spring water, this saline cold spring water has low calculated geothermometer temperatures. In this study, the temperature and depth of the thermal reservoir are evaluated along with the interrelationships between the saline cold spring water and that of the hot springs. This study was funded by the U. S. Dept. of Energy under contract DE-AC07-80ID12079.

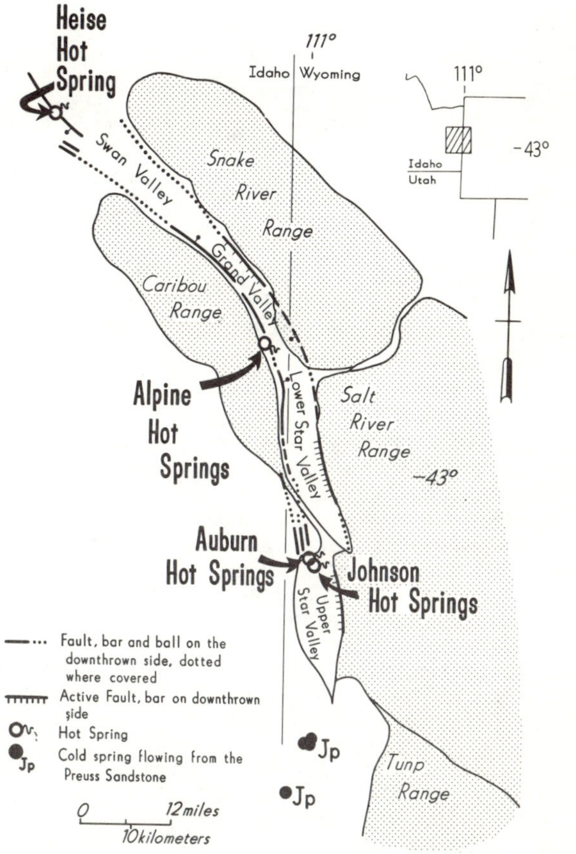

Figure 1. Location map of Swan-Star Valley trend. Fault locations from Ralston (1981).

GEOLOGY

Upper Star Valley lies near the center of the Wyoming-Idaho thrust belt. The valley is the southern extension of a concave-westward, curvilinear fault-bounded trough. The trough consists of Swan, Grand, and Lower and Upper Star Valleys (Swan-Star Valley trend) (Fig. 1). The Swan-Star Valley trend has been modeled by Stott (1974) as a series of 3000 m deep grabens.

The sedimentary strata along the Swan-Star Valley trend consists of over 15000m of limestone, sandstone, siltstone, and evaporites. These sedimentary rock units are Paleozoic to Cenozoic in

Table I - Compositional Ranges of Water and Geothermometer Temperatures[1]

	GROUP I[2]	GROUP II[3,4]	GROUP III[3]			
SOURCE		Jp	Auburn H.S.[5]	Johnson H.S.[6]	Alpine H.S.[7]	Heise H.S.[8]
Surface Temp.	4-10	9-16	55-62	43	37-56	49
TDS	102-396	672-270000	5250-5904	5962	6404-6615	6495
SiO_2	4.3-20	2.5-12	27-40	37	40-45	30
Na	0-11	150-84713	1158-1400	1284	1500-1532	1500
K	0.5-1.4	1.2-67	123-150	163	189-190	190
Ca	29-91	70-1101	400-420	447	526-560	450
Mg	5-19	12-183	66-70	72	93-100	82
HCO_3	101-268	92-276	828-860	800	880-920	1100
SO_4	2.8-130	67-2075	1100-1411	1129	1000-1052	740
CL	1-4.3	200-158000	1700-1900	2019	2412-2800	2400
F	0.0-4.3	0.1-0.4	0.2-4.0	4.2	2.7-2.8	3.1
pH	7.4-8.2	7.3-7.7	6.6-7.1	6.4	6.5	6.7
Na/K	0-12.	56-1300	9.33-10.0	7.88	8.1-8.3	7.9
CL/SO_4	0.01-0.43	.76-7,600	1.31-1.55	1.79	2.29-2.80	3.24
No. of Samples	11	4	3	1	2	1
GEOTHERMOMETERS						
Na-K-Ca(-Mg)[9]	≤ 6[11]	22-30	92-97	101	92-97	79
Quartz[10]	14-63	≤ 45[11]	75-92	88	93-94	95

1. Analyses in ppm and temperatures in °C.
2. Source of data, Lines and Glass (1975), Samples include springs and shallow wells (< 40 m).
3. Spring samples only.
4. Sources of data, Lines and Glass (1975) and this study.
5. Source of data, Lines and Glass (1975), Breckenridge and Hinckley (1977) and this study.
6. Source of data this study.
7. Sources of data, Ross (1971) and Mitchell et al. (1978).
8. Source of data, Mitchell et al. (1978).
9. Calculated by the methods of Fournier and Truesdell (1973) and Fournier and Potter (1979).
10. Calculated by the method of Fournier (1981).
11. Lowest geothermometer temperatures are below measured temperatures and therefore are not reported.

age (Mansfield, 1927). One of these units, the Jurassic Preuss sandstone, contains locally thick (50 m) accumulations of halite (Mansfield, 1927).

HOT SPRINGS

Auburn Hot Springs consists of over 100 vents covering a 12000 m^2 area with a total discharge of 2.5 liter/sec and a maximum temperature of 62°C (Hinckley and Breckenridge, 1977). Travertine deposits of gypsum, aragonite, and calcite occur throughout this area and are presently forming around active vents. Previous workers (Hinckley and Breckenridge, 1977) have suggested that native sulfur is deposited as part of the travertine to explain the yellow color. Although hydrogen sulfide evolves from the thermal fluids, X-ray examination of fresh yellow travertine did not confirm the presence of native sulfur.

Johnson Hot Spring, located 1 km south of Auburn Hot Springs, consists of a single spring with a flow of 0.13 liters/sec and a measured temperature of 46°C (Hinckley and Breckenridge, 1977). Travertine surrounds this vent and covers a 225 m^2 area.

North of the Auburn and Johnson Hot Springs study area there is other evidence of thermal activity along the Swan-Star Valley trend. This includes travertine deposits and two areas containing active hot springs. The travertine occurs sporadically along the western edge of the valley between Johnson Hot Spring and the town of Freedom, 25 km to the north. The vents related to these deposits are presently inactive. Alpine and Heise Hot Springs occur 45 and 115 km north, respectively, of Johnson Hot Spring. Alpine Hot Springs consist of numerous vents that are presently covered by the Palisades Reservoir. The water temperatures ranged from 31° to 66°C (Mitchell et al., 1978). Bradley (1885, p. 269) reported "calcareous, sulphurous and saline deposits" associated with the hot springs. Heise Hot Springs mark the northern limit of the Swan-Star Valley trend. These springs have a maximum measured temperature of 49°C and are surrounded by extensive travertine deposits (Ralston et al., 1981).

WATER CHEMISTRY

Waters from the Swan-Star Valley trend display a broad compositional range. The chemistry of these waters is listed in Table 1 and their chemical character is displayed on Figure 2. Comparison of these water analyses delineates three groups. Group I consists of cold ground water from springs and shallow wells in the vicinity of Lower and Upper Star Valleys. This water is calcium bicarbonate to calcium sulfate in character, with a dissolved solids content of less than 400 ppm, and a near-neutral pH. Hydrologic and chemical data from Lines and Glass (1975) indicate that Group I water flows from a variety of Mesozoic and Tertiary sedimentary rocks.

Group II is comprised of cold spring water that discharges exclusively just south of the Swan-Star Valley trend from the Jurassic Preuss sandstone (Fig. 1). Group II water, in contrast to the cold ground water of Group I, is sodium chloride in character and more saline, with dissolved solids content ranging from 672 to 270000 ppm. The relatively high Na and Cl contents of these fluids probably reflect interaction with halite-bearing units within the Preuss sandstone.

Group III includes all hot springs occurring along the Swan-Star Valley trend. The thermal waters are sodium chloride in character, have dissolved solids contents ranging from 5250 to 6495 ppm, and pH values ranging from slightly acidic to neutral. The similar chemical compositions and geothermometer temperatures for all four of the hot springs suggest that waters discharging from these springs have equilibrated under similar physical and chemical conditions, despite their geographic separation. In addition, this suggestion of a similar evolutionary history is further supported by the close spacial association of the springs with the faults bounding the Swan-Star Valley trend.

Groups II and III are similar in that they are both sodium chloride in character. However, they differ considerably in their element concentrations and temperatures. Group II has a wider range in dissolved solids (672 to 270000 ppm) than the relatively narrow range of Group III (5250 to 6495 ppm). Na/K and Cl/SO_4 ratios are generally higher for Group II, ranging from 56 to 1300 and 0.76 to 7600, respectively, whereas Group III Na/K and Cl/SO_4 ratios range, respectively, from 7.88 to 10.0 and from 1.3 to 3.2. The measured and predicted temperatures of Group II waters, less than 16°C and 45°C, respectively, are distinctly lower than those of Group III (Table 1). Group III has measured temperatures ranging from 49°C to 62°C, and geothermometer temperatures ranging from 75° to 101°C.

It is suggested by Mansfield (1927) that thermal waters of this area have derived their sodium chloride character from the halite beds in the Preuss formation. Nonthermal spring water flowing from the Preuss formation (Group II) is significantly enriched in Na and Cl relative to all other element concentrations in the samples. The thick halite beds present in the Preuss are the probable source for the high Na and Cl in this water and the relatively slight enrichment of the other elements. In contrast, the thermal water (Group III) is not only enriched in Na and Cl, but relative to Group II water, the thermal water is also enriched in most other elements. It is therefore suggested that the Na and Cl enrichment in the thermal water is the result of interaction of this water at elevated tempera-tures, on the order of 100°C, with the other Mesozoic and Cenozoic lithologies at depth. This does not preclude the possibility that some component of the Na and Cl in the thermal water is derived from interaction with the Preuss at depth.

SUBSURFACE TEMPERATURE

The reservoir temperatures for the hot springs occurring along the Swan-Star Valley trend were estimated using the quartz (Fournier, 1981) and the Na-K-Ca (Fournier and Truesdell, 1973) with Mg correction (Fournier and Potter, 1979) geothermometers. The temperatures derived using the two methods were concordant, ranging from 75° to 95°C for the quartz geothermometer and 79° to 101°C for the Na-K-Ca (-Mg) geothermometer.

The silica equilibrium temperatures were calculated assuming saturation with quartz rather than its polymorph, chalcedony. Frequently the chalcedony geothermometer is used to give a minimum temperature of silica equilibrium.

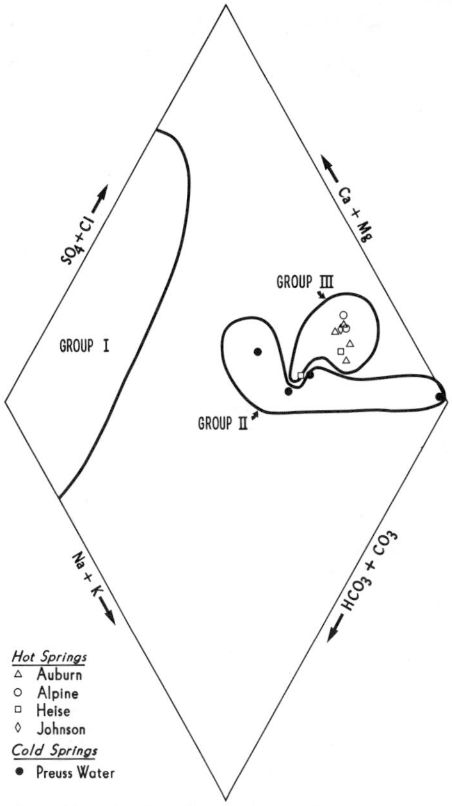

Figure 2. Chemical character of waters from the Swan-Star Valley trend. Data sources are listed in Table 1.

However, the lack of chalcedony in the sedimentary rocks and the dominance of quartz allows the assumption that quartz equilibria controls the solubility of silica in the thermal waters. Quartz geothermometers were therefore considered the most reliable for these waters.

The accuracy of the Na-K-Ca (-Mg) geothermometer is contingent on fluid-mineral equilibrium at the reservoir temperature. Fournier (1979) states that above 100° to 150°C the major control of Na/K in most natural waters is the base exchange of alkalies between coexisting feldspars. The average geothermometer temperature derived from the hot spring waters in Swan-Star Valley, 92°C, is at the lower threshhold of Fournier's generalization. Equilibrium, however, is also dependent on residence time at the reservoir temperature. It cannot be ascertained whether the thermal water was equilibrated at the elevated reservoir temperatures, although the agreement between the predicted temperatures of the quartz and Na-K-Ca (-Mg) geothermometers suggests it has.

The possible influx of sodium ions from dissolution of halite beds in the Preuss formation is an additional complication. The Na-K-Ca (-Mg) geothermometer is based on element ratios. Fournier and Truesdell (1973) indicate that while the absolute abundance of calcium is dictated by carbonate equilibrium, the ion ratios are controlled by feldspar base exchange reactions. In a similar manner, halite in the Preuss formation may provide the absolute abundance of sodium, while feldspar equilibrium dictates the ion ratios. This would be true only if the temperature is high enough or the residence time long enough for feldspar base exchange to compensate for possible sodium influx. This assumption is supported by the considerably higher Na/K ratios in waters that have interacted with halite-bearing lithologies (Group II) in comparison with the thermal waters (Group III).

The accuracy of the Na-K-Ca (-Mg) geothermometer is also affected by deposition of travertine at the surface. The hot springs along the Swan-Star Valley trend actively deposit calcium-bearing minerals around their vents. Precipitation of these minerals lowers the calcium concentration in the water, raising the derived Na-K-Ca temperature. Conversly, the magnesium correction factor is increased by calcium loss, resulting in an overall decrease in the Na-K-Ca (-Mg) geothermometer temperature. Thus the Na-K-Ca (-Mg) geothermometer temperatures for these hot springs waters are a minimum estimate with respect to mineral deposition.

SOURCE OF THERMAL WATER

The depth of the thermal reservoir was calculated using a one-dimensional conductive heat flow equation $[(Td-Ts)/\Sigma(Q/K_i)=X]$ (Gretener, 1981). For this study, Td is the estimated reservoir temperature, 92°C, Ts is the average surface temperature, estimated at 10°C, Q is the heat flow, 1.59 μ cal/cm^2 sec, derived from a thermal gradient hole near Auburn Hot Springs (H. Hesslar, pers. comm.), ΣKi is the summation of the thermal conductivities estimated from average conductivities for specific lithologies and X is the depth to the reservoir temperpature. Based on thermal conductivity of 4 to 6 mcal/cm sec, the reservoir depth for the thermal fluids is 2100 to 3100 m. Gravity data (Stott, 1974) suggests that this depth coincides with the base of the deep graben that runs the length of the Swan-Star Valley trend.

CONCLUSIONS

Auburn and Johnson Hot Springs have an estimated reservoir temperature of approximately 90°C. Their chemical compositions and geothermometer temperatures were very similar to the Heise and Alpine Hot Springs and indicate a similar evolutionary history for the fluids at depth. All of these hot springs lie along faults that bound the continuous 150 km long Swan-Star Valley trend, which consists of a series of graben structures. The depth to basement based on gravity modeling suggests that the graben are 2300 to 3400 m deep. Heat flow calculations show that the maximum temperatures recorded by the geothermometers could be reached at these depths.

REFERENCES

Bradley, F. H., 1885, Sixth Annual Report: U.S. Geol. Survey, F. V. Hayden Survey, 269 p.

Fournier, R. O. 1979, A revised equation for the Na/K geothermometer: Geotherm. Res. Council, Trans., v. 3, p. 221-224.

Fournier, R. O., 1981, Application of water chemistry to geothermal exploration and reservoir engineering: in Rybach, L., and Muffler, L. J. P.: Geothermal Systems: Principles and case histories, John Wiley and Sons, N. Y., p. 109-144.

Fournier, R. O., and Potter, R. W., II, 1979, Magnesium correction to the Na/K/Ca chemical geothermometer: Geochim. Cosmochim. Acta, v. 43, p. 1543-1550.

Fournier, R. O., and Truesdell, A. H., 1973, An empirical Na/K/Ca geothermometer for natural waters: Geochim. Cosmochim. Acta, v. 37, p. 1255-1275.

Gretener, P. E., 1981, Geothermics: Using temperature in hydrocarbon exploration: AAPG Education Course Note Series No. 17, 156 p.

Hinckley, B. S., and Breckenridge, R. M., 1977, Auburn Hot Springs, Lincoln County, Wyoming in Helsey, E. L. et al., ed Rocky Mountain Thrust Belt Geology and Resources: WY Geol. Assoc., 29th Ann. Field Conf., p. 707-710.

Lines, G. C., and Glass, W. R., 1975, Water resources of the thrust belt of western Wyoming: U. S. Geol. Survey Hydro. Inv. Atlas HA-539.

Mansfield, G. R., 1927, Geography, geology, and mineral resources of part of southeastern Idaho: U. S. Geol. Survey P. P. 152, 453 p.

Mitchell, J. C., Johnson, L. L., and Anderson, J. E., 1980, Potential for direct heat application of geothermal resources: Idaho Dept. Water Res. Water Inf. Bull., no. 30, 369 p.

Ralston, D. R., Arrigo, J. L., Baglio, J. V., Jr., Coleman, L. M., Hubbell, J. M., Souder, K., Mayo, A. L., 1981, Geothermal evaluation of the thrust zone in southeastern Idaho: Res. Tech. Completion Report, Idaho Water and Energy Res. Inst., Univ. of Idaho, 110 p.

Ross, S. H., 1971, Geothermal potential of Idaho: Idaho Bur. Mines Geol. Pamph. no. 150, 72 p.

Stott, L. R., 1974, A gravity survey of a late Cenozoic graben in the Wyoming-Idaho thrust belt: MS thesis, Univ. of Michigan, 28 p.

Geothermal Resources Council, TRANSACTIONS Vol. 6, October 1982

HYDROGEN AND OXYGEN ISOTOPE GEOCHEMISTRY OF COLD AND WARM SPRINGS FROM THE TUSCARORA, NEVADA THERMAL AREA

John R. Bowman* and David Cole**

*Dept of Geology and Geophysics, Univ. of Utah, Salt Lake City, UT
**Earth Science Laboratory, Univ. of Utah Res. Inst., Salt Lake City, UT

ABSTRACT

Eighteen cold and warm spring water samples from the Tuscarora, Nevada KGRA have been analyzed for hydrogen and oxygen isotope composition and fluid chemistry. Warm springs have δD values (-128 to -137 permil) significantly lower than those of cold springs to the north and east of the area, but similar to the δD values of cold springs to the west and south (-131 to -135 permil). The recharge area for the warm springs is unlikely to be to the immediate north, which is the local topographic highland in the area. The hydrogen isotope data would permit recharge from areas to the southwest or from high elevations to the southeast (Independence Mountains), a sector consistent with electrical resistivity evidence of fluid flow.

Warm springs are HCO_3^--rich waters, enriched by a factor of 3 to 10 in Na, HCO_3^- and SiO_2 relative to local cold springs. Average quartz (no steam loss) and Na/K/Ca geothermometer estimates suggest subsurface temperatures of 145° and 196°C, respectively. The warm springs exhibit poor correlations between either hydrogen or oxygen isotope composition and water temperature or chemistry. The absence of such correlations suggests that there is no single coherent pattern of cold water mixing or evaporation in the thermal spring system.

INTRODUCTION

The Tuscarora, Nevada Geothermal System has been an area of geological, geophysical and geochemical investigation in recent years for evaluation as a geothermal resource. Of prime importance in this evaluation is the characterization of thermal water chemistry, estimation of subsurface temperatures, and definition of recharge areas for the thermal waters. Acquisition of stable isotope and fluid chemistry is required for the quantification of these parameters. These data are also important for the evaluation of near-surface mixing and water-rock interaction in the geothermal reservoir. In order to fulfill this requirement, water samples were collected from eight warm springs and ten cold springs in the Tuscarora area (Fig. 1). These samples were analyzed for major cations and anions, and hydrogen and oxygen isotope compositions. The results of these analyses form the basis of this summary.

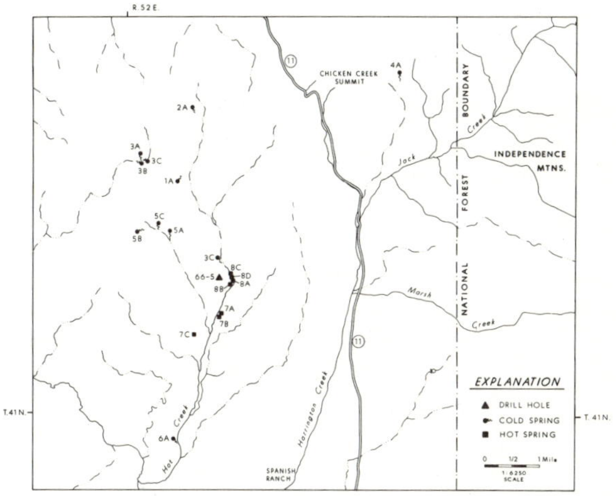

Figure 1. Location map of cold and hot springs.

STABLE ISOTOPE GEOCHEMISTRY

Hydrogen and oxygen isotope analyses of the cold and warm springs are shown in Table 1 and plotted in Figure 2. The isotope analyses were determined using standard extraction (Epstein and Mayeda, 1953; Friedman, 1953) and mass spectrometric techniques. The data are expressed in the delta (δ) notation relative to SMOW for oxygen and hydrogen. In Figure 2, cold springs north of Hot Sulfur Springs (Samples 1A through 4A) are plotted as solid circles, cold springs northwest and southwest of Hot Sulfur Springs (5A through 6A) are plotted as open circles, hot springs of Hot Sulfur Springs (7A through 7C) are plotted as solid squares and hot springs north of Hot Sulfur Springs are plotted as open squares (8A through 8D).

The cold springs can be divided into two distinct groups. The group of cold springs north of the Hot Sulfur Springs system plot, with two exceptions, close to the meteoric water line, with

TABLE 1. Hydrogen and oxygen isotope compositions of cold and warm springs.

Sample #	$\delta^{18}O$	δD	Spring Temperature (°C)
Cold Springs			
1A	-16.3	-119	10.5
2A	-16.2	-122	11.5
3A	-16.8	-125	15.0
3B	-16.7	-124	16.5
3C	-16.0	-116	10.0
3D	-15.5	-118	9.6
4A	-15.3	-123	8.0
5A	-16.9	-135	14.5
5B	-16.1	-131	18.5
5C	-16.7	-135	18.5
6A	-16.6	-134	20.0
Hot Springs			
7A	-16.1	-137	89.0
7B	-15.6	-129	82.0
7C	-16.6	-133	55.5
8A	-16.1	-128	73.0
8B	-14.0	-137	95.0
8C	-15.7	-136	59.0
8D	-15.8	-135	85.00

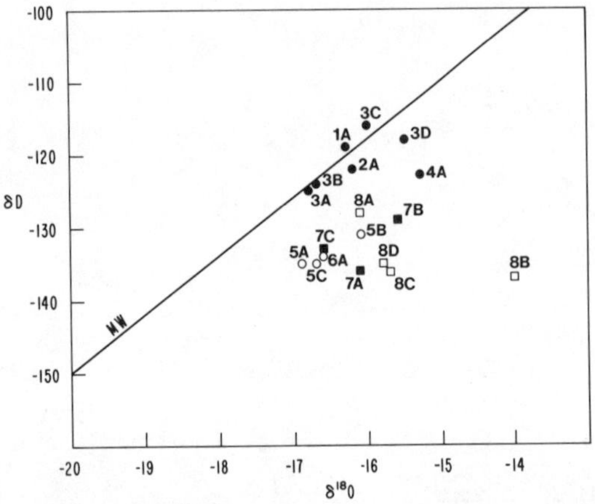

Figure 2. δD and $\delta^{18}O$ values of cold and warm springs. MW (meteoric water line).

δD values ranging from -116 to -125 permil and $\delta^{18}O$ values from -15.3 to -16.8 permil. Samples 3D and 4A have apparently experienced ^{18}O enrichment of 0.5 and 1.3 permil, respectively. Both samples come from very low-volume seeps, consistent with the possibility of evaporation. However the spring temperature of sample 4A is the coolest--8.0°C--of all cold springs sampled and is characterized by the lowest Cl, HCO_3^- and TDS contents of all springs sampled (Table 2). The cold springs from the northwest and southwest of Hot Sulfur Springs constitute the second group. These waters have significantly lower δD values (-131 to -135 permil) and plot off the meteoric water line. All these samples (5A through 5C, 6A) have experienced ^{18}O enrichments of at least 1.2 permil. These waters are significantly warmer (14.5 to 17.5°C) and contain generally higher TDS, SiO_2, Cl and $SO_4^=$ contents (Table 2). The ^{18}O enrichment and increased salinity in these samples may be in part the result of kinetic evaporative effects. However δD-$\delta^{18}O$ trends of low temperature (< 90°C) evaporation (Craig et al., 1963) would result in intersection with the meteoric water line at δD values below (<-140 permil) those of any spring water measured by us in the area or reported for this part of the Basin and Range province (Friedman, 1953). Alternatively, ^{18}O-enrichment could result from either a mixture of ^{18}O-enriched thermal fluid or from low temperature water/rock interaction in the ground water aquifer.

All seven warm spring samples plot to the right of the meteoric water line, with δD values of -128 to -137 permil and ^{18}O enrichments of 1.3 to 4.3 permil. With the exception of sample 8B ($\delta^{18}O$ = -14.0 permil), warm springs from the Hot Sulfur Springs area (solid squares) and those from the hot springs area to the north (open squares) are indistinguishable with respect to both hydrogen and oxygen isotope composition. There is considerable range in temperature of the warm springs from 55 to 95°C. $\delta^{18}O$ and δD values of both cold and hot springs are plotted versus spring temperature in Figure 3A. The warm springs and the two groups of cold springs define distinct T - δD fields as expected. The hottest spring (8B) does have the highest $\delta^{18}O$ value, perhaps reflecting more extensive interaction with the high $\delta^{18}O$ carbonate rocks present at depths greater than about 4500 feet (Sibbett, 1982). Otherwise, there is no significant difference in either hydrogen or oxygen isotope composition between the two coolest warm springs (7C and 8C) and the hotter springs. As shown in Figure 3B, a plot of δD and $\delta^{18}O$ values versus SiO_2 content (Table 2), there also are no significant differences in SiO_2 content between the two coolest warm springs and the remaining warm springs. Further inspection of the chemical data in Table 2 reveals no apparent correlation between isotope compositions, other chemical parameters (Cl^-, HCO_3^-) and temperature. The absence of trends suggests that the warm springs are not simply related by progressive dilution of thermal fluids by cool surface water or by evaporation processes. The low chloride contents of all warm springs (8 to 19 ppm) indicate that these waters are not derived from deep high temperature saline brines of formation or connate origin unless they represent steam condensates. However, the high pH and comparatively low $SO_4^=$ contents argue against such an origin.

RECHARGE AREA FOR THE THERMAL FLUIDS

All the cold springs sampled north of Hot Sulfur Springs in the Hot Creek drainage (the local topographic highland) and drainages to the east (4A) have δD values (-125 permil or heavier) significantly greater than those of the warm springs (-128 to -137 permil). If the thermal waters are derived predominantly from these springs, they must represent steam condensates to explain their lighter δD values. The thermal

TABLE 2. SUMMARY OF CHEMICAL COMPOSITIONS AND GEOTHERMOMETER TEMPERATURES FOR SELECTED COLD AND THERMAL SPRINGS AT TUSCARORA, NEVADA.

	1A	2A	3A	4A	5A	6A	7A	7C	8A	8B	DH 66-5
T°C	10.5	11.5	15	8	14.5	21	89	55.5	73	95	110
pH(20°C)	6.1	6.2	6.4	6.4	7.5	7.5	6.9	6.9	7.6	7.4	8.4
Na (ppm)	7	6	10	41	14	80	151	169	145	148	163
K (ppm)	3	4	<2.5	<2.5	6	5	15	<11	19	20	25
Ca (ppm)	5	4	6	2	6	11	10	19	17	1	14
Mg (ppm)	2	1	2	1	2	3	<0.5	3	2	<0.5	2
Fe (ppm)	0.04	0.13	0.32	0.04	0.025	<0.025	<0.025	0.09	<0.025	0.04	0.06
Sr (ppm)	0.06	0.07	0.06	0.10	0.05	0.13	0.61	0.39	0.59	0.25	1.01
B (ppm)	<0.125	<0.125	<0.125	<0.125	<0.125	0.2	0.8	0.9	0.9	0.90	0.8
SiO_2 (ppm)	43	48	40	17	61	52	129	122	103	104	109
SO_4 (ppm)	5	4	6	2	7	17	52	34	50	55	47
Cl (ppm)	6	4	5	6	7	13	18	19	16	6	26
F (ppm)	0.2	0.2	0.1	<0.1	0.2	1.6	11	8.9	8.7	8.2	8
HCO_3 (ppm)	33	47	47	22	65	232	352	484	382	345	397
Geothermometers (a)											
Qtz adia.	97	101	94	64	111	104	145	142	134	135	137
Qtz Cond.	95	100	92	57	111	104	152	149	139	139	142
Chalcedony	64	70	61	25	82	74	127	123	112	113	116
NaKCa	61	73	56	69	87	87	184	159	194	225	208
NaKCaMg	---	---	---	---	77	77	---	108	144	187	158

(a) Geothermometer temperatures (°C) were calculated from equations given in Fournier (1981), Qtz adia = Quartz adiabatic geothermometer, Qtz cond = Quartz conductive geothermometer.

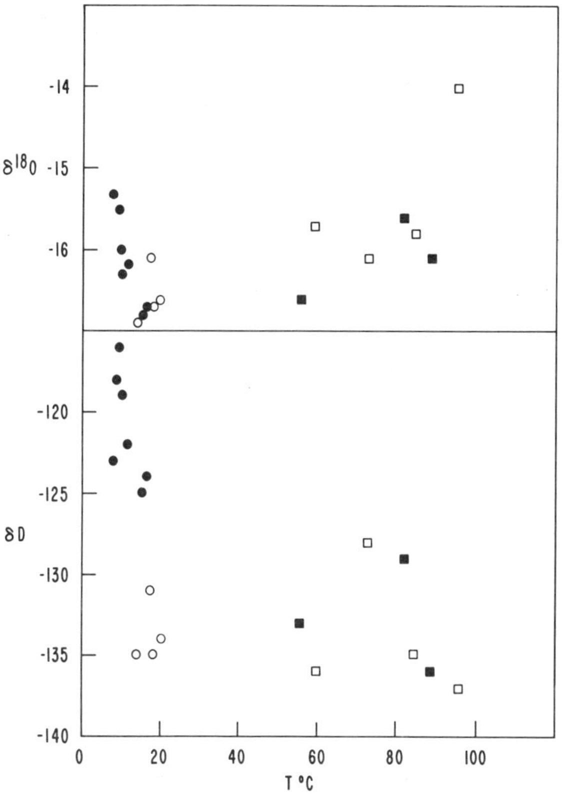

Figure 3A. δD and $\delta^{18}O$ values versus spring temperature.

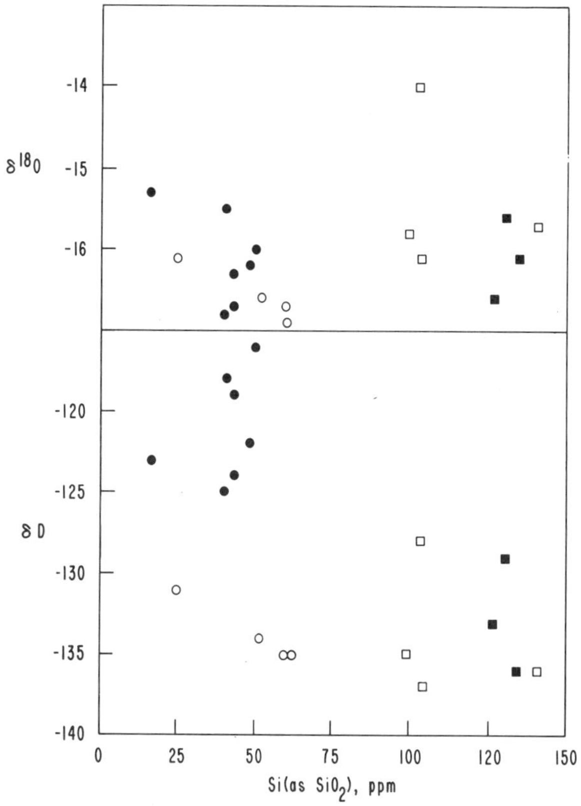

Figure 3B. δD and $\delta^{18}O$ values versus SiO_2 contents.

waters do have low chloride contents characteristic of steam condensates, but the process is not consistent with observed SiO_2 contents, pH values greater than 7.0 or comparatively low $SO_4^=$ contents (Table 2). The cold springs west-northwest (5A to 5D) and south (6A) of the thermal springs have δD values equivalent to those of the thermal waters, demonstrating that surface waters exist in the area with appropriate δD values to directly produce the thermal waters. Alternatively, these cooler, low δD fluids may represent the conductively cooled equivalents of thermal fluids that have leaked south and west-northwestward along faults.

Geophysical evidence from a resistivity survey (Mackelprang, 1982) does suggest fluid flow from southeast to northwest in the area. A 2 ohm-meter zone is thought to indicate thermal fluids issuing from aquifers within Tertiary tuffaceous sediments and Paleozoic quartzites, cherts and carbonates at depths in excess of 4000'. The fluids rise to the surface via fracture zones to the Hot Creek and Hot Sulfur Springs area where they exit as hot springs. The presence of a low resistivity area at the surface northwest of these hot springs indicates that leakage of thermal fluids persists beyond the major existing zones along Hot Creek and Hot Sulfur Creek. These near surface fluids appear to have cooled approximately 50°C or more, but have retained their low δD isotopic signature. Further sampling in the areas south and west of the present hot springs area is needed to actually locate the recharge area. The isotope data do indicate that recharge for the geothermal system is unlikely to the immediate north of Hot Sulfur Springs.

This study was funded by the Department of Energy, Division of Geothermal Energy under contract DE-AC07-80ID12079.

REFERENCES

Craig, H., Gordon, L. I., and Horibe, Y., 1963, Isotopic exchange effects in the evaporation of water, 1, Low-temperature experiment results: Jour. Geophys. Res., v. 68, p. 5979-5087.

Epstein, S. and Mayeda, T. K., 1953, Variations of oxygen-18 content of waters from natural sources: Geochim. Cosmochim. Acta, v. 4, p. 213-224.

Fournier, R. O., 1981, Application of water geochemistry to geothermal exploration and reservoir engineering: in Geothermal Systems: Principles and Case Histories, Edited by L. Rybach and L. J. P. Muffler, p. 109-143.

Friedman, I., 1953, Deuterium content of natural water and other substances: Geochim. Cosmochim. Acta, v. 4, p. 89-103.

Mackelprang, C. E., 1982, Interpretation of the dipole-dipole electrical resistivity survey, Tuscarora geothermal area, Elko County, Nevada: Univ. of Utah Res. Inst., Earth Science Laboratory Report 72, 16 p.

Sibbett, B. S., 1982, Geology of the Tuscarora geothermal prospect, Elko County, Nevada: (submitted to Bull. Geol. Soc. Am.).

Geothermal Resources Council, TRANSACTIONS Vol. 6, October 1982

CHEMICAL AND SULFUR ISOTOPE VARIATIONS IN A THERMAL SPRING SYSTEM SAMPLED THROUGH TIME

David R. Cole

Earth Science Laboratory Division
University of Utah Research Institute
Salt Lake City, Utah 84108

ABSTRACT

Major cations and anions, temperatures and sulfur isotopes of aqueous sulfates and sulfides have been monitored in a Basin and Range thermal spring system for over one year. The data indicate that significant changes in the chemistry are cyclical. Enrichments in Na, Ca, SiO_2, Cl and SO_4 are accompanied by sympathetic increases in surface temperatures and flow rates suggesting that mixing is the major process controlling the cyclic nature of these various parameters. In spite of these chemical variations, major cation ratios remain remarkably constant assuring the applicability of cation geothermometers regardless of the time of year. Major inflow of diluting groundwaters occurs on a regular basis every two to six months.

The $\delta^{34}S$ values of aqueous sulfates and sulfides indicate that these constituents are probably derived from dissolution of marine sulfates and pyrite contained in sedimentary rocks at depth. Sulfur isotopic fractionation between sulfates and sulfides is very near equilibrium (90%) with minimum estimated temperatures ranging from 211°-126°C. Using kinetic isotope data, minimum residence times of between 90 and 500 years are estimated for the interaction of these sulfur species assuming 90% equilibration at temperatures of 200°-175°C for the deep fluids.

INTRODUCTION

Natural waters, regardless of temperature, acquire their chemical characteristics through chemical reactions with solids, liquids and gases that occur during various parts of the hydrologic cycle. Percolation of near-surface waters to greater depths along steeply-dipping faults commonly results in an increase in temperature at a rate governed by the local geothermal gradient. The temperature increase with depth produces an increase in reaction rates between minerals and solutions leading to enrichments in solutions of many rock-forming elements compared to local cooler groundwaters. Typically, the direct influence of precipitation water on chemical and isotopic variations in thermal waters and the turn-over-time of meteoric precipitation are poorly known for most thermal spring systems.

This study was initiated to evaluate the chemical and stable isotopic variations of a typical Basin and Range thermal spring system over a one year period. The major aims of this study were to: (1) document the chemical and isotopic variability in a thermal spring system through time, (2) determine the influence of seasonal precipitation and runoff on the thermal regime, (3) test the reliability of applying chemical and isotopic geothermometers to the system regardless of the time of year, and (4) estimate the source and possible residences times of water in the system.

Geologic Setting

The area selected for the study was the Warm Springs Fault geothermal system, located in northern Salt Lake County (see Fig. 1). The thermal system is located immediately west of the Wasatch Mountains at the easternmost edge of the Basin and Range physiographic province within an active seimsmic zone referred to as the intermountain

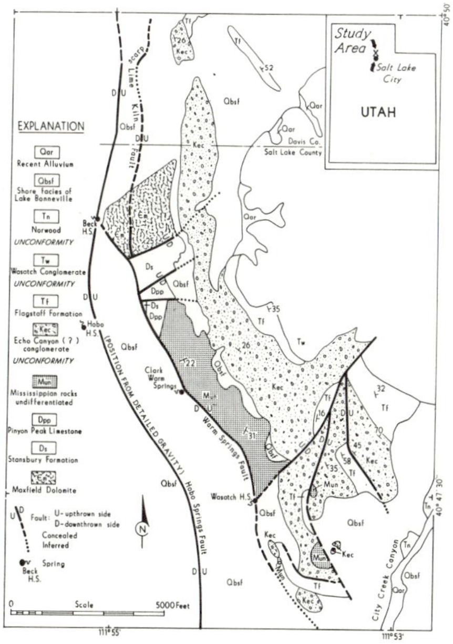

Figure 1. Geologic Map

seismic belt (Murphy and Gwynn, 1979). The thermal springs occur along normal faults that separate the Wasatch Range to the east and a deep valley graben to the west.

The graben is bounded in the thermal area by two faults. The Warm Springs fault has a minimum displacement of approximately 180 meters and the downthrown block is buried beneath approximately 120 meters of valley fill (Murphy and Gwynn, 1979). A second fault referred to as the Hobo Springs fault lies to the west and has a total displacement of approximately 1220 meters (Murphy and Gwynn, 1979). Major thermal springs appear to be located either near the intersections of these two major faults with each other (e.g. Beck Hot Spring) or with relatively minor pre-Basin and Range structures (e.g. Wasatch Hot Spring).

Basement rocks of the graben exposed in the range include Paleozoic limestones, shales, and sandstones that are unconformably overlain by a complex series of Cretaceous and Tertiary conglomerates and Tertiary lacustrine and volcanic units (Fig. 1). The conglomerates are exposed from the Wasatch Springs fault to the main Wasatch Range covering most of the pre-Cretaceous features of the area. A mantle of Lake Bonneville shoreline deposits covers the older rocks within the graben in this area.

Recharge to the system is believed to be from undefined source areas in the Wasatch Range on the east. Heat is provided to the thermal system by a normal geothermal gradient of approximately 32°C/km.

CHEMICAL VARIATIONS

Systematic chemical variations through time are characteristic of thermal springs issuing from this fault system. Figures 2 and 3 show chemical data in ppm and temperatures in °C plotted against time in weeks for two thermal springs (Wasatch and Beck) sampled over a one year period. These figures demonstrate that significant changes in chemistry are cyclical. For example, chemical enrichments in Wasatch Hot Spring are observed during the summer months and depletions occur during fall and early winter months. These chemical variations are accompanied by sympathetic changes in temperature and flow rate; during the summer months at times of maximum flow, Wasatch Hot Spring exhibits its highest surface temperatures.

Beck Hot Spring chemistry (Fig. 3) also exhibits systematic enrichments and depletions but with greater frequency. Major chemical enrichments are separated by time periods ranging from only two to three months. Temperature peaks correlate with chemical enrichments but are not as pronounced as those observed for Wasatch Hot Spring. Chemical and temperature variations in Hobo and Clark Hot Springs parallel Beck and Wasatch Hot Springs, respectively.

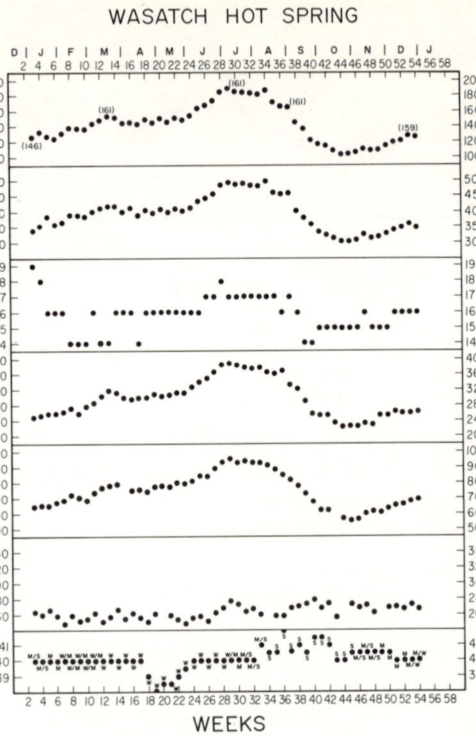

Figure 2. Chemical and temperature variations in Wasatch Hot Spring.

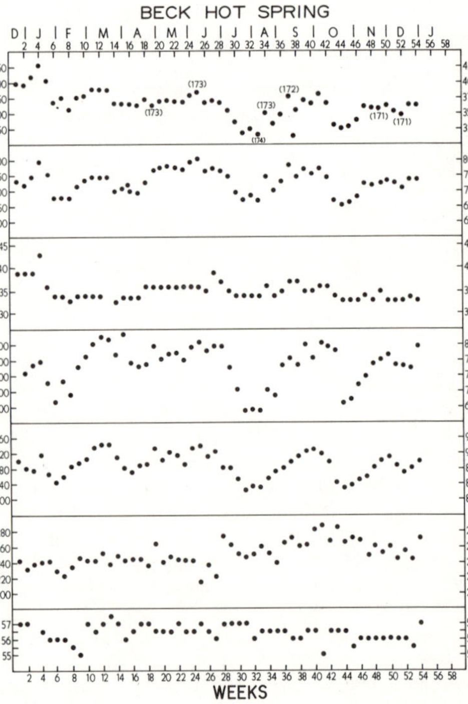

Figure 3. Chemical and temperature variations in Beck Hot Spring.

Major processes that could have produced the observed systematic changes include variations in the degree of rock-water interaction or mixing. Fluctuations in the chemistry could result from changes in temperature or in the type of mineralogy encountered by circulating fluids as the source region, residence time or water/rock mass ratio changes. Geothermometer calculations (quartz-no steam loss, Mg-corrected Na/K/Ca, Na/K, Na/K/Ca and SO_4-H_2S sulfur isotopes) indicate that subsurface temperatures for these thermal springs have remained relatively constant suggesting that the extent of rock-water interaction at depth has remained constant through the year (see examples in Table 1).

The correlation of maximum and minimum surface temperatures with chemical enrichments and depletions, respectively, strongly suggests that mixing is the major process controlling the cyclic nature of the chemistry and temperature. The high concentrations of elements occur during periods when influx of non-thermal ground water is the lowest. Significant declines in the concentrations of Na, Ca, SiO_2, Cl and SO_4 coincide with declines in surface temperatures. A temperature of between 22° and 28°C is required of the non-thermal end member assuming that it has a TDS of 500 ppm or less, which is typical of shallow-valley groundwaters adjacent to this fault zone. If this cool water is derived initially from mountain snow packs, then the duration of time from recharge to discharge is estimated to be four to six months for Wasatch Hot Spring and less than four months for Beck Hot Spring.

The major variations in chemistry and temperature observed for Wasatch Hot Spring could result from mixing of thermal fluids with a single local groundwater. Episodic chemical enrichments and depletions observed at Beck Hot Spring could, on the other hand, reflect mixing of several compositionally similar groundwaters with different sources and path lengths.

SULFUR ISOTOPE VARIATIONS

The $\delta^{34}S$ values for aqueous sulfate and sulfide species sampled from the various thermal springs are represented on a frequency histogram in Figure 4. Sulfates (with a mean of 13.3°/∘∘) are heavier isotopically than sulfide (with a mean of -21.4°/∘∘), which is consistent with the prediction of statistical thermodynamics (Tudge and Thode, 1950). In general, the $\delta^{34}S$ values of aqueous sulfates increase with increasing concentration of sulfate. Because the sulfate concentrations of most groundwaters and thermal springs are much too high to be explained by the atmospheric inflow, the sulfate must originate directly from the aquifer system through the dissolution from rock sulfate or from oxidative weathering of sulfides. The latter mechanism would produce sulfates with $\delta^{34}S$ values less than about 5°/∘∘ (Ohmoto and Rye, 1979); therefore sulfate in the thermal waters must be derived from dissolution of sulfate-bearing sediments at depth.

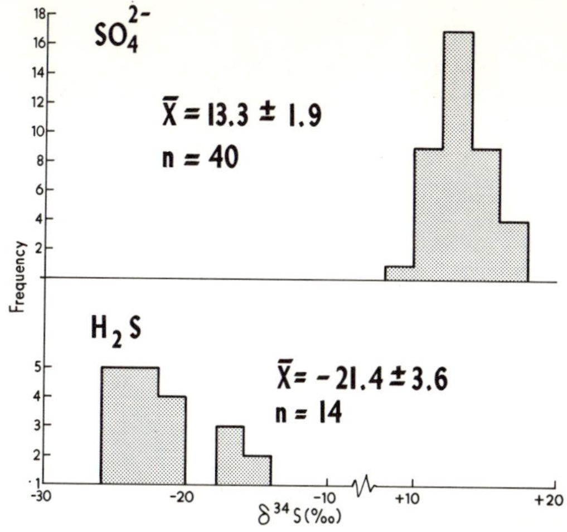

Figure 4. Frequency Histogram of Sulfur Isotope Data.

Systematic variation in the $\delta^{34}S$ within the oceanic reservoirs during Earth history have been deduced from the record preserved in marine evaporites of different geologic ages (Nielsen, 1979). Present-day seawater sulfate has a $\delta^{34}S$ value of about 20°/∘∘. The range of $\delta^{34}S$ values of 10 to 17°/∘∘ for sulfates in these thermal springs suggests that the likely source rocks for the sulfate are Mississippian, Pennsylvanian, Permian, lower Triassic and lower Cretaceous in age. Lithologies of these ages occuring at depth are dominantly carbonates and sandstones.

The $\delta^{34}S$ values of aqueous H_2S range from -25.8 to -15°/∘∘. Several major processes have been described which lead to the generation of H_2S in natural solutions; (1) bacterial reduction of sulfate, (2) organic decomposition, (3) organic reduction of sulfate, (4) inorganic reduction of sulfate, and (5) leaching of sulfides. If the initial aqueous sulfate $\delta^{34}S$ value was approximately 13°/∘∘, then mechanisms (1) through (4) could not produce the range in negative $\delta^{34}S$ values observed for aqueous H_2S (Ohmoto and Rye, 1979). Direct leaching of sedimentary sulfides by thermal fluids could produce $\delta^{34}S(H_2S)$ values ranging from +10 to -30°/∘∘. Once aqueous sulfide is generated, it is free to interact with aqueous sulfate resulting in sulfur isotopic fractionation. Evidence presented in Table 1 suggests that the sulfur isotopic fractionation is very near equilibrium. The fractionations ($\Delta_{SO_4-H_2S}$) and the temperatures estimated from them remain moderately constant through time. In most cases the estimated isotopic temperatures are 30° to 40°C less than the temperatures estimated by the Na/K/Ca geothermometer. These data suggest that reequilibration of sulfur isotopes may have occurred as the fluids ascended to the surface from deep aquifers.

Calculations using sulfur isotope kinetic data for sulfate-sulfide reactions (Ohmoto and

Table 1. Comparison of sulfur isotopic geothermometer estimates with cation and quartz geothermometer temperatures.

SPRING	WEEK	H_2S (ppm)	ΔSO_4-H_2S (o/oo)	$T_{SO_4-H_2S}$ (°C)	Na/K/Ca (°C)	Mg-Na/K/Ca (°C)	Qtz (cond) (°C)
Beck	19	34	33	173	173	87	87
Beck	25	32	37	149	173	87	87
Beck	33	39	41	126	174	88	85
Beck	34	34	31	185	173	87	87
Beck	49	45	37	148	171	85	86
Beck	54	31	36	153	172	88	84
Hobo	19	33	36	152	172	72	86
Hobo	33	30	40	133	172	72	85
Hobo	34	29	28	211	172	72	87
Hobo	49	65	37	148	169	70	84
Hobo	52	33	39	137	170	70	84
Hobo	54	34	35	158	170	70	85

Lasaga, in press) indicate a minimum time of between 90 and 525 years is required for the attainment of 90% isotopic equilibrium for thermal waters ranging in temperature from 200 to 175°C with a total sulfur content of 0.01 molal. An additional 500 to 1000 years is required for re-equilibration of sulfur isotopes at lower temperatures between 150 and 125°C. These calculations imply a very long residence time for the sulfur species in these thermal fluids.

CONCLUSIONS

Significant chemical variations observed through time for thermal springs associated with a Basin and Range fault system are produced by mixing with cooler, dilute groundwaters derived from higher elevations. In spite of these chemical variations, major cation ratios remain remarkably consistent assuring the applicability of cation geothermometers regardless of the time of year. Major inputs of diluting groundwater occur at least every six months and in some cases every three months. The volume and the timing of influx of the diluting components differs for each spring because of variations in the complexity of the fluid flow paths.

The $\delta^{34}S$ values of aqueous sulfates and sulfides indicate that these constituents are derived from dissolution of sulfates and pyrite in sedimentary rocks at depth. Specifically, the sulfate appears to come from Mississippian, Pennsylvanian, Permean and lower Triassic carbonates and sandstones. Sulfur isotope fractionation between sulfates and sulfides is very close to equilibrium with estimated temperatures ranging from 211° to 126°C. Minimum residence times of between 90 and 500 years are estimated for the interaction of these sulfur species assuming 90% equilibration at temperatures of 200° to 175°C for the deep fluids. This long-term chemical residency should not be confused with the short-term pulses of groundwater which affect the magnitudes of cations and some anions in solution.

This research was funded by the Department of Energy, Division of Geothermal Energy under contract DE-AC07-80ID12079.

References

Murphy, P. and Gwynn, J. W., 1979, Geothermal investigations of the Warm Springs fault geothermal system, Salt Lake County, Utah: Utah Geol. and Mineral Survey Report of Investigation no. 140, 29 p.

Nielsen, H., 1979, Sulfur isotopes: in Lectures in Isotope Geology, Edited by E. Jager and J. C. Hunziker, p. 283-312.

Ohmoto, H. and Rye, R. O., 1979, Isotopes of sulfur and carbon, Chapter 10: in Geochemistry of Hydrothermal Ore Deposits, Second Edition, H. L. Barnes (ed.), John Wiley and Sons, New York, p. 509-567.

Ohmoto, H. and Lasaga, A. C., in press, Kinetics of reactions between aqueous sulfates and sulfides in hydrothermal systems: (accepted to Geochim. Cosmochim. Acta.).

Tudge, A. P. and Thode, H. G., 1950, Thermodynamic properties of isotopic compounds of sulphur: Can. J. Res., v. 28b, p. 567.

AN ISOTOPIC STUDY OF THE COSO, CALIFORNIA, GEOTHERMAL AREA

Robert O. Fournier and J. M. Thompson

U.S. Geological Survey, Menlo Park, CA 94025

ABSTRACT

Thirty-nine water samples were collected from the Coso geothermal system and vicinity and were analyzed for major chemical constituents and δD and $\delta^{18}O$. Non-thermal ground waters from the Coso Range were found to be isotopically heavier than non-thermal ground waters from the Sierra Nevada to the west. The δD value for the deep thermal water at Coso is similar to that of the Sierra water, suggesting that the major recharge for the hydrothermal system comes from the Sierra Nevada rather than from local precipitation on the Coso Range. The $\delta^{18}O$ values of the thermal water are about 7°/oo heavier than those of the Sierra water. This shift in $\delta^{18}O$ is the result of water-rock reaction at high temperatures, and the magnitude of the shift indicates that the ratio of rock to total water has been large for the system up to its present stage of development. The isotopic data are compatible with the geochemical model previously proposed by Fournier, et al. (1980).

INTRODUCTION

Smith and others (1979) measured the deuterium concentrations in rain and snow collected at 26 stations in California and Nevada during the exceptionally wet 1968-69 season. Their results showed that the winter precipitation upon the Sierra Nevada was isotopically slightly lighter than the summer and fall precipitation on the nearby Mojave Desert. Most of the Sierra ground-water recharge comes from winter storms moving generally from west to east. These winter storms drop most of their moisture before reaching the Coso Range, which is directly east of the Sierra Nevada. In contrast, most of the Coso Range recharge is from large, but infrequent tropical storms that come from the south. Therefore, the isotopic composition of the normal, non-thermal ground water in the vicinity of the Coso geothermal field is slightly lighter than the isotopic composition of nearby Sierran waters. The purposes of the present study were to determine (1) whether the recharge for the Coso geothermal system comes from precipitation on the Sierra Nevada or from local precipitation at Coso, and (2) whether the (1980), the water cannot boil as it moves isotopic data are compatible with the geochemical model for the reservoir fluid proposed by Fournier and others (1980).

SAMPLES STUDIED

A total of 39 samples from 37 different sources were collected and analyzed. Eleven samples are of cold ground waters flowing from springs and wells in the Coso Range, north and east of Coso Hot Springs within the China Lake Naval Weapons Center. Eight samples are of Sierra Nevada ground water, collected through a 40-km-long region west and northwest of the Coso area; and five samples are from wells in alluvium in Rose Valley, between the Coso Range and the Sierra Nevada. Fournier and others (1980, table 1) previously reported the details of the collection and chemical analyses of two samples of thermal water from the CGEH (Coso Geothermal Exploration Hole) No. 1 well (CC77-4 and CF78-1). Two downhole samples from the Coso No. 1 well (CF79-1 at -50 m and CF79-2 at -95 m) were collected for the present study using a modified version of the Fournier and Morganstern (1971) sampling tool designed for use on wireline equipment.

RESULTS AND DISCUSSION

A plot of δD versus $\delta^{18}O$ (Fig. 1) shows that cold ground waters flowing from the Sierra Nevada and the well waters from Rose Valley all plot near the average meteoric water line of Craig (1961) and have δD values less (more negative) than -100. The δD values of the Sierra waters generally become more negative to the north (Smith and others, 1979). The cold ground waters collected from springs and wells within the Coso Range also plot near the meteoric line (Fig. 1), but they have δD values heavier than -100, averaging -94. The difference in isotopic composition between the Sierra ground water and that of the Coso Range reflects the different types of storm systems contributing the major water recharge in the two areas, as discussed above.

Water samples from the CGEH No. 1 well, CC77-4 and CF78-1, plot far to the right of the meteoric line (Fig. 1), as do thermal waters world (White, 1970). As meteoric water flows into a geothermal system and becomes heated, its oxygen exchanges with the isotopically heavy oxygen in the surrounding rock so that the oxygen in the water becomes isotopically heavier and the oxygen in the rock becomes lighter. Hydrogen reacts in a similar manner. However, because the rock contains a very large amount of oxygen and only a small amount of hydrogen, the oxygen isotopic composition of the water is changed considerably while the hydrogen isotopic composition is changed only slightly. Therefore, the amount of $\delta^{18}O$ shift away from the meteoric line gives an indication of the relative amount of meteoric water that has reacted with rock, whereas the δD value is indicative of the δD of the meteoric recharge water. The shift in $\delta^{18}O$ of 7°/oo for samples CC77-4 and CF78-1 is very large and indicates that relatively little water has moved through the system.

The δD value of the CGEH No. 1 water suggests that recharge for the hydrothermal system comes predominantly from the Sierra Nevada to the west with little or no component of recharge from the Coso Range. However, the data do not rule out the possibility that recharge is a mixture of isotopically light Sierra water from parts of the Sierra north of Coso and isotopically heavy locally derived Coso Range water. The isotopic data unambiguously show that recharge for the CGEH No. 1 thermal water is not entirely from locally derived ground water, nor could recharge be from OwensLake, which is isotopically very heavy because of extensive evaporation (Friedman and others, 1976).

The two samples from the shallow Coso No. 1 well (CF79-1 and CF79-2) also plot far to the right of the meteoric line, but at δD values of -15 and -99, respectively (Fig. 1). The sample from near the top of the water table in this well (CF79-1) at -50 m has about twice the total dissolved solids and is isotopically much heavier than the sample from near the bottom of the well (CF79-2). Evaporation from the top of a free-standing column of water in the well accounts for these differences very nicely.

The relationship of the waters entering the CGEH No. 1 and Coso No. 1 wells is of great interest. On the basis of chemical data obtained from downhole samples, Fournier and others (1980) concluded that a single parental water supplied both wells and that compositional variations in the waters collected at the wellheads were the result of (1) different amounts of boiling in the wells during upflow, and (2) a higher reservoir temperature in the vicinity of the Coso No. 1 well (245°C) than in the vicinity of the CGEH No. 1 well (205°C). These reservoirs are places in the rock where fracturing is locally more extensive than elsewhere so that permeability and the ratio of water to rock are higher than in the surrounding rock. In the model of Fournier and others (1980), the water cannot boil as it moves from most geothermal systems throughout the laterally from the 245°C reservoir to the 205°C reservoir. If the water had boiled, then the chloride concentration in the downhole samples from the two wells would have been different. The fact that the chlorides are very similar indicates very slow natural flow and conductive cooling of thermal water as the water moves from the vicinity of the Coso No. 1 well towards the CGEH No. 1 well. The flow could be slow because the permeability within the rock connecting the two reservoirs is very low. Alternatively, the permeability could be high (essentially one reservoir with small vertical extent and a horizontal temperature gradient within it) and convective flow limited by poor permeability on the outflow part of the convection system.

If the model presented by Fournier and others (1980) is correct, the isotopic composition of the water entering the two wells should be about the same prior to any boiling or evaporation during upward movement after leaving the respective local reservoirs (if little water-rock isotopic re-equilibration took place because of very slow reaction rates as the temperature changed from 245°C to 205°C). The slightly different observed isotopic compositions of the downhole sample from the Coso No. 1 well (CF79-2) and of the downhole sample from the CGEH No. 1 well (CF78-1) appears to be the result of slight contamination by evaporated water from the top of the Coso No. 1 well (CF79-1) as shown by the straight-line relationship in Figure 1 among samples from these wells.

CONCLUSIONS

The average meteoric water falling on the Coso Range is isotopically slightly heavier than the average meteoric water falling on the Sierra Nevada to the west. The deuterium concentration in the deep geothermal water is similar to that in the Sierra Nevada ground water and is different from that in the Coso Range water. Therefore, recharge into the deep part of the geothermal system probably comes predominantly from the Sierra Nevada. The main upflow in the hydrothermal system appears to be along a north-northeast-trending fault zone along which Coso Hot Springs is located. The large shift in $\delta^{18}O$ of about 7°/oo in the thermal water indicates that the rock-to-water ratio is large, suggesting very slow movement of new water into and old water out of the convection system. The isotopic data are compatible with the geochemical model of Fournier and others (1980) in which some of the chloride-rich hot water ascending along faults passing through the Coso Hot Springs area encounters other permeable zones and flows laterally toward the CGEH No. 1 well, cooling conductively and reacting chemically with the surrounding rock as it travels. The top of the chloride-rich water remains below ground and, where underground boiling occurs, fumaroles, acid-sulfate pools and acid altered rock are found at the surface.

REFERENCES

Craig, Harmon, 1961, Isotopic variations in meteoric waters: Science, v. 133, p. 1702-1703.

Fournier, R. O., and Morganstern, J. C., 1971, A device for collecting down-hole water and gas samples in geothermal wells: U.S. Geological Survey Professional Paper 750-C, p. C151-C155.

Fournier, R. O., Thompson, J. M., and Austin,C.F., 1980, Interpretation of chemical analyses of waters collected from two geothermal wells at Coso, California: Journal of Geophysical Research, v. 85, p. 2405-2410.

Friedman, Irving, Smith, G. I., and Hardcastle, K. G., 1976, Studies of Quaternary saline lakes-II. Isotopic and compositional changes during desiccation of the brines in Owens Lake, California, 1969-1971: Geochimica et Cosmochimica Acta, v. 40, p. 501-511.

Smith, G. I., Friedman, Irving, Klieforth, Harold, and Hardcastle, K. G., 1979, Areal distribution of deuterium in Eastern California precipitation, 1968-1969: Journal of Applied Meteorology, v. 18, p. 172-188.

White, D. E., 1970, Geochemistry applied to the discovery, evaluation and exploitation of geothermal energy resources: Geothermics Spec. Issue 2, v. 1, p. 58-80.

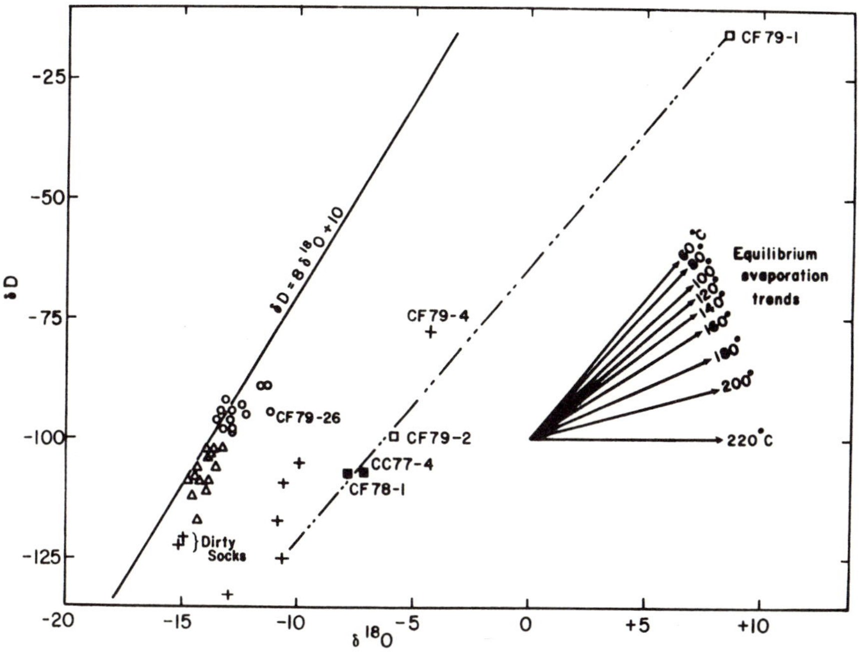

Figure 1. δD versus $\delta^{18}O$ for thermal and nonthermal waters from the Coso geothermal area. Circles, nonthermal waters from the Sierra Nevada and Rose Valley; solid squares, waters from the CEGH No. 1 well; open squares, waters from the Coso No. 1 well; crosses, other thermal waters and steam condensates.

Geothermal Resources Council, TRANSACTIONS Vol. 6, October 1982

OXYGEN AND HYDROGEN ISOTOPES IN THERMAL WATERS AT ZUNIL, GUATEMALA

Robert O. Fournier[1/], Bruce B. Hanshaw[2/], and J. F. Urrutia Sole[3/]

1/ U.S. Geological Survey, Menlo Park, Calif.
2/ U.S. Geological Survey, Reston, Va.
3/ Instituto Nacional de Electrificación, Guatemala

ABSTRACT

Enthalpy-chloride relations suggest that a deep reservoir exists at Zunil with a temperature near 300°C. Water from that reservoir moves to shallower and cooler local reservoirs, where it mixes with diluted water and then attains a new water-rock chemical equilibrium. This mixed water, in turn, generally is further diluted before being discharged from thermal springs. The stable-isotopic composition of the thermal water indicates that recharge for the deep water at Zunil comes mainly from local sources. The presence of measurable tritium, which suggests that the deep water has been underground about 20 to 30 years, also indicates a local source for the recharge.

INTRODUCTION

A geochemical model, incorporating both boiling and mixing processes, was previously developed for the Zunil geothermal system by the Guatemala Instituto Nacional de Electrificacion (INDE) from chemical analyses of waters from springs and wells (Urrutia Sole, 1979). The present study was undertaken mainly to determine whether recharge for the deep reservoir water is from a local or distant source.

The prior analytical work carried out by INDE greatly facilitated the selection of waters for isotopic analyses. At the time of the isotopic collection, the same waters were also collected for chemical analyses to guard against the possibility that a water might have changed in composition within the interval between the earlier INDE and the present work.

BOILING AND MIXING RELATIONS

Boiling and mixing relations, which must be considered in evaluating the isotopic results, are conveniently displayed on an enthalpy-chloride diagram (Fig. 1) (Fournier, 1979; Fournier et al., 1979). Most of the natural hot-spring waters in the Zunil area, which contain less than 200 mg/kg Cl, appear to be mixtures of much hotter water with cold water that has not come to a new water-rock chemical equilibrium after mixing. However, some water-rock reaction has taken place after mixing, resulting in increased amounts of magnesium and bicarbonate in the mixed water.

Of the low-chloride hot springs, those in the southern part of the area (samples 16-19, Fig. 1) show one mixing trend that is on a line extending from dilute water through points A, B, and F on the plot, and those in the northern part of the area (samples 20-22) show a slightly different but parallel mixing trend extending through points C and G. Chemical geothermometers indicate that the one high-chloride spring in the area (sample 1) issues from a reservoir in which water-rock chemical equilibrium was attained at about 210°C. That spring is vigorously boiling, and the rising water probably is cooled mainly adiabatically. The calculated position of the reservoir water feeding spring 1 is at point A (Fig. 1).

Well Z-4 is located near spring 1 but on the opposite side of the Samala River. Water from this well was collected after first separating steam at 30 psig (point S_{30}, Fig. 1) and then flashing to atmospheric pressure. The composition of the flashed water is very similar to that of spring 1, although the well water is slightly more dilute and the geothermometer temperatures are slightly higher. The difference in total salinity between the waters from well Z-4 and spring 1 may reflect slight conductive cooling of the well water because that well was discharged vigorously for only a short time before the water collection. Point B (Fig. 1) shows the condition of the reservoir feeding well Z-4, assuming no conductive cooling and a temperature equal to the maximum temperature measured in the well (205°C). The chemical composition of the water, however, suggests that this well may have intersected a fracture that brings water from a higher temperature environment of 220° to 225°C, (point C, Fig. 1).

Well Z-2 produces water that is twice as rich in Cl (after flashing) as the flashed water from well Z-4. Assuming no conductive cooling, the reservoir water would plot at point D (Fig.

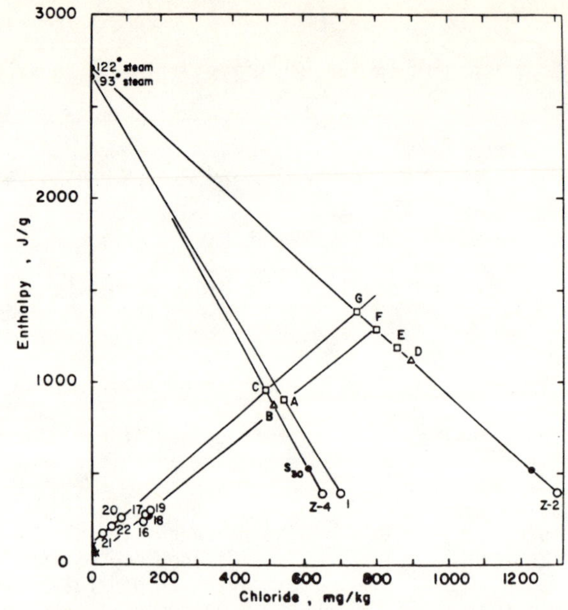

Fig. 1 Enthalpy-chloride relations for waters from Zunil area of Guatemala. Circles, thermal water at time and place of collection, triangles, calculated reservoir waters at maximum measured temperatures in wells Z-2 and Z-4; dots, conditions just after steam separation in first-stage separator; squares, reservoir conditions calculated from geochemical considerations.

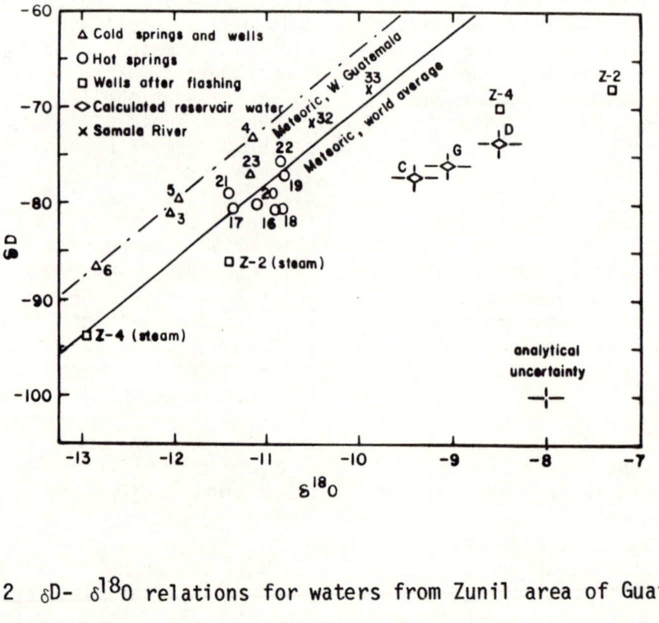

Fig. 2 δD- $\delta^{18}O$ relations for waters from Zunil area of Guatemala.

2) if that reservoir were at the maximum temperature measured in the well (255°C). The Na/K ratio of the water in well Z-2 suggests that the reservoir could be as hot as 270°C (point F, Fig. 1), and the enthalpy-chloride relations of all the thermal waters in the area suggest deeper reservoir waters of about 290° to 305°C (points F and G).

δD - δ^{18}O RELATIONS

Fig. 2 plots the δD-δ^{18}O relations for waters from the Zunil area. The solid line shows the average trend for meteoric water throughout the world (Craig, 1961). Local departures from this line occur, and the dot-dashed line shows the apparent meteoric trend for western Guatemala.

Craig et al. (1956) showed that for geothermal waters the deuterium concentration in the water remains relatively unchanged as that water circulates through silicate rocks at high temperatures, whereas the oxygen-18 content changes by water-rock reactions. Thus, it is normal to find that thermal waters plot to the right (relatively rich in oxygen-18) of the meteoric line. However, the δD value gives a good indication of the original δD of the thermal water before water-rock reaction at high temperatures. In Fig. 2, points C, D, and G have been adjusted to the compositions of the original reservoir water before steam separation. The δD values of these points fall within the range of values measured in the local cold-spring waters.

TRITIUM RELATIONS

The tritium content of cold ground waters of the area ranges from very low (0.8±0.5 T.U.) to very high (24.7±1.2 T.U.). The 24.7 T.U. value represents very young water that probably fell in a single storm. The low but measurable tritium contents (1.1, 0.7, and 1.1±0.1 T.U., respectively) of the deep thermal waters in wells Z-2 and Z-4 and spring 1 indicate that these waters have been underground about 20 to 30 years. The tritium concentrations in the well waters are about the same as those in the lower temperature "mixed water" springs (16-22). In most geothermal systems that have been studied, the deep thermal water contains no measurable tritium.

CONCLUSIONS

The stable isotopes of water indicate that recharge for the deep thermal water in the Zunil geothermal system comes mainly from local sources. The presence of measurable tritium in the water produced from well Z-2 also suggests that the deep water has not come from a distant source. If this source were many kilometers away, the water should contain no measurable tritium.

Enthalpy-chloride relations suggest that water exists deep in the system at 290° to 305°C with chloride concentrations (before flashing) of 750 to 800 mg/kg. Some of that water mixes with cold dilute water in a shallow environment and attains a new water-rock equilibrium at 210° to 225°C. The chloride concentration of the water (before flashing) is about 500 to 550 mg/kg.

REFERENCES

Craig, H., 1961, Isotopic variations in meteoric waters: Science, v. 133, p. 1702-1703.

Craig, H., Boato, G., and White, D. E., 1956, Isotopic geochemistry of thermal waters: National Academy of Sciences-National Research Council Publication 400, p. 29-38.

Fournier, R. O., 1979, Geochemical and hydrologic considerations and the use of enthalpy-chloride diagrams in the prediction of underground conditions in hot spring systems: Journal of Volcanology and Geothermal Research, v. 5, p. 1-16.

Fournier, R. O., Sorey, M. L., Mariner, R. H., and Truesdell, A. H., 1979, Chemical and isotopic prediction of aquifer temperatures in the geothermal system at Long Valley, California: Journal of Volcanology and Geothermal Research, v. 5, p. 17-34.

Urrutia Sole, J. F., 1979, Zunil geothermal system temperature estimation at depth by the chemical method: Guatemala, Instituto Nacional de Electrificacion report, 15 p.

Geothermal Resources Council, TRANSACTIONS Vol. 6, October 1982

GEOCHEMICAL AND ISOTOPIC STUDIES OF LOW TEMPERATURE GEOTHERMAL WATER, EYJAFJORDUR, NORTH ICELAND

Hrefna Kristmannsdottir and Sigfus Johnsen

National Energy Authority, Geothermal Division
Science Institute, University of Iceland

ABSTRACT

Stable isotopes and dissolved solids in the geothermal waters have been studied as a part of geothermal prospecting in the Eyjafjordur region. The recharge areas of the waters are traced by the ratios of the stable isotopes to the North western part of the glacier Vatnajokull. At least three separate reservoirs are found to feed the natural springs and drillholes. The geothermal water in the region is of a diluted type with a high pH. It can be used directly for house heating without corrosion or scaling problems.

INTRODUCTION

Geothermal activity is widespread in the Eyjafjordur region (fig. 1). The bedrocks in the area are built of Tertiary basaltic lavas, 8-10 m.y. old (see Bjornsson, 1981) with thin sedimentary layers between. The lava pile is intersected by numerous dykes, commonly with a north-south direction. The bedrock is rather impervious due to intensive secondary mineralization.

The yield of most natural springs in the area is low, mostly less than 1 l/s. The first deep drillhole in the region was drilled at Laugaland in 1975. It yielded at first 100 l/s of free flowing 96°C warm water, but during production the discharge has decreased rapidly. Now seven years later the water level in the Laugaland area has been lowered 200-300 m and about 85 l/s can be pumped out of the area.

There are now four production areas in the Eyjafjordur region, Laugaland, Ytritjarnir, Botn and Glerargil. The temperature of the water pumped out of the areas is 55-100°C. Water is still short for the needs for space heating of the area and extensive exploration is going on.

STUDY OF STABLE ISOTOPES IN THE GEOTHERMAL WATER

The $\delta D/\delta 18$ relation (fig. 2) indicates that the isotopic ratios are not affected by water-rock interaction. A best-fit line for the Eyjafjordur region falls nearer to the "meteoric line" (Dansgaard, 1964) than for the best-fit line for precipitation in Iceland (Arnason, 1976). This feature connected with the very low δ values of the water indicate a high age of the geothermal water.

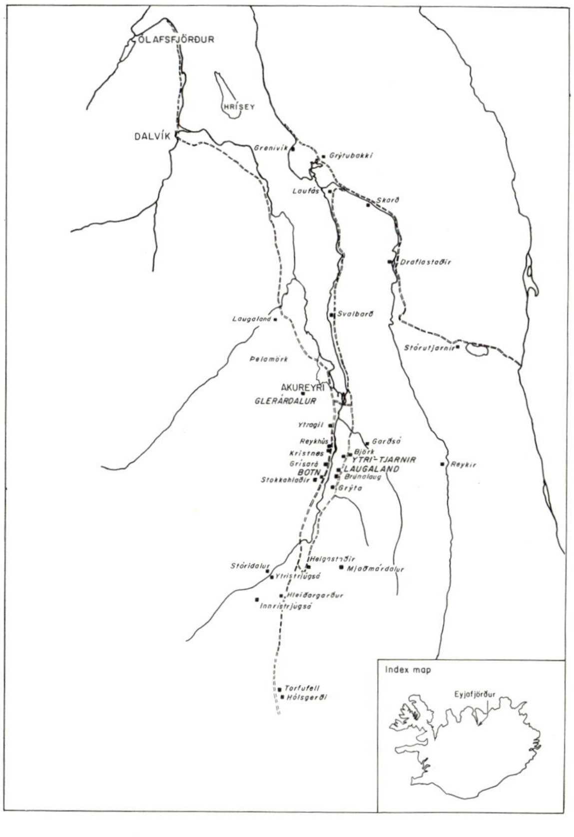

Figure 1. Location of hot springs in the Eyjafjordur region

The mean δD for the region is -99 ± 4 $^0/_{00}$ (population deviation). Local groundwater in the region has δD of about -86 $^0/_{00}$ (Arnason, 1976).

The geothermal water in northern Eyjafjordur region has its origin from North western part of the glacier Vatnajokull according to Arnason's (1976) δD map of precipitation in Iceland. A few springs show signs of mixing with local groundwater. In the southern part of the region most springs could have the same recharge area, but in one area (Hols gerdi) the springs must have a different origin. The $\delta 18$ values give a similar overall picture as the δD values. $\delta 18$ in the hot springs and drill-

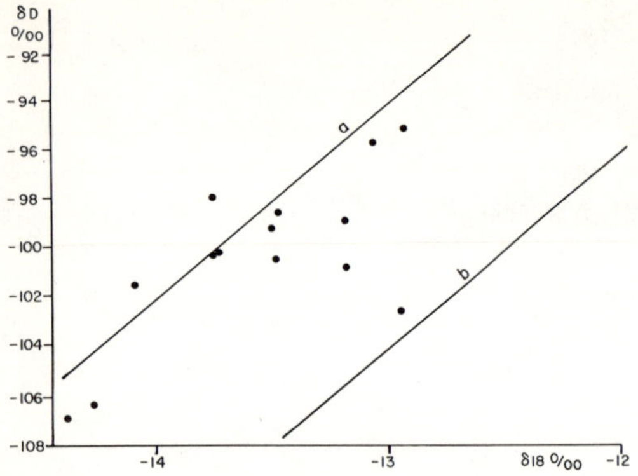

Figure 2. δD against δ18 for geothermal water in Eyjafjordur. Line a is the meteoric line (Dansgaard,1964). Line b is the "best-fit" line for precipitation in Iceland (Arnason,1976).

holes in the Eyjafjordur region is ÷13.3 ± 0.6 ‰ and ÷11.65 ‰ in local ground water. On basis of δ18 values of water samples from drillholes the geothermal water in the northern part of Eyjafjordur has been classified into two groups. The δ18 values of water from the main aquifers at Laugaland (δ18=÷13) differ significantly from Ytritjarnir, Botn and Glerargil (with δ18=÷14). The deepest, and sparsely yielding aquifer in the Laugaland area has similar δ18 value as the main aquifer in the other mentioned areas.

CHEMISTRY OF THE WATERS

Generally the waters have low contents of dissolved solids (table 1) and a high pH. The variation in contents of some substances between different aquifers is often too low for a safe detection. On basis of chloride and fluoride contents and Cl/B ratio the geothermal water in the northern part of the region could with exceptions of two springs be from the same reservoir. In the southern part of the region the water can be classified into at least three different groups on the same basis. Water from drillholes in all production areas is very similar (table 1). A good comparison can be made between measured deep temperature and calculated temperature from chemical thermometers (see Ellis and Mahon, 1977).

The chemical thermometer found to be the most applicable is a "feldspar thermometer" based on theoretical equilibrium between Na-and K-feldspar and solution (Helgeson, 1969). A plot of "feldspar temperature" against measured temperature in drillholes is shown in Figure 3 a. Chalcedony temperatures from samples in deep drillholes are either the same or lower than measured temperature.
A graph of "feldspar temperature" against chalcedony temperature is shown in Figure 3 b. A concentration of points plot on a line displaced paralel to the line connecting equal t chalcedony and t feldspar as t chalcedony is on average some 10°C lower than feldspar and also measured deep temperature.

Table 1. Selected analysis of geothermal water from the Eyjafjordur region. For location see fig. 1.

	Laugaland	Botn	Ytritjarnir	Glerargil
t°C	94	83	80	55
pH/°C	9.77/21	9.83/24	9.95/21	9.83/20
SiO_2*	102.0	81.3	92.3	78.1
Na*	50.9	48.6	53.2	48.7
K*	1.50	1.02	0.96	0.72
Ca*	2.79	3.48	3.60	2.16
Mg*	0.004	0.004	0.006	0.059
CO_2*	18.0	11.6	13.3	13.8
SO_4*	38.1	43.3	43.0	31.8
H_2S*	0.06	<0.1	0.09	<0.1
Cl*	11.6	8.3	11.4	8.8
F*	0.43	0.60	0.47	0.60
B*	0.130	0.143	0.160	0.190
Tot.diss.sol*	242	233	223	218
δD	÷99.5	÷98.0	÷102.2	÷100.3
$\delta^{18}O$	÷13.32	÷14.03	÷14.12	÷14.09
Chalced. temp °C	88	70	75	58
"Feldsp.temp °C	108	86	79	71

* in ppm

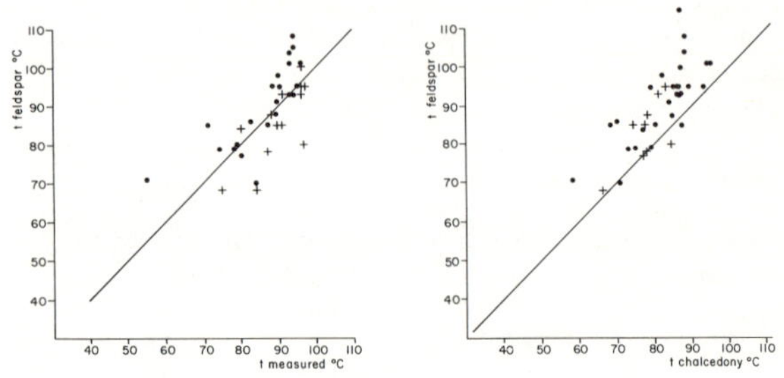

a) b)

Figure 3. a)"Feldspar temperature" against measured temperature in water from drillholes. b)"Feldspar temperature" against chalcedony temperature. The points are for water sampled at well head and crosses are samples from depth in the drillhole.

The water is quite good for space heating as there is little danger of precipitation of any kind in pipelines and it is not very corrosive. It can be used directly in preparation of food and for drinking.

REFERENCES

Arnason,B.,1976. Groundwater systems in Iceland traced by deuterium. Soc.Sci.Islandica, Occasional publ. 42, 236 pp.

Björnsson,A.,1981. Exploration and exploitation of low temperature geothermal fields for district heating in Akureyri, North Iceland. Geothermal Resource Council Transactions, 5, 495-498.

Dansgaard,W.,1964. Stable isotopes in precipitation. Tellus, 16, 4, 436-468.

Ellis,A.J. and Mahon,W.A.J.,1977. Chemistry and geothermal systems. Academic Press, N.Y.,392 pp

A SIMPLE FORMULA FOR SATURATION TEMPERATURE/PRESSURE OF NaCl SOLUTIONS

by Donald E. Michels

Republic Geothermal, Inc.
11823 East Slauson Avenue
Santa Fe Springs, California 90670

Abstract

The saturation temperatures and pressures of NaCl solutions throughout the geothermally interesting range can be calculated by:

$$\text{psia} = \exp[6.7028 - 5.52 \times 10^{-7}(\text{ppm Cl}) - 3712/(^{\circ}\text{F}+460)].$$

Results of this equation agree with tabular data generally to within ± 2.5 percent of psia for NaCl concentrations up to 15 weight percent and 300°C. At higher NaCl concentrations different sets of tabular data are themselves out of conformance while the equation yields intermediate values.

This equation can be adapted to mixed-salt brines and to dynamically flashing systems by including modifiers for just the second term in the brackets.

INTRODUCTION

Tabular data for thermodynamic properties of NaCl solutions are available (1, 2). However, as explained in (3), a mathematical form is more economical for computer usage. Polynomial equations reported in (3) aimed to imitate the tabular data in (1) and succeed within a precision of $\pm 3\%$ for enthalpy, entropy, density, viscosity, temperature, and pressure, over the range of geothermal interest--to 30% NaCl and 325°C.

The method of (3) is modestly complicated. For example, calculations for the saturation pressure and temperature involve seven equations and 32 mathematical or physical constants. It is suitable only for computer applications.

An alternative approach uses simpler logarithmic equations to provide an equally good fit to the tabular data. Saturation temperature and pressure are described by a single equation using only four mathematical constants. It is suitable for hand-held calculators.

DERIVATION

Empirically, the logarithm of pressure for a two-phase (liquid-vapor) system of one component is proportional to the reciprocal of the absolute temperature, as in (a).

$$\log P = A + \frac{B}{T_{abs}} \qquad (a)$$

Equation (a) fits the data for pure water in (1) and (4) to within better than $\pm 2.5\%$ of P when B = -3713.5 and A = 6.703. Pressure units are pounds per square inch; temperature in °F. Saturation temperature/pressure for NaCl solutions reported in (1) can also be fitted by an equation having the form of (a), but which utilizes different values of A and B, chosen according to the NaCl concentration.

The mathematical systematics of the relationship between salt content and A- and B-values can be explored by making linear least squares fits for (a) and the sets of P-T tabulations in (1) that refer to different NaCl concentrations. This approach yields sets of values for A and B which are listed in Table 1 and plotted in Figure 1.

Table 1
A- and B-values for Equation (a)

% NaCl	Molal NaCl	A	B
0.00	0.00	6.702	-3712.7
5.00	0.9006	6.688	-3712.5
10.00	1.9012	6.670	-3710.8
15.00	3.0195	6.651	-3711.0
20.00	4.277	6.632	-3714.5
22.61	5.0900	6.625	-3720.3
25.00	5.7036	6.622	-3728.8
27.53	6.500	6.622	-3742.4
29.03	7.000	6.564	-3696.2
30.00	7.3332	6.557	-2694.7
31.86	8.000	6.537	-3686.3
35.00	9.2135	6.543	-3711.1

Figure 1 shows the locus of A and B values has a discontinuity at about 28% NaCl. Such an outcome is curious, considering the smooth nature of (a) and the expectation of smooth behavior in real physical systems. The

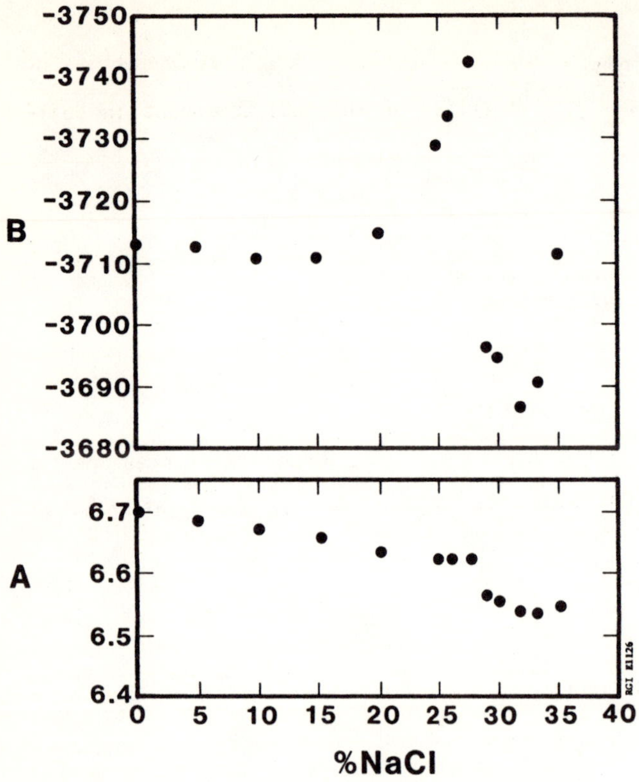

Figure 1. Percent NaCl vs coefficients for log $P=A+B/T_{abs}$

calculations used to generate the tables in (1) involve polynomial approximations to reported data. It is suspected that the discontinuity is due to artifacts in the equations in (1) that are exaggerated when the partial molal enthalpy is calculated, persuant to making the tables. Thus, the discontinuity in Figure 1 does not necessarily indicate a serious failure of logarithmic equation (a) to represent physical reality.

Accepting that the A and B values in Table 1 are approximately correct, it remains to describe their variability in terms of NaCl content in order to extend equation (a) to a more general form. For example, values of A in Table 1 can be fitted to percent NaCl by a linear least squares method, yielding (b).

$$A = 6.7028 - 0.00335\%\ NaCl \qquad (b)$$

Contrastingly, values for B, as plotted in Figure 1, appear less suited to a linear least squares fitting. Below 20% NaCl they seem to be independent of salt content whereas above 20% NaCl they seem to be strongly and wildly dependent on NaCl content. If it is accepted that such behavior is probably ficticious, as suggested above, then a nominal value of B = 3712 can be assigned for NaCl solutions.

Using these values for A and B in (a) yields equation (c), the counterpart of (a) which applies to aqueous solutions of NaCl.

$$P = \exp\left[6.7028 - 0.00335\ (\%NaCl) - \frac{3712}{^{o}F + 460}\right] \qquad (c)$$

RESULTS

Conformance of (c) with the tabular data in (1) is indicated in Table 2. Values are percent mismatch; negative signs indicate the calculated value is smaller than the table value.

Table 2
Percent Mismatch of Calculated vs Tabular Pressures

% NaCl	100°C / 212°F	150 / 302	200 / 392	250 / 482	300 / 572
0	2.8	-1.8	-1.7	0.3	2.4
5	2.0	-2.5	-2.4	-0.4	1.7
10	1.8	-2.6	-2.5	-0.6	1.7
15	2.5	-2.1	-2.0	0.00	2.4
20	4.2	-0.6	-0.6	1.4	3.7
25	7.9	2.2	1.7	3.3	5.6
30	s	6.3(1)	5.2	6.0	7.6
35	s	s	s	10.0(2)	10.0

(1) 160°C
(2) 260°C
s = saturated

The mismatch values at higher NaCl concentrations in Table 2 may not represent defective estimates about the real physical system. It has been noted (3) that references (1) and (2) contain discrepancies ranging upward beyond 8% at higher temperatures and salt contents. Specifically, values in (2) are higher than in (1); the values calculated with (c) tend to be intermediate. This suggests that values in Table 2 overestimate the mismatch between (c) and the real physical system.

Equation (c) is useful for engineering applications where nominal uncertainties are acceptable. This includes most of the geothermally interesting ranges of temperature and salt content. Rewriting (c) into the general form (d)

$$P = \exp[g - h(C) - \frac{j}{T+k}] \qquad (d)$$

makes it easy to describe sets of values for mathematical constants that are appropriate for several conventional units, as in Table 3.

Table 3
Units and Coefficients for (d)

Units for P	T	C	Values for g	h	j	k
psia	°F	%NaCl	6.7028	.00335	3712	460
psia	°F	ppmCl	6.7028	$5.52X10^{-7}$	3712	460
atm	°C	%NaCl	5.5355	.00335	2062	273
atm	°C	ppmCl	5.5355	$5.52X10^{-7}$	2062	273
kPa	°C	%NaCl	9.550	.00335	2062	273
kPa	°C	ppmCl	9.550	$5.52X10^{-7}$	2062	273

Note that the differences of P-T behavior vis-a-vis pure NaCl solutions and mixed salt brines might be described concisely by correct selection of h-values in (d), without regard to g, j, or k. Thus, expressing (C) in units of ppmCl (parts per million of chloride in the liquid phase) makes tempting the use of (d) for mixed salts such as occur in the hypersaline brines of the Imperial Valley where Cl commonly represents 99+ percent of the anions. Simply using chemical analytical results for ppmCl in (d) would account for most of the difference in P-T behavior between pure water and real brine. Some residual uncertainty would remain, of course, depending mainly on the proportions of Na:K:Ca.

An additional advantage of using (d) can be gained when considering dynamic situations, like fluid rising in wellbores where steam flashing causes progressively increasing concentrations of salts in the residual liquid. Such cases can be accommodated simply by multiplying (h) by the term 1/(1-f) where f represents the fractional mass of liquid that has flashed to steam and (h) represents the initial (pre-flash) concentration of salt.

These approaches toward describing thermodynamic behavior of brines have not been developed, apparently. This logarithmic equation appears to simplify such descriptions since some important factors can be identified with single terms in the equation.

REFERENCES

1. Haas, J. L., Jr.: "Physical Properties of the Coexisting Phases and Thermochemical Properties of the NaCl Component in Boiling NaCl Solutions," U.S. Geol. Surv. Bul. 1421-B, Government Printing Office, 70 pp., 1976.

2. Silvester, L. F. and Pitzer K. S.: "Thermodynamics of Geothermal Brines, Part I," Lawrence Berkeley Laboratory, Univ. of Calif., Rep. 4456, 1976.

3. Michaelides, E. E.: "Thermodynamic Properties of Geothermal Fluids," Geothermal Resources Council Transactions, v. 5, p. 361-364, 1981.

4. Leib, G. F.: "Thermodynamic Properties of Vapors," A.S.M.E. Transactions, Feb., 1941, reprinted as "Steam Tables by Combustion Engineering, Inc."

Geothermal Resources Council, TRANSACTIONS Vol. 6, October 1982

MERCURY-ITS ROLE IN THE EXPLORATION OF VAPOR-DOMINATED GEOTHERMAL SYSTEMS

Joseph N. Moore*, Odin Christensen**, and Robert Bamford***

*Earth Science Laboratory/Univ. of Utah Research Inst./Salt Lake City, UT
**Newmont Exploration Ltd./Sparks, NV
***Consultant-Seattle, WA

ABSTRACT

Geothermal wells drilled to depths of up to 3 km in The Geysers steam field have provided an opportunity to examine, laterally and vertically, the distribution of Hg in an active vapor-dominated hydrothermal system.

Mercury is concentrated primarily in the outer, cooler portions of the thermal system. Heating experiments indicate that Hg is contained primarily in silicate minerals and that it is heat labile at reservoir temperatures of 240°C. The relative depletion of Hg within the reservoir rocks compared to the margins of the field suggests that the distribution of this element can be used to guide exploration in poorly known portions of the field.

INTRODUCTION

Mercury frequently occurs in trace amounts in geothermally altered rocks and is currently used as a pathfinder element in the exploration of high-temperature liquid-dominated geothermal fields. Little information has, however, been published on the behavior and distribution of Hg in vapor-dominated geothermal systems. This paper illustrates the results of Hg analyses from wells drilled to depths of 3 km in the steam reservoir at The Geysers in California. The data suggest that Hg can also be used as a pathfinder element in the exploration of vapor-dominated systems. Samples were generously provided by Aminoil, U.S.A., Shell Oil Company and Union Oil Company. The work was funded by the U.S. Department of Energy, Division of Geothermal Energy, under contract DE-AC07-80ID12079.

GEOLOGIC SETTING

The Geysers steam field is located on the southwestern edge of the Clear Lake volcanic field about 175 km north of San Francisco in the structurally and stratigraphically complex Mayacmas Mountains (McLaughlin and Stanley, 1976; Hearn and others, 1976; McLaughlin, 1977). Steam is contained in faults and fractures within the regionally metamorphosed eugeosynclinal volcanic and sedimentary rocks of the late Mesozoic Franciscan assemblage.

McLaughlin and Stanley (1976) divided the Franciscan assemblage into three fault-bounded structural units on the basis of gross lithology and degree of metamorphism, which generally decreases with depth. The structurally lowest unit, Unit 1, contains most of the steam produced in the field and consists primarily of weakly metamorphosed graywacke with minor interbedded shale. Unit 2 is a highly tectonized assemblage of graywacke, conglomerate, basaltic volcanics (greenstone), chert, tectonic melange, and serpentinite, separated by imbricate thrust faults. The structurally highest unit, Unit 3, is lithologically similar to Unit 2 but is characterized by a higher metamorphic grade. Units 2 and 3 apparently form a relatively impermeable cap to the hydrothermal system over much of the area.

Temperatures within the geothermal reservoir are near 240°C and remain relatively constant with depth. James (1968) and White and others (1971), noting that the maximum enthalpy of saturated steam occurs at 236°C, have suggested that both liquid and vapor must be present in the fractures and pore spaces of the thermal reservoir. Because reservoir pressures of approximately 32 kg/cm^2 are below hydrostatic, Truesdell and White (1973) concluded that the pressure-controlling medium in the through-going fractures must be steam.

Silicic intrusions related to volcanism in the Clear Lake volcanic field were emplaced at relatively shallow depths in the Franciscan assemblage of The Geysers area during late Cenozoic time (Schriener and Suemnicht, 1980). Volcanic activity began in the Clear Lake field about 2.5 m.y. ago and continued until at least 10,000 years ago producing rhyolite to basaltic flows and pyroclastic deposits (Hearn and others, 1976). Geophysical data suggest that heat to the thermal system is supplied by a cooling silicic magma body located at a depth of approximately 10 km beneath Mount Hanna, near the northeast edge of the steam reservoir (Iyer and Hitchcock, 1975; Isherwood, 1976; Chapman, 1975).

The geothermal wells that have been studied

present a variety of physical environments within the thermal area. Eight of the wells produce commercial quantities of steam. These wells penetrated unproductive caprock consisting largely of graywacke or volcanic and ultramafic rocks before intersecting graywacke of the thermal reservoir at depths of more than 1000 m. One well is located on the high-temperature margin of the thermal system and produces subcommercial quantities of steam, and one is located outside of the hydrothermal reservoir but near known Hg mineralization.

DISTRIBUTION OF HG

Mercury mineralization is widespread in the Mayacmas Mountains (White, 1967), and in The Geysers area trace amounts of Hg are broadly distributed at depth. The Hg contents of 30 m composite whole rock samples range from less than 5 ppb to approximately 1.4 ppm.

The analytical data for Hg in two of the production wells are illustrated on downhole plot in Figure 1. Despite local variations in the Hg contents of the samples, the downhole distributions indicate a tendency for Hg to concentrate primarily in rocks overlying the steam reservoir.

Well G-4, located on the margin of the steam field, contains anomalously high Hg throughout its length (Fig. 3). Relatively high temperatures and high Hg contents in this well suggest that, in this portion of the field, the reservoir lies at depths greater than about 3 km. Although additional data are lacking, a thick caprock is consistent with the well's location in the known margin of the steam reservoir.

Mercury frequently occurs as sulfide minerals such as cinnabar, as impurities in other sulfide phases, and as Hg° in active and fossil hydrothermal systems (Christensen and others, 1980). Because of the limited stabilities of Hg-bearing minerals at elevated temperatures, thermal decomposition curves can be used to identify the various forms present (Fedorchuk, 1961; Jovanovic and Reed, 1968; Reed and others, 1972; Koksoy and others, 1967).

The nature of the Hg mineralization in The

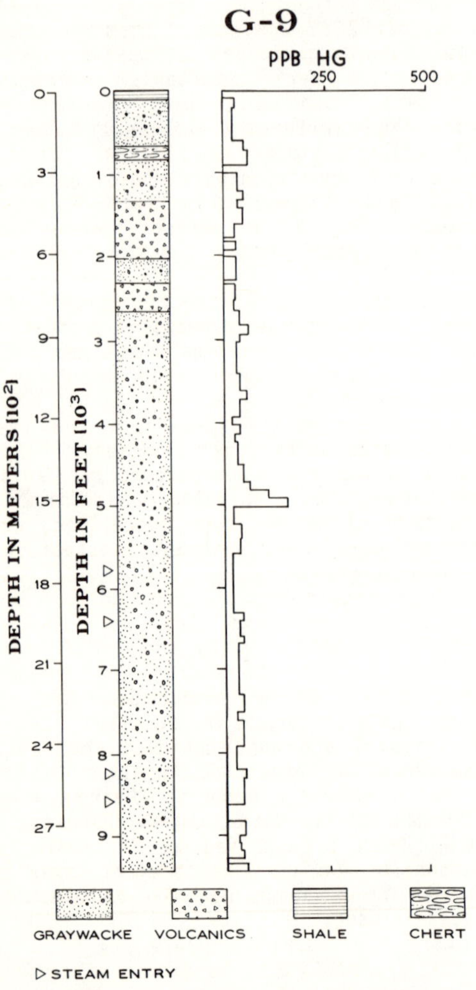

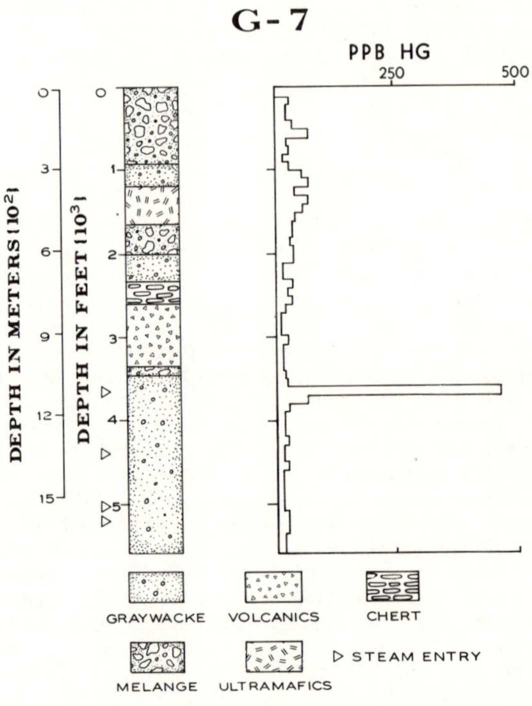

Figure 1. Distribution of Hg in production wells G-9 and G-7.

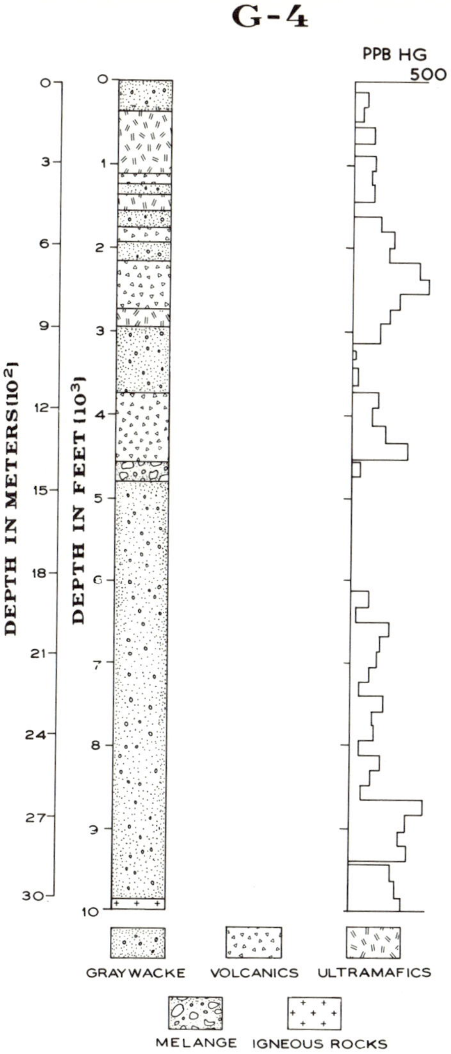

Figure 2. Distribution of Hg in well G-4, located on the margin of the field.

Geysers samples was investigated by measuring the temperature dependence of Hg liberation from drill cuttings. Pulverized samples were heated in air at one atmosphere and held at temperature for approximately one hour, and splits were removed for analysis by gold film Hg detector. The temperature of Hg release varied among samples, although the overall geometries of the Hg-loss curves are similar (Fig. 3). Mercury was released during heating of the samples in several stages, indicating that Hg occurs in more than one form. Three temperatures - 150°C, 250°C and 400°C - appear to be critical thresholds.

Petrographic examination of the samples and comparison of the release curves with published thermal decomposition curves indicate that Hg loss beginning at 150°C represents the release of Hg held within silicate minerals, possibly within chalcedony or quartz vein material. The major portion of the Hg within all samples from the wells occurs in this form. The temperature break occurring above about 250°C may represent the thermal decomposition of cinnabar even though no cinnabar was identified during examination of the cuttings. Mercury retained to 400°C probably occurs as a trace constituent of pyrite or other sulfide minerals.

Based on the heating experiments and logging of the drill cuttings, Hg in the samples studied appears to be contained mainly in silicate minerals and in sulfides other than cinnabar.

DISCUSSION

The geochemical data suggest that Hg forms a persistent and conspicuous geochemical envelope about the steam reservoir at The Geysers. Christensen and others (1980) have described similar Hg haloes around geothermal systems in the Basin and Range Province and have shown that the distribution of Hg is related to the present thermal regimes in these systems. For example, at Roosevelt Hot Springs, Utah, Hg is presently being deposited only in regions where temperatures are below about 200°C.

The heating experiments indicate that, at reservoir temperatures of 240°C, Hg within and surrounding The Geysers geothermal reservoir rocks is, in part, heat labile and is mobilized away from the high-temperature reservoir center toward the cooler fringes of the thermal field. The presence of Hg in steam discharged from some wells at The Geysers (Robertson and others, 1977) demonstrates that Hg is actively being mobilized in portions of the field. Although temperature distributions within the wells are not known, low concentrations of Hg within the reservoir rocks compared to the caprocks and high concentrations in well G-4 on the margin of the steam field are consistent with the relationships suggested by the heating experiments.

White and others (1971) suggested that vapor transport could be an important mechanism for redistributing Hg at The Geysers. The distribution of Hg we have documented, and the absence of significant alteration associated with local concentrations of Hg, make vapor transport an appealing mechanism.

REFERENCES

Chapman, R. H., 1975, Geophysical study of the Clear Lake region, California: California Division of Mines and Geology, Special Report 116, 23 p.

Christensen, O. D., Moore, J. N., Capuano, R. M., 1980, Trace element geochemical zoning in the Roosevelt Hot Springs thermal area, Utah: Geothermal Resources Council, Transactions, v. 4, p. 149-152.

Fedorchuk, V. P., 1961, Formation of aureoles of direct ore indicators around mercury

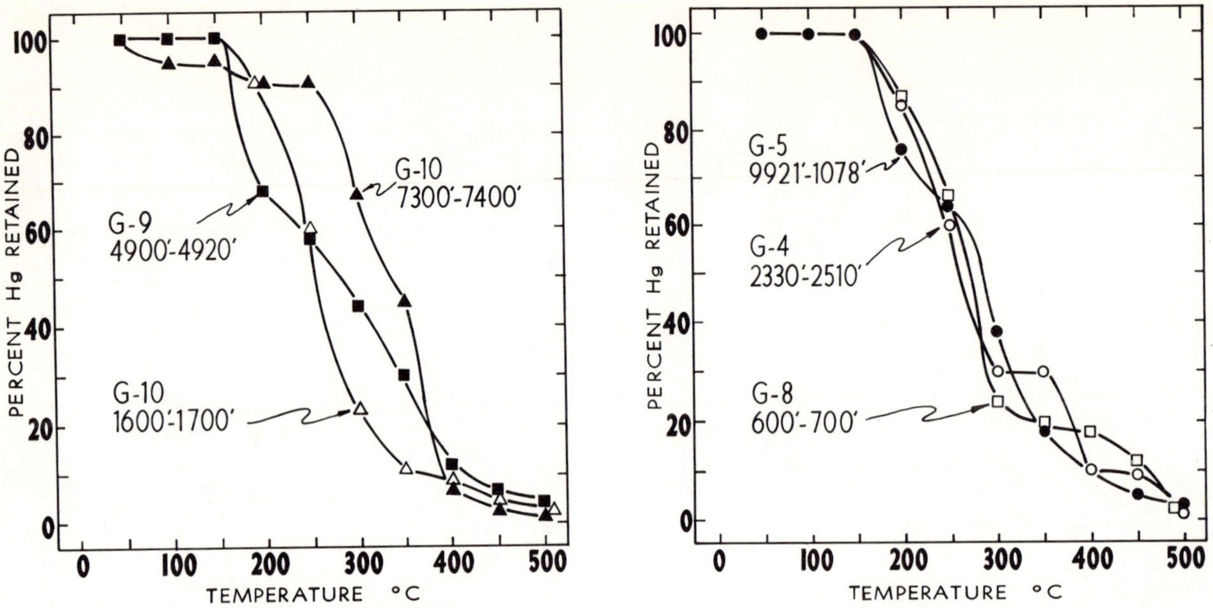

Figure 3. Temperatures of Hg liberation from the downhole samples.

deposits: Geokhimlya, no. 10, p. 911-919.

Hearn, B. C., Donnelly, J. M., and Goff, F. E., 1976, Geology and geochronology of the Clear Lake Volcanics, California: Second U. N. Symposium on the Development and Use of Geothermal Resources, San Francisco May, 1975, p. 423-428.

Isherwood, W. F, 1976, Gravity and magnetic studies of The Geysers - Clear Lake geothermal region, California: Second U. N. Symposium on the Development and use of Geothermal Resources p. 1065-1073.

Iyer, H. M., and Hitchcock, T., 1975, Teleseismic residuals at The Geysers geothermal area: EOS, Transactions of the American Geophysical Union, v. 56, p. 1020.

James, R., 1968, Wairakei and Larderello; geothermal power systems compared: New Zealand Journal Science and Technology, v. 11, p. 706-719.

Jovanovic, S., and Reed, G. W., 1968, Hg in metamorphic rocks: Geochimica et Cosmochimica Acta, v. 32, p. 341-346.

Koksoy, M., Bradshaw, P. M. D., and Tooms, J. S., 1967, Notes on the determination of mercury in geological samples: Institute of Mining and Metallurgy Transactions, v. 76, p. B121-124.

McLaughlin, R. J., 1977, The Franciscan assemblage and Great Valley sequence in The Geysers - Clear Lake region of northern California: In Field Trip Guide to The Geysers - Clear Lake area for the Cordilleran Section of the Geological Society of America, April 1977, p. 3-24.

McLaughlin, R. J., and Stanley, W. D., 1976, Pre-Tertiary geology and structural control of geothermal resources, The Geysers steam field, California: Second U. N. Symposium on the Development and Use of Geothermal Resources, San Francisco, May 1975, p. 475-485.

Reed, G. W., Jovanovic, S. and Tennissen, A. C., 1972, Concentrations and thermal release of Hg from iron and zinc sulfides: Earth and Planetary Science Letters, v. 14, p. 26-30.

Robertson, E. D., Crecelius, E. A., Fruchter, J. S., and Ludwick, J. D., 1977, Mercury emissions from geothermal power plants: Science, v. 196, p. 1094-1097.

Schriener, A., and Suemnicht, G. A., 1980, Subsurface intrusive activity at The Geysers geothermal area, California: Geological Society of America Abstracts with Programs, Corvallis, Oregon, p. 152.

Truesdell, A. H., and White, D. E., 1973, Production of superheated steam from vapor-dominated geothermal reservoirs: Geothermics, v. 2, p. 154-173.

White, D. E., 1967, Mercury and base-metal deposits with associated thermal and mineral deposits: in Geochemistry of Hydrothermal Ore deposits; ed. H. L. Barnes, Holt, Rinehart and Winston Inc., p. 575-631.

White, D. E., Muffler, L. J. P., and Truesdell, A. H., 1971, Vapor-dominated hydrothermal systems compared with hot-water systems: Economic Geology, v.66, p. 75-97.

FLUID GEOCHEMISTRY OF HOT SPRINGS BAY VALLEY, AKUTAN ISLAND, ALASKA

Roman J. Motyka*, Mary A. Moorman*, and Robert Poreda**

*Alaska Division of Geological and Geophysical Surveys, Box 80007, College, Alaska 99708; **Isotope Laboratory, Scripps Institution of Oceanography, UCSD, LaJolla, California 92093.

ABSTRACT

Several thermal springs and a fumarole field occur in Hot Springs Bay Valley on Akutan Island. The thermal springs are low in Mg, SiO_2-rich, and moderately concentrated in Na-Cl, a chemistry typically associated with hot-water hydrothermal systems. B/Cl ratios indicate the several sets of springs are derived from a common parent hot-water reservoir. Application of silica, cation, and sulfate-water oxygen-isotope geothermometry gives a range of 160-190°C for the parent reservoir. Recent geophysical investigations performed in cooperation with the Geophysical Institute, U of Alaska, identified a broad zone of apparent low resistivity underlying the lower part of the valley in the vicinity of the thermal springs. Relatively high CH_4 content and low H_2/CH_4 ratios occur in gases from both the thermal springs and from the fumarole field located at the head of the valley about 4 km southeast of the hot springs. The similarity in H_2/CH_4 suggests underlying reservoirs are similar in both areas. Both sites have a ratio of $^3He/^4He$ WRT atmospheric levels of about 7.0.

INTRODUCTION

Akutan Island is located in the eastern Aleutian Islands about 45 km northeast of Unalaska Island and 160 km southwest of the Alaska Peninsula (fig. 1). Akutan Village, the only habitation on the island, is located on the north shore of Akutan Harbor about 6 km east of Akutan Hot Springs. The present population depends on subsistence, commercial fishing, and fish processing for their economy. Several floating fish processors operate in the protected waters of Akutan Harbor, which bring in a season an influx of 200-700 nonresident workers. The nearness of thermal springs to the village prompted a detailed study of the geothermal resources of Hot Springs Bay Valley as a potential source of energy for Akutan Village and Harbor. Investigations of fluids associated with the thermal springs and with the fumarole field located at the head of the valley were undertaken to help assess the nature and extent of the underlying hydrothermal system, and to provide estimates of reservoir temperatures. The results of this study are summarized in this paper. A companion paper by Wescott and others (1982) describes the results of geophysical investigations on Hot Springs Bay Valley.

GEOLOGY

Akutan Island consists of an older sequence of volcanics deeply eroded by glaciation and a Recent volcanic pile associated with Akutan Volcano at the western end (Fig. 1) (Byers, unp. data, USGS; Romick, 1982). The older volcanic complex consists of two units: a lower member of mainly debris flow deposits overlain by a sequence of shallow-dipping basaltic and andesitic flows. Potassium-argon ages of lava flows in the upper sequence range from 1.4±0.1 m.y. for one of the oldest flows to 1.1±0.1 m.y. for one of the youngest lavas (Romick, 1982). The lower member, which is extensively exposed over much of the island, has been intruded by numerous dikes, one of which has a preliminary radiometric age of 14.3±1.1 m.y..

Akutan volcano (1,300 m), a composite shield volcano, and its satellitic vents dominate the western part of the island. A series of ring-dike intrusions exposed on the northern part of the island suggest the existence of an ancestral collapse caldera near the present site of Akutan volcano. Recent volcanic activity has concentrated at Akutan volcano with fresh lava flows, lahars, and pyroclastic debris mantling the glacially eroded surface of the older volcanic complex. The volcano is capped by a 2-km-dia caldera thought to have formed as recently as 500 years ago (Byers and Barth, 1953). Akutan volcano is one of the most active volcanoes in the Aleutian chain, having erupted more than 25 times since 1700 (Simkin and others, 1981).

Portions of the island not covered by recent volcanic flows show signs of intense glaciation: serrated ridges, cirques, hanging valleys, and broad U-shaped valleys. The east end of the island is split by Akutan Harbor, a deep 8-km-long fjord.

THERMAL FIELDS

The Akutan hot springs are located about 4 km northwest of Akutan Harbor and 10 km north-

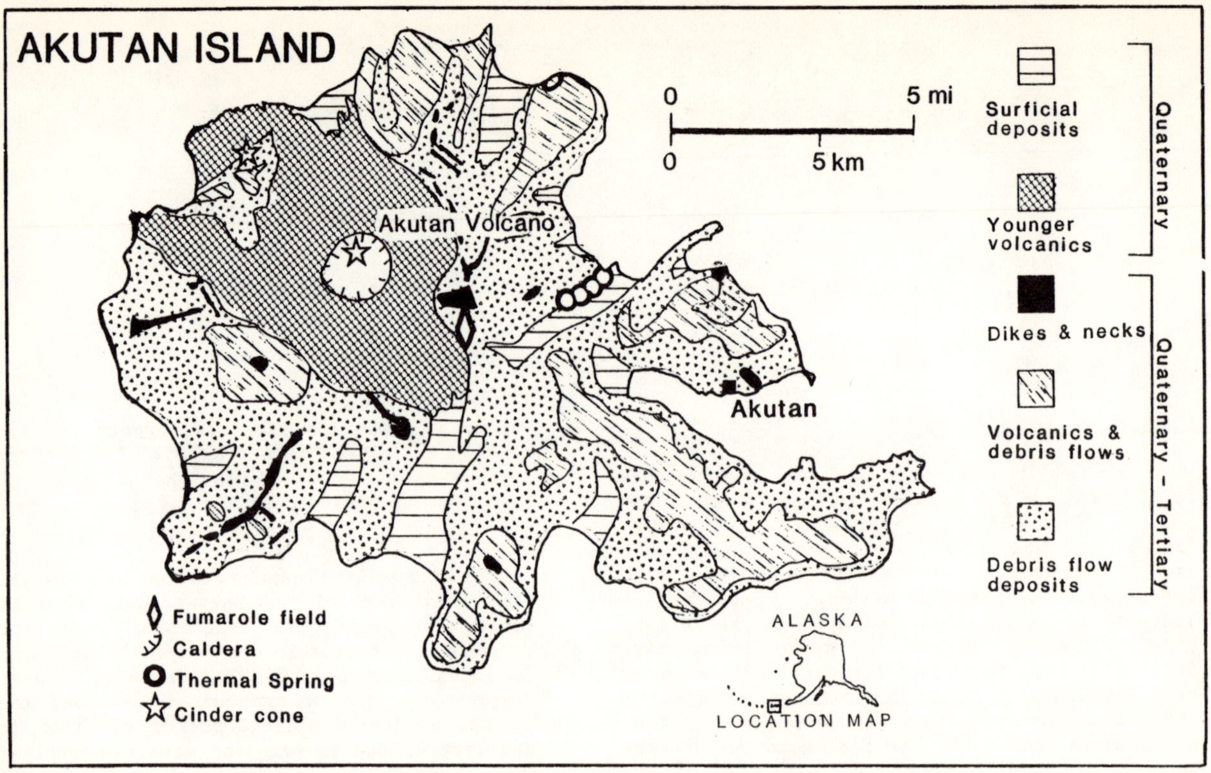

Figure 1. Generalized geologic map of Akutan Island (adapted from Byers, unp. data, USGS, and Romick, 1982)

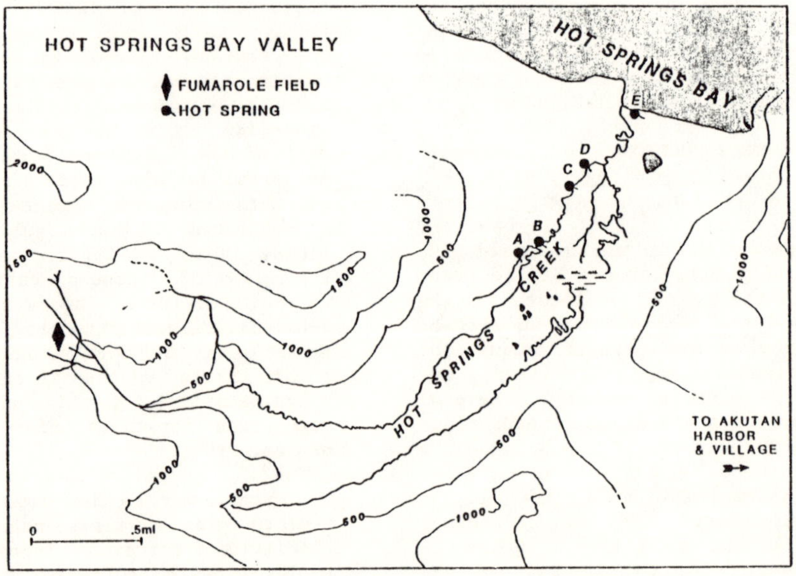

Figure 2. Location of hot springs and fumaroles in Akutan Hot Springs Bay Valley.

Table 1. Water chemistry, Akutan Bay hot springs[a]. Units are mg/l unless otherwise noted.

	1. A3	2. A3	3. B1	4. C	5. D2	6. E	Fumarole acid spr.
SiO_2	145	135	103	133	91	121	220
Fe	0.05	0.01	0.57	0.01	0.03	0.04	40.7
Ca	11.3	12.4	14.7	18.0	11.0	132	31.8
Mg	2.5	1.0	1.5	1.6	11.8	317	12.5
Na	323	328	172	207	128	1663	16.6
K	24.8	25.9	16.3	16.4	9.3	73.7	3.5
Li	1.28	1.24	0.61	0.61	0.34	1.10	0.01
Sr	0.11	0.12	0.10	0.21	0.09	1.18	0.05
HCO_3	172	nd	280	236	128	322	-
SO_4	42.7	40.6	22.4	42.7	26.4	495	1302
Cl	424	406	221	275	136	3444	5.2
F	1.1	0.94	0.64	0.97	0.88	0.54	0.1
Br	1.31	nd	1.34	0.27	nd	17.4	nd
I	0.42	nd	0.10	0.62	nd	0.35	nd
B	11.4	11.5	5.9	7.0	3.4	4.5	0.5
H_2S	0.5	nd	nd	nd	nd	nd	nd
pH, field	6.98	nd	6.39	6.47	6.82	7.34	2.55
T (°C)	84.0	83.3	47.4	73.4	58.8	67.0	92.3
Date sampled	8/7/80	7/10/81	7/9/81	7/9/81	8/8/80	7/12/81	7/7/81

[a]M. Moorman, Alaska Division of Geological & Geophysical Surveys, College, AK, Analyst
nd = not determined

east of the active Akutan volcano (fig. 1). Five sets of hot springs occur along and at the base of the west valley wall in a 1.5 km long zone that extends southwest from Hot Springs Bay (fig. 2). The thermal waters issue from fissures in hydrothermally cemented stream bank sediments, from pools in valley alluvium, and through beach sands in the intertidal zone near the mouth of the Hot Springs Creek.

The hottest springs in the valley are the southernmost, at site A, with temperatures as high as 85°C. These springs discharge into a tributary of the main valley stream. Based on the increase in chloride and silica in the tributary stream below site A compared to concentrations in the tributary above the springs, heat loss by spring discharge at site A is estimated at 1.6 MW. A similar analysis made on chloride, potassium plus sodium, and silica concentrations in the main valley stream just below site D hot springs indicates a total heat loss by spring discharge from sites A, B, C, and D of about 8-9 MW.

In addition to the hot springs, a fumarole field occurs several kilometers up valley from the springs and is located at about 350 m elevation on the flank of the active Akutan Volcano. The field consists of a series of fumaroles, mild steam venting, and boiling acid springs, covering an area of about 5,000 m^2. Temperatures of fumaroles and steam vents are at or near boiling point.

WATER CHEMISTRY

The thermal waters from spring sites A through D are moderately concentrated sodium chloride waters with a relatively high level of bicarbonate (Table 1). Springs E are much more saline because of the mixing with seawater in the intertidal zone. Site D spring waters are more dilute than thermal waters from the other sites indicating colder surface waters are mixing with the thermal waters at D. Such mixing at site D is also suggested by a preliminary enthalpy-chloride analysis and from deuterium and oxygen 18 isotope data on thermal and local meteoric waters. The similarity in the ratios of the conservative elements boron to chloride in thermal waters from sites A through D, however, suggest the spring waters are all ultimately derived from a common parent thermal reservoir.

The percentage cation content of Hot Springs Bay Valley thermal and surface meteoric waters are plotted in Fig. 3. The seawater dilution of springs E (No. 6) is readily apparent as is the dilution of thermal waters at springs D with colder surface waters (cf. No. 5 to No. 16). The Na + K trend in meteoric waters reflects the sampling of valley stream waters above, near, and below hot spring sites. Number 13 was taken from the main fork below site D, while number 7, 8, and 9 were obtained at the head of the valley. Number 14 and 15 were taken from the tributary stream above and below site A respectively.

Application of silica and cation geothermometry to site A hot springs gives a range of 160-190°C for the deep parent reservoir temperature. A sulfate-water oxygen isotope geothermometer obtained through the cooperation of N. Nehring (USGS, Menlo Park, CA) gives a concordant temperature for the deep thermal waters of 186°C.

Thermal waters associated with the fumarole field at the head of the valley are highly acidic and appear to be locally derived surface waters heated by condensing steam and volcanic gases. The constituents in these thermal waters were probably leached from local rocks by the hot acid waters. The silica content of 220 ppm in one of the acid pools reflects the breakdown of silicates and equilibration of the waters to amorphous silica.

Table 2. Chemical composition of fumarole and thermal spring gases from Akutan Hot Springs Bay valley, volume %[a].

	Fumarole Field	Hot Springs A
CO_2	95.37[b]	16.53[b]
H_2S		
H_2	0.34	0.45
CH_4	1.32	4.58
N_2	2.96	76.68
Ar	0.02	1.73
O_2	-	-
He (ppm)	7.3	12.4
$^3He/^4He$, R/Ra	7.1	6.5 $\pm$ 5
H_2/CH_4	0.26	0.10
N_2/Ar	148.	44.

[a]R. Poreda and J. Welhan, Scripps Institute of Oceanography, LaJolla, CA, analysts.
[b]Total CO_2 + sulfur gasses.

GAS CHEMISTRY

The volume percent composition of gas samples obtained from the fumarole field and from hot spring site A are given in Table 2. The much higher percentage of N_2 from hot spring A reflects a greater proportion of dissolved air in the thermal waters. Oxygen is selectively removed in oxidation reactions. Both samples contain methane and hydrogen in similar proportions. Carbon 13 isotope analyses are being made on CO_2 and CH_4 in samples to help differentiate between biogenic and abiogenic origins. Both sites show a high R/Ra ratio of $^3He/^4He$ (6.5-7.0), suggesting a magmatic influence on the hydrothermal systems in both localities. The R/Ra ratio here is typical of ratios found in gases from other island arc settings.

The fumarole field at the head of the valley and the hot spring system near the Bering Seacoast may be connected to the same thermal source of heat at depth. The similarity in H_2 to CH_4 ratio suggests that at least underlying reservoirs are similar in both localities.

DISCUSSION

Bedrock underlying the valley may be an extension of the volcanic debris flow sequence exposed along the valley walls. If capped by hydrothermal cementation, such a porous permeable host rock could house a substantial hot-water reservoir at fairly shallow depths, one that might be easily tapped. The distribution of the thermal springs suggest they are related to a subsurface zone of apparent low resistivity described in a companion paper by Wescott and others (1982).

The estimated deep-reservoir temperature of 180°C is sufficient for a variety of applications, including the generation of a modest amount of electrical power, e.g., a 1-MW well-head-driven Rankin binary system. The nearness of this hot-water system to a well-protected deep-water harbor with a population center and potential industrial users (e.g., fishing processors) make the Akutan Hot Spring site a particularly attractive one for future development.

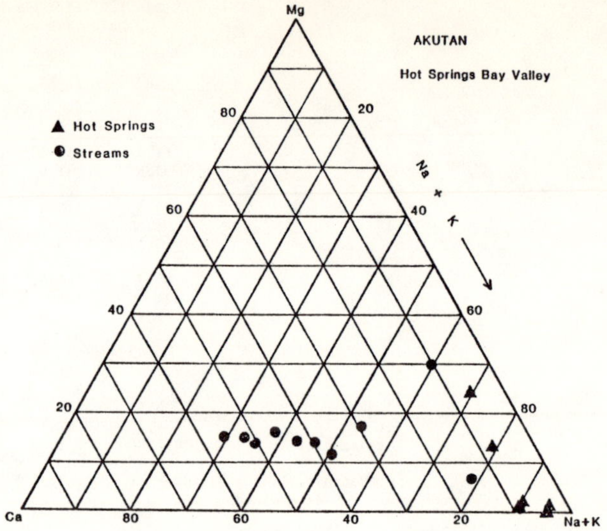

Figure 3. Cation ratios for waters from Hot Springs Bay Valley

REFERENCES

Byers, F.M., Jr., and Barth, T.F.W., 1953, Volcanic activity on Akun and Akutan Islands: Pacific Science Congress, 7th, New Zealand, 1949, Proceedings, v. 2, p. 382-397.

Romick, J., 1982, Igneous Petrology and Geochemistry of Northern Akutan Island, Alaska, Master's Thesis, submitted, University of Alaska.

Simkin, T., Siebert, L., McClelland, L., Bridge, D., Newhall, C., and Latter, J.H., 1981, Volcanoes of the World, Smithsonian Institute, Academic Press, 233 p.

Wescott, E.M., Turner, D.L., Witte, W., Petzinger, B., 1982, Geophysical investigation of Hot Springs Bay Valley, Akutan Island, Alaska, Geothermal Resources Council Transactions, this volume..

FLUID GEOCHEMISTRY OF THE MAKUSHIN GEOTHERMAL AREA, UNALASKA ISLAND, ALASKA

Roman J. Motyka*, Mary A. Moorman*, and Robert Poreda**

*Alaska Division of Geological and Geophysical Surveys, Box 80007, College, Alaska 99708; **Isotope Laboratory, Scripps Institution of Oceanography, UCSD, LaJolla, California 92093.

ABSTRACT

Several fumarole fields and numerous thermal springs occur on the eastern flanks of Makushin volcano. The thermal springs at Makushin are typically near-neutral in pH, have very low chloride levels (10 ppm), and are comparatively rich in Mg, Ca, SO_4, HCO_3 and SiO_2. Similarities in cation and isotopic compositions of Makushin thermal springs and local surface streams indicate the thermal waters are locally derived meteoric waters. These surface waters infiltrate to relatively shallow depths where they are heated by steam and gases rising from a vapor-dominated zone. Predominate fumarolic gases are CO_2, N_2, and sulfur gases. All samples showed a significant level of H_2 and a high H_2/CH_4 ratio. The ratio of $^3He/^4He$ WRT atmospheric levels range from 4.5 to 6.0. Fumaroles and hot springs in the upper parts of Glacier and Makushin Valleys form a roughly linear trend, suggesting their distribution may be structurally controlled.

INTRODUCTION

Unalaska Island, second largest in the arcuate chain of Aleutian Islands lies 200 km southwest of the Alaska Peninsula (fig. 1). Due to the large and excellent deep-water harbor located at Unalaska Bay, the village of Unalaska serves as the major base of operations for Bering Sea and northern Gulf of Alaska fisheries. The village is actively seeking an energy base to support this growing industry. Fumarole fields and thermal springs which occur on the flanks of Makushin Volcano, 20 km west of Unalaska Village, indicate the presence of a high-temperature geothermal resource that, if developed, could be used by the village. Investigations of fluids associated with these thermal fields were undertaken by DGGS to help assess the nature and extent of the underlying hydrothermal system, and to provide estimates of reservoir temperatures. This paper summarizes the results of the study as of June, 1982. Exploratory drilling funded by the State of Alaska is currently underway with Republic Geothermal of California the prime contractor. The main targets of the exploration are the hydrothermal systems located on the east flank of Makushin.

GEOLOGY

The western part of northern Unalaska is dominated by Makushin Volcano (2035 m) (fig. 1). The broad dome-shaped summit has a small caldera and is capped by a glacier with tongues that descend the larger valleys to elevations as low as 300 m (1000 ft.). Several satellitic cones and craters occur on the flanks of the volcano.

The Unalaska Formation, which is exposed over two-thirds of the island, constitutes the oldest and most extensive group of rocks on the island. The formation consists of a thick sequence of coarse and fine sedimentary and pyroclastic rocks intercalated with dacitic, andesitic, and basaltic flows and sills, cut by numerous dikes and small plutons (Drewes and others, 1961) (fig. 1). The formation has been extensively folded, faulted, and intruded by plutonic rocks, with moderate hydrothermal alteration occurring near the plutons. The batholiths and small plutons are granodiorite with border phases as mafic as gabbro. The ages of the plutons are considered to be younger than early Miocene and older than middle Pleistocene.

Basalt and andesite flows and pyroclastic rocks of the Makushin Volcanics unconformably overlie the Unalaska Formation and the plutonic rocks that intrude it (Drewes and others, 1961). Most of the Makushin Volcanics are believed to be middle to late Pleistocene. Late Wisconsin to Recent volcanic cinder cones, composite cones, and lava flows are scattered about the base of Makushin Volcano and have been collectively mapped as Eider Point Basalt (Drewes and others, 1961). Makushin Volcano is still active and is known to have erupted at least 14 times since 1760 (Simkin and others, 1981).

THERMAL AREAS

Fumaroles occur within the 2.5-km-dia Makushin caldera (fig. 2). An extensive fumarole field and a series of associated hot springs occur at the head of Glacier Valley on the south-southeast flank of Makushin Volcano. The fumaroles lie at an elevation of about 670 m.

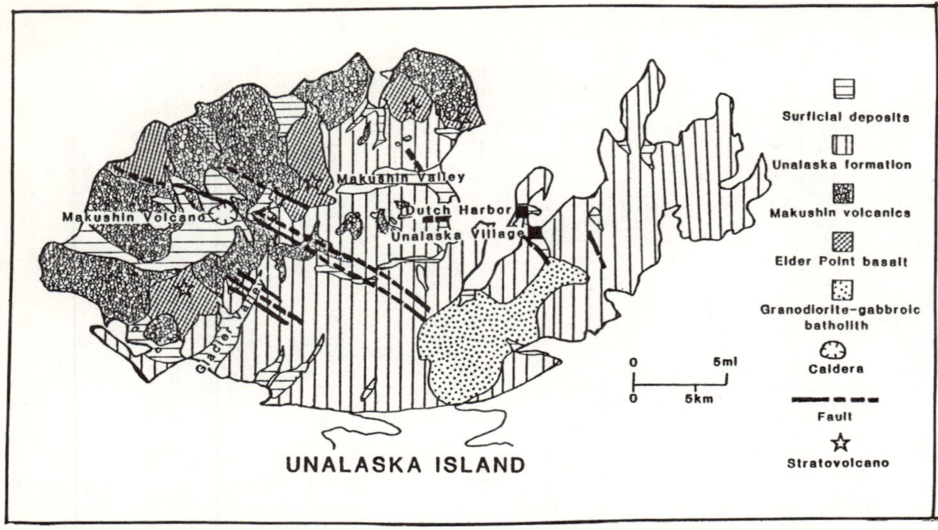

Figure 1. Generalized Geologic map of northern Unalaska Island (adapted from Drewes and others, 1961, and J. Reeder, pers. comm. ADGGS).

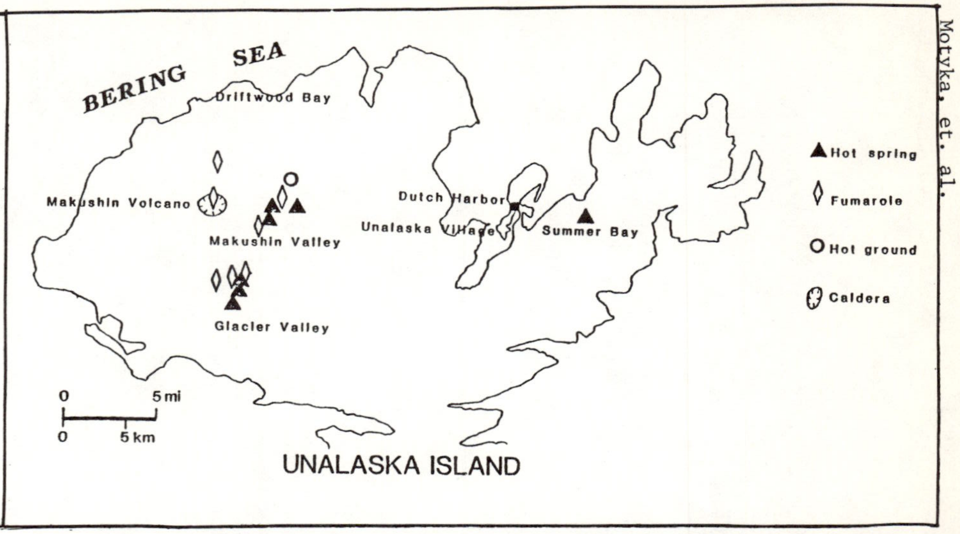

Figure 2. Location of fumaroles, hot springs, and thermal ground on northern Unalaska Island.

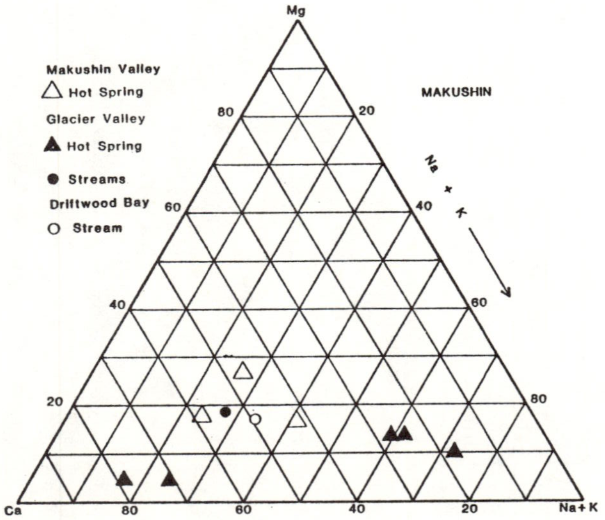

Figure 3. Cation ratios for waters from Makushin.

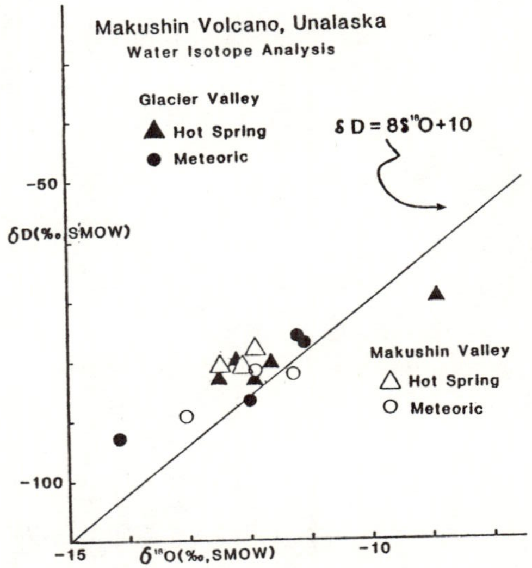

Figure 4. Isotopic analyses of waters from Makushin.

Table 1. Water chemistry, Makushin and Glacier Valley thermal springs and cold streams[a]
Units are mg/l unless otherwise noted.

	Makushin Valley			Glacier Valley						
	1. M1	2. M2	3. M3	4. G1-A	5. G1-B	6. G1-C	7. G-2	8. G-3	9. G1-Str[b]	10. D-St
SiO_2	140	88	88	94	125	120	138	142	20	4.5
Fe	0.09	0.07	0.03	0.1	0.01	nd	0.02	0.21	0.01	0.06
Ca	69.3	23.3	23.1	11.7	32.1	25.4	258	208	8.9	2.6
Mg	12.2	5.5	8.0	4.0	10.6	8.0	9.6	7.8	1.9	0.5
Na	28	24	13.9	52.0	87.2	62	61	81	4.7	2.0
K	5.6	3.2	3.4	4.8	5.7	5.2	3.3	4.8	0.8	0.06
Li	0.01	0.01	0.01	0.01	0.01	0.01	0.04	0.03	0.01	0.01
HCO_3	191	nd	116	37	288	6.0	nd	256	nd	nd
SO_4	155	25	21	129	95	218	491	476	29	3.1
Cl	5	7.8	5	10	5	6.1	2.3	7.5	5.6	2.6
F	0.12	0.13	0.11	0.14	0.28	0.1	0.26	0.24	0.1	0.1
B	0.5	0.01	0.5	0.5	0.5	0.01	0.01	0.01	0.01	0.01
Sr	0.28	0.01	0.10	0.07	0.26	0.20	1.08	1.11	0.02	0.01
pH, field	5.48	5.28	5.32	6.40	6.50	4.34	nd	6.36	nd	nd
T (°C)	87.4	57.5	67.0	96.8	82.4	77.5	68	79	4.9	3.8
Date sampled	8/13/80	7/4/81	8/13/80	8/11/80	8/11/80	7/5/81	7/5/81	7/5/81	7/5/81	7/1/81

[a]M. Moorman, Alaska Division of Geological and Geophysical Surveys, College, AK, Analyst.
[b]West fork upper Glacier Valley stream.
[c]Cold stream in Driftwood Valley.
nd = not determined.

Two fumarole fields and associated hot springs are found in the upper reaches of Makushin Valley on the northeast flank of Makushin Volcano (fig. 2). The lower and smaller field occurs at an elevation of about 360 m on a small bench located on the steep north valley wall about 25 m above a stream channel. The larger field occurs about 2 km further upstream at the head of one of the tributary valleys on the east flank of the volcano at elevations varying from 600-800 m.

Hot springs and fumaroles located in Glacier and Makushin Valleys follow a roughly linear northeast trend, suggesting their distribution may be structurally controlled. Recent satellitic cones in the area also follow a roughly northeast trend. The springs and fumaroles occur at the eastern edge of the volcanic field and emanate from faulted and intensely hydrothermally altered rocks belonging to the Tertiary-aged Unalaska formation and gabbroic plutons that intrude it. Several northwest trending faults that cut across the volcano and pass through or near the fumaroles fields may be acting as conduits for the flow of thermal fluids.

Fumaroles on the flanks of the volcano are at or near boiling point. Thermal spring temperatures in Glacier and Makushin Valleys range from about 50°C to near boiling. Discharges from individual sets of springs range from 50 to 100 lpm. Deposits of calcite were found coating channels of thermal springs in the upper part of Glacier Valley.

WATER CHEMISTRY

Key features of the thermal water chemistry in both Glacier Valley and Makushin Valley are the extremely low levels of chloride present, the relatively low cation content, and the comparatively high level of magnesium and calcium (Table 1). The sulfate and bicarbonate contents probably arise from the oxidation of H_2S and CO_2 gases associated with the fumarolic activity. Most of the thermal springs are near neutral in pH. Acid springs also occur, usually in close proximity to steam vents. The waters are similar to those that have been classified as bicarbonate-sulfate waters. Such waters occur on the flanks (or in wells drilled thereon) of active volcanoes in island-arc settings, and are often associated with vapor-dominated reservoirs (Oki and Hirano, 1970; Mahon and others, 1980). The waters are relatively high in SiO_2. The high SiO_2 may reflect equilibration of waters with quartz or chalcedony in a shallow reservoir or possibly leaching of country-rock by acid-waters formed by the reaction of volcanic gases with ground waters and equilibration with amorphous silica.

The cation and isotopic composition of Makushin Valley hot springs are similar to local meteoric water indicating the thermal waters probably originate from the near-surface mixing of steam and ground water (figs. 3 & 4). The isotopic composition of thermal waters in Glacier Valley indicates a similar situation occurs there. Thermal springs in Glacier Valley, however, are more concentrated in sodium and potassium than thermal springs in Makushin Valley indicating a greater degree of water-rock interaction occurs in Glacier Valley. Although the sodium and potassium contents of thermal springs examined in Glacier Valley are similar, the calcium concentration differs markedly: springs at lower elevation (G2 and G3) have an order of magnitude greater calcium content than springs located directly below the fumarole field (G1).

GAS CHEMISTRY

The predominate gases in all sites samples are carbon dioxide, nitrogen and the sulfur gases (Table 2). A significant percentage of hydrogen are found in both localities, the Glacier Valley fumaroles having about 2-4

Table 2. Chemical composition of fumarole and thermal spring gases from Makushin Volcano thermal fields, volume %.[a]

	M1a	M1b	M3	G1-fum.	G1-spr.
CO_2	89.14	92.28[b]	91.29	94.81[b]	88.66[b]
H_2S	2.38		2.75		
H_2	0.49	0.74	0.25	1.12	1.80
CH_4	0.002	0.021	0.031	0.006	0.020
N_2	7.91	6.87	5.61	4.04	9.41
Ar	0.08	0.06	0.07	0.04	0.11
O_2	-	0.04	-	-	-
He (ppm)	5.6	8.1	8.0	3.0	4.5
$^3He/^4He$, R/Ra	4.9	5.0	6.6	4.5	-
H_2/CH_4	274.	36.	8.	179.	88.
N_2/Ar	99.	119.	80.	104.	86.

[a]R. Poreda and J. Welhan, Scripps Institute of Oceanography, LaJolla, CA, analysts.
[b]Total CO_2 + Sulfur gasses.

times the hydrogen content of the Makushin fumaroles. Hydrogen in fumarolic gases may be derived from high-temperature interaction of water with ferrous-oxide silicates. Giggenbach (1980) indicates the percentage of hydrogen can also increase at higher temperatures by reactions such as

$$CH_4 + 2H_2O \rightarrow CO_2 + 4H_2$$

and

$$2NH_3 \rightarrow N_2 + 3H_2$$

The Makushin samples are unusual in their relatively high H_2 content, low CH_4 and high N_2/Ar. The low CH_4 contents suggest very little thermogenic gas contribution, and the high H_2/CH_4 and N_2/Ar ratios may indicate a primitive gas composition, possibly of direct volcanic origin (pers. comm., J. Welhan, Scripps Institute). Similar high ratios were found in gases from the Mud Volcano region of Yellowstone, an identified vapor-dominated system. Analyses of ^{13}C in CO_2 and CH_4 currently underway should provide additional clues regarding the origin of the gases.

D'Amore and Panichi (1980) have suggested a gas geothermometer for estimating reservoir temperatures based on the proportions of CO_2, H_2S, H_2 and CH_4. Application of this geothermometer to the Makushin gas samples for which H_2S analyses are available give reservoir temperature estimates of 232°C and 278°C.

Enrichments in ^3He WRT to atmospheric levels have been correlated with magmatic activity on a world-wide basis (Craig and Lupton, 1981). Values for the ratio of $^3He/^4He$ at Makushin fumaroles and gases compared to atmospheric ratios (R/Ra) are typical of other island arc settings (Poreda and others, 1981). The excess ^3He is thought to be derived from the mantle. A high value for the $^3He/^4He$ ratio in gases from hydrothermal systems suggests a more direct connection to magmatic sources with little crustal contamination, although it may also result from leaching of young volcanic rock (Truesdell and Hulston, 1980). Lower values indicate a crustal influence of radiogenic ^4He.

DISCUSSION

The distribution of neutral-pH, low chloride thermal springs in association with several areas of solfatara activity on the east flank of Makushin indicate the presence of at least a shallow vapor dominated system. This system must extend at least as far as the lowest thermal springs in Glacier Valley which occur at an elevation of 150 m (500'). The series of springs and fumaroles in Glacier Valley cover a distance of about 4 km. Those at the head of and along Makushin Valley cover a distance of about 5 km, if thermal ground located west of Sugarloaf cone is included. The springs and fumaroles in the upper parts of Glacier and Makushin Valleys form a roughly linear trend about 12 km in length, suggesting that their distribution may be structurally controlled. Exploratory drilling now underway should provide additional information regarding the nature of the subsurface hydrothermal system at Makushin.

REFERENCES

Craig, H., and Lupton, J.E., 1981, Helium-3 and mantle volatiles in the ocean and the oceanic crust: The Oceanic Lithosphere: Vol. 7, The Sea, pp. 391-428, John Wiley & Sons.

D'Amore, Franco, and Panichi, Costanzo, 1980, Evaluation of deep temperatures of hydrothermal systems by a new gas geothermometer: Geochemica et Cosmochimica Acta, v. 44, p. 549-556.

Drewes, H., Fraser, G.D., Snyder, G.L., and Barnett, H.F., Jr., 1961, Geology of Unalaska Island and adjacent insular shelf, Aleutian Island, Alaska: U.S. Geological Survey Open-file Report 1028-S, p. 583-676.

Giggenbach, W.F., 1980, Geothermal gas equilibria: Geochimica et Cosmachimica Acta, Vol. 44, pp. 2021-2130.

Mahon, W.A.J., Klyen, L.E., and Rhode, M., 1980, Neutral sodium/bicarbonate/ sulphate hot waters in geothermal systems: Journal of Japan Geothermal Energy Association, v. 17, p. 11-24.

Oki, Y., and Hirano, T., 1970, The Geothermal system of the Hakona Volcano, in U.N. symposium on the development and utilization of geothermal resources, Pisa, 1970, v. 2, pt. 2: Geothermics Special Issue 2, p. 1157-1166.

Poreda, R., Craig, H., and Motyka, R.J.: "Helium isotope variations along the Alaskan-Aleutian Arc", abs. EOS, vol. 62, 45, 1981, p. 1082.

Simkin, T., Siebert, L., McClelland, L., Bridge, D., Newhall, C., and Latter, J.H., 1981, Volcanoes of the World, Smithsonian Institute, Academic Press, 233 p.

Truesdell, A.H., and Hulston, J.R., 1980, Isotopic evidence on environments of geothermal systems: in Handbook of Environmental Isotope Geochemistry, pp. 179-219, Elsevier.

Geothermal Resources Council, TRANSACTIONS Vol. 6, October 1982

COMPARISON OF METHODS FOR RECOVERING METAL VALUES FROM SALTON SEA KGRA BRINES

L.E. Schultze and D.J. Bauer

Reno Research Center
Bureau of Mines, U.S. Department of the Interior
Reno, Nevada 89512

ABSTRACT

As part of its program for developing plentiful domestic resources, the Bureau of Mines, U.S. Department of the Interior, is investigating methods for recovering metal values from geobrines. Two precipitation techniques were evaluated for recovering selected metal values from Salton Sea KGRA brines. Lime addition precipitates more than 95 pct of the iron, manganese, lead, and zinc, while sulfide addition precipitates more than 90 pct of the lead and zinc and lesser amounts of iron and manganese. A comparison of these techniques and a discussion of the respective advantages and disadvantages are presented.

INTRODUCTION

As part of the goal of the Bureau of Mines to make use of plentiful domestic resources, investigations were directed toward recovering metal values from geothermal brines. Interest in this plentiful resource was prompted by the availability of large quantities of brine that will result from plans for utilizing the heat content from geothermal reservoirs to generate electricity. Methods for recovering metal values dissolved in these brines focused on treatment of brine before injection into the reservoir.

Arrangements for an evaluation of recovery techniques by contract studies were made, and samples of brine from Magmamax No. 1, which is located near the southern end of the Salton Sea, were obtained. A partial analysis of a typical brine sample (table 1) shows significant levels of several metals that could be recovered by precipitation techniques.

TABLE 1. Partial analysis of a typical brine:

Metal	Concentration, ppm
Sodium	51,000
Calcium	22,000
Potassium	11,000
Manganese	600
Strontium	460
Zinc	240
Boron	220
Iron	170
Lithium	150
Magnesium	64
Lead	43

One of the contract studies (Urbanek et al., 1979) evaluated the potential for recovering sodium, calcium, and potassium and concluded that concentration of the brine by evaporation or freeze-drying would be required. The special restrictions of the area make these approaches impractical. Two techniques that would be compatible with energy recovery were precipitation of iron, manganese, lead and zinc (Berthold et al., 1975) by addition of lime and precipitation of lead and zinc by addition of sodium sulfide (Farley et al., 1981). A method for precipitating lithium was also described (Berthold et al., 1975) which is similar to one proposed for recovering lithium from Dead Sea brines (Epstein et al., 1981) and will not be discussed.

PRECIPITATION WITH LIME

A Mineral Recovery Unit (MRU) based on an engineering design (Berthold and Stephens, 1978) was assembled and operated at the Magmamax No. 1 site. Evaluation of samples taken during operation of the MRU showed that more than 95 pct of the iron, manganese, lead, and zinc were precipitated during the addition of a 20-pct slurry of lime in water that was equal to 27 lb of lime per 1,000 gal of brine. When the addition rate was decreased to 21 lb per 1,000 gal, manganese precipitation decreased to 50-58 pct, zinc precipitation was 92-95 pct, and iron and lead precipitation was 99 pct. Increasing lime addition rates to 35 lb per 1,000 gal precipitated 95 pct of the zinc and 99 pct of the iron and manganese. Under the more basic conditions resulting from the increased lime addition, lead was measured at higher concentrations than were present in untreated brine because of redissolution from precipitated hydroxides. The main difficulty during operation of the MRU was using pH as a process control. The high temperature and salinity of the brine represent an unfavorable environment for most electrodes. Coating of electrodes with lime and/or precipitated hydroxides compounded the problem. During portions of the test in which the oxidation states were varied by injection of air or exclusion of air, the oxidation state also affected pH measurements. When the MRU was operating under oxidizing conditions, pH fluctuated less and agreed with values that would be expected at the lime addition rates. When air was excluded from the MRU, operation demonstrated characteristics of non-steady-state behavior. Measured pH and percent precipitation of metals varied over a wide range with no detectable correlation to lime addition rates. The most consistent precipitation results were obtained by adding lime slurry at set flowrates, which were calculated to provide sufficient lime to accomplish the desired precipitation of dissolved metals.

Removal of precipitated metal hydroxides presented no major difficulties. A rake thickener, used for the initial solid-liquid separation, recovered up to 98 pct of the solids. The underflow from the thickener was filtered in a plate and frame filter press and yielded filter cakes containing 35 pct solids. The solids from a typical filter cake contain 57 pct soluble salts and 43 pct precipitated hydroxides. The major monetary value was represented by zinc, which was about 2.4 pct of the filter cake.

Bench-scale experiments performed to evaluate methods of releaching zinc since were not completely satisfactory. Approximately 50 pct of the zinc was selectively leached with ammoniacal solutions. Only a few percent of the remaining zinc was recovered by releaching the resultant residues. Acidic solutions leached manganese in preference to zinc. Zinc was not leached until 20 pct of the manganese was dissolved.

The data indicate that a low-grade zinc precipitate may be recovered from the brine and that leaching of the precipitate will extract about 50 pct of the zinc and yield an overall zinc recovery of 47 pct.

PRECIPITATION WITH SODIUM SULFIDE

During the same time that the investigation of the hydroxide precipitation method was being performed, a contract research effort was in progress to selectively precipitate zinc and lead from the same brine source. By using at least a 100-pct stoichiometric excess of sulfide and a pH of higher than 3.5, more than 90 pct of the zinc and lead was precipitated. Under these conditions, less than 20 pct of the manganese and 30 pct of iron were precipitated. Addition of a larger excess of sulfide, up to four times the stoichiometric requirement for zinc and lead, increased iron precipitation to 55 pct, while manganese precipitation was less than 20 pct.

When a brine was treated with a solution containing 0.83 $\underline{M}$ Na_2S, or 3.5 times the stoichiometric requirement for zinc and lead, a precipitate containing 37 pct zinc, 7 pct lead, 15 pct manganese, and 10 pct iron was obtained. No pH adjustment was employed in the test, and the resultant pH was 5.0.

The major difficulty encountered during sulfide precipitation was in solid-liquid separation. The slurry obtained from addition of Na_2S contained only 0.08 pct solids. While the solids settled rapidly when stirring was stopped, the settled sludge contained only 2.3 pct solids. The sludge was filtered by pressure filtration to 23 pct solids, but filtration was very slow. The filter cake was washed by an equal volume displacement wash to remove salt. Filtration in a 6-inch pressure filter at a constant pressure difference of 60 psi required 38 minutes to filter 1 liter of sludge.

Characterization studies on the sludge showed that lead and zinc sulfides were crystalline and had a particle size of 0.1 micrometer or less. The iron and manganese sulfides were amorphous. The sulfides formed 1 to 2mm flocs without the aid of a flocculant.

Comparison with hydroxide precipitation shows that the sulfide precipitate represented a higher grade zinc concentrate. The precipitate would be easier to process for zinc recovery and may be suitable feed to a commercial zinc reduction facility.

Another advantage of sulfide precipitation is that treated brine may be injected without further treatment to adjust pH. Disadvantages include the difficulty in achieving solid-liquid separation and the low percent solids of the filter cake.

COMBINATION RECOVERY TECHNIQUES

A combination recovery sequence would employ sulfide precipitation of zinc and lead followed by precipitation of iron and manganese by lime addition. The brine could then be acidified to its original pH and injected. If lithium recovery is desired, the brine could be treated with aluminum chloride and lime before acidification. The sulfide and alumina precipitates would represent two high-value products. The sulfide precipitate could be sold or subjected to further hydrometallurgical processing. The alumina precipitate would be processed to recover aluminum chloride for recycle and generate a lithium carbonate or chloride product. Research on treatment of the precipitates and techniques for improving the solid-liquid separation characteristics of the sulfide precipitate is currently in progress at the Reno Research Center.

DISCUSSION AND CONCLUSIONS

The potential benefits that could be realized if all the technical problems can be solved are apparent when the potential annual recovery of metals from brines is estimated and compared with the annual U.S. consumption figures.

Estimates of the annual potential for power production from the Imperial Valley KGRAs over the next 30 years range from 3,000 to 20,000 MW. If we use the 3,000-MW figure, which corresponds to a brine flow rate of 530,000 gpm, and assume 90 pct metal recovery from a brine containing the concentrations presented in table 1 during operation for 300 24-hour days a year, the resource could provide 230,000 tons of zinc. This is 18 pct of the U.S. annual zinc consumption reported for 1980. Lead recovery would represent 42,000 tons which is 3.6 pct of the annual consumption reported for the same year. Recovery of lithium would be 160,000 tons which exceeds consumption by an order of magnitude. Lithium consumption will increase in the future because of increased usage of lithium batteries. If the use of fusion reactors becomes commercial, this would represent another new market for lithium.

While some problems remain unresolved, results of completed work indicate that techniques are available and capable of recovering significant quantities of useful products.

REFERENCES

Berthold, C.E., P. Hadzeriga, D.H. Christopher, T.A. Applegate, and D.M. Gillespie, 1975, Process Technology For Recovering Geothermal Brine Minerals. BuMines OFR 35-75. Available from NTIS, Springfield, VA., PB 867/AS.

Berthold, C.E., and F.M. Stephens, 1978, Magmamax No. 1 Geothermal Brine Bulk Solids Precipitation Pilot Plant-- Engineering Design. BuMines OFR 127-78.

Epstein, J., E. Feist, J. Zmora, and Y. Marcus, 1981, Extraction of Lithium From the Dead Sea. Hydrometallurgy, 6 pp. 269-275.

Farley, E.P., E.L. Watson, D.D. MacDonald, R.W. Bartlett, and G.N. Krishnan, 1981, Recovery of Heavy Metals From High Salinity Geothermal Brines. BuMines OFR 91-81. Available from NTIS, Springfield, VA., PB 81-222218.

Urbanek, M.W., C.D. Hornberg, and B. Lindal, 1979, Research on a Geothermal Mineral Extraction Complex, Phase 1, Preliminary Technical and Economic Assessment, BuMines OFR 51-79.

Geothermal Resources Council, TRANSACTIONS Vol. 6, October 1982

PRELIMINARY CHEMICAL STUDIES OF THERMAL WATERS IN LASSEN VOLCANIC NATIONAL PARK AND VICINITY

J. Michael Thompson

U.S. Geological Survey
Menlo Park, California

ABSTRACT

Most thermal waters in Lassen Volcanic National Park (LVNP) issue as acidic, low Cl hot springs, which are characteristic of vapor-dominated hydrothermal systems and cannot be used for liquid-chemical geothermometry. Waters from the neutral pH, low Cl hot springs at Drakesbad and in Little Hot Springs Valley may have equilibrated in shallow aquifers, and so both Na-K-Ca and Na-Li geothermometers approach the measured spring temperatures. The high Cl waters at Growler Hot Spring and Morgan Hot Springs, situated south of LVNP, are the most appropriate for liquid-chemical geothermometry and indicate subsurface temperatures between 230° and 240°C. These springs may arise either from boiling at depth and subsequent mixing with meteoric water or from conductive cooling during lateral flow. In either case, the ~230° to 240°C thermal water probably originated inside the park and flowed southward to Morgan Hot Springs.

INTRODUCTION

Interpretation of the chemistry of thermal waters issuing from the hot springs and fumaroles in Lassen Volcanic National Park (LVNP) and vicinity can provide insights about the structure of the associated hydrothermal system. Waring (1915) reported three thermal-water analyses - two from LVNP (Bumpass Hell and Devils Kitchen) and one from Morgan Hot Springs (outside LVNP approximately seven km due south of Bumpass Hell, fig. 1). Day and Allen (1925) reported analyses of waters from 23 acidic hot springs exclusively in LVNP collected during their studies of then erupting Lassen Peak and vicinity. White and others (1963) reported analyses from two hot springs near LVNP--an acidic spring in Bumpass Hell and a neutral spring at Growler Hot Spring. Ghiorso (1980) reported chemical analyses for 34 hot springs, pools, and mudpots in LVNP--the largest compilation of acidic thermal features in LVNP. Although Waring (1915), Day and Allen (1925), and Ghiorso (1980) reported that most thermal springs within the boundaries of LVNP are acidic and can be characterized by White and others' (1963) definition of "acid sulfate springs associated with volcanism," the thermal waters from Morgan Hot Springs and Growler Hot Spring were classified as "thermal waters from geyser areas in volcanic environments" by White and others (1963, p. 40, table 17). (Incidentally, springs both in Bumpass Hell and at Morgan Hot Springs were used by White and others, (1963) as examples of these two water types.)

RESULTS AND DISCUSSION

In this paper, I report on the analyses of eight hot springs, one nonthermal spring which is apparently a highly diluted thermal water, and one background, nonthermal spring (table 1). Because water samples from acidic hot springs are not usually representative of any deep thermal fluids, neutral springs are preferred for liquid-chemical geothermometry (note calculated temperatures for Bumpass Hell, table 1). These three types of spring waters were collected and analyzed by the methods of Thompson (1979a, b) and Thompson and Hutchinson (1980, 1981). Except for Growler Hot Spring and an unnamed hot spring at Morgan Hot Springs, all the thermal waters in the LVNP area contain <30 mg Cl L^{-1} (table 1).

Using the Na-K-Ca (Fournier and Truesdell, 1973) and the Na-Li geothermometers (Fouillac and Michard, 1981), the calculated temperatures for Drakesbad and Little Hot Springs Valley generally agree with the measured orifice temperatures. The rate of flow of the local ground water is sufficiently slow so that a chemical equilibrium (as defined by the two geothermometers) is established between the country rock and the water. This ground water does not appear to be associated with any deep thermal fluid, and its source is probably shallow. The shallow water may be heated by a gaseous thermal fluid and (or) by a high regional heat flow.

Morgan Hot Springs and Growler Hot Spring are the only thermal areas in the vicinity of LVNP that have thermal waters which consistently indicate a high temperature (>150°C) environment by liquid-chemical geothermometry; the adiabatic silica geothermometer (Fournier and Rowe, 1966) indicates a temperature between 176° and 187°C. The Na-K-Ca geothermometer (Fournier and Truesdell, 1973), however, indicates temperatures near 220°C for both these springs and Nehring and others (1979) reported sulfate

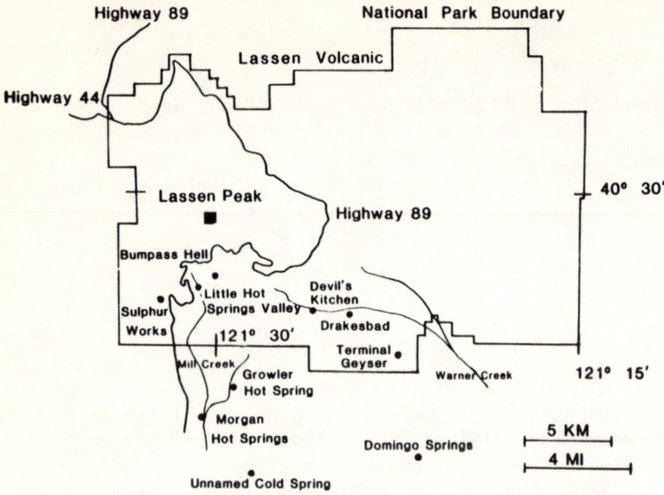

Figure 1. Location map of thermal areas near Lassen Peak and vicinity

geothermometer temperatures near 230°C for both springs. This disparity between the silica geothermometer and the Na-K-Ca and sulfate geothermometers may reflect either mixing of different waters or silica reequilibration. If mixing is assumed, the silica mixing model of Truesdell and Fournier (1977) yields temperatures of 189°C assuming steam loss at 100°C (near the boiling point at the surface), and of 220°C, assuming steam loss at 140°C.

The thermal waters from Morgan Hot Springs and Growler Hot Spring may contain some dissolved constituents from sedimentary formations. White and others (1963) suggested that any water having a Br/Cl weight ratio >0.003 is connate. From the data listed in table 1, the Br/Cl weight ratio is 0.004 at Growler Hot Springs and 0.007 for an unnamed spring at Morgan Hot Springs. Taft and others (1940) reported that Upper Cretaceous marine sandstone, shale, and conglomerate are deposited in the Chico Formation, which outcrops approximately 33 km west and southwest of Morgan Hot Springs. Additionally, the Na-Li geothermometer (Fouillac and Michard, 1981) indicates temperatures ~40°C below those determined by the Na-K-Ca geothermometer (Fournier and Truesdell, 1973). M.J. Reed (oral communication, 1982) has noted a similar pattern for geothermal areas in sediment of the Imperial Valley. Thus the Na-Li geothermometer may indicate temperatures lower than those determined by any other cation geothermometer in sedimentary environments and supports the assumption that the thermal waters from Growler Hot Spring and Morgan Hot Springs contain a sedimentary component.

It may be that hot water originating in LVNP, probably in the vicinity of Bumpass Hell, moves southward at depth toward Morgan Hot Springs. Somewhere in the vicinity of Growler Hot Spring, the water begins an upward movement passing through the Chico Formation, which may have been heated by overlying lavas. The thermal water thus gains Br and NH_3 from thermal decomposition of organisms initially buried in the oceanic sediment.

A model based on isotopic data was proposed by Truesdell and Hulston (1980). This model, which also is supported by the chemical data, assumes that the initial 230° to 240°C water cools conductively to 95°C during lateral flow, precipitating silica as it flows, to discharge at Growler Hot Spring and Morgan Hot Springs. This model also accounts for the discrepancy between the calculated quartz and Na-K-Ca temperatures. The Na-Li geothermometer may indicate lower temperatures because the sedimentary rock contains less Li than the fresher volcanic rock, and the brine contained in the sedimentary rock may also contain less Li per unit Na than the hot volcanic brine underlying Bumpass Hell. If this interpretation is correct, then the deep Terminal Geyser water, as reported by Beale (1981), could contain just slightly more Cl than reported.

Another model, also compatible with the chemistry, allows boiling that occurs as the thermal water ascends to the surface. Mixing of this boiled water with cold meteoric water then decreases the concentration of total dissolved solids. Part of this water issues at Growler Hot Spring, but most of the water moves southwesterly, mixes with more cold meteoric water, and issues at Morgan Hot Springs. The change in the Cl/SiO_2 ratio, however, indicates a loss of silica. Silica-mixing-model calculations indicate that the thermal water may have been diluted by as much as 40 percent after boiling, so that the boiled thermal water originally contained ~3500 mg Cl L^{-1} and the reservoir water ~3000 mg Cl L^{-1}.

Beale (1981) has reported as much as 2100 mg Cl L^{-1} in water from the Walker "O" No. 1 well at Terminal Geyser. Stearns and others (1937) suggested the presence of a fault running from Devils Kitchen to Terminal Geyser. Klein (1979) shows that fault in his report on the seismicity of LVNP. From the surficial hydrothermal expressions along this fault, the greatest activity occurs at Devils Kitchen and the activity decreases toward Terminal Geyser. According to Beale (1981), the highest temperatures were measured in the well at a depth of 660 to 800 m. The well had to be stimulated with N_2 to flow. No known hot springs issue south of Terminal Geyser. Domingo Springs, approximately 7 km south of Terminal Geyser, contains 10 times more Cl and 6 times more B than a typical background spring and only 0.01 times the Cl and B of Morgan Hot Springs (table 1); thus, Domingo Springs may be a heavily diluted expression of the deep Terminal Geyser thermal water. If both Growler Hot Spring - Morgan Hot Springs and the deep thermal water at Terminal Geyser originate inside LVNP and both waters have boiled, then the original LVNP thermal water may contain as much as 3000 mg Cl L^{-1}.

CONCLUSIONS

Table 1. Chemical Data from Thermal Springs in the Lassen Peak Thermal Area.

[Estimated flow in liters per minute; ne, not estimated. Temperature in degrees Celsius All values of chemical constituents in milligrams per liter, nd, not determined, nc, not calculated. All gepthermometers temperatures in degrees Celsius; chalcedony, adiabatic quartz, and conductive quartz calculated from equations of Fournier and Rowe (1966), Na-K (1) from Truesdell (1976), Na-K(2) from Fournier (1979), Na-K-Ca from Fournier and Truesdell (1973) and Na-Li from Fouillac and Michard (1981).]

	Bumpass Hell	Devils Kitchen	Sulphur Works	Little Hot Springs Valley	Drakesbad	Terminal Geyser	Growler Hot Spring	Morgan Hot Springs	Domingo Springs	Unnamed cold spring
date	Aug. 79	July 79	Sept. 76	Aug. 79	Sept. 79	July 79	Aug. 79	Aug. 79	Aug. 79	Aug. 79
est flow	25	8	12	4	6	25	20	8	High	ne
temp	52.5	93	84	93	66	95	95	95	10	10
field pH	2.1	6.8	6.9	5.65	6.8	5.20	7.45	7.25	5.9	6.5
lab pH	2.36	7.05	8.1	6.59	8.13	7.63	7.85	8.08	7.00	6.74
Chemical constituents										
SiO_2	236	60	140	123	144	64	272	225	40	39
Al	17.6	0.7	nd	.4	nd	.4	.4	.2	nd	nd
Fe	12	.41	.02	.34	.04	.1	.04	.01	nd	nd
Mn	.2	.7	.40	.15	.06	.04	.01	.03	nd	nd
As	nd	nd	nd	nd	nd	nd	11	10.6	nd	nd
Ca	34	9.2	188	26	38.6	9.25	81	94	5.3	11.6
Mg	6.6	.9	38	4.1	9.6	1.5	.07	.47	2.9	7.6
Sr	nd	nd	nd	nd	.6	nd	.9	1.1	nd	nd
Ba	nd	nd	nd	nd	.5	nd	2.5	3.0	nd	nd
Na	33	12	68	87	41	7.5	1340	1260	12	3.1
K	10	1.1	14	12	10.3	1.3	173	162	2.6	1.9
Li	.05	.05	.02	.05	.03	.05	6.3	6.1	.01	.01
Rb	nd	nd	.12	nd	.1	nd	1.4	1.2	nd	nd
Cs	nd	nd	.07	nd	.06	nd	.6	.4	nd	nd
NH_4	nd	1.0	nd	3.5	nd	nd	9.5	13	nd	nd
HCO_3	0	82	115	19	130	24	55	68	78	138
SO_4	547	78	514	301	132	16	110	123	2	3.
Cl	5.2	16	.5	5.2	1	26	2370	2210	21	2
F	.28	.22	.47	.84	.15	.29	2.3	3.0	<.01	.1
Br	nd	nd	nd	nd	nd	nd	10	14.5	nd	nd
B	1.6	.1	.5	4.2	.6	1.4	84	nd	.6	.1
H_2S	nd	.03	nd	.03	.3	nd	.02	.03	nd	nd
H_3O^+	10.6	0	0	0	0	0	0	0	0	0
Weight ratios										
Cl/F	18.6	72.7	1.06	6.19	6.0	89.6	908	936	210	20
Cl/B	3.2	160	1.0	1.24	1.5	18.6	30.7	nc	35	20
Cl/SO_4	.01	.205	.001	.017	.007	1.62	20.1	23.6	10.5	.67
Cl/SiO_2	.022	.267	.004	.042	.007	.406	8.71	9.82	.52	.05
Cl/Na	.16	1.33	.007	.06	.022	3.45	1.80	1.54	1.75	.65
Cl/K	.54	14.5	.036	.437	.087	19.5	14.3	13.1	8.08	1.05
Na/Li	656	240	3400	7.31	1370	151	177	203	1200	310
Na/Ca	.94	.82	.36	1740	1.06	.81	14.3	17.7	2.26	.27
Ca/Mg	5.30	4.32	4.95	3.39	4.02	6.2	275	782	1.83	1.53
Br/Cl	nc	nc	nc	nc	nc	nc	.004	.007	nc	nc
Calculated Geothermometer Temperatures										
Chal	172	80	131	142	133	83	184	168	nc	nc
Qtz Ad	178	111	150	158	152	114	187	176	nc	nc
Qtz Cond	191	110	157	166	158	114	202	188	nc	nc
Na-K	354	178	628	238	314	258	217	216	nc	nc
Na-K	329	210	473	253	304	267	238	238	nc	nc
Na-K-Ca	75	21	57	95	75	29	223	219	nc	nc
Na-Li	100	173	22	66	61	218	184	187	nc	nc

The thermal water discharging at Morgan Hot Springs and Growler Hot Spring, and the deep thermal water at Terminal Geyser, both probably originate inside LVNP near Bumpass Hell at ~230° to 240°C. The deep Terminal Geyser water may be diluted by a factor of as much as 100 during its ascent from ~1300 to ~1700 m above sea level. Because the Br/Cl ratio in these waters is higher than seawater and the Na-Li geothermometer indicates a lower temperature than the Na-K-Ca geothermometer, there may be a chemical contribution to the thermal waters from sedimentary material in the area, possibly from thermal decomposition of biological material contained in the Chico Formation.

ACKNOWLEDGEMENT

I thank C. K. Janik, A. H. Truesdell, and M. A. Clynne for useful discussions on the Lassen thermal system and M. L. Sorey, N. L. Nehring, and K. E. Bargar for their excellent reviews of the manuscript.

REFERENCES

Beale, J. J., 1981, A hydrologic model based on deep test data from the Walker "O" No. 1 well, Terminal Geyser, California: Geothermal Resources Council, Transactions., vol 5, p. 153-156.

Day, A. L., and Allen, E. T., 1925, The volcanic activity and hot springs of Lassen Peak: Carnegie Institution of Washington Publication no. 360, 190 p.

Fouillac, C. and Michard, G., 1981, Sodium/lithium ratio in water applied to geothermometry of geothermal reservoirs: Geothermics, v. 10, p. 55-70.

Fournier, R.O., 1979, A revised equation for the Na/K geothermometer: Geothermal Resources Council, Transactions., v. 3, p. 221-224.

Fournier, R. O., and Rowe, J. J., 1966, Estimation of underground temperaturesfrom the silica content of water from hot springs and wet steam wells: American Journal of Science, v. 264, p. 685-697.

Fournier, R. O., and Truesdell, A. H., 1973, An empirical Na-K-Ca geothermometer for natural waters: Geochimica et Cosmochimica Acta, v. 37, p. 1255-1275.

Ghiorso, M. S., 1980, Studies in natural solid-liquid equilibrium: Ph. D. thesis, University of California, Berkeley, 371 p.

Klein, F. W., 1979, Earthquakes in Lassen Volcanic National Park, California: Seismological Society of America Bulletin, v. 69, p. 867-875.

Nehring, N. L., Mariner, R. H., White, L. D., Huebner, M. A., Roberts, E. D. Harmon, K. A., Bowen, P. A., and Tanner, L., 1979, Sulfate geothermometry of the western United States: U.S. Geological Survey Open-File Report 79-1135, 11 p.

Stearns, N.D., Stearns, H.T., and Waring, G.A., 1937, Thermal springs in the United States: U.S. Geol. Survey Water Supply Paper 679-B, 205 p.

Taft, J. A., Hanna, G. D., and Cross, C. M., 1940, Type locality of the Cretaceous Chico Formation: Geological Society of America Bulletin, v. 51, p. 1311-1328.

Thompson, J. M., 1979a, A reevaluation of the geothermal potential of the Wilbur Hot Springs Area, California: Geothermal Resources Council Transactions, v. 3, p. 729-731.

Thompson, J. M., 1979b, Arsenic and fluoride in the Upper Madison River system: Firehole and Gibbon Rivers and their tributaries, Yellowstone National Park, Wyoming and southeast Montana: Environmental Geology, v. 3, p. 13-21.

Thompson, J. M., and Hutchinson, R. A., 1980, Boundary Creek Thermal Areas of Yellowstone National Park: II Thermal water analyses: Geothermal Resources Council Transactions, v. 4, p. 189-192.

Thompson, J. M., and Hutchinson, R. A., 1981, Chemical analyses of waters from the Boundary Creek Thermal Area, Yellowstone National Park, Wyoming: U.S. Geological Survey Open-File Report 81-1310, 15 p.

Truesdell, A. H., 1976, Summary of section III - geochemical techniques in exploration in Second United Nations Symposium on the Development and Use of Geothermal Resources, San Francisco, May 1975, U.S. Government Printing Office, v. 1, p. liii-lxiii.

Truesdell, A. H., and Fournier, R. O., 1977, Procedure for estimating the temperature of a hot water component in a mixed water by using a plot of dissolved silica vs enthalpy: U.S. Geological Survey Journal Research, v. 5, p. 49-52.

Truesdell, A. H., and Hulston, J. R., 1980, Isotopic evidences on environments of geothermal systems in P. Fritz and J.C. Fontes (Eds.), Handbook of Environmental Isotope Geochemistry, V, v. 1: The Terrestrial Environment, Amsterdam, Elsevier, p. 179-225.

Waring, G. A., 1915, Springs of California: U.S. Geological Survey Water-Supply Paper 338, 410 p.

White, D. E., Hem, J. D., and Waring, G. A., 1963, Chemical composition of subsurface waters, Chap F of Data of Geochemistry (M. Fleischer ed.): U.S. Geological Survey Professional Paper 440-F, p. F1-F67

EXPLORATION-III: GEOPHYSICS

Geothermal Resources Council, TRANSACTIONS Vol. 6, October 1982

THE KEMP THERMAL ANOMALY: A NEWLY DISCOVERED GEOTHERMAL RESOURCE IN PUMPERNICKEL VALLEY, NEVADA.

Thomas Flynn, Dennis T. Trexler
Division of Earth Sciences, UNLV
255 Bell Street, Reno, Nevada 89503

Brian A. Koenig
Union Oil Company, Santa Rosa, California

Abstract

Geophysical, geochemical and shallow-depth drilling surveys were successfully employed in Pumpernickel Valley, Nevada, to identify a previously unrecognized geothermal resource. The study area, located within the "Battle Mountain Heat Flow High", contains many low-temperature (<100°C) geothermal occurrences and at least one moderate temperature (100-150°C) geothermal well. The newly recognized resource is not spatially associated with any known thermal manifestation, and was first identifed on the basis of a 2-m depth temperature-probe survey. Gravity data suggest the thermal anomaly originates from a fault located 1 -2 km (.6-1.2mi.) east of a prominent northeast-trending range-bounding fault. The maximum temperature measured was 70°C at a depth of approximately 100 m (300 ft.). The areal distribution of economically recoverable geothermal fluids associated with this discovery is probably limited to one km^2 or less. Direct temperature measurement techniques were judged to be the most useful in this study.

Introduction

In May, 1981, the Division of Earth Sciences began an assessment of low and moderate temperature geothermal occurrences in Carlin and Pumpernickel Valley, Nevada. (This work was funded by the U.S. Department of Energy, Division of Geothermal Energy, (DOE/DGE) under contract number DE-AC08-81NV10220). The exploration techniques employed in this investigation are based on previous DOE/DGE funded assessment programs that were completed in Carson-Eagle Valleys, Big Smoky Valley, Caliente, Hawthorne, and Paradise Valley, Nevada. The federally-funded geothermal assessment programs were established to expand the geothermal resource data base by characterizing known geothermal areas, evaluating exploration techniques, and possibly identifying new resources.

This paper presents the results of the Pumpernickel Valley assessment. Exploration techniques that provided direct temperature measurements, including 2-m depth temperature probe surveys, and 35-meter depth and 150-meter depth temperature gradient holes, were judged to be most useful in this study. Indirect exploration techniques, including water geochemistry and gravity surveys, provided additional supporting data.

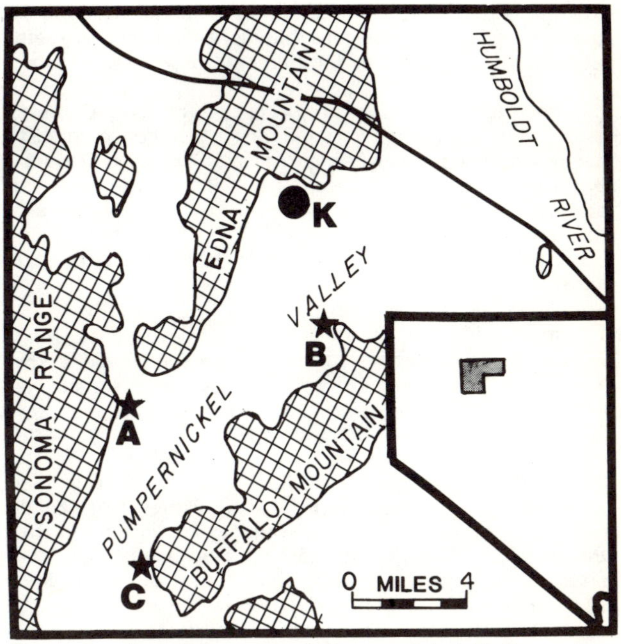

Figure 1. Location map of Pumpernickel Valley: A. Tipton Hot Springs; B. Brooks Hot Springs; C. Tipton-II Warm Spring; K. Kemp thermal anomaly.

Geology and Geothermal Occurrences

The Pumpernickel Valley study area covers approximately 500 km^2 (200 mi.^{2}) and is located in the southeast corner of Humboldt County and the northeast corner of Pershing County in north-central Nevada (fig. 1). Pumpernickel Valley trends predominantly northeast and is bounded by the Sonoma Range to the west, Edna Mountain along the northwest, Tobin Range to the south, and Buffalo Mountain on the southeast.

Rocks exposed in the mountains surrounding Pumpernickel Valley range in age from early Paleozoic to early Quaternary. Early Paleozoic rock types include both deep water off-shore and shallow water near-shore lithologies. The former is represented by the Ordovician Valmy Formation and includes chert, slate, argillite and greenstone; the latter is represented by quartzite of the Osgood Mountain Formation as well as interbedded shale and limestone of the Preble Formation. Both the Osgood Mountain and Preble Formations are Cambrian age. Chert, argillite, shale, and occasional greenstone of the Havallah sequence (Silberling and Roberts, 1962) represent the majority of late Paleozoic rocks found in the ranges surrounding the valley.

Granodiorites of Mesozoic age intrude the Paleozoic sediments and are extensively exposed in the ranges surrounding Pumpernickel Valley. Mafic rocks are observed in some areas between the granodiorites and sedimentary rocks. Limited exposures of Triassic age Koipato group rhyolitic lavas and tuffs are also present.

Tertiary age rocks are almost entirely welded to non-welded silicic ash-flow tuffs, air-fall tuffs and sedimentary rocks. Olivine-bearing basalt flows of Tertiary-Quaternary age crop out at the northern end of Pumpernickel Valley, and prominent exposures occur on Edna Mountain. These rocks represent the most recent volcanic activity in the study area. Alluvial deposits of Quaternary age include fans, playas, and stream gravels, silts and clays related to the Humboldt River.

Exposed faults in the area consist of thrust faults associated with the Permo-Triassic Golconda thrust, northeast trending high-angle normal faults related to Basin and Range style deformation and north-northwest oriented faults associated with the Oregon-Nevada lineament (Stewart and others, 1975). The latter faults have been mapped in Quaternary and Tertiary age basalts (Stewart and Carlson, 1976).

Thermal fluids are known to occur at several locations within the study area. At the Tipton Ranch along the west side of Pumpernickel Valley, several springs discharge thermal fluids from a prominent fault at 30 to 40 GPM with temperatures as high as 85°C. Steam and hot water (104°C) also issue from a 936-meter (3071 ft.) deep well drilled in the immediate vicinity of the former 85°C spring. A single spring (28°C) issues from rocks near the southwest end of Buffalo Mountain on the Pumpernickel Valley side. Brooks Hot Spring (35°C) is situated approximately along the contact between playa sediments and alluvium derived from the northeast end of Buffalo Mountain. There are no thermal springs in the immediate vicinity of the newly discovered thermal anomaly (located near the Kemp benchmark, black circle on fig. 1) although Sulfur Spring (18°C) discharges fluids that are compositionaly similar to Brooks Hot Spring fluids.

Two-Meter Depth Temperature Probe Survey

During October and November, 1981, approximately 100 holes were drilled to a depth of 2-m throughout Pumpernickel Valley. These holes were drilled along accessible roads at 1.6 km (1 mi.) intervals. Thermistor probes, encased in 1.9 cm (.75 in.) diameter PVC tubes were then emplanted and allowed to equilibrate for at least 24 hours. Measured temperatures were corrected for seasonal variations and elevation differences. This was accomplished by using control probes which were allowed to remain in place for the duration of the survey.

Corrected temperatures ranged from 17.5°C to 26°C. Background temperatures ranged from 18°C to 22°C and were affected principally by soil moisture content. Temperatures greater than 22°C were considered anomalous and were taregeted for further exploration. The corrected temperature at Kemp was 26°C.

Corrected temperatures of 23°C and 24°C occurred 1.6 km (1 mi.) north and 2.5 km (1.5 mi.) west, respectively, of Kemp.

Temperatures that were only slightly above background were recorded near Brooks Hot Spring 6.6 km (4 mi.) southeast of Kemp and Blossom Hot Pot 20 km (12 mi.) northeast of Kemp.

35-Meter Temperature Gradient Holes

Four drill sites were selected for 35 m temperature gradient holes on the basis of results obtained in the 2-meter depth temperature probe survey. An additional site was selected in an area where no other subsurface data were available. The holes were drilled with a small, rotary mud rig. Return mud temperatures were monitored continuously and cuttings were collected at 3 m (10 ft.) intervals. The holes were blind-cased with 5 cm (2 in.) diameter, water-filled PVC pipe, backfilled and allowed to equilibrate for three weeks. Lithologic logs for these holes consist of various ratios of gravel, sand, silt, and clay layers. No bedrock was encountered.

Temperature gradients were calculated from temperatures measured at 3 m (10 ft.) intervals in the equilibrated well bores. Gradients for the bottom 10 m (30 ft.) ranged from 25°C/km to a high of 600°C/km at Kemp. Gradients of 100°C/km and 220°C/km were measured at 2-m probe sites with corrected temperatures of 22°C and 19.4°C respectively.

Bottom hole temperatures for four of the wells ranged from 12°C to 20°C. At Kemp, the bottom hole temperature was 50°C.

65-145 Meter Depth Temperature Gradient Holes

Five holes, ranging in depth from 65 m (220 ft.) to 145 m (480 ft.) were drilled in the area surrounding the Kemp thermal anomaly during April and May, 1982. The objectives of this drilling program were to determine resource characteristics such as temperature and depth, to identify possible structural or stratigraphic controls, and to ascertain the areal extent of the resource.

The holes were drilled with a Gardner-Denver rotary rig with both mud and air drilling capabilities. Return mud temperatures were continuously monitored and drill chip samples were collected at 6 m (20 ft.) intervals. The holes were then blind-cased with 5 cm (2 in.) diameter water filled steel pipe and backfilled.

Temperature-depth profile measurements were completed after allowing approximately three weeks for equilibration. Lithologic logs constructed from drill chip samples indicate that alluvial material ranging in thickness from 90 to 110 m (200-350 ft.) overlies bedrock consisting of Paleozoic sedimentary rocks and Mesozoic igneous and volcanic rocks. Figure 2 illustrates the temperature-depth profile and lithologic log for the 110 m (360 ft.) hole drilled at Kemp. The diagram shows a constant gradient of 600°C/km to a depth of approximately 67 m (220 ft.). The well is isothermal to approximately 100 m (350 ft.) with a slight temperature reversal at the bottom where bedrock was encountered.

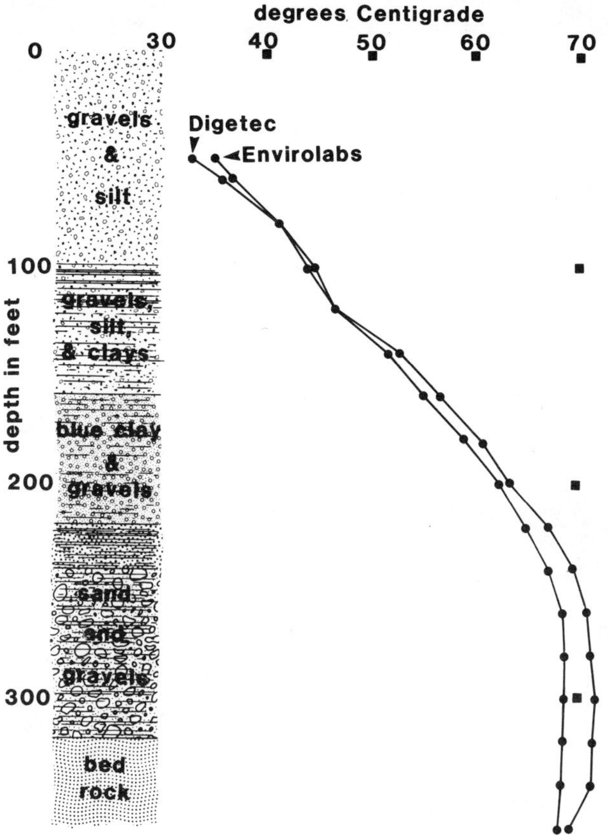

Figure 2. Lithologic log and temperature depth profile for Kemp 110 m (360 ft.) temperature gradient hole.

The zone of constant temperature is coincident with a 30 m (100 ft.) thick section of sandy gravels that lies above the sandstone bedrock and sandy blue clay. The highest temperature recorded in the Kemp well was 70°C at a depth of 76 m (250 ft.).

Temperature gradients in the other wells were significantly lower. PVTG-2, drilled 1.6 km (1 mi.) west of the Kemp well, had a gradient of 3.2°C/km. Several cold water aquifers were encountered in this hole. The gradients in the other holes ranged from 29°C/km to 32°C/km; these gradients are typical for the Battle Mountain Heat Flow High. Lithologic logs for these wells were similar to the Kemp well with the exception of the blue clay layer, which was unique to Kemp.

Gravity Survey

Proprietary gravity data from southern Pumpernickel Valley were combined with the results of a 200-station gravity survey in northern Pumpernickel Valley to construct a complete Bouguer gravity map of the entire study area. A regional gravity survey (Erwin, 1974) was available and both data sets were formatted into digitized 40 x 50 matrices that covered the entire study area. After subtracting the regional effects, a three-dimensional computer-generated graphical representation of the residual gravity was produced (fig. 3).

The view in this diagram is to the southwest with an elevation of 30° above the zero-plane horizon. This figure roughly corresponds to the basement topography and prominent features such as the Buffalo Mountain Gravity high (BM) and Pumpernickel Valley gravity low (PV) can be readily distinguished. The Kemp thermal anomaly (K) is coincident with a gravity low, which may be interpreted as an alluvial-filled asymmetric graben. The resolution of this diagram and lack of additional data in the Kemp area do not permit an exact interpretation of the small-scale structures responsible for control of the thermal fluids.

Aqueous Geochemistry

There are no surface manifestations of thermal fluids in the Kemp area. Thermal fluids were not recognized during drilling, but they are probably present in the sands and gravels that lie between bedrock and the thick layer of blue clay in the 110 m (360 ft.) hole at Kemp. Sulfur Spring, located 1.6 km (1 mi.) northeast of Kemp, discharges 18°C waters from a prominent northeast-trending fault scarp. Kerr (1940) described a hot spring with well-developed tufa deposits in the Sulfur Springs area. There are no well-developed tufa deposits at Sulfur Spring, but a fossil hot spring deposit, consisting of globular and botryoidal travertine, was discovered approximately 1.6 km (1 mi.) northeast of Sulfur Spring. The spring deposits are located in a brecciated fault zone, approximately 30 m (100 ft.) wide, in the Pumpernickel formation. The deposits suggest that fluids discharged at the surface, but it is impossible to deduce the temperature of those fluids.

Silica geothermometers for Sulfur Spring indicate an equilibrium temperature of 112°C, and magnesium-corrected cation geothermometers suggest equilibrium temperastures of 106°C. The bulk chemical composition of Sulfur Spring is similar to thermal fluids at Blossom Hot Pot (58°C), which is located 20 km (12 mi.) to the northeast. Sulfur Spring is also somewhat similar to Brooks Hot Spring (35°C), which is located 6.6 km (4 mi.) to the east. All three springs are chemically similar to Humboldt River waters.

Conclusions

Direct temperature measurement techniques were successfully employed in Pumpernickel Valley, Nevada, to locate a geothermal resource that is not associated with any surface thermal manifestations. The resource was originally identified during a 2-m depth temperature probe survey and later confirmed by 35 and 110 m (120 and 360 ft.) temperature gradient holes. Lithologic logs and temperature-depth measurements indicate that thermal fluids are present in a 30 m (100 ft.) thick layer of sand and gravels that overlies bedrock of Paleozoic sedimentary rocks. Thermal fluids do not appear at the surface and are probably trapped beneath a thick layer of blue clay.

A gravity low associated with the thermal anomaly suggests the presence of a buried fault, located 1-2 km (.6-1.2 mi.) east of a prominent northeast trending range bounding fault. The buried fault has no surface expression and is a likely source for rising thermal fluids.

Figure 4 is a cross-sectional representation of structural and stratigraphic controls at the Kemp thermal anomaly.

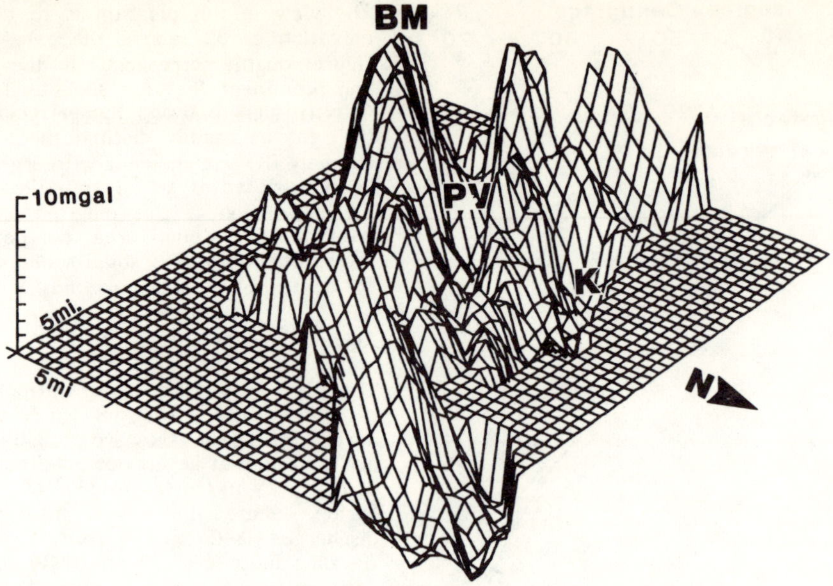

Figure 3. Three-dimensional computer-generated residual gravity map of Pumpernickel Valley: BM, Buffalo Mountain; PV, Pumpernickel Valley; K, Kemp (NCAR plot).

The diagram shows thermal fluids, rising along a buried fault zone, are channelled away from the surface by an impermeable layer of blue clay and are restricted to flow along the top of the bedrock. The ultimate destination of these fluids is unknown.

Temperature gradients for holes drilled 1-2 km north and west of Kemp ranged from 3.2°C/km to 32°C/km; the gradient at Kemp was 600°C/km. These data suggest that the resource is less than one km^2 in areal extent.

Direct temperature measurement techniques were judged to be the most effective for both resource identification and definition. Indirect methods were useful in the preparation of a structural model.

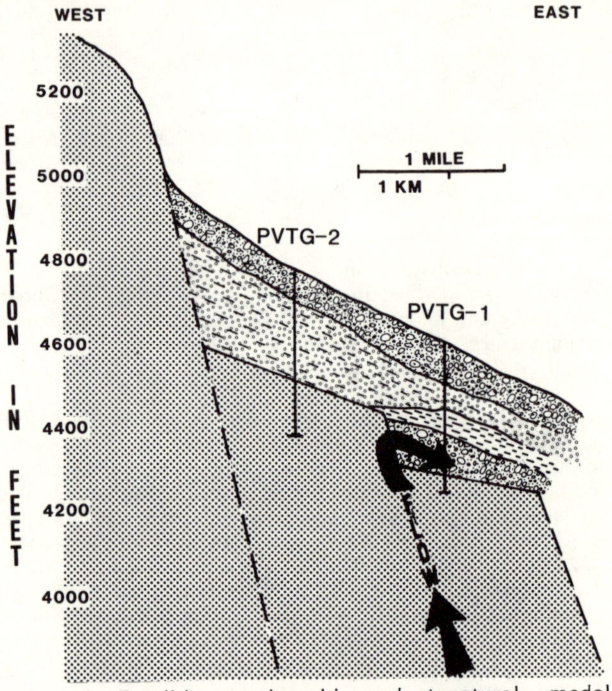

Figure 4. Possible stratigraphic and structural model through the Kemp thermal anomaly.

References

Erwin, J.W., 1974, Bouguer gravity map of Nevada, Winnemucca sheet: Nevada Bur. Mines Geol., Map 47.

Kerr, P.F., 1940, Tungsten-bearing manganese deposit at Golconda, Nevada: Geol. Soc. America Bull., v. 51, p. 1359-1389.

NCAR: National Center for Atmospheric Research, Boulder, Colorado.

Silberling, N.J., and Roberts, R.J., 1962, Preliminary stratigraphy and structure of northwest Nevada: Geol. Soc. America, Spec. Paper 72, 58 p.

Stewart, J.H., and Carlson, J.E., 1976, Geologic map of north-central Nevada: Nevada Bur. Mines Geol., Map 50, scale 1:250,000.

Stewart, J.H., Walker, G.F., and Kleinhampl, F.J., 1975, Oregon-Nevada lineament: Geology, v. 5, p. 265-268.

Acknowledgements

The authors gratefully acknowledge the assistance throughout the assessment program of technical monitor Joe Fiore, U.S. DOE, Las Vegas, Nevada. We would also like to thank Margie Johnson, Cam Covington, Steve Weiss, Elaine Bell and Patricia Flynn for help in preparing the manuscript and graphics.

Geothermal Resources Council, TRANSACTIONS Vol. 6, October 1982

STRIKE-DIP DETERMINATION OF FRACTURES IN DRILLCORES BY AN ASTATIC-MAGNETOMETER

Masao Hayashi and Naoyuki Furutani

Research Institute of Industrial Science
Kyushu University 86
Kasuga City, Fukuoka, 816 Japan

ABSTRACT

The strike and dip of fractures in drillcores from Well HT-4 drilled in the Hatchobaru geothermal field, Kyushu, Japan, have been determined using an astatic-magnetometer. Scince the drillcores consist mainly of younger andesite lavas, the measurements of the declination and inclination of remnant magnetism should yield the strike and dip of the fractures. The results show that they dip generally southward with angles from 40° to 80° (62.5 on the average), and strike NW-SE or NE-SW. The NW-SE trending fractures predominate in the Pleistocene series, which persists at depths shallower than 1000 m, while the NE-SW trending ones occur in the Neogene system at deeper levels, and are considered to be older than the former. The stress field can also be estimated by the strike-dip data and the direction of lineation on a slickenside.

INTRODUCTION

It is, of course, very important to determine the strike and dip of fractures in drillcores for the exploration of geothermal reservoirs. However, few attempts have been made to do this because of the difficulty involved. In the present paper, the strike and dip have been estimated after measuring the declination and inclination of the thermo-remnant magnetism of drillcores consisting of younger andesite lavas. This experiment has confirmed that such important fractures as slickensides and hydrothermal veins have similar tendencies at definite depth intervals. On the basis of these findings it has been possible to analyze the stress field in the rocks.

GEOLOGIC SETTING

Fig. 1 shows the location of Well HT-4 drilled to the southwest of the Hatchobaru field, where a 55 MW geothermal power station had already been constructed in 1977. The figure also indicates the existence of faults estimated by geological and geophysical (electric) surveys. Of these faults, the Komatsuike fault is the most conspicuous and is accompanied by high-temperature manifestations such as fumaroles and hot springs. The Hatchobaru fault divides the mapped area into a very active Type A field (Hayashi et al., 1981) on the northeast and a less active Type B field on the southwest. Some of the faults estimated by electrical prospecting trend obliquely to the geologically confirmed faults.

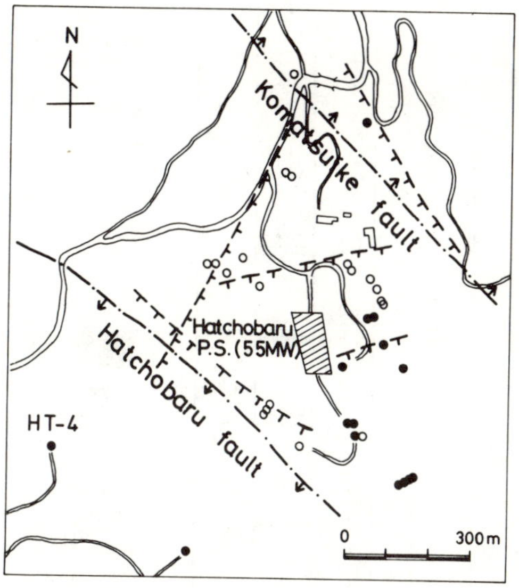

● Production well ● Exploratory well
○ Reinjection well and others
—·— Geological fault ⊤⊤ Geophysical fault

Fig. 1. Location map of Well HT-4 and faults in the Hatchobaru geothermal field.

The geology and hydrothermal alteration of Well HT-4 are illustrated in Fig. 2. From the surface to a depth of 213 m, the middle to late Pleistocene Kujyu Volcanic Rocks occur as an alternation of hornblende andesite lava flows and tuff breccias. Underlying them is a thick pile of pyroxene andesite lava flows intercalated with thin layers of tuff breccia. Known as the Hohi Volcanic Rocks, they began to erupt at the beginning of the Pleistocene and ended about 800,000 years ago. At depths greater than 1127 m, the Neogene Usa Group is found, being entirely made up of tuff breccias of andesitic composition.

The hydrothermal alteration of Well HT-4 is characterized by the presence of alunite zones of an

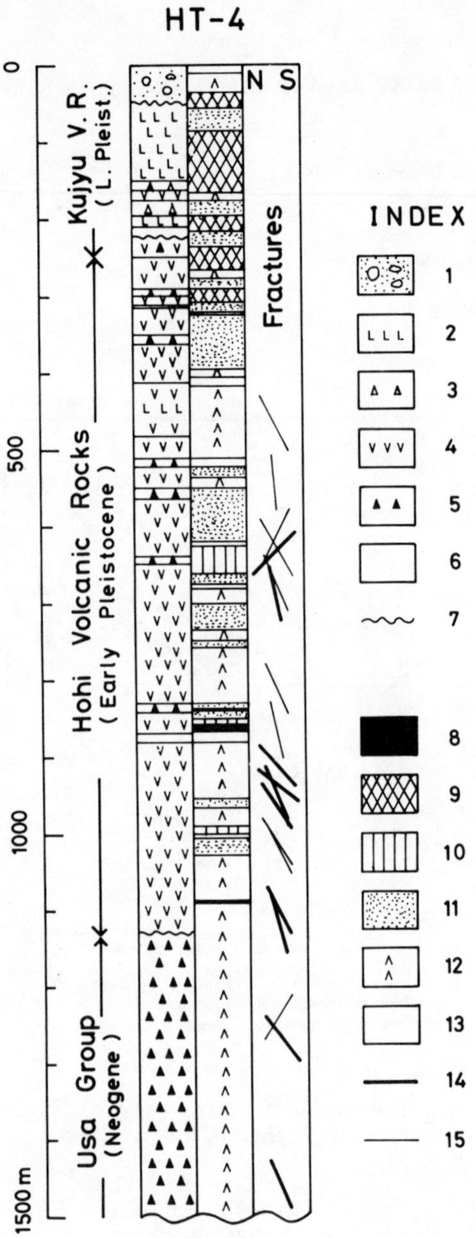

Fig. 2. Geology and hydrothermal alteration of Well HT-4 in the Hatchobaru geothermal field. 1. Tallus deposits, 2. Hornblende andesite lava, 3. Hornblende andesite tuff breccia, 4. Pyroxene andesite lava, 5. Pyroxene andesitic tuff breccia, 6. Unkown, 7. Unconformity, 8. Silicified zone, 9. Alunite zone, 10. Kaolin-pyrophyllite zone, 11. Aluminosilicate clay zone, 12. Partially altered zone, 13. Unaltered zone, 14. Slickenside, 15. Hydrothermal vein

acidic type at shallwer levels, whereas partially altered zones are predominant at deeper levels. In addition, thin silicified zones, which may correspond to conduits of geothermal fluids, are formed at three depths : 835 m, 861-872 m and 1090 m.

EXPERIMENTS

Fig. 3 illustrates an astatic magnetometer, Type NY-2, made by Kyoritsusha LTD, Japan. The device consists of two small bar magnets with nearly the same magnetic moment but opposite polarity, suspended by a thin phosphor-bronze wire. When these two magnets are so placed that the sum of their magnetic moment vanishes, the magnetic-system does not move when the magnetic field changes uniformly.

When a rock specimen is approached to the lower magnet, the magnetization of the rock specimen acts practically only upon the lower magnet. As a result of this action, the magnet-system turns around the vertical axis until it is balanced by the torsion of the suspention wire. The deflection angle can be read by the attached lamp-and-scale. Thus, the direction of magnetization of the specimen can be obtained by measuring the deflection angle, with the remote control of the attached turning table, on which the rock specimen is placed (Kyoritsusha, 1980).

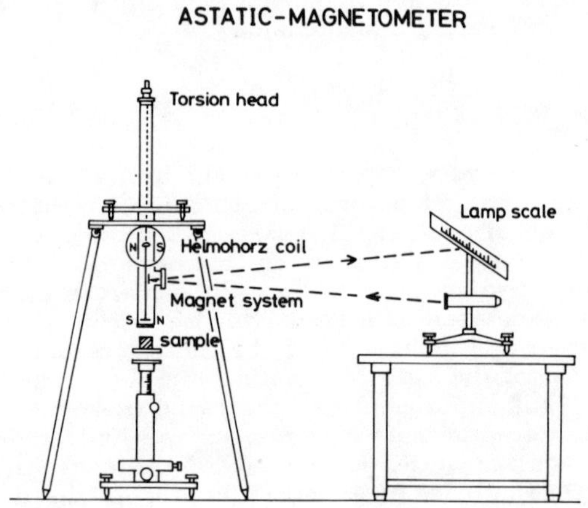

Fig. 3. An astatic-magnetometer used in this study.

RESULTS

Results of the strike-dip determination of fractures are plotted in Wulff's net of Fig. 4. They are classified into slickenside, hydrothermal vein and joint, the second of which is sibdivided into two according to the width of the veins. It is obvious from the figure that such important fractures as slickensides and hydrothermal veins dip southward in most cases. Moreover, there seem to be two groups with different trends in the fractures;

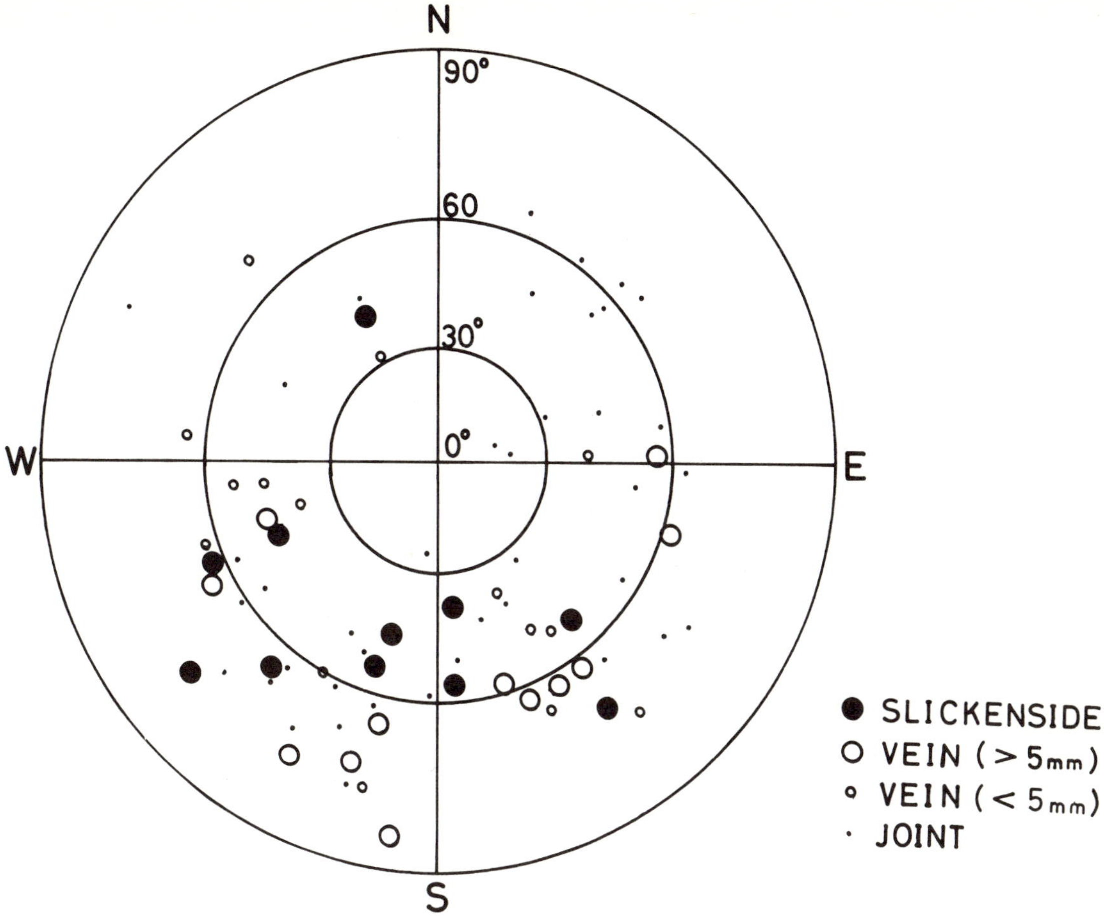

Fig. 4. Wulff's net projection of fractures in drillcores of Well HT-4 in the Hatchobaru field (The upper hemisphere is used).

one having NW-SE and the other NE-SW. The NW-SE trending fractures predominate in the Pleistocene series, which occurs at depths of less than 1000 m. On the other hand, the Neogene system at greater depths abounds in the NE-SW trending fractures, suggesting the difference of stress field between the two ages.

Fig. 5 show the frequency distribution of dip-angle of various kinds of fractures; slickensides, hydrothermal veins and joints. Dip of the fractures ranges from 15° to 85° and concentrates between 40° to 80°, with an average of 62.5°. The average value coincide with that the highest frequency of the dip-angles. Looking at the figure in detail, the hydrothermal veins seem to have steeper dips compared with the slickensides, that are in the narrowest range of the three kinds of fractures. Joints, which have no evidence of sliding or filling have almost constant frequency in a range of 40° to 75°.

The dip-angles of the fractures are also shown in Fig. 2 with the distiction of south or north in the direction of dip. As can be easyly visualized, the fractures dip south except one slickenside and two hydrothermal veins. It may be noteworthy that the slickensides are formed a little beneath the silicified zones at depths of 861-872 m and 1090 m. In other words, the strongly alterd zones are somewhat above the depths of the slickensides (fault zone). Furthermore, the fault zones themselves are made up of partially altered zone, indicating less permeability of the fault zones.

DISCUSSION

It is well-known that the North and the South Poles of the earth have moved such a short distance over the past several millions years that it has been possible to use data on rock magnetism to discuss continental drift. Thus when rocks are taken from younger formations such as the Pleistocene and the Late Neogene, the direction of thermo-remnant magnetism can be considered to be parallel to the present magnetic line of force of the earth. For this reason, the measurement of declination and inclination of drillcores reproduces their original directions in situ, resulting in the determination of the strike and dip of fractures formed in the drillcores.

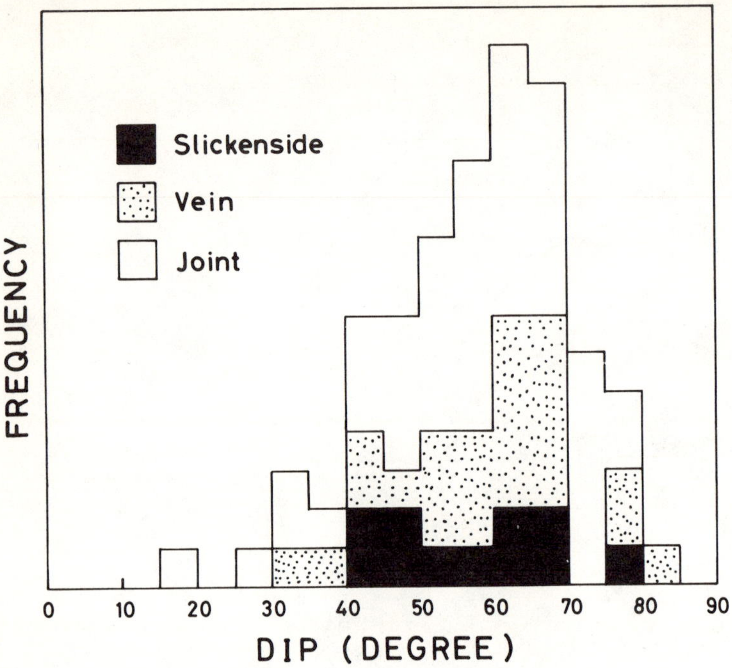

Fig. 5. Frequency distribution of the dip-angle of fractures in drillcores from Well HT-4 in the Hatchobaru geothermal field.

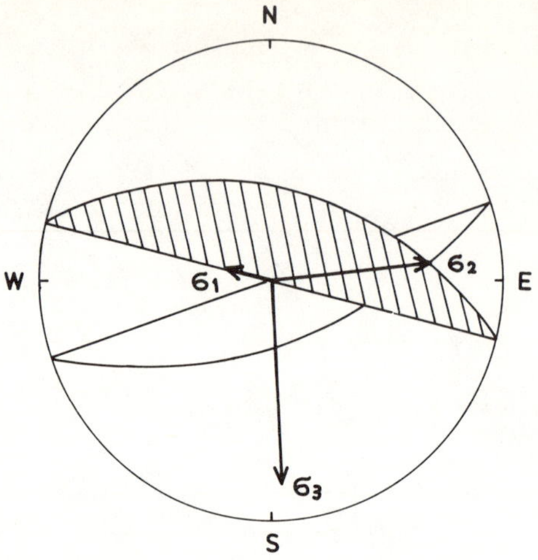

Fig. 6. Stress field estimated from the lineation on a slickenside and a joint, both being representative in the Pleistocene series in the Hatchobaru geothermal field.
Oblique lines in the great-circle indicate the direction of lineation on the slickenside.

As mentioned before, the application of this method to drillcores of Well HT-4 in the Hatchobaru field has revealed that essential fractures like slickensides and hydrothermal veins strike NW-SE or NE-SW, and dip southward with angles generally from 40° to 80°. Of these, the NW-SE trending fractures coincide in direction with the Hatchobaru fault estimated geologically (Fig. 1). At the time of the geological survey, this fault was assumed to dip south, thus meeting Well HT-4 at a depth of about 900 m , as confirmed by the present experiment. On the other hand, the NE-SW trending fractures are consistent with some of the geophysically estimated faults which seem to control the geologic structure at greater depths.

Next, the analysis of the strike and dip of fractures reveals the direction of the stress field acting on each formation. As shown in Fig. 6, if a slickenside has streaks of lineation that goes downward, the direction of the medium principal stress should be at a right angle to the streaks. As a result, the possible directions of the other two principal stresses will be limited due to the facts that the internal friction angle of rocks is about 60° and that the three principal stresses cross at right angles to each other.

For example, in the case of a slickenside at a depth of 900 m in Well HT-4 in the Hatchobaru field, a joint of N70°E and 60N can be reasonably selected from among many kinds of fractures, as the other fracture of a conjugate set. Thus, the stress field in the Pleistocene series has been estimated to be of a normal fault type with directions σ_1 N78°W-66°N, σ_2 N87°E-21°E and σ_3 N3°W-8°S, whereas that in the Neogene system is of an intermediate type whose direction of the maximum principal stress inclines moderately.

In conclusion, this method will be useful not only for younger rocks but also for older rocks if the declination and inclination of rocks of the same formation are known as a standard.

REFERENCES

Hayashi, M., Taguchi, S. and Yamasaki, T.(1981) Activity index and thermal history of geothermal systems; Geothermal Resources Council, Transactions, 5, 177-180.

Kyoritsusha, LTD (1980) An astatic-magnetometer, Type NY-2 (An instruction book).

Geothermal Resources Council, TRANSACTIONS Vol. 6, October 1982

THE THERMAL CONDUCTIVITY OF SOME CRYSTALLINE AND SEDIMENTARY ROCKS FROM SCANDINAVIA

Jan I. Kristiansen, Svend Saxov and Niels Balling

Laboratory of Geophysics
University of Aarhus
Aarhus, Denmark

ABSTRACT

Some available information on thermal conductivity of earth materials from the Scandinavian area is collected. The mean conductivities as reported from individual localities are grouped in crystalline and sedimentary rocks. Mean results are displayed in histograms and localities are mapped. The collocation of conductivity information contains new results of granites and sedimentary rocks from Sweden and of limestones and clays from Danish borings. The new values are presented as histograms of individual measurements and given as mean values with standard errors of mean. The crystalline rocks range from ~2 to ~4 W/(m K), and the sedimentary rocks range from ~0.8 to ~6 W/(m K).

INTRODUCTION

Thermal conductivity is an important parameter in the study of terrestrial heat flow and subsurface temperatures. Also in the search for areas which have potential as geothermal resources, information about the conductivities of a wide variety of earth materials is essential. The Scandinavian region mainly comprises of parts of the Precambrian Fennoscandian Shield, the Paleozoic Caledonian Orogenic Belt and the northeastern part of the Northwest European Sedimentary Basin. Thus the rock types of the region cover crystalline as well as sedimentary rocks. A fairly large number of conductivity results from Scandinavia have been reported, and the rocks which have been measured include many of those found in the region.

GEOLOGY

The Fennoscandian Shield covers the largest part of the area of interest (Fig. 1). The major parts of Sweden and Finland belong to this Precambrian shield, while only the southern and west central Norway is included. The rocks are mainly volcanics, metasediments, gneisses and granites. The Paleozoic Caledonian Orogenic Belt covers a large part of Norway and the northwestern Sweden. The rocks of these areas consist mainly of greenstones, amphibolites, metasediments, metavolcanics and various Caledonian intrusives. The Permian graben system in the Oslo area is an igneous province of minor extent. Here the rocks consist of alkaline intrusives and extrusives. The Danish area constitutes the northeastern part of the Northwest European Sedimentary Basin. Thick sequences of Paleozoic, Mesozoic and Cenozoic rocks are present in the central part of the Danish subbasin. A minor part of southern Sweden (Skåne) is covered by similar sedimentary rocks. The islands of Öland and Gotland (Sweden) are covered by ancient Paleozoic rocks, while the island of Bornholm (Denmark) consists of rocks from all four eras.

Table 1. Rock types which are included in the collocation.

No.	Rock type	No.	Rock type
1	Amphibolite	21	Porphyrie
2	" gneiss	22	Rapakivi
3	Anorthosite	23	Serpentinite
4	Carbonite	24	Sparagmite
5	Diabase	25	Syenite
6	Diorite	26	Alunshale
7	Glimmerite	27	Anhydrite
8	Gneiss	28	Clay
9	Granite	29	Claystone
10	" gneiss	30	Dolomite
11	Greenstone	31	Greenschists
12	Hällaflinta	32	Greywacke/shale
13	Ilmenite	33	Lake sediment
14	Leptite	34	Limestone
15	Liparite	35	Rocksalt
16	Mica gneiss	36	Sand
17	Mica schists	37	Sandstone
18	Migmatite	38	Siltstone
19	Pegmatite	39	Quartzite
20	Peridotite		

COLLOCATION OF THERMAL CONDUCTIVITIES

The Scandinavian thermal conductivities, which are considered here, are mainly measured by the divided bar method and the transient cylindrical probe method. In Sweden a modified probe method (Landström et al., 1980) has been used occasionally. Traditionally most measurements on crystalline rocks are made by the first method but also

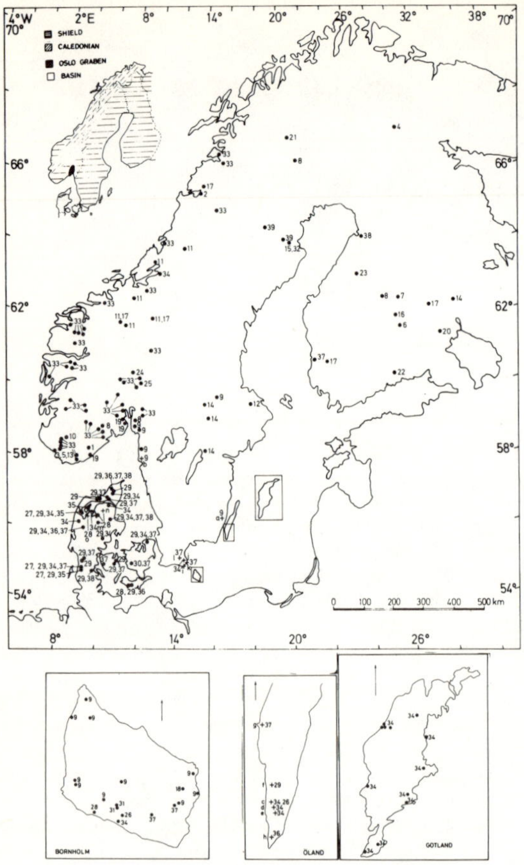

Figure 1. Sites of conductivity measurements and Tectonic map of Scandinavia. • indicates previous sites, + indicates new sites (identified by letters) and figures indicate rock type.

the probe method has been used successfully in crystalline areas. The probe method is very suitable for measurements in most sedimentary rocks. It also applies to in situ conductivity determinations on lake sediments in Norway (Hänel et al., 1974). The conductivity results as considered here consist of existing information and of new measurements. Conductivities from Norway are reported by Swanberg et al. (1974), Heier and Grønlie (1977), Hänel et al. (1979) and Grønlie et al. (1980), from Sweden by Parasnis (1975), Eriksson and Malmqvist (1979), Poulsen et al. (1981), and Malmqvist et al. (1982), from Finland by Järvimäki and Puranen (1979), and from Denmark by Balling (1979), Balling et al. (1981), and Kristiansen et al. (1982).

Not all conductivity results, which are contained in the above papers, are included in this work. Thus a few cases, where mean conductivities in the original papers are ascribed to more than one widely different rock type, are disregarded. A similar rejection of results has been used

Table 2. New sites of measurements. N: number of measurements, $\bar{k}$: mean conductivity. The figures of the rock type correspond to Table 1.

Locality	Rock type	N	$\bar{k}$(W/(m K))
a Kalmar	9	6	3.07
b Bohus	9	6	3.09
c Degerhamn 1	34	3	2.82
"	26	5	1.87
d Degerhamn 2	34	3	2.67
e Degerhamn 3	34	4	2.58
f Degerhamn 4	29	1	3.15
g Mörbylånga	37	1	4.22
h Ottenby	36	1	1.71
i Gislövhammar	34	2	2.41
j Tommarp	34	2	2.22
k Tobbisborg	37	1	3.71
l Dalby-Sandby	37	1	4.19
m Harre 1	28	138	1.08
"	28	40	0.82
n Års 2	28	45	1.89
o Selde 1	34	244	1.74

in cases where materials or results are considered not to be generally representative of a rock type. The rock types are presented in Table 1. The localities to which the conductivity samples relate are mapped in Fig. 1, and the identifying figures are given in Table 1. As the present collocation of results is based on mean values, each box of the compiled histograms contains two figures: the first relates to the rock types, and the second to the number of measurements. The histograms shown in Fig. 2 comprise the mean conductivities of crystalline and sedimentary rocks. The conductivities of crystalline rocks range from ~2 to ~4 W/(m K). The magnitude of the conductivity of a specific rock is highly dependent on the quartz content. The range of for example the values of granites may be due to this. However, the overall conductivity range ascribed to crystalline rocks is narrow compared to that of sedimentary rocks. Here values range from ~0.8 to ~6 W/(m K), and the rock types show individually much broader distributions. Limestones range from ~1.3 to ~3.2 W/(m K), and sandstones from ~2.1 to ~6.1 W/(m K). Again the content of quartz is an important factor, but also the porosity has a dominating effect. The porosity dependence of some Danish sediments has been investigated (Balling et al., 1981). For water-saturated pure limestones it is shown that the conductivity (k) varies with porosity (ϕ) as follows: $k = 0.58^{\phi}\ 3.26^{1-\phi}$. Apparently, similar formulas do not hold for other types of sediments within the same accuracy.

NEW CONDUCTIVITY VALUES

New conductivities from the Swedish

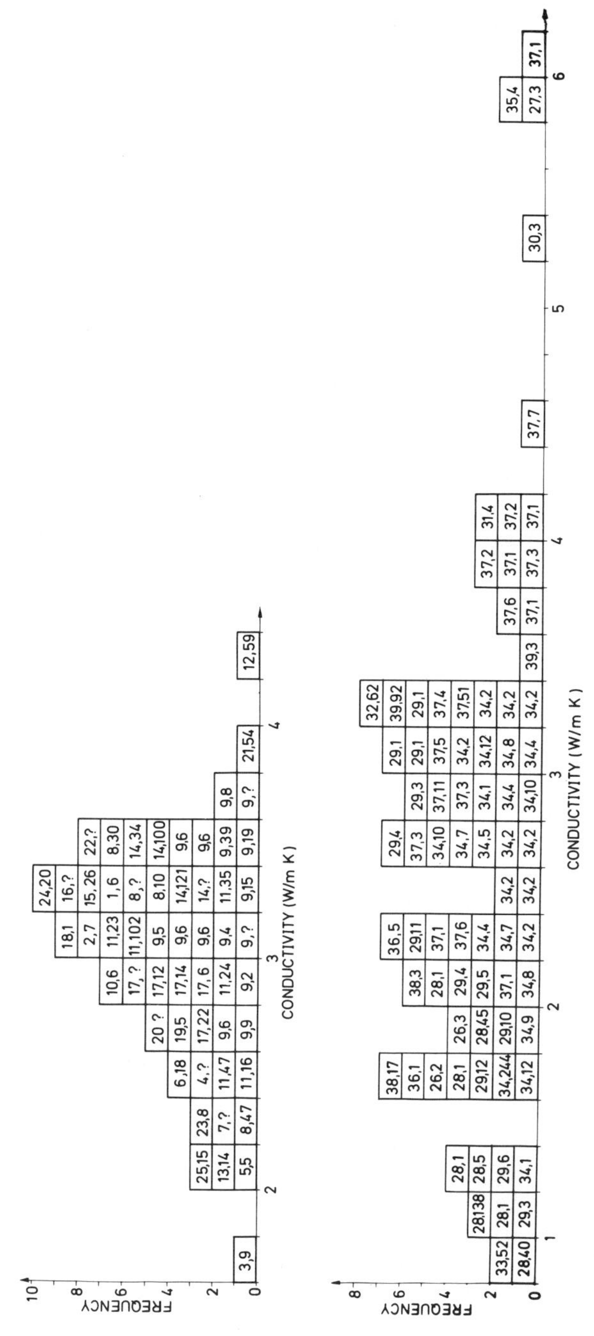

Figure 2. Histograms of Scandinavian mean conductivities. Left part: crystalline rocks. Right part: sedimentary rocks. The first figure within a box indicates rock type and the last indicates the number of measurements.

and Danish areas have been determined. The measurements have been carried out by the probe method on outcrops and on core materials. The equipments have been standardized against calibration materials. Their mode of operation as well as typical examp-

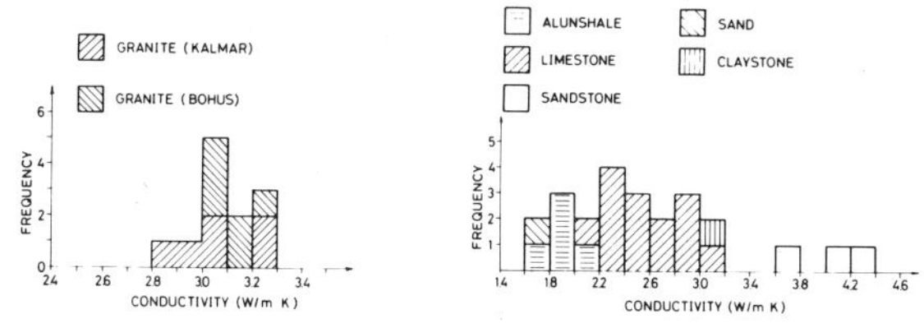

Figure 3. Conductivities of some Swedish granites and sedimentary rocks.

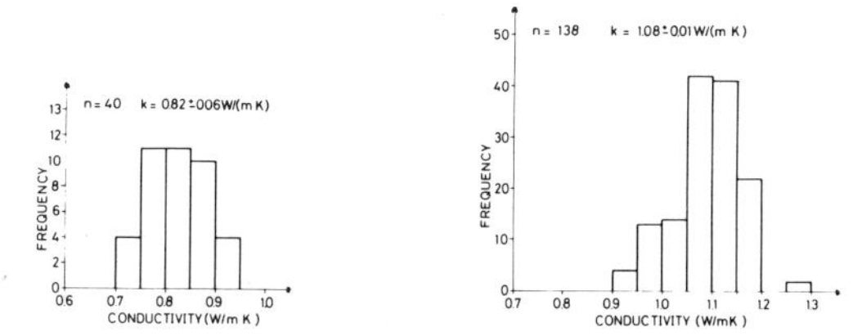

Figure 4. Conductivities of two clay sequences (Harre 1, Denmark).

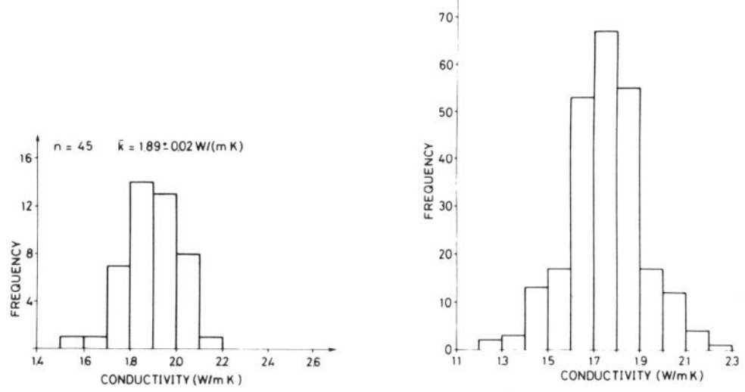

Figure 5. Conductivities of a clay sequence (Års 2, Denmark) (left part), and of a limestone sequence (Selde 1, Denmark) (right part).

les of measurements are given elsewhere (Balling et al., 1981; Poulsen et al., 1981; Kristiansen et al., 1982). The interpretations of data are based on the line source model. A major part of the results have been compared with and agree with those obtained by the more accurate integral solution combined with the iterative least squares inversion procedure (Kristiansen, 1982).

In situ measurements on two granites from the Kalmar and Bohus areas and on five sedimentary rocks from Öland and Skåne have been made. The localities and the results of the individual measurements are given in Table 2 and shown in Fig. 3. The sample sites are mapped on Fig. 1 and characterized according to Tables 1 and 2. These results show accordance with expected values for the rock types in question.

The laboratory measurements on homogeneous Danish core materials (Harre 1, Års 2, Selde 1) show accurately determined mean conductivities. The results appear in Table 2 and Figs. 4 and 5. The localities are included in Fig. 1. Results of these types have been used for accurate heat flow determinations in Denmark (Balling et. al., 1981). Our new results are included in the present collocation of data (Figs. 1 and 2).

CONCLUSION

Thermal conductivities of earth materials from Scandinavia are presented and collocated. Mean values of crystalline rocks range from ~2 to ~4 W/(m K) and show narrow individual distributions. Mean values of sedimentary rocks cover a larger range (~0.8 to ~6 W/(m K)), and the individual distributions are broader. The wide range of the sedimentary results implies that temperature gradients in sedimentary areas are not constant with depth provided constant heat flow. The new results of Danish core materials show narrow distributions. Thus mean values for clays are 0.82 ±0.06 W/(m K), 1.08 ±0.01 W/(m K), and 1.89 ±0.02 W/(m K), and for limestone 1.74 ±0.01 W/(m K). The new Swedish results which we have obtained are in accordance with expected values.

REFERENCES

Balling, N., 1979, Subsurface temperatures and heat flow estimates in Denmark in V. Cermák and L. Rybach (eds), Terrestrial Heat Flow in Europe, Springer, p. 161-171.

Balling, N., Kristiansen, J.I., Breiner, N., Poulsen, K.D., Rasmussen, R., and Saxov S., 1981, Geothermal measurements and subsurface temperature modelling in Denmark, GeoSkrifter 16, 172 p.

Eriksson, K.G. and Malmqvist, D., 1979, A review of the past and the present investigations of heat flow in Sweden in V. Čermák and L. Rybach (eds), Terrestrial Heat Flow in Europe, Spri Springer, p. 172-178.

Grønlie, G., Johansen, T.E., Karlstad, B., and Heier, K.S., 1980, Prospecting for geothermal energy in the Iddefjord granite, Østfold, Norway, Norsk Geol. Tidsk. 60, p. 263-267.

Hänel, R., Grønlie, G., and Heier, K.S., 1974, Terrestrial heat flow determinations from lakes in southern Norway, Norsk Geol. Tidsk. 54, p. 423-428.

Hänel, R., Grønlie, G., and Heier, K.S., 1979. Terrestrial heat flow determination in Norway and an attempted interpretation in V. Čermák and L. Rybach (eds), Terrestrial Heat Flow in Europe, Springer, p. 232-239.

Heier, K.S. and Grønlie, G., 1977, Heat flow - heat generation studies in Norway in S.K. Saxena and S. Bhattacharji (eds), Energetics of Geological Processes, Springer, p. 217-235.

Järvimäki, P. and Puranen, M., 1979, Heat flow measurements in Finland in V. Čermák and L. Rybach (eds), Terrestrial Heat Flow in Europe, Springer, p. 172-178.

Kristiansen, J.I., 1982, The transient cylindrical probe method for determination of thermal parameters of earth materials, GeoSkrifter (in press).

Kristiansen, J.I., Saxov, S., Balling, N., and Poulsen, K.D., 1982, In situ thermal conductivity measurements of Precambrian, Palaeozoic and Mesozoic rocks on Bornholm, Denmark, Geol. Före. i Stockholm Förh. 104 (in press).

Landstrøm, O., Larson, S.Å., Lind, G., and Malmqvist, D., 1980, Geothermal investigations in the Bohus granite area in southwestern Sweden, Tectonophysics 62, p. 131.

Malmqvist, D., Larson, S.Å., Landstrøm, O., and Lind, G., 1982, Heat flow and heat production from the Mallingsbo granite, central Sweden (personal communication - in press).

Parasnis, D.S., 1975, Temperature phenomena and heat flow estimates in two Precambrian ore-bearing areas in north Sweden, Geophys. J. R. astr. Soc. 43, p. 531-554.

Poulsen, K.D., Saxov, S., Balling, N., and Kristiansen, J.I., 1981, Thermal conductivity measurements on Silurian limestones from the island of Gotland, Sweden. Geol. Före. i Stockholm Förh. 103, p. 349-356.

Swanberg, C.A., Chessman, M.D., Simmons, G., Smithson, S.B., Grønlie, G., and Heier, K.S., 1974, Heat-flow - heat generation studies in Norway, Tectonophysics 23, p. 31-48.

Geothermal Resources Council, TRANSACTIONS Vol. 6, October 1982

COMPARATIVE STUDIES OF GEOTHERMAL SURVEYS IN 3-METER AND TEMPERATURE-GRADIENT HOLES

Arthur L. Lange**, H.D. Pilkington* & John Deymonaz*

*AMAX Exploration Inc.
1707 Cole Blvd
Golden, CO 80401

** Mincomp Exploration Resources
4175 Harlan St.
Wheat Ridge, CO 80033

ABSTRACT

The reliability of conducting temperature surveys within the upper 3 meters of the surface to map geothermal anomalies is demonstrated in experiments at two prospects in which deeper gradient hole data were obtained. The 3m temperatures faithfully outlined the thermal anomaly at McCoy, Nevada; and in Dixie Valley, NV 3m surveys reproduced and detailed patterns derived from 40m data. These encouraging results led to the development of multi-thermistor strings for logging the seasonal wave within the upper 3 meters. From many such logs, diffusivity variations can be detected, which might otherwise be misconstrued as thermal anomalies. The technique is demonstrated by a typical Basin-Range reconnaissance project.

As many as 10 or more 3m holes can be emplaced in the time required for a conventional gradient well, and with considerably less impact on the environment.

INTRODUCTION

Temperature surveys made within a few meters of the surface have been widely used as a prospecting tool for the mapping of groundwater aquifers, upwelling cool and warm waters, and for the detection of geothermal anomalies and mineral deposits. Kappelmeyer & Haenel (1974) treat in considerable detail these divers applications of near-surface temperature surveys. Birman (1969) reports on the thermal mapping of aquifers; Lee (1977), Kintzinger (1956) and Le Shack (1978) have applied shallow surveys to the mapping of geothermal anomalies.

This paper summarizes experiments conducted by the authors within the upper 3 meters of the surface, which led to the development of an effective system for mapping geothermal anomalies and recognizing the perturbing effects of soil diffusivity variations.

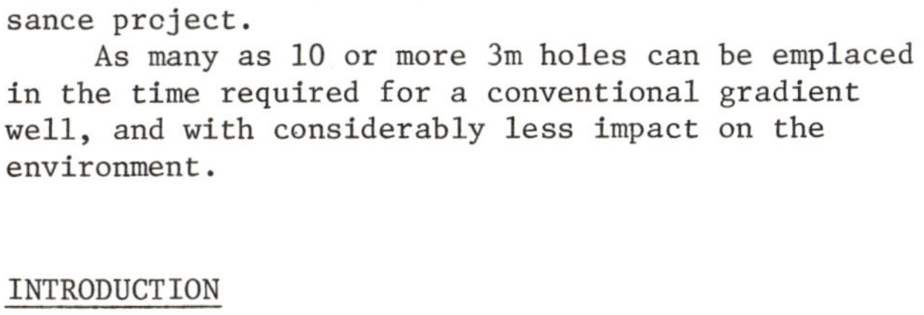

Figure 1. Daily range and mean of temperatures recorded at Milford, Utah airport.

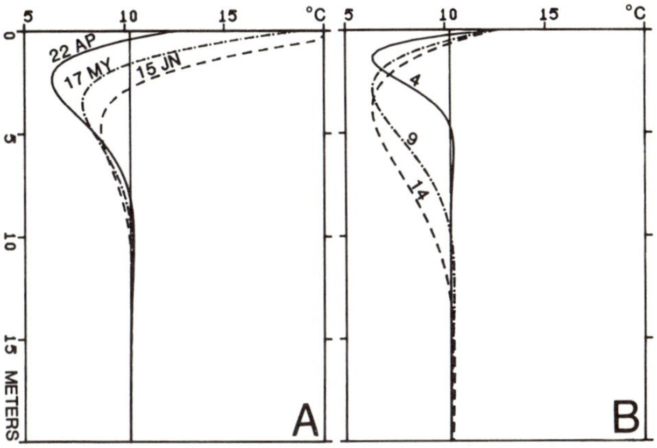

Figure 2. Seasonal wave effects: A) Downward progress of the wave at a diffusivity of $0.006 cm^2/sec$ at 4-week intervals beginning 22 April; B) Effect of different diffusivities (4-, 9-, and $14 \times 10^{-3} cm^2/sec$) on the seasonal wave. Extremum amplitude is independent of diffusivity.

BASIC PRINCIPLES

The steady state equation governing temperature T at depth z in a uniform medium is

$$T(z) = T_o + z \frac{dT}{dz}, \tag{1}$$

where T_o is the mean annual surface temperature (a function of altitude) and dT/dz is the geothermal gradient. Near the surface two additional terms must be added to the right-hand side of Equation 1; namely, a diurnal temperature variation T_d and a seasonal variation T_s. The former generally does not affect the subsurface temperature regime in alluvium below a depth of 1.0m and will be omitted here. The seasonal effect, however, is superimposed on the geothermal gradient and distorts the temperature log to a depth exceeding 10m in alluvium and as great as 20m in rock. The seasonal variation at the surface is a quasi-sinusoid (Figure 1) oscillating around the mean annual temperature of the site. The annual cycle penetrates the ground as a thermal wave of amplitude T_a attenuating exponentially with depth. Ideally the seasonal term is given by

$$T_s = T_a e^{-mz} \cos(\omega \Delta t - mz), \tag{2}$$

where $m = \sqrt{\pi/\tau\alpha}$ and $\omega = 2\pi/\tau$. Here τ is the period

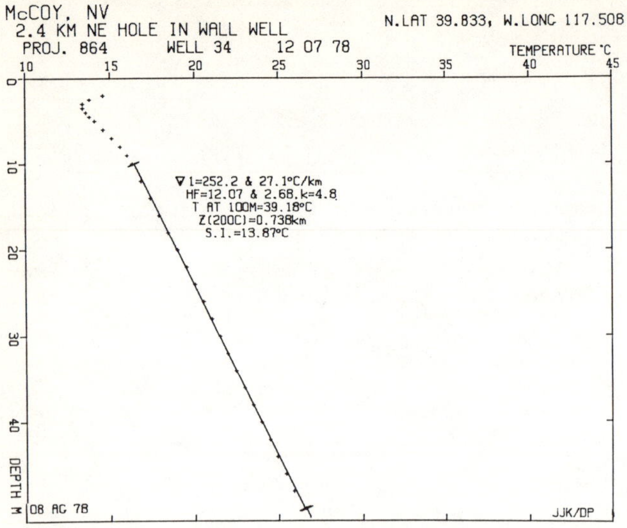

Figure 3. Computer plot of well log showing a seasonal wave's winter low superimposed on a geothermal gradient of 252°/km (12 July 1978).

(one year); Δt is the time elapsed since the summer temperature high; and α is thermal diffusivity, related to thermal conductivity by

$$\alpha = \kappa / c\delta, \tag{3}$$

where c is specific heat and δ is density.

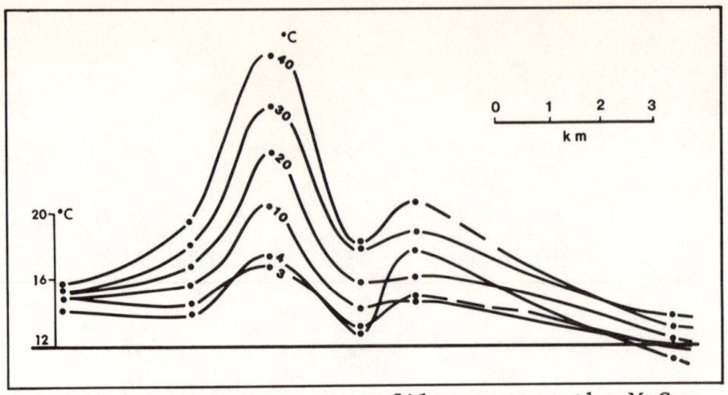

Figure 4. Temperature profiles across the McCoy, Nevada, thermal anomaly at depths from 3 to 40m (11 July 1978).

Figure 2A illustrates the progress of the seasonal wave as it penetrates alluvium of diffusivity 6 ($\times 10^{-3}$). Here the extreme temperature of the previous winter's low migrates downward while dwindling in amplitude. By the time it reaches 10m depth it virtually disappears. In Figure 2B we see the effect of different diffusivities on the log for a particular spring day. The depth of the previous winter's low is a direct effect of the diffusivity; however, the amplitude is the same for all diffusivities. Thus, by determining the depth of the extremum z_x on a given day, we can compute the mean diffusivity between that depth and the surface by the relation-

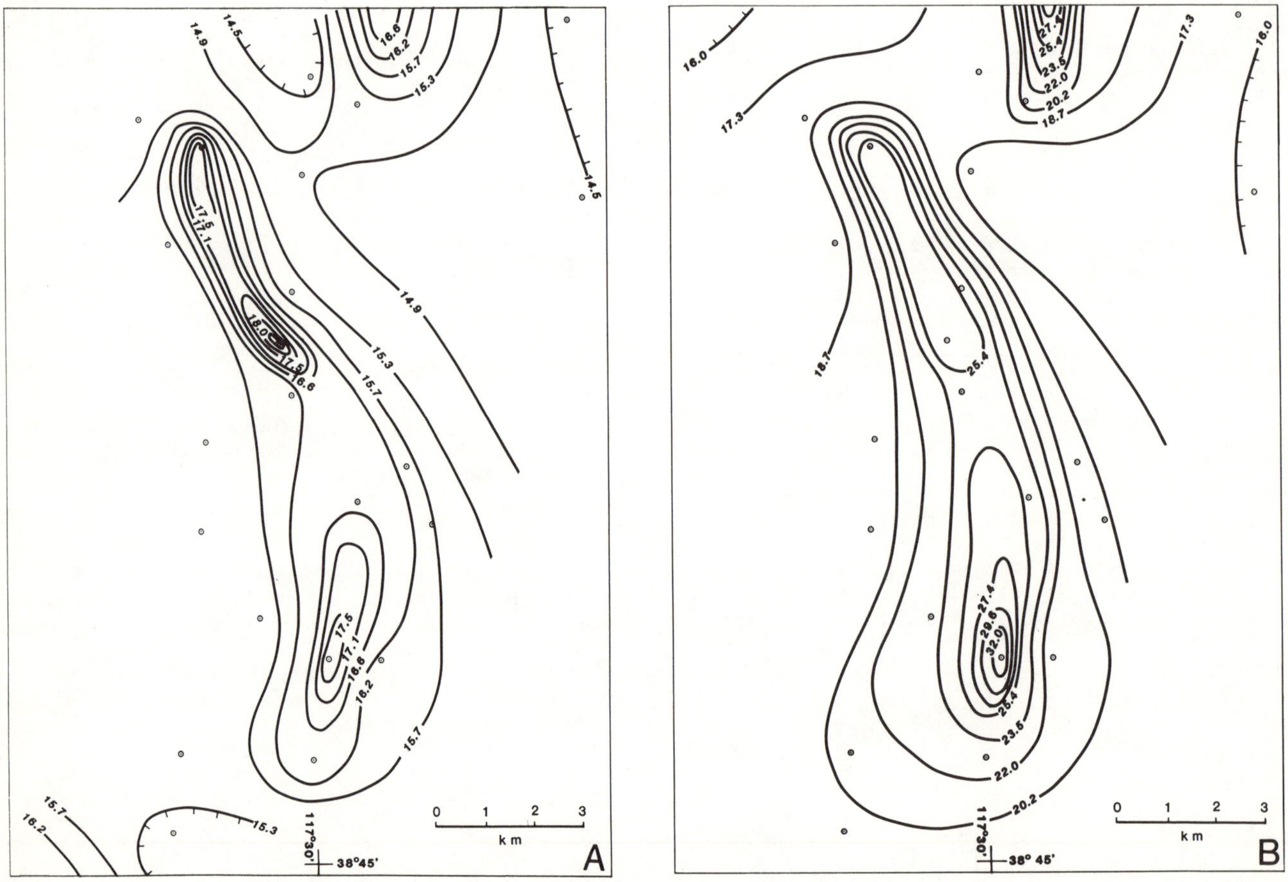

Figure 5. Plan views of the thermal anomaly at McCoy, Nevada, 11 July 1978; A) Temperatures at 3m; B) Temperatures at 30m (both adjusted for altitude).

ship

$$\alpha = \frac{\tau z_x^2}{4\pi(\Delta t - \tau/8)^2}. \qquad (4)$$

When the seasonal wave of Equation 2 is superimposed on the temperature gradient of Equation 1, the result--taken from an actual log--appears as in Figure 3.

EXPERIMENT AT McCOY NEVADA

In order to test the 3m methods, the authors

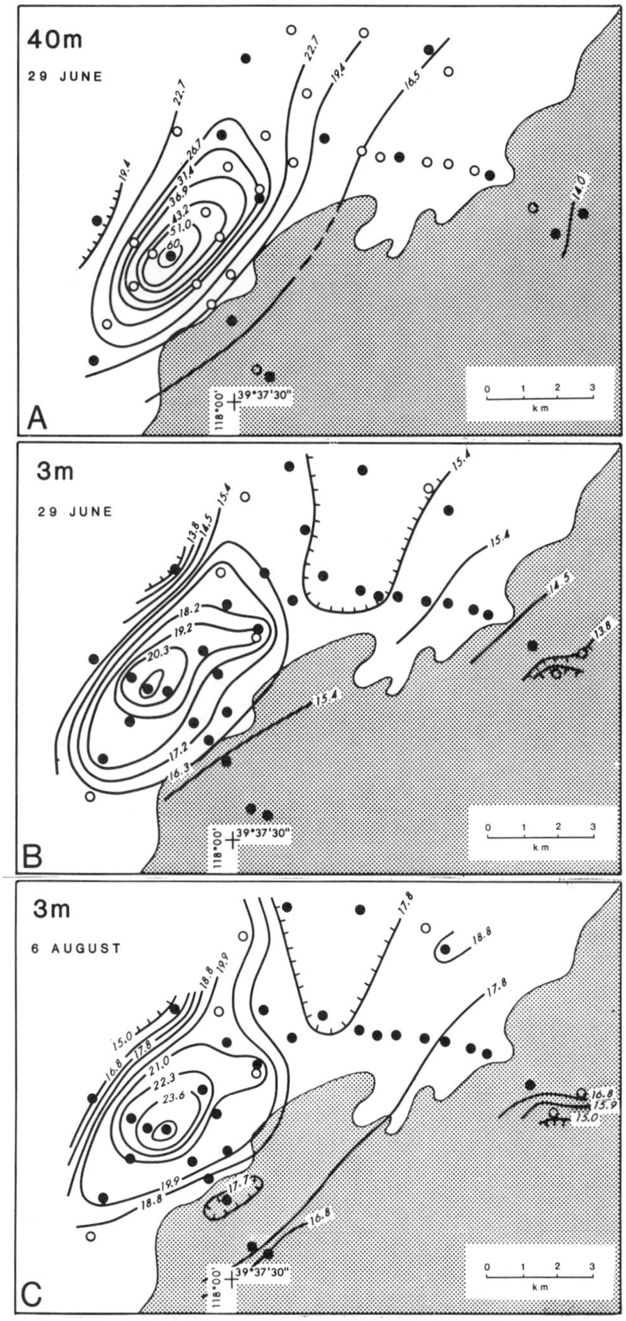

Figure 6. Plan views of a thermal anomaly at Dixie Valley, Nevada: A) Temperatures at 40m, in gradient holes (solid circles) 29 June 1979; B) Temperatures at 2.9m, 29 June 1979; C) Repeat of 2.9m temperatures 38 days later (6 August 1979). Hills and outcrop shaded.

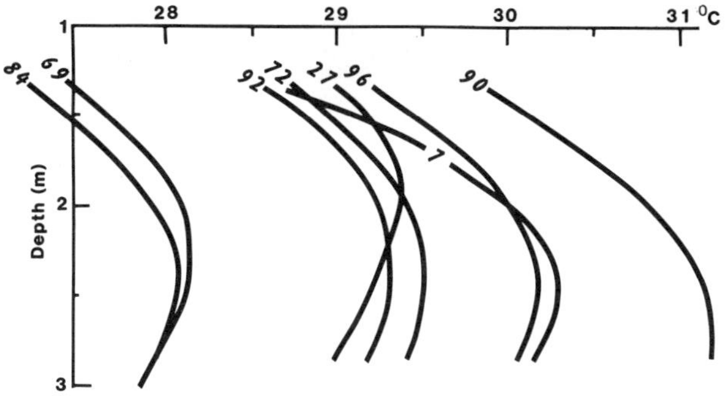

Figure 7. Desert Center, California: Typical logs of the seasonal wave obtained using thermistor strings in 8 holes, November 1980. The vertical shift of the curves is due to diffusivity changes from well-to-well; lateral shift results from different temperature gradients and site altitudes.

logged 24 gradient holes within AMAX' McCoy geothermal prospect, 65km NW of Austin, Nevada. In Figure 4, a profile of temperatures at successive depths across the McCoy anomaly, demonstrates that even within the water column of gradient holes the thermal features were mappable at 3m (Figure 5).

EXPERIMENT IN DIXIE VALLEY, NEVADA

In order to test further the applicability of shallow temperature measurements, single thermistors were installed in 3m PVC pipes adjacent to and in-between gradient holes in a geothermal prospect on the east side of Dixie Valley, Nevada. The thermistors were covered with dry vermiculite in order to insulate them from the outside air.

Results of the 40m temperature survey are shown in Figure 6A. The corresponding plot obtained on 29 June 1979 from the 2.9m thermistors appears in 6B wherein the thermal anomaly here is faithfully depicted and refined, owing to the higher sampling density. In 6C, we see that the pattern remained essentially unchanged 38 days later.

INSTRUMENTATION AND APPLICATION

Encouraged by the above results, we fabricated thermistor strings, consisting of seven transducers spaced 0.25m apart between 1½ and 3m, sealed in PVC pipe. Multiple thermistors permit the logging of a significant portion of the seasonal wave, so that diffusivities might be evaluated using Equation 4. The strings were installed in sleeves in pre-drilled holes during a reconnaissance in the Basin-Range. Usually 20 to 30 strings were logged each day using a digital ohmmeter. Calibration coefficients for over 400 thermistors were stored on magnetic cards of an HP-97 hand-computer, so that temperatures could be rapidly computed on-site. Typical thermal logs obtained during a particular day in November 1980 show the divers depths reached by the summer maximum in 8 wells, and the temperature shifts due to differing thermal gradients and altitudes (Figure 7). One 3m temperature map produced is shown in Figure 8, after correcting temperatures (T_o) for elevation using an empirical adiabatic lapse rate of 8.5°C/km. An almost identical map resulted from measurements made during the previous July. From the 3m measurements, one is lead to identify osten-

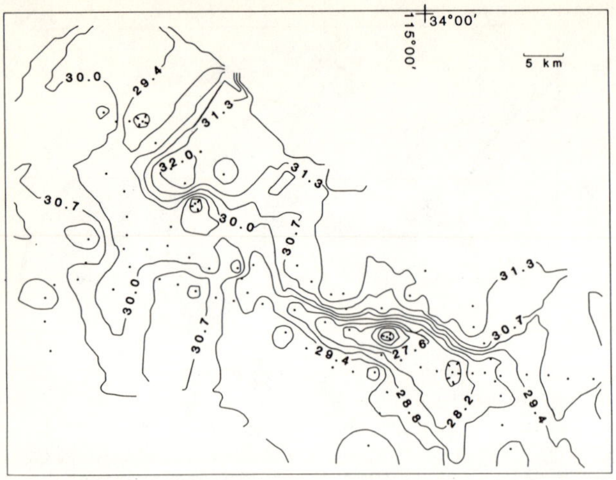

Figure 8. Desert Center: Computer map of 3m temperatures around 7 November 1980, adjusted for altitude.

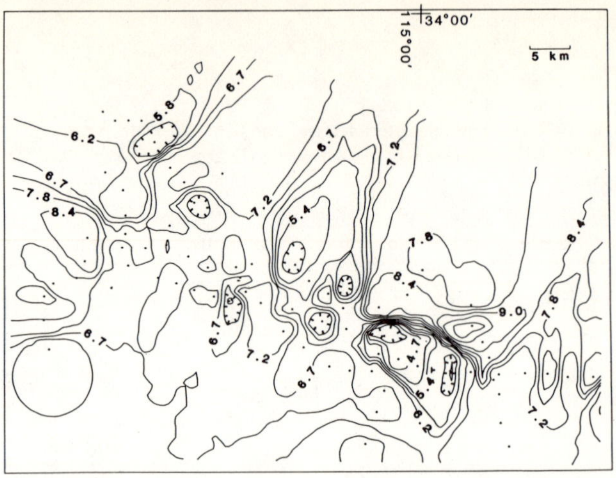

Figure 9. Desert Center: Diffusivity map (values x 1000), showing the correspondence between a diffusivity low to the southeast and the thermal low of Figure 8.

sibly warm and cold anomalies; however, when soil diffusivities are computed a map of diffusivities results that displays considerable coincidence between low 3m temperatures and low diffusivities (Figure 9). Indeed the thermal low in the southeast corner of the map occupies a playa region in which soil moisture, by raising the specific heat of the medium, likely lowers the diffusivity (cf, Equation 3) giving rise to a near-surface thermal low. Birman has suggested that evaporation may also play a role in lowering the shallow temperatures at this site (Personal communication).

In such cases, where soil changes are likely, it is desirable to compute diffusivities at each site by utilizing either the extremum depth or fitting the seasonal log to an appropriate model curve (Lee, 1977). One may then recompute the logs for a common diffusivity, and in effect produce a residual temperature map, independent of soil variations.

PRACTICAL CONSIDERATIONS

In order to drill 3m holes effectively in typical basin fanglomerates, we found it expedient to employ a very mobile rubber-tired truck-mounted drill rig. By this means, approximately 10 or more 3m holes could be drilled in the time that a conventional 50-100m gradient hole could be put down, resulting in about 10 data points for each gradient hole. Furthermore, it is desirable that 3m holes be tied to occasional deeper wells to insure that aquifers do not intervene between the upper and lower thermal regimes.

CONCLUSIONS

Temperature surveys conducted using single thermistors installed in 3m holes reproduced known geothermal anomalies in the absence of severe soil variations and intervening aquifers. If diffusivity changes occur among the well sites they can be detected and evaluated by employing 3m-thermistor strings consisting of a series of closely spaced tranducers individually calibrated. By logging an array of strings on a given day, one can produce not only a 3m-temperature map but also logs of a portion of the seasonal wave superimposed on the soil temperature gradient. Mean diffusivities can be determined at each site and a corresponding map of diffusivity anomalies produced.

The 3m holes are most effectively drilled using a mobile truck-mounted rig capable of auguring in fanglomerate. As many as 10 or more 3m holes can be drilled in the time required to drill a single 50- to 100m well, with a much lesser impact on the environment.

ACKNOWLEDGEMENTS

The authors thank AMAX Exploration, Inc. for permission to present these survey results, and Mincomp Corporation for generating the computer plots exhibited above.

REFERENCES

BIRMAN, JOSEPH H. Geothermal Exploration for ground water. Geological Society of America, Bulletin 80: 617-630.

KAPPELMEYER, O. & R. HAENEL (1974). Geothermics, with Special Reference to Application. Gebrüder, Borntraeger, Stuttgart, 239p., 123 fig.

KINTZINGER, P.R. (1956). Geothermal survey of hot ground near Lordsburg, New Mexico. Science 124: 629-630.

LEE, TIEN-CHANG (1977). On shallow-hole temperature measurements--A test study in the Salton Sea geothermal field. Geophysics 42: 572-583.

LeSHACK, LEONARD A., et al. (1978). Shallo-Temp™ Surveys--A Rapid Reconnaissance Technique for Geothermal Prospects. Paper presented at 48th Annual Meeting, Society of Exploration Geophysics, San Francisco.

Geothermal Resources Council, TRANSACTIONS Vol. 6, October 1982

A METHOD FOR FILTERING HOT SPRING NOISE FROM SHALLOW TEMPERATURE GRADIENT DATA

Todd M. C. Li
Chandler A. Swanberg
John F. Ferguson

TELEDYNE GEOTECH
Division of Geophysical Research
3401 Shiloh Road
Garland, Texas 75041

ABSTRACT

A technique for separating shallow heat source effects from temperature gradient data is presented. The technique makes use of the depth-dependent information available in the wave number spectrum of the gradient data. The effectiveness of the technique is demonstrated on a two-dimensional numerical model of a geothermal system containing a deep geothermal reservoir which is masked by a warm, shallow aquifer and a thermal spring. This geothermal system is representative of those found throughout the Basin and Range province. The resulting filtered gradients produce an excellent prediction of the temperatures in the modeled geothermal reservoir.

I. Statement of the Problem

The thermal method for locating and evaluating commercial geothermal fields consists of using shallow (~500 foot) temperature gradient test holes. A recent industry survey (Ward et al, 1981) has shown that the thermal method is the preferred exploration technique in all regions of the western United States. Despite this industry endorsement, the thermal method is not without its limitations, the most severe of which involves the adverse effects caused by the convection of near-surface hydrothermal fluids (hot spring noise). For example, the technique of downward continuation, which yields reliable estimates of subsurface temperatures in strictly conductive regimes, tends to go unstable and predict unreliable temperatures in the presence of hot spring noise.

Because of these problems, and the rather substantial cost of drilling and logging temperature gradient test holes, it seems desirable to develop a practical technique for filtering out the effects of "hot spring noise". In this paper, we present a technique for filtering hot spring noise that makes use of the depth-dependent information contained in the wave number spectrum of the thermal gradient data. The depth-dependent information is used to distinguish between the thermal effects resulting from shallow "hot spring noise" and the effects resulting from the deeper, potentially commercial geothermal reservoirs. In addition, the technique is designed to incorporate constraints from other geothermal studies such as gravity surveys, passive or active electrical surveys, geologic mapping, and geochemical studies.

II. Mathematical Basis of the Wave Number Filtering Technique

The concept of using wave number spectra to delineate source depth parameters has appeared previously in the context of gravity and magnetics (e.g., Odegard and Berg, 1965; Spector and Grant, 1970). The mathematical relationship between source depth and surface temperature-gradient data becomes relatively simple when written in the wave number domain rather than in the space domain.

For example, consider the gradient due to a line source located at (x_s, z_s) in a homogeneous isotropic halfspace. The vertical temperature gradient at the surface is given by:

$$g(x) = \frac{dT}{dz}(x, z=0) = \frac{Q}{\pi K} \frac{z_s}{(x-x_s)^2+z_s^2} \qquad (1)$$

where Q (heat/unit length/unit time) is the source strength, K (heat/unit length/unit time/degree) is the thermal conductivity of the medium, T is temperature, and z is the vertical coordinate, postive downward. This expression can be obtained from the Green's function solution to Poisson's equation, which is given by Carslaw and Jaeger (1959, pg. 422), and using the method of images with T (x, z = 0) = 0.

The complex wave number spectrum of the gradient is the Fourier transform of g(x) (see Erdelyi et al, 1954):

$$G(\omega) = \frac{Q}{2\pi K} \left(e^{-i\omega x_s}\right) \left(e^{-\omega z_s}\right) \qquad (2)$$

where ω = wave number (radians/distance).

This expression of the wave number spectrum permits a factorization of the linear operator into several components, one of which expresses the depth dependence.

The Fourier transform is also a linear operator, and the wave number spectrum of the gradient due to M line sources Q_j, $j=1,\ldots,M$, can be written as a linear combination of source strengths:

$$G(\omega) = \sum_{j=1}^{M} A_j Q_j \tag{3}$$

where, from Equation (2),

$$A_j = \frac{1}{2\pi K} \left(e^{-i\omega x_{sj}}\right) \left(e^{-\omega z_{sj}}\right) \tag{4}$$

The wave number filtering process consists of three basic steps. These steps are (1) parameterization of the model in terms of a discrete source distribution, (2) least-squares inversion for the source strengths, and (3) the filtering of the data based on the source solution.

The first step, parameterization of the model, consists of specifying a distribution of line sources to represent the shallow and deep parts of the geothermal system. This parameterization of the model involves choosing source locations such that an effective separation of shallow and deep components can be accomplished upon inversion for the source strengths. There is an inherent nonuniqueness in the source distribution that is based upon the fact that the observations are made on a two-dimensional surface, while the sources exist in a three-dimensional configuration (Roy, 1962). Some of this ambiguity can be removed if there is independent evidence for the locations of the sources. Such information can be obtained from gravity or seismic surveys, electrical or electromagnetic resistivity surveys, geologic mapping, and from disturbances in the temperature-depth profiles themselves.

The second step in the filtering process is the inversion of the measured data to obtain the source strengths at the specified parameter locations. The inversion is obtained by a least-squares fit of the predicted wave number spectrum to the observed wave number spectrum, subject to a set of constraints on the source strengths. Each of the predicted wave number spectral data points, $\hat{G}_k$, $k=1,\ldots,N$, can be expressed as a linear combination of the M source strengths, Q_j, $j=1,\ldots,M$. This can be written in matrix form as

$$\underset{N\times 1}{\underset{\sim}{\hat{G}}} = \underset{N\times M}{\underset{\approx}{A}} \; \underset{M\times 1}{\underset{\sim}{Q}} \tag{5}$$

The least-squares problem can be stated as

$$\text{Minimize} \quad R = [\,(\underset{\sim}{\hat{G}}-\underset{\sim}{G})^* \, (\underset{\sim}{\hat{G}}-\underset{\sim}{G})\,]^{1/2} \tag{6}$$

subject to L constraints

$$\underset{L\times M}{\underset{\approx}{H}} \; \underset{M\times 1}{\underset{\sim}{Q}} \geq \underset{L\times 1}{\underset{\sim}{b}} \tag{7}$$

where $\underset{\sim}{G}$ is the observed wave number spectrum, and $\underset{\sim}{G}^*$ is the complex conjugate of $\underset{\sim}{G}$. The inequality in Equation (7) is a mathematical representation of the constraints on source strength that arise from the known physical properties or behavior of the geothermal system. The geologic effectiveness of the filter may be strongly influenced by the addition of appropriate independent constraints. The constrained least-squares problem expressed by Equations (6) and (7) is essentially a quadratic programming problem that can be solved by a number of standard techniques (see, for example, Wilde and Beightler, 1967; Land and Powell, 1973; Lawson and Hanson, 1974).

The third step is the construction of the matched filter, which is based on the source strength distribution (Green's functions) obtained in the separation model. The solution for the deeper source regime can be used to generate a set of filtered gradients that are relatively free from shallow source effects and can then be interpreted with regard to the existence of the deep geothermal field.

III. Application of the Technique to a Numerical Model of a Geothermal System

The purpose of this example is to demonstrate that the filtering technique can be successfully applied to a temperature gradient profile as might be observed over a typical "basin and range" geothermal system. The computer model of a geothermal system and the gradients generated by each of the model components are shown schematically in Figure 1. The essential features of the system are a commercially viable geothermal reservoir, an active fault which permits vertical migration of geothermal fluids to a surface hot spring, and a shallow aquifer which permits horizontal migration of the geothermal fluids in such a way as to "mask" the geothermal reservoir.

Figure 2 (top) shows the observed surface gradient as it might be measured in the field. It has been constructed by combining the individual effects shown in the top of Figure 1. This is the gradient profile to be filtered in order to reproduce the plot labeled "Geothermal Reservoir Contribution" in Figure 1. The effect of the deep geothermal reservoir can be seen in the observed gradient profile shown in Figure 2, although it is certainly not obvious.

The technique of downward continuation has been applied to the gradient contribution from the geothermal reservoir, and the resulting subsurface temperature estimates are shown in Figure 3. While these temperatures are not the true temperatures within the model, they do represent the geothermal exploration target and can also be used to compare the results from downward continuation of the filtered gradients.

Figure 2 (bottom) shows the results of downward continuation of the observed temperature gradient profile. This represents the current "state-of-the-art" of processing tem-

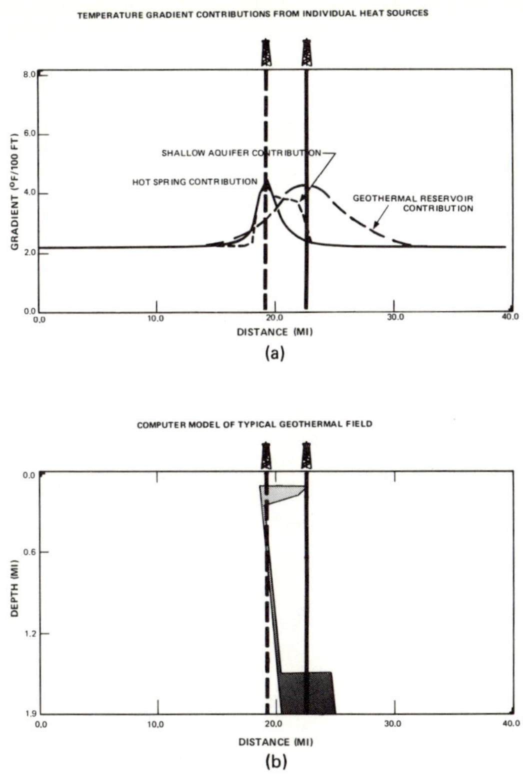

Figure 1.

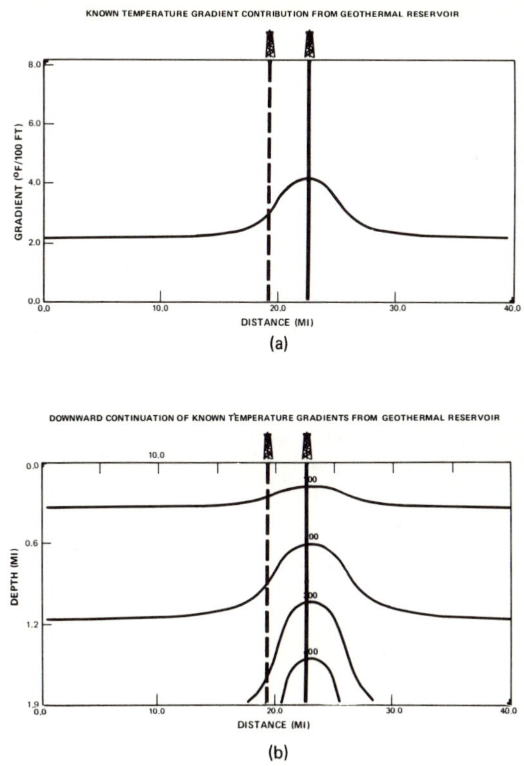

Figure 3.

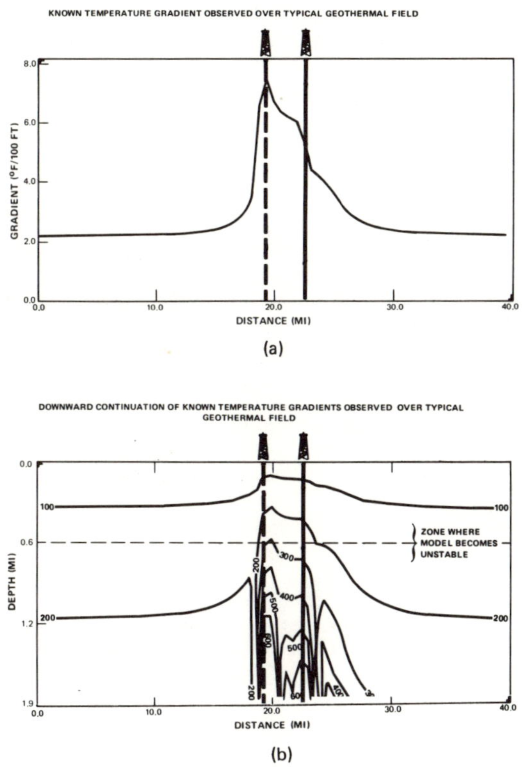

Figure 2.

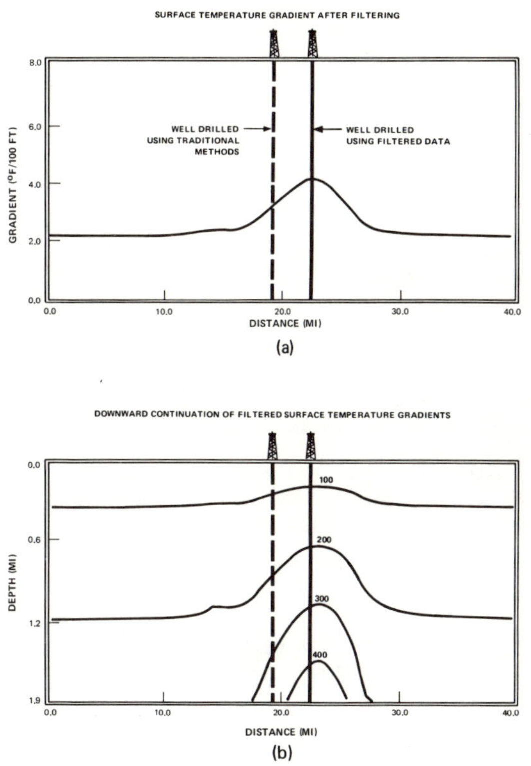

Figure 4.

perature gradient data and shows the problems inherent in processing gradient data in the presence of hot spring noise. There are four significant features revealed by the results shown in Figure 2: (1) The model starts to become unstable at shallow depths and has not adequately located or resolved the temperatures of the deep geothermal source. (2) There is some ambiguity regarding the precise depth at which the model goes unstable. This ambiguity raises the possibility that the downward continuation (or direct extrapolation) process may be extended to too great a depth, with subsequent overestimation of subsurface temperatures. (3) The maximum subsurface temperatures are located at the fault rather than at either the shallow aquifer or the deep reservoir. This was anticipated since the maximum gradients are located over the fault. (4) A deep geothermal well drilled on the basis of the observed temperature gradients will not encounter the deep geothermal source. Rather, the well will undergo a temperature reversal when the fault is encountered, and the gradient will never recover.

The observed temperature gradient profile shown in Figure 2 has been filtered using the wave number technique represented by Equations (5)-(7). For this example, the sources have been placed at 1640 ft (500 m) and 6560 ft (2km). The source strengths were constrained to be nonnegative. In addition, the deep layer was constrained to be smooth by limiting the magnitude of the first derivative of the source strengths with respect to horizontal distance. The resulting gradient profile with the shallow information removed is shown in Figure 4, along with the subsurface temperatures predicted by downward continuation of the filtered gradients. By comparing Figure 4 to Figures 1 and 2, the following conclusions can be drawn: (1) The estimated subsurface temperatures in Figure 4 are now stable at all depths throughout the model. That is, the filtered gradients can be continued through the hot spring noise to resolve the temperatures at the deep geothermal source. (2) The model accurately predicts the temperatures known to exist at the deep geothermal source (Figure 3). (3) The maximum temperature gradient in the filtered profile is located over the deep geothermal source and is not displaced towards the hot spring (compare Figure 1 and Figure 4). (4) A deep geothermal well drilled on the basis of the processed temperature gradients <u>will</u> encounter the deep geothermal source (Figure 1). The well will undergo a temperature reversal below the shallow aquifers, but the gradient will recover.

CONCLUSION

We have applied a wave number filtering technique to a two-dimensional numerical model of a basin-and-range type geothermal system, and the results are quite good. In addition, we have applied the technique to a set of field data from the Valles Caldera (Swanberg and Li, 1982).

This wave number filtering technique has great theoretical promise, but additional work needs to be performed to make it practical. Extension of the technique to three dimensions should be relatively straight-forward, and limited only by the capabilities of the computer. More importantly, the sensitivity of the technique to the presence of errors must be studied. These include the effects of uneven and sparse data sampling, the effects of variations in thermal conductivity, the effects of random or systematic measurement errors, and the effects of various constraints. We are currently working on these problems.

REFERENCES

Carslaw, H. S., and J. C. Jaeger, 1959, <u>Conduction of Heat in Solids</u>, second edition, Oxford University Press, London, 510 pp.

Erdelyi, A., W. Magnus, F. Oberhettinger, and F. G. Tricomi, 1954, <u>Tables of Integral Transforms</u>, Vol. I, Bateman Manuscript Project, California Institute of Technology, McGraw-Hill, New York, pg. 118.

Land, A. H., and S. Powell, 1973, <u>FORTRAN Codes for Mathematical Programming: Linear, Quadratic and Discrete</u>, John Wiley & Sons, Chichester, 249 pp.

Lawson, C. L., and R. J. Hanson, 1974, <u>Solving Least Squares Problems</u>, Prentice-Hall, Englewood Cliffs, New Jersey, 340 pp.

Odegard, M. E., and J. W. Berg, Jr., 1965, Gravity interpretation using the Fourier integral, Geophysics, <u>30</u> (3), 424-438.

Spector, A., and F. S. Grant, 1970, Statistical models for interpreting aeromagnetic data, Geophysics, <u>35</u> (2), 293-302.

Swanberg, C. A., and T. M. C. Li, 1982, Wave number filtering of thermal data from the Valles Caldera, New Mexico, Geothermal Resources Council Transactions, in press.

Roy, A., 1962, Ambiguity in geophysical interpretation, Geophysics, <u>27</u> (1), 90-99.

Ward, S. H., H. P. Ross, and D. L. Nielson, 1981, Exploration strategy for high-temperature hydrothermal systems in Basin and Range province, Amer. Assoc. Pet. Geol., <u>65</u> (1), 86-102.

Wilde, D. J., and C. S. Beightler, 1967, <u>Foundations of Optimization</u>, Prentice-Hall, Englewood Cliffs, New Jersey, 480 pp.

Geothermal Resources Council, TRANSACTIONS Vol. 6, October 1982

TEMPERATURE GRADIENT DRILLING IN THE LAS CRUCES EAST MESA GEOTHERMAL FIELD

Richard L. Lohse and Larry Icerman

New Mexico State University Energy Institute
Las Cruces, New Mexico 88003

ABSTRACT

Thirty-four shallow temperature gradient holes were drilled on the Las Cruces east mesa adjacent to the Las Alturas geothermal anomaly. Temperature and heat flow data indicate that the anomaly, now named the Las Cruces East Mesa Geothermal Field, is at least 75 km^2 in size. These data together with geological and geophysical data collected previously suggest that the hydrothermal system is fault controlled and is composed of NW-SE and N-S components, which give the system an apparent NNW-SSE trend.

INTRODUCTION

Thirty-four shallow temperature gradient holes were drilled on the Las Cruces east mesa in order to further delineate a geothermal anomaly which was first confirmed in the Las Alturas area in 1979. University-based researchers had previously defined the Las Alturas geothermal anomaly (see Fig. 1) by conducting geophysical surveys and temperature gradient well drilling programs. Electrical resistivity data (Hohmann and Jiracek 1979) were interpreted as suggesting that the anomaly was of limited extent to the north. In 1980, the City of Las Cruces drilled potable water wells north of the Las Alturas geothermal anomaly, which encountered warm temperatures and suggested that the anomaly was much larger and hotter than first anticipated. The results of the current temperature gradient well drilling program support this conjecture.

TEMPERATURE DATA

The locations of the temperature gradient holes, identified by the label TG, are shown in Figure 1. Hole depths range from 35 to 91 meters. All of the holes encountered Quaternary deposits ranging in compaction from unconsolidated sands and clays, with drilling rates exceeding 40 ft/hr, to well-cemented and hard sands and gravels, with drilling rates less than 3 ft/hr, except holes TG-31 and TG-32 which were drilled into limestone formations. Measured temperature gradients range from 40 to 400°C/km, with a temperature as high as 54°C observed at 90 meters (TG-43) and a temperature of 48°C measured at a depth as shallow as 55 meters (TG-31).

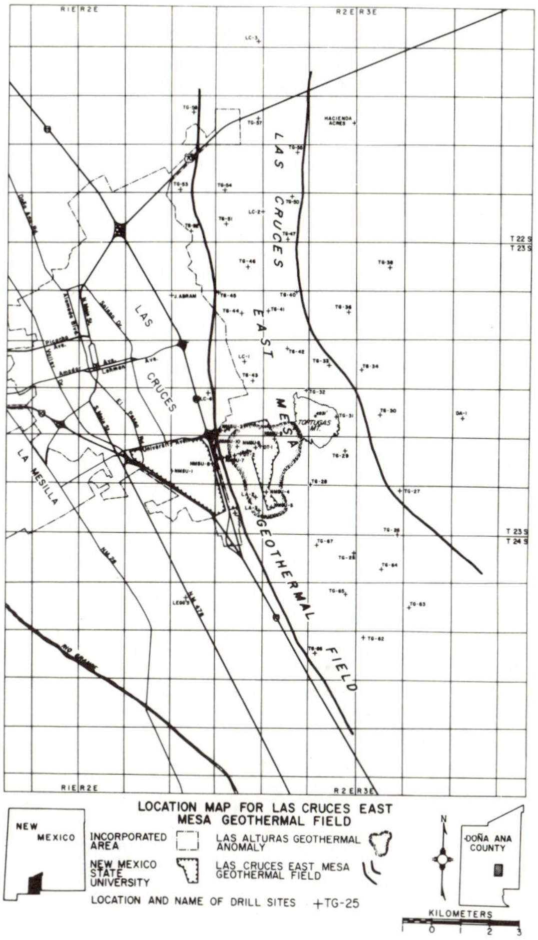

Figure 1. Location map of the study area.

Figures 2 and 3 show the hole locations and temperatures at 30 meters and 90 meters,

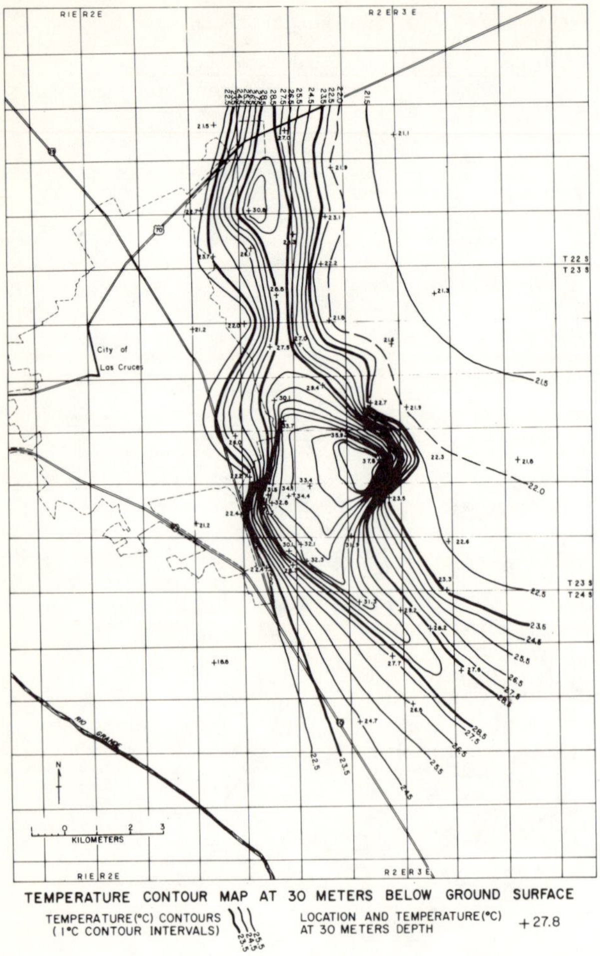

Figure 2. Temperature contours at 30 meters depth.

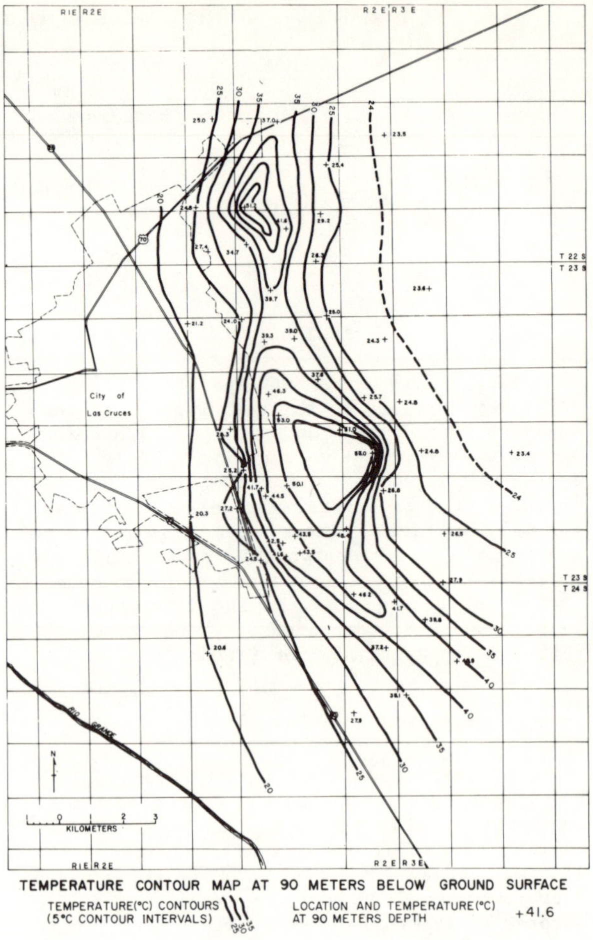

Figure 3. Temperature contours at 90 meters depth.

respectively, of new and existing drill sites within the study area. Temperatures indicated in Figure 3 for those temperature gradient holes drilled to depths less than 90 meters were estimated by extrapolating the measured temperature-depth data. A linear extrapolation was used for the holes that had nearly linear measured temperature gradients, while a gradual decrease in the temperature gradient with depth was assumed for holes that were drilled into zones believed to be convection-dominated. The magnitude of the assumed decrease in temperature gradient is proportional to the curvature of the actual temperature-depth profile.

Because the 30-m data are contoured at 1°C intervals rather than 5°C intervals, more contour lines are presented and, thus, one may be misled visually to believe that a larger temperature increase over the width of the anomaly exists at a depth of 30 meters than at 90 meters (compare Figs. 2 and 3). In fact, the temperature regime at 90 meters is better defined and more detailed than the one at 30 meters (i.e., a variation of 30°C across the width of the anomaly compared to 17°C).

HEAT FLOW DATA

The hole locations and best estimates for heat flow values for the current and existing well sites within the study area are shown in Figure 4. Heat flow values for specified depth intervals were calculated for each of the holes drilled under the current program based on measured temperature gradients and estimated thermal conductivities. The estimated best heat flow value represents a simple average of the heat flow values for each hole. These best estimates range from 2 to 19 HFU within the anomaly.

Figure 4 also shows the generalized

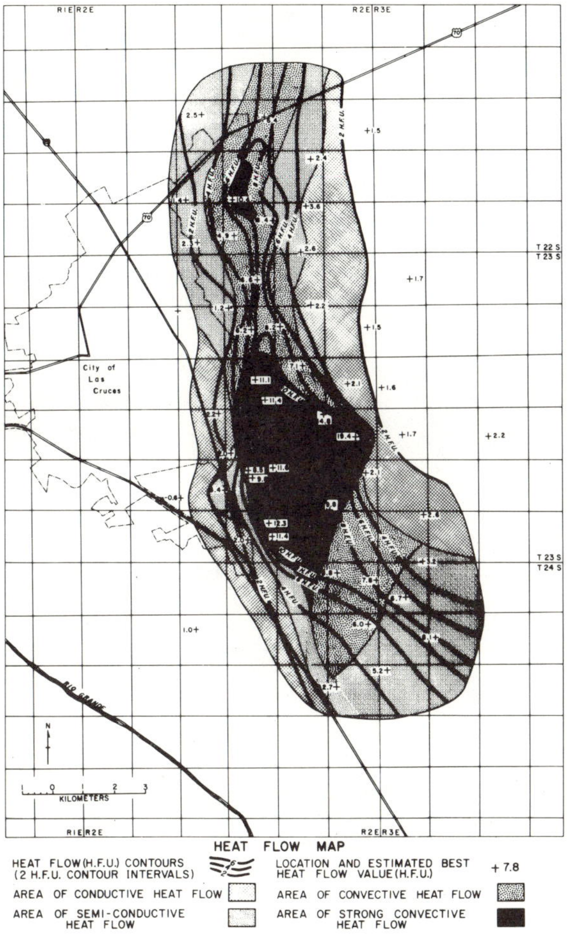

Figure 4. Heat flow map of the study area.

results of an analysis of the heat flow data versus depth for all of the current and some of the existing drill sites. Based on the magnitude and consistency of the decrease in heat flow with depth, the sites are assigned to a heat flow regime characterized as: (1) conduction dominated; (2) semi-conductive (i.e., approximately equal components of conduction and convection); (3) convection dominated; or (4) very strongly convection dominated.

GEOLOGICAL AND GEOPHYSICAL DATA

Tectonic Features

Several miles to the east of the study area, Seager (1981) has mapped pre-Quaternary faults which trend in a NW-SE direction. The Massey Tank Fault Zone and the Pena Blanca Fault in the southern part of the Organ Mountains, approximately 10 km to the southeast of Tortugas Mountain, are on line with the NW-SE trending northern boundary fault of Tortugas Mountain. These faults are also shown on a geologic map of the area (Kottlowski 1960). An area in the Organ Mountains near Squaw Mountain, approximately 13 km to the southeast of the study area, shows some NW-SE structural control (Seager 1981) and lies on a line with a suspected NW-SE trend in the northern part of the study area evident in Figures 2 through 4.

Seager et al. (1981) show late Quaternary N-S faulting throughout Dona Ana County. This observation is supported by geologic cross sections developed by Wilson et al. (1981), which are based on electrical resistivity data collected by Jackson (1976) and on lithologies of drill holes.

Lineament Data

A lineament study of New Mexico (Lepley 1982) and LANDSAT and SKYLAB satellite imagery clearly show a NW-SE fracture zone extending from the northern boundary of Tortugas Mountain to the northwest across Dona Ana County. Two main trends of structural weakness in NW-SE and N-S directions are believed to intersect in Dona Ana County (Lepley 1982; Lohse 1982).

Aeromagnetic and Gravity Data

The aeromagnetic data of Keller (1979) show NW-SE and N-S structural trends. These trends, defined by high magnetic gradients, are predominately NW-SE and N-S in the southern and northern parts of the study area, respectively, and may identify the positions of basement faults. Because of a limited density, the discrete fault structure is not easily recognized in the residual Bouguer gravity data of Aiken et al. (1978).

DISCUSSION

The heat flow data (see Fig. 4) show a similar pattern for the thermal anomaly as the 90-m temperature data (see Fig. 3). Although a strong convective area lies adjacent to a conductive area in Figure 4, this condition should not be interpreted as a strict physical model but instead as a measure of the contrast between the heat flow character of two adjacent drill sites. Sharp contrasts in bottom-hole temperatures (i.e., 20 to 54°C), heat flow values (i.e., 2 to 19 HFU), and in the character of the heat flow (i.e., conductive to strongly convective) over distances on the order of 1 km within the study area indicate the existence of a near-surface, fault-controlled hydrothermal system.

Where the contrast is not so sharp, the fault-controlled hydrothermal system is probably somewhat deeper, hidden below the Quaternary sediments. Because the sharpest contrast occurs where the electrical basement (Permo-Penn) is well exposed (Tortugas Mountain), the basement fault structure may well control the hydrothermal system. However, because large lateral hot water flows lead to high heat flow values, are of a convective nature, and probably occur following the ascent of the geothermal fluids through the basement (Permo-Penn) and into the overlying Quaternary sediments, there may be large deviations between actual fault positions and the positions inferred from sharp contrasts in temperatures, heat flow values, and modes of heat transfer. In particular, this condition may be true for the high heat flow area west of Tortugas Mountain (see Fig. 4). This strongly convective zone may be caused by: (1) hot water flowing in a southernly direction following its ascent up the NW-SE trending northern boundary fault of Tortugas Mountain or (2) vertical flow of hot water up through an extremely fractured basement caprock overlying a very large convective system and geothermal reservoir.

CONCLUSIONS

It is now known, based on shallow temperature gradient data, that the Las Cruces East Mesa Geothermal Field, which includes the approximately 5-km^2 Las Alturas geothermal anomaly, extends at least 20 km in length and 3 to 5 km in width. Temperature, heat flow, geologic, and geophysical data suggest that the fault-controlled hydrothermal system is composed of NW-SE and N-S components, which give the system an apparent NNW-SSE trend. There is also a strong spatial correlation between the apparent NNW-SSE trending thermal anomaly and deeper NNW-SSE trends of structural control defined by gravity and aeromagnetic maxima. These maxima and the thermal anomaly closely coincide in the northwest and southeast portions of the study area, while the gravity and aeromagnetic maxima trend slightly to the east of the thermal anomaly in the central portion of the study area. Both the gravity and aeromagnetic data show very large maxima in the southeast portion of the study area where analyses of temperature gradient and heat flow data show an increase in the temperature gradient with depth and suggest a semi-conductive to conductive mode of heat transfer, respectively. The source of the gravity and aeromagnetic maxima may well be the ultimate source of heat in the study area.

Temperature gradient and heat flow values in the southeast part of the study area are as high as 200°C/km and 7 to 8 HFU, respectively, with bottom-hole temperatures of 41°C at 90 meters. If temperature gradients of this magnitude persist with depth, temperatures of 100°C could exist at depths as shallow as 400 meters.

ACKNOWLEDGMENT

This work was supported by the New Mexico Energy Research and Development Institute, Contract 2-69-2202.

REFERENCES

Aiken, C.L.V., Laughlin, A.W. and West, F.G., 1978, Residual Bouguer gravity anomaly map of New Mexico: Los Alamos National Laboratory, scale 1:500,000.

Hohmann, G.W. and Jiracek, G.R., 1979, Bipole-dipole interpretation with three dimensional models (including field example from Las Alturas Estates, New Mexico): University of Utah Research Institute, Earth Science Laboratory, Report ESL-20, 20 pp.

Jackson, D.B., 1976, Schlumberger soundings in the Las Cruces, New Mexico area: U.S. Geological Survey, Open-File Report 76-231, 170 pp.

Keller, G.R., 1979, Aeromagnetic map of Tortugas Mountain, Dona Ana County, New Mexico: New Mexico Bureau of Mines and Mineral Resources, Open-File Report 110.

Kottlowski, F.E., 1960, Reconnaissance geologic map of the Las Cruces 30 minute quadrangle: New Mexico Bureau of Mines and Mineral Resources, Geologic Map 14.

Lepley, L., 1982, Lineament map of the state of New Mexico, in State-coupled low temperature geothermal resource assessment program, fiscal year 1981: New Mexico Energy Institute at New Mexico State University, p. 4/1-4/21.

Lohse, R.L., 1982, Temperature studies in southwestern New Mexico, in State-coupled low temperature geothermal resource assessment program, fiscal year 1981: New Mexico Energy Institute at New Mexico State University, p. 7/37-7/82.

Seager, W.R., 1981, Geology of Organ Mountains and southern San Andres Mountains, New Mexico: New Mexico Bureau of Mines and Mineral Resources, Memoir 36, 97 pp.

Seager, W.R., Shafiquallah, M. and Clemons, R. E., 1981, Active fault analysis and radiometric dating of young basalts in southern New Mexico, in State-coupled low temperature geothermal resource assessment program, fiscal year 1980: New Mexico Energy Institute at New Mexico State University, p. 6/1-6/17.

Wilson, C., White, R., Orr, B. and Roybal, G. R., 1981, Water resources of the Mesilla and Rincon valleys and adjacent areas, New Mexico: U.S. Geological Survey, Open-File Report 81-487.

CLASSIFICATION OF SCHLUMBERGER RESISTIVITY SOUNDING CURVES IN GEOTHERMAL FIELD

Seibe ONODERA

Department of Mining
Faculty of Engineering
Kyushu University

ABSTRACT

The classification of Schlumberger resistivity sounding curves in various geothermal fields is treated in the application of a multiple-layer resistivity problem. In comparison of the result of interpretation of resistivity sounding curves with near exploratory or production wells, it is demonstrated that only some type curves, which are satisfied by geothermal requirement, indicate the existence of geothermal reservoir. As a result, the distribution map of type curves in Kirishima geothermal field is presented as a basic data to determine the drilling target. Finally the result of interpretation of Schlumberger resistivity sounding curve observed at the station of I-23 is correlated with the drilling logs of the exploratory well KT-4.

TYPE CURVES IN GEOTHERMAL FIELD

If we classify a total of 597 Schlumberger RS (resistivity sounding) curves obtained in such geothermal fields as Otake, Hatchobaru, Sensui, Kirishima, Onikobe, Dieng in Central Java, and El Tatio in Chile by use of 2- to 6-layer type symbols, and if we sum up the number of similar types, then we have Table 1, according to the classification of RS curves in a multiple-layer resistivity problem (Onodera, 1961).

Table 1. The type of resistivity sounding curves in geothermal areas

2 layer		3 layer		4 layer		5 layer		6 layer	
Type	No.	Type	No.	Type	No.	Type	No.	Type	No.
1(1)	0	1(11)	0	1(111)	0	1(1111)	0	1(11111)	0
2(2)	5	2(12)	15	2(112)	2	2(1112)	0	2(11112)	0
		3(21)	29	3(121)	68	3(1121)	0	3(11121)	0
		4(22)	27	4(122)	97	4(1122)	1	4(11122)	0
				5(211)	10	5(1211)	1	5(11211)	0
				6(212)	22	6(1212)	23	6(11212)	0
				7(221)	69	7(1221)	64	7(11221)	0
				8(222)	13	8(1222)	16	8(11222)	0
						9(2111)	1	9(12111)	0
						10(2112)	1	10(12112)	0
						11(2121)	37	11(12121)	7
						12(2122)	8	12(12122)	1
						13(2211)	7	13(12211)	0
						14(2212)	1	14(12212)	0
						15(2221)	26	15(12221)	13
						16(2222)	0	16(12222)	0
								17(21111)	0
								18(21112)	0
								19(21121)	0
								20(21122)	0
								21(21211)	0
								22(21212)	4
								23(21221)	10
								24(21222)	0
								25(22111)	1
								26(22112)	0
								27(22121)	1
								28(22122)	0
								29(22211)	0
								30(22212)	0
								31(22221)	0
								32(22222)	0

Table 2.
The distribution of wells for various type curves

No. of wells	Layer	Type curves
1	3	2(12)
2		3(21)
1		4(22)
5	4	3(121)
2		5(211)
11		7(221)
4	5	7(1221)
2		11(2121)
1		15(2221)
29		

In practice, the classification of these curves extends to several categories of 2- to 8-layer types.

Of these, only Schlumberger RS curves, which are satisfied by the geothermal requirement (Onodera, 1980), indicate the existence of fracture type geothermal reservoir, which are proved by exploratory and producing wells, as listed in Table 2. It is known from the Table that particularly such types of 7(221) for the four-layer curve, 3(121) for the for the four-layer curve, and 7(1221) for the five-layer curve demonstrate possible existence of the fracture type geothermal reservoir.

This statistical data teaches us the fact that, in selecting the site of geothermal drilling based on the geothermal requirement for Schlumberger RS curves, at first, observed RS curves should be classified into the type of curve groups, and the zoning map of type curves should be drawn.

THE GEOTHERMAL REQUIREMENT

The geothermal requirement is defined as the combination of resistivity values of sequence layers, which are correlated to an altered formation and a fracture type geothermal reservoir, respectively, and a depth of their interface from the surface, whose value must be greater than 350 m.

Table 3 shows some numerical values for the geothermal requriement, and also is called necessary condition for the Schlumberger RS curve indicating the existence of the fracture type geothermal reservoir.

Fig. 1 shows the solution for the iterative least-squares interpretation of Schlumberger RS curve observed near the production well H-7 in Hatchobaru geothermal field, and Fig. 2 illustrates the correlation of the interpreted result of the same curve with the temperature log in well H-7.

Table 3. Some numerical values for Geothermal Requirement

Well	Station	Type curves	A in ohm-m	B in meters	C in ohm-m
O-15	LOS13	3(21)	20	480	100
H-16	LOS24	3(121)	77	1,000	270
H-12	LOS23	3(121)	40	700	250
O-6	LOS12	3(121)	7.4	430	140
O-7	LOS14	7(221)	10	550	300
H-1	C-13	7(221)	8	996.8	106
H-7	A-10	7(1221)	3.6	702	198.7
H-12	B-12	7(1221)	2	694.3	198.7
U-1	ES-17	11(2121)	12	750	200
7	7-EW	13(2211)	4.6	478.3	207.3

A: Resistivity of altered formation; B: Depth of interface; C: Resistivity of fracture type geothermal reservoir; O: Otake; H:Hatchobaru; U: Onikobe; 7: El Tatio.

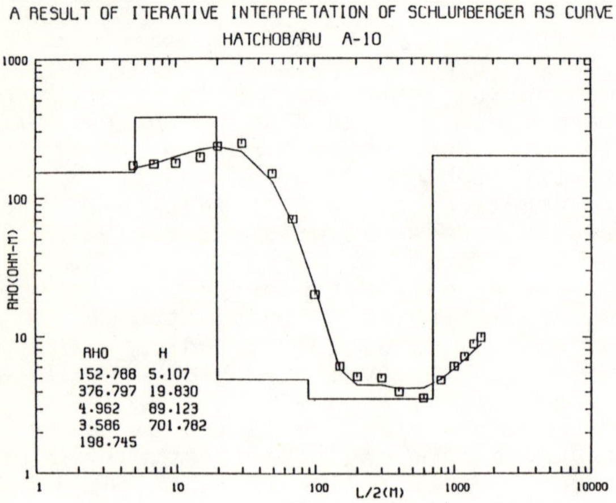

Fig. 1.

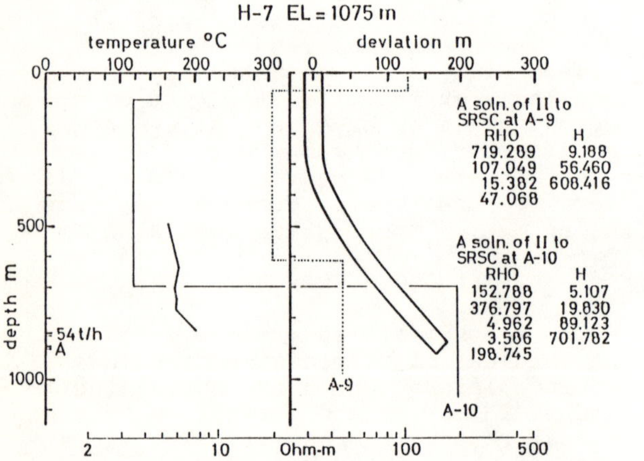

Fig. 2. Correlation of the result of iterative interpretation of RS curves observed at stations A-10 and A-9 with the temperature log in well H-7.

In the solution of RS curve for A-10, the fourth resistivity layer (resistivity of 3.6 ohm-m) corresponds to an altered formation, say Hohi volcanic complex, and the fifth resistivity layer is correlated to the fracture type geothermal reservoir, called Usa formation. And the depth of their interface is given by 702 m. These values show a typical geothermal requirement. The flow rate of steam from H-7 was 21.8 t/h dated March 3, 1980. Geothermal fluid was ejected from cracks at a depth of

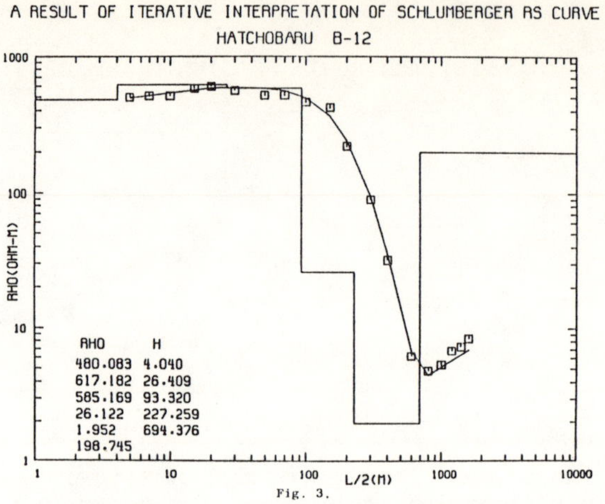

Fig. 3.

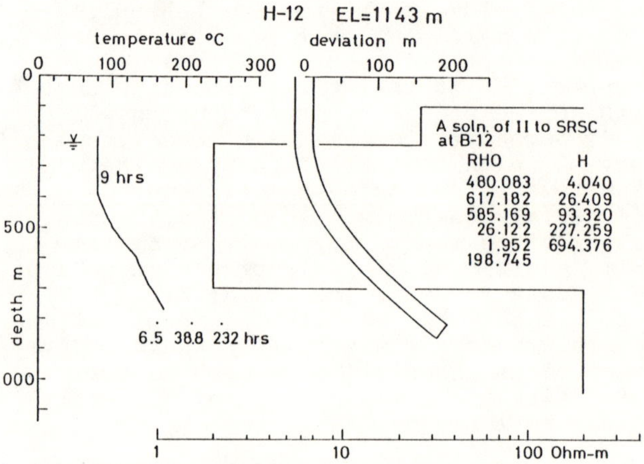

Fig. 4. Correlation of the result of iterative interpretation of RS curve observed at the station B-12 with the temperature log in well H-12.

850 m, which is formed in the fifth resistivity layer of 198.7 ohm-m in resistivity.

Similar example is also shown in Figs. 3 and 4. The well H-12 produced the steam flow rate of 45.7 t/h under pressure of 6 atg. dated March 4, 1980.

APPLICATION

Fig. 5 shows the distribution map of type curves for the resistivity sounding curves observed in Kirishima geothermal field now under geothermal development by the Geothermal Section of Thermal Power Department in the Kyushu Electric Power Co., Inc. This zoning map shows an area, which is occupied by the type curves at the resistivity sounding stations given by color visions.

A total number of 85 belong to the four-layer type 4(122) of the Schlumberger RS curves, which do not indicate the existence of the fracture type geothermal reservoir, are observed in Kirishima geothermal field. Painting out these areas by making use of pink, the residual area corresponds to a promising geothermal development area, which is supported by such substantial statistical data as the four-layer type 7(221)

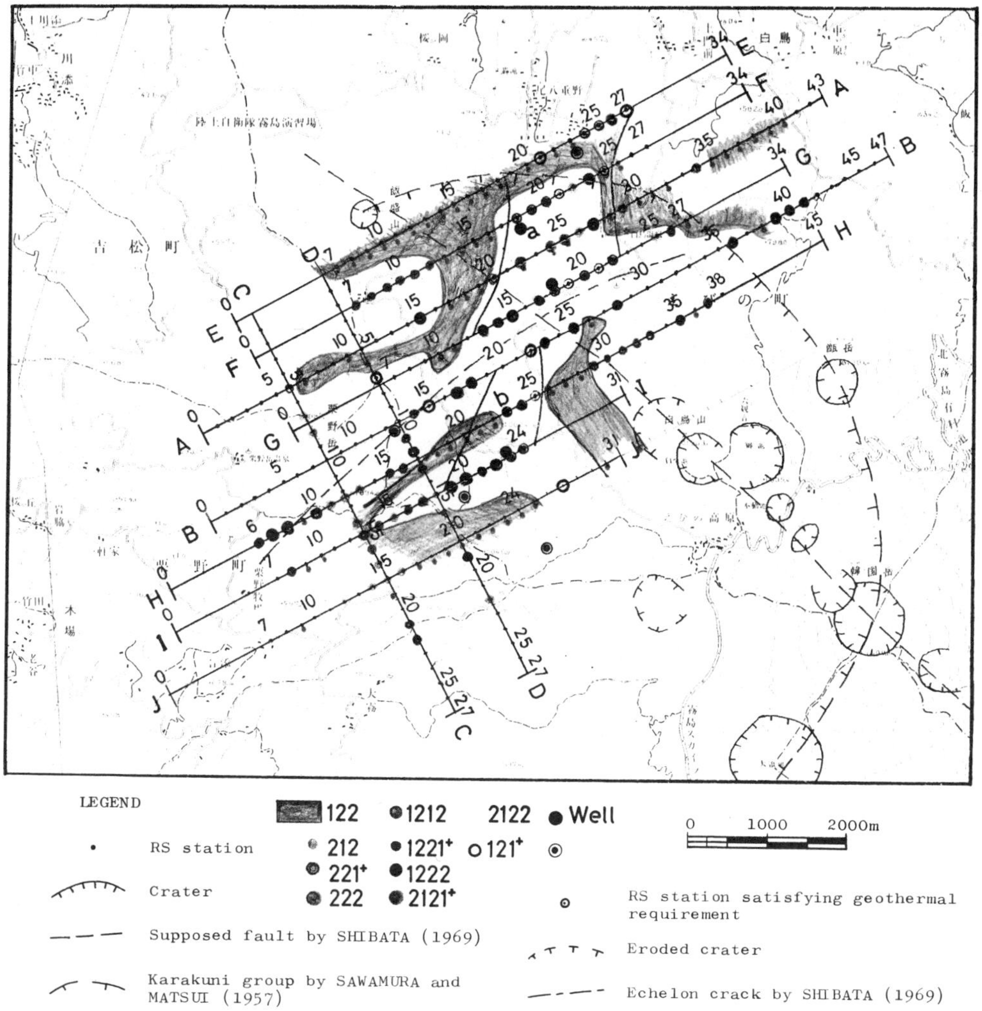

Fig. 5. Map showing the distribution of types of resistivity sounding curves in Kirishima geothermal field

and 3(121) and the five-layer type 7(1221) and 11(2121).

These can be divided into three areas, namely (1) West area of Mt. Shiratoriyama-Shiratori spa., (2) area of Kurino spa.-Kurinodake-Shiratori spa., and (3) central part of the area investigated by resistivity sounding exploration.

As for the types of RS curves indicating statistically the existence of the fracture type geothermal reservoir, the numbers of type curves are 7 for the type 3(121) and 7 for the type 7(221) of the four-layer RS curves, and are 26 for the type 7(1221), including the RS curve at the station of I-23 near the exploratory well of KT-4, and 13 for the 11(2121) of the five-layer RS curves. Consequently, from a view point of the classification of RS curves, it is naturally concluded that the areas numbered by (1), (2), and (3) are fully expected to develop Kirishima geothermal resources.

CORRLATION OF THE RESULT OF INTERPRETATION OF RS CURVES OBSERVED AT THE STATION, I-23 WITH THE DRILLING LOG OF KT-4

Fig. 6 shows the correlation of the result of interpretation of Schlumberger RS curve observed at the station, I-23 with the drilling log of the exploratory well, KT-4.

From the RS curve, geothermal requirement is given by the combination of the resistivity values of 20 ohm-m for the fourth resistivity layer, which corresponds to the altered formation, called the Saga-riyama lava, the resistivity value of 180 ohm-m for the fifth resistivity layer, which is correlated to the fracture type geothermal reservoir, called the Makizono lava, and the depth of these interface of 780 m. Especially, the resistivity sudden fall for the half electrode spacing of 900 m indicates the existence of a hydrothermal system.

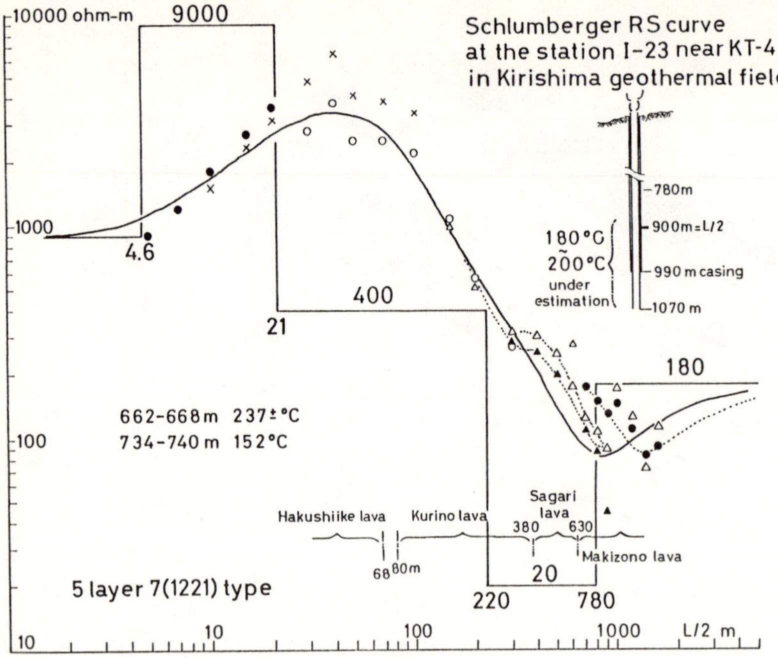

Fig. 6. Five layer 7(1221) type of the Schlumberger resistivity sounding curve indicating the existence of fracture type geothermal reservoir, which was proved by the exploratory well KT-4 in Kirishima geothermal field.

On the other hand, as shown in the right upper corner of Fig. 6, the casing with 7 inches in diameter was inserted at a depth of 990 m. The bottom temperature distribution in the well gives approximately 180 °C to 200 °C at a depth range from 900 m to 1,070 m. Such temperature distribution satisfies a condition of swabbing. Within the interval from 990 m to 1,070 m with 6 inches in diameter, the portion of escaping the drilling mud is existent at a depth of 990 m, and the escaping rate was observed by 60 t/h over. From here, the flow rate of steam of 15 t/h approximately was observed under the well pressure of 1.5 atg. Although this exploratory well does not correspond to an ordinary geothermal producing well, it is worth while to prove the existence of geothermal resources at a depth of 1,000 m, approximately.

As a result, the error of the interpreted result by making use of the ordinary curve matching method is given by 22 per cent, which is too large compared with another cases.

In addition, two temperature data, that is, 237 °C± for depths from 662 m to 668 m and 152 °C for 734 m to 740 m, which are shown in the left middle part of Fig. 6, are obtained by the estimation from the fluid filling temperature of core samples at laboratory after Assistant Prof. M. Hayashi, Division of Geothermal Geology, Research Institute of Industrial Science, Kyushu University.

DRILLING TARGET IN FUTURE

It is suggested from geoelectrical point of view that the most promising area as drilling target should be pointed out to the ridge of Mt.Shiratoriyama.

CONCLUSION

The research on the classification of resistivity sounding curves goes back to the academic year of 1961. In its application, it was presented here that the zoning map of the type for Schlumberger RS curves should be treated and discussed before writting the report of resistivity exploration for geothermal field. At present, the statistical probability of resistivity sounding curves, which were proved by the production wells, is of few magnitude. However, this distribution map not only presents and delineates an area of geothermal development, but also ought to have a close relation to a geological map in geothermal field.

ACKNOWLEDGEMENT

The author would like to express his sincere appreciation to the Kyushu Electric Power Co., Inc. for the offer of the drilling logs of KT-4 in Kirishima geothermal field.

REFERENCES

Onodera, S., 1961, A contribution of the theory of a multiple-layer resistivity problem, BUTSURI-TANKO (GEOPHYSICAL EXPLORATION), vol. 14, no. 4, 165-175.

Onodera, S., 1980, Geothermal indicators of resistivity and self-potential explorations, BUTSURI-TANKO (GEOPHYSICAL EXPLORATION), vol. 33, no. 5, 255-262.

Geothermal Resources Council, TRANSACTIONS Vol. 6, October 1982

DETERMINATION OF DRILLING TARGET IN ONIKOBE GEOTHERMAL AREA

Seibe ONODERA

Department of Mining
Faculty of Engineering
Kyushu University

ABSTRACT

Based on the definition of geothermal requirement, that is, the combination of resistivities of an altered formation and a fracture type geothermal reservoir and a depth of their interface from the surface, a drilling target for geothermal development at Onikobe was determined by the interpreted result of resistivity sounding curves together with seismic reflection data of lacking reflectors. And this was proved by the exploratory well, U-1.

INTRODUCTION

The geothermal development in Onikobe area began with geological survey and geochemical and geophysical explorations in 1962. In the report of resistivity survey at Onikobe area worked by CGG from May 13 to July 29, 1963, it was concluded that "Faults F3 and F4 bordering a central horst, and this horst itself, should be obviously drilled to investigate geothermal conditions," and also "Extended exploration of the area delineated by faults F2 EN and F6 SW should be determined only after obtaining production from the central horst."

According to the suggestion, the Electric Power Development Co., Inc. (EPDC) executed steam well drillings in large diameter. Without necessary and sufficient evaluations of steam flow rate ejected from production wells, the geothermal power plant with a capacity of 25 MW was constructed in March, 1975. For the duration of 16 years from 1962 to 1978, more than 50 wells were drilled in the existing geothermal area, which is delineated by faults F3 and F4, but it was unexpected to obtain necessary steam flow rate to generate elec-

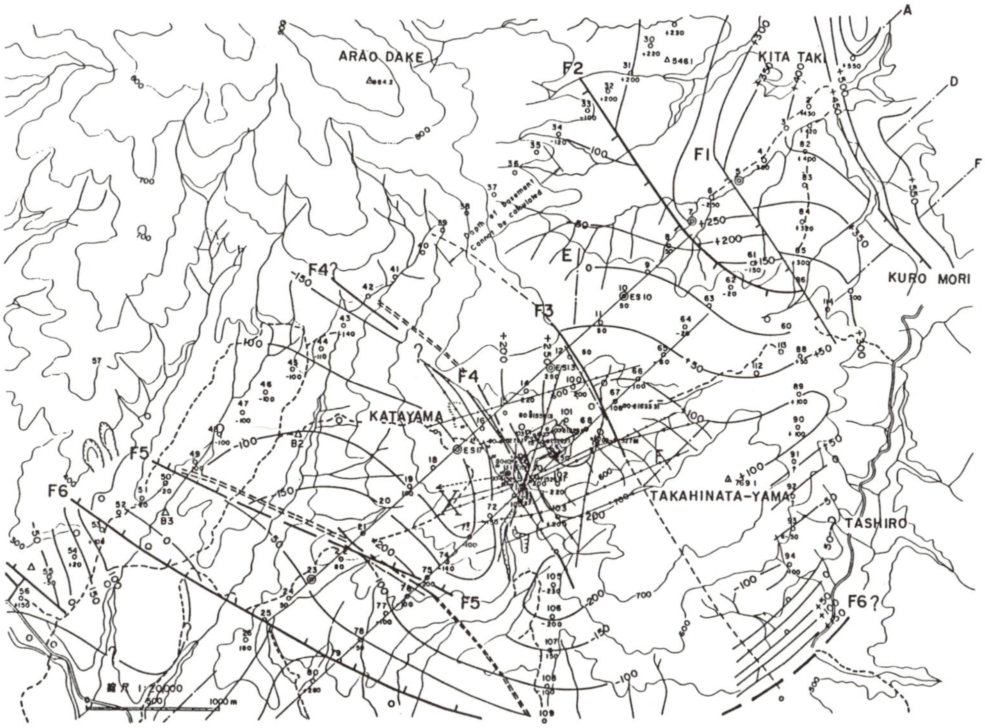

Fig. 1. Arrangements of resistivity surveying lines and the spread lines of seismic reflection and the area of X surrounded by faults lines F4 and F5 is promising for geothermal development

tricity of 25 MWH. At present, the power plant is operated for a fall of 6 MWH by feeding about 60 t/h of steam from 11 production wells at a depth of 300 m into the turbine generator.

On the other hand, geothermal wells at a depth of 1,300 m ejected acidic fluid of 3.0 in pH, therefore it was impossible to use. Is there a good measure to overcome this difficulty? Yes, the answer can be found by the reinterpretation of geophysical data.

In November of 1978, the author had an opportunity to discuss the data of geothermal exploration in Onikobe at the request of Mr. Hiroshi TAKAGI, Manager of Onikobe power plant.

After the examination of various data, the drilling target was determined from the result of interpretation of 6 resistivity sounding curves together with the section of seismic reflection of lacking reflectors. As for discovering a new fracture type geothermal reservoir, the suggestion will be given in this paper in connection with geophysical indications.

INDICATION OF RESISTIVITY SOUNDING CURVES

Fig. 1 shows the arrangement of resistivity surveying lines and stations set up by CGG. Only 6 Schlumberger RS (Resistivity Sounding) curves observed at stations ES-5, -7, -10, -13, -17, and -23 are illustrated here, as it is stated that 109 RS curves of the remainder are lost. Values of observed apparent resistivities corresponding to the half electrode separations, denote L/2, for the Schlumberger array are plotted by small circles as shown in Fig. 2. So far as 6 RS curves are concerned, the indications are classified into 4-layer type 7(221) for ES-5, 3-layer type 3(21) for ES-7, and 5-layer type 11(2121) for ES-10, -13, -17, and -23, respectively, which fall under a category of typical curves of curve groups (Onodera, 1973) in geothermal areas of Otake, Hatchobaru and Kirishima. Of these ES-17 RS curve is of interest in searching geothermal reservoir by use of its geoelectrical indication due to high resistivity contrast.

RESULTS OF INTERPRETATION OF RS CURVES

Applying the simulative interpretation of field RS curves (Onodera et al., 1980), we have the following results: As shown in Fig. 2, (a) values of resistivity for sequence layers and depth of interface for them are plotted in two rows print type at the left of the same Figure, (b) each step-resistivity distribution is drawn by (2N-1) partial straight lines, where N is the number of layers, and (c) each computed RS curve is plotted by a solid line.

At present, the geothermal requirement is defined rationally such that resistivities of resistivity layer corresponding to altered formation are in the range from several ohm-m to 77 ohm-m, those correlating to a fracture type geothermal reservoir from one hundred ohm-m to 300 ohm-m, and a depth of their interface is greater than 350 m. These will be changed as the research makes progress.

The initial screening for interpreted results, especially the maximum depth of interface, removes those for ES-5 and ES-7 from our consideration.

Resistivity values of the 4-th layer for ES-10, -13, -17 and -23 are in the range from 7 ohm-m to 18 ohm-m, those of the 5-th layers for them from 100 ohm-m to 200 ohm-m, and the maximum depth of interface for them are in the range from 500 m to 900 m, which satisfy the geothermal requirement. Thus, the former corresponds to an altered formation and the latter is correlated to a fracture type geothermal reservoir.

Since the evidence is that acidic geothermal fluid is ejected from wells located in the existing geothermal area,

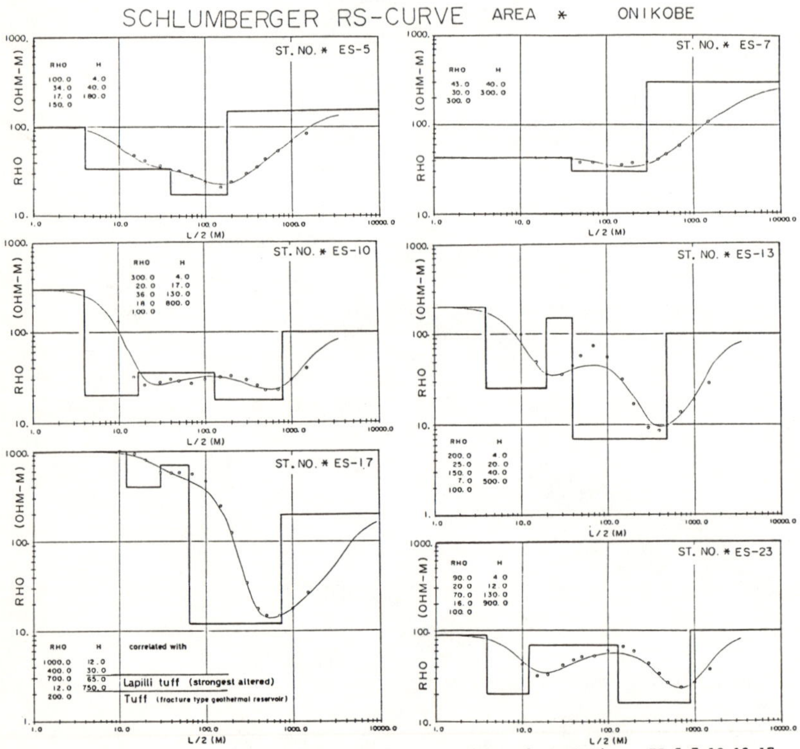

Fig. 2. Results of interpreting Schlumberger RS curves observed at stations, ES-5,7,10,13,17 and 23 belong to the A-Line. All curves fall under the category of the typical curves. Of these ES-13 is located in the existing geothermal development area. The ES-17 shows a drilling target to get geothermal fluid for generating electricity.

where a reservoir at a depth of 1,300 m originates in the formation of 100 ohm-m, it seems naturally that the same fluid will be discharged from subsurfaces at a depth of 800 m and 900 m for ES-10 and -23, respectively.

To get more good quality of steam, it needs to choose an area having more high resistivity formation. The place satisfying such a condition is given by ES-17, because that the resistivity formation with 200 ohm-m is distributed at a depth of 750 m from the surface. So the final screening tells us ES-17 as a possible drilling target.

In short, the 4-th resistivity layer for ES-17 (resistivity of 12 ohm-m; layer thickness 685 m) can be correlated with the strongest altered formation, say Lapilli tuff; and the 5-th resistivity layer (200 ohm-m, depth from the surface of 750 m) is correlated with the weak altered formation, that is to say, Tuff, which forms the so-called fracture type geothermal reservoir. And also the 4-th and 5-th resistivity layers and the depth of interface between them show sufficient values as the geothermal requirement.

SECTION OF RESISTIVITY LAYERS

If we plot results of interpretation of RS curves on the section of resistivity layer after CGG, then we obtain Fig. 3.

Comparing this with these, we may find that depths of the second and fourth interfaces, resistivity values of the second resistivity layer for ES-7, and resistivity values of the fourth resistivity layer for ES-10, ES-13, and ES-17 are equal to those given by CGG. Particularly, from geothermal point of view, the result of interpretation of the RS curve for ES-17 is better than that of CGG.

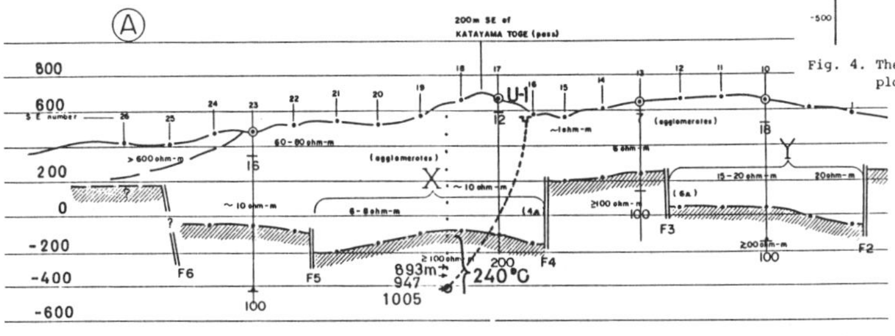

Fig. 3. The resistivity section for the A-Line after CGG together with the plot of results for ES's 5 to 23 and projection of directional well,U-1 to the A-Line

In the same Figure, as far as the part (which is marked with a Roman capital letter of X and is delineated by fault lines F4 and F5) is concerned, the section of resistivity layers after CGG gives an excellent geoelectrical information.

Further, it is only natural for this part to continue with the X-part of D-Line. This X-area is a site proposed for geothermal development.

DISTRIBUTION MAP OF RESISTIVITY LAYERS

The existing geothermal development area consists of the up-lift geoelectrical structure, which is delineated by fault lines of F3 and F4, as shown in Figs. 1 and 3.

In practice, geothermal development in hydrothermal system is effective in the graben geoelectrical structure. From this point of view, geothermal exploratory wells should be drilled in the X-area delineated by fault lines of F4 and F5, especially near ES-17.

The X-area, that is proposed as the site for geothermal drilling, is located in western part of the existing geothermal area.

CHECKING THE X-AREA BY SEISMIC REFLECTION DATA

Fig. 4 shows the section of seismic

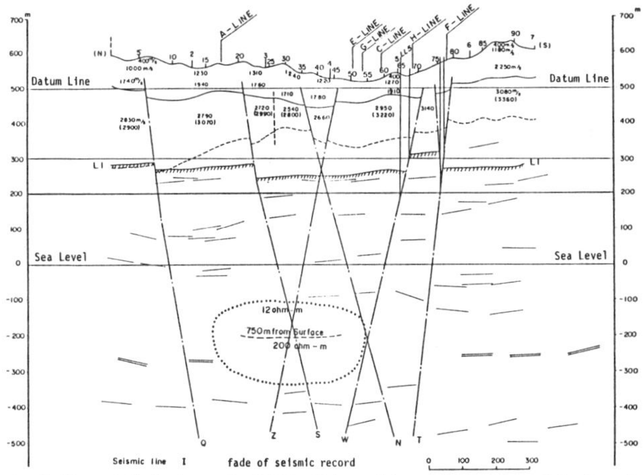

Fig. 4. The resistivity layers for ES-17, which are interested in geothermal development, plotted on the section of seismic reflection after the Ube Industrial Co., Ltd.

refraction and reflection for I-Line. The numbers of seismic stations close to the X-area are in the range from 20 to 50. Remarkable reflectors exist in both ends (North and South parts) of the spread at sea level of -200 m to -300 m. If we thus project the result of interpretation of the RS curve for ES-17 on the lower part of station numbers from 20 to 50, then we have an explicit Figure as shown in Fig. 4. Clearly, reflectors do not exist in the dotted section, because of a fade of seismic record due to the existence of fracturing formations.

Such a distribution of reflectors enables still more the suggestion of resistivity exploration for geothermal field, that is, the analysis of resistivity sounding curve indicates the existence of the fracture type geothermal reservoir at a depth of 750 m from the surface of the X-area.

DRILLING TARGET

At first, the site of geothermal exploratory wells should be chosen near the resistivity station ES-17. In this case the drilling should be aimed at a depth of 750 m from the surface of the area.

When an exploratory well is drilled through the fourth resistivity layer into the fifth resistivity layer and encountered with a fracture type geothermal reservoir within the fifth resistivity layer, the escape of drilling circulating fluid will occur. Then a hole slightly smaller than a casing must be drilled for a few meters below a shoe, and a string of casing must be put down and landed on the bottom of the hole, and finally the well completion can be executed.

In general, accuracy of the interpretation of resistivity sounding curves is approximately -13 per cent in depth of interfaces. Adding this, an actual drilling depth becomes 850 m approximately.

THE RECORD OF EXPLORATORY WELL, U-1

At the end of June in 1980, the suggestion was proved by the directional drilling U-1, which was designed by EDPC as shown in Fig. 5. The temperature distribution in the well was observed as 240 °C for the depth range from 800 m to 1,150 m along the inclination as shown in the projection chart of the directional well of U-1 (see Fig. 3), which is interesting phenomenon of temperature distribution.

From this Figure, it is well-known that the fourth depth of interface between fourth and fifth resistivity layers is nearly

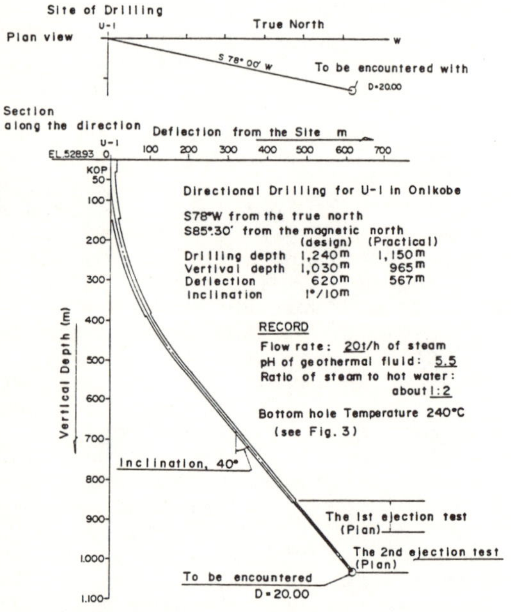

Fig. 5. Design of directional drilling of U-1 in Onikobe

equal to the depth showing the temperature of 240 °C, and three cracks exist in the fifth resistivity layer at a depth of 893 m, 947 m, and 1,005 m along the inclination of the well.

The principal well data are as follows:

pH of geothermal fluid .... 5.5 (possible use)
Flow rate of steam for well size of 6 1/2 in. ... 20 t/h about
The ratio of steam and hot water .. 1:2
Temperature in the well at depth range from 800 m to 1,150 m along the inclination of the well .... 240 °C

Assuming the flow rate of steam for 8 5/8 in. of the production well instead of 6 1/2 in, the well will produce steam of 40 t/h approximately.

CONCLUSION

A new possible geothermal development area, say the X-area, was presented here by the discussion of geophysical data in Onikobe, and the result was proved by the exploratory well, U-1.

Geophysical indications caused by the existence of the fracture type geothermal reservoir are composed of two categories, that is, one is the geothermal requirement, which is the combination of resistivity values for a series of resistivity layers, which corresponds to altered formations and fracture type geothermal reservoir, and the depth of interface of them, and the other is given by the place where no reflectors exist in seismic reflection data.

ACKNOWLEDGEMENTS

The author would like to extend his deep appreciation to the Electric Power Development Co., Inc. for the design of the directionally exploratory drilling of U-1.

REFERENCES

Onodera, Seibe, 1973, Application of resistivity method to searching geothermal resources, Chinetsu, vol. 10, no. 3 (Ser. No. 38), 47-57.

Onodera, Seibe, Michihiro Fukuda, and Yutaka Sasaki, 1980, Interpretation and graphic representation of resistivity data using a curve plotter connected to a computer, BUTSURI-TANKO (GEOPHYSICAL EXPLORATION), vol. 33, no. 3, 116-126.

Mattice, Mark D. and Barry R. Lienert, 1980, Schlumberger survey of Maui Island, State of Hawaii, Geothermal Resources Council, TRANSACTIONS vol. 4, 85-88.

Geothermal Resources Council, TRANSACTIONS Vol. 6, October 1982

TYPE CURVES FOR RESISTIVITY SOUNDING WITH TWO-ELECTRODE, WENNER, SCHLUMBERGER, AND DIPOLE-DIPOLE ARRANGEMENTS FOR A TWO-, THREE-, FOUR-, AND FIVE-LAYERED MODEL

Seibe ONODERA

Department of Mining
Faculty of Engineering
Kyushu University

ABSTRACT

Type curves for resistivity sounding with the two-electrode, Wenner, Schlumberger, and Dipole-Dipole arrays are presented for a two-, three-, four-, and five-layered earth, based on the theory of classification of relative resistivity sounding curves for an n-layered resistivity problem.

These graphs are useful to get an elementary knowledge of the interpretation of resistivity sounding curves for various electrode arrangements.

EXACT RELATIVE RESISTIVITY COMPUTATION FORMULAS FOR AN N-LAYER RESISTIVITY PROBLEM

Suppose an n-layer resistivity problem. Equations for computing the relative resistivity, which is defined as the ratio of apparent resistivity and resistivity of the first layer, at the surface of the model involved are:

for the two-electrode array

$$\rho^*_2(a) = 1 + F(a) \quad \text{(1)}$$

for the Wenner array

$$\rho^*_w(a) = 1 + 2F(a) - F(2a) \quad \text{(2)}$$

for the Schlumberger array

$$\rho^*_s(a,ma) = 1 + \frac{m}{m-1}F(a) - \frac{1}{m-1}F(ma) \quad \text{(3)}$$

where $m = (L+\ell)/(L-\ell)$

and for the Dipole-Dipole array

$$\rho^*_D(x,mx) = 1 + \frac{1}{2}\{(m+1)(m+2)F(mx) - 2m(m+2)F((m+1)x) + m(m+1)F((m+2)x)\} \quad \text{(4)}$$

where x is the Dipole length and mx is the Dipole separation,

where $F(a) = 2a\int_0^\infty K_n(\lambda)J_0(\lambda a)d\lambda$,

a is the electrode separation, $K_n(\lambda)$ is the kernel function, J_0 is the Bessel function of the first kind of zero order, and λ is a parameter.

KINDS OF RELATIVE RESISTIVITY SOUNDING CURVES

The relative resistivity sounding curves for a two-layer earth can be grouped into two kinds of type curves depending on the condition such that $\rho_1<\rho_2$ and $\rho_1>\rho_2$. This is valid in the master curves by Tagg (1934) and Roman (1934). The procedure of classification is not affected by various electrode arrangements. Basically, the number of type curves involved can be expressed by

$$N_2 = 2 \quad \text{(5)}$$

The kernel function for a two-layer case is given in the form

$$K_2(\lambda) = \frac{k_1e^{-2\lambda h_1}}{1-k_1e^{-2\lambda h_1}} \quad \text{(6)}$$

The values of reflection coefficients k_1 take possitive and negative such that $1\geq k_1>0$ for $\rho_1<\rho_2$ and $-1\leq k_1<0$ for $\rho_1>\rho_2$. Thus, such values of k_1 are substituted into (6), and carrying out the numerical computation of (1) to (4), we have two groups or types of relative resistivity sounding curves.

In general, the reflection coefficients $(k_1,k_2,\ldots,k_{n-1})$ included in the kernel function for an n-layer resistivity problem consist of n-1 kinds. In accordance with possitive and negative in each value, the number of combination of 2^{n-1} exists. This permits the classification of resistivity sounding curves. Thus, the number of type curves in the relative resistivity sounding curves for an n-layer resistivity problem can be expressed as

$$N_n = 2^{n-1} \quad \text{(7)}$$

NOTATION IN CLASSIFICATION

As illustarted in the foregoing section, fundamentally, RS (Resistivity Sounding) curves for a two-layer model can be grouped into two types: one is for resistivity constrast such that $\rho_1<\rho_2$, for which the curve groups are denoted by a numeral 1 with (1), and the other is for $\rho_1>\rho_2$, for

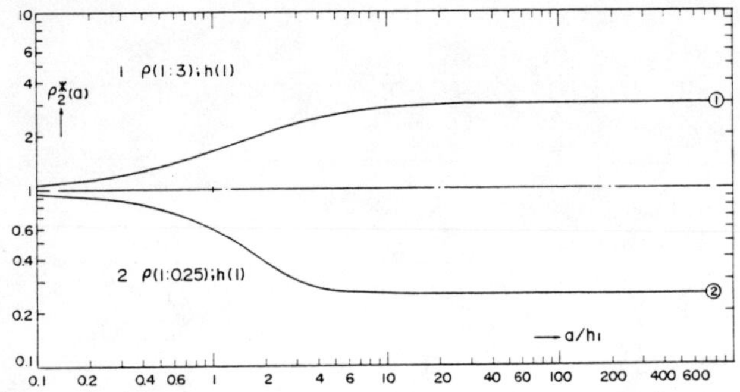

Fig. 1(a). Two types of curve groups for the two-layer model for the two electrode array

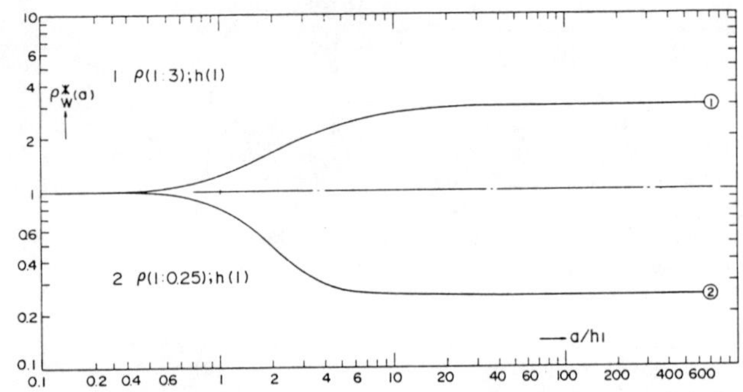

Fig. 1(b). Two types of curve groups for the two-layer model for the Wenner array

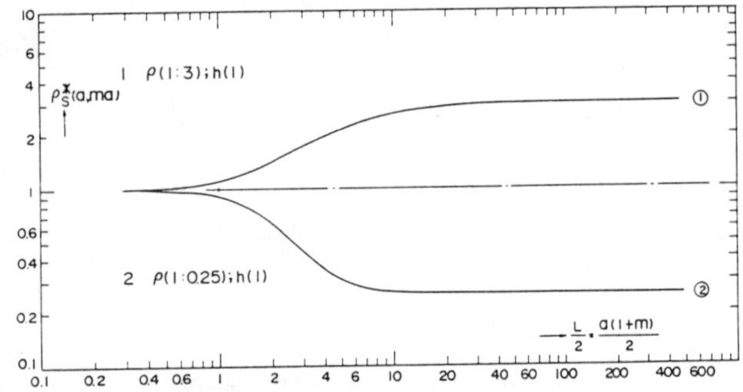

Fig. 1(c). Two types of curve groups for the two-layer model for the Schlumberger array

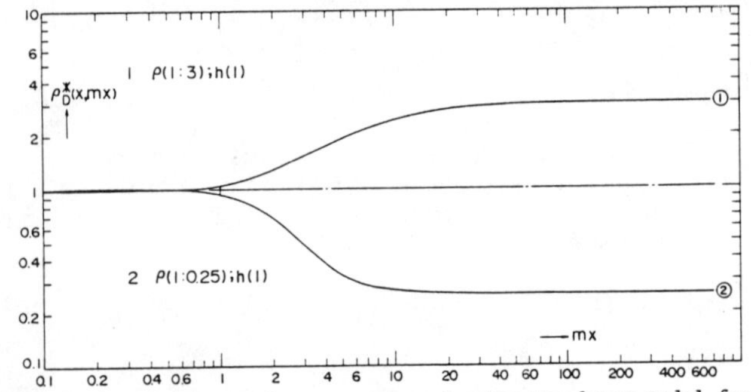

Fig. 1(d). Two types of curve groups for the two-layer model for the Dipole-Dipole array

which those are represented by a numeral 2 with (2). In short, relative RS curves for a two-layer can be grouped into

Type 1(1) for $\rho_1 < \rho_2$, and Type 2(2) for $\rho_1 > \rho_2$.

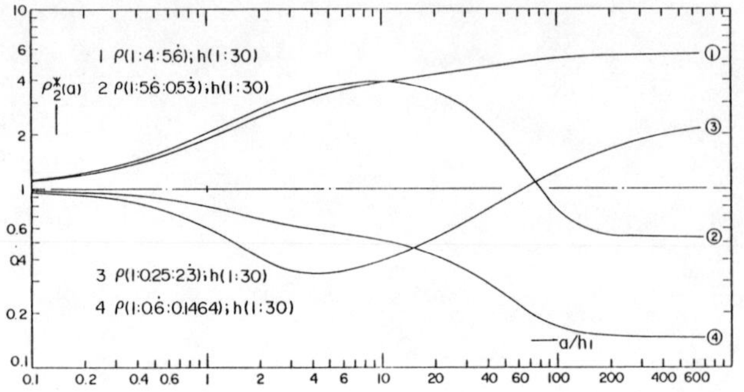

Fig. 2(a). Four type of curve groups for the three-layer model for the two electrode array

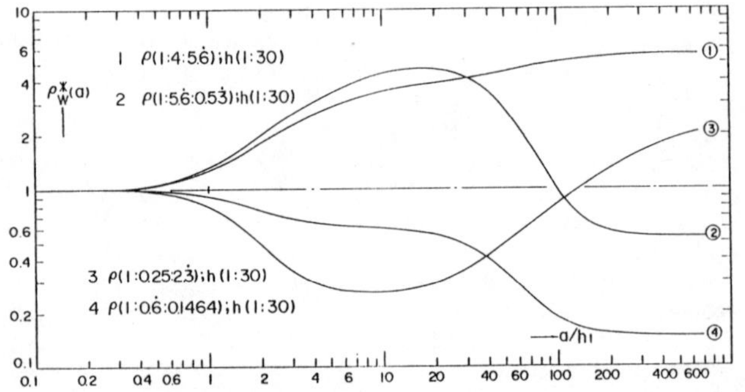

Fig. 2(b). Four types of curve groups for the three-layer model for the Wenner array

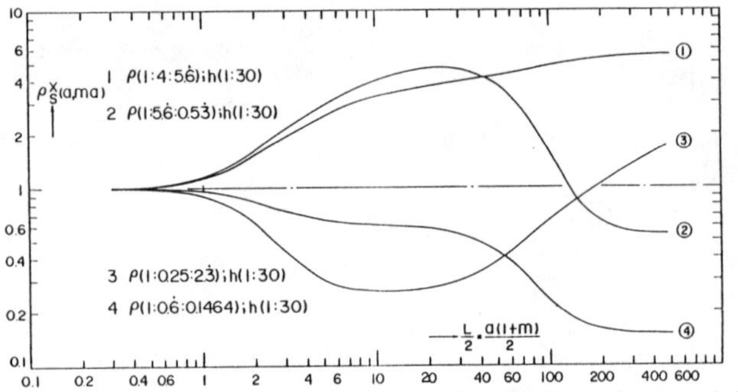

Fig. 2(c). Four types of curve groups for the three-layer model for the Schlumberger array

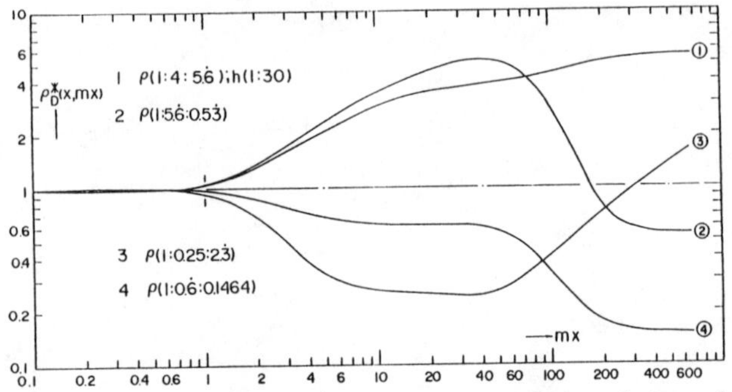

Fig. 2(d). Four types of curve groups for the three-layer model for the Dipole-Dipole array

For three layer RS curves it follows that 4 types can be grouped such that

Type 1(11) for $\rho_1 < \rho_2 < \rho_3$,

Type 2(12) for $\rho_1 < \rho_2 > \rho_3$,

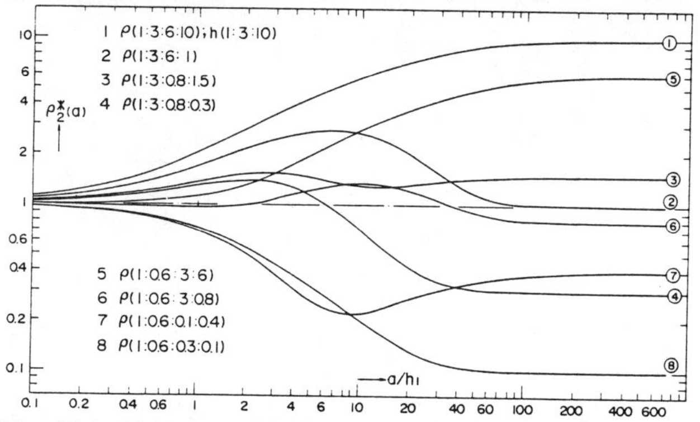

Fig. 3(a). Eight types of curve groups for the four-layer model for the two electrode array

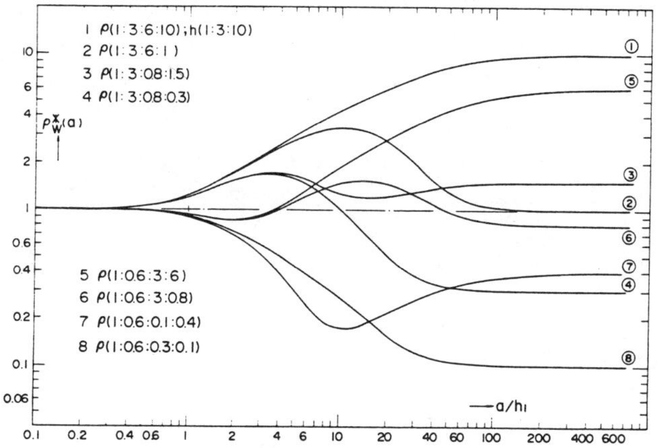

Fig. 3(b). Eight types of curve groups for the four-layer model for the Wenner array

Type 3(21) for $\rho_1 > \rho_2 < \rho_3$, and

Type 4(22) for $\rho_1 > \rho_2 > \rho_3$.

For four-layer RS curves there exists 8 types such that

Type 1(111) for $\rho_1 < \rho_2 < \rho_3 < \rho_4$,
Type 2(112) for $\rho_1 < \rho_2 < \rho_3 > \rho_4$,
Type 3(121) for $\rho_1 < \rho_2 > \rho_3 < \rho_4$,
Type 4(122) for $\rho_1 < \rho_2 > \rho_3 > \rho_4$,
Type 5(211) for $\rho_1 > \rho_2 < \rho_3 < \rho_4$,
Type 6(212) for $\rho_1 > \rho_2 < \rho_3 > \rho_4$,
Type 7(221) for $\rho_1 > \rho_2 > \rho_3 < \rho_4$, and
Type 8(222) for $\rho_1 > \rho_2 > \rho_3 > \rho_4$.

Similarly, five-layer RS curves can be classified into 16 types such that

Type 1(1111) for $\rho_1 < \rho_2 < \rho_3 < \rho_4 < \rho_5$,
Tyep 2(1112) for $\rho_1 < \rho_2 < \rho_3 < \rho_4 > \rho_5$,
Tyep 3(1121) for $\rho_1 < \rho_2 < \rho_3 > \rho_4 < \rho_5$,
Type 4(1122) for $\rho_1 < \rho_2 < \rho_3 > \rho_4 > \rho_5$,
Type 5(1211) for $\rho_1 < \rho_2 > \rho_3 < \rho_4 < \rho_5$,
Type 6(1212) for $\rho_1 < \rho_2 > \rho_3 < \rho_4 > \rho_5$,
Type 7(1221) for $\rho_1 < \rho_2 > \rho_3 > \rho_4 < \rho_5$,

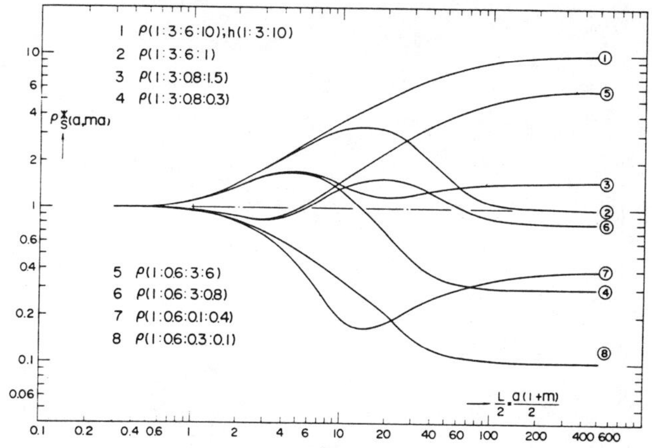

Fig. 3(c). Eight types of curve groups for the four-layer model for the Schlumberger array

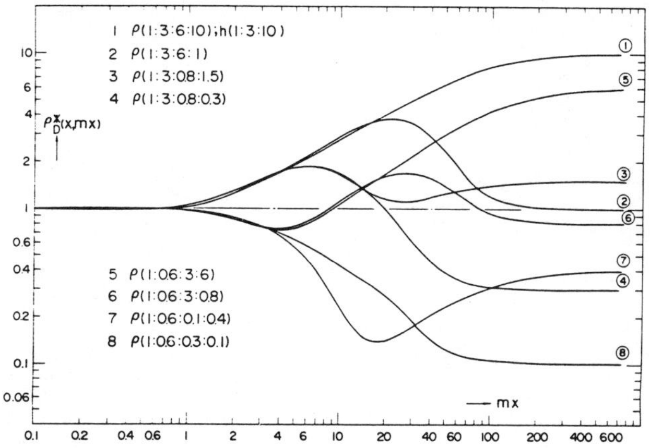

Fig. 3(d). Eight types of curve groups for the four-layer model for the Dipole-Dipole array

Type 8(1222) for $\rho_1 < \rho_2 > \rho_3 > \rho_4 > \rho_5$,
Type 9(2111) for $\rho_1 > \rho_2 < \rho_3 < \rho_4 < \rho_5$,
Type 10(2112) for $\rho_1 > \rho_2 < \rho_3 < \rho_4 > \rho_5$,
Type 11(2121) for $\rho_1 > \rho_2 < \rho_3 > \rho_4 < \rho_5$,
Type 12(2122) for $\rho_1 > \rho_2 < \rho_3 > \rho_4 > \rho_5$,
Type 13(2211) for $\rho_1 > \rho_2 > \rho_3 < \rho_4 < \rho_5$,
Type 14(2212) for $\rho_1 > \rho_2 > \rho_3 < \rho_4 > \rho_5$,
Type 15(2221) for $\rho_1 > \rho_2 > \rho_3 > \rho_4 < \rho_5$, and
Type 16(2222) for $\rho_1 > \rho_2 > \rho_3 > \rho_4 > \rho_5$.

GRAPHS OF RELATIVE RS CURVES FOR A TWO-, THREE-, FOUR-, AND FIVE-LAYER MODELS

The typical curves of curve groups for a two-, three-, four-, and five-layer cases are shown in Figs. 1, 2, 3, and 4 for two electrode, Wenner, Schlumberger, and Dipole-Dipole arrays with (a), (b), (c), and (d), respectively, and no difficulty is encountered in ascertaining their general feature. These calculations are due to a polynomial approximation of the kernel function (Onodera, 1963).

Depths of interfaces are selected as increased geometrical progression from upper to lower, so that each resistivity

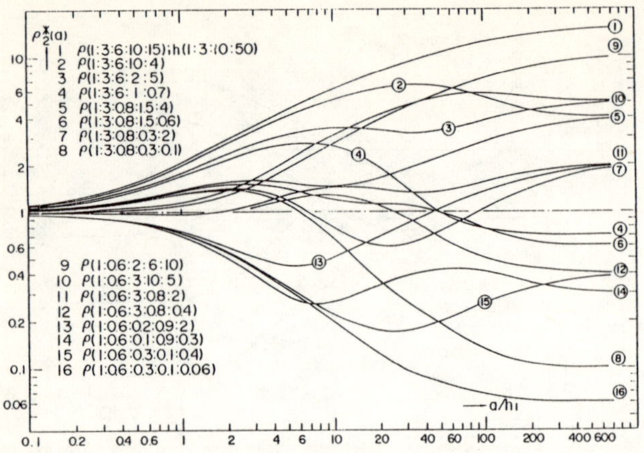

Fig. 4(a). Sixteen types of curve groups for the five-layer model for the two electrode array

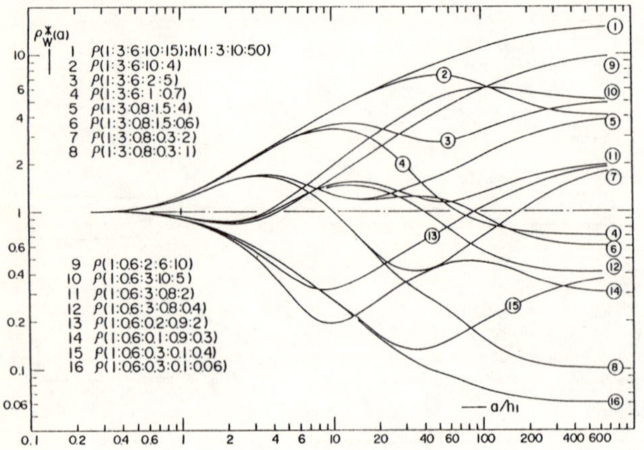

Fig. 4(b). Sixteen types of curve groups for the five-layer model for the Wenner array

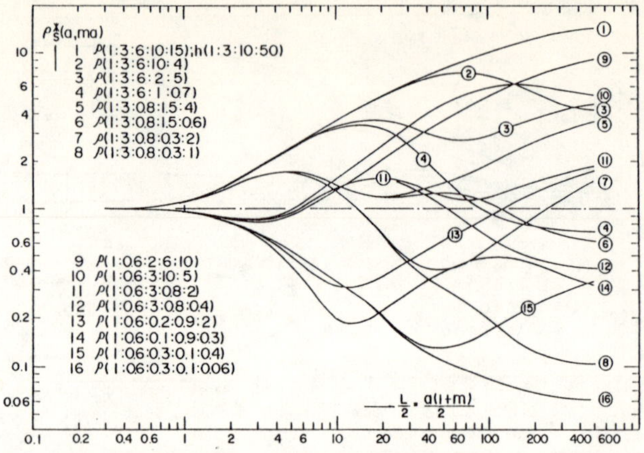

Fig. 4(c). Sixteen types of curve groups for the five-layer model for the Schlumberger array

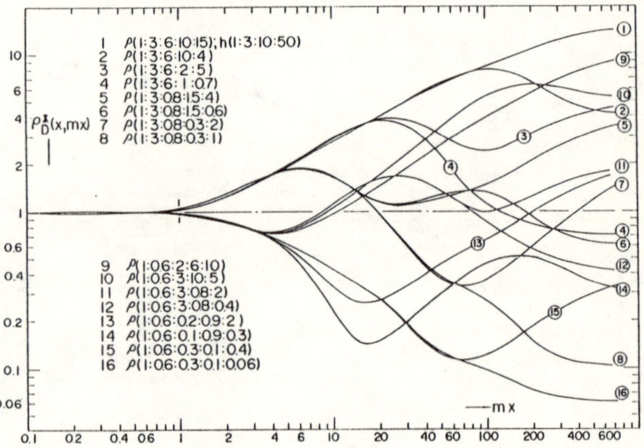

Fig. 4(d). Sixteen types of curve groups for the five-layer model for the Dipole-Dipole array

layer leflects clearly upon each relative resistivity sounding curve. For convenience, such a curve is called the type curve of the relative resistivity sounding curves, of which the interpretation is very easy.

EFFECT OF DEPTHS

If each of depths for a given model is close with each other, then it is impossible to classify the type curve for a given RS curve. This makes the interpretation of RS curve still worse, particularly, in the case of containing a thin layer.

CONCLUSION

The type curves for the relative resistivity sounding for a two-, three-, four-, and five-layer models were presented for the two electrode, Wenner, Schlumberger, and Dipole-Dipole arrangements.

These type curves furnish an elementary knowledge of the interpretation of resistivity sounding curves to us.

REFERENCES

Onodera, S., 1961, A contribution to the theory of a multiple-layer resistivity problem, BUTSURI-TANKO (GEOPHYSICAL EXPLORATION), vol. 14, no. 4, 165-175.

Onodera, S., 1963, An evaluation of the relative resistivity for a multiple layered earth, GEOPHYSICAL PAPERS DEDICATED TO PROFESSOR KENZO SASSA, 393-402.

Orellana, E. and H.M. Mooney, 1966, Master tables and curves for vertical electrical sounding over layered structure, Interciencia.

Roman, I., 1934, Some interpretations of earth resistivity data, Am. Inst. Mining & Metallurgical Engrs., 110, 183-197.

Roman, I., 1960, Apparent resistivity of a single uniform overburden, GEOLOGICAL SURVEY PROFESSIONAL PAPER 365.

Tagg, G.F., 1934, Interpretation of resistivity measurements, Trans. A.I.M.E., 110,Geophys. Prospecting, 135-147.

ROCK FAILURE DURING MASSIVE HYDRAULIC STIMULATION OF THE BACA LOCATION GEOTHERMAL RESERVOIR

Chris Pearson, Hans Keppler,* James Albright, and Robert Potter

Earth and Space Sciences Division, Los Alamos National Laboratory, Los Alamos, New Mexico 87545
*Visiting Scientist from Institute for Geosciences and Natural Resources (BGR), Hannover, Germany

ABSTRACT

The analyses of microearthquake signals occurring during hydraulic stimulation provide an estimate of the size and location of the fractures thus produced. We report studies of microearthquakes occurring during two large ($>10^3 m^3$) hydraulic stimulations of the hydrothermal reservoir at the Baca Location in the Jemez Mountains of Northeastern New Mexico. Both stimulations consisted of water, viscosity enhancer, and proppant. The microearthquake event rate was low but variable throughout most of the treatment. Rock failure as indicated by the distribution of the microearthquakes' foci appeared restricted to a nearly vertical NE striking zone. This orientation is in good agreement with the local earth stresses inferred from geological considerations. The second stimulation which occurred in a neighboring well was similar to the first except for a larger injected volume. The lateral extent of the detected fracture system was 600 m in both stimulations.

INTRODUCTION

Determining the orientation and size of hydraulic fractures produced by hydraulic stimulation of geothermal wells remains an important unsolved problem in geothermal engineering. This paper describes a technique for using the location of very small (magnitude -5 to -2) microearthquakes that often accompany hydraulic stimulation experiments in geothermal wells. Since these events, which are induced by a localized zone of high pore pressure, occur in a narrow band about the hydraulic fracture, their locations can be used to infer the fracture size and orientation.

Microearthquake observation techniques were originally developed to locate hydraulic fractures in low permeability, low porosity crystalline rocks for Los Alamos National Laboratory Hot Dry Rock Geothermal Project (Albright and Hanold, 1976). However, in this paper, we describe experiments that successfully used microearthquake signals to locate hydraulic fractures produced during stimulations of hydrothermal wells on the Union Geothermal Baca lease in NE New Mexico.

PHYSICAL SETTING

The Baca Geothermal lease is located in the Jemez Mountains, a well known Miocene to Quaternary volcanic center in Northwestern New Mexico. Volcanic activity here culminated in the formation of the Valles Caldera 1.1 Myr ago (Smith and Bailey, 1980). The two wells (Baca No. 23 and No. 20), which were stimulated during this program are located in Redondo Creek valley, a prominent graben structure slightly northwest of Mt. Redondo, the resurgent dome.

The graben structure is filled with 1200-1800 m of tuff which is underlain by 300-600 m of andesite. All of the tuff has been hydrothermally altered resulting in quite low matrix permeabilities (<1 md). The upper part of the tuff is highly silicified forming a nearly impermeable caprock for the reservoir. Figure 1 shows a generalized stratigraphy. Because of the low permeability production is from fractures principally in the lower 300 m of the tuff. Formation temperatures in this zone can be as high as 260°C (Dondanville, 1978).

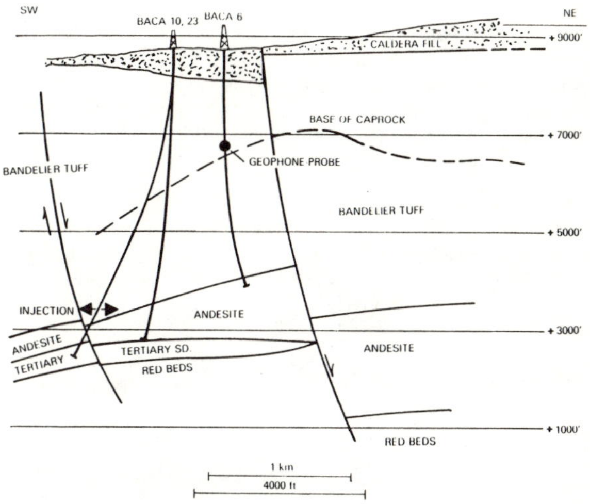

Figure 1. Lithology near Union's Baca Well No. 23. Well No. 6, our observation well and Well No. 10 in which vertical seismic profile measurements were made.

When locating these events we assumed that the P wave velocity was 5.3 km/s. This value came from a preliminary analysis of a vertical seismic profile, measured in cooperation with Union Geothermal Co. in well No. 10, which is within a few meters of well No. 23. During this survey 10 explosive charges were fired in 10 m boreholes located approximately 100 m from the well to avoid generating tube waves.

Our analysis of this survey suggests two layers, the upper one having a velocity of 3.9 km/s (12,900 ft/s) and the lower one having a velocity of 5.3 km/s (17,300 ft/s). The interface between the two layers is at 610 m (2000 ft) depth in well No. 10. As shown in Fig. 1 this depth correlates with the base of the caprock in well No. 10 so the 5.3 km/s layer represents the saturated tuff and the 3.9 km/s layer represents the dry caprock.

INSTRUMENTATION

The microearthquakes were recorded using a removable three-component geophone package (Albright and Pearson, 1982) which we lowered down a shut-in geothermal well to a depth of 305 m during the first experiment and 1000 m during the second. This downhole system allows us to record events within a few hundred meters of the source and avoid the excessive attenuation in the overburden which would be expected, if detection by surface instruments were attempted. Good coupling between the rock and the geophone package is ensured by extending the retractable arm, which locks the package in the well.

EVENTS

An example of the microearthquake signals detected during each stimulation is shown in Fig. 2. The records usually show two clear arrivals, the first of which is the compressional (P) wave and the second is the shear (S) wave. These characteristics allowed us to distinguish between microseismic signals and electronic noise. In practice, surface pumping noise was not a problem because the geophone was located more than 300 m below the surface in a shut-in well.

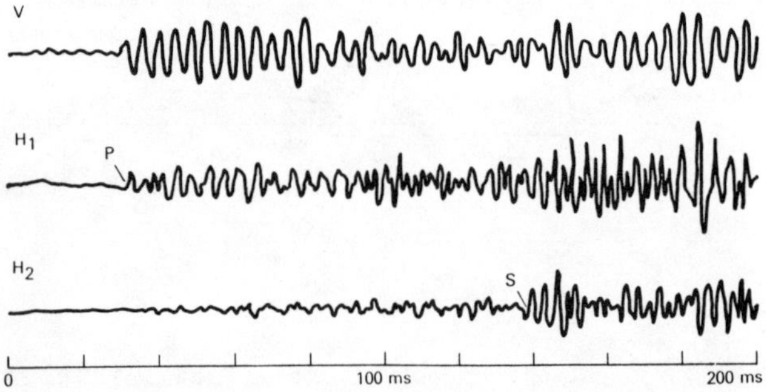

Figure 2. Microearthquake signal from the Baca No. 23 stimulation.

These events have magnitudes ranging between -5 and -3, approximately four orders of magnitude smaller in energy than earthquakes which would be noticed or felt on the surface. The magnitudes were estimated using the signal duration or coda length method described by Real and Teng (1973). In calculating the magnitude we used an empirical relation between magnitude and duration developed by Newton et al. (1976) for the Jemez region.

The direction and distance to microearthquake foci were determined using P-wave onset polarization and the S- and P-wave arrival time difference, respectively (Albright and Pearson, 1982).

Because the orientation of microearthquake radiation patterns is not known _a priori_, there are two possible locations for the origin of the microearthquake. By design, during the hydraulic fracturing experiments the geophone tool was stationed substantially above the stimulated reservoir volume. Thus, one can assume that the event locations can be properly placed below the geophone station.

The first measurements were conducted during a hydraulic stimulation of Baca Well No. 23. 1300 m^3 water, viscosity enhancer, and proppant were injected at up to 0.20 m^3/s (76 bpm). Microearthquake signals were detected within 18 min after the start of pumping. The event rate was low but variable throughout most of the treatment. Bursts of activity, which included the largest events detected occurred both after injecting 120 m^3 of untreated water and after having pumped 140 m^3 of the proppant schedule. No change in event rate was observed on injection of the viscosity-enhancing gel. Figure 3 shows the rate of energy release during this experiment. The first cluster of events, labeled A in Figure 3, represent a natural earthquake which occurred 1.5 hr before the hydraulic stimulation. Fortunately this swarm died out before the beginning of pumping.

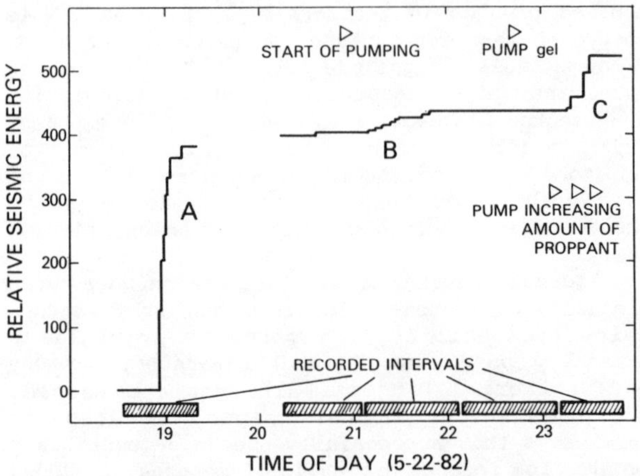

Figure 3. Seismic energy release as a function of time for the Baca No. 23 stimulation. Focal energies are plotted on a scale normalized to the smallest detectable event of $M_L \sim -5$.

The second stimulation, which occurred in Baca Well No. 20, a well 2 km up Redondo Creek from Baca Well No. 23, was similar to the first except for a larger injected volume of both water and proppant as well as a higher injection rate. During this experiment most of the activity occurred within one hour after the start of pumping. There was no burst of activity similar to C in Figure 3 at the start of the proppant schedule.

EVENT LOCATIONS

During the first experiment rock failure appeared restricted to a nearly vertical NE striking zone approximately 600 m in length. This orientation is roughly perpendicular to the least principal horizontal earth stress as inferred from geological findings and regional earthquake fault-plane solutions (Aldrich and Laughlin, 1982).

Most events clustered in a narrow band 100 m high near the injection point which may represent the main fracture opened during this experiment. However, three events occurred as much as 400 m higher, suggesting hydraulic communication over that great a distance in the reservoir. Figure 4 shows the event locations both in plan view and projected to a plane close to the probable fractured zone. The injection zone in the stimulated well falls slightly off the fracture plane in the plan view projection but this is probably caused by small errors in determining the orientation of the geophone package in the wellbore.

During the second stimulation (see Figure 5) the resulting fracture was somewhat larger, nearly 400 m high and almost 600 m long. An equipment failure prevented us from determining the orientation of our geophone package in the wellbore so we were not able to measure the strike of this fracture.

During both experiments the seismic zones were approximately 150 m thick. Some but not all of this thickness is due to errors in our location technique (Pearson, 1981). Albright and Pearson (1982) show that the events are caused by shear

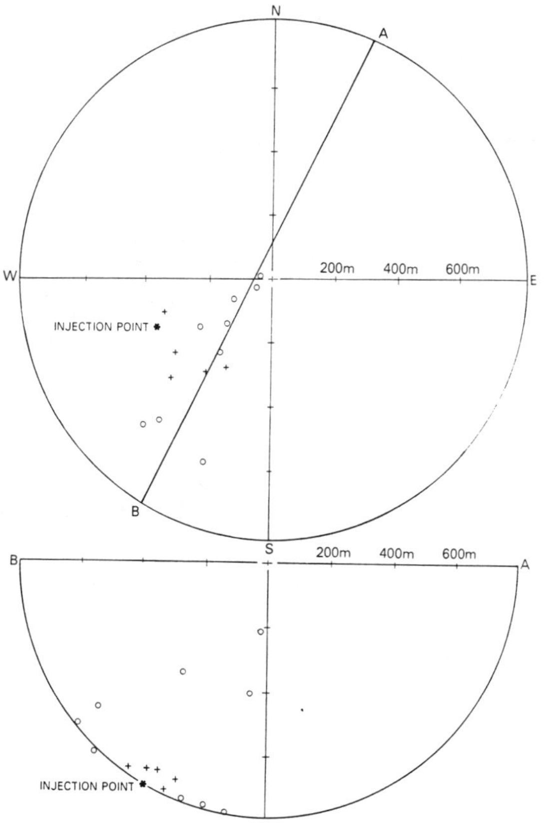

Figure 4. Location of microearthquakes that occurred during the stimulation of Baca Well No. 23 — plan view (upper) and projection to the vertical AB plane. Crosses indicate positive first motions, circles indicate negative first motions. The coordinate origin is the geophone location and the injection point is indicated by an asterisk.

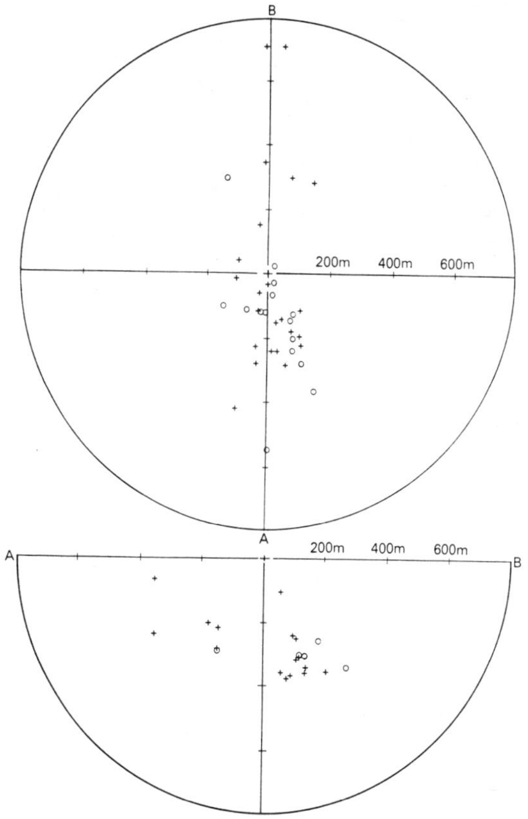

Figure 5. Location of microearthquakes that occurred during the stimulation of Baca Well No. 20 — plan view (upper) and projection to the vertical AB plane. Crosses indicate positive first motions, circles indicate negative first motions. The coordinate origin is the geophone location.

failure on secondary joints and fractures that intersect the main hydraulic fracture. Failure thus occurs in a potentially wide zone where leak-off has locally increased the pore water pressure.

CONCLUSION

Both stimulation experiments turned unproductive wells into producers (Verity and Morris, 1981). Large fractures or fracture zones with dimensions on the order of several hundred meters opened up during the hydraulic stimulation. Since part of the fractured rock may have low hydraulic conductivity, the effective area accessible during production may be smaller.

The results of these measurements suggest that microearthquake observation methods are a potentially useful technique for evaluating hydraulic stimulations of wells in other geothermal fields. However for this technique to work, reservoir rock must transmit both the P- and S-waves without excessive attenuation. In many geothermal reservoirs, particularly if unconsolidated sand or partially saturated rock is present, this may not be possible (Nur et al., 1980).

ACKNOWLEDGMENTS

The authors gratefully acknowledge Gerhard Suhr of Parkla Seismos, William Cumming of Union Geothermal, Bob Hanold, and the Earth Science Instrumentation Group of Los Alamos National Laboratory for their help during the experiments described herein. The authors also thank Union Geothermal Co. for permission to publish some of this data. This work was supported by the U.S. Department of Energy.

REFERENCES

Albright, J. N. and Hanold, R. J., 1976, Seismic mapping of hydraulic fractures made in basement rocks, Second Ann. ERDA Enhanced Oil and Gas Recovery Symposium, 9-10 Sept. 1976, Tulsa, OK.

Albright, J. N. and Pearson, C. F., 1982, Acoustic emissions as a tool for hydraulic fracture location: the experience at the Fenton Hill Hot Dry Rock Site, Soc. Petrol. Eng. J. (in press).

Aldrich, M. J. and Laughlin, A. W., 1982, Orientation of least-principal horizontal stress: Arizona, New Mexico, and the Trans-Pecos Area of West Texas, Los Alamos National Laboratory Map LA-9158-MAP.

Dondanville, R. F., 1978, Geologic characteristics of the Valles Caldera geothermal system, New Mexico Geotherm. Res. Council Transactions 2.

Newton, C., Cash, D., Olsen, K., and Homuth, E., 1976, Los Alamos Scientific Laboratory Seismic programs in the vicinity of Los Alamos, New Mexico, LASL Report, LA-6406-MS.

Nur, A. M., Walls, J. D., Winkler, K., DeWilbiss, J., 1980, Effect of fluid saturation on waves in porous rock and relations to hydraulic permeability, Soc. Petrol. Eng. J., 20, 450-458.

Pearson, C. F., 1981, The relationship between microseismicity and high pore pressures during hydraulic stimulation experiments in low permeability granitic rocks, J. Geophys. Res. 86, 7855-7864.

Real, C. D. and Teng, T. L., 1973, Local Richter magnitude and total signal duration in southern California, Bulletin of the Seismological Society of America, 63, 1809-1827.

Smith, R. L. and Bailey, R. A., 1968, Resurgent cauldrons _in_ R. R. Coats, R. L. Hay, and Anderson (eds), Studies in Volcanology Geol. Soc. Am. Mem. 116, 613-662.

Verity, R. V. and Morris, C. W., 1981, Fracture stimulation of Union's Baca Well 23, Geotherm. Res. Coun. Transactions 5, 315-318.

Geothermal Resources Council, TRANSACTIONS Vol. 6, October 1982

DEVELOPMENT OF DOWNHOLE GEOTHERMAL HEAT FLUX AND THERMAL CONDUCTIVITY TRANSDUCERS

H. F. Poppendiek
D. J. Connelly
A. J. Sellers

Geoscience Ltd
Solana Beach, California 92075

ABSTRACT

Two new downhole geothermal transducers were developed for the Department of Energy* for the purpose of measuring the geothermal heat flux and thermal conductivity in exploration holes without the necessity of taking core samples. One transducer system was based on the principle of quasi steady state two dimensional heat conduction; the transducers consisted of cylindrical rods having thermal conductivities different from the surrounding earth. A second cylindrical transducer system based on transient conduction generated a step function heat flux that was transferred into the earth. Both systems required that the annular space between the transducer and the hole wall be filled with water or a drilling fluid.

This paper describes the new measurement techniques, the field tests and the results.

INTRODUCTION

The common method of determining the geothermal heat flux in exploration holes consists of measuring the thermal conductivity of core or drill chip samples in the laboratory and using in-place vertical temperature profile measurements to obtain the product. The conductivity measurement is generally made in a "divided bar apparatus," which involves the comparison of the rock sample with a specimen whose thermal conductivity is known. This method has some disadvantages. One is the difficulty in acquiring samples for measurement. Further, if drill chips are used, the reconstituted samples may be different from the original rock. The water content of the rock samples may also change during the collection and transportation processes as a result of evaporation, which may affect the measured thermal conductivity.(1) There are more accurate methods of measuring the thermal conductivity of rocks in the laboratory than the divided bar method; (2) however, most of the disadvantages noted still apply.

There are a number of different methods being used by the geothermal community to assess geothermal reserves. Included are the geochemical, magnetic, electrical resistivity, microseismic, acoustic, infrared imagery, and the heat flux methods.(3) Staff members of exploration companies and the U. S. Geological Survey working with thermal techniques feel that heat flux is a logical and appropriate index of the strength of a geothermal reserve. It is felt that the in-situ measurement of heat flux and thermal conductivity using the two transducers described here has several advantages.

In the following sections, descriptions of the new transducers, field test results and interpretations of the results are presented.

DESCRIPTION OF THE MEASUREMENT TECHNIQUES

A. Rod Heat Flux Transducers

Consider the idealized rod heat flux transducer system shown in Figure 1. A steady state heat transfer analysis of this boundary value problem was made which relates the pertinent system parameters (4) (see the Appendix for the analytical closed form functions). The solution contains two unknowns, namely, the thermal conductivity of the earth and the vertical earth temperature gradient. These two unknowns are evaluated by making steady state thermopile voltage measurements with two rod transducers of different but known thermal conductivities. The solution also accounts for the effect of a fluid annulus between the transducer and the borehole wall.

In addition to the closed form solution, a two dimensional cylindrical finite difference solution was performed for a range of system parameters.

The equations for the temperature distribution along the rod heat flux transducers are functions only of the thermal conductivity of the earth, the temperature gradient, and the physical characteristics of the transducers. The equation for one transducer can be divided by the equation for the other. The resulting ratio is then only a function of the earth thermal conductivity because the unknown earth temperature gradient cancels out.

Each rod transducer consisted of a sealed tube (one having a thick aluminum wall and the other a thin stainless steel wall). Both had long internal thermopiles with packing. The output leads

*Contract EG-77-C-03-1318

connected with a two conductor cable that also deployed the sensors. The typical output signals from these transducers range from 0.1 mv to 0.5 mv.

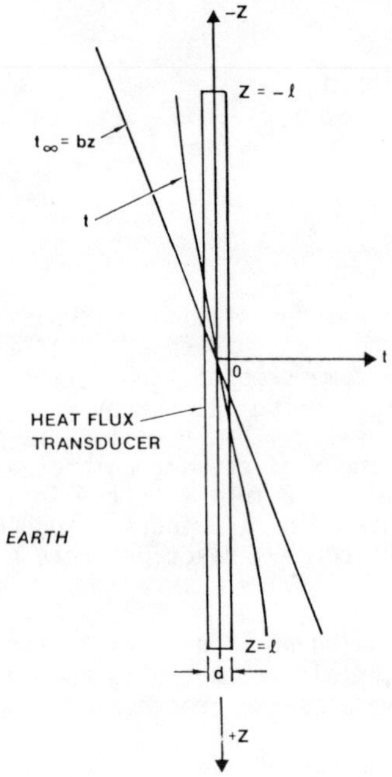

Figure 1. Idealized thin rod transducer system.

B. Thermal Conductivity Probe

Consider the idealized cylindrical thermal conductivity probe system shown in Figure 2. Mathematical heat transfer analyses were used to relate the earth thermal conductivities (and the thermal capacities per unit volume) to the time-temperature measurements, the electrical surface heat addition and the system geometry. Also included was the effect of a fluid annulus between the probe and the borehole wall. The analytical functions are given in the Appendix. (5) (6)

The thermal conductivity probe consisted of a long cylindrical section containing a surface heater over part of its length and a thermopile that measured the temperature rise of the heated surface relative to the unheated surface. Typically, the heater was operated at 0.64 amps and 64 volts. This power level was maintained for a period of about one hour. From a time-temperature recording, while power is applied to the transducer heater, the thermal conductivity and heat capacity per unit volume values can be extracted. At short time periods after the beginning of the constant heating process, the thermal capacity per unit volume controls; for longer time periods after heating initiation, the thermal conductivity controls.

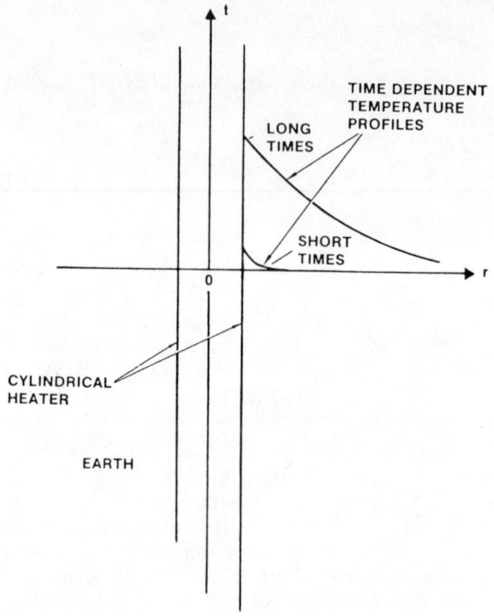

Figure 2. Idealized thermal conductivity probe system.

TEST SITE DESCRIPTION

The well used by Geoscience for these tests was located near Middletown, California. It had been drilled by the Phillips Petroleum Company as a temperature observation well in June of 1977. The drilling log shows that clay and volcanics were the types of earth encountered to a depth of about 122 meters. Deeper than 122 meters, the rock consisted of clay and greenstone to about 305 meters and then principally gabbro to the bottom of the hole. It was almost completely filled with water. The well contained a steel liner surrounded by a thin cement annulus to a depth of about 91 meters; the hole was uncased from 91 meters to the bottom. Both transducer types had an outside diameter of 0.152 meters and a length of about 2.75 meters. The inside diameter of the steel liner of the well was 0.165 meters and the hole diameter of the uncased region was about 0.159 meters.

Geothermal measurements were made at three primary depths. Efforts were made to precondition the two sensor types prior to hole insertion so that equilibration times (to quasi steady state) were not excessive.

RESULTS

A. Rod Heat Flux Transducer

The rod heat flux transducer data were reduced using the equations given in the Appendix of this paper. The thermal conductivities and geothermal heat fluxes at the three primary depths that resulted are given in Table I.

TABLE I

Rod Heat Flux Transducer Results

r_o = 0.91 meter

Depth m	k_∞ W/m·K	q/A (HFU) μ-cal/sec cm^2
52	1.1	2.6
76	1.2	2.3
110	5.1	2.4

B. Thermal Conductivity Probe

The data derived from the thermal conductivity probe were evaluated using the methods given in the Appendix. The data obtained from the thermal conductivity probe were corrected for the small temperature drop across the stainless steel wall of the probe. The results are shown in Table II. The thermal capacity per unit volume ($\rho_\infty c_{p_\infty}$) is shown in Table II, as it is also a parameter in the analysis.

TABLE II

Thermal Conductivity Probe Results

Depth m	$\rho_\infty c_{p_\infty}$ J/m^3·K	k_∞ W/m·K	q/A (HFU) μ-cal/sec cm^2
52	2.0×10^6	1.2	2.6
76	2.3×10^6	1.2	2.3
110	1.1×10^6	3.6	1.9

SOME INTERPRETATIONS

The literature thermal conductivity range for the type of rock described by Phillips Petroleum for its well ranged from approximately $0.5 < k_\infty < 4$ W/m·K; (7) the measured values presented in this document fall within the range of $1 < k_\infty < 5$ W/m·K.

Both sets of data presented indicate that the heat flux did not change significantly for the three depths (as would be expected) yet the thermal conductivity and the thermal capacity per unit volume did change significantly at the 110 meter level. These changes are believed to be related to the different lithology that occurred at that level (as pointed out by Phillips Petroleum staff). Some rock cuttings were obtained from Phillips Petroleum for the 110 meter level and thermal conductivity, specific heat and density measurements were made in the laboratory; the thermal conductivity was found to be approximately 3.3 W/m·K and the specific heat was found to be 750 J/Kg·K. The thermal properties of a reconstituted rock test sample would not be the same as the in-situ rock; thus, the agreement between laboratory and field thermal conductivities and thermal capacities per unit volume found in this study is believed to be reasonable.

The comparison of thermal conductivity and heat flux results as measured by the two different transducers (Table I and Table II) is encouraging at this stage of the transducer development. Both transducers tested can play a role in downhole thermal conductivity and heat flux measurement efforts in future geothermal exploration evaluations.

NOMENCLATURE

- A , cross sectional area of the rod
- a , thickness of the thin slab
- b , undisturbed vertical temperature gradient in the earth
- c_{p_∞} , specific heat of the earth
- D_1 , thermal diffusivity of the thin slab
- D_2 , thermal diffusivity of the semi-infinite solid
- k , thermal conductivity of the rod
- k_1 , thermal conductivity of the thin slab
- k_2 , thermal conductivity of the semi-infinite solid
- k_∞ , thermal conductivity of the infinite solid; thermal conductivity of the earth
- ℓ , half length of the rod
- P , perimeter of the rod
- q , the constant heat release per unit length
- $\left(\frac{q}{A}\right)_o$, constant heat flux addition at x = 0
- r , radial distance from the cylinder centerline
- r_i , radius of the rod
- r_o , radius at which the undisturbed linear temperature field in solid exists; radius of the cylinder
- R_e , equivalent end thermal resistance of the solid surrounding the rod
- R_∞ , equivalent radial thermal resistance of the solid surrounding the rod
- t , temperature; rod temperature (above the rod midpoint temperature datum); temperature at distance r
- t_∞ , the linear lateral temperature variation (above the rod midpoint temperature datum) in the solid at a radial distance sufficiently great so that the presence of the rod does not effect it
- x , distance into the semi-infinite solid measured from the outer surface of the thin slab

z , distance along rod from rod midpoint

θ , time

ρ_∞ , density of the earth

APPENDIX

Recapitulation of the Mathematical Models for the Heat Flux and Thermal Conductivity Transducers

A. Rod Heat Flux Transducers

A heat balance was made on the transducer shown in Figure 1, accounting for vertical conduction along the rod and heat loss to and from its cylindrical surface and its ends, yielding an ordinary second order differential equation. The vertical temperature profile of the undisturbed earth at some distance from the rod is a linear function. Solution of this boundary value problem in terms of the temperature distribution along the rod was (4)

$$t = c_1 \left(e^{-\sqrt{C}z} - e^{\sqrt{C}z} \right) + \frac{B}{C} z$$

where

$$C = \frac{P}{R_\infty kA} \quad \text{and} \quad B = \frac{Pb}{R_\infty kA}$$

$$c_1 = \frac{-b}{-\sqrt{C}\left(e^{-\sqrt{C}\ell} + e^{\sqrt{C}\ell}\right) + \frac{1}{kR_e}\left(e^{-\sqrt{C}\ell} - e^{\sqrt{C}\ell}\right)}$$

$$R_\infty = \frac{r_i}{k_\infty} \ln \frac{r_o}{r_i} \quad \text{and} \quad R_e = \frac{r_i^2\left(\frac{1}{r_i} - \frac{1}{r_o}\right)}{k_\infty}$$

When a transducer is used in a test hole, an annulus of liquid surrounds the transducer. In order to account in the equations for the presence of this layer, the thermal resistance of the fluid annulus, R_{an}, is added to the cylindrical earth resistance to obtain an increased value for R (i.e., $R' = R + R_{an}$). Similarly, the effect of fluid at the end of the transducer is also included, resulting in a modified R'_e (end heat conduction into a two region system).

B. Thermal Conductivity Probe

The transient heat transfer performance of the thermal conductivity probe can be described by the classical step function surface heating boundary condition for a long cylinder (see Figure 2), namely, (5)

$$\frac{4\pi k_\infty}{2.30q} t = \int_{-\infty}^{\log \frac{k_\infty}{\rho_\infty c_{p_\infty}}\theta} e^{-\frac{(r_o^2 + r^2)}{4\frac{k_\infty}{\rho_\infty c_{p_\infty}}\theta}} J_o\left(\frac{i r_o r}{2\frac{k_\infty}{\rho_\infty c_{p_\infty}}\theta}\right) d\left(\log \frac{k_\infty}{\rho_\infty c_{p_\infty}}\theta\right)$$

The addition of a fluid annulus between the thermal conductivity probe and the borehole changes the conduction system into a two region problem. This boundary value problem was first solved by considering the annulus to be a thin slab located adjacent to a semi-infinite solid (the surrounding earth). The transient temperature solution for the annulus region is: (5)

$$t = \frac{\left(\frac{q}{A}\right)_0 \sqrt{D_1 D_2}}{\sqrt{D_1}k_2 + \sqrt{D_2}k_1} \sum_0^n \lambda^n \left[2\sqrt{\frac{\theta}{\pi}}\, e^{-\frac{(2an+x)^2}{4D_1\theta}} - \frac{(2an+x)}{\sqrt{D_1}} \operatorname{erfc}\left(\frac{2an+x}{2\sqrt{D_1\theta}}\right) \right.$$

$$\left. + 2\sqrt{\frac{\theta}{\pi}}\, e^{-\frac{(2an+2a-x)^2}{4D_1\theta}} - \frac{(2an+2a-x)}{\sqrt{D_1}} \operatorname{erfc}\left(\frac{2an+2a-x}{2\sqrt{D_1\theta}}\right) \right]$$

$$+ \frac{\left(\frac{q}{A}\right)_0 k_2}{\sqrt{D_1}k_2 + \sqrt{D_2}k_1} \left(\frac{D_1}{k_1}\right) \sum_0^n \lambda^n \left[2\sqrt{\frac{\theta}{\pi}}\, e^{-\frac{(2an+x)^2}{4D_1\theta}} - \frac{(2an+x)}{\sqrt{D_1}} \operatorname{erfc}\left(\frac{2an+x}{2\sqrt{D_1\theta}}\right) \right.$$

$$\left. - 2\sqrt{\frac{\theta}{\pi}}\, e^{-\frac{(2an+2a-x)^2}{4D_1\theta}} + \frac{(2an+2a-x)}{\sqrt{D_1}} \operatorname{erfc}\left(\frac{2an+2a-x}{2\sqrt{D_1\theta}}\right) \right]$$

$$\text{where} \quad \lambda = \frac{\sqrt{D_2}k_1 - \sqrt{D_1}k_2}{\sqrt{D_2}k_1 + \sqrt{D_1}k_2} \quad \text{and } n = 0, 1, 2 \ldots$$

From the evaluation of this two-slab system it was shown that the temperature difference across the annulus is small in comparison to the transducer surface temperature rise above the initial temperature datum for times greater than the annulus time constant (which is of the order of 0.05 hours for a 0.5 cm thick water layer). Therefore, in the cylindrical coordinate system consisting of the fluid annulus and the surrounding semi-infinite earth, the annulus time constants and temperature drops will likewise be small compared to transducer values. If the water annulus thickness is no more than 15 percent of the probe radius, then one can neglect the annulus effect (for times greater than the annulus time constant).

ACKNOWLEDGMENTS

Support of the following people is acknowledged: A. F. Veneruso, J. A. Coquat, G. W. Crosby, S. D. Johnson, C. M. Sabin, A. C. Reynard, E. M. Boughton, and E. W. Fowler.

REFERENCES

1 Gross, R., and Combs, J., 1975, "Thermal Conductivity Measurement and Prediction from Geophysical Well Log Parameters with Borehole Application," in Proc. Sec. U N Sym. on Devel. and Use of Geoth. Res., Vol 2, p 1019.

2 Kappelmeyer, O., and Haenel, R., 1974, Geothermics, with Special Ref. to Application, Gebruder Borntraeger.

3 Proceedings, Sec. U N Sym. on Devel. and Use of Geoth. Res., 1975, U. S. Gov't. Printing Office.

4 Poppendiek, H. F., and Connelly, D. J., 1977, "A Thin Rod Heat Flux Transducer Positioned in the Earth Having A Uniform Temperature Gradient," GLR-178.

5 Poppendiek, H. F., and Connelly, D. J., 1977, "Cylindrical and Spherical Transient Heat Transfer Solutions and Their Application to Thermal Conductivity Measurements in the Earth," GLR-185.

6 Poppendiek, H. F., and Connelly, D. J., 1977, "The Effects of a Fluid Annulus on the Performance of Thermal Transducers in a Borehole," GLR-189.

7 Jumikis, A. F., Thermal Soil Mechanics, 1966, Rutgers University Press.

Geothermal Resources Council, TRANSACTIONS Vol. 6, October 1982

THEORETICAL NUCLEAR LOG RESPONSES OF THE COMPONENTS OF THE GEYSERS GEOTHERMAL RESERVOIR, CALIFORNIA

S. K. SANYAL, M. CHE, RICHARD E. DUNLAP*
and MARY K. TWICHELL*

GeothermEx, Inc., Berkeley, California
*Aminoil USA, Santa Rosa, California

ABSTRACT

This paper presents theoretically calculated matrix response parameters for nuclear logs for the common minerals encountered at The Geysers geothermal field, California. From fundamental equations, matrix responses were derived for 15 minerals for neutron, density and thermal neutron capture cross-section logs. Most of these matrix parameters are not available in the literature and cannot be readily obtained by other means. These parameters are invaluable in analysis of geothermal well logs from The Geysers.

INTRODUCTION

A knowledge of the matrix properties of various minerals and rocks is essential in well log analysis. Such data are available from the logging manuals only for a limited number of lithologies. Data are insufficient or nonexistent for most of the minerals encountered in geothermal reservoirs.

There are essentially four ways of estimating matrix properties: (1) Calculation from theoretical considerations; (2) laboratory measurement; (3) derivation from log response in unfractured, "tight" zones; (4) calculation from multi-log suites.

Laboratory-derived data reported in the literature are usually obtained under ambient conditions, whereas reservoir conditions are different: the pressure and temperature are higher and the stress condition is not the same. The mastrix parameters are often reported as a range of values, the exact value for a specific reservoir being difficult to know without site-specific data. For most minerals at The Geysers, matrix parameters are not available in the literature. Matrix properties can sometimes be derived from logs through "tight" zones. If a well section is known to be nonporous ("tight"), its log response is the same as the matrix response for the lithology around the well bore. Unfortunately, it is often impossible to know independently whether a zone is tight or not. When a multi-log suite is available, one can calculate the matrix responses by simultaneous solution of the response equations. However, usually there are too few equations (limited to the number of logs run) compared to the number of unknown matrix responses.

Matrix properties can be estimated by theoretical means also. Such an estimate is usually an approximation because tool response cannot usually be rigorously taken into account. For example, Edmundson and Raymer (1979) presented theoretical equations for the calculation of matrix properties for several logging tools. Such calculated responses can be valuable guides in geothermal well log analysis.

CALCULATION OF MATRIX RESPONSES

Neutron Log

Neutron logs are used in geothermal wells for geologic correlation, for evaluation of porosity and lithology, and for detection of rock alteration. The modern neutron devices virtually eliminate undesirable borehole effects, and some have increased lateral investigation. In neutron logging, a high-energy neutron source such as plutonium-beryllium, radium-beryllium or americium-beryllium is attached to the lower end of the downhole tool to bombard the formations with neutrons. After emission from the source, the neutrons are slowed to thermal velocities (thermalized) mainly by hydrogen atoms. Then they may be captured by other atoms, which then emit gamma rays of capture. The gamma rays of capture and/or thermal neutron or epithermal neutrons are detected by one or two detectors and telemetered to the surface as electrical pulses. The number of these pulses is logged as an indication of the hydrogen density in the formation surrounding the borehole.

A compensated neutron tool has two detectors. The long-detector spacing provides greater depths of investigation outside the borehole, and the short-detector spacing provides for investigation of the formation near the source. The neutron log is available as the compensated or as the sidewall variety. The neutron log response is expressed either in API neutron units or in percent porosity (with reference to either limestone or sandstone) based on calibration in API test pit. If there is no hydrogen in the rock matrix, the higher the true porosity, the lower the neutron response in API units or higher the neutron response in apparent neutron porosity units.

The hydrogen index (HI) of a material is defined as:

$$HI = \frac{\text{Hydrogen Density of the material (gms/cc)}}{\text{Hydrogen Density of Water at STP (gms/cc)}} \quad (1)$$

or,

$$HI = 9 \text{ (Hydrogen Density of the material in gms/cc)} \quad (2)$$

The neutron response of a material is a function of HI of the material and the tool design. For a rock of porosity ϕ,

$$HI = \phi(HI)_f + (1-\phi)\ (HI)_{ma} \quad (3)$$

where $(HI)_f$ = HI of the fluid in the porespaces and
$(HI)_{ma}$ = HI of the rock matrix.

If there is no hydrogen in the rock matrix, $(HI)_{ma} = 0$, and if the pores are filled with water, $(HI)_f = 1$. Under these conditions, equation (3) reduces to

$$HI = \phi \quad (4)$$

and the porosity of the formation can be calculated from HI.

In reality, however, neutron response is not completely controlled by HI only; the formation exhibits a lithology effect which is related to the thermalizing ability of atoms other than hydrogen. Minerals should thermalize neutrons according to their molecular thermalizing coefficient (K_T). It can be shown that:

$$K_T = \sum_{i=1}^{n} \frac{1}{A_i} \quad (5)$$

where A_i is the atomic weight of the ith atom in a molecule of the mineral composed of n atoms. In order to estimate the relative thermalizing effect of various minerals on neutron log response, one can define a "thermalizing index" (TI) as:

$$TI = \frac{\text{(KT} \cdot \text{Molecular Density) of Mineral}}{\text{(KT} \cdot \text{Molecular Density) of Quartz}} \quad (6)$$

$$\text{where Molecular Density} = \frac{N\rho}{M} \quad (7)$$

N = Avogadro's Number
ρ = Bulk density
M = Molecular weight

A quartz standard is chosen here for convenience because at The Geysers we are dealing with graywackes.

Another factor that can affect neutron log response is the thermal neutron capture cross-section (Σ) of the material in the formation. A high capture cross-section implies that a higher proportion of thermal neutrons can be captured. Σ of a mineral can be calculated approximately from theoretical considerations, as shown later.

Density Log

Density logs are used in geothermal wells for correlation, for porosity evaluation and for lithology identification. A source of gamma rays and a gamma ray detector (or two detectors in case of compensated density tools) are mounted in the lower part of the instrument, with radioactivity shielding between them. The density tool is a pad-mounted device kept pressed against the borehole wall during logging. The gamma rays from the source radiate into the formation where they are scattered and deflected by collisions with electrons in the formation. Some of the scattered gamma rays are deflected toward the detector. The number of gamma rays that reach the detector is proportional to the number of electrons encountered in the formation, and the number of electrons per unit volume of the formation (electron density) is proportional to the bulk density (ρ) of the formation.

The density log, therefore, consists of the sum of the electrical pulses generated by the detector when a gamma ray passes through it. The output of this log is a measurement of the formation density in grams/cc.

The density log usually provides an apparent bulk density of the formation, which is a simple function of the "electron density index" of the formation material, and is approximately equal to the true bulk density for most rocks. The "electron density index" (ρ_e) of a chemical compound is given by:

$$\rho_e = \frac{(2\Sigma Z)\rho}{M} \quad (8)$$

where ΣZ = sum of the atomic numbers of the atoms comprising the molecule, and
M = molecular weight.

For most rocks $2\ \Sigma\ Z/M \simeq 1$ and hence $\rho_e = \rho$. For minerals with $2\ \Sigma\ Z/M$ substantially different from unity, apparent bulk density is different from true bulk density. The bulk density a formation of porosity ϕ is given by:

$$\rho = \phi\ \rho_f + (1-\phi)\ \rho_{ma} \quad (9)$$

where ρ_f = bulk density of the fluid, and
ρ_{ma} = bulk density of the matrix.

Thus knowing ρ_f and ρ_{ma} one can calculate ϕ from measured bulk density using equation (9). If a mineral has a $2\ \Sigma\ Z/M$ substantially different from 1, the apparent formation density read from density log needs correction.

In a compensated density log, the response is presented either as the counts per second from each of the two detectors, or as corrected density in gms/cc. The calculation of density (ρ) in gms/cc from the long-spacing detector reading (LS) and the short-spaced detector reading (SS) is achieved from a functional relation of the form:

$$\rho = \rho_0 - \alpha \sqrt{\left(\log \frac{LS}{LS_0}\right)^2 + \left(\log \frac{SS}{SS_0}\right)^2} \qquad (10)$$

where LS_0 and SS_0 respectively are the long-spacing and short-spacing detector reading for a calibration standard of known density ρ_0, and α is a known parameter. Equation (10) assumes no borehole effect. If there is significant borehole effect, ρ is calculated from LS and SS by using an empirical plot.

Thermal Neutron Capture Cross-Section Log

This log has not been used much in the geothermal industry, presumably because of the temperature limitation of the present tool. But this log has potential of becoming a useful geothermal well log, particularly in correlation, detection of minerals containing high capture cross-section atoms (such as B or Li) and detection of porous and permeable zones. This log measures the cross-section for thermal neutron capture in the formation around the borehole.

The capture cross-section (Σ) thermal neutron shown in Table 2 is calculated theoretically from:

$$\Sigma = \frac{602.5\rho}{M} \sigma_M \qquad (11)$$

where σ_M is the molecular capture cross-section of the mineral (in Barns). Here,

$$\sigma_M = \sum_{i=1}^{n} A_i \sigma_i \qquad (12)$$

where A_i is the atomic weight and σ_i is the capture cross-section of the ith mineral (i = 1, 2, ... n).

Gamma Ray Log

All rocks contain some radioactive material, primarily potassium 40, the uranium series and the thorium series. Among sediments, shales have the highest concentrations of radioactive material, while sands and carbonates usually have the lowest. Among igneous rocks, usually the more acidic the rock, the higher the radioactive material content. For metamorphic rocks, radioactive material content depends on the composition of the parent rock and the process of metamorphism the rock has undergone. However, hydrothermal alteration often changes the radioactive material content of geothermal reservoir rocks.

When gamma rays emitted by the formations penetrate suitable detectors in boreholes, an electrical pulse is generated and sent to a surface recorder. The pulses are counted systematically and expressed on the logs in analog form, in terms of gamma ray intensity on a standard quantitative scale devised by the American Petroleum Institute (API). The gamma ray tool is empirically calibrated in API units based on the API test pits in Houston. In geothermal wells, the log is used for correlation of subsurface formations, and often for identification of lithology or diagnosis of rock alteration. The gamma ray log response of a formation is a function of the amounts of minute quantities of radiactive minerals in the formation. It is difficult to assign a typical gamma ray response to any particular lithologic component, except potassium-bearing minerals, which are characterized by very high gamma ray intensity. The only potassium bearing minerals present at The Geysers are orthoclase and sometimes biotite. These two have high gamma ray intersities; the response for other minerals cannot be calculated theoretically.

Table 1 lists some of the matrix parameters of the most frequent minerals and rocks encountered at The Geysers. The parameters in table 1 were calculated from the chemical formulas and true bulk densities of the minerals (obtained from the literature) using basic principles of physics and equations discussed.

Table 1 also includes fluid parameters for water and steam. If the influence of thermal neutron capture cross-section on neutron log response is neglected, the apparent neutron porosity (ϕ_N) of a mineral or rock should be proportional to its thermalizing index (TI). Figure 1 shows a plot of the approximate ϕ_N versus TI for various minerals listed in table 1. The straight line in figure 1 was obtained by joining the points for quartz (TI = 1, ϕ_N = 0) and water (TI = 15.92, ϕ_N = 1). Figure 2 is a plot of approximate ϕ_N versus apparent bulk density (ρ_a) of the various minerals considered at The Geysers. The ϕ_N values were obtained from figure 1 and ρ_a values from table 1.

Table 1 and figure 2 can be used as a guide in analyzing geothermal well logs from The Geysers. Some minerals stand out in table 1 and figure 2. For example, tourmaline and datolite have extreme neutron capture cross-sections. Pyrite and pyrrhotite have high densities and large capture cross-sections. Clay minerals, such as kaolinite and chlorite show high neutron porosities. To analyze response for other minerals, crossplot analysis pattern recognition can be used.

TABLE 1. Calculated Log Responses

Mineral/Fluid	Formula	Apparent Bulk Density (gms/cc)	Hydrogen Index	Thermal-izing Index	Capture Cross-sections (c.u.)
Water (at STP)	H_2O	1.0	1.0	15.92	22.1
Steam (of Density s)	H_2O	s	s	15.92	
Quartz	SiO_2	2.644	0	1.00	4.4
Pyrite	FeS_2	4.845	0	0.47	91.2
Pyrrhotite	Fe_7S_8-FeS	4.405	0	0.37	90.5
Tourmaline (Dravite)	$NaMg_3Al_6B_3Si_6O_{27}(OH)_4$	3.005	0.114	3.01	4338.2
Epidote	$CaFeAl_2Si_3O_{12}(OH)$ $CaCO_3$	3.40	0.070	2.21	19.9
Calcite	$CaCO_3$	2.710	0	1.13	7.1
Axinite	$(Ca,Mn,Fe)_3Al_2BO_3[Si_4O_{12}](OH)$	3.25	0.051	1.88	2647.4
Datolite	$CaB[SiO_4](OH)$	2.98	0.168	3.81	8527.0
Actinolite	$Ca_2Fe_5[Si_8O_{22}](OH)_2$	3.337	0.063	1.91	32.6
Albite	$NaAlSi_3O_8$	2.595	0	0.97	7.7
Chlorite	$(Mg,Al,Fe)_{12}[(Si_1Al)_8O_{20}](OH)_{16}$	2.93	0.342	6.27	24.3
Sphene	$CaTiSiO_5$	3.447	0	1.00	75.9
Anorthite	$CaAl_2Si_{12}O_8$	2.734	0	0.93	7.5
Orthoclase	$KAISi_3O_8$	2.55	0	0.87	16.0
Biotite	$H_2K(Mg_1Fe)_3Al(SiO_4)_3$	2.87	0.116	1.37	24.0

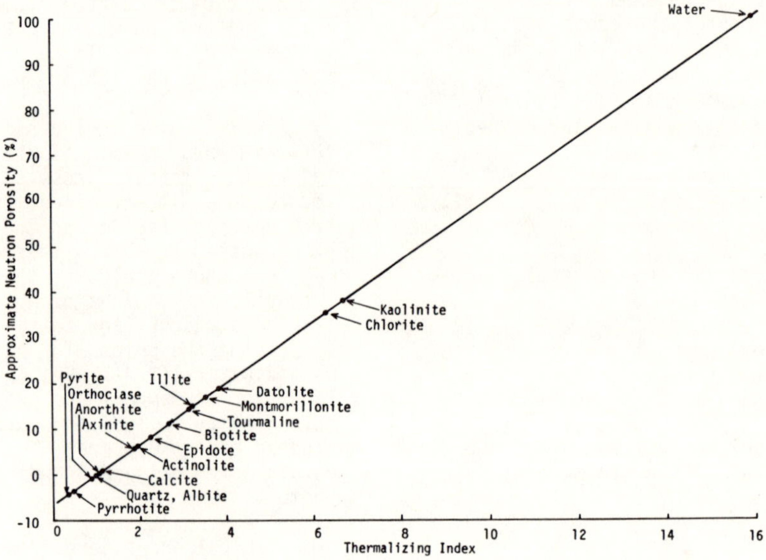

FIGURE 1. Approximate Neutron Porosity vs. Thermalizing Index (assuming no influence of capture cross-section).

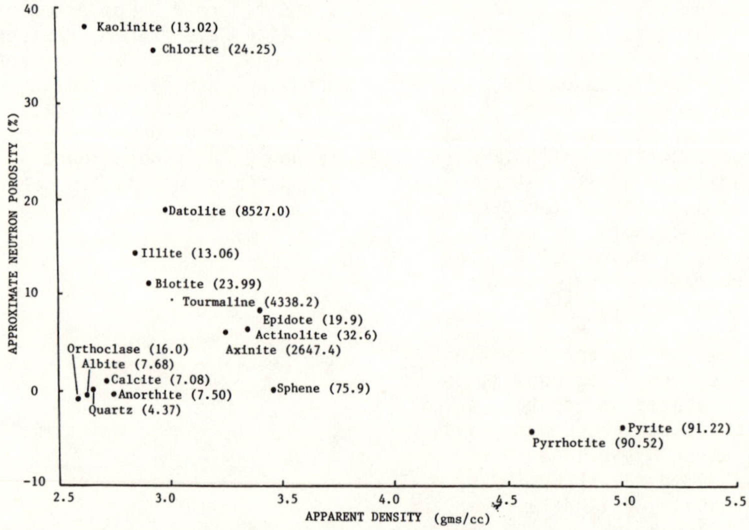

FIGURE 2. Calculated ϕ_N vs. ρ_b for Relevant Minerals (Thermal Neutron Capture Cross-section in capture units in parentheses).

Geothermal Resources Council, TRANSACTIONS Vol. 6, October 1982

THE CIRCULAR VERTICAL SOUNDING METHOD APPLIED TO THE EXPLORATION OF THE DHAMAR-RADA'A (Y.A.R.) GEOTHERMAL AREA

G.STANGALINO+, F.AUMENTO+, ABDELLATIF AL MASRI++, TOWFIK NOAMAN++

\+ ELC-Electroconsult, Milano, Italy
++ YOMINCO, Sana'a, Yemen Arab Republic

ABSTRACT

We illustrate the use of a new geophysical prospecting method in the detailed investigations of a geothermal field. The method, called " Circular Vertical Soundings" (C.V.S.) is useful in determining the predominant orientation of deep fracturing of rocks and of their schistosity. In geothermal studies it is particularly useful to know the direction or directions of the predominant basement fracture systems since these could be the conduits responsible for the flow of hot fluids. The method is based on the Quadripole Schlumberger Technique, and in practice one carries out 4 to 6 vertical electrical soundings (V.E.S.) with the same centre but different azimuth (say every 30-45°), thereby permitting one to measure the relationship between apparent resistivity and azimuth; results are plotted on polar diagrams. The validity of the method is demonstrated through its application to the Dhamar-Rada'a (Y.A.R.) geothermal area, where four C.V.S. were carried out together with 83 conventional V.E.S.

INTRODUCTION

Geoelectrical resistivity exploration has proved to be one of the most effective methods in geothermal exploration, being a method which measures directly parameters which can be affected by the presence of geothermal fluids. In geothermal studies it is particularly important to know if there exist fractures which affect the basement; these fractures need not necessarily have surface expressions; recent volcanism, for example, often covers the fault/fracture system which was responsible for feeding its eruptive centres, and later for the circulation of geothermal fluids. There exists the possibility of detecting these fractures and their directions through geoelectrical exploration. The technique relies on the fact that rocks with well developed and preferentially oriented fissures also exhibit electrical anisotropy. The existence of this anisotropy can be explained as follows:in all media, be they isotropic or anisotropic, electrical energy radiates outwards from a point source; the current density in the direction of least resistance (i.e. major conductance) will be greater than in a direction perpendicular. The increase in current density in a particular direction will cause a corresponding decrease in the potential difference between receptor electrodes when placed in the ground in a direction parallel to this direction; placing the electrodes in any other direction will cause them to detect greater potential differences. Therefore, if one determines the apparent resistivity on the ground over such anisotropic rocks with one's equipment oriented perpendicular to the fractures, one obtains higher values than with the equipment parallel to the fractures. This phenomenon is known as the "anisotropy index". The diversity of values of apparent resistivities can be used to determine the predominant direction of fracturing, of schistosity, and in igneous rocks, even of magma flow or crystal layering. In practice one carries out a minimum of four standard vertical soundings with the same centre but different azimuth, permitting one to measure the relationship between apparent resistivity and azimuth. Typical results are shown in Figure 4. The major axis of the ellipse so described represents the direction of major conductivity, which will correspond to that of the predominant fracturing, fissuring or schistosity of the rock. Ellipses can be drawn for the apparent resistivities at different AB/2 dipole separations, giving anisotropies at different apparent depths.

As part of the preliminary investigations of the Dhamar-Rada'a geothermal area of the Yemen Arab Republic, ELC-Electroconsult of Milan, Italy, and YOMINCO, Yemen Oil and Minerals Resources Corporation of Sana'a, Y.A.R., undertook a geolectrical exploration program using both conventional vertical soundings (V.E.S.) and the circular vertical soundings (C.V.S.) described here with the purpose of: a) Locating low resistivity zones which might be related to geothermal activity; b) Collecting structural information including the detection of predominant fracture systems, through the recognition of lateral variations and discontinuities in resistivity; c) Outlining the stratigraphy through the interpretation of the different electrostratigraphic units encountered.

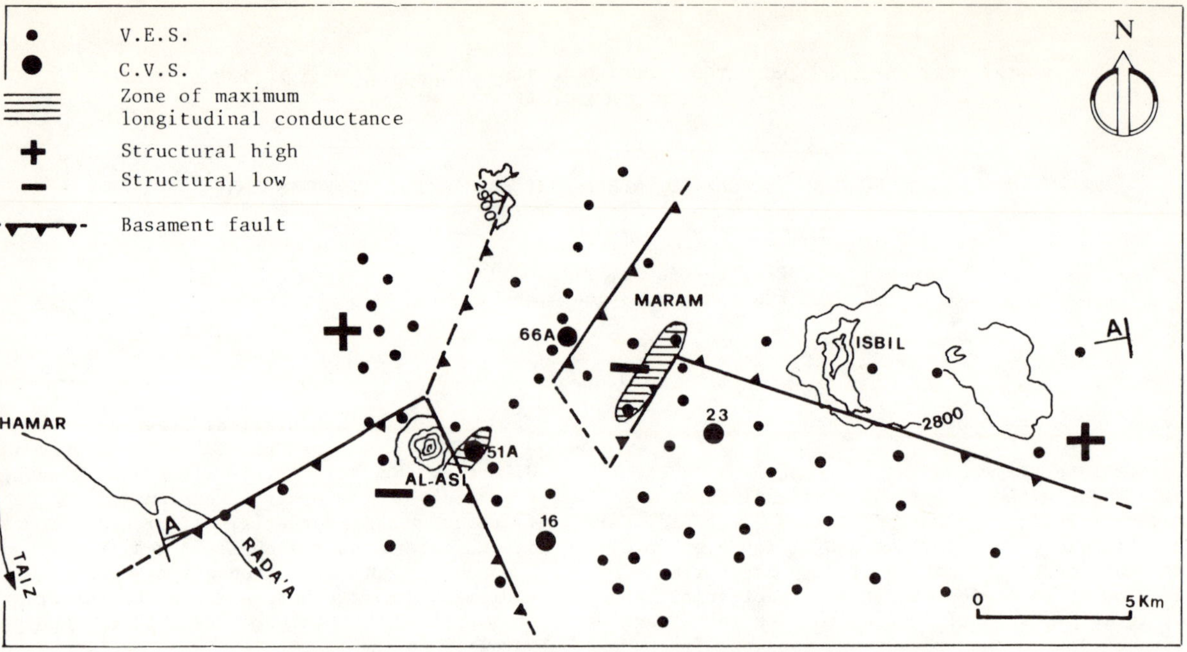

Figure 1. The Dhamar-Rada'a Geothermal Area, showing V.E.S. and C.V.S. locations.

METHODOLOGY

In the Dhamar-Rada'a area 83 standard quadripole Schlumberger V.E.S. with AB electrodic separations of 4,000-6,000 m were accomplished, located as per the map of Figure 1. Four circular soundings, the results of which are represented as "polar diagrams of the apparent resistivity", were made only after the exploratory phase has been completed and preliminary interpretations had been made.

RESULTS

Electrostratigraphis Sections (Quantitative Interpretations) Quantitative interpretations attempt to determine the spatial distribution of the true resistivities below ground, using the apparent resistivity data. The procedure is to compare the field -obtained curves with synthetic curves until a match is obtained. When this occurs, one has one correct solution for the resistivity/thickness combinations giving the curve measured. This solution is by no means unique, especially in complex layered cases. The quantitative interpretation of all V.E.S. is summarized in electrostratigraphic sections, in which the results of similar resistivity are correlated. These have revealed four distinct geoelectric units which are clearly visible, for example, in Figure 2. Those zones with resistivity below 12.5 ohm.m have been highlighted, since they may be of geothermal interest.

Map of the Roof of the Basal Resistive Layer
The contours of the elevation of the roof the basal resistive layer define both the structure of said unit and that of the overlying conductive layer (see Figure 1). The faults shown indicate the presence of strong lateral discontinuities and vertical displacements of the resistive layer. Soundings close to these faults are always affected by them, such that the exact position of the faults cannot be determined exactly. This problem was later resolved using the results of the circular vertical soundings, so that it is believed that the structural reconstruction presented is a reasonable representation of the situation. The map reveals two horsts outlined by important faults, to the NW of the Al Asi volcano and E of the Isbil volcano respectively. Between these two volcanoes there is an enormous graben, inside which there are two minor structures, with associated depressions well outlined by faults, the first being on the Al Asi volcano and the second around the village of Maram.

Longitudinal Conductivity Map
An electric sounding curve can be interpreted different ways when, as in the case here, a conductive layer is sandwiched between two resistive layers. In this case the real value of resistivity of the conductive layer cannot be measured exactly, but can only be inferred. However, the relationship h/ϱ, where h is the thickness and ϱ the actual resistivity of the conductive layer, can be measured objectively from the curve of a sounding.

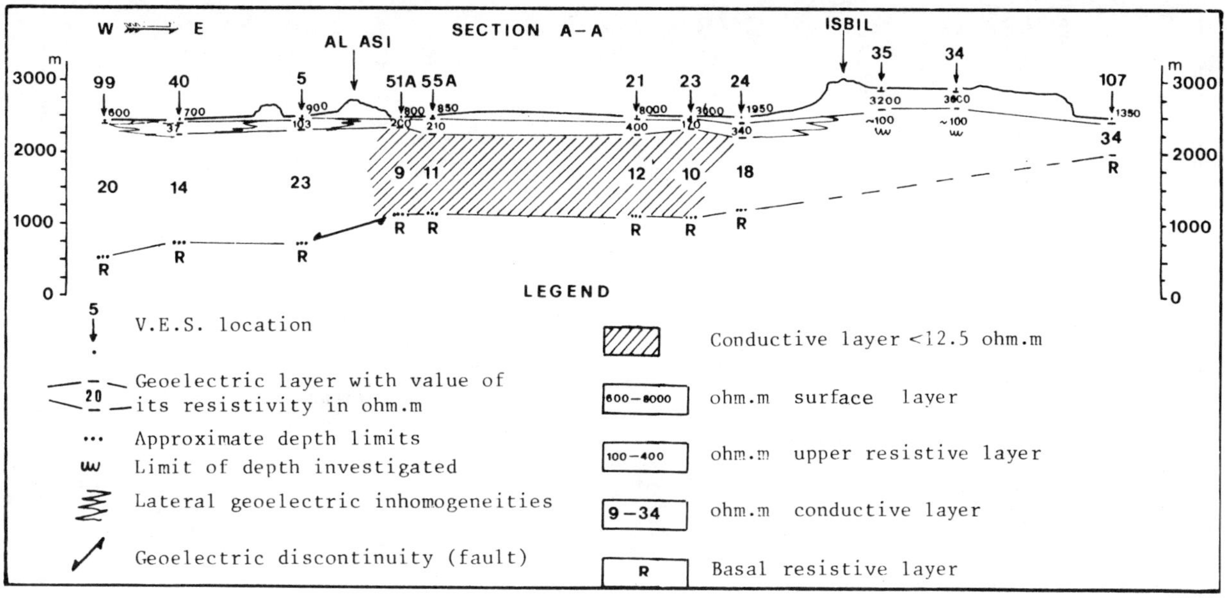

Figure 2. Electrostratigraphic Section A - A

This relationship is called the "Longitudinal Conductivity"; representing a quantity which will increase in size with both thickness and conductivity of a given layer, from a geothermal point of view may indicate more favourable zones. Furthermore, subjective evaluations do not enter into the measurement of h/ρ . The zones of maximum h/ρ are shown Figure 1. Comparing these locations with the "roof" structure of the basal resistive layer, one notices that the conductivity maxima coincide with the structural lows near the Al Asi volcano and the village of Maram.

Circular Vertical Soundings

Results of the four C.V.S. carried out are plotted in Figure 4: only values for AB/2 of 1,500 and 2,000 m are given, these separations ensuring that one is looking at the upper crystalline basement and the overlying conductive layer, which geothermally speaking are of more interest to us. Two C.V.S., Nos 16 and 51A were positioned to the SE of the Al Asi volcano, No. 66A to the SW of the village of Maram, and 23 to the SW of the Isbil volcano, all being areas of structural and/or geochemical interest. C.V.S. Nos 16, 51A and 66A show strong anisotropy, with a maximum of conductace oriented NE-SW. This direction coincides with the NE-SW basement block faulting direction visible in the map of the roof of the basal resistive layer; the other visible on the map, at NW-SE, does not appear. Even C.V.S. No. 51A, less than 500 m from a major NW-SE fault, shows no effect of said fault, but still bear the inprint of the predominant NE-SW trend. Surface geology, which shows both trends, would have suggested that the NW-SE was the more important. C.V.S. No. 23, to the SW of the Isbil Volcano, shows a different trend, about E-W. This direction is close to that of a scissor fault detected south of Isbil both in the basement roof map and by surface geology. This area is obviously differently affected tectonically.

DISCUSSION

The geoelectric investigations have permitted the reconstruction, even if only in a simplified way, of the local geological/structural situation especially of the basement. Different deep fracture trends have been detected, namely NE-SW, NW-SE and E-W. All directions are visible on surface geology but have different importance. One of the major

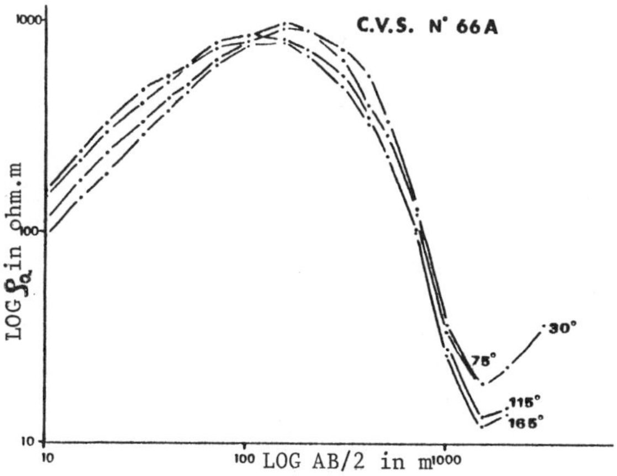

Figure 3. Log-Log C.V.S. curves, showing differences in apparent minimum resistivity at AB/2 of 1,500 and 2,000 m.

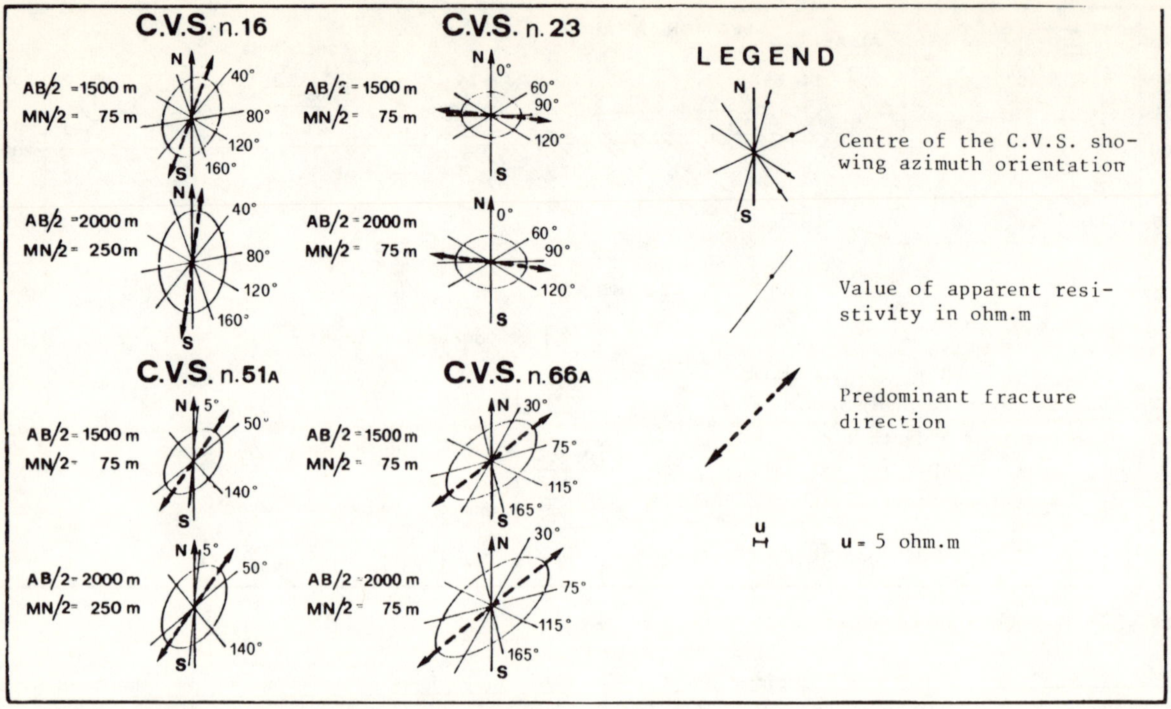

Figure 4.. Polar Diagrams of Apparent Resistivity

surface lineaments, namely NW-SE, also related to basement fracturing, is not apparent in the electric anisotropy diagrams. It is significant that zones affected by the NE-SW anisotropy and fault directions, parallel to a major trend visible in satellite photos, are the zones containing other parameters of geothermal interest (ie. low resistivity, trace element, geothermometer and water chemistry anomalies). The other trend detected E-W, lies in the geothermally barren area south of the Isbil volcano. Therefore, the application of the C.V.S. method permitted to us: a) Identify the predominant trends of those fracture which affect the basement; b) Verify whether the conductive zones identified lie beneath the central point of the soundings or are displaced laterally; c) Rely on the quantitative interpretations of those V.E.S. containing the C.V.S., and to use these interpretations as fixed points from which to interpret and correlate other nearby V.E.S.; with one C.V.S., equivalent to four V.E.S. with the same centre and different azimuths, lateral and surface effects can be identified and controlled. The use of C.V.S. should be considered economically viable since, even given difficult terrain conditions, it is possible to carry out one C.V.S. consisting of 4 V.E.S. with AB/2 of 2,000 m every two working days; the use of the same centre for each V.E.S. facilitates operations considerably.

REFERENCES

ALFANO, L., 1950, Introduction to the interpretations of resistivity measurements for complicated structural conditions. Geoph.Prosp.VIII, p.311-366

ASTIER, J.L., 1971, Geophisique appliquèe à l'hidrogeologie. Masson, Paris, 277 pags

IAKUBOVSKII, IU.V.and Liajov L.L., 1975, Exploracion electrica. Editorial Revertè S.A., Barcelona, 380 pags

ORELLANA, E., 1982, Prospecion geoelectrica en corriente continua. Paranifo Madrid 578 pags.

Geothermal Resources Council, TRANSACTIONS Vol. 6, October 1982

CALCULATED GEOTHERMAL GRADIENT FOR THE WESTERN OUACHITA MOUNTAINS, ARKANSAS

Kenneth F. Steele, George H. Wagner [1]
John B. Sharp [2]

Dept. of Geology, University of AR, Fayetteville, AR 72701 [1]
Texas Oil and Gas, Ft. Smith, AR [2]

ABSTRACT

Based on silica concentration of ground water for a 130 x 140 km area of the Ouachita Mountains, Arkansas a mean reservoir temperature of 44.3°C was calculated utilizing quartz solubility. Additional calculations obtain a mean heat flow value of 46.4 mW/m^2, geothermal gradient of 20.2°C/km and depth of water circulation of 1.4 km. This calculated geothermal gradient (20.2°C/km) compares to other limited values ranging from 18.2 to 45.6°C/km.

INTRODUCTION

Conventional heat flow data for the Ouachita Mountains are very sparse; therefore, it is of interest to calculate heat flow and associated parameters for this area. This can be done by using the following equation of Swanberg and Morgan (1978) which relates the mean regional heat flow and mean silica temperatures based on quartz thermometry.

$$(1) \quad q = (T_{SiO_2} - b)/m$$

where q is heat flow in milliwatts (mW) per square meter (m^2)
T_{SiO_2} is the subsurface temperature in °C calculated from silica concentration of ground water (see equation 2).
b is a constant equal to 13.2°C
m is a constant equal to 0.67°C m^2/mW

Thus, it is possible to calculate a mean regional heat flow, and then geothermal gradient and depth of water circulation for a significant portion of the Ouachita Mountains (Fig. 1) from existing silica data for ground water. The results presented here differ appreciably from preliminary calculations (Sharp et al., 1979), because actual conductivity measurements rather than estimates have been used. Steele and Wagner (1981) and Wagner and Steele (1979) have reported on the anomalous surface and calculated subsurface ground water temperatures for a 130 x 140 km area of the western Ouachita Mountains, Arkansas and their data for 92 springs and 9 wells (Fig. 1) can be utilized to calculate a mean regional heat flow for the area.

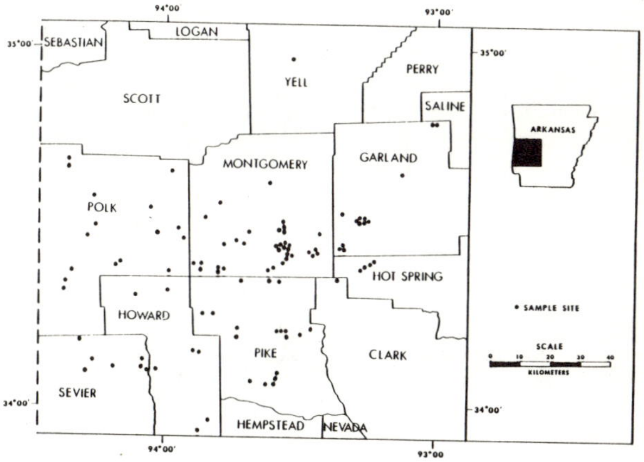

Figure 1. Location of study area and distribution of ground water sites.

GEOLOGY

The western Ouachita Mountains, Arkansas (Fig. 1) are comprised primarily of Paleozoic sedimentary rocks in the northern part and Cretaceous sands and gravels in the south. The northern part of the area is located in the core of the Ouachita Mountains which has been intensely folded and faulted; whereas the southern part is less intensely deformed, yet there are some faults present.

QUARTZ GEOTHERMOMETRY

The relationship between quartz solubility and temperature is given by the equation of Truesdell (Swanberg and Morgan, 1978):

$$(2) \quad T_{SiO_2} = -273.15 + 1315/(5.205 - \log_{10} SiO_2)$$

where T_{SiO_2} is the quartz geotemperature in degrees Celius
SiO_2 is the analytically determined silica concentration of the ground water in parts per million

Several investigators have shown the usefulness of this equation in determining the reservoir temperature of springs (Mahon, 1966; Fournier et al., 1974; Mariner and Willey, 1976).

A histogram of the silica concentrations of the ground water sites in the study area (Fig. 1) is shown in Figure 2. The principle population is in the 5-15 ppm range with a mean of 12.4 ppm which yields a calculated reservoir temperature of 44.3°C from equation 2.

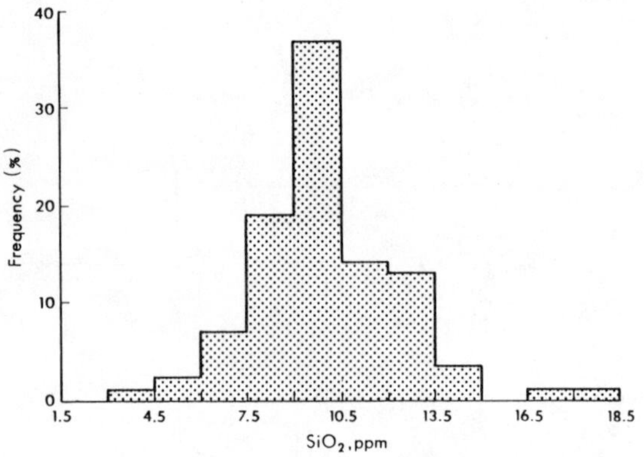

Figure 2. Histogram of silica concentrations for groundwater in study area.

RESULTS

The mean heat flow calculated from equation 1 (using the mean silica geotemperature of 44.3°C) is 46.4 mW/m^2 for the western Ouachita Mountains, Arkansas. The mean geothermal gradient and depth of groundwater circulation can be calculated from equations 3 and 4, respectively.

(3) Geothermal Gradient = Heat Flow (q)/Thermal Conductivity

(4) Depth of Circulation = $(T_{SiO_2} - T_a)$/Geothermal Gradient

where T_a is the mean air temperature

Thermal conductivities averaging 2.3 x 10^3 mW/m/°C have been measured in the center of the study area (Bruce Taylor, written communication, 1980), and have been used to calculate the mean geothermal gradient of 20.2 °C/km from equation 3. The depth of water circulation is then calculated as 1.4 km from equation 4 using T_a = 15.8°C (average data for Ft. Smith and Little Rock, NOAA, 1978).

DISCUSSION

The geothermal gradient (20.2 °C/km) calculated for the western Ouachita Mountains, Arkansas is generally lower than the sparse data available for the area. The Geothermal Map for Arkansas (AAPG, no date) presents only rough extrapolations for the Ouachita Mountains because of the lack of wells drilled in the area. The AAPG average value for the center of the area is 30°C/km.

A well in the center of the area has yielded a measured geothermal gradient of 45.6 °C/km (Bruce Taylor, written communication, 1978). However, it should be noted that this well is located within 24 to 32 km of three warm springs with surface temperatures of over 29°C and thus should be expected to yield a higher geothermal gradient than the entire area. Using the silica content of another well within 2 km as a "regional" value, a geothermal gradient of 43.5 °C/km is obtained.

Geothermal gradient values for the central part of the United States read from a heat flow map constructed from silica geotemperatures (Swanberg and Morgan, 1978) yield values of 18.2 to 27.3 °C/km. Thus the calculated regional geothermal gradient values obtained for this area of the Ouachita Mountains compares reasonably well with the limited data available; howbeit, somewhat lower. The calculated value, 20.2 °C/km, presented here is based on more sites and/or more closely spaced sites than the other available data, and thus is thought to better represent the geothermal conditions of this portion of the Ouachita Mountain Region.

Although the calculated values for this area compare reasonably well with the limited data available, refinement of equation 1 could possibly improve the comparison. Equation 1 is based on quartz solubility and undoubtedly in many regions, such as the Ouachita Mountains, chalcedony solubility relationships (Steele and Wagner, 1981) should be utilized. Also ground water sites with significant supersaturation, e.g. those above the dashed line in Figure 3, should be eliminated as the supersaturation is probably from weathering of silicates (Steele and Wagner, 1981 and Hobba et al., 1979). However, because mean silica concentrations for a region are utilized and equation 1 is an empirical equation, problems of these types are minimized, and calculated heat flow values, such as these for the Ouachita Mountains, offer valuable information which is often otherwise missing. In conclusion the potential for commerical geothermal sites in this area of Arkansas (Fig. 1) is very poor based on the available regional data.

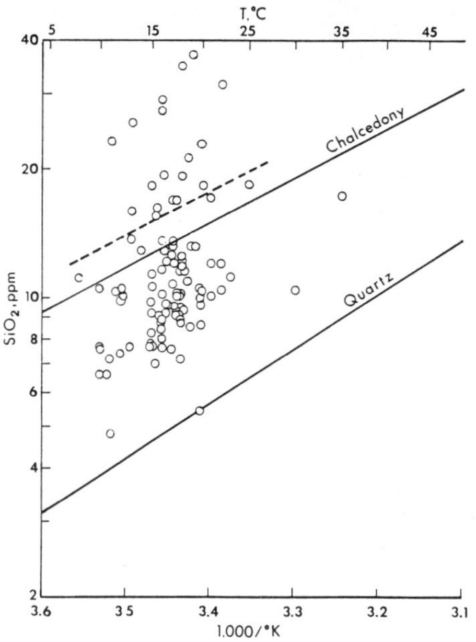

Figure 3. Relation of silica concentration with temperature. Quartz and chalcedony solubility from Bedinger et al., (1979). See text for meaning of dashed line.

ACKNOWLEDGEMENTS

The work upon which this publication is based was supported in part by funds provided by the Office of Water Research and Technology, U.S. Department of the Interior, through the Water Resources Research Center at the University of Arkansas under Project B-055-ARK as authorized by the Water Research and Development Act of 1978.

REFERENCES

Bedinger, M.S. Pearson, F.J. Jr., Reed, J.E., Sniegocki, R.T. and Stone, C.G., 1979, The waters of Hot Springs National Park, Arkansas - their nature and origin: Geol. Sur. Prof. Paper 1044-C, 33 p.

Fournier, R.O., White, D.E. and Truesdell, A.H., 1974, Geochemical indicators of subsurface temperature. I. Basic assumptions, Jour. Res. U.S. Geol. Sur., Vol. 2, p. 259-262.

Hobba, W.A. Jr., Fisher, D.W., Pearson, F.J. Jr., and Chemerys, J.C., 1979, Hydrology and geochemistry of thermal springs of the Appalachians, U.S. Geol. Sur. Prof. Paper 1044E, 36 p.

Mahon, W.A.J., 1966, Silica in hot water discharged from drill waters at Wairakel, New Zealand, New Zealand Jour. Sci., Vol. 9, p. 135-144.

Mariner, R.H. and Willey, L.M., 1976, Geochemistry of thermal waters in Long Valley, Mono County, California, Jour. Geophys. Res., Vol. 81, p. 792-800.

NOAA, (Natural Oceanic and Atmospheric Administration) 1978, Climatological data, Annual Summary, Vol. 29, 37 p., Environmental Data and Information Service, National Climatic Center, Ashville, NC.

Sharp, J.B., Steele, K.F. and Wagner, G.H., 1979, Geothermal reconnaissance survey of the Ouachita Mtns. AR, GSA, Ab., Vol. 11, p. 165.

Steele, K.F. and Wagner, G.H., 1981, Warm Springs of the western Ouachita Mountains, Arkansas: Geothermal Res. Council, Trans., Vol. 5, p. 137-140.

Swanberg, C.A. and Morgan, P., 1978, The linear relation between temperature based on the silica content of groundwater, and regional heat flow: a new heat flow map of the United States: Pure and Applied Geophysics, Vol. 117, p. 227-241.

Wagner, G.H. and Steele, K.F., 1980, Chemistry of the spring waters of the Ouachita Mountains - excluding Hot Springs, Arkansas: Arkansas Water Resources Research Center, Publication Number 69, Univ. of Arkansas, Fayetteville, 130 p.

DOWNWARD CONTINUATION OF TEMPERATURE GRADIENTS AT MACFARLANE'S HOT SPRING, NORTHERN NEVADA

Chandler A. Swanberg*
Roger L. Bowers**

*Teledyne Geotech
Garland, Texas
**Hunt Energy Corporation
Dallas, Texas

ABSTRACT

MacFarlane's Hot Spring is located on the eastern margin of the Black Rock Desert of northwest Nevada. Detailed temperature logs from thirty-eight shallow boreholes (500 feet) and six intermediate depth boreholes (1500-2000 feet) have been used to construct a temperature gradient contour map covering approximately 144 square miles, both within and adjacent to the geothermal area. These temperature gradients were then continued downward through a detailed conductivity model to complete the three-dimensional thermal picture. The principal results are as follows: (1) The maximum measured temperature is 178°F at 2,000 feet, and the maximum projected temperatures at greater depths are not likely to exceed the 250-350°F range. (2) The area of hydrothermal activity is confined to the western front of a structural platform bounded by two roughly parallel normal faults. (3) The anomaly is best explained in terms of a simple groundwater flow model. The groundwater flows west through the structural platform and ascends when it intersects the conduit provided by the fault. The faults on the eastern side of the platform permit recharge to the system.

INTRODUCTION

The MacFarlane's Hot Spring geothermal prospect is located in the basin and range province in Northwest Nevada (Figure 1). It is within the "Battle Mountain High" heat flow province of Sass et al (1981) and is located about sixty miles west of Winnemucca and twenty miles east of the Double Hot Springs KGRA. The geology of the area is typical of that found throughout the Basin and Range and has been described in detail by Sibbett et al (1982). To summarize, the study area lies on an alluvial fan to the west of the Jackson Mountains (Figure 1). These mountains are lithologically characterized by Permian volcanics (the Happy Creek Volcanic series) which are overlain by sedimentary and volcanic rocks of Triassic and Jurassic age. A single, major thrust sheet containing Cretaceous and younger rocks overlies this Mesozoic sequence. Several high-angle, normal faults are present in the area and juxtapose formations of contrasting lighology and age.

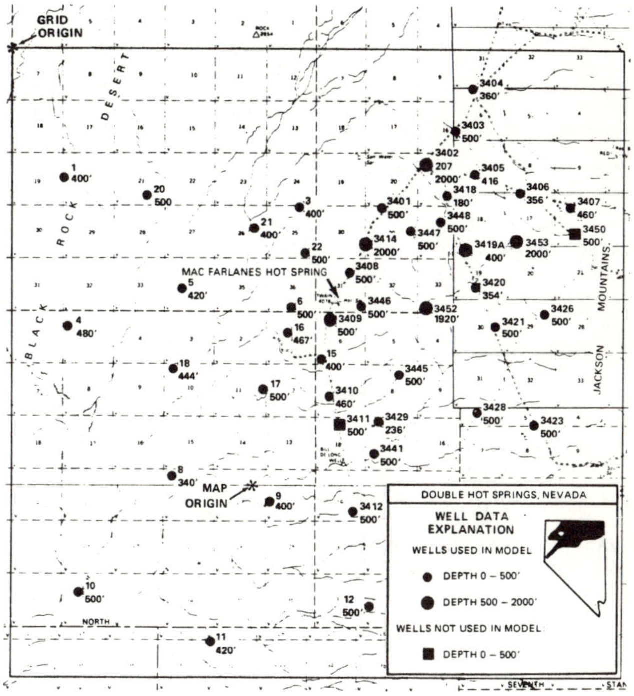

Figure 1. Location of MacFarlane's Hot Spring geothermal prospect, Nevada. Map origin is at 41° N. Lat. and 118° 45' W. Long. The shaded area in the inset is the Battle Mountain high heat flow province.

The fan, which includes most of the study area, is composed of Quaternary alluvial deposits. Two long, normal faults cross the study area in a northeast to southwest direction and define the structural platform mentioned above. Between these two faults, an inselberg of Tertiary basalt is exposed (Figure 2).

DATA PREPARATION

The technique of downward continuation as applied to temperature gradient data has been described by Brott et al (1981). The estimation of subsurface temperatures by this technique requires two basic sets of information. The

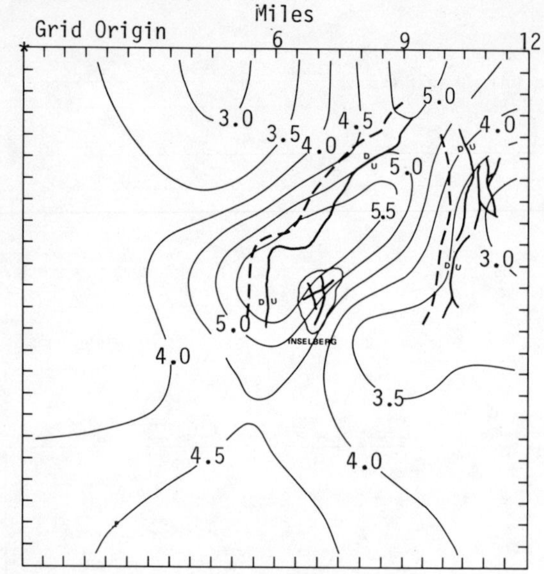

Figure 2. Low-order polynomial fit (trend surface) to temperature gradients (°F/100'). Also shown are surface faults (solid lines) and subsurface extension of these faults (dashed lines).

first is a set of near-surface temperature and thermal gradient data corrected to a common datum and contoured over the region of interest. The second set of information is a three-dimensional thermal conductivity model underlying the prospect area. The former data set is based on temperatures measured in shallow boreholes and, for the present study, corrected for topographic effects. The conductivity model is based on laboratory measurements of well cuttings and information about the lithology and structure of the region.

The locations of the forty-four boreholes used in this analysis are shown in Figure 1. Two shallow boreholes (3411 and 3450) were not included in the analysis because the temperatures were clearly disturbed by groundwater movement and provide no reliable estimate of regional heat flow (see Figure 3). The surface temperatures and gradients were taken as the midpoint of the deepest zone which exhibited clear evidence of conductive heat flow. In addition to the surface temperatures and gradients which were taken from the individual well logs, background values were assigned to represent those areas which were outside the data set yet within the perimeter of the grid model. Proceeding clockwise from the grid location (Figure 1), the assigned corner gradients are 3.7, 4.6, 4.4, and 4.4°F/100', and the assigned temperatures are 58.5, 64.1, 87.3 and 65.4°F.

All of the data from the MacFarlane's Hot Spring prospect have been corrected for topographic effects out to a distance of 1.25 miles or roughly ten times the average hole

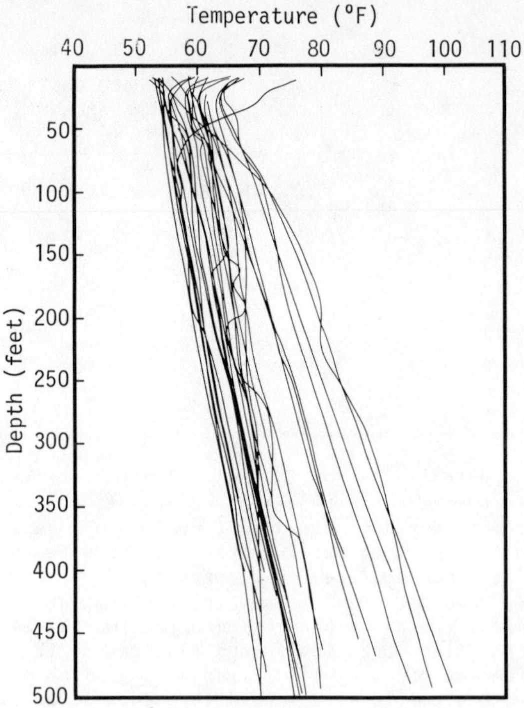

Figure 3a.

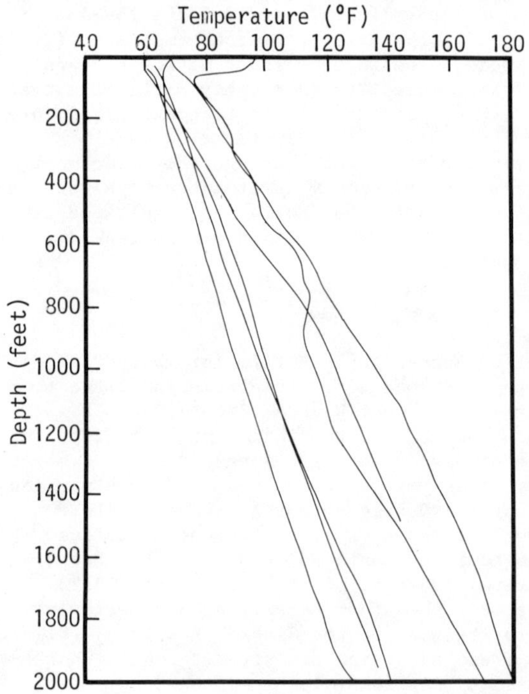

Figure 3b.

Figure 3. Composite diagram showing temperature measured in the boreholes used in the analysis.

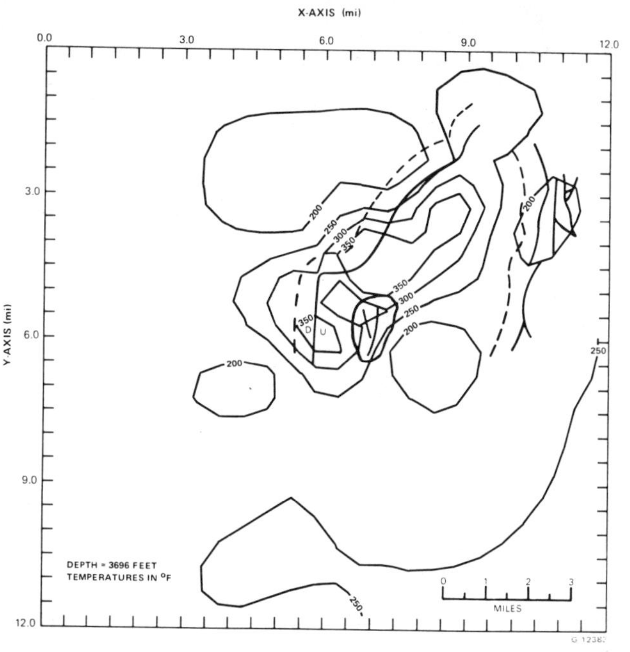

Figure 4. Temperatures continued downward to a depth of 3696 feet, the maximum depth for stable continuation. Also shown are the surface faults (solid lines) and subsurface extension of these faults (dashed lines).

depth. A lapse rate of -0.44°F/100' (-8.00°C/km) was used in the calculations. This lapse rate, which describes the linear decrease in surface temperature with increasing elevation elevation, was obtained by a least-squares approximation to the elevations (well collar) and surface temperatures (extrapolated from the temperature-depth data) of the forty-four boreholes used in this analysis. The topographic effects are small. The mean correction is 0.19 $\pm$ 0.61% with a range of -0.87 to 2.33%. The near-zero average means that the boreholes are evenly distributed throughout the topography.

The conductivity values are generally assigned in such a way as to match the known geology as closely as possbile while still maintaining a relatively simple model. In the present case, however, all of the rock units encountered in the boreholes, including Tertiary basalts as well as Quaternary alluvium, have very nearly the same thermal conductivity (3.75 mcal/cm-sec-°C). This conclusion is based on almost 200 individual conductivity values obtained using standard, divided bar techniques and well cuttings from the boreholes. No attempt, therefore, was made to vary the conductivity in the lateral direction. The conductivity values were varied with depth, however, due to the progressive decrease in porosity resulting from overburden pressure. The conductivity values have been systematically varied from 3.75 mcal/cm-sec-°C at the surface to 5.4 mcal/cm-sec-°C at a depth of 7,000 feet.

RESULTS

Most of the temperature logs available for study show strong evidence of conductive heat transfer (Figure 3) making this area an ideal place for applying the principles of downward continuation. Only two boreholes show evidence of groundwater movement (Figures 1 and 3), and both were deleted from the data set prior to the analysis. Of the undisturbed boreholes, the measured temperature gradients ranged from 2.85 to 10.36°F/100' (52-189°C/km) with an average gradient of 4.7°F/100' (86°C/km). The lowest gradients are found in the recharge areas on the flank of the Jackson Mountains where the average gradient is 2.97°F/100' (54°C/km, including the disturbed borehole 3450), (Figure 1). The maximum gradient of 10.36°F/100' represents borehole 3408 located one-half mile north of MacFarlane's Hot Spring. The average gradient for the non-geothermal portion of the Black Rock Desert on the western side of the study area is 3.97°F/100' (72°C/km). The geothermal area as defined by the 5.5°F/100' contour includes 11 square miles.

Figure 2 shows a low-order polynomial fit to the measured surface temperature gradients (trend surface) and the principal mappable structural features within the study area. The purpose of this figure is to demonstrate the close association between the geothermal anomaly and the regional structural setting. This correlation almost demands a cause-effect relationship. The region of elevated temperature gradients is confined to the western front of the structural platform located between the two northeast trending fault systems, and the maximum gradients are consistently located over the westernmost fault. Another interesting aspect of Figure 2 is that the gradients return to normal at the southern end of the westernmost fault. These observations, coupled with the low gradients observed over the fault system of the eastern margin of the platform, form the basis of our hydrology model to explain the origin of the MacFarlane's Hot Spring geothermal system.

Figure 4 shows the results of the downward continuation analysis. The downward continuation was terminated at 3696 feet (0.70 miles). Below this depth, oscillations started to occur indicating that the source depths had been encountered. As shown in Figure 4, the maximum temperature that is likely to be encountered is 350°F, and a value of 300°F is probably more realistic. The maximum temperatures predicted by chemical geothermometry are near 280°F (Sibbett et al, 1982).

The origin of the MacFarlane's Hot Spring geothermal area is most easily explained in terms of a simple groundwater flow model. This is shown schematically in Figures 5 and 6. The high-elevation region near the eastern fault complex is an area of groundwater recharge as

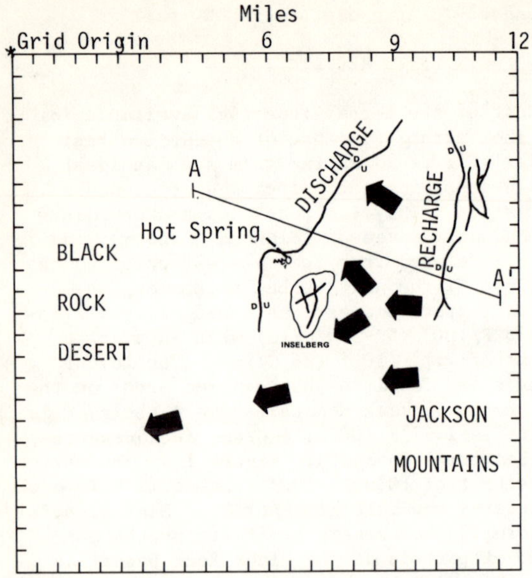

Figure 5. Proposed model for MacFarlane's Hot Spring geothermal system. Arrows refer to direction of groundwater flow.

evidenced by the low surface gradients (Figure 2) and the presence of the isothermal borehole 3450 (Figure 1). Meteoric water enters the hydrologic system via these faults and flows westward through aquifers which are comprised mostly of porphyritic and vesicular basalts (Sibbett et al, 1982). At the inselberg, a topographic high of several hundred feet, the flow is diverted to the northwest and southwest. To the northwest, the flow intersects the conduit provided by the fault and is free to ascend to the surface at MacFarlane's Hot Spring and forms the shallow geothermal system. To the southwest, the fault is not encountered, and the groundwater continues into the Black Rock Desert without forming a geothermal system.

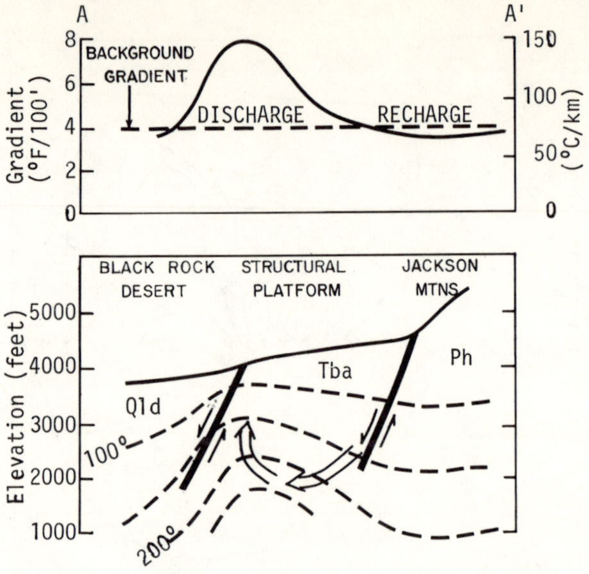

Figure 6. Proposed model for MacFarlane's Hot Spring geothermal system. (Top) Temperature gradients. (Bottom) Isotherms, faults, and assumed groundwater flow paths. Qld, Tba, Ph are respectively Lake Lohantan deposits, vesicular basalt flows, and the Happy Creek group (Sibbett et al, 1982).

ACKNOWLEDGEMENTS

The present study represents temperature gradient evaluation as it is currently practiced by the geothermal industry. We would like to acknowledge the following companies for their respective roles in the data acquisition and analysis. Geothermal drilling, temperature logs, and thermal conductivity analysis were performed by Geothermal Services, Inc., and Microgeophysics Corporation. Geological information was provided by Earth Science Laboratories, University of Utah Research Institute, and Texas Instruments Geophoto Services. Downward continuation analysis was provided by Teledyne Geotech. We especially would like to acknowledge Hunt Energy Corporation of Dallas, Texas, for their overall management of the program and their permission to release the information.

REFERENCES

Brott, C. A., Blackwell, D. D., and Morgan, P., 1981, Continuation of heat flow data: A method to construct isotherms in geothermal areas; Geophysics, Vol. 46, p. 1732-1744.

Sibbett, B. S., Zeisloft, J., and Bowers, R. L., 1982, Geology of MacFarlane's spring thermal area, Nevada; Trans, Geoth. Res. Co., Vol. 6, in press.

Sass, J. H., Blackwell, D. D., Chapman, D. S., Costain, J. K., Decker, E. R., Lawver, L. A., and Swanberg, C. A., 1981, Heat flow from the crust of the United States; in Touloukian, Y.S., Judd, W.R., and Roy, R. F., Eds., Physical Properties of Rocks and Minerals, McGraw-Hill, New York, p. 503-548.

Geothermal Resources Council, TRANSACTIONS Vol. 6, October 1982

WAVE NUMBER FILTERING OF THERMAL DATA FROM THE VALLES CALDERA, NEW MEXICO

Chandler A. Swanberg
Todd M. C. Li

TELEDYNE GEOTECH
Garland, Texas
P.O. Box 28277
Dallas, Texas 75228

ABSTRACT

The Valles Caldera is located in northern New Mexico and is one of the major silicic volcanic centers of the United States. Detailed temperature logs from seventy shallow boreholes (250 feet) and fifteen deep geothermal wells (5000 - 8000 feet) constitute the data base for the present study.

The temperature gradient data have been processed using a newly developed wave number filtering technique with the following results: (1) The technique is very effective in locating the top of the geothermal reservoir where the temperature gradient begins to decrease. This cannot be accomplished using traditional techniques of processing thermal data such as direct extrapolation or downward continuation. (2) The technique predicts realistic temperatures to a depth of at least 4000 feet (1300 m). Since the data base generally represents only the upper 250 feet (75 m) of section, this represents a large extrapolation in a very thermally complex regime. (3) The data base appears to contain a component of heat representing the transient effect of a cooling magma body. The wavelength of this anomaly may be too long to be accurately resolved with the available data base.

INTRODUCTION

In a previous paper (Li et al, 1982), a technique was outlined for filtering hot spring noise from shallow temperature gradient data. The technique makes use of the depth information contained in the wave number spectrum. This technique has been applied to a two-dimensional temperature gradient profile at the Baca location #1 KGRA located in northern New Mexico (Figure 1). This KGRA contains the Valles Caldera, which is one of the major silicic volcanic centers in the United States. The geothermal system associated with the Valles Caldera can be characterized as follows: (1) It is a high-temperature system (400-600°F: 200-300°C) with most of the geothermal wells reaching 400°F (200°C) by 2000 feet (600 m). (2) It is an areally large system with the 11°F/100 ft (200°C/km) temperature gradient contour enclosing sixty square miles (150 km^2). (3) The similarity between the temperature gradient contours and the position of the caldera's main ring fracture system implies the presence of a magma body at shallow depths.

DATA PREPROCESSING

Detailed temperature logs from seventy shallow boreholes have been taken from Goupillaud et al (1981) and from Reiter et al (1976), and temperature gradients over the depth range 50-250 feet have been calculated by the method of least squares. These gradients have been corrected for topographic effects by standard techniques and contoured using a hermite cubic interpolation routine to generate the temperature gradient contour map shown in Figure 2. The contouring method has been designed to be a very accurate reflection of the measured surface data. The average error between the contours shown in Figure 2 and the temperature gradients measured in the boreholes is 0.38°F/100 ft or 2.66% of the average temperature gradient. The effect that this type of contouring may have on the wave number filtering process, and the more general problem of aliasing (the ambiguity that may result from inadequate distribution of borehole data), is being examined in a separate study.

A two-dimensional temperature gradient profile across the contoured anomaly was taken at line A-A' (Figure 2). The profile (Figure 3) was constructed by reading temperature gradients along A-A' at one-half kilometer intervals. The profile was designed to pass through the greatest concentration of shallow boreholes in order to minimize the problem of aliasing and also to pass through the area containing the geothermal wells so that the temperatures from these wells could be used to evaluate the results of the wave number filtering process. The temperature gradient data shown in Figure 3 (42 points) constitute the "observed gradients" to be used in the two-dimensional wave number filtering analysis.

WAVE NUMBER FILTERING

The wave number filtering technique was described for the case of two dimensions by Li et al (1982). Basically, the technique consists of three steps. The first step consists of parameterizing the heat source model into a discrete set of line sources located to allow separation of shallow and deep contributions. The second step consists of solving for the model source strengths that minimize the error in a least-squares sense between the observed gradients and the gradients predicted by the model sources. The third step consists of constructing what amounts to a matched filter for the shallow source effects based on the model sources. In addition, the model sources can be used to estimate subsurface temperatures.

To illustrate this latter point, the temperature gradient profile A-A' was processed in a series of steps that can be summarized as follows:

1. The heat sources that give rise to the surface temperature gradients are assumed to lie along either of two horizontal planes, one of which represents the shallow "hot spring noise" to be removed and the other representing the deep geothermal reservoir to be evaluated.

2. The two source planes are separated by a distance of ~2000 feet (600 m), and the source strengths are calculated. An example of source planes at 656 and 2624 feet (200 and 800 m) is shown in Figure 4 (top).

3. The shallow sources, representing the "hot spring noise", are removed.

4. The sources in the deep-source plane are used to calculate the temperatures along a plane midway between the two source planes.

5. The temperatures at that depth which fall within the area covered by the geothermal wells (see Figure 2) are replotted for comparison with the observed temperatures (Figure 5).

6. The two source planes are changed to new depths, and the process is repeated. Figure 4 (bottom) shows the source strengths for planes at depths of 2296 and 4265 feet (700 and 1300 m).

RESULTS OF WAVE NUMBER FILTERING

Figure 4 shows the separation of sources between the shallow and deep planes for cases when the planes are placed at shallow depths (656 and 2624 feet) and at deeper depths (2296 and 4265 feet). There are two significant features that are apparent in the data shown in Figure 4. First is the notable decrease in source strength for the sources placed in the lower plane as the two planes are shifted to greater depths. Correspondingly, there is an increase in the shallow sources for the same shift. This is a direct consequence of passing through some of the sources as the planes are shifted to greater depths. Secondly there is a clear tendency for the strongest deep-plane sources to migrate to the northeast along profile A-A' as the planes are shifted to greater depths. This systematic migration has been observed in all test runs as the deep-source layer is increased from 2296 to 5413 feet (700-1650 meters) and is also apparent from the two examples presented in Figure 4.

The temperatures calculated by the wave number filtering technique over the depth range 1312-4424 feet (400-1350 meters) are shown in Figure 5 as well as the temperatures actually measured in the boreholes. Both sets of temperatures represent the same portion of the geothermal field (shaded area of Figure 2). In general, the agreement between calculated and measured temperatures is excellent. The two temperature fields overlap at all depths tested, and the average temperatures agree to within 15% (~50°F) for seven of the eight depths tested. Even the results at 3936 feet (1200 meters) are probably acceptable. Additionally, there are several specific features that are worth special mention.

1. The wave number filtering technique accurately predicts the temperature rollback at depths around 2000 feet. Physically, this rollback results from the temperatures having reached the flash point with temperatures at greater depths being controlled by the thermodynamics of water. However, the filtering module has not been programmed to treat this phenomenon. Rather, the module has independently reached the same conclusion on the basis of spectral information contained in the shallow temperature data.

2. The wave number filtering technique has predicted realistic temperatures at depths at least as great as 4000 feet. However, the "observed gradients" are generally taken from depths less than 250 feet. This represents a meaningful extrapolation over a considerable depth range in a highly complex thermal environment.

3. The present experiment was discontinued when the deeper source plane exceeded 5413 feet (1650 meters). The reason for this termination is the reluctance of the filtering module to assign sources to the deeper layer at these depths. Mathematically, this means that sources in excess of 5400 feet are spectrally incompatible with the observed temperature gradient data. Physically, this means that, without deeper sources, it is unlikely that continued drilling will result in significantly higher temperatures, and such drilling should probably not be attempted unless suitable temperatures have already been encountered and drilling is continued for the purpose of locating production horizons.

4. Finally, it is constructive to compare the temperatures estimated by the wavelength filtering technique to those obtained by direct extrapolation of surface temperature

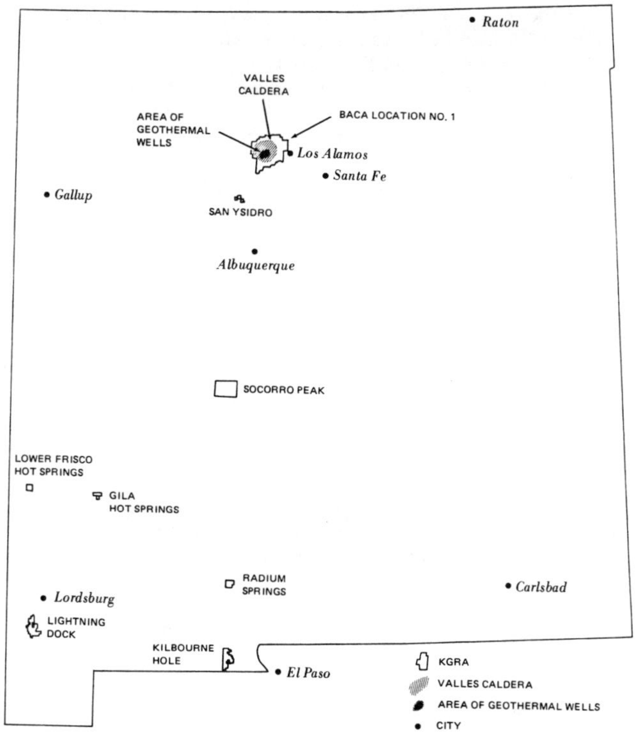

Figure 1. Location of the Valles Caldera and KGRA's in New Mexico

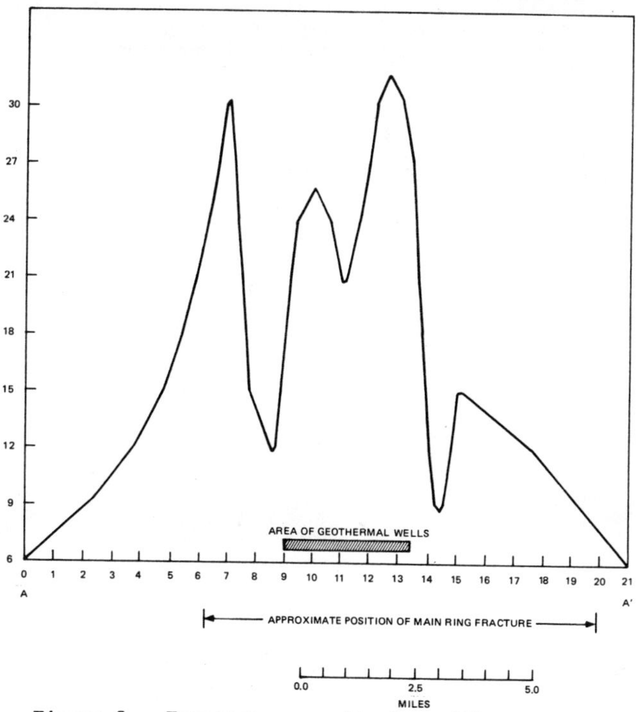

Figure 3. Temperature gradient profile across Valles Caldera

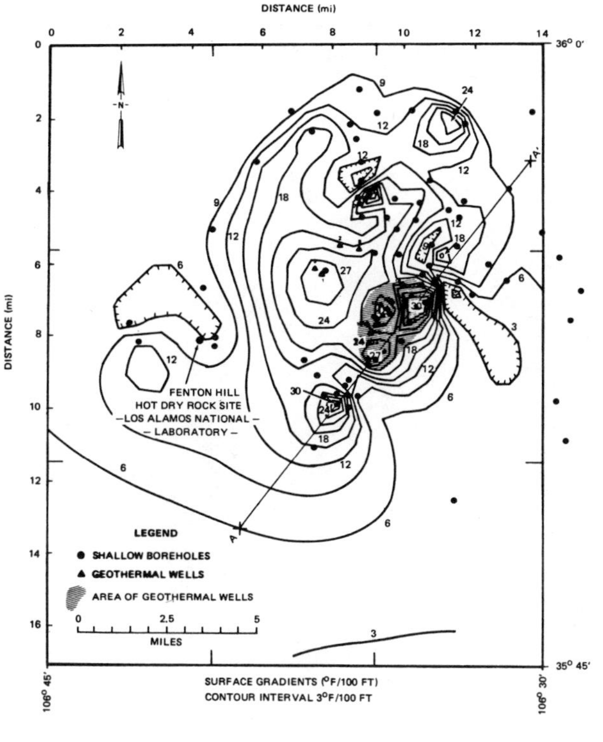

Figure 2. Temperature gradient contours at Valles Caldera

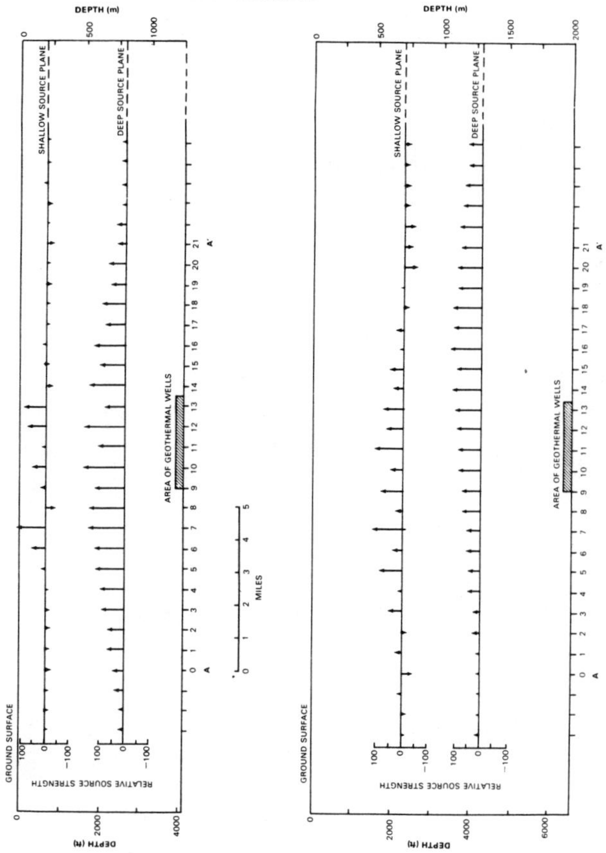

Figure 4. Separation of sources between the shallow and deep source plane for two different depth regions.

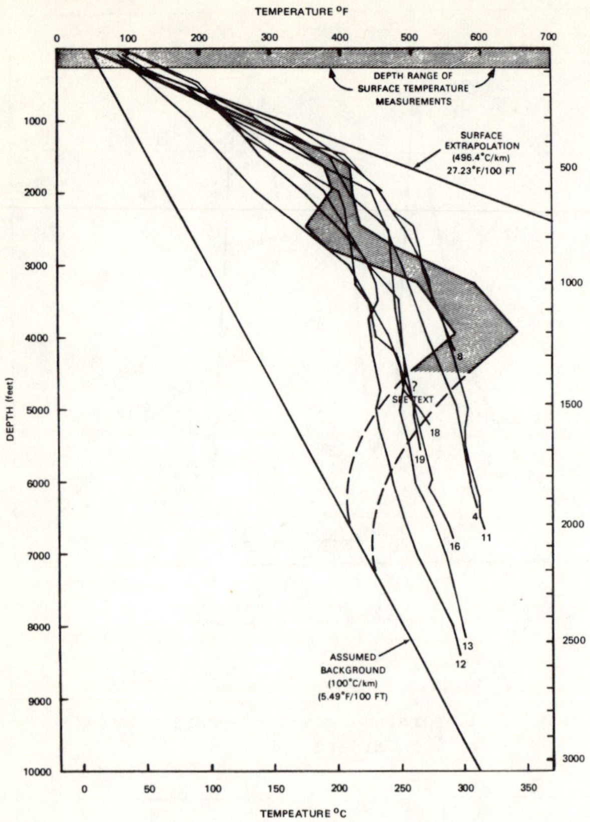

Figure 5. Temperatures, Valles Caldera. Estimated (shaded) and measured subsurface temperatures, Valles Caldera. Well locations shown by the triangles in Figure 2.

gradients. The average gradient of the six wells located within the shaded area of Figure 2 is 27.23°F/100 feet, and this gradient has been reproduced on Figure 5. Direct extrapolation of this gradient results in significant overestimates (~ 100°F) at a depth of 1500 feet, and the results are completely untenable by a depth of 2000 feet.

SUMMARY AND CONCLUSION

We have applied a two-dimensional technique for wave number filtering of hot spring noise to a set of shallow temperature gradient data at the Valles Caldera, New Mexico. The results of the filtering were used to obtain estimates of subsurface temperatures for comparison to the temperature-depth profiles observed in several geothermal production wells at the site. These estimated temperatures agree reasonably well with the observed temperatures, particularly in the identification of the turnover in the temperature-depth profile, which could not have been predicted using direct extrapolation or downward continuation of the shallow gradients.

REFERENCES

Goupillaud, P. L., L. F. Rice, and B. F. Mason, BACA geothermal demonstration project; Geological and Geophysical Data File, Topical Report, WESTEC Service, Inc., Albuquerque, New Mexico, 1981.

Li, T. M. C., C. A. Swanberg, and J. F. Ferguson, 1982. A method for filtering hot spring noise from shallow temperature gradient data, Geothermal Resources Council Transactions, in press, 1982.

Reiter, M., C. Weidman, C. L. Edwards, and N. Hartman, Subsurface temperature data in Jemez Mountains, New Mexico; New Mexico Bureau of Mines and Mineral Resources, CIVC, 151, 16pp., 1976.

Geothermal Resources Council, TRANSACTIONS Vol. 6, October 1982

A GEOPHYSICAL SURVEY OF HOT SPRINGS BAY VALLEY, AKUTAN ISLAND, ALASKA

Eugene M. Wescott, Donald L. Turner, William Witte and Becky Petzinger

Geophysical Institute, University of Alaska, Fairbanks, Alaska 99701

ABSTRACT

Geothermometry of hot springs in Hot Springs Bay Valley on Akutan Island indicate a common reservoir temperature of 160-190°C. Geophysical exploration of the valley has revealed a probable hot water reservoir 1.4 km x 0.5 km in area, 40-75 m thick and capped by 40-70 m of impermeable volcanic debris flows. Resistivity of the reservoir rock ranges from 2.2-12 Ω-m. High seismic velocity and high resistivity of the basement rock may be due to mineral cementation.

INTRODUCTION

A series of hot springs occur along a 1.5 km distance at the NW edge of Hot Springs Bay Valley, Akutan Island, Alaska. The island has an active volcano - Akutan. Motyka et al. (1982) have reported that the geothermometry gives an estimated reservoir temperature range of 160°-190°C. The heat loss by the springs is estimated at 1.6 MW (Motyka et al., 1981). In the summer of 1981 geologic mapping and geophysical surveys were carried out in the valley to locate and evaluate geothermal energy resources. Geophysical techniques included near surface conductivity profiling, seismic refraction pro-

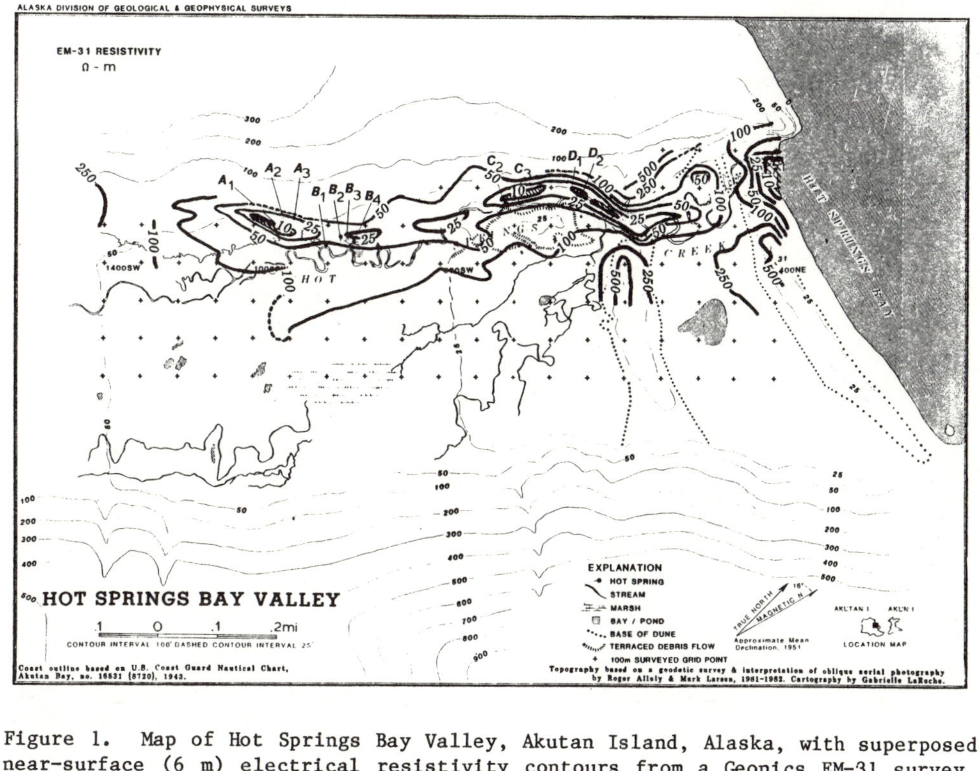

Figure 1. Map of Hot Springs Bay Valley, Akutan Island, Alaska, with superposed near-surface (6 m) electrical resistivity contours from a Geonics EM-31 survey. The resistivity values appear to be inversely related to anomalously high ground temperatures.

filing, Schlumberger vertical electrical soundings, and dipole-dipole resistivity profiling.

GEOLOGY

The Hot Springs Bay Valley has steep side walls composed of an upper section of lava flows and minor mudflows of 1.4 m.y. age lying unconformably on a thick section of well indurated volcanic mudflows cut by abundant fine-grained dikes dated at 14 m.y. by the ^{40}K - ^{40}Ar method.

The valleys on the island appear to have been scoured to U shapes by glaciers. Hot Springs Bay Valley has a flat floor formed by volcanic debris flows. The uppermost unit is a lahar which is nonporous and impermeable where exposed.

GEOPHYSICAL SURVEYS

Electrical resistivities of the waters from two of the springs with temperatures of 84 and 59°C are 2.39 and 8.03 Ω-m respectively. The resistivities of the valley sediments are generally much higher, so the near surface resistivity should be well correlated with temperature anomalies where hot water is closest to the surface. We used a Geonics EM-31 electromagnetic conductivity meter to measure the near surface (6 m depth) resistivity over a 1.26 square km area. The contoured data are shown in Figure 1. The low resistivity zones form a narrow sinuous pattern along the NW side of the valley. No other low resistivity zones were located. The pattern suggests that an ancient

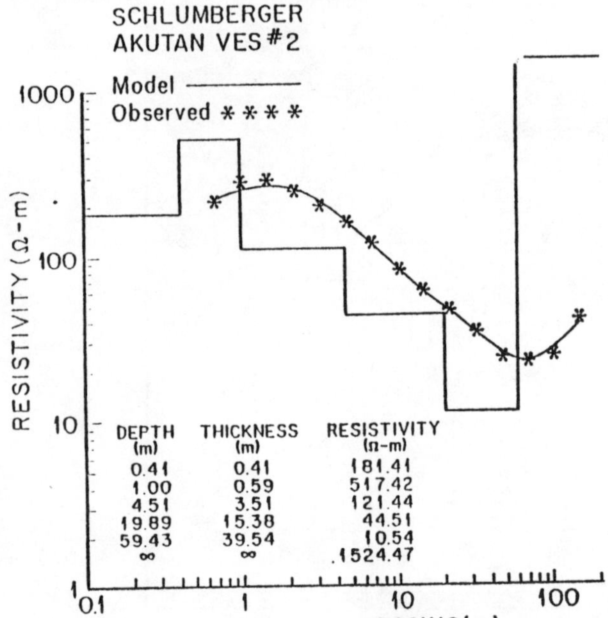

Figure 2. Plot of vertical electric sounding #2 on a lahar terrace 100 SE of the C springs (see Figure 1). True resistivity vs. depth of the model is also plotted. The lowest resistivity layer is interpreted as a geothermal reservoir of porosity 45-82%.

buried stream channel forms a permeable zone through the lahar. The near surface resistivity is about 100 Ω-m over the rest of the valley except for two sand dune areas where it is 500 Ω-m.

Four Schlumberger vertical electrical soundings were made: one 900 m inland parallel to the valley and 200 m from the NW side, a second parallel to it on a terraced debris flow, a third between hot springs C and D and a fourth near hot springs A (Figure 1). The data were interpreted by an automatic curve fitting program (Zhody, 1974). Figure 2 shows the vertical electric sounding (VES) curve and interpretation run about 150 m from the NW edge of the valley on a lahar terrace. All appear to indicate a geothermal reservoir of 2.2 to 12 Ω-m resistivity, 23 to 42.5 m thickness and a depth of 13 m near the NW edge of the valley to 52 m towards the center.

Three 100 m dipole-dipole pseudo sections were run along and perpendicular to the valley. The results are plotted in Figure 3 to show where they intersect. They all suggest a cap rock, presumably an impermeable lahar, of thickness 40-70 m underlain by a reservoir thicker towards the center of the valley and towards the NE. The SE-NW dipole-dipole no. 1 section was modeled using a two-dimensional program. The result suggests a basement rock of 1500 Ω-m perhaps 150 m deep towards the valley center. Archie's law suggests a porosity of 45-82% for the reservoir and 4-7% for the basement rocks (Archie, 1942).

Seismic profiling results agree with the resistivity data. There are three basic units: the top or lahar unit has a velocity ranging between 1630-1960 m/sec and a thickness of 30-75 m; the geothermal reservoir unit has a velocity of 3240-3505 m/s, a thickness of 40-75 m and a base which slopes steeply towards the center of the valley and down the valley towards the ocean. The basement rocks have a velocity of 4900 m/s, and probably correspond to the ≥ 14 m.y. old rocks forming the valley walls, although they could also be lava flows, intrusives, or hydrothermally cemented sediments. Figure 4 shows the seismic profile parallel to the valley. The reservoir apparently coincides with the medium velocity layer.

CONCLUSION

Geophysical surveys have located a probable geothermal reservoir at least 1.5 x 0.5 km^2 in area with a thickness ranging between 40-75 m. Figure 5 shows a composite profile parallel to the valley about 200 m from the NW side. The reservoir probably extends further to the NE as evidenced by hot springs E (Figure 1) at the ocean shoreline. It may be thicker towards the NE end of the valley which was not fully explored. Deep resistivity and seismic profiling agree on the general shape of the reservoir which probably rests on a glaciated volcanic bedrock surface. High seismic velocity of

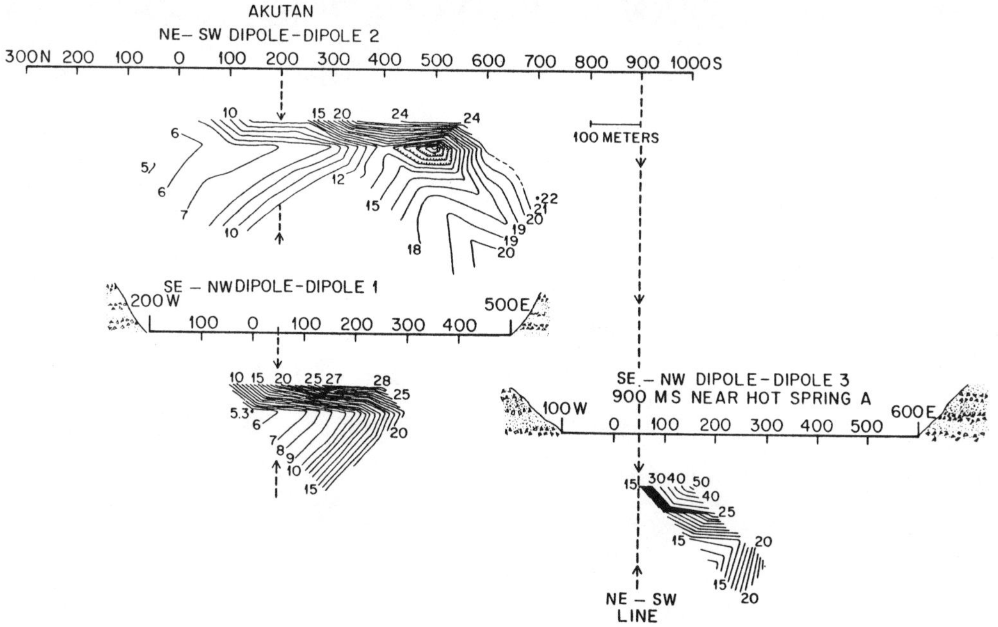

Figure 3. 100 m dipole-dipole resistivity pseudo-sections plotted to show where the NW-SE sections intersect the NE-SW section contours in Ω-m. The sections generally show a medium resistivity cap layer 40-70 m thick overlying a geothermal reservoir layer 40-75 m thick underlain by high resistivity bedrock.

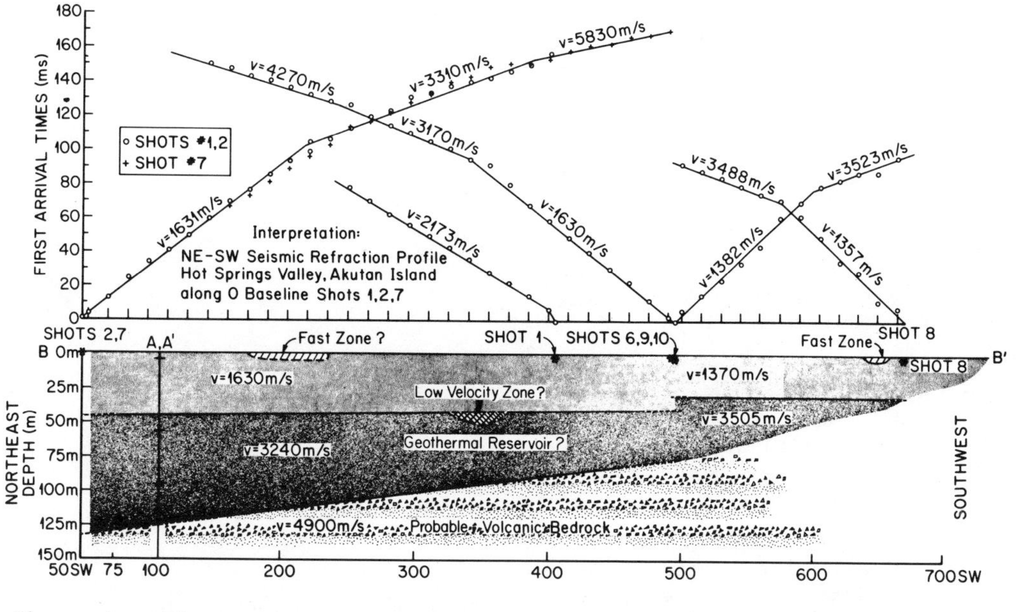

Figure 4. First arrival seismic travel time curves for profile parallel to valley. Bottom interpreted cross section. The interface where dashed indicates zones of no seismic information.

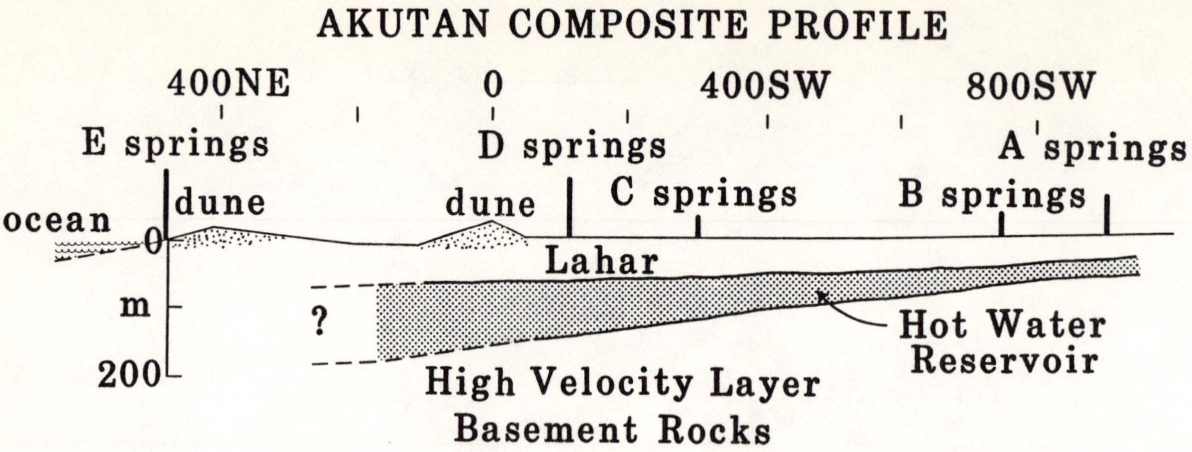

Figure 5. Composite cross section of lower Akutan Hot Springs Bay Valley taken parallel to axis of valley.

bedrock could be a result of silica or other mineral deposition from a deeper, hotter reservoir. The resource outlined is sufficient to supply hot water and power to the town of Akutan and an expanded fish processing industry 5 km distant.

ACKNOWLEDGMENTS

The work was supported by the USDOE under cooperative agreement DE-FC07-79-ET-27034 and by the Alaska Division of Geological and Geophysical Surveys. Dr. John Reeder managed the very difficult field logistics and Roman Motyka and others did preliminary exploration of Akutan and provided the impetus for the geophysical exploration.

REFERENCES

Archie, G. E., 1942, The electrical resistivity log as an aid in determining some reservoir characteristics, Trans. AIME 146, p.54-62.

Motyka, R. J., Moorman, M. A. and Liss, S. A., 1981, Assessment of thermal spring sites, Aleutian Islands, Atka Is. to Beckerof Lake: Preliminary results and evaluation, Alaska Div. of Geol. and Geoph. Surv. Open-File Report 1A4, 175 p.

Motyka, R. J., Moorman, M. A. and Poreda, R., 1982, Fluid geochemistry of Hot Springs Bay Valley, Akutan Island, Alaska, Geothermal Resources Council Transactions, 1982 Annual Meeting.

Zhody, A. R., 1974, A computer program for the automatic interpretation of Schlumberger sounding curves over horizontally stratified media, U.S. Geol. Surv. Report, GD-74-017.

DRILLING/STIMULATION TECHNOLOGY

PRECISION DIRECTIONAL DRILLLING OF HOT DRY ROCK GEOTHERMAL PRODUCTION WELL EE-3*

Richard S. Carden,[1] John C. Rowley, and Catherine Helmick

[1]Grace, Shursen, Moore and Associates, Oklahoma City, Oklahoma 73112
Earth and Space Sciences Division, Los Alamos National Laboratory, Los Alamos, New Mexico 87545

ABSTRACT

The deviated directional drilling of the hot dry rock (HDR) geothermal production well EE-3 (Energy Extraction No. 3) was successfully completed on August 1981. The injection well, EE-2, previously had been drilled with its lower part at an inclination of 35° to the vertical. It reached an on-line depth of 15,292 feet and its bottom-hole temperature was 608°F (320°C). The production well EE-3 was required to be drilled 1200 feet (370 m) above and parallel to the injection well. This necessitated high precision, controlled-trajectory directional drilling operations. The directional drilling of EE-3 was accomplished within the required tolerances at a depth of 13,933 feet and a bottom-hole temperature of 580°F (280°C).

INTRODUCTION

Hot dry rock geothermal reservoir analyses (Cummings, et al., 1979) indicate that more effective and optimum heat extraction systems will result when multiple, parallel, vertically oriented fractures are produced in inclined well-bores. This concept is illustrated in Figure 1; a cross-sectional view of the EE-2/EE-3 wellbores.

The injection well, EE-2, was completed May 12, 1980 (Brittenham et al., 1980). As this report indicates, EE-2 was directionally drilled with downhole motors in the NNE octant and inclined at 35° to the vertical by a 10,500 ft (3200 m) depth. This placed the wellbore approximately perpendicular to the fracture planes determined by the previous successful reservoir formation efforts (Dash et al., 1980). Well EE-2 was drilled with packed (stiff) bottom-hole assemblies that held the 35° inclination, but permitted the borehole to turn in azimuth to N88°E at its total depth (TD) of 15,292 ft (4461 m).** The bottom-hole static temperature was 320°C (608°F) (Pettitt et al., 1982). The directional-drilling operations and equipment used in well EE-2 had established the basic approach and procedures required to direct and control the trajectory of well EE-3 (Pettitt et al., 1982; Neudecker et al., 1982; Brittenham et al., 1980; Williams et al. 1979).

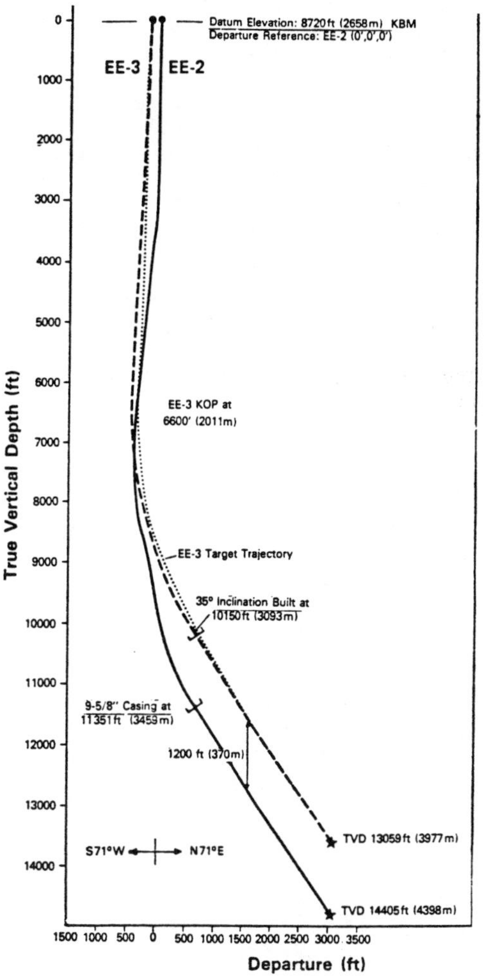

Figure 1. Cross section of EE-2/EE-3 wellbores.

*Work supported by the U.S. DOE, Division of Geothermal and Hydropower Technologies.

**All depths are measured drill string depths.

DIRECTIONAL DRILLING PARAMETERS FOR WELL EE-3

Well EE-3 required very precise directional drilling because the borehole trajectory had to be drilled within specified tolerances with respect to the injection well EE-2. The trajectory parameters set as targets for the reservoir portion of EE-3 were:

- 1200 ± 50 ft vertically above EE-2,
- ±100 ft maximum lateral deviation from the horizontal projection of EE-2.

These objectives were met within practical drilling limits and resource constraints. The maximum deviations observed were a slightly greater vertical spacing of about 1300 ft over the upper part of the reservoir which decreased to 1200 ± 50 ft within the reservoir section. The widest lateral deviation occurred at the TD of EE-3, i.e., 13,933 ft (4247 m) measured depth, or 13,048 ft (3977 m) TVD, where the EE-3 wellbore had drifted to the north 180 ft as evaluated from the magnetic single-shot survey data (Rowley et al., 1982). The final relationship between EE-2 and EE-3 and the HDR reservoir volume is illustrated in Figures 2 and 3.

DIRECTIONAL DRILLING OPERATIONS

The drilling of production well EE-3 was finished August 21, 1981. The planned controlled-trajectory drilling required a high-precision application of directional-drilling technology. The directional-drilling strategy for EE-3 used downhole-motor runs, guided by steering tools for azimuth change. The motor runs were alternated with rotary drilling assemblies intended to hold or build the inclination angle of the borehole, and to drill hole.

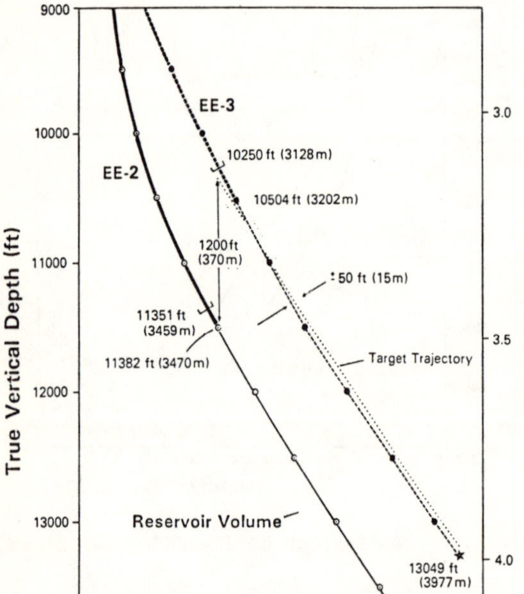

Figure 2. Section view of EE-2 & EE-3 in reservoir section.

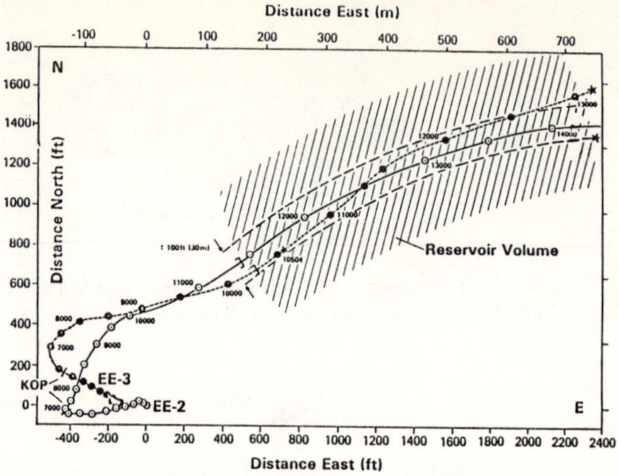

Figure 3. Horizontal projection of EE-3 wellbore trajectories.

The 53 directional downhole-motor runs are summarized in Table 1.

Various equipment items had been tested in EE-2 and two definite temperature limits were established. At temperatures less than about 400°F (200°C), corresponding to Fenton Hill depths less than 10,000 ft (3.0 km), positive displacement motors (PDM) could be used. These downhole drilling motors are constant-volume-displacement devices, so the bit rpm is controlled directly by the flow rate of the drilling fluid.

At temperatures above 400°F (200°C) two sizes of all-metal, high-temperature rated turbodrills were used (Neudecker et al., 1982). The two sizes of high-temperature turbodrills from Maurer Engineering Inc. (MEI) generally operated very satisfactorily. The 7-3/4-in.-diam tools were used in the 12-1/4-in.-diam hole. Fourteen of the 17 runs were successful. The three directional run failures were due to two cases of float-valve failures that resulted in plugged bearings, and one incident where the motor stalled because of tight hole conditions.

The eight directional-drilling runs with the 5-3/8-in.-diam MEI turbodrills were conducted in the 8-3/4-in.-diam reservoir section of the borehole. These tools had been tested at TD in EE-2 and in the laboratory. The last three runs had to be aborted because of steering tool failures. The smaller MEI turbodrills performed very well. Their operations were stable and reliable once it was learned that a high load on the bit, i.e., 20,000 to 25,000 lb_f, was required. Excessive bit

Table 1. Summary of directional drilling runs and results for well EE-3.

Directional Drill Motor Run No.	Drill Motor[a]	Steering Tool Service[b]	Bent Sub Angle	Measured Depth[c] m (ft)	Borehole Trajectory	Distance Drilled m (ft)	Remarks
1	BPDM	SS	2°	1981 (6520)	9 1/2°N,53°W	23 (76)	Problems with motor
2	BPDM	SS	2°	2027 (6649)	14 3/4°N,54°W	39 (129)	Angle building excessively
3	BPDM	EYE	2°	2075 (6809)	14 3/4°N,37°W	30 (99)	Replace bit
4	DDPDM	EYE	2°	2083 (6868)	14 1/2°N,20°W	17.9 (59)	Plugged bit on connection (no float)
5	DDPDM	EYE	2°	2168 (7114)	13°N,23°E	64.3 (211)	Motor quit, bit very worn, 2.5 cm (1 in.) undergauge
6	BPDM	EYE	2°	2216 (7269)	10 1/2°N,43°E	27.4 (90)	Motor stalled
7	BPDM	EYE	2°	2216 (7269)	---	0 (0)	Bit pinched and motor bent, hit ledge going into hole
8	NPDM	EYE	2°	2236 (7337)	10 1/2°N,43°E	20.7 (68)	Bit worn
9	NPDM	EYE	2°	2264 (7427)	8 3/4°N,62 1/2°E	27.4 (90)	Bit worn
10	NPDM	EYE	2°	2281 (7482)	7°N,47°E	16.8 (55)	Bit worn, bit 13 mm (1/2 in.) undergauge
11	DDT7D	EYE	1 1/2°	2395 (7856)	---	0.3 (1)	Steering tool failed, test run of turbine
12	DDPDM	EYE	1 1/2°	2402 (7883)	15 3/4°N,51°E	8.2 (27)	Dull bit
13	DDPDM	EYE	1 1/2°	2409 (7905)	15°N,47°E	6.7 (22)	EYE quit, motor and bit worn out
14	DDPDM	EYE	1 1/2°	2420 (7940)	16°N,51°E	10.7 (35)	Motor quit
15	NPDM	EYE	1 1/2°	2454 (8052)	16 1/4°N,59°E	34.1 (112)	Bit locked up
16	NPDM	EYE	1 1/2°	2457 (8062)	---	3.0 (10)	Could not orient due to torque
17	NPDM	EYE	1 1/2°	2486 (8159)	---	28.0 (92)	Bit cones loose
18	NPDM	EYE	1 1/2°	2490 (8170)	---	3.4 (11)	Motor quit
19	NPDM	EYE	1 1/2°	2519 (8265)	16 1/4°N,87°E	28.9 (95)	Bit locked up
20	DDT7S	---	---	2562 (8407)	17 1/2°N,84°E	0 (0)	Tachometer not operative, motor not rotating, straight hole tool with stabilizer, test run of turbine
21	DDT7S	---	---	2562 (8407)	17 1/2°N,84°E	0 (0)	Could not get to bottom, reamed 27 m (90 in.), straight hole tool with stabilizer, test run of turbine
22	NPDM	EYE	1 1/2°	3011 (9879)	25 1/2°N,66°E	8.8 (29)	Bit dull
23	NPDM	EYE	1 1/2°	3011 (9879)	25 1/2°N,66°E	0 (0)	Motor wouldn't rotate, reached temperature limit of PDMs
24	DDT7D	EYE	1 1/2°	3053 (10,017)	24°N,72°E	0 (0)	Washout in drill string
25	DDT7D	EYE	1 1/2°	3053 (10,017)	24°N,72°E	0 (0)	Turbine would not rotate
26	MEIT7	EYE	1 1/2°	3083 (10,116)	27°N,70°E	30.4 (99)	Bit undergauge, build angle attempt
27	MEIT7	EYE	1 1/2°	3085 (10,123)	27°N,70°E	2.1 (7)	Turbodrill quit, build angle attempt
28	MEIT7	EYE	1 1/2°	3094 (10,150)	27°N,67°E	8.2 (27)	Bit locked up, build angle attempt
29	MEIT7	EYE	1 1/2°	3110 (10,204)	27 3/4°N,68°E	16.5 (54)	Bit undergauge, build angle attempt
30	MEIT7	EYE	2°	3150 (10,334)	27 1/4°N,51°E	9.4 (31)	Motor quit, build angle attempt
31	MEIT7	EYE	2°	3163 (10,378)	24 3/4°N,66°E	13.4 (44)	Dropping angle, steering tool seated 170° from key
32	MEIT7	EYE	2°	3175 (10,417)	24 3/4°N,66°E	11.9 (39)	EYE failed
33	MEIT7	EYE	2°	3189 (10,463)	24 3/4°N,66°E	14.0 (46)	Motor quit, radial bearing failed, build angle attempt
34	MEIT7	EYE	1 1/2°	3197 (10,489)	24 3/4°N,66°E	7.9 (26)	Bit locked up, undergauge, 0.9 cm (3/8-in.)
35	MEIT7	EYE	1 1/2°	3209 (10,528)	26 1/4°N,66°E	0 (0)	Motor would not rotate, tight hole
				(9396)		(86)	
36	DDPDM	EYE	1 1/2°	2913 (9559)	24°N,73°E	0.3 (1)	Motor failed, plugged bit
37	DDPDM	EYE	1 1/2°	2936 (9631)	24°N,75°E	21.9 (72)	Turn hole to east, avoid old hole
38	DDPDM	EYE	2°	2978 (9771)	24 1/2°N,78°E	12.9 (42)	Turn hole to east, avoid old hole
39/40	MEIT7	EYE	1 1/2°	3117 (10,226)	31 1/2°N,74°E	0 (0)	Float failed, plugged two turbodrills[d]
41	MEIT7	EYE	1 1/2°	3123 (10,245)	31 1/2°N,72°E	5.6 (19)	Eye wireline failed[e]
42	MEIT7	EYE	1 1/2°	3155 (10,352)	28°N,67°E	29.6 (97)	Pulled due to drop in inclination
43	MEIT7	EYE	2°	3189 (10,462)	30 3/4°N,57°E	18.3 (60)	Light weight on assembly, good run
44	MEIT7	EYE	1 1/2°	3245 (10,654)	32°N,56°E	24.7 (81)	Good run
45	MEIT7	EYE	2°	3289 (10,791)	34°N,55°E	17.4 (57)	Both floats failed
					Directional Control 22.2 cm (8-3/4 in.) Hole		
46	MEIT5	EYE	1 1/2°	3643 (11,951)	35 1/4°N,53°E	7.6 (25)	Dull bit
47	MEIT5	EYE	1 1/2°	3695 (12,123)	36°N,53°E	7.3 (23)	Dull bit
48	MEIT5	EYE	1 1/2°	3714 (12,186)	35 1/2°N,59°E	19.2 (63)	Pulled for survey
49	MEIT5	EYE	1 1/2°	3752 (12,310)	35°N,72°E	36.3 (119)	Bit locked up, good run
50	MEIT5	EYE	1 1/2°	3950 (12,961)	36°N,73°E	20.1 (66)	Radial bearing cage twisted
51	MEIT5	EYE	1 1/2°	3950 (12,961	36°N,73°E	0 (0)	Bit pinched
52	MEIT5	EYE	1 1/2°	3950 (12,961)	36°N,73°E	0 (0)	Eye tool failed
53	MEIT5	EYE	1 1/2°	3950 (13,365)	36°N,73°E	0 (0)	Eye tool failed, motor plugged while waiting

[a]Key for downhole motion types:
BPDM = Baker Service Tools.
DDPDM = Dyna-Drill (Smith International).
NPDM = Navi-Drill (Christensen Oil Tools).
DDT7D = Dyna-Drill, 7 in. directional turbine.
DDT7S = Dyna-Drill, 7 in. straight-hole turbine.
MEIT7 = Maurer Engineering Inc., 7 in. turbine.
MEIT5 = Maurer Engineering Inc., 5 3/8 in. turbine.

[b]OSS = Single-shot orientation; EYE = Scientific Drilling Controls steering tool.

[c]Measured depth at end of run.

[d]Started running tandem float valves.

[e]Used high temperature float seals for remainder of runs.

wear (Figure 4), bit lift-off, and turbine-speed run-away were symptoms of too light bit loadings.

Two types of commercial turbine motors were tested during the EE-3 drilling program. Two 7-in.-diam Dyna-Drill (Smith International) directional motor tests were conducted (see Table 1). Also, two straight-hole high-temperature Dyna-Drill turbines were run. The Dyna-Drill turbine runs were supported by the Sandia National Laboratories Geothermal Drilling and Completions R&D Program. These motors appear to have excellent high-temperature capabilities (Rowley et al., 1982).

The performance of the downhole-motor assemblies is summarized in Table 2 and is dependent on the proper functioning of all components of the directional system. The major reasons for directional runs to be aborted were when the orienting or steering tool failed, bit dulled, float valve leaked, or motor stalled.

Figure 4. Photo of cone wear due to high rpm on bottom.

Table 2. Downhole-motor directional-drilling-assembly performance; EE-3 — excluding sidetracking operations.

Motor Type	Total No. of Runs Attempted	No. of Runs that Drilled	Average[c] Distance per Run[d] m	(ft)	Average[c] ROP per Run m/h	(ft/h)	Average[c] Duration of Run, h	Diam of Motor cm (in.)
MEIT 7[a]	17	14	13.2	(43.3)	3.9	(12.7)	3.4	19.7(7-3/4)
MEIT 5[a]	8	5	18.1	(59.3)	8.6	(28.2)	2.1	13.7(5-3/8)
NPDM	10	9	19.0	(62.2)	4.5	(14.8)	4.0	20.3(8)
DDPDM	8[b]	8	18.2	(59.6)	3.3	(10.8)	5.5	19.7(7-3/4)
BPDM	5	4	30.0	(98.5)	2.1	(7.0)	14.0	17.1(6-3/4)

[a] High temperature turbodrills.
[b] Excluding those used for sidetracking.
[c] Based upon successful runs; i.e. those that drilled.
[d] Note that most directional runs were set up with a 20-m (60-ft) length, 2 joints, of drill pipe initially in the derrick.

CONCLUSIONS

Although the directional-drilling program for EE-3 consumed only 2.9 percent of the total time invested in drilling EE-3, it constituted the most important part of that program because of the required alignment with the injection well EE-2. With the successful directional drilling of EE-3 within the desired trajectory tolerances, the EE-2/EE-3 reservoir volume was ready for fracturing operations.

ACKNOWLEDGMENT AND DISCLAIMER

The authors thank the many individuals who supported the EE-3 drilling campaign with their expertise, and acknowledge the firms that provided the equipment, instruments, services, and supporting personnel. The supervisors and crew members of the Brinkerhoff-Signal Rig No. 56 contributed significantly to the success of these operations. We especially note with appreciation those companies that supplied developmental high-temperature-rated hardware and tools for evaluation and use.

However, specific references to a company, product name, service, tool, or equipment item does not imply approval or recommendation of the product, service or hardware by the University of California (Los Alamos National Laboratory) or the U.S. Department of Energy, to the exclusion of others that may be suitable.

REFERENCES

Brittenham, T. L., Williams, R. E., Rowley, J. C., and Neudecker, J. W., 1980, Directional drilling operations, hot dry rock well EE-2, Geotherm. Resour. Coun. Trans., v. 4, p. 273-276.

Cummings, R. G., 1979, Mining the Earth's heat: Hot dry rock geothermal energy, Technology Review, MIT Press, v. 81, no. 4, p. 2-19.

Dash, Z. V., Murphy, H. D., and Cremer, G. M., 1981, Hot dry rock geothermal reservoir testing: 1978-1980, Los Alamos National Laboratory Report LA-9080-SR (November 1981).

Neudecker, J. W. and Rowley, J. C., 1982, High temperature directional drilling turbodrill, Los Alamos National Laboratory Report LA-9231-MS.

Pettitt, R. A., Koczan, S., Helmick, C., and Williams, R. E., 1982, Drilling of hot dry rock geothermal well EE-2, Los Alamos National Laboratory report LA-9302-HDR.

Williams, R. E., Rowley, J. C., Neudecker, J. W., and Brittenham, T. L., 1979, Equipment for drilling and fracturing hot granite wells, Soc. Pet. Eng. Paper No. 8268, Las Vegas, NV.

Geothermal Resources Council, TRANSACTIONS Vol. 6, October 1982

THE IMPACT OF COMMON PROBLEMS IN GEOTHERMAL DRILLING AND COMPLETION

Charles C. Carson and Y. T. Lin

Geothermal Technology Development Division
Sandia National Laboratories
Albuquerque, New Mexico 87185

ABSTRACT

Problems that arise in geothermal drilling and completion account for a significant portion of geothermal well costs. In order to evaluate new technologies for combatting these problems, the relative frequencies and severities of different problems have been estimated. The estimates were based on both subjective judgements and analysis of available drilling records. The most common problems include lost circulation, stuck pipe and cementing, and their impact is to increase well cost by an average of at least 15%.

INTRODUCTION

Sandia National Laboratories manages the Department of Energy program in Geothermal Technology Development. A major portion of this program has been the development of new drilling and completion technologies specifically for geothermal applications. (1) A common measure for determining which technologies to pursue has been the potential impacts of proposed technologies on well cost. One useful tool for evaluating cost impacts has been a family of representative well models that describe in detail the costs of the activities required in drilling and completing geothermal wells. (2,3,4) However, comparison of the costs determined from the models to actual well costs indicates that the money and time spent solving unplanned-for problems comprise a significant part of the total drilling and completion cost. Because of their importance, the program has emphasized technologies for solving common drilling and completion problems; and in turn, a portion of the well cost analysis has been directed toward understanding and estimating the costs associated with those problems. This paper presents results from that analysis.

Drilling and completion problems play a larger part in geothermal drilling than they do in drilling for oil and gas. To a certain extent this can be attributed to the relative immaturity of geothermal exploitation. Due to lack of experience in most geothermal areas, geothermal drilling tends to be more like wildcat drilling than field development drilling. However, the primary reason that problems are so important arises from the nature of the resource and the fact that most geothermal drilling and completion methods have been directly adapted from petroleum drilling. Geothermal resources are typically found in hard, fractured formations that often are underpressured. Geothermal fluids are generally quite corrosive, and by its nature, geothermal drilling is into hot formations. These and other inherent characteristics introduce problems to conventional oil drilling technology.

THE GENERAL IMPACT OF TROUBLE

In general, drilling and completion troubles increase the time and cost of making wells. In addition, they introduce significant scatter or dispersion to total well costs. Other statistical variables that affect time and cost, such as rates of penetration, introduce much smaller dispersions. In their extreme, drilling troubles can increase well cost or time sufficiently to cause abandonment of a well. In an economic sense, the effect of drilling trouble is increased investment risk.

It is difficult to find much published data on the costs of geothermal wells, and so much of the analysis of trouble has been based on consideration of its impact on drilling time. Figure 1 presents total rig time as a function of depth for 123 wells drilled at the Geysers, California. For comparison of the modeled and actual wells, the average "trouble-free" drilling times given by the representative, or generic, well models are indicated. The scatter in the drilling time data is obvious. Much of the extreme scatter is attributed to encountered drilling problems. Charts similar to Figure 1 have been constructed for all U.S. areas in which there has been significant drilling, and they show the same general pattern of scatter. The charts graphically demonstrate the general impact of trouble and indicate that trouble costs can dominate other costs. They also indicate the other extreme as well. Cases that are trouble-free and in which rates of penetration are high can be drilled much faster than might be expected.

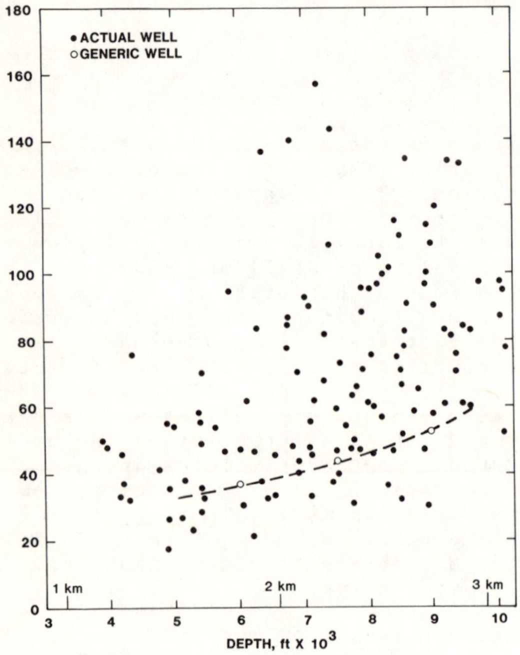

Figure 1. Total Drilling Time vs. Depth--The Geysers

SUBJECTIVE ESTIMATES OF TROUBLE PROBABILITIES

Results such as those shown in Figure 1 illustrate the general impact of trouble, but they do little to identify specific drilling problems or to quantify their effects. Very little information exists in the proper form to help evaluate the impacts of drilling troubles, and so a structure was developed to generate such data from the experiences and judgements of people familiar with geothermal drilling. The probability tree in Figure 2 illustrates the procedure used. For a given interval of drilling at a given resource area, the probabilities of encountering specific troubles were estimated. In addition, estimates were made for the probabilities of encountering different severities of problems.

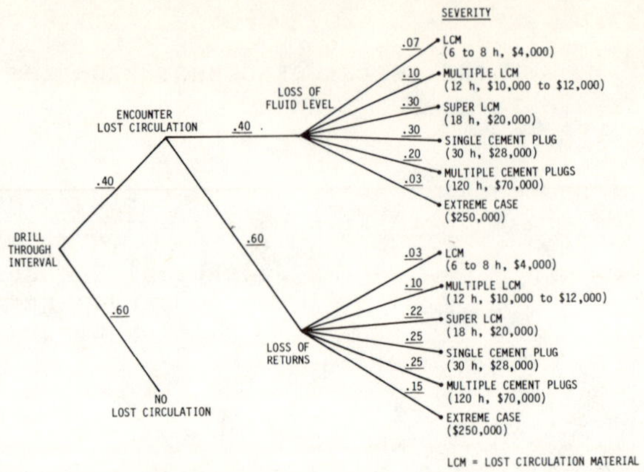

Figure 2. Sample Probability Tree Used for Subjective Analysis

This sampling procedure for generating subjective estimates of trouble probabilities is continuing. Early results are presented in Table 1, for three potential problems: lost circulation, stuck pipe, and cementing. The cementing problems included are only those that can be recognized during the cementing operation and do not include many problems that arise after completion (due to poor cementing). The results in Table 1 are preliminary and will be updated, but effort in the trouble characterization area is moving toward the analysis of historical data as described below.

Table 1. Preliminary results from the subjective analysis.

Region	Problem Area	Avg. Time (hrs)	Avg. Cost ($1000)
The Geysers	Lost Circulation	41	53
	Stuck Pipe	24	14
	Cementing	17	20
	totals	82	87
Baca	Lost Circulation	75	125
	Stuck Pipe	39	20
	Cementing	26	27
	totals	140	172
Imperial Valley	Lost Circulation	6	10
	Stuck Pipe	15	9
	Cementing	45	48
	totals	66	67

ANALYSIS OF HISTORICAL DATA

A primary source of the actual drilling-time data used to analyze the frequency and severity of drilling and completion problems has been the drilling records filed with governmental units for geothermal wells. For wells in many geothermal resource areas, these records are the only drilling data available in the open literature. The trouble information recorded on them is generally quite sketchy and includes only major problems, and then only what the operator chooses to include. For example, the drilling records are not likely to show the lesser levels of lost circulation shown in the probability tree of Figure 2. However, the data contained in them are interesting and do provide useful results.

Baca

Data from the records on early Baca wells were discussed in Ref. 5. The most frequently cited problem was lost circulation and the second most common problem was stuck pipe. 29% of the trouble references were to lost circulation and 17% were to stuck pipe. Lost circulation occurrences cost an average of 2-1/2 days, while delays from the stuck pipe incidents averaged 8 days.

The Geysers

Newly analyzed data from 49 Geysers wells drilled in the early 1970s display strikingly similar characteristics to the Baca wells. (The drilling records analyzed were obtained from the California State Division of Oil and Gas open files.) Lost circulation was again the most common problem, with 30% of troubles being lost circulation and each occurrence costing roughly 2-1/4 days. 16% of the trouble references were to stuck pipe, but for these wells, the stuck pipe incidents cost an average of 3-1/4 days.

Table 2. Trouble frequency and severity--the Geysers.

Problem	Portion of Wells (%)	Average Lost Time per Occurrence (days)
Cementing Problems	54	2
Lost Circulation	46	2-1/4
Stuck Pipe	38	3-1/4
Fishing	26	3
Side Tracking	26	4
Twist Off	24	4-3/4
Sloughed Hole	16	2-1/4
Casing Problems	10	1-3/4
Rig Problems	10	2-1/4

Roughly 90% of the Geysers wells studied encountered a problem severe enough to be noted on the drilling record. Table 2 shows that more than half of the wells had cementing problems and that nearly half encountered lost circulation. Though fewer wells suffered from lost circulation, it was the most common problem--nearly twice as as common as cementing problems--because it repeated in the same well much more than did cement troubles. However, Figure 3 shows that there was no great tendency for lost circulation problems to repeat.

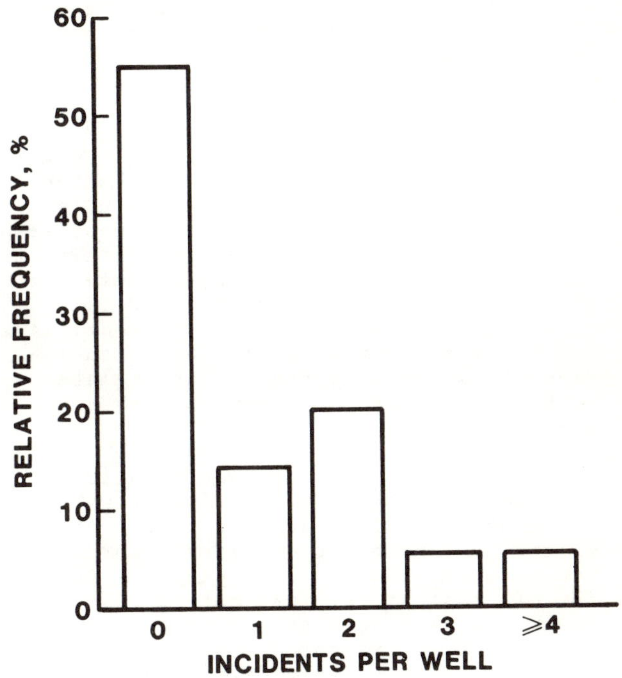

Figure 3. The Geysers Lost Circulation Experience

In terms of the total impact of trouble, the nine problems identified on the Geysers drilling records added an average of 8 days to the time required to drill a well. They increased the average well cost by an estimated 15%. Because of the nature of the drilling records, this constitutes a lower-bound estimate for the Geysers trouble severity and likelihood.

The overall figures for lost circulation, stuck pipe, and cementing problems (the three most common problems) agreed surprisingly well with the subjective estimates described above. However, they differed significantly in detail. In the subjective estimates, the problems, while greatest at the middle depths, occurred at comparable rates for the entire depth range of the well. For the historical data, the problems were concentrated in the intermediate casing portion of the well. It appears from the drilling records that drilling and completion problems at the Geysers are most severe at the middle depths, above the producing reservoir. The time lost between 2000 ft and 5000 ft was nearly ten times greater than for the shallower depths and five times greater than for the deeper region. This conclusion must be tempered by the fact that other problems such as erosion, corrosion, bit failure, casing failure, etc., are generally more severe at greater depths.

It should be noted that the historical trouble data from the drilling records do not entirely account for the wide spread in drilling time displayed in Figure 1. However, they do explain a significant portion of it.

SUMMARY

Geothermal drilling and completion rpoblems vary from region to region, primarily because of differences in geothermal resources. However, there are certain inherent similarities among resources that give rise to common problems. Drilling records indicate that the largest drilling and completions problem is lost circulation. Also significant are the related problems of incomplete cement jobs and stuck pipe.

Drilling and completion problems give rise to wide dispersion in drilling times and costs. Indications are that problems of the type noted on drilling records account for at least a 15% (average) increase in well cost at the most mature U.S. geothermal area. Accounting for the many problems that are not normally recorded would raise this estimate. Clearly, developing technologies to avoid or combat geothermal drilling and completion problems could have a significant impact on well costs.

REFERENCES

1. Varnado, S. G., editor, Geothermal Drilling and Completion Technology Development Program, SAND80-2179, November 1980.

2. Carson, C. C. et al., Representative Well Models for Eight Geothermal Resource Areas, SAND81-2202, to be published.

3. Carson, C. C. et al., Well Descriptions for Geothermal Drilling, Geothermal Resources Council Transactions, Vol. 5, October 1981.

4. Lin, Y. T., The Impact of Bit Performance on Geothermal Well Cost, Geothermal Resources Council Transactions, Vol. 5, October 1981.

5. Carson, C. C. et al., Geothermal Well Costs and Their Sensitivities to Changes in Drilling and Completion Operations, Proceedings of the International Conference on Geothermal Drilling and Completion Technology, Albuquerque, NM, SAND81-0036C, January 1981.

Geothermal Resources Council, TRANSACTIONS Vol. 6, October 1982

DESIGN OF A DIESEL EXHAUST GAS PURIFICATION SYSTEM FOR INERT GAS DRILLING*

Billy C. Caskey

Geothermal Technology Development Division
Sandia National Laboratories
Albuquerque, New Mexico 87185

ABSTRACT

To combat the serious oxygen corrosion of drill pipe when a low density drilling fluid (air or mist) is used in geothermal drilling, a system has been designed that produces an inert gas (essentially nitrogen) to be substituted for air. The system fits on three flatbed trailers, is roadable and produces 2000 scfm of gas. The projected cost for gas is slightly less than \$2.00 per thousand standard cubic feet. Detailed design information is available in a Sandia report written by Foster-Miller Associates, Inc.

INTRODUCTION

Sandia National Laboratories manages the U.S. Department of Energy's Geothermal Technology Development Program. A portion of the program has been aimed at helping solve the serious drill pipe corrosion problem that occurs when underpressured geothermal formations are drilled with a low density drilling fluid (air or mist). Even with corrosion control chemical costs of \$2000 to \$7000 per day, drill pipe corrosion can reduce drill pipe life to a few months (typical oil field life is a few years) resulting in drill pipe replacement costs of more than \$1000 per day.

One approach to reducing drill pipe corrosion involves removing oxygen from the drilling fluid. To assess the effectiveness of this method under actual field conditions, a test was planned (Caskey 1980) and conducted. The results (Caskey 1981) showed more than a ten-fold reduction in corrosion rates when nitrogen was used rather than air. However, as known beforehand, the cost of trucking and vaporizing liquid nitrogen is prohibitive for drilling operations (\$15,000-\$20,000) per day).

Sandia also performed laboratory tests and a study to determine appropriate gas composition specifications (Hinkebein 1981) that would apply to purifying diesel available "free" from drill rig engines. Based on these specifications, Sandia contracted with Foster-Miller Associates, Inc. for the design of a portable diesel exhaust gas purification system.

This paper describes the design concept (Dougherty 1982).

DESIGN PARAMETERS

In addition to the requirements defined in the Sandia contract, discussions with diesel engine manufacturers, compressor manufacturers, air drilling services, drilling contractors and geothermal operators resulted in final specifications for the design. The basic features and specifications are listed in Table 1. Requirements that most impacted the final design were the lack of cooling water and the removal of water from the product gas. A smaller, less costly system would result if these requirements could be relaxed. Unfortunately, water for cooling is generally not available at drilling sites and the air compressors presently used for drilling operations require that no condensation occurs during compression.

Table 1. Diesel exhaust purification system design features and specifications.

Basic features:
- Portable/roadable
- No cooling water available
- No electrical power available
- Ambient temperature of -50 to +110°F
- Long Life
- Easy setup and operation
- Reliable
- Low Cost

Inlet (diesel exhaust) specifications:
- Oxygen 6-12 percent
- Carbon dioxide ∿6 percent
- Oxides of nitrogen ∿0.1 percent
- Allowable back pressure <25 inches of water

*Work performed under the auspices of U.S. Department of Energy Contract No. DE-AC04-76PK00789.

Table 1. cont.

Outlet (product gas) specificiations:
Oxygen <50 ppm
Carbon dioxide <1 percent
Oxides of nitrogen <2 ppm
Water -55^oF dewpoint
Nitrogen 98+ percent

BASIC DESIGN

The system consists of three subsystems--O_2/NO_x removal, CO_2 removal and H_2O removal. Both CO_2 and H_2O removal are accomplished with minimal incremental expense since the thermal energy generated at the catalytic converters is in excess of that required by the regeneration of monoethanolamine and desiccant.

O_2/NO_x Removal -- O_2/NO_x removal is accomplished with standard catalytic converters based on automotive technology-operating characteristics and life expectancy are well understood. An electric preheater maintains the diesel exhaust temperature at 650^oF to ensure proper reaction rates in the first catalyst bed (CB-1 as shown in Figure 1.); fuel flow (65 gph for 11% O_2) is proportioned to each catalyst bed in order to remove the correct percentage of oxygen. NO_x is removed in the fourth stage which runs slightly rich. Therminol heat exchangers cool the gas stream after each stage to control catalyst and gas temperatures. As much as 8 million BTU per hour is generated during the catalytic reaction. Finally, the gas is cooled to 130^oF in a large air-cooled heat exchanger. Also, a centrifugal blower increases the gas pressure, limiting the back pressure on the diesel engine to 25 inches H_2O.

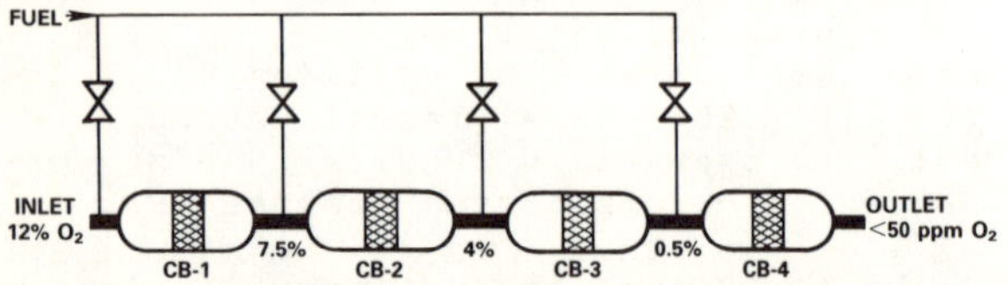

Figure 1. Diesel exhaust purification system four stage O_2/NO_x removal scheme. Oxygen levels are shown; NO_x is removed in CB-4 to less than 2 ppm. The product gas is cooled after each stage.

CO_2 Removal -- As shown in Figure 2, a high efficiency packed column scrubber using aqueous monoethanolamine (MEA) at 140 gpm is used to scrub the CO_2. MEA absorbs CO_2 and is regenerated (CO_2 boiled off) with Therminol previously heated by the catalytic reaction. Water saturated nitrogen exits the top of the scrubber, still at about 130^oF.

H_2O Removal -- Desiccant (activated alumina) is used to remove water from the nitrogen stream as shown in Figure 3. Three large (>6 foot diameter) dryers are required. Each will 1) absorb moisture, 2) be regenerated with Therminol-heated air, and 3) be cooled with dry air in a closed loop. Each step will be automatically selected by dryer valving. Step 1 will take about an hour and steps 2 and 3 about 45 minutes.

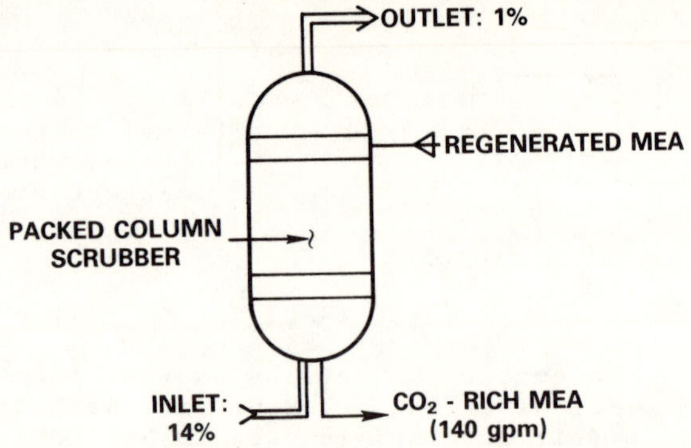

Figure 2. Diesel exhaust purification system carbon dioxide removal scheme. The monoethanolamine (MEA) is only 15% of the MEA stream; the rest is water.

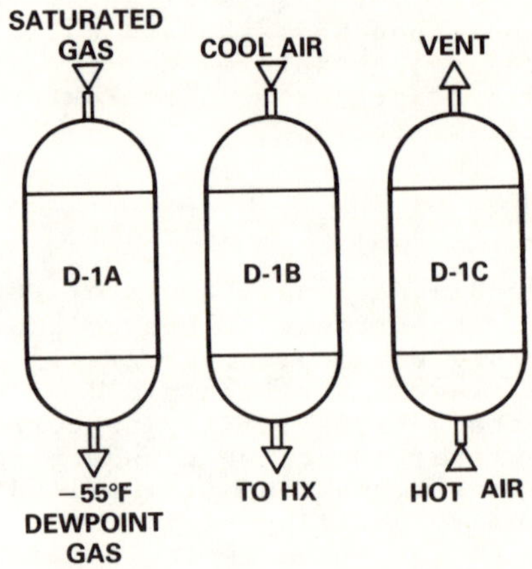

Figure 3. Diesel exhaust purification system desiccant dryers. D-1A drying product gas; D-1C desiccant being dried with hot air; D-1B being cooled with dry air in closed loop.

Ancillary Equipment -- The entire process flow diagram is presented in Figure 4. It shows, in addition to the three subsystems described above, the complete Therminol loop, the pumps, heat exchangers and valving required. Pressures and temperatures at various points are also shown.

Physical Layout -- Three thirty-six foot trailers are required to hold the system. The arrangements of each trailer are shown in Figures 5, 6, and 7. Components on the trailers may be located on Figure 4 by their identification code. A 150 kW diesel electric generator is placed on the second trailer; it supplies electric power for the entire system.

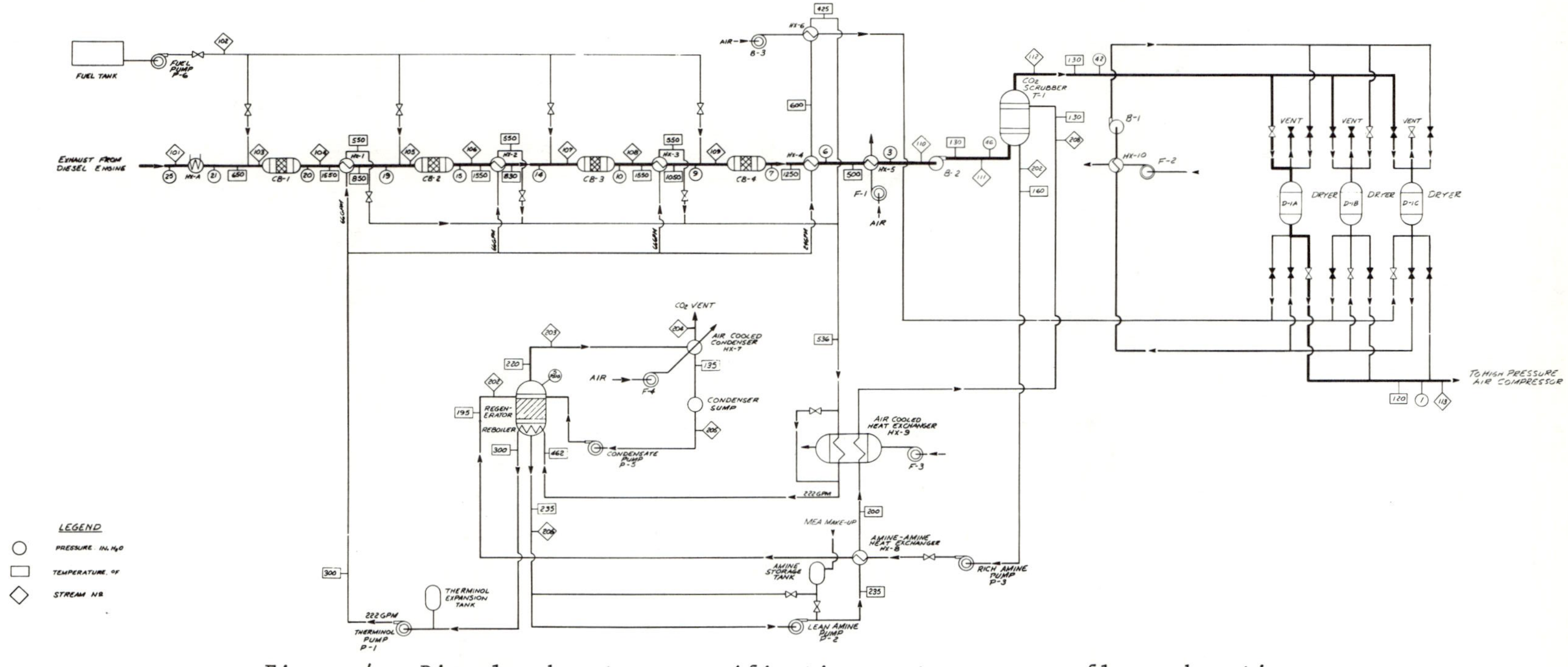

Figure 4. Diesel exhaust gas purification system process flow schematic.

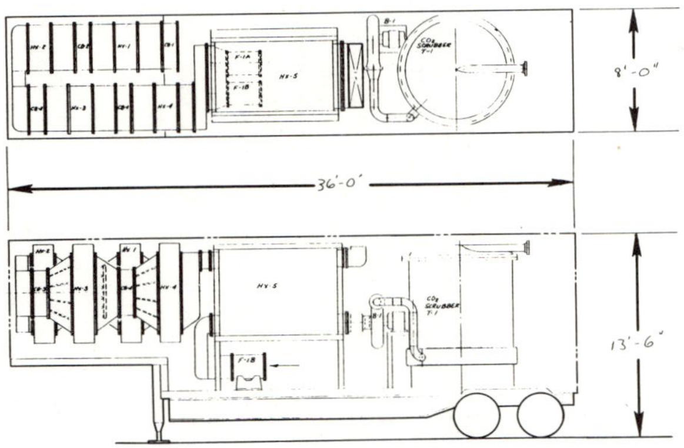

Figure 5. Diesel exhaust purification system trailer containing the catalytic combustors and carbon dioxide scrubber.

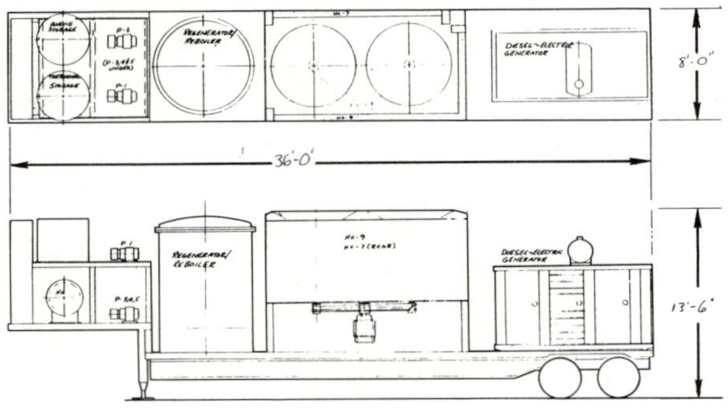

Figure 6. Diesel exhaust purification system trailer containing ancillary equipment.

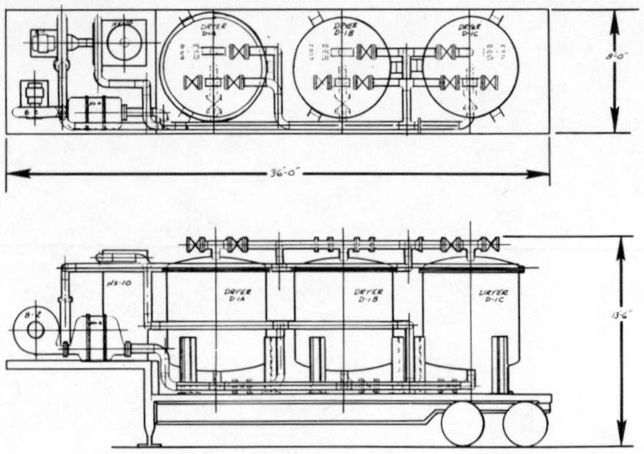

Figure 7. Diesel exhaust purification system trailer containing the desiccant dryers.

COST INFORMATION

Based on the economic assumptions presented in Table 2, the cost figures in Table 3 were calculated. The annual capital cost covers the payment required for a $1,126,860 loan to buy the system. Operating costs include fuel and labor; one operator is in attendance when the system is operating. Maintenance costs cover catalyst replacement at the end of its life, amine makeup, desiccant replacement every three years and 3 man weeks per year for in-shop cleaning and major component replacement.

Depending on the amount of oxygen to be removed, the resulting cost for nitrogen is between $1.47 and $1.73 per thousand standard cubic feet. Assuming the price was $2.00/1000 scf, the daily bill for this service would be about $4000 per day.

CONCLUSIONS

A design is available for a portable diesel exhaust gas purification system that produces an inert low density drilling fluid at an acceptable cost. Construction and use of this system would solve geothermal drill pipe corrosion problems--potentially lowering the cost of drilling by reducing drill pipe failures and reducing chemical costs. More importantly, use of this technique may open some geothermal areas to development--areas where the corrosion problems are so severe that drilling is impractical.

Table 2. Assumptions used in cost determinations.

Capital costs
- G & A 10 percent
- Warranty 5 percent
- Profit 20 percent
- Interest 18 percent
- Loan duration 8 years

Operating cost
- Fuel $1.10 per gallon
- Utilization 40 percent (16 hours per day for 18 days per month)
- Labor $98,112 per year ($14/hr and 100% overhead)

Maintenance cost
- 6000 hour catalyst life with 11 percent O_2 in exhaust
- 17,000 hour catalyst life with 6.5 percent O_2 in exhaust
- Amine $0.64 per pound
- Desiccant $1.00 per pound

Table 3. Cost figures.

	Oxygen in diesel exhaust 11% O_2 (2 cycle engines)	6.5% O_2 (4 cycle engines)
Capital cost (per year)	$276,080	$276,080
Operating cost (per year)	370,232	278,112
Maintenance cost (per year)	71,900	53,900
Totals	$718,212	$608,092
Nitrogen cost (per 1000 scf)	$1.73	$1.47

REFERENCES

Caskey, B. C. and Copass, K. S., Geothermal Drill Pipe Corrosion Test Plan, Sandia National Laboratories Report SAND80-1090, December 1980.

Caskey, B. C. and Copass, K. S., Drill Pipe Corrosion Control Using an Inert Drilling Fluid, published in the Proceedings of the International Conference on Geothermal Drilling and Completion Technology, Jan. 21-23, 1981, SAND81-0036C, p.28-1 thru 28-28.

Hinkebein, T. E. and Snyder, T. L., Corrosion Inhibition by Control of Gas Composition During Mist Drilling, Sandia National Laboratories Report SAND81-0973, May 1981

Dougherty, B. J., Foster-Miller Associates, Diesel Exhaust Gas Purification System, Sandia National Laboratories Report SAND81-7027, July 1982.

Geothermal Resources Council, TRANSACTIONS Vol. 6, October 1982

GEOTHERMAL COMPLETION TECHNOLOGY LIFE-CYCLE COST MODEL (GEOCOM)

A. J. Mansure and C. C. Carson

The BDM Corporation
1801 Randolph Rd. S.E.
Albuquerque, New Mexico 87106

Sandia National Laboratories
Division 4741
Albuquerque, New Mexico 87185

ABSTRACT

GEOCOM is a model developed to evaluate the cost effectiveness of alternative technologies used in the completion, production, and maintenance of geothermal wells. The model calculates the ratio of life-cycle cost to life-cycle production or injection and thus is appropriate for evaluating the cost effectiveness of a geothermal well even when the most economically profitable well completion strategies do not result in lowest capital costs.

The project to develop the GEOCOM model included the establishment of a data base for studying geothermal completions and preliminary case/sensitivity studies. The code has the data base built into its structure as default parameters. These parameters include geothermal resource characteristics; costs of geothermal wells, workovers, and equipment; and other data.

The GEOCOM model has been written in ANSI (American National Standard Institute) FORTRAN 1966 version. Documentation of the model will be available in a Sandia National Laboratories report (Anderson et. al.).

INTRODUCTION

In the development of technologies for completing geothermal wells it is necessary to evaluate the impacts of proposed new completion technologies. The Geothermal Completion Technology Life Cycle Cost Model (GEOCOM) was designed to aid this evaluation. It uses estimates of costs for completion operations and estimates of well performance to derive a measure of effectiveness for completion procedures. It facilitates comparisons of various completion activities and sensitivity analysis of individual technologies.

USE OF THE MODEL

The model includes the basic data required to analyze completions as default parameters. These include data on performance and characteristics of the major resources in the U.S., information on the cost of performing workovers, data on the scheduling of workovers, and general information on the economics and use of geothermal wells. The user of the model can generate a new case or study by modifying the appropriate parameters to change the values from default. This is done by use of a simple input processor designed to ease implementation of the model. Cases can be stacked one behind the other for convenience of parameter or sensitivity studies. The code itself was written using program design language and is explained with frequent comments making it easy for a person not already familiar with the code to understand its operation and make any modifications that might be desired. The code is divided into functional modules which reflect the various steps that must be done in calculating life-cycle costs of completion technologies. These modules include elements that initialize and allow for input of the parameters, determine the workover schedule, compute total production or injection of the well, compute the cost of the well, and output the results.

The GEOCOM model is an analytical tool designed to assist the engineer in evaluating the cost effectiveness of a geothermal well. Specifically, it will aid in integrating engineering economics, operations, resource characteristics, etc. into a single measure of the value of a well. It automatically accounts for such things as time-value-of-money, reservoir depletion, etc., but it is not intended to replace the engineer's understanding of geothermal completions. It is possible, for example, to ask the model to determine the cost effectiveness of configuraitons the engineer would automatically reject, such as a submersible pump in a dry stream reservoir.

For a single initial analysis, use of the model would likely require more effort than doing the work without a computer; however, many workovers have already been analyzed and are included in the model. Thus, for new technologies or uses, the model provides a quick method of comparison to baseline conditions and a structure for consistent analysis. Futhermore, "new" technologies or uses are often variations of existing methods already included in the model. The strength of the model lies in its ability to perform repeated calculations of cost effectiveness for multiple values of the parameters. It can thus be used for sensitivity studies or optimization.

SCOPE OF THE MODEL

The model includes sufficient detail to allow comparisons between various completion and usage alternatives for both production and injection wells. It combines capital cost of the well and other equipment; continuing cost for workovers, the cost of routine O&M, electricity, and chemicals; and analysis of productivity. To adequately analyze productivity, the model uses different data to characterize different geothermal resources and associates the appropriate well drilling cost with each resource. Since the scheduling of workovers can greatly impact productivity, the model allows freedom in scheduling workovers. The user can select which workovers he wants and when they are to be performed. The user can also select how the well is to be produced (artesian flow, submersible pump, line shaft pump, etc.). In order for the model to be appropriately sensitive to completion technologies, it could not use costs based on gross historical averages, such as annual workover costs equal to a percentage of capital cost. Instead, the model calculates cost for each workover performed and determines production as a function of time.

Geothermal Reservoirs

The differences in well performance and completion technology between the Geysers, with dry steam at 365°F, and East Mesa, with hot water at 340°F, are so great that it would be unreasonable to use average parameter values in GEOCOM. Thus, each important reservoir was considered separately and described by representative values for the basic reservoir-related parameters. The reservoirs included in the model are Brawley, Heber, The Geysers, Baca Ranch, Roosevelt Hot Springs, and East Mesa.

The parameters used to define the characteristics for each of these resources are well depth, length of completion interval, well head temperature, well life, initial well flow rate, total dissolved solids fraction, steam fraction, and reservoir performance decline rate. The decline rate is important in determining how long the well will flow at a usable rate. It includes such elements as drawdown, interference, formation plugging, etc.; all effects outside of the well. In contrast, the well lifetime is a physical or mechanical parameter determining how long until a failure, such as casing collapse, destroys the well.

Technologies

Many of the completion technologies that have potential to reduce life-cycle costs are new ones for which there are no historical data on cost or performance. These new technologies are often composed of, or substituted for, existing technologies, operations, or equipment. Jet descaling is an example of a potential technology to improve the cost effectiveness of geothermal wells. It would supplant mechanical descaling (done with bit and reamer), but would utilize many of the same support equipment items, such as a rig. The difference between mechanical descaling and jet descaling would depend on such things as the rate of scale removal, effectiveness of scale removal, and any extra costs for additional or different equipment.

The GEOCOM model does not have sufficient data base to analyze all potential completion technologies. For many new ideas, cost and performance data do not exist. GEOCOM can, however, help evaluate such ideas. First, it is used to analyze the present completion method to which the new method will be compared. Second, the parameter changes which determine the cost difference between the new and old technologies are identified. Last, GEOCOM is used to estimate the impacts of the parameter changes that define the new completion method.

The basic technologies that have been included in the GEOCOM model are:

(1) Logging
(2) Perforating
(3) Mechanical Descaling
(4) Hydrojet Descaling
(5) Chemical Scale Inhibition
(6) Injection Pumps
(7) Submersible Pumps
(8) Line Shaft Pumps
(9) Remedial Cementing
(10) Underreaming and Gravel Pack
(11) Slotted Liner Replacement
(12) Well Repair with a Liner

MODEL STRUCTURE AND PARAMETERS

The parameters used in the calculation of an individual GEOCOM model case can be grouped according to function into the following categories: capital cost, continuing cost, productivity, schedule, and financial. A list of these parameters is given in table 1. For many parameters a default value is given. For others the value depends on which geothermal reservoir is picked or upon which workover type or method of utilitization is chosen, and this is indicated. In addition to the parameters directly used in calculation, the GEOCOM model also uses parameters of three types: case selection, general, and special. The case selection parameters determine the geothermal resource and method of using the well (injection or production well). The general parameters include parameters such as the rejection temperature (important in converting from pounds of production to Btus). Special parameters are used in the costing of the individual workovers. The model automatically calculates the base case cost for each workover type but allows the user to change the basis for cost calculations. For example, by use of the special parameter the user can change the number of shots per foot in cemeted perforated completions.

Use of the model is not limited to those workovers already programmed into it. Engineering calculations for the cost, frequency, duration, and effectiveness of a new workover can be input directly into the model.

TABLE 1. MODEL PARAMETERS

PARAMETER	VALUE	TYPE OF PARAMETER
INITIAL WELL FLOW	1	PRODUCTIVITY
RESERVOIR DECLINE RATE	1	PRODUCTIVITY
DISCOUNT FACTOR	0.00878/MO	FINANCIAL
INFLATION FACTOR	0.00695/MO	FINANCIAL
WELL LIFE	1	SCHEDULE
STUDY PERIOD	360.MO	SCHEDULE
WELL FLOW LOSS	0.125/MO	PRODUCTIVITY
INITIAL DELAY	0.0 MO	SCHEDULE
CAPITAL COSTS		CAPITAL COSTS
- WELL	1	
- OTHER	2	
WORKOVER		
- COSTS	2	CONTINUING COSTS
- FREQUENCY	2	SCHEDULE
- DURATION	2	SCHEDULE
- EFFECTIVENESS	2	PRODUCTIVITY
ROUTINE O&M		CONTINUING COSTS
- WELL	1	
- OTHER	2	
PER POUND PRODUCTION COSTS	2	CONTINUING COSTS
REPAIR		
- COST	$121,000.	CONTINUING COSTS
- TIME	999. MO	SCHEDULE
- DURATION	0.33 MO	SCHEDULE
- EFFECTIVENESS	40%	PRODUCTIVITY
ABANDONMENT COST	$18,000.	CONTINUING COST

1 RESOURCE DEPENDENT VALUE

2 CONFIGURATION AND USAGE DEPENDENT VALUE

Continuing Cost Parameters

There are five basic continuing cost parameters: workover costs, routine O&M costs, costs per pound production, repair costs, and abandonment costs.

Workover costs are the costs to workover a well such as those for bringing in a workover rig to mechanically drill and ream out scale. Only the cost of the specific workovers selected by the user are included in the total cost of the well. Costs of the workovers are put into the cost stream at the times determined by the workover schedule. Routine O&M costs include labor to operate the well, materials and supplies, and maintenance. Routine O&M includes a flat yearly amount ($54,000/yr), plus a percentage of the well capital cost (3%), plus a percentage of other capital cost (varying depending upon the usage of the well).

Per pound production costs are the operating costs that depend on the amount of production. Examples are the electricity cost for downhole pumps and chemical costs for chemical inhibition.

The repair of a well is a one-time event such as the repair of a casing failure by cementing in a liner. This operation connot be repeated and changes the profile and productivity of the well.

Productivity and Schedule Parameters

There are five basic productivity parameters: initial well flow, reservoir decline rate, well flow loss, workover effectiveness, and repair effectiveness. Figure 1 shows the productivity and schedule parameters.

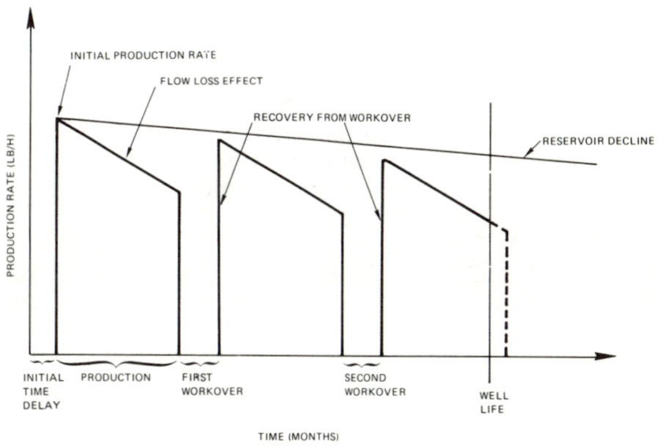

Figure 1. Workover and Production History Illustration

There are six schedule parameters: well life, study period, workover frequency, workover duration, repair time, repair duration, and initial delay. The GEOCOM model allows up to three types of workovers to be scheduled. The frequencies of the workovers are mutually independent. The model calculates the combined workover schedule and determines cost and productivity appropriately.

The initial delay parameter allows for inclusion of the effect of the delay between when the well is drilled and when the power plant is first on-line. The reservoir decline rate parameter is used to take into account the production or injection decline of the entire reservoir. The flow of individual wells may also decline because of damage to the well or scaling of the well. These effects are taken into account by the flow loss parameter. When a workover is performed the flow rate of the well can recover only well loss, not reservoir decline. In cases where the workover is not 100% effective, recovery is not complete.

If the study period, e.g., power plant lifetime, is more than the well life, the model combines the appropriate number of completed well lives plus a partial one. For the partial well, it prorates the initial capital cost according to the fraction of production during the partial well life divided by production of the whole well life. Continuing costs and benefit are taken as the values accumulated up to the end of the partial well life. All values for the wells needed after the initial ones are discounted and inflated to present value when the first well was drilled.

Mansure, Carson

SUMMARY

The results of the GEOCOM project are three fold: the establishment of a data base for studying geothermal completions, the development of the GEOCOM model, and preliminary case/sensitivity studies. The data base is included in the model in the form of default parameters. The GEOCOM code is documented in a Sandia report. The preliminary case/sensitivity studies have been reported elsewhere and are also in the Sandia documentation (Carson et. al.).

The model uses the ratio between life-cycle cost and life-cycle production or injection as a measure of effectiveness and provides sufficient flexibility to allow comparisons between the various completion and workover alternatives for wells in order to aid in management direction of a technology development program.

The GEOCOM model is an analytical tool designed to assist the evaluation of the cost effectiveness of a geothermal well. Specifically, it will aid in integrating engineering economics, operations, resource characteristics, etc. into a single measure of the value of a well. It will automatically account for such things as time-value-of-money, reservoir depletion, etc. The model provides a quick method of comparison to base-line conditions and a structure for consistent analysis. The strength of the model lies in its ability to perform calculations of cost effectiveness for multiple values of parameters. It can thus be used for sensitivity studies or optimization.

BIBLIOGRAPHY

Anderson, E. R., Hoessel, W. C., Mansure, A. J., and McKisson, P., Geothermal Completion Technology Life Cycle Cost Model (GEOCOM), Volume I - Final Report, Volume II - User Instruction Manual, SAND82-7006, 1982.

Brown, G. L., Mansure, A. J., and Miewald, J. N., Geothermal Wells: A Forecast of Drilling Activity, SAND81-7127, 1981.

Carson, C. C. and Mansure, A. J., The Impact of Common Completion and Workover Activities on the Effective Costs of Geothermal Wells, SPE11155, 1982.

Geothermal Resources Council, TRANSACTIONS Vol. 6, October 1982

A NEW BENTONITE SOURCE WITH UNUSUAL THERMAL STABILITY

William J. Miles

Miles Industrial Mineral Research
12940 West 16th Drive
Golden, CO 80401

ABSTRACT

A new bentonite source is being developed in Queensland, Australia that meets all current API specifications and appears to have unusual geothermal stability. It does not undergo thermal flocculation that is characteristic of Wyoming bentonite until approximately 500°F. Viscosity of this Australian bentonite is essentially constant over the temperature range studied. Its response to polymers and thinners is similar to Wyoming bentonite. When used in geothermal drilling fluids such as HTM-1 and several variations of it, filtration control is superior to sepiolite and similar to Wyoming bentonite. Hot rolling at 500°F for extensive periods indicate that this Australian bentonite is durable.

INTRODUCTION

Development of geothermal resources is highly dependent on drilling technology and drilling fluid systems. Drilling fluids are required that maintain the required properties of relatively constant viscosity, filtration control to prevent reduction of the porosity of the formation, and freedom from temperature flocculation with resulting gellation when exposed to formation conditions.

In search of improved geothermal drilling muds, the D.O.E. has sponsored the development of the HTM-1 geothermal mud. It is based on sepiolite for controlled viscosity, some bentonite and polymer for filtration control, brown coal for thinning the mud, and caustic. Sepiolite possesses the required thermal stability but lacks filtration control due to its lath-like morphology, although a platey morphology develops eventually as sepiolite alters to stevensite. Wyoming bentonite lacks the thermal stability and eventually undergoes temperature flocculation.

A new bentonite deposit is being developed that has much greater thermal stability in drilling systems than Wyoming bentonite, and much greater filtration control than sepiolite. This bentonite is produced by Cudgen R.Z., Ltd., in Queensland, Australia under the trade name TRUGEL RED. It is an API approved drilling mud product.

THERMAL STABILITY

Rheology at elevated temperatures and pressure was acquired with a Fann 50 C viscometer for TRUGEL RED and an API grade Wyoming bentonite. The clays were prepared by API procedures with aging of the slurries for completion of hydration of the bentonite prior to rheology measurement. The rheology curves were obtained at 60 rpm. since this corresponds to typical drilling conditions.

TRUGEL RED exhibits an almost flat viscosity profile to 500°F, the limit of the Fann viscometer. During the temperature program, the viscosity drifts between 20 and 42 cps., finally increasing to about 65 cps. upon cooling. The temperature profile is illustrated by the second curve.

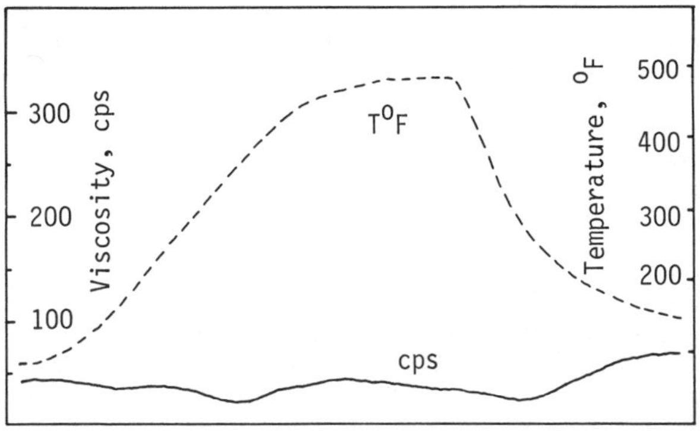

Figure 1. TRUGEL RED viscosity at elevated temperatures and pressure. 25 ppb bentonite concentration.

The corresponding Wyoming bentonite slurry exhibits temperature flocculation beginning at about 250°F, as illustrated by the rapidly increasing viscosity. The temperature program was terminated at 375°F due to extremely high viscosity of 320 cps. Upon cooling, a portion of the temperature flocculation was reversible; however, the viscosity decreased to 170 cps, much greater than its initial value. This residual viscosity indicates a permanent change of the clay due to heating.

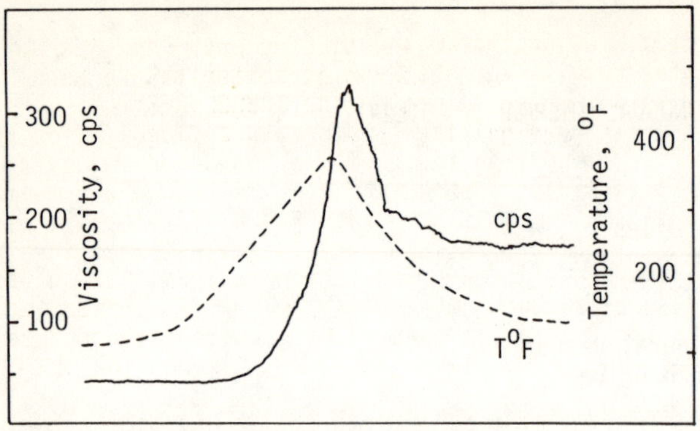

Figure 2. Wyoming bentonite viscosity at 60 rpm at elevated temperatures and pressure. 25 ppb bentonite concentration.

HOT ROLLING TESTS

Since the heating of the bentonite is relatively short in the Fann 50 C viscometer, both TRUGEL RED and Wyoming bentonite were hot rolled for 24 hours at temperatures ranging from 265°F to 500°F. Slurry concentrations were 22.5 ppb clay. After cooling, the slurries were resheared for 5 minutes to break down gel structure, followed by measurement of viscosity and API filtration. Residual viscosity increases after heating are interpreted as an indication of temperature flocculation, based upon the previous Fann 50 C data.

The TRUGEL RED slurries were all flowable muds when removed from the roller cells. The rheology results indicate that viscosity characteristics are essentially constant to 450°F. Some viscosity increase is observed at 500°F; however, it appears to retain functional properties. Filtrates increase with increasing temperature, but remain comparable to Wyoming bentonite and superior to sepiolite.

Table 1.

TRUGEL RED slurry properties after 48 hour hot rolling at 500°F.

T°F	Plastic Viscosity	Yield Point	Gel Strengths	API Filtrates
API	10	18	6-19	10.4
265	10	10	3-20	12.6
300	14	10	4- 9	14.5
350	13	11	3- 4	17.2
400	12	12	3- 3	17.5
450	12	12	3- 4	17.5
475	15	12	3- 4	16.4
500	17	16	3- 3	16.0

The Wyoming bentonite slurries were all found to be flocculated and gelled into pastes. These pastes did thin with reshearing; however, the rheology data shows considerably higher residual viscosity indicating temperature flocculation. Above 300°F, the rheology characteristics were not stable and the values reported are averages of the high and low viscosities observed. API filtrates increase through the temperature profile, similarly to TRUGEL RED.

Table 2.

Wyoming bentonite slurry properties after 48 hour hot rolling at 500°F.

T°F	Plastic Viscosity	Yield Point	Gel Strengths	API Filtrates
API	14	12	3-14	10.4
265	25	28	11-40	10.7
300	33	43	12-37	11.6
350	30	50	9-26	12.0
400	30	45	7-14	13.4
450	30	50	14-30	13.8
475	30	60		16.6
500	27	89		18.3

These tests were repeated for 96 hour hot rolling periods through the temperature profile. Viscosity and filtrate results observed were similar to the 24 hour results for each bentonite.

EFFECT OF THINNERS AT ELEVATED TEMPERATURE

Since drilling mud systems are more complex than the simple systems already discussed, these clays were hot rolled with sodium hydroxide for pH adjustment to over 10, and 2 ppb lignite or ferrochrome lignosulfonate. These 22.5 ppb clay slurries were hot rolled overnight at temperatures from 300°F to 500°F.

When cooled, TRUGEL RED slurries were all flowable muds. After shearing, the rheology results for the lignite thinned slurries show essentially constant plastic viscosity and yield point over the temperature range, while filtrates increase with temperature. With lignosulfonate, the plastic viscosity and yield point are constant to 400°F, with a noticeable increase in both values at 500°F. API filtrates increase with temperature, but remain similar to those of Wyoming bentonite.

Table 3.

Properties of TRUGEL RED and thinner slurries after hot rolling. 22.5 ppb bentonite, 2 ppb thinner, pH 10, overnight hot rolling.

T°F	Plastic Viscosity	Yield Point	Gel Strengths	API Filtrates
Lignite				
300	9	11	4-30	15.3
400	10	3	2- 3	16.6
500	12	8	2- 2	18.4

Table 3 continued

Ferrochrome lignosulfonate				
300	8	4	2-10	16.1
400	8	4	2- 4	18.2
500	16	24	2- 4	21.5

The Wyoming bentonite exhibits relatively high residual viscosities with lignite after heating at all of the temperatures studied. It does not show a significant increase in plastic viscosity or yield point at 500°F. With lignosulfonate, plastic viscosity and yield point are higher than TRUGEL RED, but constant to 400°F. At 500°F, its viscosity characteristics increase dramatically. API filtrates increase consistently with increasing temperature.

Table 4.

Properties of an API grade Wyoming bentonite and thinner slurries after hot rolling. 22.5 ppb bentonite, 2 ppb thinner, pH 10, overnight hot rolling.

T°F	Plastic Viscosity	Yield Point	Gel Strengths	API Filtrates
Lignite				
300	24	24	4-10	11.4
400	29	19	3- 4	12.4
500	29	17	3- 4	13.9
Ferrochrome Lignosulfonate				
300	23	6	3- 3	14.2
400	22	6	2- 3	14.8
500	40	22	3- 3	16.0

Both lignite and ferrochrome lignosulfonate thin TRUGEL RED more efficiently than Wyoming bentonite at all of the temperatures studied. This has also been observed in practice in Australia, although formation temperatures have been limited to 265°F to date. Gel strengths remain comparable for both clays and thinner combinations. Filtrates are higher for TRUGEL RED, but basically similar.

GEOTHERMAL DRILLING FLUID FORMULATIONS

The HTM-1 drilling fluid formulation contains 15 ppb sepiolite for rheology control, 5 ppb bentonite for some filtration control, 2.5 ppb polymer for additional filtration control, 20 ppb brown coal for thinning and inhibition of temperature flocculation, and 2 ppb caustic to solubilize the humates and maintain pH. Brown coal was chosen for this formulation instead of typical drilling grade lignite because of its greater thermal stability.

Both TRUGEL RED and Wyoming bentonite were tested in this formulation as the bentonite component by hot rolling for 24 hours at 500°F. Shearometer tests were carried out on the cooled muds within the roller cells to quantify the flocculation thickening that occurs. The shearometer results for the Wyoming bentonite are more than twice the value for the TRUGEL RED mud, indicating significant temperature flocculation. After reshearing the drilling fluids, rheology and API filtrates were determined. Although the muds were similar in viscosity, TRUGEL RED viscosities are significantly lower and filtrates are slightly higher than the Wyoming bentonite. Filtration was also carried out at 500 psi and 300°F, indicating better filtration control at elevated temperature for TRUGEL RED.

Since many geothermal drilling fluids presently used do not contain the quantities of sepiolite present in HTM-1, the formulation was modified to replace all of the sepiolite with bentonite. After hot rolling 24 hours at 500°F, shearometer results indicate much weaker gel structure for TRUGEL RED. After shearing, rheology results are quite similar for both mud systems. In this formulation, high temperature filtration is superior for TRUGEL RED.

Table 4.

Evaluation of physical properties of TRUGEL RED and Wyoming bentonite in the HTM-1 formulation after hot rolling for 24 hours at 500°F.

ppb sepiolite	15	15		
ppb W. bentonite		5		20
ppb TRUGEL RED	5		20	
ppb brown coal	20	20	20	20
ppb polymer	2.5	2.5	2.5	2.5
ppb caustic	2	2	2	2
shearometer $lb/100ft^2$	28	60	5	60
Plastic Viscosity	14	17	11	13
Yield Point	8	12	7	5
Gel strengths	4-12	3-5	4-14	3-4
API Filtrate, cc.	7.7	6.0	5.7	4.5
Filtrate, 55 psi, 300°F	26.6	35.0	22.4	27.2

Additional 120 hour hot rolling tests were carried out for TRUGEL RED, and results show essentially no deterioration of the system. The brown coal was reduced to 10 ppb in a 20 ppb TRUGEL RED formulation. Results for this system also indicate preservation of rheology and filtration characteristics with 24 and 96 hour heating.

As an alternate to brown coal, causticized lignite and another high temperature filtration control polymer (DFE-105) were substituted in the HTM-1 formulation. These slurries were hot rolled at 500°F for 48 hours, followed by cooling and measurement of physical properties. Again, shearometer results indicate much stronger residual flocculation gel structure for the Wyoming bentonite formulations. After reshearing, the Wyoming bentonite muds retained significantly more viscosity, indicating permanent effects from the heat treatment. API filtrates for the Wyoming bentonite slurries were lower, but the high temperature filtrates were poorer.

Table 5.

Evaluation of physical properties of TRUGEL RED and Wyoming bentonite in a geothermal drilling fluid formulation after hot rolling for 48 hours at 500°F.

ppb TRUGEL RED	20		20	
ppb W. bentonite		20		20
ppb causticized lignite	10	10	5	5
ppb DFE-105	5	5	3	3
ppb caustic	2	2	2	2
shearometer lb/100 ft^2	3	40	5	37
Plastic Viscosity	16	15	14	14
Yield Point	7	19	5	20
Gel strengths, sec.	4-6	5-6	3-7	6-6
API Filtrate, cc.	5.1	4.3	7.4	5.8
Filtrate, 500 psi, 300°F	42.0	47.6	25.0	34.2

CONCLUSION

After testing in simple systems and in geothermal drilling fluids under simulated rigorous conditions, an Australian bentonite deposit appears to have unusual thermal stability. Testing for extended periods at 500°F indicates that TRUGEL RED can be used to formulate durable drilling fluids which compare favorably to Wyoming bentonite based drilling fluids. This bentonite has not been used to date in a geothermal well, but it has been used successfully in a variety of conditions in Australia.

ACKNOWLEDGEMENTS

The author would like to thank Mr. Orien Van Dyke and Fann Instrument Operations for providing the Fann 50 C viscosity data. This work was sponsored by Cudgen R.Z., Ltd. located in Brisbane, Q., Australia.

REFERENCES

Carney, L. L., and Meyer, R. L., A new approach to high temperature drilling fluids, 1976, Society of Petroleum Engineers of AIME, paper no. SPE 6025.

Kelsey, J. R., 1980, Recent developments in geothermal drilling fluids, SAND-80-2702C.

Remont, L. J., 1979, Application and utilization of improved geothermal drilling fluid technology, SAND-7120.

Remont, L. J., 1979, Field trial of HTM-1, SAND-79-7090C.

Remont, L. J., Gray, G. R., Chenevert, M. E., Carwile, C. E., 1977, Temperature limitations of drilling fluids, CONF-771153-P1.

Remont, L. J., Rehm, W. A., McDonald, W. J., Maurer, W. C., 1976, Evaluation of commercially available gelthermal drilling fluids, SAND-77-7001.

Geothermal Resources Council, TRANSACTIONS Vol. 6, October 1982

FRACTURE STIMULATION EXPERIMENT IN BACA 20

by C. W. Morris, R. V. Verity, and C. L. Ritz

Republic Geothermal, Inc.
11823 East Slauson Avenue
Santa Fe Springs, California 90670

Abstract

The Geothermal Reservoir Well Stimulation Program group performed a large hydraulic fracture treatment in the well Baca 20, located in Union's Redondo Creek Project Area in north-central New Mexico. A nonproductive 240-foot openhole interval, with a formation temperature of 520°F, was selected and isolated for the job. The treatment utilized a large cooling water pre-pad followed by a high viscosity frac fluid carrying sintered bauxite as the proppant. Post-stimulation tests and analyses indicate a fracture was created with a vertical height of about 100 feet at the wellbore. This is the highest temperature well in the world to be successfully prop-fractured to date.

A number of fracture evaluation techniques were used to analyze the job, including some that are quite unique. Among these techniques were fracture mapping by monitoring the miscroseismic events caused by the fracture job, downhole pressure data analyses, temperature profiles, electric log surveys, production tests, and chemical tracer monitoring. The post-stimulation fluid productivity of the well is sub-commercial, probably because of low formation permeability surrounding the artificially created fracture.

Introduction

The U.S. Department of Energy-sponsored Geothermal Reservoir Well Stimulation Program (GRWSP) was initiated in February 1979 to pursue industry interest in geothermal well stimulation work and to develop technical expertise in areas directly related to geothermal well stimulation activities. Republic Geothermal, Inc. and its principal subcontractors (Maurer Engineering, Inc. and Vetter Research) have completed seven full-scale field experiments to date. One of these experiments, performed on October 5, 1981 in the Baca 20 well located in New Mexico, is described herein where unique diagnostic data were obtained during the treatment. These measurements, along with conventional techniques, were used to evaluate the fracture treatment.

The Baca reservoir lies within the Jemez Crater, Valles Caldera, and is defined by more than 20 wells completed to date in the Redondo Creek area by Union Geothermal Company of New Mexico (Union). The main reservoir, 4,000 to 6,000 feet in thickness, is composed of volcanic tuffs with low permeability and a primary flow system of open fracture channels. In the Redondo Creek area, wells have encountered a high temperature (500°F+) liquid-dominated geothermal reservoir, but a number of wells have not been of commercial capacity, primarily because of the absence of productive natural fractures at the wellbore. The large hydraulic fracture job was intended to create the high conductivity flow channel needed to make Baca 20 a commercial well. This experiment was cost-shared with Union.

Description of Stimulation Treatment

Baca 20 was originally completed as shown in Figure 1A with a 9-5/8 inch liner cemented at 2,505 feet and a 7 inch slotted liner to 5,812 feet. In preparation for the treatment, the well was recompleted as shown in Figure 1B to isolate 240 feet of openhole interval. This particular 240-foot interval, from 4,880 to 5,120 feet, was chosen primarily because the best production in the area has been found in comparable intervals near the bottom of the Bandelier Tuff and because of its high reservoir temperature (520°F).

The hydraulic fracture treatment was accomplished in the eleven stages defined in Table 1. The high formation temperature (520°F) dictated special design and materials selection. A 3,000 bbl fresh water pre-pad (Stages 1-3) was used to cool the wellbore and fracture. In the following Stages 4-10, 5,600 bbl of crosslinked polymer gel frac fluid were injected, carrying 119,700 lb of 16/20-mesh and 119,700 lb of 12/20-mesh sintered bauxite proppant in the last four of the stages. The frac fluid was a 60 lb/1,000 gal hydroxypropyl guar polymer gel mixed in fresh water and crosslinked as it was pumped. This fluid was a new high-pH system (Western HT FF-60) designed for improved stability at high temperature. Sintered bauxite proppant was used because of its laboratory demonstrated ability (GRWSP-Phase 1) to withstand high temperature and stress. The majority of the treatment

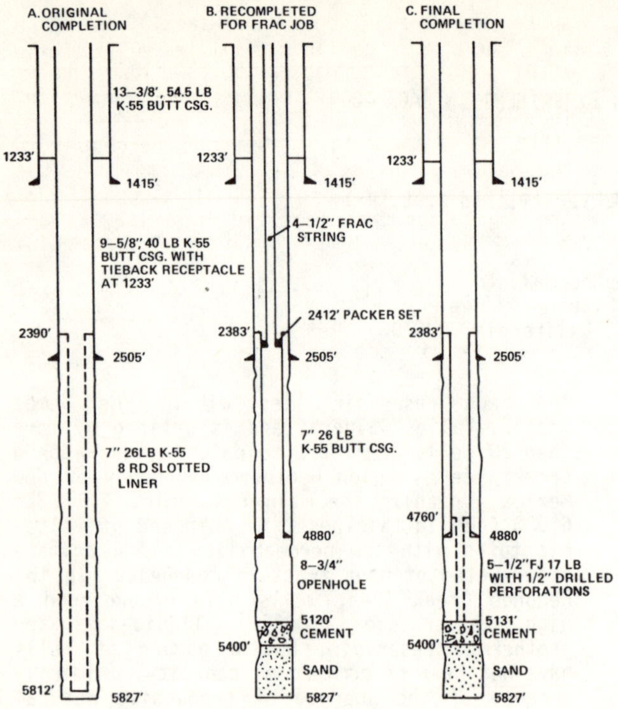

Figure 1. Baca 20 completion details.

Table 1

BACA TREATING SCHEDULE

Stage No.	Planned Size (bbl)	Actual Size (bbl)	Proppant (lb/gal)	Size	Fluid
1.	2000	2000			FRESH WATER WITH FLUID LOSS ADDITIVE (FLA)
2.	500	639	0.39	100-MESH $CaCO_3$ (10,500 LB)	FRESH WATER WITH FLA
3.	500	350			FRESH WATER WITH FLA
4.	1500	1400			POLYMER GEL WITH FLA
5.	500	566	1.33	100-MESH $CaCO_3$ (31,500 LB)	POLYMER GEL WITH FLA
6.	500	500			POLYMER GEL WITH FLA
7.	1150	1168	0.46	16/20-MESH BAUXITE	POLYMER GEL
8. a	850	682	1.85	16/20-MESH BAUXITE	POLYMER GEL
b		378	2.77	16/20-MESH BAUXITE	POLYMER GEL
9.	300	450	2.11	12/20-MESH BAUXITE	POLYMER GEL
10.	750	451	4.21	12/20-MESH BAUXITE	POLYMER GEL
11.	150	151			FRESH WATER
	8700	8735			

fluid was pumped at approximately 80 BPM. The rate was slowed to 40 BPM in Stage 10 when the proppant concentration was increased to 4.2 lb/gal.

Fracture Treatment Evaluation

During the fracture treatment of Baca 20, Los Alamos National Laboratory (LANL) performed a fracture mapping experiment using Baca 22, located approximately 1,000 feet from Baca 20, as an observation well. A triaxial geophone system was placed in the Baca 22 well at a depth of about 3,000 feet and, using techniques developed for the Hot Dry Rock Geothermal Energy Program, microseismic activity caused by the fracture job was mapped. A large number of discrete events (38) were recorded during the frac job; however, the geomagnetic orientation measurement of the tool was lost. The microseismic activity, illustrated in Figures 2 and 3, occurred in a broad zone which was roughly 2,300 feet long, 700 feet wide, and 2,000 feet high. Each mapped event location is probably known within 150 feet in relation to other rock failure locations, and thus, clearly suggests the stimulation treatment did not result in the creation of a singular monolithic fracture. Also the microseismic events would be expected to proceed in advance of any significantly widened fracture and would not necessarily define a propped flow path to the wellbore at Baca 20.

The 240-foot isolated open interval was nonproductive prior to the treatment, although there was a small rate of fluid loss during the well completion operations. After the frac job, several temperature surveys were obtained in the lower part of the well. In general, these surveys showed the heat-up of the wellbore following the frac job and the same trends as the pre-frac survey except for the bottom zone of the open interval. These temperature surveys confirmed that only the bottom interval of about 100 feet in height (below 5,000 feet) accepted all or most of the frac fluid.

On October 10-11, 1981, a 6-hour modified drillstem test (DST) was performed which was a unique combination of conventional DST methods and gas lift to maintain steady, single-phase flow to the wellbore. Transient pressure and temperature data were obtained downhole during the production period and the subsequent pressure buildup period. Conventional transient pressure analysis techniques were used to analyze the data which yielded a reservoir permeability-thickness of 1,000 md-ft.

Theoretical calculations were made to estimate the fracture geometry resulting from the actual stimulation treatment pumping schedule. Assuming a large vertical fracture geometry to approximate a penny-shape fracture (i.e., a 600-foot fracture height), the calculated fracture length and width are 340 feet and 0.432 inches, respectively. Fracture conductivity calculations also suggest the artifi-

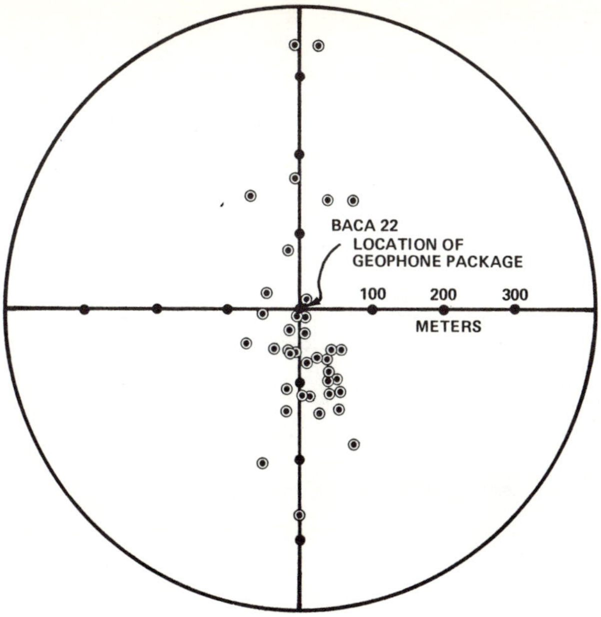

Figure 2. Baca 20 microseismic event locations.
Shown in plan view (No geomagnetic orientation)

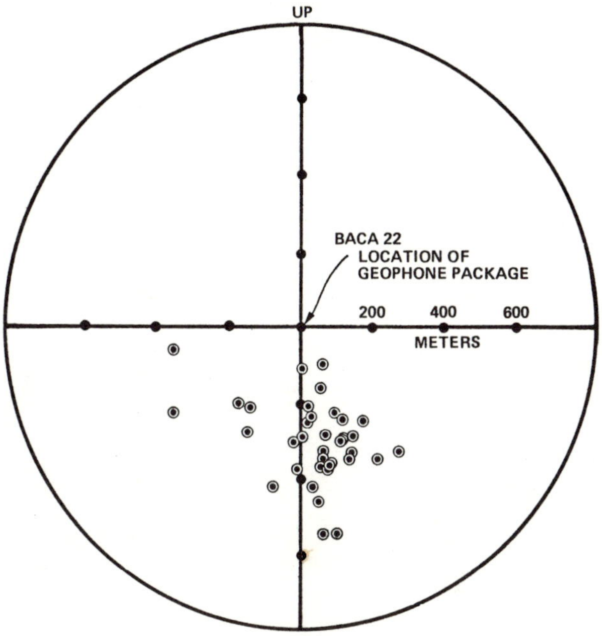

Figure 3. Baca 20 microseismic event locations.
Shown in vertical projection (No geomagnetic orientation)

cially created fracture has relatively high conductivity (i.e., in the range of 500 darcies, which corresponds to laboratory data for proppant packs using this size material).

Following the modified DST, a 14-day flow test was performed by Union to determine the well's productive capacity. The well produced approximately 110,000 lb/hr total mass flow initially, but declined rapidly to a final stabilized rate of about 50,000 lb/hr under two-phase flow conditions in the formation. The measured steam fraction increased from about 50 percent to a final 86 percent with the wellhead pressure declining from 50 psig to a final 24 psig. These data suggest that the production test resulted in a large pressure decline in the well and two-phase flow conditions were occurring in the formation.

The most probable cause of the low productivity is the relative permeability reduction associated with two-phase flow effects in the formation. These effects result principally from the low reservoir permeability conditions in this area, combined with the sub-hydrostatic reservoir pressure.

In order to determine the fraction of injected frac fluid that was returned when production commenced, several chemical tracers were employed. The tracers were methanol, n-propanol, Tinopal CBS-X, and the polymer itself. All the chemicals, except the Tinopal CBS-X, are included in the normal composition of the fracture fluid.

During the production testing done after the stimulation treatment, fluids were sampled and analyzed for their tracer content. Figure 4 shows the analyses results for the methanol and n-propanol concentrations in the returned fluids versus the cumulative flow volume. The other tracers exhibited similar trends.

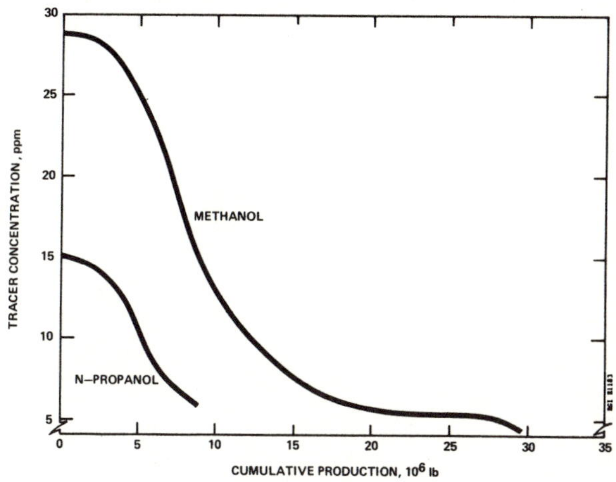

Figure 4. Concentration of methanol and N-propanol tracer versus cumulative production.

Material balance calculations show that 26 percent of the methanol, 19 percent of the n-propanol, and 21 percent of the polymer were returned to the surface. Tinopal CBS-X, a fluorescent, water soluble tracer was added to the gelled fluid in Stage 4. Only one percent of the Tinopal CBS-X was produced during the post-stimulation flow tests. The low recovery of the Tinopal CBS-X is most likely related to the fact that, unlike the other tracers, it was injected as a slug early in the treatment. The frac fluids injected in the early stages may have been pushed far enough into the formation to intersect the natural fracture system and be diluted by the reservoir fluids and/or trapped in the formation. The total frac fluid volume, therefore, does not appear to have been contained within a fracture system which was swept back to the wellbore as a slug by the reservoir fluids.

Conclusions

1. A large hydraulic fracture treatment was successfully performed at Baca 20 in a 540°F interval. Production tests indicated that high conductivity fractures were propped near the wellbore and communication with the reservoir system was established. This is the highest temperature well in the world to be successfully prop-fractured to date.

2. The productivity of Baca 20 has declined to a noncommercial level since the fracture treatment. The probable cause is relative permeability reduction associated with two-phase flow effects in the formation. This, in turn, probably results from restricted inflow because of the low permeability formation surrounding the fracture.

3. Although the stimulation treatment did not result in a commercial well, the hydraulic fracturing technique shows promise for future stimulation operations and for being a valid alternative to redrilling and/or new well drilling in high temperature geothermal reservoirs.

Geothermal Resources Council, TRANSACTIONS Vol. 6, October 1982

GEOFRAC™--AN EXPLOSIVE STIMULATION TECHNIQUE FOR A GEOTHERMAL WELL

Douglas M. Mumma, Physics International Company; Foy McCullough, Jr., and Eckart W. Schmidt, Rocket Research Company; D. Stephen Pye, Wm. Christopher Allen, and Delbert Pyle, Union Oil Company; Robert J. Hanold, Los Alamos National Laboratory

ABSTRACT

The first known attempt using explosives to stimulate a geothermal well was successfully conducted in December 1981 with a process called GEOFRAC™. The 260°C well was located at the Union Oil Company's Geysers Field in Northern California. For the initial test, 364 kg of a new explosive called HITEX™ II was placed at a depth of 2256 meters and detonated to verify techniques. The explosive was contained in an aluminum canister to separate it from the well fluids. In the second test, 5000 kg of explosive was used, representing a column length of approximately 191 meters. The explosive was detonated at a depth of 1697 meters in the same well. The results of these tests show that HITEX™ II can be safely emplaced and successfully detonated in a hot geothermal well without causing damage to the well bore or casing.

INTRODUCTION

Using explosives to stimulate a geothermal well has long been considered an attractive method by many researchers to enhance production (Ref. 1-3). However, no known field demonstrations have been carried out prior to this time because of the high cost and limited availability of heat resistant explosives. This paper describes a process called GEOFRAC™ in which a new, low-cost heat-resistant explosive called HITEX™ II was used in a field demonstration to show that it could be safely emplaced and detonated in a geothermal well.

HITEX II™

Rocket Research Company recently developed a new heat-resistant explosive called HITEX™ II* for use in geothermal well stimulation (Ref. 4). This explosive consists of a mixture of nitrate salts. When compounded as solid grains the material, which has an off-white color, cannot be detonated at room temperature. It begins to melt at a temperature near 178°C and becomes an almost colorless liquid that can be detonated. Further heating to 260°C shows no occurrence of incipient exotherms. Weight loss for a 24-hour period at 260°C was less than 1%. The detonation velocity at this temperature was measured as 6394 m/s. The properties of HITEX™ II are summarized in Table 1.

Table 1. Physical and chemical characteristics of HITEX II.

Density	
Solid Grain (21°C)	1.63 g/cm^3
Liquid (185° to 205°C)	1.53 g/cm^3
Liquid (232° to 260°C)	1.50 g/cm^3 (contains bubbles)
Melting Temperature Range	178°C (350°F) to 191°C (375°F)
Color	
Solid (21°C)	White
Liquid (260°C)	Clear
Detonation Velocity (260°C)	6394 m/s (measured) 6459 m/s (calculated)
Heat of Explosion	-2844 J/g
Energy Release	864 J/mol gas

GEOFRAC™ TECHNIQUE

GEOFRAC™ involves the process of loading solid cylindrical grains of HITEX™ II explosive into a container and emplacing the container in a geothermal well with a detonator package; the HITEX™ II melts when heated by the hot well temperature and can then be detonated to create additional fractures around the well bore. This explosive stimulation concept was evaluated in a dry steam well provided by Union Oil Company at the Geysers Field in Northern California. Two tests were planned. The initial test was designed to use a small amount of explosive at the bottom of the well to verify the operational techniques prior to conducting a full stimulation test in a higher section of the well with a large amount of explosive. The two configurations tested are described below.

*U.S. Patent No. 4,274,893

INITIAL TEST

For the initial test, a relatively small amount of explosive, 367 kg, was placed in sections of aluminum pipe joined with steel couplings coated with Teflon™. This canister was then lowered into the well on X-line casing and a detonator package placed on top of the explosive by wireline. The canister was then sealed with a bridge plug to prevent intrusion of well fluids. A shaped charge cutter was used to sever the casing from the canister after difficulty occurred with the "J" slot releasing tool. Retrieval of the casing was made with no further difficulties.

To ensure that the explosive did not melt during emplacement, the steam production was killed by injecting water into the well, thus lowering the temperature of the well below the melt temperature of HITEX™ II. When all equipment was safely secured, the water flow was turned off, the well was pressurized with air to drive the water into the surrounding formation, and the well was allowed to produce steam, thus heating the HITEX™ II to liquid form. The sequence of operations for this test is shown in Figure 1. For these tests no sand or water tamp was used.

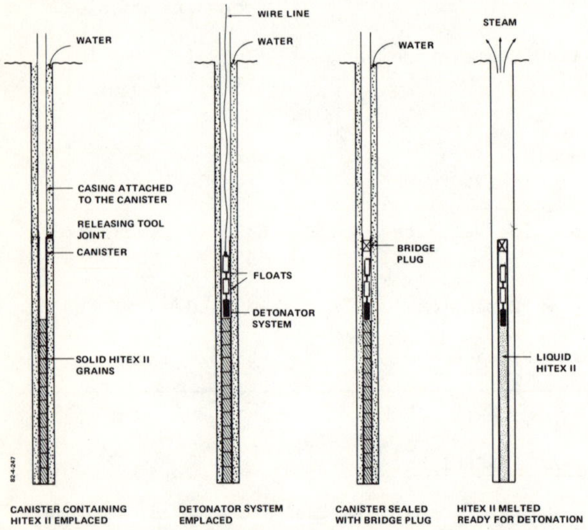

Figure 1. GEOFRAC emplacement concept: detonating the top of the HITEX II explosive column.

The specially designed detonator package consisted of a shaped charge with electronic timing and firing circuits. These elements were protected from the hostile environment of the well by placing them inside a protective pressure vessel with a thermal shield. These components are illustrated in Figure 2. Buoyancy floats were attached to the detonator package so it would remain at the top of the liquid explosive column after melt occurred. The well and canister configuration after emplacement are shown in Figure 3. The explosives were successfully detonated at the preset time of 48 hours after emplacement.

Cleaning out the well proved to be difficult. A short piece of X-line, the canister lengths above the liquid explosive level, and the remains of the detonator/float package were lodged at various locations in the well. After retrieving most of these pieces of junk, a solid debris plug was located at a depth of 1698 meters; consequently, further cleanout of the well was discontinued. This plug unfortunately shut off the lower two steam entries in the well.

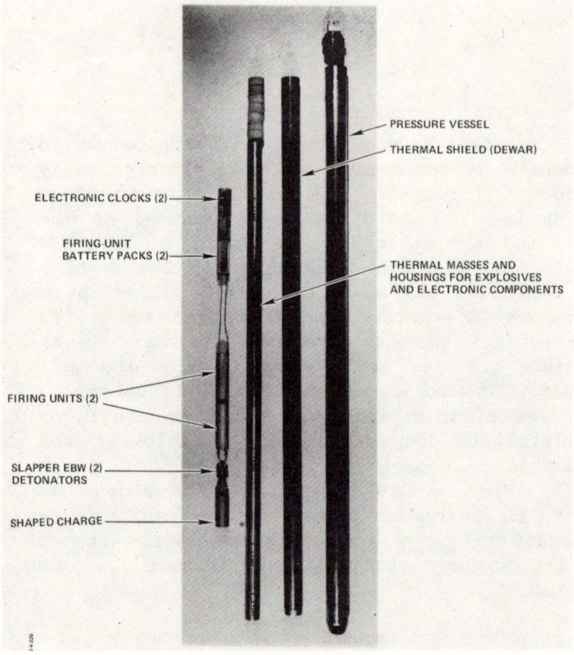

Figure 2. Photograph of the detonator-system components.

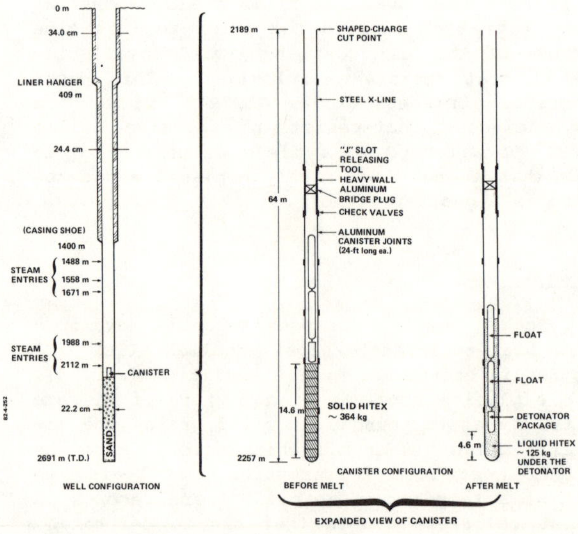

Figure 3. Configuration of the canister emplaced in the LF-30 well for the initial test. (Fired on December 4, 1981 at 3:00 a.m.).

STIMULATION TEST

To prevent a similar problem of well plugging during the 5000 kg stimulation test, an alternate emplacement configuration was selected. In this configuration, the detonator package was placed at the bottom of the canister, eliminating the need for the floats and void space for the detonator package. A safety joint was selected instead of the "J" slot tool to improve the method of canister disconnect. The main operational difference using this technique was that the detonator package was placed close to the bottom of the explosive column, the canister could be sealed during assembly at the surface thus eliminating the need for running a bridge plug, and the canister could be lowered into the hole in a shorter time on drill pipe. The configuration of the well and canister for this test is shown in Figure 4.

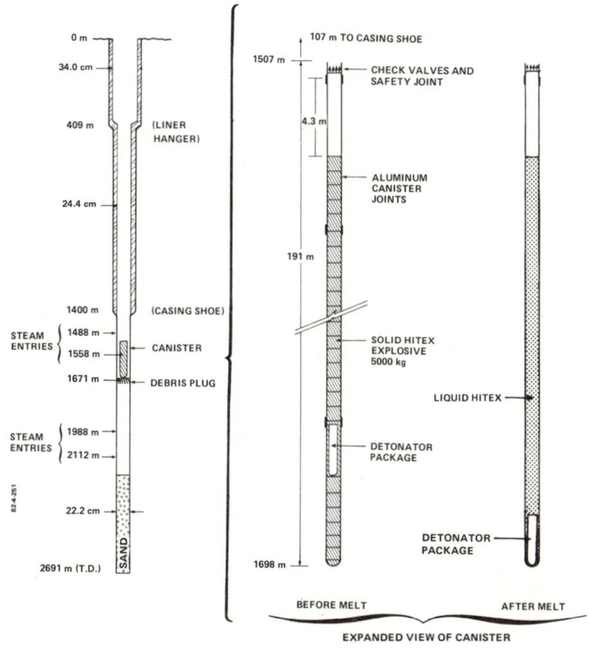

Figure 4. Configuration of the canister emplaced in the LF-30 well for the simulation test. (Fired at 9:45 p.m. on December 14, 1981).

No operational problems were encountered during the emplacement of the explosives. The canister was run in the hole and rested on top of the debris plug left in the well from the initial test. The total column length of explosives was 168 meters. The detonator system caused the explosives to detonate again on schedule 48 hours after emplacement. The well vented rocks and small pieces of metal for a short time after the detonation. Well cleanup proved to be quite minimal and, in fact, the solid debris plug on which the canister assembly was placed was driven 202 meters deeper in the well, where again it became solidly lodged. Complete cleanout of the well below this point was not attempted.

PRODUCTION TEST RESULTS

Production test results from the well prior to, between, and after the two explosive tests are shown in Figure 5. The data show that the steam production decreased from about 22,200 kg/hr before the tests to 14,100 kg/hr afterward. This decrease is probably a result of the debris plug that formed after the initial test, effectively sealing off the lower two steam entries. The measured skin factor changed from -0.6 to -2.4, which could be attributable to a change in the flow conditions, the loss of the lower portion of the well, or an improvement in the permeability of the stimulated section. The fact that the skin factor did not become more positive would suggest that there was no damage to the well bore caused by the creation of a residual "stress cage," compaction of the formation, or damage to well casing.

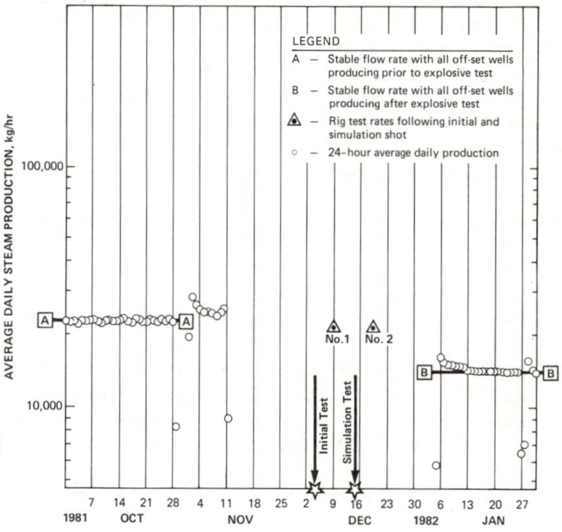

Figure 5. Production data (LF-30).

CONCLUSIONS

It was demonstrated that the unique properties of HITEX™ II can be used in the explosive stimulation of geothermal wells. A shaped-charge detonator package was developed that can survive temperatures of 260°C and pressures of 1000 psi for 48 hours, and then cause molten HITEX™ II to detonate at a preset time. No damage to the well casing was observed even though the explosive column was placed within 107 meters of the casing shoe.

A successful technique was developed to place the explosives and detonator package in a geothermal well so that personnel hazards were minimized during emplacement and well cleanup after detonation was not a big problem. The primary objective of developing the GEOFRAC™ technique was met; however, no improvement in well productivity was observed over the wellbore interval that was treated. A number of additional wells will need to be tested before the full economic utility of the concept can be developed and demonstrated.

ACKNOWLEDGEMENTS

The development of the detonator package by Physics International Company was sponsored by the Los Alamos National Laboratory under Contract No. XP9-1358F-1. Union Oil Company developed the canister assembly used to contain the explosives and the detonator package. In addition, Union Oil provided the well and necessary services to conduct these field tests. Rocket Research Company characterized and provided the explosives under contract No. LP9-1359F-1 to the Los Alamos National Laboratory.

REFERENCES

(1) Carl F. Austin and Guy William Leonard, "Chemical Explosive Stimulation of Geothermal Wells," Geothermal Energy, ed. Paul Kruger and Carel Otte, Stanford University Press, Stanford, California (1973), 269-292.

(2) A. E. Sherwood, "Explosive Stimulation of a Geothermal Steam Reservoir," UCRL-51303 Lawrence Livermore Laboratory, Livermore, California (November 1972).

(3) Robert J. Hanold and H. L. Overton, "Geothermal Reservoir Categorization and Stimulation Study," Los Alamos Scientific Laboratory Informal Report LA-6889-HS (July 1977).

(4) Eckart W. Schmidt and Arthur M. Spencer," HITEX™ Qualification and Manufacture for Explosive Stimulation of a Geothermal Well at the Geysers," Final Report No. RRC-80-R-705, Rocket Research Company, Redmond, Washington (March 1978).

Geothermal Resources Council, TRANSACTIONS Vol. 6, October 1982

PRODUCTION CASING FOR HOT DRY ROCK WELLS EE-2 AND EE-3*

R. W. Nicholson**, R. Pettitt+ and J. Sims+

**Well Production Testing, Inc., Carlsbad, CA.
+Earth and Space Sciences Division
Los Alamos National Laboratory, Los Alamos, NM 87545

ABSTRACT

The production casing for a pair of hot dry rock (HDR) energy extraction wells had to be designed for unique conditions. Two hot dry rock wells (EE-2 and EE-3) were drilled and production casing installed at Fenton Hill, NM for the Los Alamos National Laboratory HDR program. The design of the production casing and subsequent completion operations in these wells revealed that thermal cycling, anticipated operating pressures, and wear during downhole operations are major considerations for both casing specifications and installation procedures. The first well (Energy Extraction No. 2; EE-2) is intended to be the injection well and EE-3 the production well. The top joint strain in EE-3 was monitored during installation, cementing and tensioning.

INTRODUCTION

The production casing for HDR wells is anticipated to undergo extreme changes in temperature. The wells are cased and exposed to the geothermal gradient until production operations commence. Figure 1 is a schematic of the drilled and cased wells, EE-2 and EE-3. Figure 2 shows the static geothermal gradient in EE-2. The production casing in EE-3 is anticipated to undergo a large increase in temperature during future heat production. The injection well casing string also will undergo a large decrease in temperature during injection. The order of magnitude of these changes has been estimated, and for design purposes the maximum extremes are being used.

The primary casing problem thus far was wear of the 9-5/8 in., 40 lb/ft, N-80 casing in EE-2. The reduction in pipe wall has caused a limitation on application of internal pressure. This wear occurred primarily during drilling of the 8-3/4 in. reservoir section.

EE-2 PRODUCTION CASING

The original design concept for cementing of the 9-5/8 in. production casing called for running the casing to approximately 11,600 ft and then cementing to the surface in four stages, because

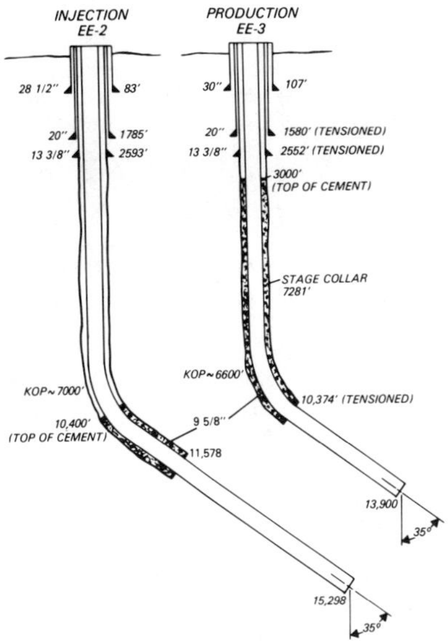

Figure 1. Schematic of the pair of Fenton Hill, HDR wells.

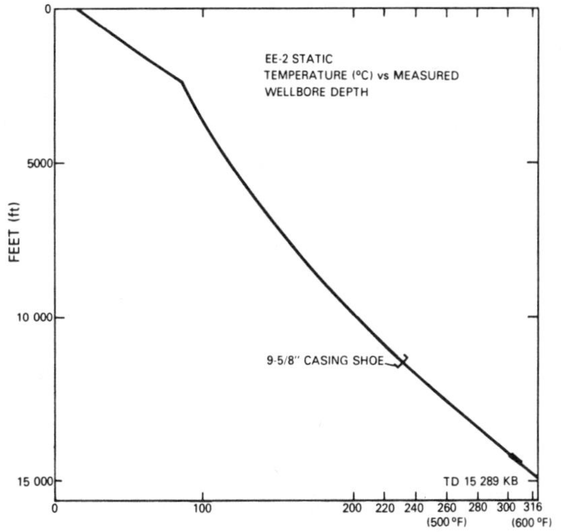

Figure 2. EE-2 static temperature (^{o}C) vs measured wellbore depth.

of FO stage collars at 8000, 5000, and 2000 ft (Pettitt et al. 1982). The requirement for cementing to the surface was largely determined by

---

*This work was supported by U.S. Dept. of Energy, Div. of Geothermal and Hydropower Technologies.

the necessity to reduce the severe axial stresses in the casing string to an injection water temperature of 40°C (104°F) during circulation of the underground loop. Also, because of this cool operation mode, it was planned to cool the hole before cementing, and to maintain a temperature of 120°C (250°F) in the casing for 48 h after cementing, to allow the cement to set around a cooled and contracted casing string.

However, due to last minute developments, the procedure was revised to cement only the bottom 1000 ft. The criteria that influenced this cementing design were as follows:

- Necessity to install the casing in the least amount of time, while pumping without returns.
- Probable failure of the FO stage collars to function above temperature of 163°C (325°F) as experienced in the Cerro Prieto geothermal field in Mexico (Snyder, 1979).
- Pilot test data on cement slurries indicated a tensile strength of 387 psi after curing for 72 h at 60°C (140°F).
- Intervals of high dogleg severity below 9000 ft which might cause nonuniform circumferential loading of casing and result in buckling.
- Even though a 200-ft length of cement sheath was adequate, a 1000-ft length was selected as desirable to resist potential water leakage.

Originally it was planned that the 9-5/8 in. production casing would be installed after drilling to total well depth. However, problems with the 13-3/8 in. intermediate casing dictated that the 9-5/8 in. casing be installed prior to completing the drilling of EE-2.

The precementing cool-down of the hole took four days. Running of the 9-5/8 in. production casing was completed to the design depth of 11,578 ft (see Table 1). Shortly afterwards, the cool-down flow through the casing was started. Three days later, the bottom 1000 ft of the production casing was cemented with 75 bbls of cement consisting of 237 sacks Class H with 80% silica flour. Following the cementing operation, circulation was maintained in the casing through open-ended drill pipe to produce a slow cure of the cement.

Approximately 28 days of drill pipe and wireline operations were made through the 9-5/8 in. production casing. Two areas of severe wear were detected at approximately 675 and 6500 ft.

Table 1. EE-2 production casing 9-5/8-in. buttress thread, N-80 #40, 395 wall, 8.835 I.D.

No. of joints	Size	Weight	Grade	Length	SLM Set from	KB Set to
1	9-5/8	Shoe	Float	2.38	11,578.02	11,576.64
1	9-5/8	40 lb	N-80	43.95	11,576.64	11,531.69
1	9-5/8	Collar	Hi-Temp	2.21	11,531.69	11,529.48
1	9-5/8	40 lb	N-80	38.25	11,529.48	11,491.23
1	9-5/8	Collar	Float	1.83	11,491.23	11,489.40
270	9-5/8	40 lb	N-80	11,467.2	11,489.40	22.20

EE-3 PRODUCTION CASING DESIGN

The following production casing plan for well EE-3 was developed to alleviate or minimize axial movement, axial stress and any potential collapse problems (Edwards et al., 1982):

Bottom (ft)	Length (ft)	Size (in.)	Wt. (lb/ft)	Steel Grade	Thread
200	200	9-5/8	47	S-95	Buttress
6300	6100	9-5/8	43.5	S-95	Buttress
10300	4000	9-5/8	47	S-95	Buttress

(Or 9-5/8 in., 43.5 lb/ft, S-95, Buttress Thread for entire well length.)

The string was to be cemented from 10,300 ft to 7300 ft, tensioned and then cemented from 7300 ft to 3000 ft.

Efforts were made to obtain 47 lb/ft and 43.5 lb/ft S-95 casing. Limited availability of 9-5/8 in. casing during the spring and summer of 1981 hampered attempts to obtain the desired casing. However, a source of 9-5/8 in. 47 lb/ft P-110 grade with Vallorec VAM connections was found and although the P-110 grade steel and premium VAM threads exceeded the requirements, it was deemed satisfactory and purchased.

Since availability of a reliable high-temperature (500°F) wellhead packing was uncertain it was decided to design the top joint of the string so that it could become part of a welded seal. This required that it be made of an easily field-welded grade of steel. A 53.5 lb/ft L-80 steel grade buttress threaded joint was selected as the landing joint to meet the weldability criteria with another 53.5 lb/ft L-80 buttress thread joint placed under it to form a gage section.

EE-3 PRODUCTION CASING INSTALLATION

The 8-3/4 in. reservoir section of the well was drilled to measured depth of 13,933 ft (13,048 ft TVD). The hole was then plugged back with cement from 10,900 ft to 10,400 ft, and casing tensioning jacks were installed around the wellhead (Fig. 3) (Rowley et al., 1982).

The float shoe, float collar, stage tool (DV tool), 9-5/8 in. diam. 47 lb/ft P-110 grade and 53.5 lb/ft L-80 grade casing was picked up and run in the hole to 10,374 ft. The float shoe, float collar and stage tool (DV tool) were inspected. All of the 9-5/8 in. dia. casing used in the string received Analog IV, special end-area and full-length drifting inspections.

The casing string was run in the hole in the following sequence (Table 2):

1. High-temperature aluminum jet down float shoe with buttress box thread.
2. Full-length joint of P-110 9-5/8 in. dia. 47 lb/ft casing with buttress pin and VAM box threads.

3. Full-length joint of P-110 9-5/8 in. dia. 47 lb/ft casing with mill scale removed by sandblasting, a rigid centralizer attached, and VAM pin and box threads.
4. Half-length joint of P-110 9-5/8 in. dia. 47 lb/ft casing with VAM pin and buttress box threads.
5. Halliburton conventional float collar with a buttress pin and box.
6. Full-length joint of P-110 9-5/8 in. dia. 47 lb/ft casing with a rigid centralizer attached and a buttress thread pin and VAM thread box.
7. 75 full-length joints of P-110 9-5/8 in. dia. 47 lb/ft casing with the mill scale removed by sandblasting and with VAM thread pins and boxes; rigid centralizers were attached to the first 13 joints (one per joint) and then on every other joint.
8. Half-length joint of P-110 9-5/8 in. dia. 47 lb/ft casing with a rigid centralizer attached and a VAM thread pin and buttress thread box.
9. Stage collar (DV tool) with buttress thread pin and box.
10. Full-length joint of P-110 9-5/8 in. dia. 47 lb/ft casing with a rigid centralizer attached and a buttress thread pin and a VAM thread box.
11. 179 full-length joints of P-110 9-5/8 in. dia. 47 lb/ft casing with a rigid centralizer attached to every fourth joint for the first 107 joints.
12. Full-length joint P-110 9-5/8 in. dia. 53.5 lb/ft casing with a buttress thread pin and box (8.500 in. drifted I.D.).
13. Full-length joint of L-80 9-5/8 in. dia. 53.5 lb/ft casing with a buttress thread pin and box (8.500 in. drifted I.D.).
14. Full-length joint of L-80 9-5/8 in. dia. 53.5 lb/ft casing (8.500 in. drifted) with 4 mounted strain gages (to monitor casing strain and resulting stress) and a buttress thread pin and box. This joint was landed in the casing landing spool (8.500 in. drifted I.D.).
15. Half-length joint of P-110 9-5/8 in. dia. 47 lb/ft casing with a buttress thread pin and box. This joint was used when landing casing on the rig floor.

Once the casing was run in the hole a stretch test on the casing string was conducted. An initial 380,000 lb load on the rig's drawworks was increased to 434,000 lb which resulted in a 10-1/2 in. movement of the casing out of the hole, a 500,000 lb pull resulted in a total casing movement of 24 in. When the pull force was lowered to 380,000 lb the casing moved back down hole 28 in. or 4 in. further than it had been before the stretch test was started.

The first stage of cementing was accomplished without major problems. The stage collar opened and excess cement above the stage collar in the annulus was circulated out.

Approximately 72 h after finishing the first stage of the cementing the casing tensioning and

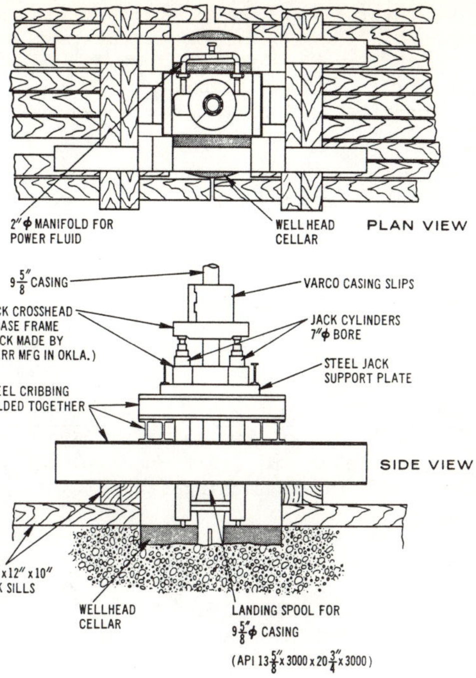

Figure 3. Sketch of casing tensioning jack system.

second stage cementing began. After testing, the tensioning jacks were cycled up and down 4 times to stretch the casing and work tension down the casing. These cycles resulted in the following casing stretch (see Fig. 4):

1. stretched casing 2.25 ft w/7000 psi (529,000 lbs)
2. stretched casing 1.9 ft w/6475 psi (498,600 lbs)
3. stretched casing 4.1 ft w/8350 psi (643,000 lbs)
4. stretched casing 4.1 ft w/8300 psi (639,100 lbs)

Upon completion of cycling, tension resulting from the jack cycling had been worked down to approximately 6440 ft and 3.0 ft of casing stretch was retained. The casing head slips were then set. Next the casing was cemented from 7280 ft (DV collar) to 3000 ft and the DV collar closed.

The jacks were then cycled up and down three times resulting in the casing being stretched 5.1 more feet for a total (including 3.0 ft of pre-second stage cementing casing stretch) of 8.1 ft.

Table 2. EE-3 production casing 9-5/8 in.

No. of joints	Size	Weight	Grade	Length	KB Depth
1	9-5/8	Shoe	Float	2.33	10,374
3	9-5/8	47	P-110	99.81	10,272
1	9-5/8	Float	Collar	1.75	10,270
77	9-5/8	47	P-110	2,985.76	7,284
1	9-5/8	Stage	Collar	3.40	7,280
181	9-5/8	47	P-110	71 93.83	86
2	9-5/8	53.5	L-80	76.33	10

Note: Float collar, float shoe, stage collar and top two joints were Buttress Threads. All other joints were VAM threads.

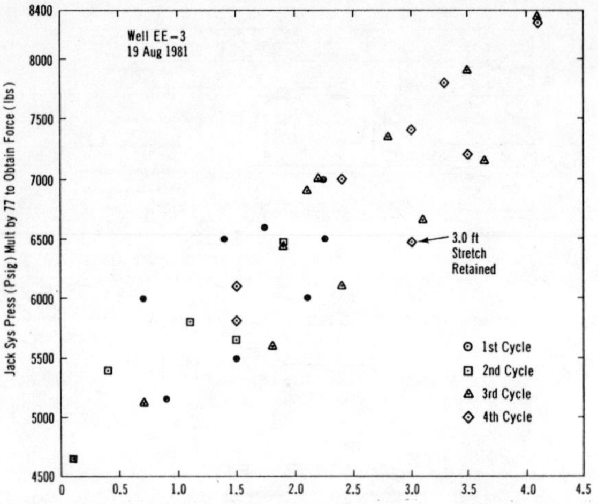

Figure 4. Stretch of 9-5/8 in. casing during initial tensioning (ft).

A pressure of 11,500 psi was applied to jacks resulting in a tensioning force of 885,000 lb which produced the 5.1 ft of casing stretch (see Fig. 5). The 9-5/8 in. casing head slips were then set in the casing landing spool. The jacks were lowered and 0.1 ft of casing stretch was lost. The 9-5/8 in. dia. casing was then cut off 5 ft above the top of the casing landing spool flange.

During the installation cementing and tensioning of the production casing in EE-3 the strain in the top joint of the string was measured using electrical resistance strain gages. Four axial strain measuring gages were welded onto the outside of the casing at equal intervals. The strain gage data (Table 3) shows that the 9-5/8 in. L-80 strain gage joint was stretched in tension on the 20th of August when the casing string was tensioned. A calculation using the change in length or strain from gages 1 and 3 (between the 19th and 20th of August) of approximately 550 μin./in. gives a change in tensile stress of 16,500 psi ($E\varepsilon = \sigma$, where $E = 30 \times 10^6$ psi, ε = strain in μin./in. and σ = stress in psi). Using an area of 15.5 in.^{2} for the 53.5 lb/ft 9-5/8 in. casing gives a tensile load of 256,000 lb which is only about half of the 580,000 lb load expected

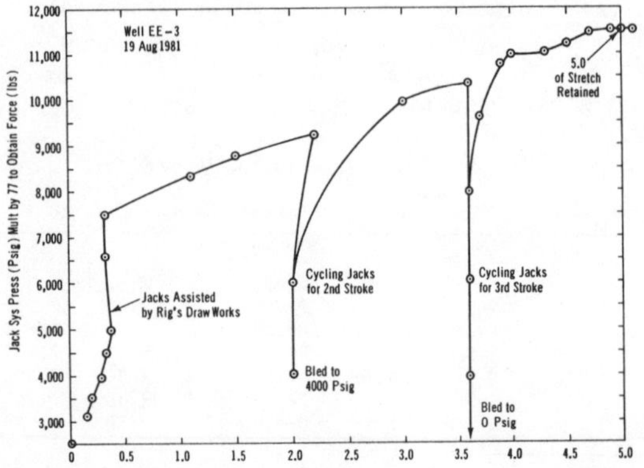

Figure 5. Stretch of 9-5/8 in. casing during final tensioning (ft).

Table 3. EE-3 9-5/8 in., L-80 strain gage data, μin./in.

Gage No.	19 Aug 81	20 Aug 81	22 Aug 81	22 Sept 81	19 Jan 82
1	550	1090	800	840	475
2	-770	open	-	-	-
3	300	857	490	790	962
4	500	375	-	-	-

from the estimated free point of 6400 ft and 8.0 ft of retained casing stretch. The explanation of this discrepancy probably is changes in apparent strain due to temperature changes. The gages were not compensated for temperature and the temperature of the L-80 strain gage joint probably varied by over 70°F between strain readings due to hot cementing fluids flowing over gages. A dummy gage mounted on unstrained casing material located near the strain gages (in the same temperature environment) should be used in future installations in order to compensate fully for temperature induced apparent strain.

Values of strain read after casing installation show changes which are again probably temperature induced apparent strain caused by the gages cooling, however some of the changes may reflect stress changes in the casing and still other changes may be due to the strain gages deteriorating in their wet environment.

SUMMARY AND CONCLUSIONS

Production casing strings that meet the requirements for HDR injection and production wells can be designed and installed with existing technology. The stress and strain in the top joint of a casing string can be monitored during cementing and tensioning.

DISCLAIMER

Reference to a company, product, name, service, tool, or equipment does not imply approval or recommendation of the product, service, or tool by the University of California (Los Alamos) or the U.S. Department of Energy to the exclusion of others that may be suitable.

REFERENCES

Edwards, L. M. et al., eds., 1982, Handbook of Geothermal Energy, Chapter 6, Casing and Tubular Design Concepts, H. H. Rieke and G. V. Chillingar; Gulf Pub. Co.

Pettitt, R. A., Koczan, S., Helmick, C., and Williams, R. E., 1982, Drilling of Hot Dry Rock Geothermal Well EE-2, Los Alamos National Laboratory report LA-9302-HDR.

Rowley, J. C. and Carden, R. S., (in preparation), Drilling of HDR Geothermal Energy Extraction Well EE-3, Los Alamos National Laboratory report.

Snyder, R. E., 1979, Geothermal Well Completions: A Critical Review of Downhole Problems and Specialized Technology Needs, SPE Paper 8211, 54th Ann. SPE Meeting, Las Vegas, NM, 6 p.

Geothermal Resources Council, TRANSACTIONS Vol. 6, October 1982

DOWNHOLE CEMENT TEST IN A VERY HOT HOLE

Roland A. Pettitt, George G. Cocks, Donald S. Dreesen, James R. Sims, Robert W. Nicholson,* and Billy G. Boevers**

Earth and Space Sciences Division, Los Alamos National Laboratory, Los Alamos, New Mexico 87545
*Well Production Testing, Carlsbad, California
**Dowell Division of The Dow Chemical, Tulsa, Oklahoma

ABSTRACT

Completion of the commercial-sized Hot Dry Rock Geothermal Energy Project requires that hydraulic fractures be created between two inclined wellbores at a depth of about 4 km (15,000 ft). Isolation of a section of the open wellbore is necessary for pressurization to achieve the fracture connections. A cemented-in liner/PBR assembly is one of the methods used for zone isolation near the bottom of the injection well. A downhole, pumped cement test was first conducted at a wellbore temperature of 275°C (525°F) to determine if a suitable slurry could be designed, pumped, and later recovered to assure the success of the cemented-in liner operation.

INTRODUCTION

In the development of the first Hot Dry Rock (HDR) geothermal energy extraction system at Fenton Hill, west of Los Alamos, New Mexico, man-made reservoirs were created by connecting two holes in hot, impermeable crystalline rock with hydraulically-produced fractures. This initial system consists of two near-vertical 9-5/8-in. diameter holes approximately 3 km (10,000 ft) deep in Precambrian basement rock, at a bottom-hole temperature of 200°C (400°F).

Drilling of the first hole for the deeper, commercial-sized reservoir was begun in 1979 at the same site (Pettitt et al., 1982). The second hole was completed in 1981 (Rowley et al., 1982). This reservoir system consists of two boreholes drilled to approximately 4 km (15,000 ft) at an angle of 35° from the vertical, which will be connected by multiple, hydraulic fractures. The bottom-hole temperature of the deeper injection well (EE-2) is 335°C (635°F) (Figure 1). Both wells are cased with 9-5/8-in. o.d. casing to within approximately 1220 m (4000 ft) of total depth; the 8-3/4-in. drilled diameter open hole sections will constitute the reservoir area.

The first connecting fractures were planned to be created from the bottom portion of EE-2, extending upward to EE-3. Methods of isolating sections of the borehole for fracturing operations included the use of open-hole packers (both straddle and single), and a cemented-in scab liner/PBR assembly.

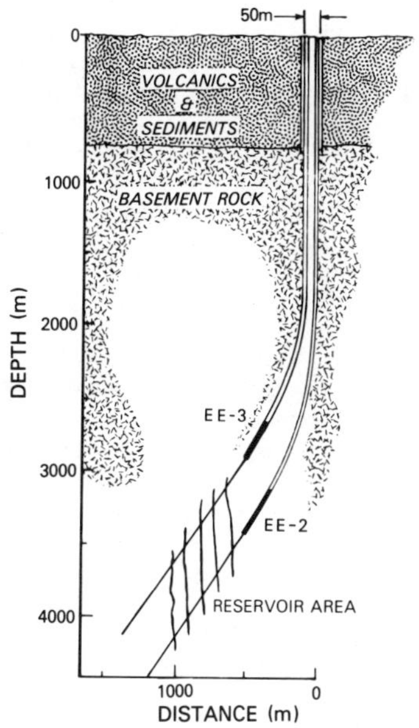

Figure 1. Schematic cross section of wellbores EE-2 and EE-3, showing area of the commercial-sized reservoir.

Since suitable cement, cement placement procedures, and elastomeric materials are required to successfully cement-in and fracture below the cemented-in scab liner, an investigative program was undertaken to (a) test cement mixtures in the laboratory and (b) to test the "best" slurry and a special wiper plug under downhole conditions at the extreme temperature to be encountered.

TEST OBJECTIVES

The use of a cemented-in permanent scab liner/PBR assembly, set at about the 15,000 ft depth in EE-2 for zone isolation for hydraulic fracturing below it, was decided in December 1981. At that time, investigations were also started to determine suitable cement mixes, elastomeric seals and wiper plugs, and an appropriate method to adequately test them. In January 1982, a cement test procedure was designed to test the proposed slurry under real downhole conditions at the depth, temperature, and pressure that the scab liner would encounter in EE-2. Concurrently, the Dowell Research Laboratory in Tulsa, Oklahoma was requested to start conducting tests on these three proposed slurries and downhole conditions:

A. Cement Type and Silica Content
   1. Class H cement, with 40% silica
   2. Class H cement, with 60% silica
   3. Class J cement, with 60% silica equivalent

B. Properties:
   48 hour compressive strength = >4000 psi
   Slurry density = >15.0 ppg

C. Placement Conditions:
   BHCT = 275°C (525°F)
   Pumping time: 4 hours
   Depth: 14,500 ft
   Pipe: 3-1/2-in. drill pipe, 13.3 lb/ft
   Fluid in the hole: water

D. Possible Contaminants in the Hole:
   Big Cat commercial detergent: 2% solution
   Friction reducer: polyacrylamide, 1/40% solution.

CEMENT AND ELASTOMER TESTING PROCEDURES

The Dowell station in Farmington, New Mexico (200 miles from Fenton Hill) reserved 200 sacks of Amarillo Class H cement, plus appropriate amounts of admixtures, to assure that the testing and placement mixes would consist of the same ingredients. Samples of each were sent to the Dowell Regional Lab in Casper, Wyoming, to the Dowell Research Lab in Tulsa, and to Los Alamos for independent testing. A silo of Class J cement was also reserved in Waco, Texas, with samples being obtained and sent to the three laboratories.

In order to keep the three laboratories informed of testing progress, a weekly progress update memo was written in Los Alamos, and sent to the Dowell Farmington station, the Denver office, and to the two Dowell laboratories. After a suitable mix design had been obtained by the Tulsa Laboratory, replicate tests were planned in Casper and Los Alamos to check for consistency of results.

Testing in Tulsa continued through February and March, 1982. Trial mixes were first made with an experimental retarder/dispersant admixture (designated XE67565) in a powder form. The efficiency of the powder was later tested by being added to the mixing water, with the result that reduced amounts of retarder/dispersant were required when added in this manner.

By the end of March, testing was completed by the Tulsa Laboratory on the three mixes. The cement systems had a thickening time of 4 to 5 hours at a BHCT of 275°C (525°F), and a minimum compressive strength of 4000 psi in 48 hours at 315°C (600°F). Cement testing was accomplished by standard API procedures, except that the heat-up times were accelerated to more closely simulate placement times. The cement test results are listed below and in Tables 1 and 2.

LABORATORY TEST RESULTS

The final mixes of the three cement systems were:

1. Amarillo Class H cement
   40% D66 (silica flour)
   5.5% Retarder/Dispersant (XE67565)*
   0.2% D46 (defoamer)

2. Amarillo Class H cement
   60% D66 (silica flour)
   6.0% Retarder/Dispersant (XE67565)*
   0.2% D46 (defoamer)

3. Unadeep Class J cement
   20% D66 (silica flour)
   2.5% Retarder/Dispersant (XE67565)*
   0.2% D46 (defoamer)

*The XE67565 is an experimental Dowell retarder/dispersant which was premixed in the mix water.

Table 1. Cement Properties

System	Density PPG	Yield cu ft/sk	Mix Water gal/sk	TT* hrs:min
1	16.6	1.44	5.03	4:10
2	16.3	1.69	6.0	3:40
3	16.2	1.27	4.39	4:37

*Samples for these thickening time tests were heated as fast as possible in the Dowell consistometer. Normal heat up time from 80 to 525°F was 65 to 75 minutes with a maximum pressure of 6300 psi.

Table 2. Compressive Strengths, PSI

System	Temp °F	48 Hours	48 hours Contaminated*
1	600	4805	4099
2	600	5068	4304
3	600	5699	4928

*The samples were contaminated with 2% by volume of a detergent (Calgon Big Cat) supplied by Los Alamos.

Dowell wiper plugs (made of EPDM Y267, formulated by L'Garde, Inc.) and Brown Oil Tool (BOT) darts were statically tested at 300°C (572°F) and 4400 psi in the Los Alamos autoclave. BOT molyglass chevron seals for the permanent scab liner/PBR assembly were also statically and dynamically tested at 300°C (572°F), withstanding 3300 psi differential pressure before failure, in the Los Alamos autoclave. (Only two seals were dynamically tested, whereas a large number of seals will be used in the actual downhole assembly.)

DOWNHOLE PLACEMENT

The bottom-hole assembly (BHA) for the downhole cement test is shown schematically in Figure 2. Basically, this BHA consisted of 742 ft of 2-7/8-in. tubing, containing a dart catcher sub, and a pressure and temperature (Kuster) gauge holder (Figure 3).

Run-in of the cement test assembly was completed on April 18, with the pressure and temperature gauge sub being located at a depth of 14,580 ft, where the in-situ rock temperature is 305°C (580°F) (Figure 4). As the design thickening time temperature of the mix to be used (System No. 1, containing 40% silica flour) was 275°C (525°F), the hole was cooled by circulating through the BHA for 3 hours with the rig pumps at 5.4 bbls/min.

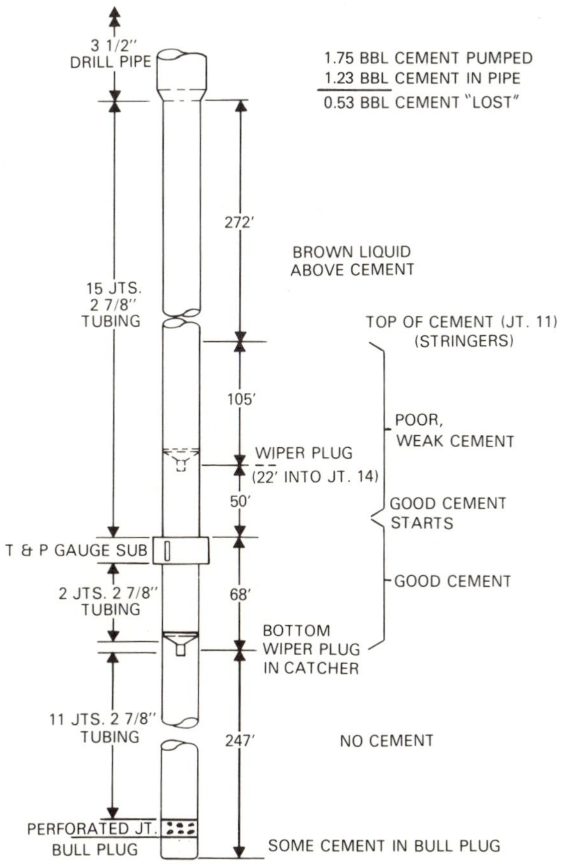

Figure 2. Bottom-hole assembly for the downhole cement test, indicating occurrence of cement upon retrieval.

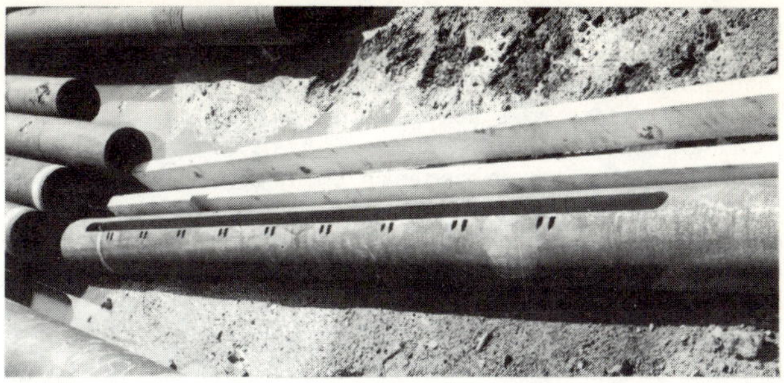

Figure 3. Photo of Kuster gauge holder sub. A 6-ft 8-in.-long slot, 1-1/2-in. wide by 1-3/4-in. deep, was cut in a 6-in. diameter drill collar. The Kuster gauge was held in place by set screws, tapped into the side of the slot, on 8-in. centers.

A representative from Brown Oil Tool supervised the loading and release of the darts (wiper plugs), as this same type of dart would be used in the permanent liner/PBR assembly. After the lower dart was released, 1.75 bbls of cement slurry were pumped downhole. The upper plug was then released, followed by 100 bbls of displacement water. Pumping pressure was 500 psi; after the lower dart bumped into the dart catcher, the pressure raised to 1500 psi. The pressure was held without leakage for 5 minutes; then the cementing head was rigged down, and the pipe string pulled up one stand. During the 48-hr waiting period for the cement to set, the pipe string was reciprocated for 15 ft at 30 minute intervals for the first six hours. Later, during the waiting period, the reciprocating interval was increased to 60 minutes and then to 90 minutes.

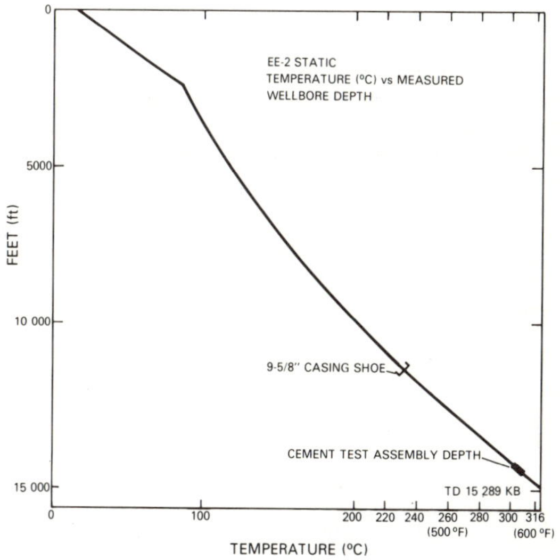

Figure 4. EE-2 static temperature vs measured wellbore depth.

DOWNHOLE TEST RESULTS

On May 20, the cement test assembly was started out of the hole, after having set at depth for 48 hours. The first 10 joints of 2-7/8-in. tubing contained no cement; the rest of the joints down to the dart catcher contained cement which was progressively harder with depth. There was no cement below the dart catcher, except for a small amount in the bull plug. The Kuster temperature gauge apparently leaked, giving no data after 24 hours. A maximum reading thermometer, also included in the Kuster package, indicated 293°C (558°F) at 14,580 ft. Figure 2 shows the occurrence of the cement in the cement test assembly. Approximately 1.23 bbl of cement remained in the pipe, leaving 0.53 bbl unaccounted for, presumably leaking by the plug in the dart catcher. However, there was no indication of cement in the bottom of the hole when it was later cleaned out to bottom. The top plug, expected to be recovered at the top of the cement column, was actually found 105 ft down from the top of the cement.

Sledge hammers were required to break the tubing connections of the joints containing hard cement, during removal from the hole. It was noted that small bubbles appeared immediately on the fresh broken surface of the hard cement at the tubing connections, indicating that the cement still contained water at pressure and that the cement-to-tubing bond was extremely good.

The condition of the darts after retrieval was very revealing. The bottom dart still retained some resiliency, but the top dart (when it was cut out of the tubing) was extremely brittle and decomposed.

Compressive strength of the hardened cement retrieved from the tubing was 5270 psi.

CONCLUSIONS

The downhole cement test successfully answered questions that had worried the designers of the liner/PBR assembly operation. These points were definitely and positively made:

- A cement slurry was designed, laboratory tested and pumped downhole.
- The retarder acted within the temperature and time constraints.
- The cement set and gained strength under the downhole conditions.
- The test provided confidence in the darts and dart catcher.
- An indication of the amount of cement slurry leaking by the darts was provided.
- It is possible to verify the actual downhole temperature and pressure that the cement slurry encountered.
- The cement test gave assurance that a satisfactory cementing operation could be performed for the liner/PBR assembly.

ACKNOWLEDGMENTS AND DISCLAIMER

Reference to a company, product, name, service, tool, or equipment item does not imply approval or recommendation of the product, service, or tool by the University of California (Los Alamos) or the U.S. Department of Energy to the exclusion of others that may be suitable. The project described in this article was funded by the U.S. Department of Energy, Division of Geothermal Energy.

Thanks are due the Los Alamos workover managers, WPT consultants, Dowell, Brown Oil Tool, and Brinkerhoff-Signal Company people who contributed so much personal effort to these operations. Special appreciation is also due these Dowell individuals: Ron Root/Denver, Richard Jones/Tulsa, Sid Christensen and Mark Stewart/Farmington, and Don Mitchell/Casper.

REFERENCES

Pettitt, R., Helmich, C., Koczan, S., 1982, Planning and drilling geothermal energy extraction hole EE-2, Los Alamos National Laboratory Report LA-9302-HDR.

Rowley, J., Carden, R., 1982, Drilling geothermal energy extraction hole EE-3, Los Alamos National Laboratory Report, in preparation.

Geothermal Resources Council, TRANSACTIONS Vol. 6, October 1982

DRILLING PROBLEMS AND THE LEARNING CURVE -- AN EXAMPLE FROM PALIMPINON GEOTHERMAL FIELD, THE PHILIPPINES

S.K. SANYAL*, M. CHE*, J.R. McNITT*, N. VASQUEZ
B.S. TOLENTINO**, A. ALCARAZ**, AND R. DATUIN***

* GEOTHERMEX, INC., BERKELEY, CALIFORNIA
** PHILIPPINES NATIONAL OIL COMPANY, PHILIPPINES
*** NATIONAL POWER CORPORATION, PHILIPPINES

ABSTRACT

This paper discusses the drilling problems encountered in developing the Palimpinon geothermal field in the Southern Negros, Philippines. The effective drilling rate has been used as a criterion of drilling efficiency and it is shown that this criterion shows the learning curve effect. Two different learning curves, corresponding to two different parts of the field were identified and their implication on the overall field development program is discussed.

INTRODUCTION

The "learning curve" concept has been used in analyzing drilling efficiency in the oil industry. However, not much is published about the applicability of this concept in geothermal industry. This paper discusses a case history of the application of this concept in the Palimpinon geothermal field in the Southern Negros island in The Philippines. The location of the field is about 20 kilometers west of the coastal city of Dumaguete. The field has been drilled for the development of the geothermal resource since 1976 by the Philippines National Oil Company. Twenty-one wells have been drilled to date and five are being drilled at this time. Figure 1 shows the locations of all the wells except one (Well N-2). The developed area is divided into three "blocks", each with different reservoir characteristics. Table 1 presents the well names, casing head flange elevation (meters above sea level), date the well was spudded, number of days required to drill the well and total measured drilled depth (meters) for each of the 21 existing wells. Wells are listed in chronological order of their drilling. The letter "D" after a well number indicates that the well is directional. A 112.5 MWe electrical power plant is under construction in this field and the successful wells among the 21 already drilled have a total electrical power capacity of 88.6 MWe. Development of further power plant capacity is being contemplated.

DRILLING PROBLEMS

Every geothermal field has its drilling problems; and Palimpinon is no exception. As a part of the review of the current development program at Palimpinon, the drilling problems in the area were studied. The aim was to analyze the exact nature of the drilling problems and to evaluate the drilling program vis a vis overall development strategy. This paper reports a part of this study.

After a review of the drilling records of the existing wells and interviews with the field and office personnel involved in the drilling programs, three basic areas of drilling problems were identified:

1. Access problem
2. Directional drilling problem
3. Hard rock problem.

The access problem is the major impediment. The area of drilling is in an extremely rugged, forested terrain in a tropical climate. Very few flat sites are available for drilling-pad construction. Building of access roads is difficult. Both drilling sites and access roads are subject to flooding and mudslides during the rainy season. Difficult access has increased drilling cost. It was concluded that the access problem cannot be avoided unless the reservoir is shown to extend towards less rugged parts of the terrain. This is possible as reservoir boundaries are yet to be fully defined.

Because of the limited number of available drilling-pad sites, Philippines National Oil Company adopted a program of directional drilling of several wells from each drilling pad. Horizontal offsets between wellhead and bottom hole locations are up to 1500 m. for a measured total depth of 3000 m. Directional drilling significantly increased drilling cost because more footage needed to be drilled for the same amount of vertical penetration. Well logging and workover operations are difficult in deviated wells. It was concluded that directional drilling is necessary unless the reservoir is shown to extend towards the less rugged parts of the terrain.

The developed part of the field has been

In estimating stable formation temperatures from transient temperature profiles, allowances were made for the effect of: (a) conductive cooling during mud circulation; (b) cooling due to mud loss; (c) cooling due to injection of cold water into the well; (d) cooling due to pressure drop caused by production from two-phase zones; (e) heating of the upper part of the hole due to production; and (f) interflow between two or more aquifers open to the well and existing at different potentials.

In most wells, the survey preceded by the longest period of shut-in condition was assumed to most closely approximate stable formation temperatures. In those cases where the temperature profile in the upper part of the well is above a boiling-point curve originating at the water surface in the well due to heating from production, estimation of initial formation temperatures were adjusted downward. Conversely, if static temperatures were lower than downhole production temperatures measured at the same elevation, estimates of initial formation temperature were adjusted upward. Formation temperatures between interflow zones were assumed to be on a straight line or a boiling-point curve connecting the permeable zones.

After completing the interpretation of the temperature profiles, temperatures were plotted on constant elevation surfaces at 200 m intervals of elevation starting at the lowest water level recorded in the wells (+200 msl), and continuing downward to -2,000 msl, which is the deepest datum for which information is available. Next, temperature contours were drawn within the area enclosed by data points on each elevation map. This area becomes larger with depth because of the high angle of deviation from vertical in the directional wells and the fact that the drilling directions tend to radiate outward from the center to the periphery of the drilled area. To take advantage of this vertical variation in the distribution of data points, once the general patterns of temperature distribution were known, additional temperature contours were extrapolated upward from directionally drilled wells and down-ward from shallow wells, thereby significantly enlarging the contoured areas. Figure 1 shows such contours at -1,400 msl level, as well as the location of the wells and the major faults.

To help visualize the temperature distribution defined by these constant elevation maps, maps showing the elevation of constant temperature surfaces were constructed at temperature intervals of 20°C from 180°C through 300°C. In addition, a SW-NE temperature cross-section across the field was constructed (figure 2, which also shows the faults and the main stratigraphic units).

These illustrations clarify several important aspects of temperature distribution within the Palimpinon Field:

1. Above an elevation of -200 msl, temperature distribution is dominated by vertical gradients caused by horizontal flow. The large lobe flowing to the NE across the NE part of the field is particularly apparent. This flow, beginning at an elevation of -350 msl and at a temperature of 260°C, rises to sea level over a horizontal distance of almost 3 km (figure 2). A large temperature reversal is apparent beneath the lobe at the NE end of the cross-section.

2. Horizontal gradients, caused by vertical circulation, occur from -300 msl to -1,200 msl. From its position, the zone appears to be related to the Ticala Fault and its north-trending branches (figure 2). Because of this association, it is proposed that the vertical flow is occurring along the Ticala Fault, or on closely associated but unmapped faults.

3. The main deep feature of the temperature distribution is a NW-trending high temperature nose, closed to the NW and open to the SE (figure 1). The top of the feature, as seen in temperature cross-section (figure 2), rises to an elevation of -500 msl, where its temperature is 250°C. On the SW it is cut off abruptly by the Sogongon Fault, which juxtaposes diorite on the SW against the Okoy Sedimentary Formation on the NE. In contrast to its abrupt termination on the SW, the high temperature feature decends gradually to the NE along temperature contours that dip at about 45° beneath the zone of vertical flow described above. Except for its furthest NE extension below the upflow zone, the high temperature nose is developed almost entirely in the Okoy Sedimentary Formation, terminating upward against the contact with the overlying Southern Negros Formation. The temperature gradient in the Southern Negros Formation is about 140°C/km as compared to about 60°C/km in the Okoy Sedimentary Formation. It is not clear if this difference in gradient between the two Formations is due to a difference in thermal conductivity or to a greater amount of internal fracturing within the Okoy Sedimentary Formation allowing greater convective circulation.

4. Horizontal gradients appear again in the SW part of the drilled area below -500 msl in the diorite block. Because of the extensive width of this zone, and its restriction to the diorite block, it is assumed that these gradients are not caused by vertical flow, but rather by heat being conducted away from the deep high temperature nose through the relatively impermeable diorite.

Two important conclusions can be drawn from the temperature distribution patterns described above: First, the Palimpinon wells have not penetrated the main center of convective upflow, but rather, they have been drilled NW of the main upflow zone which is probably located below the extrusion centers of the young Cuernos de Negros volcano approximately 3 to 4 km to the SE. Second, the stratigraphy and structure of the area exert a strong influence over the detailed pattern of temperature distribution.

Deep outflow from the convection center beneath the Cuernos volcano appears to be concentrated in the structurally highest part of

TABLE 1: SUMMARY OF WELL DATA

Well Name	Block No.	Casing Head Flange Elevation (m)	Date Spudded	No. of Days to Drill	Total Drilled Depth (m)
N-1	1	300.7	9-4-76	37	603.3
N-2	1	132.2	10-20-76	30	610.5
N-3	1	417.5	11-8-76	53	970.9
OK-1	1	587.9	1-25-78	84	1,978.9
OK-2	2	704.4	4-28-78	32	1,164.4
OK-3	2	468.3	6-7-78	42	1,521.8
OK-4	2	847.3	8-27-78	53	2,130.0
OK-5	2	932.2	10-20-78	43	1,975.2
OK-6	3	1,105.6	9-26-79	110	2,770.8
OK-7	2	759.7	7-26-80	71	2,822.8
OK-8	2*	1,004.1	4-11-80	119	2,981.8
OK-9D	2	772.7	8-13-80	73	2,889.9
OK-10D	2	759.9	10-27-80	68	3,018.9
OK-11D	3	1,190.7	12-4-80	100	2,817.9
OK-12D	2	644.8	1-7-81	106	2,797.5
PN-13D	2	756.2	3-23-81	107	2,855.2
SG-1	3	947.7	3-29-81	97	2,762.8
PN-15D	2	715.8	7-7-81	58	2,806.9
SG-2	3	1,002.1	7-11-81	92	2,945.0
PN-17D	2	715.8	9-7-81	77	3,042.2
NJ-1D	3	1,105.7	10-19-81	105	2,952.6

* Although it is in Block 2, it drills into diorite.

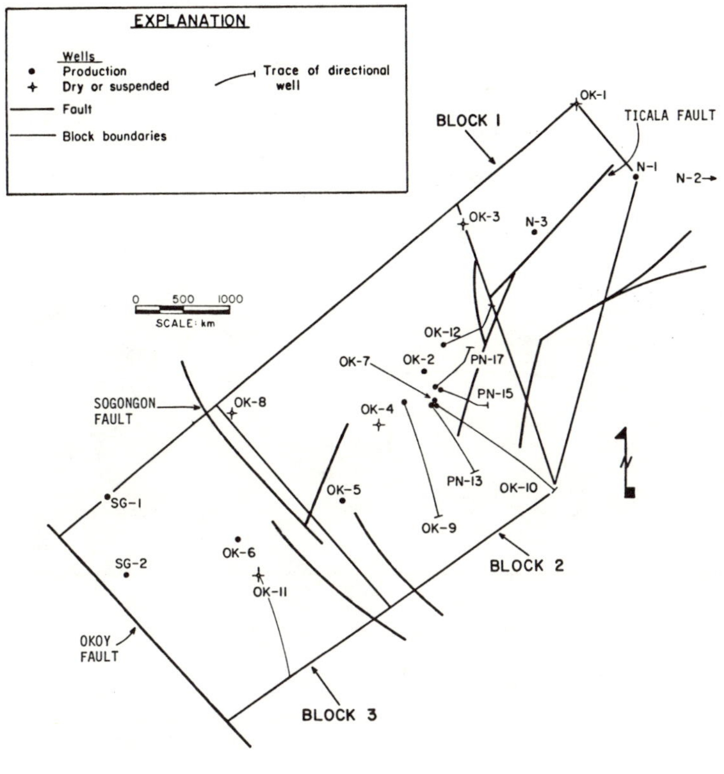

FIGURE 1. Well and block locations, Palimpinon Field, Negros, Philippines

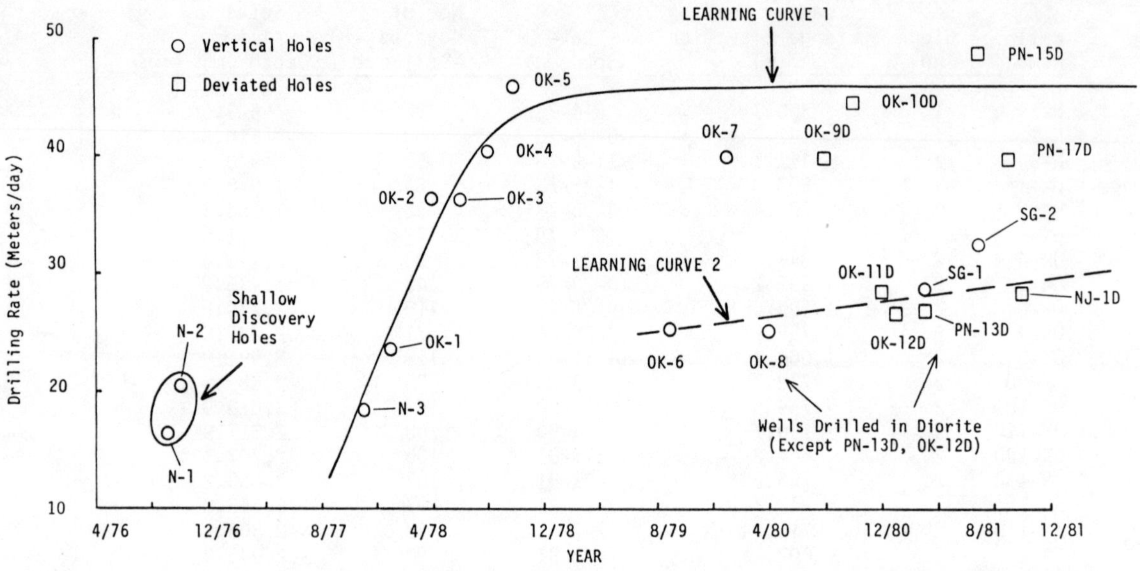

FIGURE 2 . Analysis of Drilling Rates of Palimpinon Wells

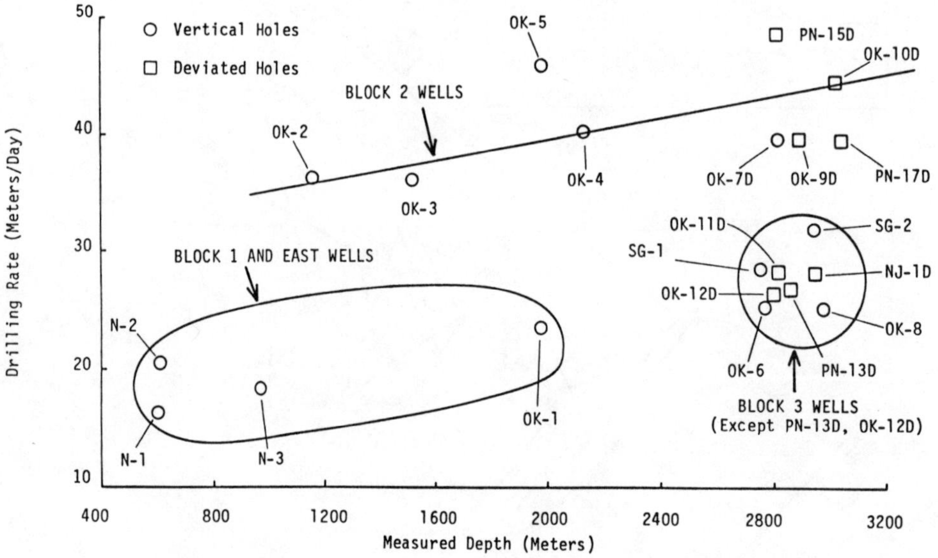

FIGURE 3 . Correlation of Drilling Rate vs. Depth Drilled for Palimpinon Wells

Geothermal Resources Council, TRANSACTIONS Vol. 6, October 1982

AN INTRODUCTION TO REVERSE CIRCULATION DRILLING FOR GEOTHERMAL EXPLORATION

John A. Thurston and Marc T. Eckels

TerraDrill, Inc. and Materi Exploration, Inc.
Denver, Colorado

ABSTRACT

Reverse circulation drilling techniques, in which returns are brought to the surface inside the drill pipe, offer significant advantages over conventional drilling for geothermal exploration in terranes where adverse drilling conditions are anticipated. Such techniques require special equipment and skills, making them more expensive than conventional drilling in uncomplicated environments. However, the special capabilities that allow the reverse circulation completion of holes in zones of severe lost circulation or formation instability typically result in substantial relative cost reductions for drilling programs conducted under such adverse conditions.

INTRODUCTION

An irony of geothermal exploration drilling is that many of the geologic characteristics sought by the explorationist condemn the driller to operations under extremely adverse conditions. Severe lost circulation and hole stability problems have been obstacles to the cost effective completion of many geothermal drilling programs, and have probably caused the premature termination of work on promising prospects.

The traditional remedies for lost circulation and hole instability are massive doses of lost circulation materials and frequent cement jobs. These approaches are not often wholly successful in the worst cases encountered in geothermal exploration and are very costly in terms of rigtime and materials.

Reverse circulation drilling techniques have been employed in mineral exploration, oil field, and water well drilling for many years to help to improve performance in areas of difficult drilling, to facilitate the drilling of relatively large diameter holes, and to optimize sample recovery and quality. This paper is specifically aimed at those explorationists wishing to conduct thermal gradient drilling programs to depths of 2000' or less in difficult drilling environments. However, many of the principles and techniques discussed are applicable to the drilling of deeper tests and production wells.

FUNDAMENTALS OF REVERSE CIRCULATION

Reverse circulation drilling differs from conventional rotary drilling in that all cuttings and drilling fluids (air, foam, water, mud) are returned to the surface through the inside of the drill pipe. There are two general catagories of reverse circulation drilling: static head/air lift, and dual tube rotary. These two techniques require somewhat different equipment and approaches and one may offer advantages over the other in any given drilling environment. Ideally, a drill rig would be able to change from one method to the other routinely. Any rig equipped for dual tube rotary drilling can also perform static head/air lift reverse circulation drilling, but the converse is not necessarily true.

STATIC HEAD/AIR LIFT REVERSE CIRCULATION DRILLING

Static head/air lift systems are appropriate for use when drilling will be conducted below the static water level, or where the hole can be kept relatively full of fluid. By injecting compressed air into the fluid column inside the drill pipe at a point halfway or more from the static fluid level to the bottom of the hole, the fluid column will be aerated and lightened enough to create air lift flow to the surface inside the drill pipe. The drill cuttings are separated from the fluid at the surface and the fluid reintroduced into the hole anulus under gravity feed. In operation, air is injected through a drop pipe inserted down the drill pipe through the rig swivel. In the case of dual tube equipped rigs, the air is injected through the drill pipe anulus into the inner tube.

The chief advantage that static head/air lift drilling enjoys over conventional fluid circulation is in its inherent ability to prevent or to remedy lost circulation problems. Lost circulation occurs when the hydrostatic pressure of the drilling fluid column exceeds the formation pressure. Aeration of the fluid column lightens the fluid weight, lessening the hydrostatic pressure to the point at which the drill fluid pressure is less than the formation pressure and circulation is maintained or regained.

A secondary advantage of maintaining circulation is in the capacity to treat unstable formations. Formation instability may be due to mechanical weakness (caving alluvium, fractured rock,etc.)

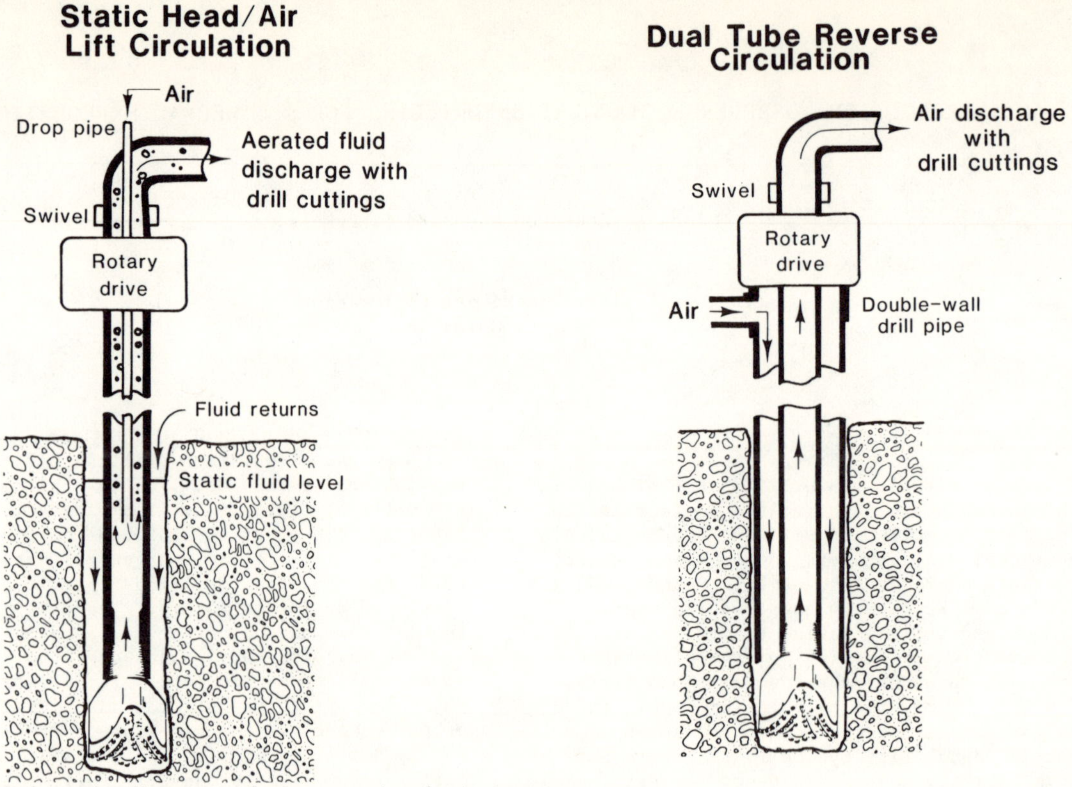

FIGURE 1. Schematic diagrams of reverse circulation systems.

or water sensivity (clays, etc.). Mechanical instability is treated best with the maintenance of some hydrostatic pressure on the formation, which is prohibitively expensive or impossible in lost circulation areas. Moreover, mechanical instability is aggravated by erosion caused by turbulent flow in the hole anulus. Flow turbulence in the hole is always at a minimum in reverse circulation drilling.

Water sensitive formations must be treated with hydration inhibiting chemicals which are too expensive to sacrifice in quantity to lost circulation zones.

The maintenance of circulation in either case of instability contributes significantly to treatment of the problem(s) at hand and reduces material costs to manageable levels.

DUAL TUBE ROTARY REVERSE CIRCULATION DRILLING

Dual tube systems employ a special drill string made of two concentric tubes. Circulation may be with air, foam, mud or water, but is most often with dry air or air mist. The circulating fluid is pumped down the anular space between the concentric tubes, accross the bit, and returns to the surface inside the inner tube. The air and the cuttings/fluid fractions of the returns are separated in a de-aerating cyclone and the cuttings collected.

Most drill rigs can be adapted for dual tube reverse circulation systems, however the top-head drive or power swivel systems are the most readily adapted.

Dual tube reverse circulation drilling is appropriate for use in nearly all drilling conditions. Circulation is usually done with air, which maximizes penetration rates and allows sampling of formation fluids. Reasonably large water flows may be handled with air/foam circulation. In extremely high water flows booster compressors may be used or dual tube air lift drilling may be employed.

Dual tube drilling allows circulation to be maintained even in the most severe lost circulation environments. For example, in the process of drilling through a large underground void, such as a covern or an old mine working, circulation has been lost momentarily but regained upon penetration of the floor. This is because all circulation occurs inside the drill pipe and the only exposure of the drilling fluid to the formation is at the bit.

Schematic diagrams of both types of reverse circulation systems are presented in figure 1.

REVERSE CIRCULATION ECONOMICS

The economic advantages of reverse circulation drilling are entirely dependant on the drilling characteristics of the prospect. For comparable holes, reverse circulation footage prices and hourly rates are commonly 20% to 50% higher than those for conventional drilling. However, unit prices rarely reflect actual costs for holes where significant problems have occurred, because most drilling contracts revert to time and materials charges in adverse drilling conditions. It is therefore conceivable that a determined operator might pay several times the quoted footage prices to complete a hole under adverse conditions. It is also conceivable that the hole might not be completed at all.

The use of reverse circulation is appropriate where experience or intuition indicate that rigtime and materials (drilling fluids, lost circulation materials, casing, etc.) are likely to increase actual costs by 50% or more over quoted footage rates for conventional drilling, or where time and materials charges account for 50% or more of the gross cost per foot of hole.

In most cases where such cost conditions prevail, there will be additional economic advantages favoring reverse circulation drilling. Such advantages may include increased production rates, reduced project overhead, a broader data base from which to gauge later exploration decisions, and fewer abandoned holes.

CONCLUSIONS

The use of reverse circulation drilling procedures has significant economic advantages in geothermal exploration drilling, particularly in difficult drilling environments. Reverse circulation systems have proven their worth for more than 25 years and over millions of feet of mineral exploration, water well, and oil field drilling. The required hardware is readily adaptable to the types of drilling equipment and drilling environments common in geothermal exploration. The advantages enjoyed by these systems in controlling lost circulation and hole instability yield significant savings in money, time, and materials expended on exploration drilling programs, as well as improving data quality from each hole.

Exploration managers should realize that reverse circulation drilling systems are available and should be aware of the cost and technical considerations involved with the use of the systems.

BIBLIOGRAPHY

1) Balanced pressure techniques applied to geothermal drilling. Sandia Government Contract #62-7201 April '81.

2) Bobo,Roy A., "New Aerated Mud Equipment Can Control Lost Circulation" Oil and Gas Journal April 29, 1968 pp. 64-69.

3) Binkley, J. F., "Concentric Drill Pipe/Air Lift: New Way to Curb Lost Circulation." World Oil, June 1968, p. 64.

4) Rehm, Bill, "Don't Overlook Aerated Mud" Oil & Gas Journal, December 2, 1963, pp. 150-153.

5) Dareing, Don W., and Kelsey, James R., "Balanced Pressure Techniques Applied to Geothermal Drilling" Geothermal Resources Council Transactions Vol. 5, October 1981. pp. 233-236

6) White, Robert J., "Bottom Hole Pressure Reduction Due to Gas Cut Mud" Journal of Petroleum Technology, July 1957, pp. 49-50.

RESERVOIR TECHNOLOGY

Geothermal Resources Council, TRANSACTIONS Vol. 6, October 1982

Well Test Data Analysis from a Naturally Fractured Liquid-Dominated Hydrothermal System

S. M. Benson
Earth Sciences Division
Lawrence Berkeley Laboratory
University of California
Berkeley, California 94720

Abstract

Production test data from a moderate-temperature geothermal well in the Basin and Range Province have been analyzed. The well is completed in granitic basement rock. Both the pressure transient and spinner data confirm that fractures provide the major component of the reservoir permeability. The productivity index of the well decreases and the apparent skin factor increases with increasing flow rate. This behavior is attributed to non-Darcy flow in the fractures near the well bore. A mathematical relation between flow rate and drawdown has been established that includes the non-Darcy and Darcy flow components.

Introduction

Well WEN-1 was drilled by GeoProducts Corporation under the Department of Energy's User Coupled Drilling Program. In March 1982 the well was tested for a period of approximately one week. The geothermal group at Lawrence Berkeley Laboratory was invited to participate in the test by collecting downhole pressure data with a Hewlett Packard Downhole Pressure/Temperature gauge. In this report the analysis of the production test data is discussed.

The well, WEN-1, is located on the eastern side of the Honey Lake valley, California, near the Wendel and Amedee Hot Springs. It was drilled to a depth of 5837' and cased 5068' with a 9 5/8 in. casing. From a depth of 5068' to 5837' the well is open hole. The entire open interval of the well is completed in granitic basement rock. Pre-test temperature and spinner surveys were conducted, but space limitations do not permit their discussion here. It will suffice to say that the measured bottom-hole temperature was ~ 248°F and that ~ 80% of the flow comes from one major fracture zone.

Test Description

Figure 1 shows a plot of all the test data.

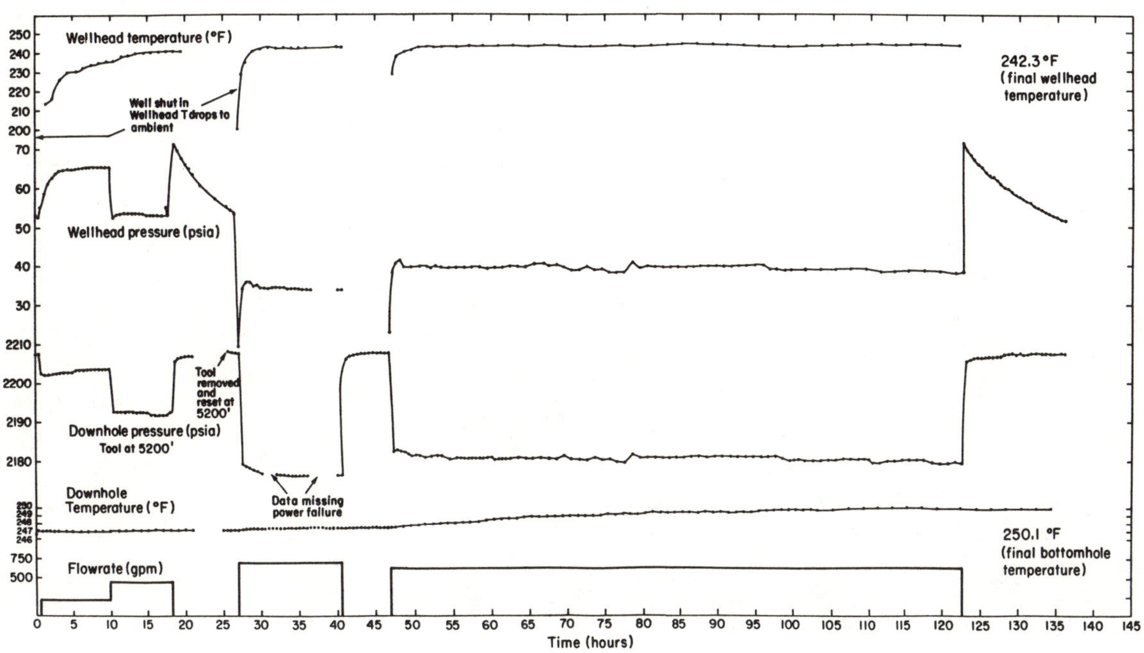

Figure 1. Test data from WEN-1 well test; wellhead pressure and temperature, downhole pressure and temperature, and flowrate.

Downhole pressure data were obtained with a Hewlett Packard quartz crystal gauge. Wellhead pressure was measured with a Paroscientific gauge. Flow rates were metered by measuring pressure differential across an orifice with Paroscientific gauges. Downhole temperature was measured with a G.O. temperature tool. Wellhead temperature was measured with a thermocouple.

The well was flowed artesian at four rates. The first three rates (220 gpm, 440 gpm, and 680 gpm) were held at nearly constant values for a duration of approximatley 12 hours each. The fourth flow rate (620 gpm) was held nearly constant for a period of 75 hours. At each flow rate the downhole pressure quickly stabilized and showed little or no change for the duration of the flow period. Both wellhead and downhole temperature rose steadily throughout the test. The wellhead temperature rose from 240°F to 242.3°F, reflecting heating of the wellbore. The bottomhole temperatures increased from 246.8°F to 250.1°F over the test period.

WEN-1 Productivity Analysis

Table 1 summarizes the downhole productivity data from WEN-1. At 220 gpm the Productivity Index is ~ 49 gpm/psi as compared to ~ 22 gpm/psi at 680 gpm. In Figure 2 the downhole pressure change vs. flow rate is plotted. The relationship between the flow rate and drawdown, shown in Figure 4, can be expressed by Equation 1:

$$\Delta P(q) = 1.13 \times 10^{-2} q + 5.16 \times 10^{-5} q^2 . \quad (1)$$

Table 1. Downhole productivity data for WEN-1.

Flow Rate (gpm)	Downhole ΔP (psia)	q/ΔP (gpm/psi)
220	4.5	48.9
440	15 0	29.3
620	26.5	23.4
680	31.5	21.6

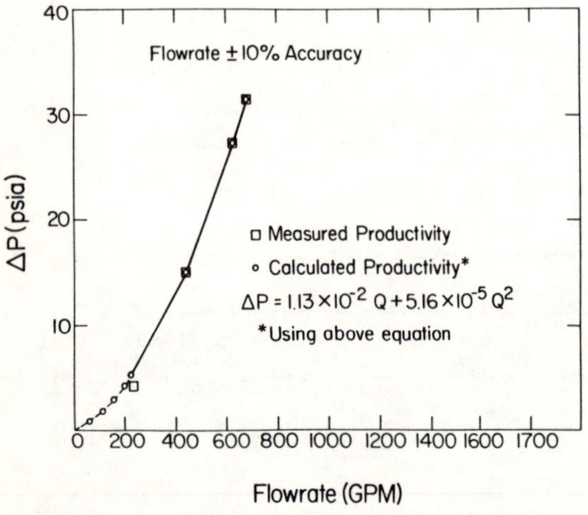

Figure 2. Downhole pressure change vs. flowrate for WEN-1.

Typically, the drawdown component proportional to q^2 is the result of well losses created by the pressure drop through the production liner and flow up the well bore (Jacob, 1947). However, since the well is completed with an open hole, and since downhole pressure data are used, the proportionality to q^2 must be a result of flow in the reservoir. This is discussed further in the following sections.

Pressure Transient Analysis

The pressure transient data were analyzed by the Horner and Miller-Dyes-Hutchinson techniques to obtain values for the reservoir transmisivity and skin factor (Mathews and Russell, 1967). Since the pressure drawdown quickly stabilizied and an absolutely constant flow rate was difficult to maintain, emphasis was placed on the analysis of the buildup data. A Horner plot of the buildup data following step 4 (620 gpm) is shown in Figure 3.

Each of the Horner plots has the same distinctive character demonstrated in Figure 3. When the well is shutin, the pressure immediately increases by nearly 95% of the total pressure drop. (Note that only the last 3 psi of pressure buildup are shown.) Then, a semilog straight line is apparent. After approximately one log cycle (10 min), the rate of pressure buildup decreases. Approximately half a log cycle later, the rate of buildup increases and once again a semilog straight line is apparent. In each case, the final semilog straight line gives nearly the same transmisivity, 3.3 - 3.5 x 10^6 md-ft/cp; moreover, the final semilog straight line extrapolates to the correct reservoir pressure. The later portion of the pressure transient data can be indicative of a double porosity or naturally fractured reservoir. Assuming a viscosity of 0.24 cp, the reservoir has a permeability-thickness of ~ 8.4 x 10^5 md-ft. Pressure data from these tests do not appear to be influenced by any reservoir boundaries. However, the relatively small drawdown and difficulty in analyzing the drawdown data may obscure boundary effects.

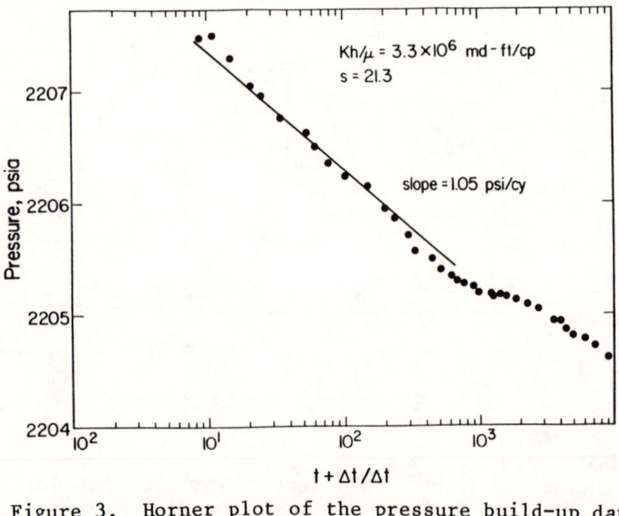

Figure 3. Horner plot of the pressure build-up data following step 4 (620 gpm).

The skin value, indicative of well bore damage or enhancement, was calculated for each of the pressure buildups. A storativity (ϕch) of 4.5 x 10^{-4} ft/psi was used in the skin value calculations. Even if this value is off by an order of magnitude, only a small error is introduced into the calculation because of the logarithmic dependence on ϕch. Table 2 lists the results of the analysis for each test segment. The skin value increases from 16 at 440 gpm to 24.1 at 680 gpm. These very high skin values are unusual for a geothermal well and very surprising, considering that the reservoir is fractured. Drilling fluid (water) and cuttings were lost in the production interval, but it is unlikely that this caused such a large permeability reduction. In addition, the skin value of a well should not be a function of the flow rate if the conventional definition of skin were applicable.

Table 2. Results of pressure buildup data analysis for WEN-1.

Segment	Flow Rate (gpm)	kh/μ (md-ft/cp)	Skin
2	440	3.3 x 10^6	16
3	680	3.5 x 10^6	24.1
4	620	3.3 x 10^6	21.3

Ramey (1965) discussed the concept of a skin effect in gas wells that included non-Darcy flow in the reservoir. The formulation he proposed is shown in Equations 2 and 3. Note that this formula assumes that q is measured in gallons/minute:

$$s' = s + Dq \, , \quad (2)$$

where

$$\Delta P(s') = 4840 \times qs'\mu/kH \, . \quad (3)$$

D can be calculated from the slope of a plot of s' vs. q. In turn, s can be calculated from the intercept at q = 0. In Figure 4, s' vs. q is plotted. As shown, a single straight line with a slope of 3.54 x 10^{-2} gpm^{-1} can be drawn through the points. The q = 0 intercept is nearly zero, indicating a zero true skin value for the well.

To summarize, kH/μ ≈ 3.5 x 10^6 md-ft/cp and D = 3.54 x 10^{-2} gpm^{-1}. The drawdown (at a single constant flow rate) can be calculated by

$$\Delta P(t) = \frac{4840q}{3.5 \times 10^6} P_D + \frac{4840q^2}{3.5 \times 10^6} \times 3.54 \times 10^{-2} . \quad (4)$$

The match of the observed and calculated pressure, based on Equation 4, is shown in Figure 5. The calculated and observed pressure drops are in excellent agreement.

Discussion

Non-Darcy flow in gas wells is quite common due to the very high near-well-bore velocity (Smith, 1961; Swift and Kiel, 1962). Similar phenomena, related to well losses, are also discussed in the literature on groundwater hydrology (Rorabaugh, 1953). However, in the groundwater literature, the q^n ($n \sim 2$) drawdown component is usually due to pressure losses resulting from flow through the production liner and friction losses up the well bore. As mentioned earlier, the q^2 dependency discussed here is due strictly to flow in the formation.

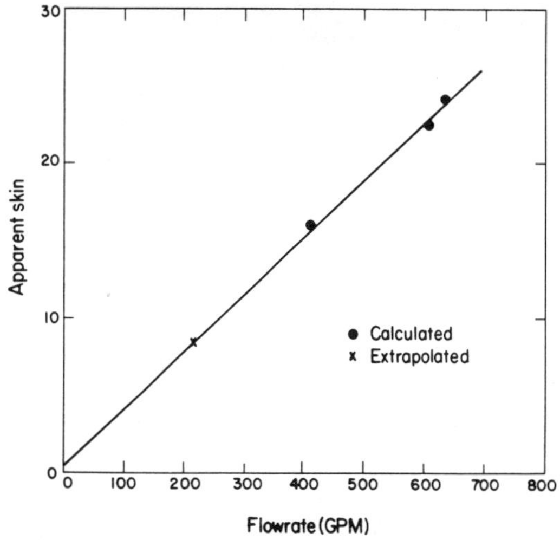

Figure 4. Plot of s' vs. q.

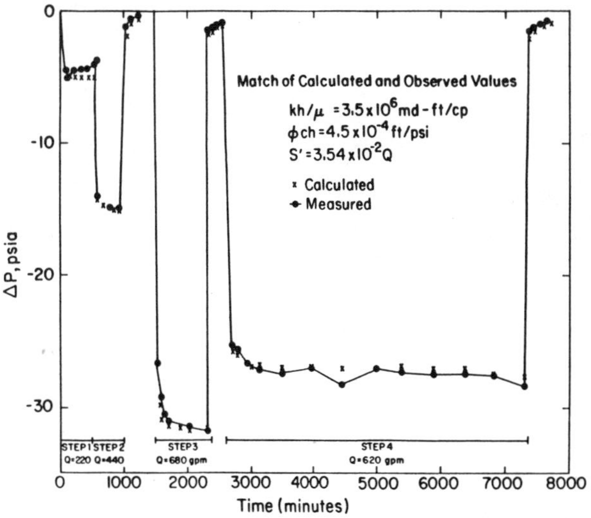

Figure 5. Match of observed and calculated data.

To obtain more insight into the problem, the velocity distribution in the formation was calculated. Because little is known of the precise fracture spacing, distribution, and aperture: velocity estimates were made for several cases. Knowing the permeability-thickness of the reservoir, the fracture aperture can be calculated from the cubic law (Snow, 1968):

$$kH = \frac{b^3}{12} \, , \quad (5)$$

where b = fracture aperture (m).

Estimates of b were made assuming that 1, 10, and 100 equally sized fractures intersect the well bore. Table 3 lists the fracture apertures, equivalent fracture mobility-thickness (per fracture), and number of fractures for a total of kH of 8.4 x 10^5 md-ft (2.56 x 10^{-10} m^3).

Table 3. Fracture aperture, equivalent kH, and number of fractures for a total kH of 8.4 x 10^5 md-ft.

Number of Fractures	Aperture (mm)	kH (md-ft)
1	1.5	8.4 x 10^5
10	0.67	8.4 x 10^4
100	0.31	8.4 x 10^3

Assuming a fracture aperture, the flow velocity in the fractures (as a function of radial distance from the axis of the well bore) for each of the step rates was calculated. Table 4 summarize the results at the highest flow rate, 680 gpm.

One of the primary assumptions in Darcy's law is that a laminar flow regime exists in the reservoir. The flow regime in the reservoir is governed by the Reynolds number (R) and the roughness (ε) of the surface. Usually, for flow between parallel plates, at Reynolds numbers greater than 2000, the flow is no longer laminar. At Reynolds numbers greater than 2000, the flow regime is either transitional or turbulent (Knudsen and Katz, 1956). The Reynolds number for flow between parallel plates is calculated by

$$R = \frac{b\bar{v}\rho}{\mu} . \quad (6)$$

Calculation of the Reynolds numbers for the flow velocities listed in Table 4 indicate that; at 680 gpm, turbulent flow will exist to more than 5 m into the reservoir if a single fracture feeds the well, up to 1 m into the formation if 10 fractures feed the well and most likely, will not exist if 100 fractures feed the well. A unique analysis of the fracture spacing and size is not possible with current fracture flow theory. However, the above analysis indicates that non-Darcy flow can significantly contribute to the drawdown if only a few major fractures feed the wellbore.

Conclusions

Both pressure transient analysis and productivity data indicate that the pressure drawdown has a component that is proportional to q^2. Analysis of the velocity distribution in the formation and the resultant Reynolds numbers indicate that turbulent flow in the near-well-bore formation may very well be the cause of the q^2 dependent component of the drawdown. Because the fracture aperture, roughness and distribution are unknown, it is not possible to fully evaluate the flow regime in the reservoir. However, an empirical relationship between flow rate and drawdown, with a very good correlation to the observed data, can be derived from this analysis. The analysis indicates a reservoir transmissivity of 3.5 x 10^6 md-ft/cp and a non-Darcy coefficient of 3.54 x 10^{-2} gpm^{-1}.

Table 4. Average fluid velocity in the fractures (m/sec)

No. of Fractures	Radial Distance from Well Axis (m) 0.2	0.6	1.	2.	5.
1	23.54	7.84	4.70	2.35	0.94
10	5.09	1.70	1.02	0.51	0.20
100	1.10	0.36	0.22	0.11	0.04

References

Jacob, C.E., 1947, Drawdown test to determine effective radius of artesian wells: American Society of Civil Engineers Transactions, v. 112, p. 1047-1070.

Knudsen, J.G., and Katz, D.L., 1956, Fluid dynamics and heat transfer: McGraw-Hill Chemical Engineering Series, p. 106-129.

Matthews, C.S., and Russell, D.G., 1967, Pressure build-up and flow tests in wells: Dallas, Society of Petroleum Engineering Monograph series (Vol. 1).

Ramey, H.J., Jr., 1965, Non-Darcy flow and wellbore storage effects in pressure build-up and drawdown of gas wells: Journal of Petroleum Technology, v. 17, no. 2, p. 223-233.

Rorabaugh, M.I., 1953, Graphical and theoretical analysis of step-drawdown test of artesian well: American Society Civil Engineers Proceedings, v. 79, sec. 362, p. 23.

Smith, R.V., 1961, Unsteady-state gas flow into gas wells: Journal of Petroleum Technology, November, p. 1151-1159.

Snow, D.T., 1968, Rock fracture spacings, openings and porosities: J. Soil Mech. Fdns. Div. Am. Soc. Civ. Engineers, v. 94, 73-91.

Swift, G.W., and Kiel, O.G., 1962, The prediction of gas-well performance including the effect of non-Darcy flow: Journal of Petroleum Technology, v. 14, no. 7, 791-798.

Nomenclature:

b	-	fracture aperture (m)
c	-	total compressibility psi^{-1} (Pa^{-1})
D	-	Non-Darcy coefficient gpm^{-1} (sec/m^3)
H	-	formation thickness ft (m)
k	-	permeability md (m^2)
P_D	-	dimensionless pressure
P	-	pressure psia (Pa)
R	-	Reynolds number
q	-	flowrate gpm (m^3/sec)
s	-	infinitesimal skin
s'	-	apparent skin = s + Dq
v	-	average velocity (m/s)
ϕ	-	porosity
μ	-	dynamic viscosity cp (Pa-s)
ρ	-	density (kg/m^3)

Acknowledgements

This work was supported by the Assistant Secretary for Conservation and Renewable Energy, Office of Renewable Technology, Division of Geothermal and Hydropower Technologies of the U.S. Department of Energy under Contract No. DE-AC03-76SF00098. The assistance of Ray Solbau and Don Lippert in obtaining this data is gratefully acknowledged. I would like to thank GeoProducts Corporation, Russell Juncal, and Gus Gertsch for providing the opportunity to obtain the data discussion in this in this paper.

Geothermal Resources Council, TRANSACTIONS Vol. 6, October 1982

CHARACTERISTICS OF THE ZUNIL GEOTHERMAL FIELD (WESTERN GUATEMALA)

H.R. Bethancourt*, E. Dominco**

* INDE, Instituto Nacional de Electrificación, Guatemala, C.A.
** ELC-Electroconsult, Milan, Italy

ABSTRACT

The Zunil geothermal field represents the marginal, shallow expression of a vast geothermal complex buried beneath active volcanic edifices (Cerro Quemado, Volcán Santa Maria, Western Guatemala) some kilometers to the west. The area lies at the edge of a tecto-volcanic depression where some 1,000 m of Tertiary and Quaternary volcanics are underlain by a granodioritic basement.

High temperature geothermal fluids (over 280ºC) reach the field from the west, upflowing along the inclined contact between the granodioritic and the overlying volcanics, and along fractures in the basement itself. A conglomeratic layer at the volcanics/basement contact, and the underlying weathered cap of the basement form the only permeable horizon of the succession; this horizon forms the local reservoir tapped by the productive wells. Its reduced thickness (around 50 m) allows for a limited fluid storage such that field production relies on external recharge along the permeable horizon and underlying fractures in the granodiorite.

Production testing and simulation models indicate a fairly rapid evolution of reservoir conditions from the liquid to the steam phase, due to pressure drawdown, in its turn due to a restricted inflow. The phenomenon determines an upgrading of the fluid enthalpy, but a decline of mass output. Simulated reinjection into the reservoir proved to be an effective measure to slow down such an evolution and optimize the field exploitation.

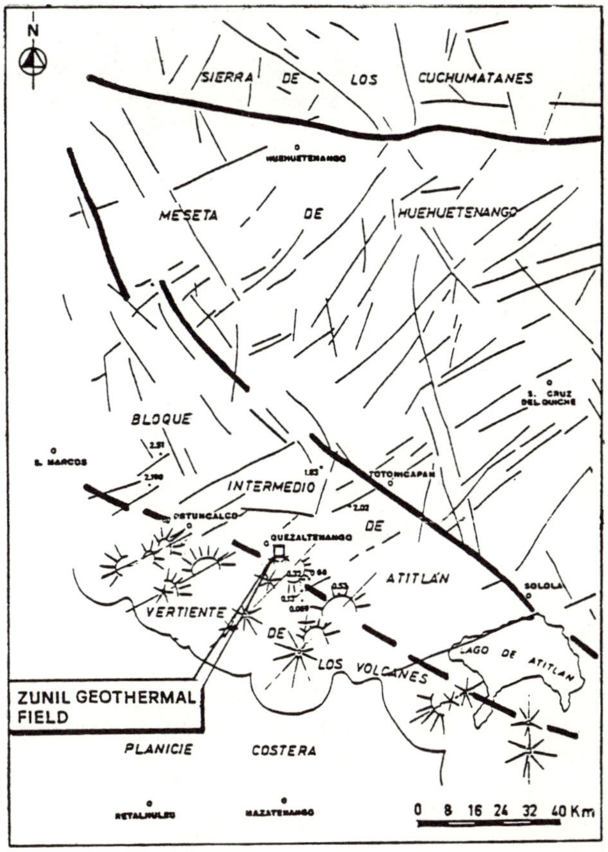

FIG. 1 TECTONIC SCHEME OF WESTERN GUATEMALA

INTRODUCTION

Active volcanism and widespread hydrothermal manifestations instigated geothermal exploration in the Zunil area. Systematic investigations carried out by INDE (Instituto Nacional de Electrificación) in the period 1975-79 produced favourable indications on the existence of an exploitable geothermal field. Exploratory and evaluation drilling, executed in the years

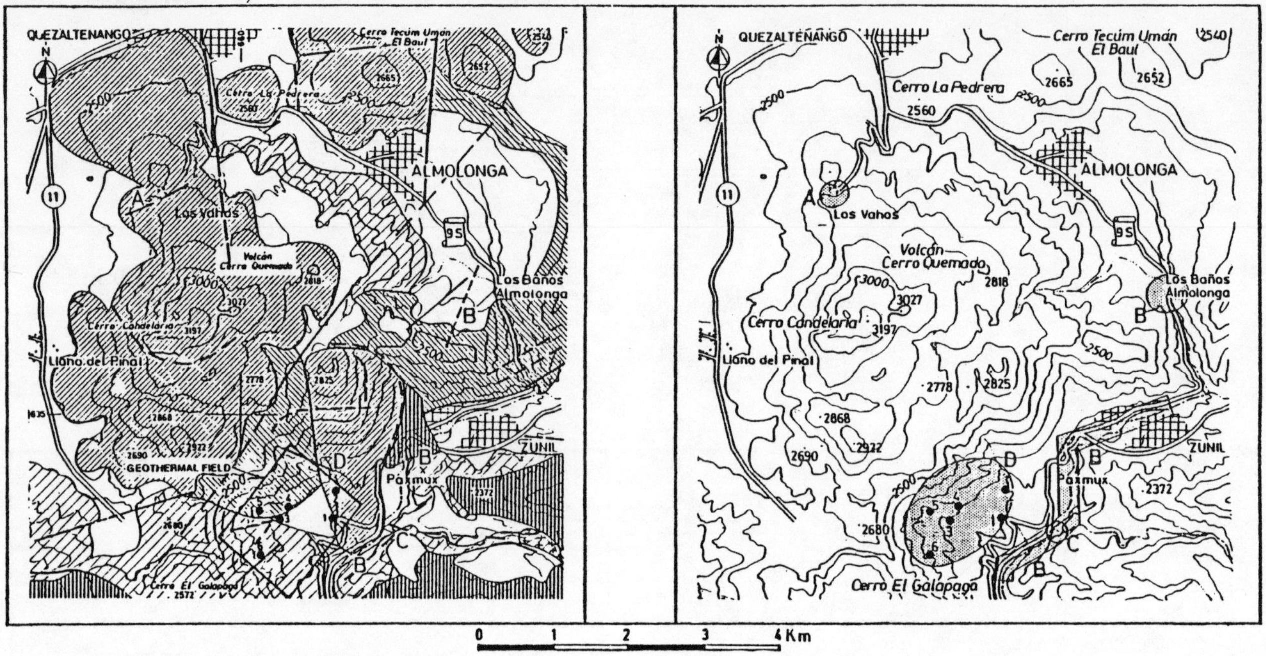

FIG. 2 GEOLOGIC MAP

- Alluvial deposits and pyroclasts
- Cerro Quemado dome complex
- Upper Quaternary volcanic edifices
- Almolonga lavas
- Lower Quaternary volcanic edifices
- Cretaceous granodioritie basement
- Fault
- 5 Deep geothermal well

FIG. 3 ZUNIL HYDROTHERMAL MANIFESTATIONS

A Low temperature springs, heated by conduction

B Medium to high temperature springs, essentially heated by conduction

C Boiling springs, partly fed by the deep geothermal aquifer

D Steam manifestations

1980-81, proved the occurrence of a geothermal reservoir, suitable for power generation.

GEOLOGIC OUTLINE

As shown in Fig.1, the Zunil geothermal field is located on the southern edge of a fault-block ("Bloque de Atitlán"), which occupies an intermediate position between the large inland Paleozoic plateau ("Meseta de Huehuetenango") and the Quaternary volcanic belt ("Vertiente de los Volcanes"). The echelon fault arrangement of the region, downstepping towards the Pacific Ocean, is the result of NW-SE and E-W tensional faulting which was active since Tertiary and which also gave origin to the volcanic activity. During Tertiary, such an activity was mainly of the fissural type, with deposition of extended andesitic lava flows on the top of the Cretaceous basement. It evolved into central type activity in the Pleistocene, with growth of large edifices, now deeply eroded. The most recent volcanic centers are aligned along NE-SW axes, intersecting the older regional tectonic trends. During Quaternary, the volcanism evolved towards more acidic products, and the last activity generated mainly dome-type volcanoes, such as the Cerro Quemado complex (active until late last century) and the Santiaguito, a lateral cone of the large extinct Santa Maria strato-volcano, which shows an almost continuous explosive activity.

As shown in Fig.2, the explored geothermal field is located on the southeastern margin of the Cerro Quemado dome complex which occupies the center of a volcano-tectonic depression. Exploratory wells confirmed that the granodioritic basement

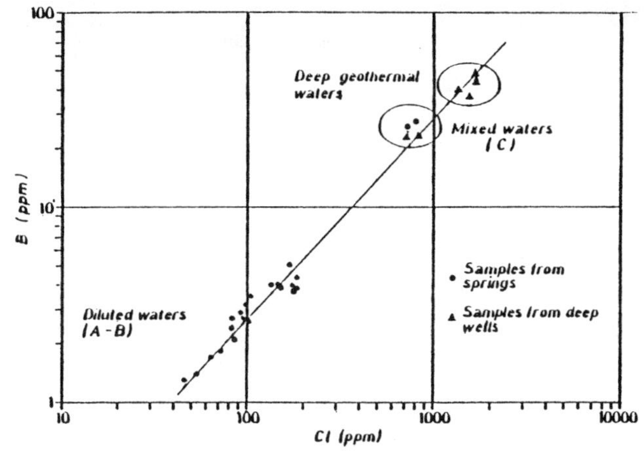

FIG. 4 B/CL RATIO OF ZUNIL SPRINGS AND WELL WATERS

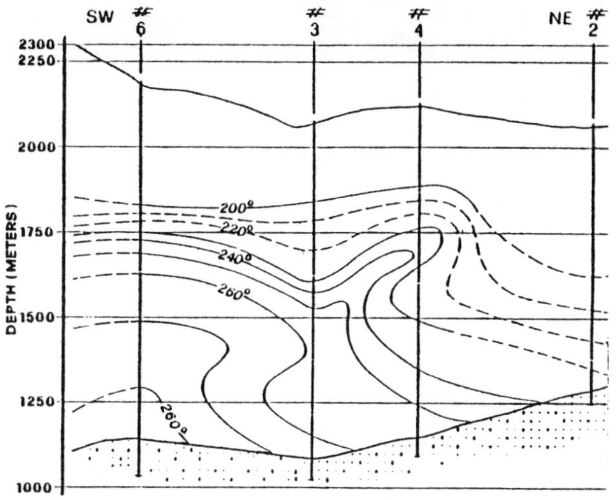

FIG. 5 Cross section of the Zunil field with isotherms, reflecting hot water flow eastwards. The nose between well 3 and 4 corresponds to an oblique upflow along a fractured zone.

gradually rises towards the east of the field, outcropping 2.5 km further east. Following the attitude of the basement, the overlying volcanic cover thins in the same direction, its thickness ranging between 1,000 and 800 m. Schematically, the stratigraphic sequence is the following:

- Quaternary volcanics (dacitic and rhyodacitic lavas, pyroclastics): 500-600 m;
- Tertiary lava flows, mainly andesitic: 300-400 m;
- Basal conglomeratic deposits, made of volcanic and granitic fragments: 0-50 m;
- Cretaceous granodioritic basement.

The field appears to be weakly affected by faulting and fracturing, which may be partly explained by the presence of the rigid granodioritic mass at shallow depth.

GEOTHERMAL MANIFESTATIONS AND GEOCHEMISTRY OF FLUIDS

The Zunil geothermal system manifests itself in the form of steam manifestations and hot and warm springs. Most of the hot springs are located on the eastern side of the field, along the deep cut of the Samalá River, while steam vents occur within the drilled area (see Fig.3).

With few exceptions, the chemical composition of the springs indicates an origin from shallow, fresh water aquifers heated by conduction, with a possible minor contribution of deep geothermal fluid. The geothermal wells showed that the latter is constituted by a NaCl water. Its concentration (TDS = 3,500 ppm), appreciably lower than in other known geothermal fields of Central America (Ahuachapán: 19,000 ppm; Momotombo: 8,000 ppm), may be explained by

TABLE 1

CHEMICAL COMPOSITION OF SPRING AND WELL WATERS (ppm)

	diluted springs	mixed waters	deep geoth. waters
t°C	62	89	-
pH	7.6	9.0	7.8
Na	150	545	960
K	13	51	146
Ca	16	7	32
Mg	5	0.3	0.3
Cl	101	728	1,660
SO_4	116	210	78
HCO_3	260	96	67
B	2.6	26	48
SiO_2	153	404	476

a relatively fast circulation in contact with the granodioritic basement, less susceptible to leaching than volcanic rocks.

The type C springs, located along one of the few deep faults, discharge a mixture of deep geothermal water and fresh water (see Fig.4).

THE GEOTHERMAL SYSTEM

The combined data from surface and well studies resulted in the interpretative scheme for the field shown in Fig.6.

The Zunil field receives its fluid supply by lateral migration from a source area which can be identified with the volcanic axis Cerro Quemado-Volcán Santa Maria and its associated system of faults and fractures, which have created the conditions for an active vertical circulation and a considerable accumulation of fluids in the underlying formations.

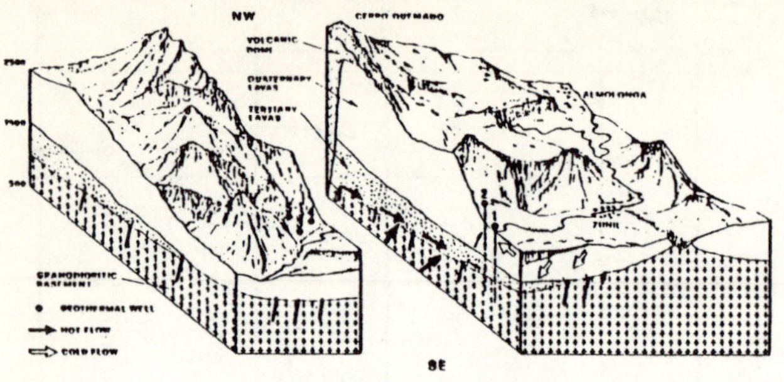

FIG. 6 ZUNIL CIRCULATION SCHEME

The migration of fluids from the source area towards the Zunil field occurs along the contact between the basement and the volcanic cover, where the basal conglomerate and the weathered cap of the granodiorite constitute a continuous permeable horizon, and alongopen fractures within the granodiorite itself.

The circulation scheme, based on geological and hydrogeological considerations, is also supported by the following facts: the isotherms shape across the field (see Fig.5), the increasing temperatures westwards, the secondary minerals suites suggesting a hot inflow from the west, the temperature inversion in the upper basement, indicating a hot flow on the top of a cooler surface. Such a feeding mechanism has an important bearing over the exploitability of the field: the fluid extraction rate is directly controlled by the velocity of the recharge from the source area, the storage capacity of the Zunil reservoir being rather limited.

PRODUCTION CHARACTERISTICS

Six wells were drilled to evaluate the geothermal aquifer. Due to topographic restraints, an area of only 2 km^2 was accessible for drilling. All the wells reached the granodioritic basement, and one penetrated deep into it. Five wells resulted productive with different characteristics. Two wells (4 and 5) produce dry saturated steam, while the others, a mixture of steam and water (see Table 2).

The wells confirmed the existence of a thin permeable layer coincident with the basal clastics, while the overlying volcanic cover resulted almost totally impervious.

Reservoir testing indicated a fast and pronounced pressure drawdown caused by production, which denoted both a reduced local availability of fluid and a restricted inflow. As a consequence of the strong

TABLE 2

PRODUCTION CHARACTERISTICS OF THE ZUNIL WELLS

	WELL n.1	WELL n.2	WELL n.3	WELL n.4	WELL n.5	WELL n.6
Reservoir Temp. (°C)	250	256	271	237	240	280
Transmissivity (m^3/Pa.s)	2.5E-9	1.5E-8	1.0E-8	1.5E-8	not determ.	2.0E-8
Total flow at 8 bar (kg/s)	-	15	39	7	8	25
Enthalpy (kJ/kg)	-	1,000	1,000	2,700	2,700	1,000

pressure drawdown (of the order of tens of bars), there appeared the formation of steam in the reservoir, which introduced an additional obstacle to fluid circulation, due to the lower ϱ/μ ratio of steam with respect to water. This is the case of well N° 5, which immediately reached dry steam production, due to the limited influx of fluid from the surrounding area. On the contrary, well N° 4 produces dry steam from an upflow zone, intersected at 600 m depth, where, due to the reduced hydrostatic pressure, flashing occurs in the formation.

The upflow is clearly reflected by the isotherm "nose" visible in well N° 4 (Fig.5). The analysis of the reservoir conditions and the simulation of its future behaviour by means of mathematical models showed that the field would evolve towards a rapid increase in enthalpy and a progressive decline of production, which would limit its power potential. Reinjection into the reservoir of the residual water looked as an effective measure to control vaporization in the formation and restore pressure to the field.

Reservoir behaviour was simulated at an extraction rate of 125 l/s and simultaneous reinjection of 100 l/s. The results showed that reinjection appreciably delays vaporization in the formation, assuring a higher margin of "exploitability" to the field.

SELECTED REFERENCES

BETHANCOURT, H.R. (1978): Exploración Geotérmica en Guatemala. Documento OLADE No.2.

ELC-ELECTROCONSULT (1980): Evaluación del Potencial del Campo, Petrografía Ignea y Mineralogía de Alteración. Proyecto Zunil. Unpublished Report.

ELC-ELECTROCONSULT (1981): Evaluación del Potencial del Campo, Estudio Geo-Volcanológico de la Region Zunil - Proyecto Zunil Cerro Quemado. Unpublished Report.

INDE (1978): Informe de Factibilidad Preliminar. Proyecto Zunil. Unpublished Report.

Geothermal Resources Council, TRANSACTIONS Vol. 6, October 1982

STUDIES OF INJECTION INTO NATURALLY FRACTURED RESERVOIRS

Gudmundur S. Bödvarsson and Cheng Hsien Lai

Earth Sciences Divisison
Lawrence Berkeley Laboratory
University of California
Berkeley, California 94720

ABSTRACT

A semi-analytical model for studies of cold water injection into naturally fractured reservoirs has been developed. The model can be used to design the flow rates and location of injection wells in such systems. The results obtained using the model show that initially the cold water will move very rapidly through the fracture system away from the well. Later on, conductive heat transfer from the rock matrix blocks will retard the advancement of the cold water front, and eventually uniform energy sweep conditions will prevail. Where uniform energy sweep conditions are reached the cold water movement away from the injection well will be identical to that in a porous medium; consequently maximum energy recovery from the rock matrix will be attained. The time of uniform energy sweep and the radial distance from the injection well where it occurs are greatly dependent upon the fracture spacing, but independent of the fracture aperture.

INTRODUCTION

Various theoretical studies have shown that injection of water into geothermal reservoirs during exploitation can greatly enhance the energy recovery from the resource. Injection will help maintain reservoir pressure and provide water that will extract energy from the reservoir rocks. However, experience gained through commercial scale reinjection has shown that the injection operation must be carefully designed. Horne (1981) reports enthalpy decline of produced fluids at several Japanese fields due to injection. This interference is attributed to rapid flow of the injected "cold" water through fractures, as evidenced by high tracer velocities. On the other hand commercial scale injection has been successful at other geothermal fields such as the Geysers, U.S. and Ahuachapán, El Salvador. This indicates that fundamental studies of cold water movement in fractured geothermal reservoirs are needed, before confidence in the design of commercial reinjection operations can be established.

In this study the problem of cold water injection into naturally fractured reservoirs is considered. The basis model used is that of Warren and Root (1963), where three orthogonal fracture sets of equal spacing are used. Similar work on non-isothermal flow in horizontal fractures are reported by Lauwerier (1955), Bodvarsson (1969), and Bodvarsson and Tsang (1982). The objective of the present work is to extend their work to include the effects of vertical fractures and to develop a methodology for the design of injection schemes for naturally fractured geothermal reservoirs.

BASIC MODEL

The model used in this study is shown in Figure 1. It consists of rectangular matrix blocks bounded by three sets of orthogonal fractures. Steady state fluid flow is assumed in the fractures, but fully transient conductive heat transfer between the impermeable rock matrix and the fractures is considered. Thus, the cold water will flow from the injection well into the fracture network, and as it moves away from the well, it will gradually get heated up due to the heat transfer from adjacent matrix blocks.

The equation for conductive heat transfer between the matrix blocks and the fractures is derived based on the basic element shown in Figure 2. The basic element represents 1/6 of a single matrix block (cube), i.e. only one face of the cube is considered (1-Dimensional approach). In this approach we assume that the thermal gradients are much smaller within the fracture network, than in the rock matrix. Thus, if the temperature in the fractures bounding a matrix block is rather

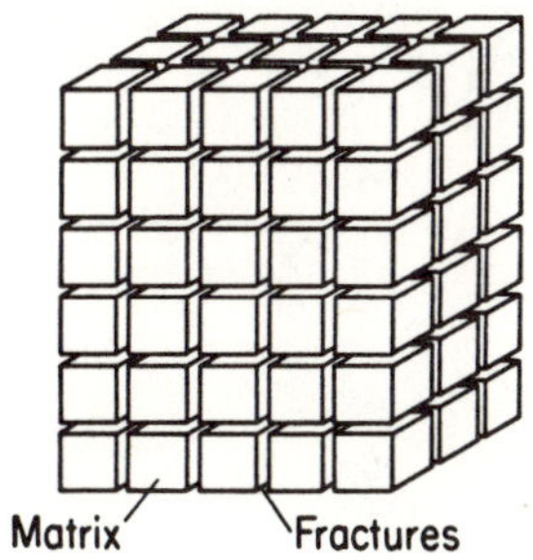

Figure 1. Idealized model of naturally fractured reservoirs (after Warren and Root (1963).

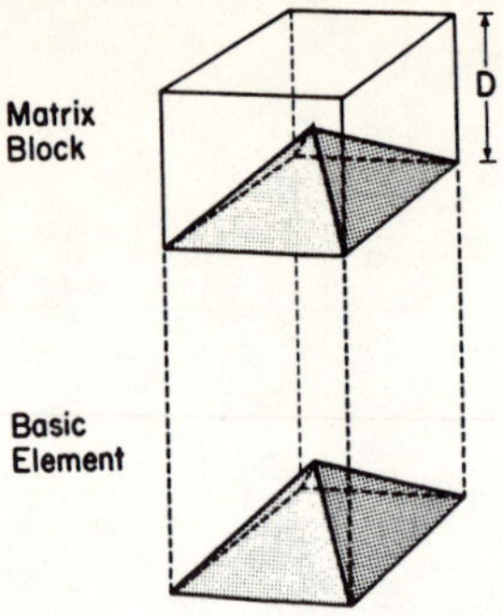

Figure 2. 1-Dimensional approximation of heat conduction in rock matrix blocks.

uniform, a heat conduction equation based on the basic element (Fig. 2) is a reasonable approximation. We have verified this approach by comparing our model to a 3-dimensional solution (Lai and Bodvarsson, 1982). The approach described above is quite similar to the one employed in the "Multiple Interacting Continuum" Method (MINC) for numerical modeling of heat and fluid flow in fractured porous media (Pruess and Narasimhan, 1982).

MATHEMATICAL DEVELOPMENT

In dimensionless form the governing equations for the temperatures in the fractures (T_2) and the rock matrix (T_1) are:

Fractures ($\eta = 1$):

$$-\frac{\delta T_{D_2}}{\delta \xi} - 12 \left.\frac{\delta T_{D_1}}{\delta \eta}\right|_{\eta=1} = \theta \frac{\delta T_{D_2}}{\delta \tau} \qquad (1)$$

Rock matrix ($\eta < 1$):

$$\frac{\delta^2 T_{D_1}}{\delta \eta^2} + \frac{2}{\eta}\frac{\delta T_{D_1}}{\delta \eta} = \frac{\delta T_{D_1}}{\delta \tau} \qquad (2)$$

The intial and boundary conditions are

$$T_{D_1}(\xi,\eta,0) = T_{D_2}(\xi,0) = 0 \qquad (3a)$$

$$T_{D_2}(0,\tau) = \begin{cases} 0 & \tau < 0 \\ 1 & \tau \geq 0 \end{cases} \qquad (3b)$$

$$T_{D_1}(\xi,1,\tau) = T_{D_2}(\xi,\tau) \qquad (3c)$$

$$T_{D_1}(\xi,0,\tau) < \infty \qquad (3d)$$

The dimensionless parameters in equations (1-3) are defined as:

$$T_D = \frac{T - T_o}{T_i - T_o} \quad (4a) \qquad \tau = \frac{4\lambda t}{\rho_r c_r D^2} \quad (4b)$$

$$\eta = \frac{2z}{D} \quad (4c) \qquad \xi = \frac{\pi r^2 H \lambda}{\rho_w c_w q D^2} \quad (4d)$$

$$\theta = \frac{4\rho_f c_f b}{\rho_r c_r H} \qquad (4e)$$

The real parameters in equations 4 are defined in the Nomenclature.

The solution of equations (1-3) can be obtained in the Laplace domain as (Bodvarsson and Lai, 1982):

$$u = \frac{1}{p} \exp - \left[\theta p + 12\sqrt{p}\,\frac{I_{3/2}(\sqrt{p})}{I_{1/2}(\sqrt{p})}\right] \xi \qquad (5)$$

$$v = \frac{u}{\sqrt{\eta}} \frac{I_{1/2}(\sqrt{p}\eta)}{I_{1/2}(\sqrt{p})} \qquad (6)$$

Equations (5) and (6) are inverted using a numerical method developed by Stehfest (1970).

RESULTS

It is of primary interest in this study to examine the rate with which the "cold" water front advances away from the injection well during injection. This information is useful in the design of the safe location and rates of injection wells in relation to the production wells. We will define the thermal front (cold water front) as the locus of points with temperature being the average of the initial temperature of the reservoir (T_o) and, the temperature of the injected water T_i ($T_{TF} = 1/2[T_o+T_i]$). In Figure 3 the dimensionless radial distance ξ of the thermal front from the well is plotted against dimensionless time τ for various values of θ. The parameter θ represents the ratio of the average fracture aperture (b) to the reservoir thickness (H). In most cases realistic values of θ lie in the

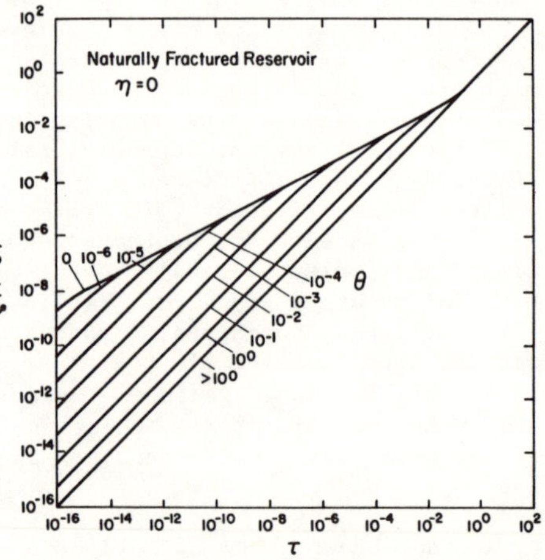

Figure 3. Type curves for thermal front movement in naturally fractured reservoirs.

range of 10^{-5} - 10^{-8}. The figure actually shows for a given value of θ, the radial location of the thermal front (cold water front in the fractures) away from the injection well, at any given time. If one follows the advancement of the thermal front for one value of θ, say θ = 10^{-5}, one can see three different rates of advancement. At early times when conduction heat transfer from the rock matrix is negligible, the front moves as r^2/t away from the injection well. During this period the thermal front moves in the fractures only, in an analogous manner to a single radial system with insulated upper (caprock) and lower (bedrock) boundaries, and a thickness corresponding to the average fracture aperture. Bodvarsson (1972) has derived an expression for the movement of the thermal front in this case. At intermediate times, the slope in Figure 3 decreases by half, and consequently the advancement of the thermal front is proportional to r^4/t. During this period the conductive heat transfer between the rock matrix and the fractures dominates, resulting in a much slower movement of the thermal front away from the injection well. The large heat transfer causes a very slow movement of the thermal front in the fractures, but rapid extraction of heat from the rock matrix.

Finally, at very late times ($\tau \geq 1.0$), the thermal front again advances at a rate proportional to r^2/t. At this time quasi-steady state heat transfer between rock matrix and the fractures has been reached, and consequently the thermal front will move as if only a porous medium was present i.e. independent of the fracture nature of the reservoir. However, in contrast to the early time behavior, the thermal front now moves at the same rate in the fractures as in the rock matrix.

In order to explain this more fully Figures 4-5 were constructed. The figures show the time sequence of the dimensionless temperature profiles away from the injection wells in the fractures and the rock matrix for a given value of θ. The dimensionless temperature of T_D = 1.0 represents the temperature of the injected water, whereas the dimensionless temperature T_D = 0.0 corresponds to the initial reservoir temperature. Temperature profiles are given for the fractures (η = 1.0), the center of the cubes (η = 0.0) and two intermediate values (η = 0.3, 0.7). The figures show

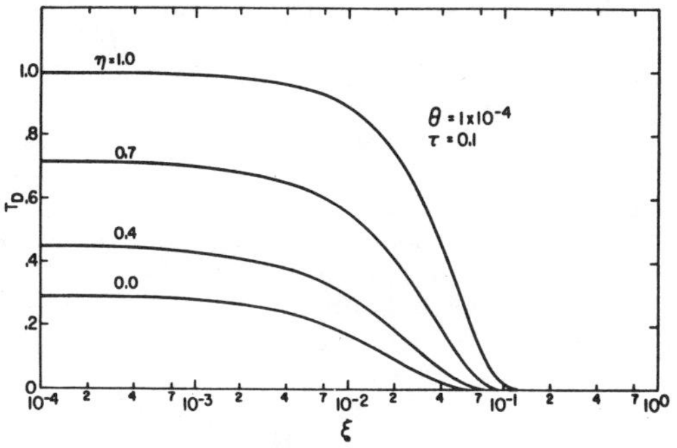

Figure 4. Temperature profiles in fractures and rock matrix; τ = 0.1.

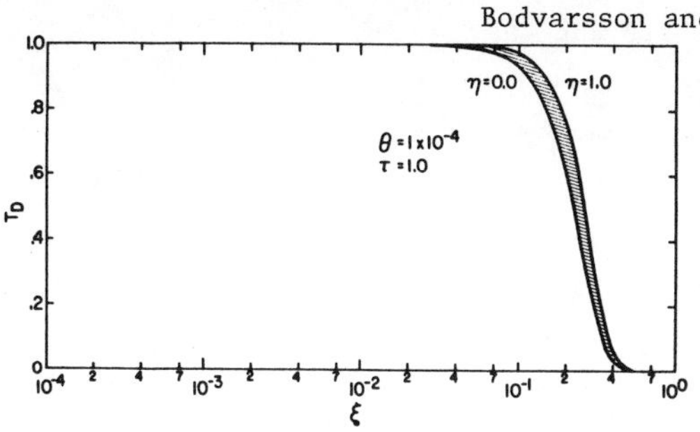

Figure 5. Temperature profiles in fractures and rock matrix; τ = 1.0.

that at early times (τ = 0.1) there is a considerable difference between the temperature profiles in the fractures and the rock matrix. At later times the curves start to converge, although the cold water front is constantly moving away from the well. At a dimensionless time of τ = 1.0 the temperature profiles are practically identical in the fractures and the rock matrix. This can be shown analytically by considering an asymptotic solution for the late time behavior of equation (6).

The reason for this phenomenon is that at early time the cold water shoots rapidly through the fractures, increasing the surface area for conductive heat transfer between the fractures and the rock matrix. The large surface area enhances energy transfer from the rock matrix to the fracture fluids, thus retarding the advancement of the cold water front along the fractures. This in turn, tends to equilibrate the temperatures in the fractures and the rock matrix so that eventually the temperature profiles away from the well are identical for the fractures and the rock matrix.

PRACTICAL CONSIDERATIONS

The most interesting aspect of the results obtained is that even for fractured reservoirs, uniform energy sweep of the reservoir can be obtained. A uniform energy sweep will maximize the amount of recoverable energy from the resource. A necessary requirement for such conditions is that the injection wells be appropriately located with respect to the production wells. Similar conclusions were obtained by Bodvarsson and Tsang (1982) for the case of horizontal fractures only; however in that case the criteria for proper siting of the injection wells are different from what we propose here for naturally fractured reservoirs.

For the design of an injection system for naturally fractured reservoirs mathematical expressions that can be used to calculate the time and radial distance from the injection wells where uniform sweep conditions prevail are quite useful. Figure 4 shows that uniform energy sweep condition will prevail when:

$$\xi\ (4 + \theta) = \tau \simeq 1.0 \qquad (7)$$

In general 4 >> θ so that equation (7) can be written in terms of real parameters as:

Bodvarsson and Lai

$$r_c = \left(\frac{\rho_w c_w q D^2}{4\pi\lambda H}\right)^{1/2} \quad (8a)$$

$$t_c = \frac{\delta_r c_r D^2}{4\lambda} \quad (8b)$$

Inspection of equations (8a) and (8b) shows that both the time and radial distance of the uniform energy sweep condition depend greatly on the fracture spacing D. However, both quantities are independent of the fracture aperture b. Fractures possessing small apertures will contain very small amounts of fluids, so that even though fluid velocities are high very little energy is needed to increase the temperature.

Now let us consider an injection well in a naturally fractured reservoir using the parameters shown in Table 1. If the average fracture spacing is not known the following expressions can be calculated.

$$r_c = 2.6 \times D \text{ (meters)} \quad (9a)$$

$$t_c = 0.01 \times D^2 \text{ (years)} \quad (9b)$$

Thus, for an average fracture spacing of 50 meters, uniform energy sweep conditions will prevail 130 m away from the well after 25 years of injection.

Table 1: Parameters Used in Example

Injection rate, q	:	20 kg/s
Fluid density, ρ_w	:	1000 kg/m^3
Fluid heat capacity, c_w	:	4200 $^J/kg.^\circ C$
Thermal conductivity, λ	:	2.0 $^J/m.s.^\circ C$
Reservoir thickness, H	:	500 m
Rock density, ρ_r	:	2700 kg/m^3
Rock heat capacity, c_r	:	1000 $^J/kg.^\circ C$

In summary, we have developed a model that can be used to determine proper locations and flow rates of injection wells in naturally fractured reservoirs. The model can help in optimizing an injection operation so that maximum energy recovery from the resource is obtained.

NOMENCLATURE

b: Fracture aperture (m)
D: Fracture spacing (m)
H: Reservoir thickness (m)
$I_{1/2}$, $I_{3/2}$: modified spherical Bessel functions of the first kind
p: Laplace parameter
ϕ: Porosity (-)
q: Injection rate (m^3/s)
r: Radial coordinate (m)
r_c: Radial distance from injection well to the location of uniform energy sweep
t: Time (seconds)
t_c: Time of uniform energy sweep (seconds)
T: Temperature (°C)
T_i: Injection temperature (°C)
T_o: Initial reservoir temperature (°C)
T_{TF}: The isotherm defined as the "thermal front", $T_{TF} = (T_i + T_o)/2$
u: Temperature in fracture in Laplace domain
v: Temperature in rock matrix in Laplace domain
z: Rock matrix coordinate (m)
λ: Thermal conductivity of rock matrix (J/m.s.°C)
ρc: Volumetric heat capacity ($J/m^3.^\circ C$)

Subscripts:

f: Fracture
r: Rock matrix
w: Liquid water

ACKNOWLEDGEMENT

Critical review of this manuscript by S. Benson, C. Doughty, M. Lippmann and K. Pruess is most appreciated. This work was supported by the Assistant Secretary for Conservation and Renewable Energy, Office of Renewable Technology, Division of Geothermal and Hydropower Technologies of the U.S. Department of Energy under Contract No. DE-AC03-76SF00098.

REFERENCES

Bodvarsson, G., On the temperature of water flowing through fractures: J. Geophys. Res., Vol. 74, No. 8, 1969.

Bodvarsson, G., Thermal problems in the siting of reinjection wells: Geothermics, Vol. 1, No. 2, 1972.

Bodvarsson, G.S., and Tsang, C.F., Injection and thermal breakthrough in fractured geothermal reservoirs: J. Geophys. Res., Vol. 87, No. B2, pages 1031-1048, 1982.

Bodvarsson, G.S., and Lai, C.H., Injection into naturally fractured reservoirs; paper in preparation, 1982.

Horne, R., Geothermal reinjection experience in Japan: paper presented at the California Regional SPE meeting, Bakersfield, California, SPE-9925, (March 25-26) 1981.

Lai, C.H., and Bodvarsson, G.S.,: Verification of the MINC method, in preparation, 1982.

Lauwerier, H.A., The transport of heat in an oil layer caused by the injection of hot fluid: Appl. Sci. Res. Scat. A., 5, 145, 1955.

Pruess, K., and Narasimhan, T.N., A practical method for modeling fluid and heat flow in fractured porous media: paper presented at the 6th annual meeting of SPE in New Orleans, SPE-20509, (Jan. 31-Feb. 2), 1982.

Stehfest, H., Numerical inversion of Laplace transforms: Communications of ACM, Vol. 13, p. 44-49, 1970.

Warren, J.E., and Root, P.J., The behavior of naturally fractured reservoirs: Society of Petroleum Engineers Journal, September 1963, pp. 245-255.

Geothermal Resources Council, TRANSACTIONS Vol. 6, October 1982

TEMPERATURE CHANGES WITH TIME IN THE SLOTTED INTERVAL OF A DEEP, SHUT-IN GEOTHERMAL WELL NEAR THERMAL EQUILIBRIUM: EAST MESA WELL 31-1, IMPERIAL COUNTY, CALIFORNIA, 1977-1982

William H. Diment and Thomas C. Urban

U.S. Geological Survey
Federal Center, MS 966, Box 25046
Denver, CO 80225

ABSTRACT

Five precision temperature logs were obtained over a five year period (1977-1982) beginning three years after drilling and one year after last significant testing in East Mesa well 31-1 (32°48.6'N, 115°15.7'W). A sensitivity of measurement of 0.001°C was maintained to the bottom of the well (165°C, 180 bars). Useable precision was limited by convective motions within the hole (0.01-0.3°C depending on ambient geothermal gradient), fluid leakage through the lubricator at the well head (4 bars) and, occasionally, by electrical or electronic noise. Comparison of these continuous logs indicates a general warming in and just above the slotted interval (1647-1877 m) which is attributed to flow into the lower levels of the slotted interval and up the casing and out into permeable zones at higher levels. Some flow continued upward through an uncemented interval of the annulus between casing and hole and out into the formation where it is blocked by cement (~1594 m). Some points of entrance or egress of fluid are marked by small, sharp temperature anomalies which have persisted over the 5-year period. Apparently, the uncemented part of the annulus has partly filled in, and flow into the bottom of slotted interval has increased (1978-1982). Both of these changes may have been induced by the nearby (~30 km) Imperial Valley earthquake of 10/15/79, or by the production or testing of nearby wells.

INTRODUCTION

Opportunities to log deep, shut-in, geothermal wells over a long period of time are rare. Operational and legal difficulties are usually formidable. Fortunately, East Mesa well 31-1 has been maintained as a test facility, first by the U.S. Bureau of Reclamation (USBR, 1977) and subsequently by the U.S. Department of Energy (e.g. Mathews, et al., 1979). This well provided us with a first opportunity to test our equipment in a high-temperature environment (Urban et al., 1978), and an opportunity to extend our studies of thermal convection (Diment, 1967; Urban and Diment, 1980; Diment, et al., 1981) in cased, water-filled holes to a large diameter hole of high temperature and high thermal gradient.

Although continuous temperature logs of high precision have been made routinely for several decades in deep holes of moderate temperatures, the conventional logging techniques tend to fail at temperatures above about 140°C. For an example, Mathews et al. (1979, p. 5) note that problems have been encountered in the comparison of temperature logs obtained in well 31-1 by four different groups in 1978-1979 and conclude that: "it appears that recording a correct temperature log in a geothermal well is still a difficult logging problem."

If conventional continuous logs cannot be obtained, clock-driven thermometers are lowered on a wire line. Thus far, such logs have not provided sufficient precision to examine the phenomenon considered here (points of influx and egress of small quantities of fluid, leakage of fluid along the annulus between casing and hole, and significance of long-term variations in temperature).

EXPERIMENTAL PROCEDURES

Temperature measurements were made using a glass-enclosed bead thermistor (50,000 ohms at 25°C) attached to a steel-armored Teflon-insulated cable. The probe containing the thermistor was at the lower end of an assembly consisting of several sinker bars (20-40 kg) and sometimes one or two centralizers. Resistances and depths were sampled at intervals of 0.3 or 0.6 m and recorded on punch-paper tape for later computer processing. The 18 cm (I.D.) cased hole was logged downward at 6 m/min through a lubricator (well-head pressure 4 bars). The measurements reported here are relative to ground level which is 5 meters below the Kelley bushing to which earlier measurements are referenced (USBR, 1977).

UNCERTAINTIES

In comparing temperature logs obtained at diffrent times, as well as judging the precision of an individual log, several kinds of uncertainties arise. Although individual, glass-encapsulated thermistors do not drift much (~0.01°C/yr judging from repeated calibrations on the probes used in well 31-1), trace quantities of water in the probe, cable-head, or cable tend to produce noisy logs and to increase the time constant of the system. Not all noise can be traced to this source, however. In one log noise will occur in certain depth intervals. In a subsequent log obtained under virtually identical conditions, the noise will occur in different intervals. Logs obtained with the probe centered in the hole are usually less noisy than those obtained by permitting the probe to swing free as it is lowered. Temperature-time recordings at fixed depths very rarely exhibit bursts of noise. Clearly, some of the noise is somehow related to lowering the probe.

Convective motions in a water-filled hole may cause significant temperature variations (R) at a given depth if the ambient geothermal gradient (G) is large. $R \approx GaA$, where a is the radius of the hole and A is the aspect ratio (height/radius) of the convective eddies. The aspect ratio usually falls between 6 and 10 for motions of periods of less than 10 minutes (Urban and Diment, 1980). In this hole (a = 8.9 cm), the maximum gradient is about 300°C/km; thus, temperature variations as high as 0.3°C might be expected from short period convective motions. Temperature-time recordings at several depths (Fig. 1) illustrate the nature of the variations at different gradients. Note that there are longer period variations than accounted for in our simple parameter A. Their significance is unknown but thought to be small on the basis of a single long record in another hole (Diment, 1967).

Escape of the fluid through the lubricator at the well head raises the fluid column in the drill hole above the point of entry of the fluid somewhere in the slotted intervals near the bottom of the hole. A crude estimate of this loss for our operation is about 0.1 ℓ/min which translates into a rise in the fluid column of about 0.3 m/hr or about 2 m over the 6 hours required to log the hole at 6 m/min. A rise of 2 meters at gradient of 100°C/km corresponds to an temperature rise of 0.2°C. Clearly, some care must be taken to minimize leakage if optimum results are to be achieved.

INTERPRETATION OF TEMPERATURE LOGS

To a first approximation the temperature profile (Fig. 2) can be divided into two parts at level D. The upper, hight-gradient, part corresponds to the "conductive cap" of the geothermal system. The lower, low-gradient, part is the "convective reservoir." This general configuration is found in many wells throughout this geothermal field (e.g. USBR, 1977; Swanberg, 1976; Riney, et al., 1979). However, the gradients in the reservoir are considerably less, or even negative, in some wells.

The sediments of the uppermost few hundred meters are comprised of a variety of deltaic, lacustrine, and aeolian sands and shales of high porosity and a wide range in thermal conductivity which partially accounts for the high range in geothermal gradients in this interval. Temperatures and gradients in this interval are also influenced by leakage from large, unlined canals (Diment, et al., 1977). This influence may extend to level A and possibly to level B which is also the approximate lower limit of water production in adjacent agricultural lands. Although the relatively large differences in temperature between 1978 and 1982 (Fig. 2) above level A are in the direction to be expected for canal leakage, the uncertainties due to upward movement of the water column during logging are also large because of the high gradient.

In general the hole cooled from 1978 to 1982 above 1000 m and warmed below it (Fig. 2). The amount of cooling is consistent with the conductive decay of the drilling (1974) and the testing (1976) disturbances each of which lasted on the order of ten days (USBR, 1977). This conclusion was reached using a procedure developed by Lachenbruch and Brewer (1959). The agreement could be entirely fortuitous, of course, but it does illustrate that large temperature perturbations of significant duration are slow to decay.

The warming of the lower part of the hole is primarily due to migration of fluid upward from various levels in the slotted intervals (Fig. 3) and outward into permeable zones above. The anomaly near level E, which is at the bottom of a cemented interval as determined by cement bond logs (Mathews, 1979), is a case in point. Judging from the residual temperature logs (Fig. 3), hot water rose along the uncemented annulus and then out into a permeable zone near the base of the cement. From sometime before the first log (4/17/77) until sometime after the fourth log (6/7/78), the anomaly continued to grow. Then sometime between the fourth and the fifth log (2/8/82), the annulus closed or was greatly restricted, and the invaded zone began to cool. This scenario would account for the change in form of the anomaly evident in figure 3, as well as the marked cooling trend at this level (E) in the difference log (Fig. 2).

Level F marks the top of the slotted intervals in the casing. It is also the location of a persistent temperature anomaly whose sharpness indicates active flow of water out of the casing into permeable formation or up the uncemented annulus. Temperatures increased at this level from 1977 to 1978; but they remained constant between 1978 and 1982 (Fig. 2). This is consistent with the notion expressed above that flow through the uncemented annulus has been cut off. The interval below level F has warmed suggesting that outflow of some of the rising hot water has been diverted to permeable horizons just below level F.

The large and persistent temperature anomaly at level G as well as lesser but still persistent ones below it cannot be so easily explained. Some of the smaller anomalies are clearly related to the slotted and non-slotted intervals. However, the anomaly to G occurs within a slotted interval, and there is no obvious reason in the construction or the condition of the well that it should occur where it does, other than that it occurs close to a casing joint and that cement bond logs suggest various degrees of in-filling in the annulus. A fault at 1699 m was inferred from a dipmeter survey (USBR, 1977). Perhaps there are other faults or permeable zones that could account for the anomaly. Even more puzzling is the behavior of temperature with time of the whole interval below level G. From 1977 to 1978 the temperatures remained rather constant; but from 1978 to 1982 they increased significantly (Figs. 2 and 3). This could be explained by a slow influx of hot water into the lowermost parts of casing, or through the bottom of the casing, but we are not sure. Could our system of measurement have failed in mode that is unfamiliar from past experience? Could the change at this level (G) as well as that at level F been the result of a catastrophic event, such as the Imperial Valley earthquake of 10/15/79? Clearly, more frequent logs would help resolve such problems.

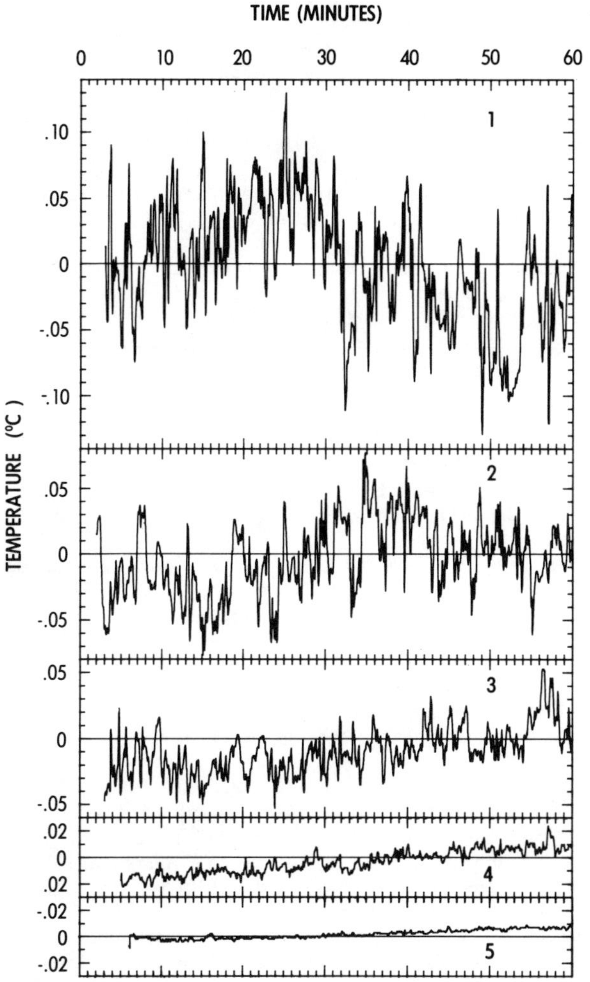

Figure 1. Selected temperature-time recordings at fixed depths in East Mesa well 31-1. Temperature and time scales are the same for each recording. The recordings are arranged in order of decreasing geothermal gradient (1) 278°C/km at 98 m; (2) 178°C/km at 1597 m; (3) 103°C/km at 1588 m; (4) 37°C/km at 1033 m; (5) 8°C/km at 1722 m. The recordings were made on 6/5/78 and 6/6/78. Zero-time is the time at which the probe reached depth. The time constant (1/e) of the probe is about 2 seconds. The probe was centered in the hole for each recording.

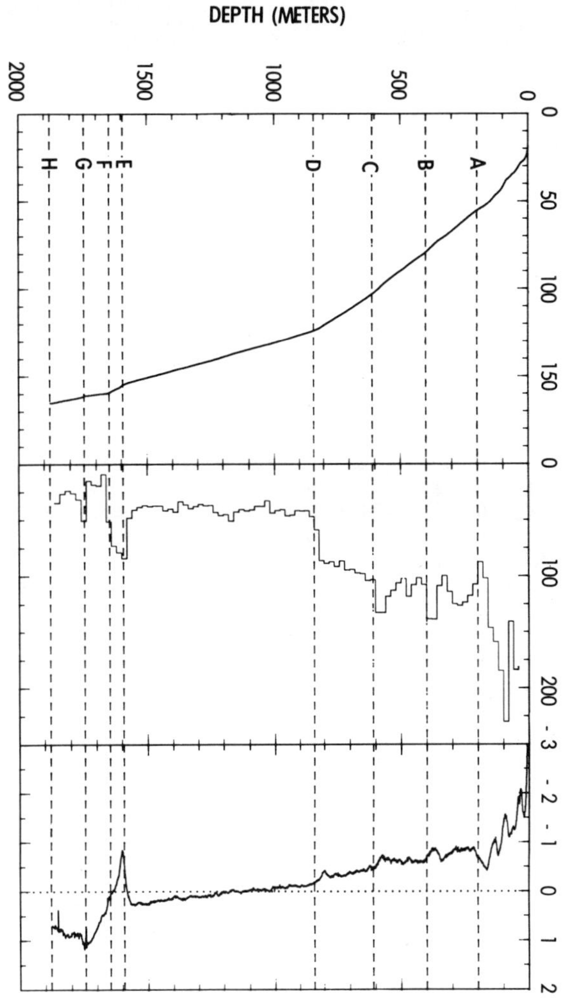

Figure 2. Temperature and temperature gradient (over 20 meter intervals) logs observed on 2/8/82 in East Mesa well 31-1, and the difference between this temperature log and one obtained on 6/7/78. Both logs were obtained with a probe centered in the hole. The observation and plotting interval for the 1982 log is 0.3 m; it is 0.6 m for the difference log. The horizontal dashed lines are levels of special interest or reference: A is the approximate lower limit of the more obvious thermal effects of canal leakage. B is the approximate lower limit of water production for agriculturual purposes in adjacent areas. C marks a change in geothermal gradient which was used in an analysis by Urban et al. (1978). D is close to the level generally regarded as the bottom of the "conductive" cap or top of "convective" geothermal reservoir. E is the bottom of cement in the annulus between hole and casing. F is the top of the slotted intervals in the casing. See figure 3 for details. G is at the level of a prominent anomaly within a slotted interval. H is the bottom of the casing.

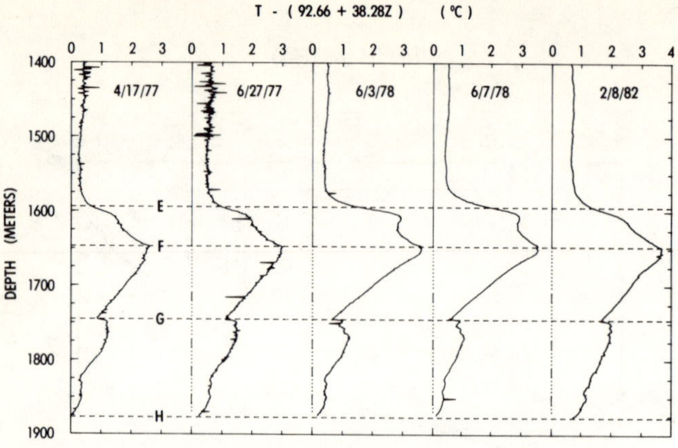

Figure 3. Residual temperature logs obtained in the lower part of East Mesa well 31-1 on the five dates indicated. Temperatures shown are observed temperatures minus a linear temperature distribution (92.66 + 38.28 z), which has a slope close to the gradient above the slotted intevals and an intercept arbitrarily chosen so all of the residual temperatures are positive and small. No attempt has been made to remove electronic/electrical noise which appears as sharp spikes on some of the profiles. The last log was recorded at 0.3 m-intervals, the other four at 0.6m-intervals. The probe was mechanically centered in the hole in the last two logs; it was allowed to swing free in the others. The horizontal dashed lines have the same meaning as in figure 2. Dotted intervals in the vertical lines separating the plots indicate slotted intervals in the casing (USBR, 1977, p. 38).

CONCLUSIONS

Continuous temperature logs can be made with precision in deep, high temperature wells. Improved cables and instruments are available, and reliable results could be obtained routinely if demand were sufficient. Such logs can reveal points of entrance and egress of fluid and movement of fluid behind the casing. Logs repeated at appropriate times in the same hole may provide additional information on flow rates, provided proper attention is given to instrument calibration and the logging operation. Unlike most other geophysical logs, the maximum utilization of temperature logs requires a complete history of the construction of a well and of thermal disturbances in it. This fact is commonly overlooked in planning and executing a logging program.

ACKNOWLEDGEMENTS

We thank K. E. Mathias, R. L. Littleton, and Robert Somes for arranging access and support at the site, and for sharing their knowledge of the history of the drilling and testing of the well. Manuel Nathenson, Mark Bond, and Benny Loebner provided significant help in the investigation. The manuscript was read and improved by W. B. Joyner and A. H. Lachenbruch.

REFERENCES

Diment, W. H., 1967, Thermal regime of a large diameter borehole - Instability of the water column and comparison of air- and water-filled conditions: Geophysics, v. 32, p. 720-726.

Diment, W. H., Urban, T. C., Nathenson, M., and Mathias, K. E., 1977, East Mesa geothermal anomaly, Imperial County, California: Effects of canal leakage on shallow thermal regime: EOS, American Geophysical Union Transactions, v. 58, p. 1241.

Diment, W. H., Urban, T. C., Nathenson, M., Nehring, N. L., and Shaeffer, M. H., 1981, Thermal convection in cased water-filled drill holes - Effect of small quantities of gas: EOS, American Geophysical Union Transactions, v. 62, p. 392.

Lachenbruch, A. H. and Brewer, M. C., 1959, Dissipation of the temperature effect of drilling a well in arctic Alaska: U.S. Geological Survey, Bulletin 1083-C, p. 73-109.

Mathews, Mark, Arney, Barbara, and Sayer, Suzanne, 1979, Log comparison from geothermal calibration/test well C/T-1: Society of Professional Well Log Analysts, 20th Annual Symposium, 3-6 June 1979, p. 1-25.

Riney, T. D., Pritchett, J. W., and Rice, L. F., 1979, Three-dimensional model of East Mesa hydrothermal systems: Geothermal Resources Council, Transactions, v. 4, p. 467-470.

Swanberg, C. A., 1976, The Mesa geothermal anomaly, Imperial Valley, California - A comparison and evaluation of results obtained from surface geophysics and deep drilling: Second United Nations Symposium on the Development and Use of Geothermal Resources, Washington, D. C., U.S. Government Printing Office, p. 1217-1228.

Urban, T. C., and Diment, W. H., 1980, Thermal convection in cased, water-filled drill holes - Observations over a wide range in conditions: EOS, American Geophysical Union Transactions, v. 61, p. 1130.

Urban, T. C., Diment, W. H., and Nathenson, M., 1978, East Mesa geothermal anomaly, Imperial County California - Significance of temperatures in a deep hole near thermal equilibrium: Geothermal Resources Council Transactions, v. 2, p. 667-670.

U.S. Bureau of Reclamation, 1977, Geothermal resource investigations, East Mesa test site, Imperial Valley, California, status report, April, 1977, Boulder City, Nevada, 99 p.

TEMPORAL CHANGES IN PHYSICO-CHEMICAL PARAMETERS OF FLUIDS DURING A PRODUCTION TEST AT MOMOTOMBO, NICARAGUA

Yoram Eckstein and Carlos V. Lopez

Department of Geology, Kent State University
Kent, Ohio 44242

ABSTRACT

Evolution of physico-chemical characteristics of the fluids discharged during a monitored six-month production test at Momotombo points to rapid transformation of the reservoir from water-, to steam-dominated. The transformation followed an invasion of the peripheries of the hot reservoir by cold ground water from the pyroclastic and lacustrine formations underneath Lake Managua. Rapid cooling of the periphery of the geothermal reservoir led to intraformational precipitation of silica, sealing off the reservoir to any fresh recharge. The process was reflected in radical changes in pressure, temperature and production rates, accompanied by significant evolution of chemical parameters of the discharged fluids, such as B/Cl, Na/K and Na/Cl ratios, and concentrations of silica, chloride, sulfate and boron.

INTRODUCTION

The Momotombo geothermal field is located on the southern flank of an active volcano, Momotombo, which is one of many Quaternary volcanoes of the Central-American island-arc-type volcanic chain (Fig.1). Since the "discovery drilling" in 1974 the field was explored by over 30 wells drilled in an area of approximately 5 sq. km. The subsurface temperature distribution and structure of the geothermal reservoir at Momotombo was described by Lopez et al. (1980). They concluded that the reservoir is supplied with steam and/or hot water from a narrow, NNE-trending sub-vertical fracture-zone, located in the western part of the geothermal field (Fig.1). The steam and/or hot water ascends along the feeder-fracture to zone of a sub-horizontal fractures, or of high permeability, at a level between -200 and -400 m (m.s.l.) east of the feeder-fracture, where it spreads laterally as a hot plume. The feeder-fracture is bounded on the west by a zone of impermeable hot-dry rock. Consequently, the wells located to the west of the fracture encountered purely conductive thermal regime, while the wells drilled to the east

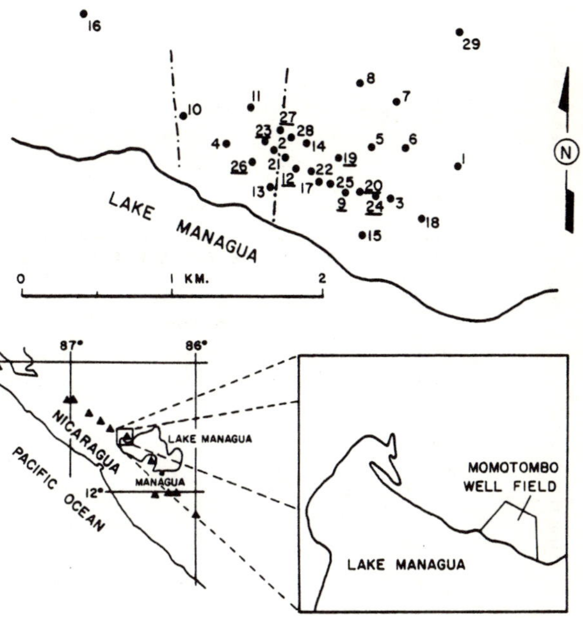

Fig. 1. Index map. Numbers correspond with the well identification MT#. Identification numbers of the wells used in the flow-test are underlined. Two dash-and-dot lines designate an approximate location of the feeder fracture. Triangles in the lower left hand inset represent Quaternary volcanic centers.

have distinct temperature inversion on their depth/temperature profiles, with the amplitude of the inversion increasing eastwards (Lopez et al., 1980).

In order to evaluate the overall potential of the field in general, and the feasibility of production from the wells with temperature inversion in particular, questioned by Eckstein (1979), a long-term simultaneous flow-test from eight wells was initiated in May 1979. The flow-test, monitored for the well-head temperature (WHT), well-head pressure (WHP) and well-lip pressure (WLP), with periodic sampling of the discharge-fluids from a bleeder valve, extended beyond the scheduled 60-days, through December 1979. Preliminary

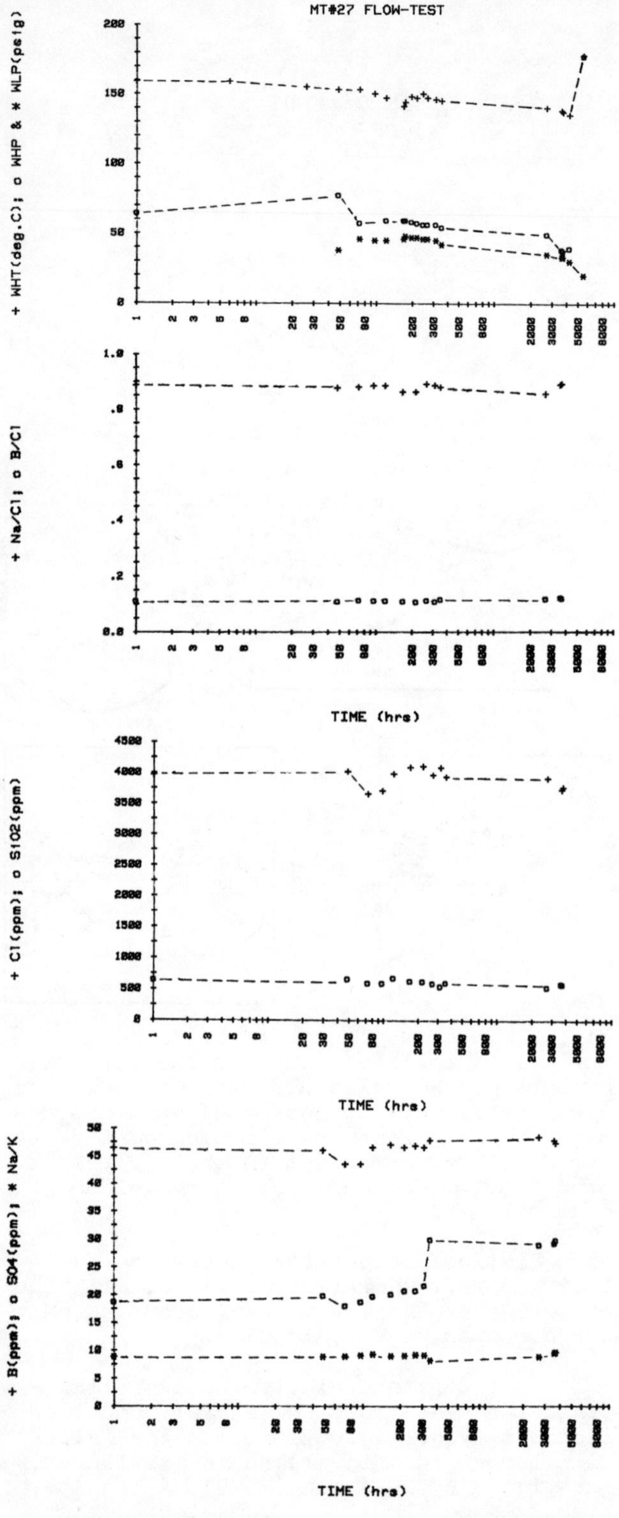

Fig. 2. Temporal changes in physico-chemical parameters of fluids discharged during flow-test from MT#27.

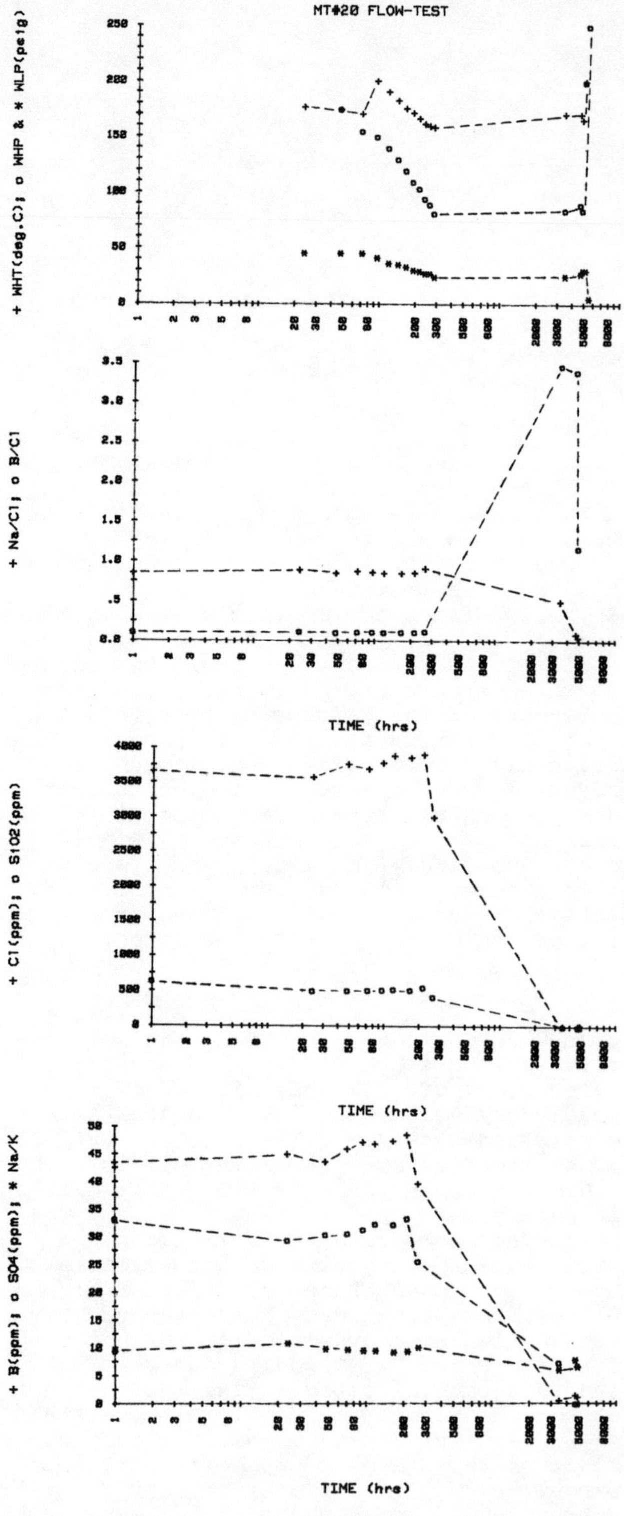

Fig. 3. Temporal changes in physico-chemical parameters of fluids discharged during flow-test from MT#20.

results and interpretation of the temporal changes in the WHT, WHP and WLP during the flow-test were presented by Lopez and Eckstein (1980). They suggested that during the flow-test the reservoir was penetrated at it's peripheries by cold water. "Cooling

Table 1. Chemical composition of the fluids discharged during 72-nd hour of the flow-test at Momotombo, May 1979.

Well #	WHT °C	WHP (psig.)	WLP (psig.)	pH	Na (ppm)	K (ppm)	Ca (ppm)	Mg (ppm)	SiO_2 (ppm)	Cl (ppm)	SO_4 (ppm)	HCO_3 (ppm)	B (ppm)
26	154	51	30	7.96	2185.	400.	70.0	0.072	706.	2963.	19.0	48.2	46.4
23	185	168	68	7.50	1850.	375.	60.0	0.036	571.	3388.	24.4	35.9	40.5
27	152	58	45	7.80	2080.	400.	67.0	0.039	612.	3654.	18.1	44.7	43.4
12	198	198	60	7.61	1970.	400.	80.0	0.026	561.	3885.	23.5	42.5	49.0
9	188	175	55	7.89	2150.	370.	84.5	0.026	539.	3911.	27.5	37.9	45.7
19	195	188	62	7.84	2050.	370.	83.0	0.016	578.	3687.	28.5	41.0	43.5
24	208	175	110	7.87	2015.	335.	96.0	0.018	471.	3839.	40.0	37.7	40.0
20	170	170	45	7.69	2080.	370.	81.5	0.024	520.	3690.	30.5	40.3	45.7

of the periphery of the reservoir during the flow-test resulted in an abrupt precipitation of silica and consequent sealing of the reservoir to any fresh recharge. The process was reflected in the dramatic increase in the temperature and well-head pressure, the drastic decline in the production (lip pressure) and the decline in concentration of chloride and silica as the vapor mixture gradually 'dried up'..." (Lopez and Eckstein, 1980). In the ensuing discussion questions were risen whether the changes in the physical properties and rate of the discharge could be attributed to precipitation of silica in the piping of the discharge system and pressure gauges. The present paper summarizes correlation between the temporal changes in the physical properties of the discharge (WHT, WHP and WLP) with the evolution of some of the hydrochemical parameters, indicative of the changes that occurred during the flow-test within the geothermal reservoir of Momotombo.

RESULTS

The wells were selected for the flow-test so that the production would be effected simultaneously from the entire reservoir. Wells MT#23, #26 and #27 are located in the immediate vicinity of the feeder-fracture (Fig.1). The remaining five wells are located to the east-south-east off the feeder-fracture, with MT#20 and #24 at the far end of the hot plume.

The Momotombo geothermal wells produce near-neutral, to slightly alkaline waters, with pH ranging between 6.8 and 8.0. Chemical composition of the water is dominated by alkali-chlorides, with characteristically low concentrations of magnesium, typical for high-enthalpy geothermal reservoirs in volcanic regions. Representative analyses are shown in Table 1.

Significant changes in the WHT, WHP and WLP, that occurred during the latter stages of the flow-test, were accompanied by parallel changes in the physics and chemistry of the discharged fluids. Decline in the production rates, slowest at the wells located near the feeder-fracture, and fastest at the east end of the field, was coincidental with increasing "dryness", or dominance of steam in the discharge. Smallest changes in the physics, as well as in the chemistry of the discharge were observed in wells MT#26 and #27 (Fig.2). The two wells discharged fluids with very simillar chemical composition (Table 1). In both wells there was slight increase in the concentrations of sulfate and boron, paralleling slow decline in WHT, WHP and WLP (Fig. 2a,d). In the final stages of the flow-test a marked decline in WLP was accompanied by increase in Na/Cl and Na/K ratios, along with decline in WHT, WHP and the concentrations of boron and chloride. Unfortunately no fluid samples were collected neither from MT#26, nor from MT#27 with the last measurements, when the most dramatic decline in WLP, and increase in WHP and WHT was recorded (Fig. 2a).

The most dramatic changes in the physics, as well as in the chemistry of the discharge occurred in the wells located farthermost to the east from the feeder-fracture, i.e. MT#20 and #24. Relatively rapid decline in WHT, WHP and WLP during the earlier stages of the flow-test was paralleled by little change in the chemical parameters. During the latter stage of the flow-test, however, sharp increase in WHP and WHT, and abrupt decline in WLP can be correlated with significant increase in B/Cl ratio, decline in Na/Cl, Na/K and concentrations of boron, sulfate, chloride and silica (Fig. 3).

Wells located in the middle of the geothermal field: MT#9, #12, #19 and well #23 located close to the feeder-fracture, all behaved during the flow-test in quite similar manner. Slight to moderate decline in WHT, WHP and WLP (Fig. 4a) throughout the early stages of the flow-test was accompanied by gradual decline in the concentrations of chloride and silica (Fig. 4c), and of boron and sulfate (Fig. 4d). In the latter stages of the flow-test, however, accelerated decline in the WLP and WHP can be correlated with sharp increase in WHT, B/Cl and Na/Cl ratios, synchronous with

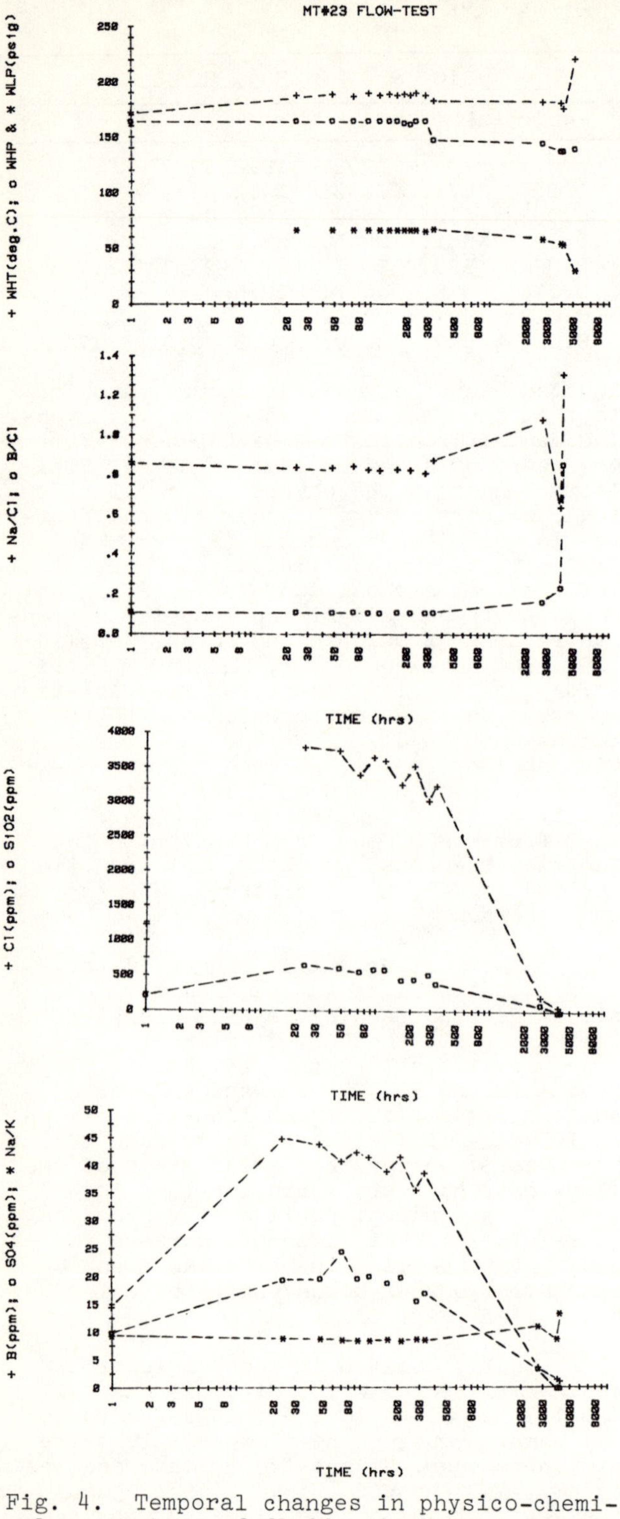

Fig. 4. Temporal changes in physico-chemical parameters of fluids discharged during flow-test from Mt#23.

substantial decline in Na/K ratio and concentrations of chloride, silica, sulfate and boron (Fig. 4).

DISCUSSION

B/Cl index may be used to assess qualitatively thermodynamic changes occurring within the reservoir. During flashing boron fractionates preferentially in relation to chloride into the steam phase. Thus an increase in the steam proportion in the discharge of a geothermal well is usually accompanied by an increase in B/Cl ratio. B/Cl index varied during the flow-test at MT#26 and #27 between 0.037 and 0.039 suggesting a continuing supply of steam and water to those wells at unchanging proportions. Slight increase in the concentration of sulfate, boron and chloride during the early stages of the flow from MT#26 and #27 followed by an increase in Na/K and Na/Cl ratios, along with decline in the concentrations of chloride and boron in the latter stages of the flow may, however, signal an incipient shift in the water/steam equilibrium.

Despite geographic proximity, physico-chemical characteristics of the fluids discharged at MT#23 evolved differently from that of MT#26 and #27. Gradual decline in concentrations of chloride and silica (Fig. 4c) along with the decline in concentration of sulfate and boron, and an increase in Na/K index (Fig. 4d) may reflect a cooling trend triggered by invasion of cold-fresh ground water into the hot fracture system, when the pressure dropped in the vicinity of a flowing well. The ensuing intraformational precipitation of silica led to flashing of the declining quantities of water available in the reservoir, producing during the latter stages of the flow-test predominantly "dry steam" with high B/Cl index (Fig. 4b). B/Cl rose in Mt#23 discharge-fluids from 0.038 in the early stages, to 0.280 at the end of the flow-test, whereas Na/K rose at the same time from 8.4 to 13.5. Similar trends were observed in the chemical composition of the fluids discharged during the flow-test from all wells located to the east of the feeder-fracture. Demonstrated close correlation between the trends in evolution of chemical and physical parameters of fluids discharged at Momotombo confirms the earlier conclusions postulated by Lopez and Eckstein (1980).

REFERENCES

Eckstein, Y. 1979 - Hydrogeochemistry of the Momotombo geothermal system; Technical Report; pp. 60.

Lopez, C.V., Hoyt, B.R. and Eckstein, Y. 1980 - Subsurface temperature distribution and structure of the geothermal reservoir at Momotombo; G.R.C., v. 4, p. 459-462.

Lopez, C.V. and Eckstein, Y. 1980 - Six month production test at Momotombo, Nicaragua; G.R.C. Trans., v. 4, p. 357-360.

Dept. of Geology, KSU, Contr. No. 243

Geothermal Resources Council, TRANSACTIONS Vol. 6, October 1982

BEOWAWE, NEVADA, WELL TESTING: HISTORY AND RESULTS

I. J. Epperson

Chevron Resources Company
San Francisco, CA

ABSTRACT

Since the initial drilling in 1959 within the Beowawe Geothermal Field in north-central Nevada, a total of sixteen wells have been drilled. Numerous flow tests, build-ups, interference tests and injectivity tests have been conducted on eight of these wells. All data indicate that the geologically complicated Beowawe system is charged from a deep carbonate reservoir. The wells exhibit high productivities from a very large reservoir.

INTRODUCTION

The Beowawe Geothermal Field is in north-central Nevada, a few miles south of Interstate Highway 80 and approximately thirty miles southeast of Battle Mountain. The field is in the Basin and Range province at the boundary between a plateau of volcanics to the south and the downfaulted Whirlwind Valley to the north. Geysers, fumaroles and boiling springs have deposited a large sinter terrace approximately 300 feet high at the northern base of the plateau. These numerous geothermal surface manifestations led Magma Power Company and Sierra Pacific Power Company to drill eleven shallow wells from 1959 thru 1965 (Figure 1). Chevron acquired the leases in 1973-1975 and began an extensive geologic study of the area. Chevron's first well, Ginn 1-13, was drilled in 1974 to a depth of 9,600 feet. Since then, Chevron has drilled three more wells (plus numerous shallow temperature holes). This paper presents an overview of the Beowawe testing history, an example of the type of test analyses used and a summary of the testing results to date.

GEOLOGY

No attempt will be made to describe in detail the complicated geology of the Beowawe Area. However, at least a conceptual description is necessary to provide an understanding of the well test environment. Simply stated, the reservoir is thought to be in lower Paleozoic carbonates at depths approaching 20,000 to 30,000 feet. Indication of this is provided by the $Na \cdot HCO_3$ nature of the geothermal fluids. The Beowawe brine is exceptionally low salinity with only 1,100 - 1,200 ppm total dissolved solids. The geothermal brine flows by way of a complicated series of faults, charges permeable stratigraphic horizons along the way and finally discharges at the surface. None of the wells drilled to date have reached the carbonate reservoir, but a number have tapped and tested various charged units and fluid conducting faults.

MAGMA TESTING HISTORY

The first Beowawe well testing was conducted by Magma on four of their wells located on the sinter terrace (Beowawe 2, Vulcan 1, Vulcan 2, Vulcan 3). They flowed each of the wells through separator facilities and measured surface pressures, temperatures, flowrates, etc. No bottomhole data were measured, but the flowing bottomhole temperatures were estimated by Magma's consultant to be near 365°F. The three Vulcan wells attained rates of 1,500,000 lbs/hr at wellhead pressures of ±110 psia. Beowawe 2 produced only 500,000 lbs/hr at ±20 psia wellhead pressure. After testing, the wells were allowed to flow for about a year, during which time visual checks of the discharge appeared to indicate a decrease in flowrate, although no flow measurements were made.

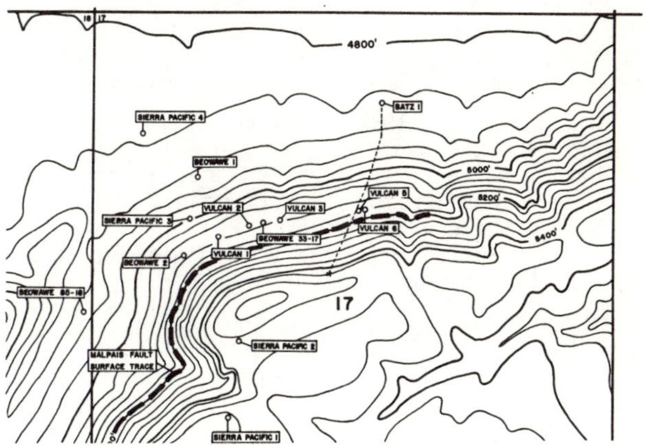

Figure 1: Surface Contours and Well Locations

Temperature surveys made in 1965 indicated maximum temperatures of 340 - 345°F within the completion intervals of two of the tested wells. Based on this data, it was concluded that the Beowawe reservoir was depleting and being invaded by cold water.

Recent Chevron data indicates that the geology of the sinter terrace consists of a fractured stratigraphic interval at 650 - 900 feet, below which is a highly-permeable fault containing +320°F brine. Computer simulation of the Magma test data indicates the three high rate wells probably had flowing temperatures close to 380°F. These wells were only completed in the fractured interval. Beowawe 2 exposed the fault as well as the fractured interval. Its lower test flowrate is explained by the cooler fluid contribution from the fault. Additionally, as the other terrace wells continued to flow, this well acted as a flue and allowed the cooler brine to migrate into the fractured interval, thus causing the observed loss of temperature. Once the flowing wells were shut-in, pressure equalization was quickly reached, stopping the cooler brine influx and allowing the fractured interval temperatures to return to normal.

CHEVRON WELL TESTING

Extensive well testing by Chevron commenced with the completion in late 1979 of Beowawe 33-17 on the terrace between two of the old Magma wells. Initially, the tests were conducted using a simple blooie line and a flowing temperature survey to estimate flowrate by the James Method. For later tests, the full James instrumentation was installed so that both flowrate and enthalpy could be monitored as a function of time. Bottomhole pressures were measured and recorded in all tests using nitrogen-filled capillary tubing with a quartz transducer at the surface.

In the first half of 1980, a number of static and flowing temperature-pressure surveys were made in Beowawe 33-17. In addition, four flow tests were also made, varying in length from one day to sixteen days. The well was also flowed as a source of drilling water for three months during this period. Each successive flow period showed improvement in wellhead pressures and flowrates as the well cleaned up and the wellbore heated up.

By the first of 1981, Beowawe 85-18 had been drilled, and the old Magma terrace well, Vulcan 2, had been cleaned out. At this time, interference tests were conducted to determine the degree of continuity between the wells. Pressure responses in shut-in wells were observed within one hour, even at distances up to 3/4 miles.

As is common in well testing, the testing supplied many answers but also raised additional questions. Consequently, in late 1981, additional testing was conducted. The primary purpose was to test the injectivity of two intervals in each of two wells. The opportunity was also taken to instrument additional wells to check for interference. Once again, pressure responses were observed in less than one hour in shut-in wells up to 1-1/4 miles away.

TYPICAL TEST DATA ANALYSIS

The field data is, of course, monitored for accuracy and consistancy as it is taken. Once in the office, the data is once again reviewed. An important data review technique is the use of a computer wellbore flow simulator to make sure bottomhole and surface data are consistent.

The first step in analyzing a flow test build-up is to construct a log-log plot to determine wellbore storage effects (Figure 2). Even though all flow tests are controlled at the surface, wellbore storage effects are never significant for more than 5 - 6 minutes. Although not obvious in Figure 2, in most cases, a linear flow period is observed for another 1 - 5 minutes (indicated by a half slope in the log-log plot). The Horner straight line normally starts within ten to thirty minutes after shut-in.

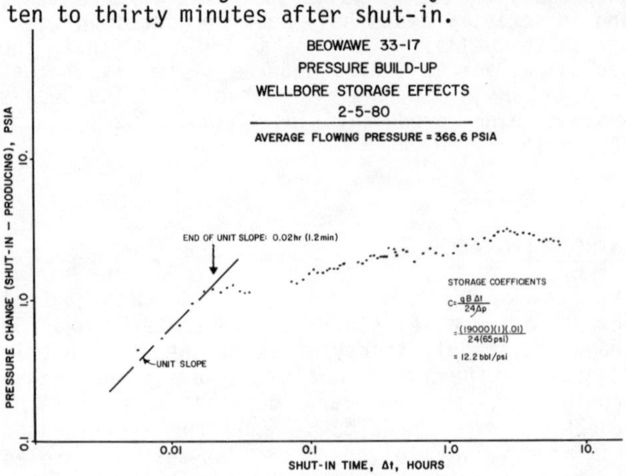

Figure 2: Wellbore Storage Effects

A typical Horner analysis is presented in Figure 3. There are several things worth noting concerning Beowawe build-ups. First, the data generally conforms closely to a straight line. Flowing drawdowns are very small - varying from about three to forty psi. Also note, that even though the test illustrated in Figure 3 flowed for over three weeks, the pressure had nearly built up to static within one hour. A standard kh calculation on this particular well indicates an extremely permeable formation with a kh of 750,000 md-ft. The kh for most Beowawe wells is in the range of 200,000 to 800,000 md-ft.

With the mechanical set-up normally used on Beowawe well tests, the bottomhole pressures can be recorded as often as at thirty-second intervals. Consequently, an early transient analysis can often be made to verify the build-up results. The analysis shown in Figure 4 indicates a kh of 850,000 md-ft, which for all practical purposes is equivalent to that calculated from the build-up; thus, confirming the extremely prolific nature of the Beowawe wells.

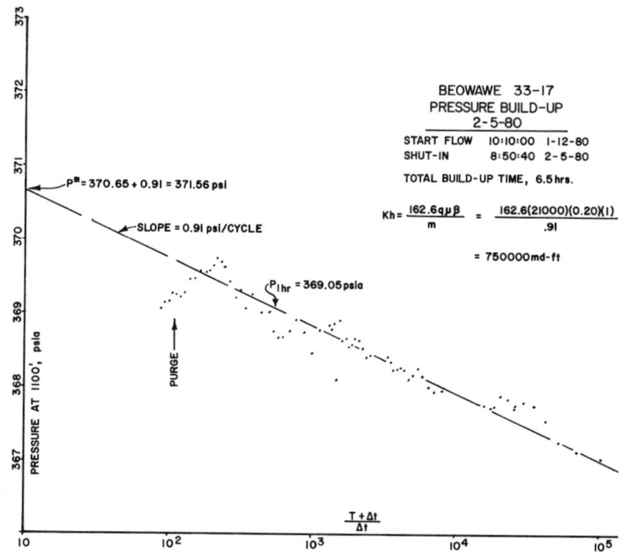

Figure 3: Pressure Build-Up Analysis

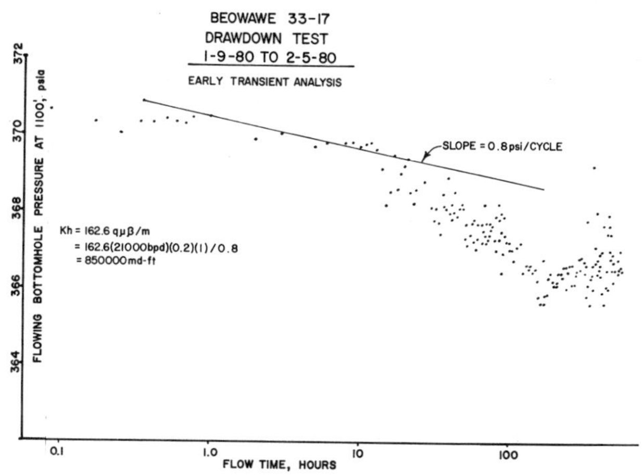

Figure 4: Early Transient Analysis

Another important question to be answered as soon as possible is an estimate of the size of the reservoir. Due to the geologic complexity of Beowawe, this question can only be partially answered at this point in time. A semi-steady state analysis of the long-term flow test data (Figure 5) indicates an extremely large fluid volume in excess of 10^{10} bbls. The steady state performance after $\pm$200 hours is indicative of a steady-state infinite reservoir or perhaps a constant pressure "boundary". Flow data from other Beowawe wells indicate fluid volumes in excess of 10^{12} bbls; in other words, a volume larger than the Prudhoe Bay Oil Field. None of the tests have ever indicated a flow barrier or reservoir boundary, so these fluid volume estimates are only minimum values.

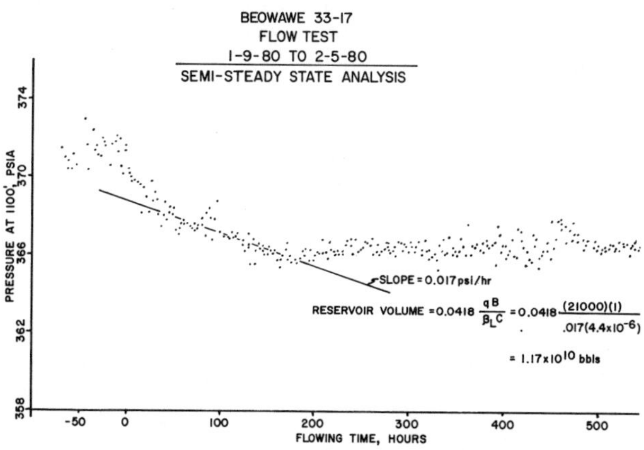

Figure 5: Reservoir Volume Determination

The numerous interference tests conducted at Beowawe show pressure communication between all the wells. Response times are less than one hour, irrespective of the distance between wells. Analyses of the interference data verifies the calculated permeability-thicknesses from the flow and build-up analyses.

A summary of the Chevron well tests is presented in Table 1. Note that all wells flowed at rates of about 300,000 lbs/hr, total mass. These relatively lower flowrates are due to mechanical completion restrictions. For instance, the Ginn well ($\pm$420°F) was completed with about 9,500' of 7" casing to the surface. Beowawe 33-17 (368°F) produces through a 6-5/8" blank and slotted liner. Using Chevron's wellbore flow simulator and the Ginn flow test data, it was found that a "Ginn type" well with 13-3/8" casing to 1,500' and 9-5/8" casing to TD ($\pm$9,500') will flow two phase brine at about 1,000,000 lbs/hr with $\pm$120 psig wellhead pressure.

TABLE 1

Beowawe Flow Test Summary

	Ginn	33-17	85-18	Vulcan 2
Number of Flow Tests	1	6	4	1
Max. Static Temp. °F	420	365	360+	360+
Flowrate Mlbs/hr	285	305	320	240
kh, md-ft	232,000	800,000	550,000	780,000
PI, lbs/hr/psi	8,900	200,000	7,800	24,000

CONCLUSIONS

1. Several shallow charged productive intervals underlying the Beowawe Area have been identified as commercially productive with temperatures near 370°F.

2. Deeper fault production (8,000' to 10,000') with +420°F temperatures has also been identified as commercially productive with potential two-phase rates exceeding 1,000,000 lbs/hr.

3. Interference tests indicate pressure communication between all wells, even those completed in different geologic horizons.

4. The Beowawe system is a complicated combination of fractured and faulted horizons which are fed from a deep carbonate reservoir.

5. The Beowawe system is very large and prolific.

ACKNOWLEDGEMENT

The author would like to thank Chevron Resources for permission to publish this paper.

BIBLIOGRAPHY

Keonig, J. B., 1970, "Geothermal Exploration in the Western United States", U.N. Symposium on the Development and Utilization of Geothermal Resources, Pisa, Vol. 2, Part 1.

Middleton, W. M., 1961, "Report on Beowawe, Nevada, Geothermal Steam Wells", (unpublished).

Nolan, T. B. and Anderson, G. H., 1934, "The Geyser Area Near Beowawe, Eureka Co., Nevada", Am. Jour. Sci., Fifth Ser., Vol. XXVII, No. 159, p. 215-229.

Oesterling, W. A., 1962, "Geothermal Power Potential of Northern Nevada", Text of paper presented at the 1962 Pacific Southwest Mineral Industry Conference of the AIME.

Smith, C., Struhsacker, E. M. and Struhsacker, D. W., 1979, "Structural Inferences from Geologic and Geophysical Data at the Beowawe KGRA, North-Central Nevada", GRC Trans., Vol. 3, p. 659-662.

Struhsacker, E. M. and Smith, C., 1980, "Model for a Deep Conduit to the Beowawe Geothermal System, Eureka and Lander Counties, Nevada", GRC Trans., Vol. 4, p. 249-251.

Swift, C. M., 1979 "Geophysical Data Beowawe Geothermal Area, Nevada", GRC Trans., Vol. 3, p. 701-703.

Zoback, M. L., 1979, "Geologic and Geophysical Investigation of the Beowawe Geothrmal Area, North-Central Nevada", Stanford University, School of Earth Sciences, Geological Sciences, Geological Sciences Series, Vol. XVI.

Geothermal Resources Council, TRANSACTIONS Vol. 6, October 1982

INTERPRETATION OF TRACER RETURN PROFILES AT WAIRAKEI GEOTHERMAL FIELD USING FRACTURE ANAYSIS

Martin P. Fossum and Roland N. Horne

Department of Petroleum Engineering
Stanford University, CA 94305

ABSTRACT

Tracer tests performed in geothermal reservoirs have indicated inter-well flows at speeds up to 100m/hr. Such rapid movement of injected water underground indicates a clear danger of premature thermal breakthrough during reinjection. Early thermal drawdown has in fact been observed in several fields. The very rapid returns suggest that fractures are the principal conduits for tracer transport and thus a new method of tracer profile interpretation has been formulated. This method is based on flow through individual fractures rather than through porous medium as has been the case in standard methods. Applications of the method to tracer tests performed at the Wairakei geothermal field show excellent matching of the observed tracer profiles based on a model of only two fractures joining the injection and production wells.

INTRODUCTION

Tracer analysis is important in determining geothermal reservoir characteristics. Most importantly, it provides a means of identifying cold reinjection water flowing to producer wells. In Japan, rapid interference from reinjection wells has caused thermal drawdown in four of five liquid-dominated fields (Horne, 1982). Although maintaining reservoir pressure is one motive for total injection of waste water, the reduction in enthalpy and subsequent decrease in steam discharge rates may imply that maintenance of discharge enthalpy is more important.

Tracer analysis will be instrumental in identifying the relative importance of the flow paths which make enthalpic reduction possible. Unfortunately, models currently used for oil reservoirs cannot be applied to geothermal systems. Oil field models assume flow through porous media. Geothermal reservoirs, however, are usually highly fractured. Since the preferred flow channels of reinjected water will be through intervening fractures between producer and injector, a method of determining the extent and type of fracturing is needed. A tracer model to characterize flow in a fractured geothermal system is developed in this paper. It can be used to: (a) Investigate possible damage from premature reservoir cooling; (b) Determine natural flow patterns within the reservoir; (c) Predict interference from current injection wells; and (d) Plan reinjection programs to minimize thermal drawdown.

Recent field test data from tracer tests in Wairakei, New Zealand, are analyzed and compared. This study illustrates how tracers can be used to understand not only the nature of fracture systems, but also their relative influence as preferred pathways for well-to-well flow. It is a vital first step in designing a reinjection program.

The recent reinjection results in Japan (Horne, 1982) have shown that tracer flow is principally within fractures. Drawing on this experience in highly fractured fields Horne and Rodriguez (1981) developed a preliminary model to characterize flow mechanisms in fractures. The model yields information on the effective tracer dispersion in the fractured system.

INTERPRETATION MODEL

Inter-well tracer surveys are often used in the oil industry. In order to quantify these survey results Brigham and Smith (1965) formulated a model to predict the time of tracer breakthrough, peak concentration of the tracer, and the general form of the breakthrough curve in a five-spot flood. Baldwin (1966) developed a similar model by dividing the five-spot pattern into a series of diverging-converging radial elements between the injector and the producer. Yuen, Brigham and Cinco (1979) then constructed a computer algorithm to test for the degree of heterogeneity between different reservoir layers. Recent work by Abbaszadeh-Dehghani and Brigham (1981) has extended and refined this approach.

However, since most geothermal reservoirs are highly fractured, the above-mentioned models are not applicable. The five-spot model assumes flow through a porous medium. Tester, Bivins and Potter (1979) used tracer profiles and residence time distributions to analyze a hydraulically fractured granitic geothermal reservoir. This was the first attempt at formulating an approach to the problem of flow in fractures. A multizone model assuming independent, one-dimensional dispersion in several separate zones was used. This, however, still assumed flow in porous media.

Tester, Bivens and Potter (1979) proposed a method to characterize the flow paths between injector and producer by analyzing N measured values of the exit tracer concentration C_i. By minimizing the objective function R,

$$R = \sum_{i=1}^{N} \left(\tilde{C} - C_i\right)^2 \tag{1}$$

the parameters of the proposed transfer function $\tilde{C}$ can be estimated. These parameters are dependent on the characteristics of the flow channels in the fracture system, and in this case are the dimensionless time relative to mean arrival $(\bar{u}t/L)$ and the Peclet number, $Pe=uL/\eta$.
Here $\bar{u}$ is the mean flow speed of tracer through a fracture of length L, and η is the effective dispersion coefficient found by Horne and Rodriguez (1981) to be

$$\eta = \frac{2}{105}\frac{(b\bar{u})^2}{D} \tag{2}$$

where D is the molecular diffusivity of the tracer in the geothermal fluid and b is the fracture aperture. The function $\tilde{C}$ may be made up of contributions from several different flow paths, but in this work it was found that in every case at most two were indicated.
Thus, the transfer function can be written

$$\tilde{C} = \varepsilon_1\tilde{C}_1\left((L/\bar{u})_1, Pe_1\right) + \varepsilon_2\tilde{C}_2\left((L/\bar{u})_2, Pe_2\right) \tag{3}$$

where ε_1 and ε_2 are the flow fractions in paths 1 and 2.

The two-component signal approach has been successfully applied to match observed tracer profile peaks. It can also be used to analyze the shape of a single peak tracer response. An important point to be stressed is that a single peak profile may not be a result of only one flow path, and therefore may be better modeled assuming multiple flow channels.

The form of $\tilde{C}$ for each of the paths for a mass s of tracer concentrated at point x=0 at time t=0 is

$$\tilde{C} = \frac{s}{2\sqrt{\pi\left(\frac{\eta}{\bar{u}L}\right)\left(\frac{\bar{u}}{L}\right)t}}$$

$$\exp\left\{-\left[1-\left(\frac{\bar{u}}{L}\right)t\right]^2 \Big/ \left[4\left(\frac{\eta}{\bar{u}L}\right)\left(\frac{\bar{u}}{L}\right)t\right]\right\} \tag{4}$$

The minimization defined by equation (1) will then provide values for ε_1 and ε_2 (the relative rate of tracer flow in each of the two fractures), $(L/\bar{u})_1$ and $(L/\bar{u})_2$ (the mean arrival time in each of the two fractures) and $(\bar{u}L/\eta)_1$ and $(\bar{u}L/\eta)_2$ (the Peclet numbers of each of the two fractures). The minimization can be achieved using non-linear regression techniques.

APPLICATION TO WAIRAKEI TRACER TESTS

In 1979 and 1980, several tracer tests using radioactive Iodine-131 and Bromine-82 were performed at Wairakei (McCabe, Manning, and Barry, 1980). Several of these tests have been interpreted in this work and some illustrative examples are shown.

Figures 1, 2 and 3 show the observed tracer returns in producing wells WK116, WK76 and WK108, all from the same tracer injection into well WK80. It is noted that the first arrival of tracer is from three to five days after injection even though WK116 is 500m from WK80, WK76 is 145m distant and WK108 is 230m distant. The speed of tracer movement through the reservoir is therefore of order 1-5 m/hr in these examples, a quite moderate speed compared to some other returns in the field and in other fractured fields (see Horne, 1982).

Table 1 shows the flow fractions, peak arrival times and Peclet numbers calculated for the three tests. Also listed are the inferred fracture apertures assuming the molecular diffusivity D of the tracer to be $10^{-10}m^2/sec$. This parameter is not well known, so these values are shown mainly for example - it is difficult to imagine open fractures of such a wide aperture extending over such distances, as this would not be consistent with observed reservoir permeabilities. Fractures of such a size (and larger) have been observed in the wellbores however.

Tracer return profiles calculated using the inferred parameters are shown in Figures 1, 2, and 3. The precision of the fit to the observed data is encouraging and lends plausibility to the model. As another check, the peak arrival times can be compared to those observed in the data. The peak arrival time is a function of both Peclet number and mean arrival time and can be calculated by finding the maximum of equation (4):

$$\bar{t} = \left\{\left(\bar{u}/L\right)\left(\left[\eta/\bar{u}L\right]^2 + \eta/\bar{u}L + 1\right)^{1/2}\right\}^{-1} \tag{5}$$

The comparison of calculated to observed peak arrivals is shown in Table 2. Only the first peak is listed since determining the second peak in the observed data is difficult.

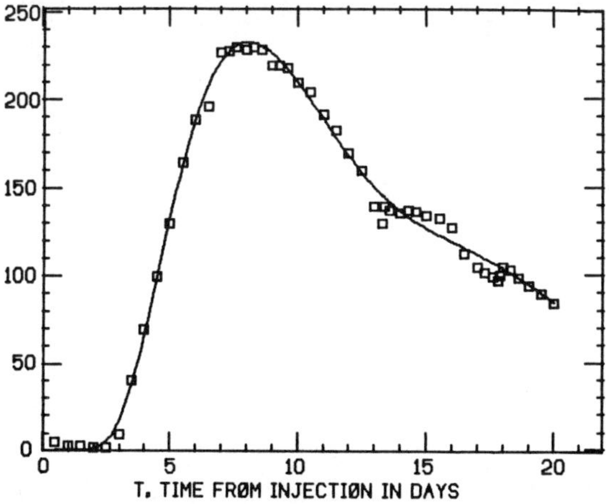

Figure 1. Tracer breakthrough concentration in WK116 from injection into WK80. Solid lines are calculated result, open squares are observed result.

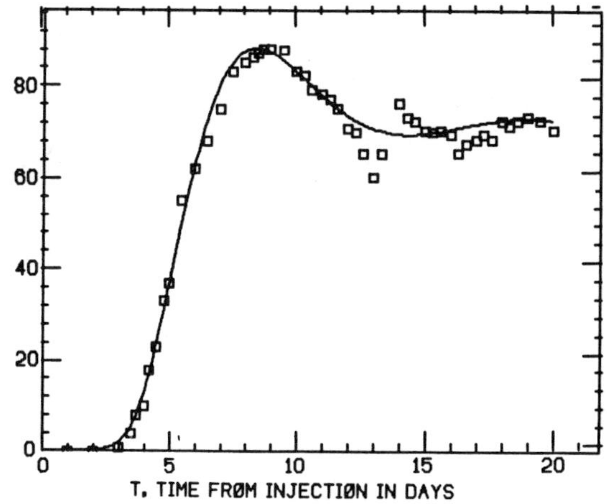

Figure 2. Tracer breakthrough concentration in WK76 from injection into WK80. Solid lines are calculated result, open squares are observed result.

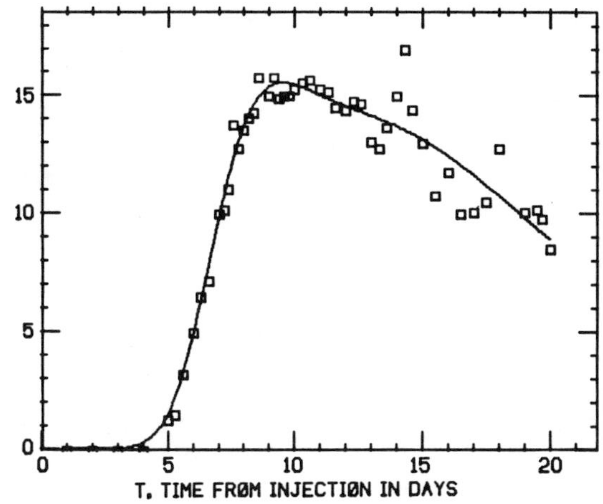

Figure 3. Tracer breakthrough concentration in WK108 from injection into WK80. Solid lines are calculated result, open squares are observed result.

Table 1

Well	Flow Fraction	$\bar{t}$ Calculated (days)	Peclet Number	Inferred Aperture (mm)
WK116	.87	8.5	9.3	21
	.13	18.9	41.6	14
WK76	.42	21.5	15.80	25
	.58	8.6	13.15	18
WK108	.80	8.30	9.9	20
	.20	13.95	43.4	12

Table 2

Well	$\bar{t}_1$, Calculated (days)	$\bar{t}_1$, Observed (days)	% Error in t_1
WK116	8.5	7.6	9
WK76	8.6	8.7	1
WK108	8.3	9.0	8

CONCLUSIONS

1. A double flowpath model gives a more accurate data match than a single component model in most instances. However, larger numbers of flowpaths do not appear to be necessary to obtain a very good fit of the data observed at Wairakei.

2. Effective dispersion in the models is larger than that predicted for one-dimensional flow in fractures unless the fracture aperture is around 2cm. This is perhaps because flow is affected by mixing or multi-dimensional flow within the faults.

3. Analysis of a tracer profile's components yields a preliminary idea of what type of inter-well connection exists. This model indicates the type of flow paths and can also say a little about their configuration.

4. The interpreted results of the Wairakei tracer tests show that the observed behavior is consistent with inter-well flow over extremely long distances in only a very few open fissures.

REFERENCES

Abbaszadeh-Dehghani, M., and Brigham, W.E.: "Analysis of Well-to-Well Tracer Flow to Determine Reservoir Heterogeneity," SPE 10760, presented at the 1982 California Regional Meeting of the SPE, San Francisco, CA., March 24-26, 1982.

Baldwin, D.E., Jr.: "Prediction of Tracer Performance in a Five-Spot Pattern," J. Petroleum Technology, Vol. 18 (April 1966), 513-517.

Brigham, W.E., and Smith, D.H.: "Prediction of Tracer Behavior in Five-Spot Flow," SPE 1130, presented at the SPE of AIME 40th Annual Fall Meeting, Denver, CO., Oct. 1965.

Horne, R.N.: "Geothermal Reinjection Experience in Japan," J. Petroleum Technology, Vol. 34 (1982), 495-503.

Horne, R.N., and Rodriguez, F.: "Dispersion in Tracer Flow in Fractured Geothermal Systems," Proceedings, 7th Annual Stanford University Geothermal Workshop (1981).

McCabe, W.J., Manning, M.R., and Barry, B.J.: "Tracer Tests - Wairakei," Institute of Nuclear Sciences Report INS-R-275, Dept. of Scientific and Industrial Research, Lower Hutt, New Zealand, July 1980. Geothermal Circular WJMcC 2.

Tester, J.N., Bivins, R.L., and Potter, R.M.: "Inter-well Tracer Analysis of a Hydraulically Fractured Granite Geothermal Reservoir," SPE 8270, presented at the SPE of AIME 54th Annual Fall Meeting, Las Vegas, NV, Sept. 1979.

Yuen, D.C., Brigham, W.E., and Cinco-L., H.: "Analysis of Five-Spot Tracer Tests to Determine Reservoir Layers," U.S. Dept. of Energy Report SAN-1265, Feb. 1979.

Geothermal Resources Council, TRANSACTIONS Vol. 6, October 1982

IDENTIFICATION OF FLUID FLOW PATHS IN THE CERRO PRIETO GEOTHERMAL FIELD

Susan E. Halfman*, Marcelo J. Lippmann*, Rubén Zelwer*, John H. Howard**

*Lawrence Berkeley Laboratory, University of California,
Berkeley, California 94720
**Lawrence Livermore National Laboratory, University of California,
Livermore, California 94550

ABSTRACT

A hydrogeologic model of the Cerro Prieto geothermal field has been developed based on geophysical and lithologic well logs, downhole temperature, and well completion data from about 90 deep wells. The hot brines seem to originate in the eastern part of the field, flowing in a westward direction and rising through gaps in the shaly layers which otherwise act as partial caprocks to the geothermal resource.

INTRODUCTION

The Cerro Prieto geothermal field, located about 20 miles south of the US-Mexico border in Baja California, Mexico (Figure 1), has approximately 100 wells, some as deep as 11,600 ft. These wells provide a vast amount of data for the study of the field's subsurface geology (Figure 2).

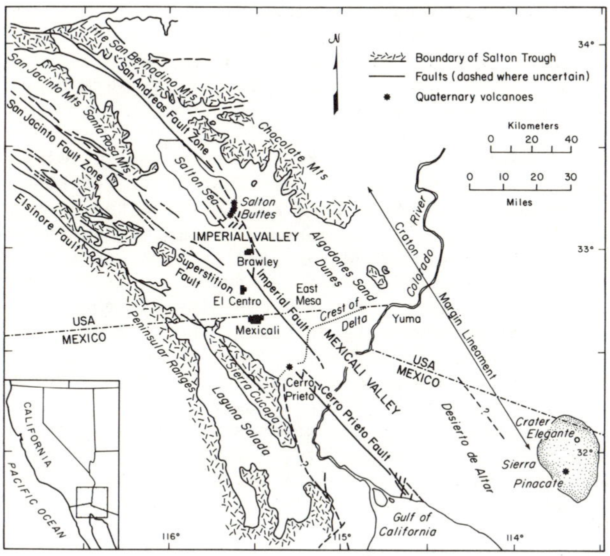

Figure 1. Regional geology of the Salton Trough and location of the Cerro Prieto area.

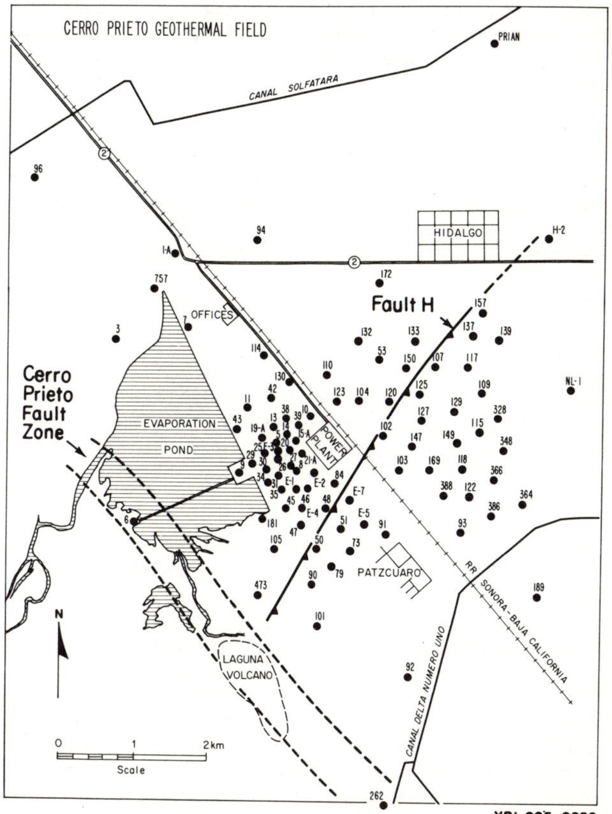

Figure 2. Location of wells and of the two faults which have the greatest impact on geothermal fluid flow at Cerro Prieto.

Based mainly on the analysis of geophysical and lithologic well log data, a geologic model of the field has been developed. By integrating downhole temperature data and depth of producing intervals into this model, flow paths for the geothermal fluids have been identified. The results presented in this paper are consistent with those of other geological, geophysical, and geochemical studies carried out in this area (Lawrence Berkeley Laboratory, 1981).

GEOLOGICAL SETTING

The Cerro Prieto field is located in a sedimentary environment resulting from the southwesternly progradation of the Colorado Delta into the Salton Trough (Lyons and van de Kamp, 1980). In this area, the thick sedimentary column consists essentially of Pliocene and Pleistocene materials deposited in alluvial, deltaic, estuarine, and shallow marine environments (Ingle, 1982). The stratigraphy is dominated by intertonguing sandstone-shale sequences which have been grouped into two main units (Units A and B) by Puente and de la Peña (1978) based on drilling data. Essentially, Unit A consists of unconsolidated sediments, while Unit B consists of consolidated or indurated sediments. Elders et al. (1978) consider that the transition from unconsolidated to consolidated sediments may occur in different ways in different parts of the system (i.e., compaction, cementation, and methamorphic reactions). They suggest that the transitional contact between Units A and B is post-depositional cautioning on its use for stratigraphic correlations or for structural interpretations.

Two liquid-dominated reservoirs have been identified in Cerro Prieto. The shallower one, designated A by Prian (1979) and α by Sánchez and de la Peña (1981), is restricted to the western part of the field. The deeper and hotter reservoir, designated B (β) by the same authors, is found at different depths throughout the field.

Structurally, the Cerro Prieto field is located between two major NW-SE strike-slip faults, the Imperial and Cerro Prieto faults (Figure 1). The field itself has been postulated to be located over a pull-apart basin (e.g. Lomnitz et al., 1970).

GEOLOGIC MODEL

A geologic model of the field has been developed by Halfman et al. (1982) based mainly on the interpretation of gamma-ray, spontaneous potential, deep induction resistivity, compensated formation density gamma-gamma, and lithologic well logs.

Various faults were identified based on visual well log correlations. The location of the two faults believed to have the greatest impact on geothermal fluid flow are shown in Figure 2. The buried fault, tentatively named "H", exhibits a 1500-foot downthrow toward the southeast (Figures 3 and 4). It plays a significant role in providing a path for geothermal fluids to flow upward. The other major fault, the Cerro Prieto strike-slip fault, corresponds to the western boundary of the field and does not seem to carry fluids from great depths.

Based on the logs and depths of geothermal fluid production, the stratigraphic column has been subdivided into a number of formations having distinct lithologic and hydrologic characteristics. Figure 3 presents a northwest-southeast section across the eastern part of the field illustrating the correlations used to establish the various formations and to identify faults. This figure shows the gamma-ray logs of three wells (M-150, M-107, and T-366), although other types of logs were also used in the correlations.

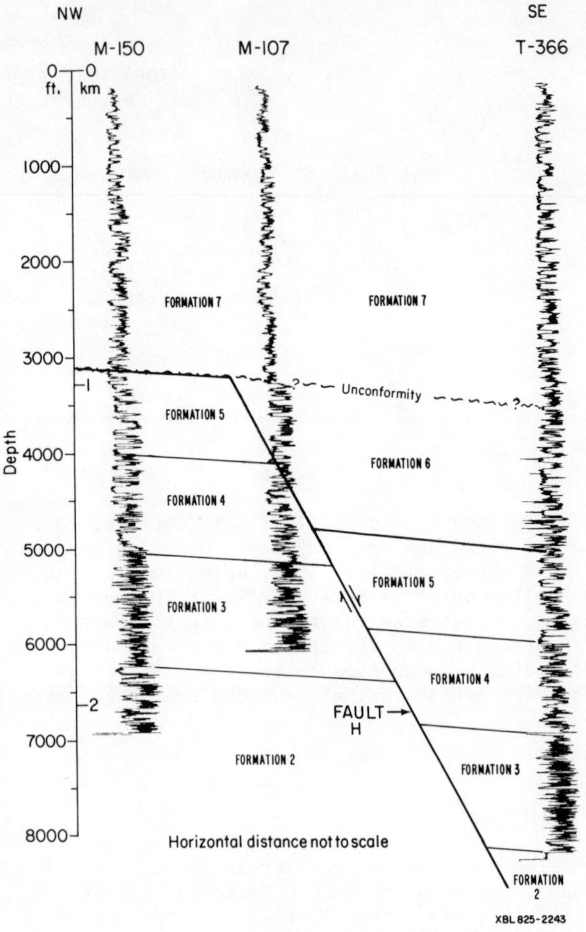

Figure 3. Simplified stratigraphy of the eastern part of the Cerro Prieto field. The traces correspond to gamma-ray logs.

Formation 2 is comprised of some well-developed sandstone beds (some more than 70 ft thick) interbedded with shale beds and contains the deeper reservoir (B or β). Formation 3, comprised of very thinly-bedded sandstones and shales, acts as a local caprock to the lower reservoir. West of the railroad tracks, Formation 4 contains the shallower geothermal reservoir (A or α). In that region, this formation consists of rather thick shales (up to 100 ft thick) and sandstone-shale sequences. Formation 5, comprised of alternating sandstones and shale units (average thicknesses: 30-60 ft), overlies most of the producing intervals of the wells completed in the Reservoir A (or α). In the western part of the field, Formation 7, which has several thick shale units near its base, acts as the local caprock to the shallower reservoir.

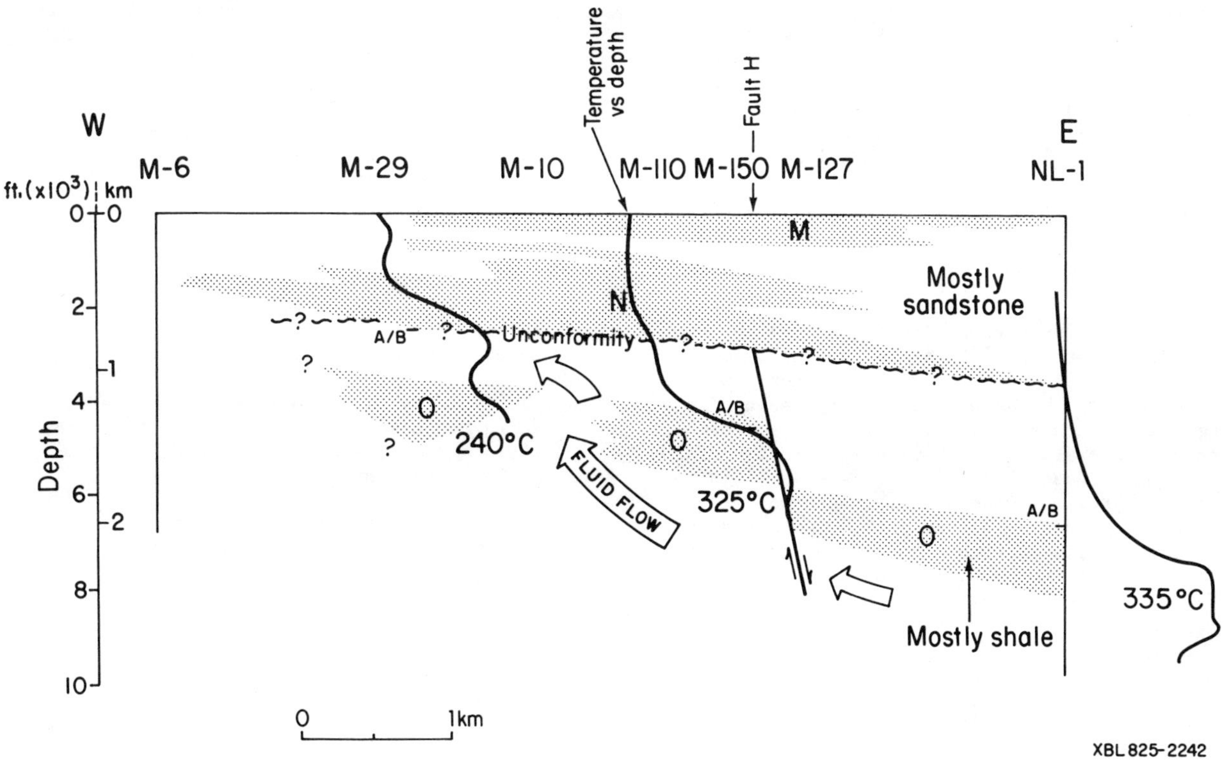

Figure 4. West-east cross-section of the Cerro Prieto field showing a schematic diagram of geothermal fluid flow through the field (indicated by arrows).

GEOTHERMAL FLUID PATHS

Figure 4 shows a west-east section across the entire field. It is a schematic version of one of the cross-sections presented by Lyons and van de Kamp (1980), who defined various lithofacies classes at Cerro Prieto using geophysical well log data.

The simplified lithology shown in Figure 4 was obtained by grouping these authors' lithofacies classes into mostly sandstone and mostly shale units, hereafter referred to as sand and shale units. Shale Unit O approximately correlates with Formations 3 and 4. Shale Units N and M are within Formation 7. Superimposed on this section are the temperature profiles for wells M-29, M-110, and NL-1 (from Bermejo et al., 1979, and personal communication, 1982), and the locations of the contact between Units A and B (A/B contact) for wells M-29, M-150, and NL-1. The producing intervals (not shown) are located within Shale Unit O in the vicinity of M-29. These intervals straddle and lie below the base of Unit O in the region east of M-10.

Some important features are noticeable in Figure 4: (1) The A/B contact is located at the top of Shale Unit O in wells M-150 and NL-1, and at the bottom of Shale Unit N in M-29, (2) The producing intervals are deepest near NL-1, and gradually become shallower toward the west, and (3) The temperature profiles show sudden increases in temperature gradients at depth near NL-1; these sharp inflections in the temperature profiles appear at shallower depths toward the west.

By comparing the temperature profiles at M-110 and NL-1 with the lithology, it can be noted that the increase in temperature gradients occur near the boundary between Shale Unit O and the overlying sand unit. On the other hand, in M-29 this increase is observed near the boundary between Shale Unit N and the underlying sand unit. The increase of thermal gradient suggests that the shale units must be barriers to convective heat transport, i.e., they essentially are acting as local caprocks. Furthermore, the maximum temperatures measured in these wells are highest near NL-1 (335°C), and decrease gradually toward the west (240°C, in M-29).

Based on the distribution of sand and shale units, location of producing intervals, and temperature profiles, a flow path for the geothermal fluids can be postulated. The fluids are believed to be heated at depth by a swarm of dikes intruded in the eastern portion of the field (W. Elders, personal communication, 1982). The fluids are thought to enter the field from that direction through the sands underlying Shale Unit O, then moving westward toward Fault H (the fluid flow path is indicated with arrows in Figure 4). As the fluids flow up the fault, encountering once again the bottom of Shale Unit O, they continue to flow westward below this shale unit. This flow continues until it reaches the gap in Shale Unit O between wells M-10 and M-110. There, the geothermal fluids flow upward through the gap and westward under Shale Unit N, which acts as a shallower local caprock. Eventually, some of the fluids leak to the surface through the Cerro Prieto fault zone, and the rest of the fluids mix with the colder waters which surround the geothermal anomaly.

CONCLUSIONS

The flow of geothermal fluids through the subsurface of the Cerro Prieto field is controlled by both stratigraphic and structural features. Many of the mineralogic, thermal, reservoir engineering, and geochemical characteristics of this geothermal system can be explained with reference to the hydrogeologic model described here. This model is a refinement of previous studies (e.g. Mercado, 1976; Elders et al., 1981). As new data is gathered and analyzed it will be possible to further extend the model to reflect the complex lithology and structure of the Cerro Prieto field.

ACKNOWLEDGEMENTS

We would like to thank the Coordinadora Ejecutiva de Cerro Prieto of the Comisión Federal de Electricidad for providing the data used in this study. This work was supported by the Assistant Secretary for Conservation and Renewable Energy, Office of Renewable Technology, Division of Geothermal and Hydropower Technologies of the U.S. Department of Energy under Contract No. DE-AC03-76SF00098.

REFERENCES

Bermejo M., F.J., Navarro O., F.X., Castillo B., F., Esquer, C.A., and Cortez A., C., 1979. Pressure variation at the Cerro Prieto reservoir during production, in Proceedings, Second Symposium on the Cerro Prieto Geothermal Field, October 17-19, 1979: Mexicali, Comisión Federal de Electricidad, pp. 473-496.

Elders, W.A., Hoagland, J.R., McDowell, S.D., and Cobo R., J.M., 1978. Hydrothermal mineral zones in the geothermal reservoir of Cerro Prieto, in Proceedings, First Symposium on the Cerro Prieto Geothermal Field, September 20-22, 1978: Berkeley, Lawrence Berkeley Laboratory Report LBL-7098, pp. 68-75.

Elders, W.A., Williams A.E., and Hoagland, J.R., 1981. An integrated model for the natural flow regime in the Cerro Prieto hydrothermal system based upon petrological and isotope geochemical data, in Proceedings, Third Symposium on the Cerro Prieto geothermal field, March 21-28, 1981: Berkeley, Lawrence Berkeley Laboratory Report LBL-11967, pp. 102-113.

Halfman, S.E., Lippmann, M.J. and Zelwer, R., 1982. Hydrogeological model of the Cerro Prieto field, Lawrence Berkeley Laboratory Report (in preparation).

Ingle, J.C., Jr., 1982. Microfaunal evidence of age and depositional environments of the Cerro Prieto section (Plio-Pleistocene), Baja California, Mexico: Berkeley, Lawrence Berkeley Laboratory Report LBL-13897.

Lawrence Berkeley Laboratory, 1981. Proceedings Third Symposium on the Cerro Prieto Geothermal Field, March 24-26, 1981: Berkeley, Lawrence Berkeley Laboratory Report LBL-11967.

Lomnitz, C., Mooser, F., Allen C., Brune J., and Thatcher, W., 1970. Seismicity and tectonics of the northern Gulf of California region, Mexico: Preliminary results: Geofísica Internacional, 10, pp. 37-48.

Lyons, D.J., and van de Kamp, P.C., 1980. Subsurface geological and geophysical study of the Cerro Prieto geothermal field, Baja California, Mexico: Berkeley, Lawrence Berkeley Laboratory Report LBL-10540.

Mercado G., S., 1976. Movement of geothermal fluids and temperature distribution in the Cerro Prieto geothermal field, in Proceedings Second United Nations Symposium on the Development and Use of Geothermal Resources, May 20-29, 1975, pp. 498-494.

Prian C., R., 1979. Development possibilities at the Cerro Prieto geothermal area, Baja California Norte, in Proceedings Second Symposium on the Cerro Prieto field, October 17-19, 1979: Mexicali, Comisión Federal de Electricidad, pp. 146-161.

Puente C., I., and de la Peña L., A., 1978. Geology of the Cerro Prieto geothermal field, in Proceedings, First Symposium on the Cerro Prieto Geothermal Field, September 20-22, 1978: Berkeley, Lawrence Berkeley Laboratory, LBL-7098, pp. 17-40.

Sánchez R., J., and de la Peña L., A., 1981. Geohydrology of the Cerro Prieto geothermal aquifer, in Proceedings Third Symposium on the Cerro Prieto Field, March 21-28, 1981: Berkeley, Lawrence Berkeley Laboratory Report LBL-11967, pp. 309-327.

STEAM FLOW RATE CALCULATIONS BY A VERY SIMPLE EQUATION

Gerardo Hiriart

Departamento de Proyectos de Geotermia
Comisión Federal de Electricidad
Morelia, Mich. México

ABSTRACT

Assuming that dry saturated steam behaves like a perfect gas, the steam flow rate discharging at sonic velocity is calculated. The comparison of the calculated flow rate with the well known experiments of Russell James indicates that a constant factor of 1.11 makes a perfect match of both methods.
Then, a very simple equation is proposed for the dry steam wells of Los Azufres using only lip pressure. The agreement with experimental values is excellent. The equation is extended to other dry steam fields.
For wells having steam and water, the same equation can be used to calculate steam flow rate with reasonably good accuracy. However, for better precision, a very simple correction factor renders the equation exact.
The proposed equations are extremely simple and as accurate as the traditional Russell James correlations.

INTRODUCTION

In the geothermal field of Los Azufres, in central México, more than 30 wells have been drilled (Ref.1). Several of them produce dry steam and many others produce high enthalpy mixture.
Up to now, the well known Russell James correlation (Ref.2) have been used to calculate the steam production based on lip pressure of the steam pipe that discharges to the silencer and the water level over a weir that indicates the water flow rate. The correlation gives excellent results as demonstrated by Russell James with a series of well performed experiments.
Unfortunately, to calculate the steam flow rate in a well that also produces water, because of the odd exponents of the equation and the implicit form in which the steam flow rate appears, makes mandatory the use of a small computer program to solve it by iterations.
In this paper, starting from purely theoretical considerations, an equation is deduced for dry saturated steam. The flow rate is found to be proportional to the lip pressure and inversely proportional to the square root of the enthalpy calculated at lip pressure. The comparison of these results with those obtained by Russell James shows that a consistent 10 % error appears.
If one attributes this error to the non - ideal conditions in which the steam flows and correct the predicted values by a factor of 1.11 an excellent agreement is obtained for a wide rage of lip pressure.
The derived equation is then simplified to eliminate the enthalpy and let the steam flow rate be a function of the lip pressure only. This simplified formula is applied to the Los Azufres field with an extraordinary good agreement with Russell James predictions.
Then, by realizing that the correlation between steam flow rate and lip pressure is not seriously affected by the presence of water, a simple correction factor is proposed to calculate exactly the steam flow rate, for mixtures with up to 60 % of water, starting from the dry steam formula.
Eventhough the Russell James equation is accurate enough for all the practical purposes in geothermics, we feel that the formulas proposed in this paper have the following advantages.

- The simplicity is extraordinary.
- The accuracy is as good as Russell James correlation.
- The derivation starts from basic considerations and the assumptions made are clearly stated.
- The above mentioned simplicity permits to obtain direct readings of the steam flow rate in the field by looking only at the lip pressure manometer.

It is the author's belief that the proposed formulas may simplify the task of evaluation of the production of the wells in many geothermal fields.

THEORETICAL CONSIDERATIONS.

It is well known (Ref.3) that steam discharging to the atmosphere, will leave the pipe with a lip pressure equal to the atmospheric if the speed is bellow sonic or with a lip pressure greather than atmospheric if the steam has a sonic velocity at the exit.
If we consider for the moment only the case of dry saturated steam, at lip pressure conditions the following will hold.

$$V = \sqrt{k\,p\,v} \qquad (1)$$

$$G = \frac{\sqrt{k\,p\,v}}{v} \qquad (2)$$

Where V is the velocity of the steam a few millimeters upstream of the end of the tube; k is the ratio of the specif heats at constant pressure (Cp) and constant volume (Cv), p the lip pressure; v the specific volume at pressure p, and G is the mass flow rate per unit area.
In the region of interest of p, the following expression can be used.

$$h = Cp\ T + \text{Constant} \tag{3}$$

Where h is the enthalpy; Cp specific heat at constant pressure and T the absolute temperature.
Assuming the dry saturated steam to behave as a perfect gas, one has

$$pv = RT \tag{4}$$

where $R = Cp - Cv$ (5)

and $k = Cp/Cv$ (6)

Replacing 4,5 and 6 in 3 one obtains

$$h = \frac{k}{k-1}\ pv + \text{Constant} \tag{7}$$

The values of k and the constant that best fit equation 7 are k = 1.33, Constant = 2000
Therefore

$$h = 4\ Pv\ 100 + 2000 \tag{8}$$

(h in Kj/Kg; P in bar; v in m^3/Kg)

In Table 1 are shown the values of h calculated from Equation 8 and those obtained from the steam tables (Ref.4)

TABLE 1

P bar	1	2	3	4	5
h Ref 4	2675.4	2706.3	2724.7	2737.6	2747.5
h Eq 8	2677.6	2708.3	2726.7	2739.5	2749.4

h in (Kj/Kg)

Rearranging equation 2 one has

$$G = \sqrt{\frac{k}{pv}}\, p \tag{9}$$

Replacing pv from equation (7) and the constant by 2000

$$G = \frac{k}{\sqrt{k-1}}\ \frac{P\ 10^5}{\sqrt{(h - 2000)10^3}} \tag{10}$$

where P is in bar; h in Kj/Kg and G in Kg/s m^2.
Replacing k by 1.33 one obtains

$$G = 7321\ \frac{P}{\sqrt{h - 2000}} \tag{11}$$

for practical purpose, if one expresses G in (Ton/hr m^2) and P in psia, equation 11 become

$$G\ \ (\text{Ton/hr } m^2) = 1818\ \frac{P}{\sqrt{h - 2000}} \tag{12}$$

Equation (12) has been obtained from purely theoretical considerations. It is obvious to find some minor discrepancy from experiments, eventhough one will expect that after applying a constant correction factor, that theoretical prediction will agree with experiments in the whole range.
To compare our predictions we will take the values predicted by Russell James'formula, Ref 2, as the exact values.
To make the comparison more evident, both cases will be taken in (Ton/hr) with a pipe of 8" nominal diameter, schedule 60 (7.813 "ID), the lip pressure in psia, and the atmospheric pressure as that of Los Azufres (10.5 psia).
Therefore equation (12) becomes.

$$Qs = F\ 56\ \frac{P}{\sqrt{h - 2000}} \tag{13}$$

Qs (Ton/hr) through a 7.813 "ID pipe
P lip pressure (psia)
h (Kj/Kg) evaluated at pressure P.
F correction factor

For a lip pressure of 30 psig one has

From Eq 13 Qs = F 64.156
From Ref 2 Qs = 71.17
Therefore F = 1.11, and (13) becomes.

$$Qs = 62\ \frac{P}{\sqrt{h - 2000}} \tag{14}$$

Table 2 compares the results obtained by using Equation (14) and those predicted (and checked experimentally) by Russell James, Ref 2.

TABLE 2

P lip	10	20	30	40	50	60
Qs Eq 14	48.3	71.2	93.4	115.8	138.0	160.0
Qs Ref 2	48.6	71.2	93.4	115.5	137.3	159.1
% Error	0.6	0.0	-0.2	-0.3	-0.5	-0.6

Qs in (Ton/hr); P in psig

Equation (14) can be considered as the proper deduction of the flow rate equation for dry saturated steam through a 7.813 "ID pipe.

SIMPLIFICATION OF THE PROPOSED FORMULA.

Equation (14), eventhough its simplicity, has the inconvenient of having to look up from the tables the value of the enthalpy "h" for every lip pressure P. The following aproximation, greatly simplifies the equation without lost of accuracy.

$$Qs = 27 + 2.2\ P' \tag{15}$$

Where P' is the manometric lip pressure (psig). Table 3 campares for Los Azufres, with a 7.813 "ID pipe, the flow rate obtained from equation 15 and the predictions from Ref 2

TABLE 3

P lip	5	10	20	30	40	50	60
Qs Eq 15	38	49	71	93	115	137	159
Qs Ref 2	37.2	48.6	71.2	93.4	115.5	137.3	159.1

Qs in (Ton/hr); P in psig

For a 6" pipe, schedule 40, 6.065 "ID,equation 15 becomes

$$Qs = 16 + 4/3\ P'$$

In general

$$Qs = (0.066 + 0.036\ P)\ D^2 \qquad (16)$$

Qs steam flow rate (Ton/hr)
P lip absolute pressure (psia)
D inner pipe diameter (in)

EXTENSION TO WATER-STEAM MIXTURES.

So far, the agreement is excellent for the cases of dry saturated steam.
Examining closely the changes in steam flow rate, for a given lip pressure for cases when there is water present, one finds, as shown in table 4, that water doesn't affect significantly the value of the steam flow rate calculated for dry steam.
Table 4 shows the value of the steam flow rate calculated with the exact Russell James'formula (Ref 2) and the error that one will find if the equations for dry steam were used.

TABLE 4

P psig		10	20	30	40	50	60
steam 100 %	Qs	48.6	71.2	93.4	115.4	137.3	159.1
	Error	.0	.0	.0	.0	.0	.0
80 %	Qs	47.8	70.1	92.0	113.7	135.2	156.7
	Error	1.6	1.5	1.5	1.5	1.5	1.5
60 %	Qs	46.3	67.8	89.0	110.1	131.0	151.7
	Error	4.7	4.7	4.7	4.6	4.6	4.6
40 %	Qs	43.2	63.2	82.8	102.7	121.8	141.2
	Error	11.1	11.2	11.3	11.3	11.3	11.2

Qs steam flow rate (Ton/hr) from Ref 2
Error (%) if the presence of water were neglected
The consistency of the error at every % steam quality permits to propose the following formula to correct the values of Qs calculated by equation (15) for Los Azufres.

$$Q's = Qs\ \left(1 - \frac{Qw}{Qs}\ 0.07 \right) \qquad (17)$$

where Q's corrected steam flow rate (Ton/hr)
Qs steam flow rate from equation 15 (Ton/hr)
Qw water flow rate (Ton/hr) measured

Table 5 shows the results of equation 17 compared with those of Ref 2 for the steam flow rate.

TABLE 5

Qw/Qs	P psig	10	20	30	40	50	60
0.25	Ref 2	47.8	70.1	92.0	113.7	135.2	156.7
	Eq 17	47.7	70.0	91.7	113.4	134.9	156.3
0.66	Ref 2	46.3	67.8	89.0	110.1	131.0	151.7
	Eq 17	46.3	67.9	89.1	110.1	131.0	151.7
1.5	Ref 2	43.2	63.2	82.8	102.7	121.8	141.2
	Eq 17	43.5	63.7	83.6	103.2	122.8	142.4

As observed, the agreement of equation 17 and the values calculated using Ref 2, is excellent.

CONCLUSIONS.

For the case of sonic velocity at the dis-- charge of dry steam to the atmosphere, assuming perfect gas behavior, the steam flow rate (Qs) expressed in (Ton/hr) was found to be given by

$$Qs = F\ 29.8\ \frac{P\ D^2}{\sqrt{h - 2000}}$$

Where P is the lip absolute pressure in (psia) and h is the enthalpy of the saturated-dry (Kj/Kg), steam at pressure P, and D is the inner diameter of the pipe in inches. The factor F was found to be 1.11.
For dry steam, discharging through an 7.813 "ID it was found for Los Azufres, where the atmospheric pressure is 10.5 psia, that

$$Qs = 27 + 2.2\ P'$$

where P' is the manometric lip pressure (psig). In general the steam flow rate Qs (Ton/hr) for dry steam is given

$$Qs = (0.066 + 0.036\ P)\ D^2$$

P absolute lip pressure (psia)
For the cases where a mixture of steam water flows,it was found that the previous equation has an accuracy better than 90 % up to 50 % of water. This equation, when the water flow rate (Qw) is known, can be corrected to give exact results by means of

$$Q's = Qs\ \left(1 - \frac{Qw}{Qs}\ 0.07\right)$$

Where Q's is the exact result of the steam flow rate.
In all the cases the results obtained by means of the well known Russell James correlation were used as the true, exact results to be compared with the proposed equations.

ACKNOWLEDGMENT.

The author would like to tank Ing. Carlos Miranda and Héctor Gutiérrez Puente for their help in preparing this paper.

REFERENCES.

1.- Hiriart, G., 1981; Alternativas del desarrollo del campo Geotérmico de Los Azufres, CFE.
2.- James, R., 1962; Steam - water critical flow through pipes, Proc. Inst. Mech. Engr. Vol. 176 No. 26
3.- Shapiro, A., 1953; Compressible fluid flow, John Wiley, New York.
4.- Grigull, U., 1979; Properties of water and steam in SI - Units, Springer-Verlag, Berlin.

Geothermal Resources Council, TRANSACTIONS Vol. 6, October 1982

A PRESUMPTION OF STEAM AND WATER FLOW CHARACTERISTICS BASED ON THE FLOW CONDITIONS AT THE WELL INLET

Ryuichi Itoi and Michihiro Fukuda

Research Institute of Industrial Science, Kyushu University, Kasuga City, Fukuoka, Japan

ABSTRACT

This study is concerned with an estimation of the productivity of the geothermal well. Steam and water flow rates under a fixed well head pressure of 8 ata are calculated for analytical well models by the equation dealing with performance of steam-water two-phase mixture flashing up in a vertical pipe. As the result, relationships between fluid pressure at the well inlet and the depth of the inlet,and between water flow rate and the depth of the inlet, with steam flow rates as a parameter, are obtained, and cosequently, the minimum fluid pressure to obtain any quantity of steam flow rate correlative with the depth, and the upper limit of water flow rate, are calculated.

INTRODUCTION

On planning to develop a geothermal field, it is frequently important to presume steam and water flow rates, and to settle appropriate numbers and depths of wells, for efficient development of geothermal resources.

Flow characteristics observed at the well head are strongly dependent on the well parameters and flow conditions at the well inlet, such as fluid pressure and the dryness fraction, which are directly affected by reservoir parameters, such as transmissibility, fluid pressure and the dryness fraction. Therefore, in order to presume steam and water flow rates,these reservoir parameters should be grasped. However, generally speaking, it takes much time to collect systematic informations from the underground and to clarify complicated reservoir structure.

In this paper, standing on the present state of geothermal development, we carried out numerical calculations of steam and water flow characteristics for the analytical well models, and from the results, discussions on correlations of steam and water flow rates with fluid conditions at the inlet, were performed.

BASIC EQUATIONS AND CALCULATION PROCEDURE

Taking into account the relative velocity between steam and water,the momentum equation and the equation of continuity of an upward flashing flow in a vertical well, can be respectively expressed by

$$dP + \frac{1}{2\,g\,v}\,dw^2 + \frac{\lambda}{D}\frac{w^2}{2\,g\,v}\,dH + \frac{1}{v_\alpha}dH = 0 \qquad (1)$$

and

$$w = \frac{G}{F}\,v_x \qquad (2)$$

From equations (1) and (2), we obtain

$$dP + \left(\frac{G}{F}\right)^2\frac{dv_x}{g} + \frac{\lambda}{2gD}\left(\frac{G}{F}\right)^2 v_x\,dH + \frac{1}{v_\alpha}\,dH = 0 \qquad (3)$$

where P is pressure, v is the specific volume, D is well diameter, w is flow velocity, H is well depth, F is cross sectional area of well, G is total flow rate, λ is the friction coefficient, g is the acceleration due to gravity. v_x is defined by

$$v_x = v_g\,x + v_\ell\,(1 - x) \qquad (4)$$

where v_g and v_ℓ are the specific volumes of steam and water respectively. v_α is defined by

$$v_\alpha = 1 / \gamma_\alpha \qquad (5)$$

and

$$\gamma_\alpha = \gamma_g\alpha + \gamma_\ell\,(1 - \alpha) \qquad (6)$$

where γ_g and γ_ℓ are the densities of steam and water respectively, α is the void fraction.

Integrating equation (3) from the well inlet, 1, to the well head, 2 , we obtain

$$\frac{1}{k}\int_{P_1}^{P_2}\frac{dP}{kv_x + \frac{1}{kv_\alpha}} + \frac{2gD}{\lambda}\int_1^2\frac{dv_x}{kv_x + \frac{1}{kv_\alpha}} + H_{12} = 0 \qquad (7)$$

where H_{12} is the depth of the well inlet, and

$$k = \frac{G}{F}\sqrt{\frac{\lambda}{2gD}} \qquad (8)$$

As such the second term of equation (7) is rather small compared to others,v_x can be substituted for v_α in the second term. Thus equation (7) can be rewritten as

$$\frac{1}{k}\int_{P_1}^{P_2}\frac{dP}{kv_x + \frac{1}{kv_\alpha}} + \frac{D}{\lambda}\ln\frac{k^2v_{x2}^2 + 1}{k^2v_{x1}^2 + 1} + H_{12} = 0 \qquad (9)$$

This is the basic equation dealing with a steam-water two-phase flashing flow in a vertical well. The void fraction α can be evaluated by the following empirical correlations;
the first is

$$\alpha = \left[1+\frac{\gamma_g}{\gamma_\ell}\psi\left(\frac{1}{x}-1\right)+\frac{\gamma_g}{\gamma_\ell}(1-\psi)\left(\frac{1}{x}-1\right)\left\{\frac{\frac{\gamma_\ell}{\gamma_g}+\psi\left(\frac{1}{x}-1\right)}{1+\psi\left(\frac{1}{x}-1\right)}\right\}^{\frac{1}{2}}\right]^{-1} \quad [1] \tag{10}$$

where $\psi = 0.4$ is adopted,
and the second is

$$\alpha = \left\{1-\frac{1}{\varepsilon}\left(1-\frac{1}{x}\right)\right\}^{-1} \quad [2] \tag{11}$$

where

$$\varepsilon = 1.5\left(\frac{\gamma_\ell}{\gamma_g}\right)^{0.692} - 0.5 \tag{12}$$

Assuming that the steam-water mixture flashes in an isenthalpic expansion, the dryness fraction, x, which is necessary to calculate v_x and α, can be calculated by

$$x = \frac{i - i_\ell}{i_g - i_\ell} \tag{13}$$

where i, i_g and i_ℓ are the enthalpies of the mixture, steam and water respectively.

The calculation procedure for flow characteristics is as follows; By inserting the values of D, λ, P, x and H_{12}, into equation (8), k can be numerically calculated correlative with the well head pressure, P_2. Using k obtained above, G is calculated by equation (8), and G_g and G_ℓ are obtained by

$$G_g = G\,x_2 \tag{14}$$

and

$$G_\ell = G\,(1 - x_2) \tag{15}$$

where x_2 is the dryness fraction at P_2.

RESULTS OF THE CALCULATIONS

The calculations were made under the following conditions; $D = 0.2$ m, $\lambda = 0.015$($Re > 2\times10^6$ for wrought iron pipe), $P_2 = 8$ ata(average well head pressure adopted in the Hatchobaru area), $x_1 = 0$ and 0.2, H_{12} = from 100 m to 3500 m and α is evaluated by equation (10).

Figure 1 shows the relationship between P_1 and H_{12} when $x_1 = 0$, with G_g as parameter of steam flow rates. Curves A and B are the pressure profiles for the case where the water level is at the groundsurface. A represents the pressure of saturated water and B the constant water density of γ=1 g/cm^3 respectively. In the case where $x_1 = 0$, as vaporization begins just at the well inlet, the water temperature in the reservoir, T_R, is nearly equal to the saturated temperature, T_0, for P_1. Therefore, the minimal pressure at the well inlet and the reservoir temperature for any quantity of steam flow rate, can be determined from this figure. For example, in order to obtain 100 t/h of steam from the well inlet of H_{12} = 1000 m, the corresponding values of P_1 and T_R should be at least 77 ata and 291°C respectively. It is obvious that the above values of P_1 and T_R increase on the curves of G_g, with increase in H_{12}. The maximum steam flow rates from given depths coincide with A, because P_1 never exceeds A when $x_1 = 0$. When geothermal water bellow the saturated temperature flows into a well,

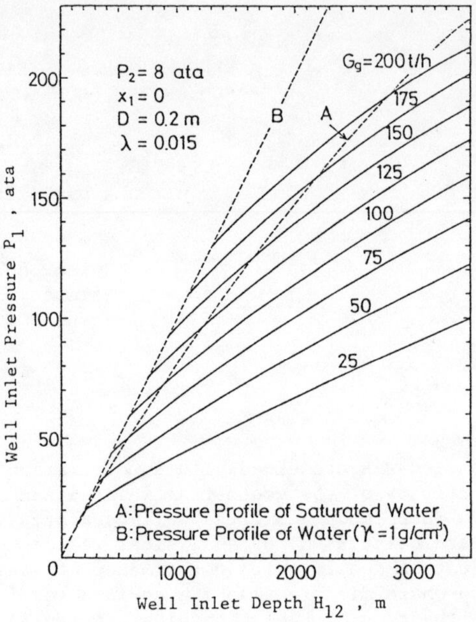

Figure 1. The relationship between P_1 and H_{12}

the water flows up in a one-phase state with decreasing pressure, and begins vaporization at the depth of H_f ($< H_{12}$) and pressure P_f ($< P_1$). In other words, the region between A and B is only applicable for the case where the flash inception is recognizable in a well. Besides, because of the fact that the analytical model (a), in figure 2, is equivalent to (b), P_f and H_f can be substi-

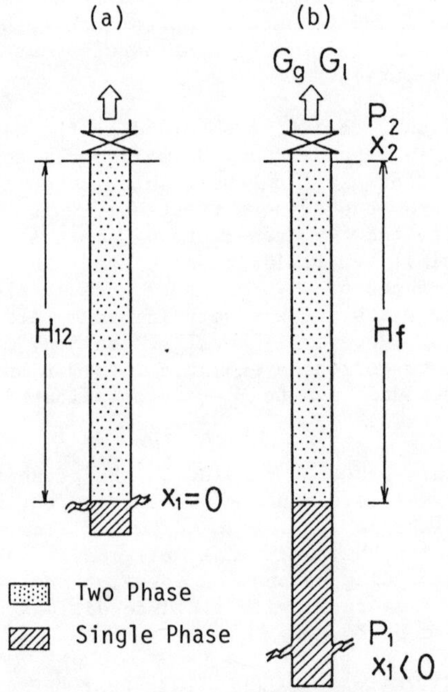

Figure 2. Sketch of analytical well models, (a) and (b) are equivalent

tuted for P_1 and H_{12}, and then figure 1 can be applied for any kind of analytical well model except the case where $x_1 > 0$.

Figure 3 shows the relationship between G_ℓ and H_{12} under the same well inlet condition of $x_1 = 0$ as in figure 2, where curves (a) and (b) correspond respectively with curves A and B. As seen in the figure, the values of G_ℓ decrease with increase in H_{12}, and the larger the value of G_g the smaller the quantity of increase in G_ℓ is. It is noticeable that

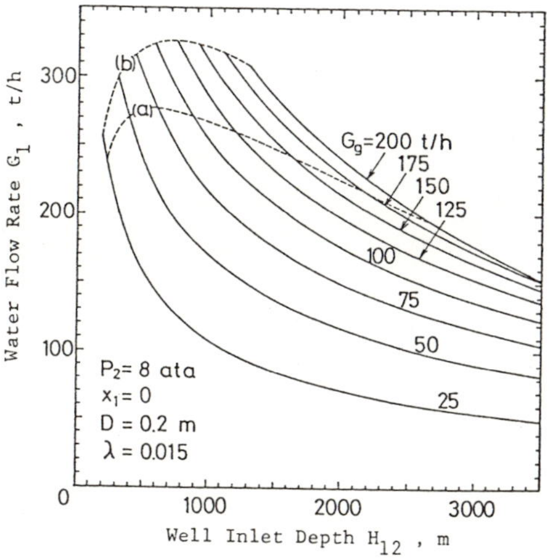

Figure 3. The relationship between G_1 and H_{12}

both (a) and (b) have the maximum values, such as G_ℓ = 280 t/h at H_{12} = 600 m and G_ℓ = 330 t/h at H_{12} = 800 m, and these maximum values of G_ℓ are the upper limits of water flow rate obtainable under the conditions of x_1 = 0 and $x_1 > 0$ respectively.

Figure 4 shows the relationship between P_1 and H_{12} under the condition of x_1 = 0.2, the upper limits of P_1 at given depths are assumed to be the same as the pressure profile A in figure 1. Comparing with figure 1, the values of P_1 required to obtain any desired quantity of steam flow rate, are rather lower. For example, the value of P_1 for G_g = 100 t/h at H_{12} = 2000 m, are 114 ata in figure 1, and 92 ata in figure 4.

Figure 5 shows the relationship between G_ℓ and H_{12} under the same condition of x_1 = 0.2 as in figure 4. Comparing with figure 3, water flow rates are generally smaller, and the inclination is more pronounced at shallower depths. The maximum flow rate is evaluated from curve (a) as G_ℓ = 190 t/h.

CONCLUSION

1) In the case where the dryness fraction, x_1, at the well inlet is zero, the minimal values of the well inlet pressure, P_1, and the reservoir temperature, T_R, required to obtain any quantity of steam flow rate, can be determined from figure 1.
2) In the case where x_1 is positive, only P_1 can be determined from figure 4, and the larger the value of x_1, the lower the value of P_1 is.
3) The maximum water flow rate in the case where x_1 is negative, (i.e., G_ℓ = 330 t/h) is the upper limit, because the value of G_ℓ has a tendency to decrease as x_1 increases.

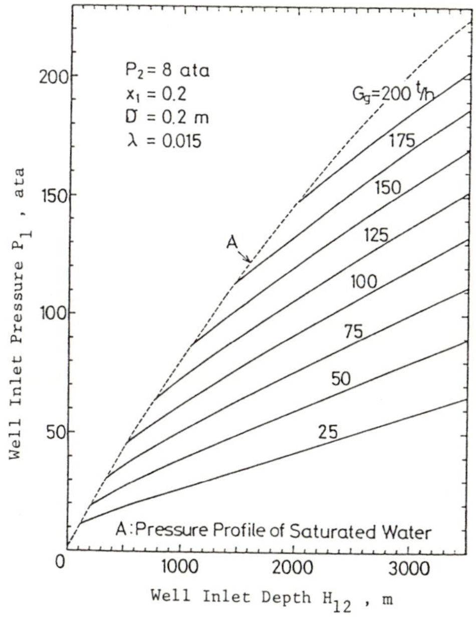

Figure 4. The relationship between P_1 and H_{12}

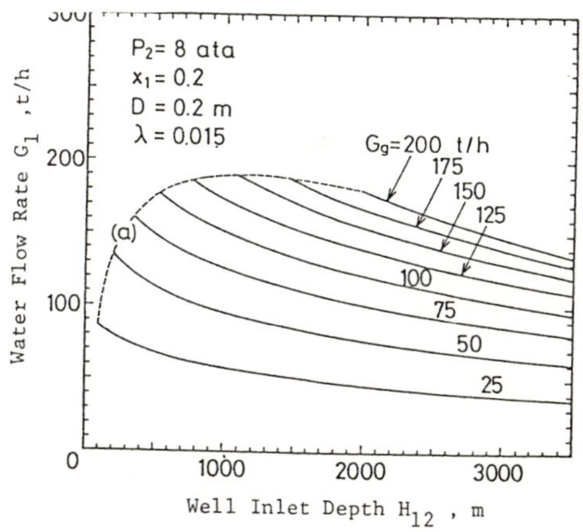

Figure 5. The relationship between G_1 and H_{12}

REFERENCES

1) Smith, S. L., 1969-70, Inst. Mech. Engrs. 184, Part 1, p. 647-664.
2) Sekoguchi, K., 1973, Aspect of Advanced Heat Transfer, 1, p.202.

Geothermal Resources Council, TRANSACTIONS Vol. 6, October 1982

WELL INLET PRESSURE AND STEAM QUALITY OF HATCHOBARU PRODUCTION WELLS

Ryuichi Itoi and Michihiro Fukuda

Research Institute of Industrial Science , Kyushu University 86
Sakamoto Kasuga City 816 , Japan

ABSTRACT

Steam-water two-phase flow analysis in a wellbore were carried out to estimate pressure and steam quality of geothermal fluid at well inlet, using the field data of Hatchobaru production wells observed in the past two years.

The calculated results suggest that two-phase condition is dominant at the well inlet which means the geothermal fluid starts flashing in the reservoir and enter the wellbore as two-phase mixture. Well inlet steam quality are not uniform but differes widely in the range from 0.1 to 0.7. Two wells intersecting thr fracture at shallow zone of the reservoir show the steam of high quality more than 0.5 and increase with time. Largr increase in well inlet pressure explains that flash point moves into the wellbore from the reservoir due the decrease in saturation pressure which may account for the effects of mixing of reinjection water.

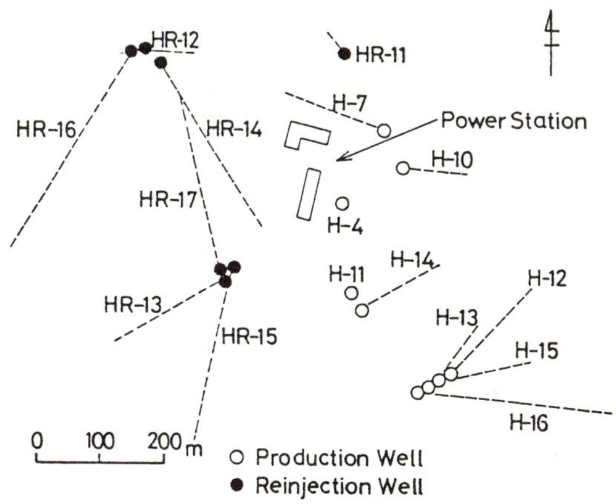

Figure 1. Location of production and reinjection wells in Hatchobaru geothermal field. Dotted line indicates the directional well.

INTRODUCTION

The Hatchobaru geothermal field is situated 1100 m above sea level at central part of Kyushu, where the power station of 55 MW installed capacity is operating. There are nine wells which produce 300 t/h of steam and 700t/h of water in total. Most production wells in Hatchobaru intersect a few fractures at short intervals where geothermal fluid enters the wellbore, although the fracture depth of each well varies from -40 m to +340 m above sea level. Waste water is reinjected into the formation at the same level as the production zone through seven reinjection wells drilled west part of this field as shown in Fig.1.

It is important to observe the change in pressure and steam quality at well inlet in order to understand the reservoir behavior such as the growth of two-phse region in the reservoir due to pressure decrease and the effects of reinjection water mixing on production.

Downhole pressure and temperature survey in a flowing well is available to obtain the information on fluid properties at well inlet. This method, however, tends to involve many technical problems in actual situations in spite of the simplicity of principle, for example, many weights must be used in wells of large mass flowrate to set probe at a given depth. No downhole measurement in the flowing well has been performed so far with success in Hatchobaru field.

In this paper, we present the pressure and steam quality at well inlet which are estimated by steam-water two-phase flow analysis in a wellbore, using the field data of Hatchobaru production wells obtained in the past two years.

TWO-PHASE FLOW IN WELLBORE

Momentum equation of steam-water two-phase flow in a vertical wellbore is expressed by

$$dp + \frac{G^2}{F^2}\frac{dv}{g}x + \frac{\lambda}{2gD}\frac{G^2}{F^2}v_x dH + (\rho''\alpha + \rho'(1-\alpha))dH=0 \quad (1)$$

where p=pressure (ata)

v_x=average specific volume (m^3/kg)

G=total mass flowrate (kg/s)

D=inner diameter of wellbore (m)

F=cross-sectional area of wellbore (m^2)

H=well length (m)

λ=friction factor (-)

g=gravitational acceleration (m/s^2)

ρ''=specific weight of steam (kg/m^3)

ρ'=specific weight of water (kg/m^3)

α=void fraction (-)

v_x is calculated from

$$v_x = v''x + v'(1-x) \quad (2)$$

where v''=specific volume of steam (m^3/kg)

v'=specific volume of water (m^3/kg)

x=steam quality (-)

Integrating eq.(1) between well inlet 1 and wellhead 2 yields

$$\frac{1}{k}\int_1^2 \frac{dp}{kv_x + \frac{1}{kv_\alpha}} + \frac{2kD}{\lambda}\int_1^2 \frac{dv_x}{kv_x + \frac{1}{kv_\alpha}} + H_{12} = 0 \quad (3)$$

where H_{12}=well length between well inlet and wellhead (m)

$v_\alpha = 1/(\rho''\alpha + \rho'(1-\alpha))$

$k = \frac{G}{F}\sqrt{\frac{\lambda}{2gD}}$ (kg/m^3)

For calculating void fraction, the formula proposed by Smith is used

$$\alpha = \left[1 + \frac{\rho''}{\rho'}\psi(\frac{1}{x}-1) + \frac{\rho''}{\rho'}(1-\psi)(\frac{1}{x}-1)\left\{\frac{\frac{\rho'}{\rho''}+\psi(\frac{1}{x}-1)}{1+\psi(\frac{1}{x}-1)}\right\}^{\frac{1}{2}}\right]^{-1} \quad (4)$$

where ψ=0.4 is given.
Steam quality x is necessary to calculate v_α and v_x. We assume that there is no heat transfer from fluid to the surroundings because it gives small effect on the total energy of fluid for large flowrate and long period of production and also ignore the kinetic and potential energy which are negligible compared with fluid enthalpy. Then, steam quality is given by

$$x = (i-i')/(i''-i') \quad (5)$$

where i=enthalpy of fluid in wellbore (kcal/kg)

i''=enthalpy of steam (kcal/kg)

i'=enthalpy of water (kcal/kg)

Pressure and steam quality at well inlet are calculated using eq.(3) by giving pressure,total mass flowrate,steam quality at wellhead and diameter,friction factor of wellbore and well length. Eq.(3) is solved numericaly in trial-and error manner using Simpson's rule for the calculation of integral terms.

Thermodynamic properties such as specific weight and enthalpy of steam and water are expressed as a polynominal function of pressure, which are obtained by least-square method using the data of steam table with subdivided pressure range in order to gain enough accuracy. In case flashing occurrs in a wellbore, specific weight and enthalpy of water at the flash pressure are used for the calculation for single-phase(water) flow from flash point down to well inlet.

For the calculation, we use λ=0.015 for friction factor and D=0.201 m for diameter of wellbore.

Table 1. Field data of Hatchobaru production wells

Well Name	Well Depth m	Wellhead Pressure ata	Total Flowrate t/h (kg/s)	Wellhead Steam Quality -
H-4	1055	3.3	92.4 (25.7)	0.14
H-7	880	7.7	120.5 (33.5)	0.29
H-10	758	7.4	63.8 (17.7)	0.69
H-11	1046	8.5	76.1 (21.2)	0.48
H-12	850	8.7	59.6 (16.6)	0.75
H-13	1014	8.3	98.8 (27.4)	0.28
H-14	1169	8.7	248.2 (68.9)	0.15
H-15	1020	6.9	146.8 (40.8)	0.27
H-16	1080	8.7	101.3 (28.1)	0.30

FIELD DATA OF HATCHOBARU WELLS

Table 1 shows an example of the data of flowrate and pressure at the wellhead measured in August 1980 and the well depth which is the well length between the well inlet and the wellhead. Wellhead pressure of H-4 is a half that of other production wells, thus only the steam separated at the wellhead separator is transported to the secondary flash tank and the water to reinjection wells.

CALCULATED RESULTS AND DISCUSSION

Fig.2 shows the calculated pressure and steam quality at the well inlet versus the elevation above sea level using the data in Table 1. Well inlet elevation is pointed by arrow with well names. Steam quality of wells except H-4 and H-14 implies that the geothermal fluid enters the wellbore as two-phase mixture. On the other hand, zero fraction in steam quality of well H-4 and H-14 indicates that water enters the wellbore. However, the saturation temperature of water determined by the produced fluid enthalpy gives 206 °C and 241 °C respectively. Low water temperature accounts for significant effect of reinjection water mixing as this is confirmed by tracer test.

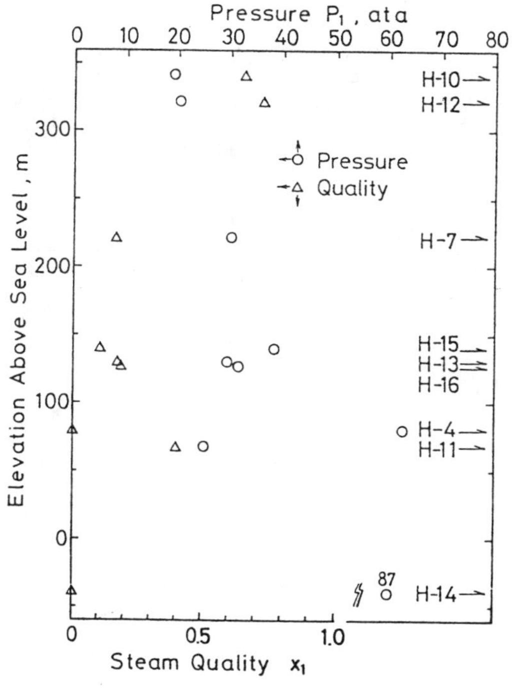

Figure 2. Vertical profile of calculated well inlet pressure and steam quality

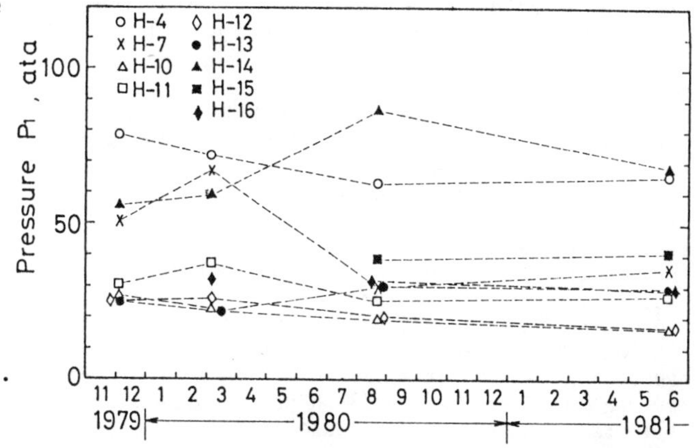

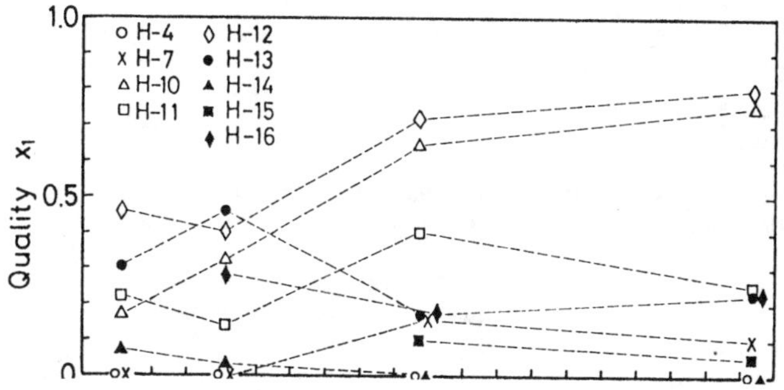

Figure 3. Calculated well inlet pressure and steam quality of Hatchobaru production wells with time.

Fig.3 shows the yearly change of calculated pressure and steam quality at well inlet using the field data measured in the past two years. Both pressure and steam quality vary in a wide range. During this period, flash point was formed in a wellbore of H-14 sometime after the flowrate measurement in March 1980 which gave pressure rise by approximately 30 at and it has not retreated into the reservoir untill the latest measurement. On the other hand, flash point of well H-7 moved into the reservoir with 37 at decrease in pressure while it remained in the wellbore during first two measurement.

Two wells of H-10 and H-12 has increased in well inlet steam quality to a great extent with time although other wells indicate constant or slight decrease in quality. These two well show similar tendency in pressure and steam quality at well inlet which suggests that these wells intersect the identical fracture while they are separated by 200 m in holizontal distance at well inlet.

Fig.4 shows well H-4 history of calculated well inlet pressure and saturation pressure determined from the enthalpy of produced fluids. The steam production of H-4 once reached more than 110 t/h which is the largest in the steam flowrates recorded in Hatchobaru geothermal field. In 1974, the well inlet pressure increased by 20 at abruptly with such a great decrease in saturation pressure as its value came lower than well inlet pressure. As well inlet pressure increased, the water flowrate increased from 200 t/h to 290 t/h, while the steam flowrate decreased from 100 t/h to 90 t/h. This is attributed to the effect of reinjection water mixing since it started at the same level of production zone in 1974. From this fact it is concluded that there is strong connection between production well inlet and reinjection well outlet through the fracture and flash point moves into the wellbore following the increase in well inlet pressure and water flowrate.

The well inlet pressure remained constant for two years after the abrupt increase, then it decreased gradually from 1977 as the saturation pressure decreased more rapidly

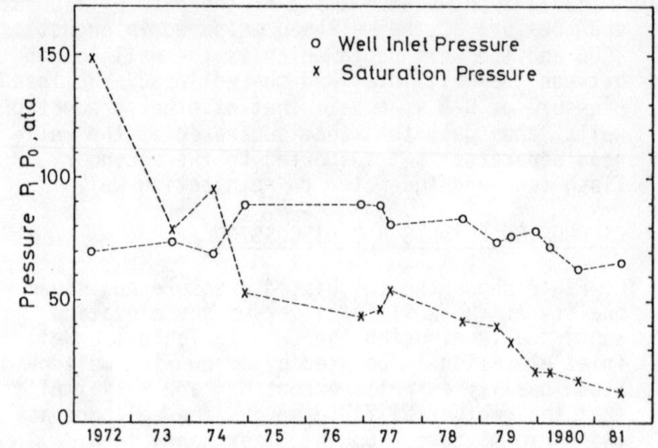

Figure 4. History of calculated well inlet pressure and saturation pressure of well H-4

CONCLUSION

Pressure and steam quality at well inlet of Hatchobaru production wells were calculated by steam-water two-phase flow analysis in a wellbore, using the field data obtained at the wellhead. It is concluded that steam-water mixture enters the wellbore in all wells except two wells throughout the period of measurement. Abrupt increase in well inlet pressure explains that the flash point is formed in a wellbore as a result of decrease in saturation pressure of water due to mainly the mixing of reinjection water.

ACKNOWLEDGEMENT

The authers would like to thank Kyushu Electric Power Company who provided the field data used in this study.

REFERENCES

Gould, T. L., 1974, Vertical Two-Phase Steam-Water Flow In Geothermal Wells, J.P.T., pp 833-842

Fukuda, M. et.al., 1975, Some Considerations on Well Characteristics at Otake and Hatchobaru Geothermal Areas,Japan, Proceedings Second UN Symp. on the Development & Use of Geothermal Resources. San Francisco, pp 1642-1650

Smith, S. L., 1969, Void Fractions in Two-Phase Flow: A Correlation Based Upon an Equal Velocity Head Model, Instn. Mech. Engrs. 184 Pt.1 36, pp 647-664

Geothermal Resources Council, TRANSACTIONS Vol. 6, October 1982

WEN-1: FIRST SUCCESSFUL WELL IN THE WENDEL-AMEDEE KGRA

R.W. Juncal, W.D. Gertsch and K.R. Johnson

Geologic Consultant, Reno, NV
Consultant, Idaho Falls, ID
GeoProducts Corp. Oakland, CA

ABSTRACT

Field exploration activities initiated by GeoProducts Corp. in 1977 culminated in the successful completion of a deep production well in Sept. 1981. The well was drilled to a depth of 5823' and completed in a fractured, dominantly granodiorite lithology. A broad fault zone with associated fracturing occurs at approximately 5300'-5335' and serves as the main production zone. Well testing has indicated a highly productive well with no immediate boundaries. Long term testing was conducted at a constant artesian flow rate of 670 gpm with a maximum downhole temperature during flow of 251 F. The thermal fluid is a Na-SO_4-Cl type of low salinity. Various geothermometers suggest that the fluid originates from a reservoir in the 275-310 F range. Further study and refinement of the system model is ongoing in preparation for future deep drilling.

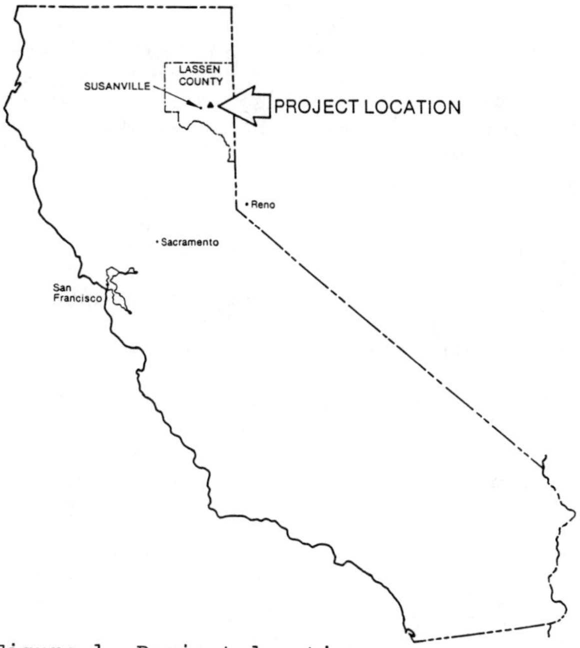

Figure 1. Project location.

INTRODUCTION

WEN-1, the first successful deep well in the Wendel-Amedee KGRA, was drilled as part of the feasibility phase of a project which will be the first commercial plant to combine the use of geothermal energy and wood fuel for power production. The project, centered near Susanville in Lassen County CA (fig.1), has been sponsered jointly by the GeoProducts Corp. of Oakland, CA, the U.S. Dept. of Agriculture-Forest Service, the California Dept. of Water Resources and the U.S. Dept of Energy. Field exploration to outline the resource potential for the project began in 1977 with reconnaissance and detailed geophysical surveys which included gravity, ground magnetic, scalar audio magneto tellurics, tensor MT, as well as dc soundings and time domain electro-magnetic profiles (see Pritchard and Zebal, GRC, 1978). The initial site selection was based on a synthesis of all available data on the area, including geologic and geophysical logs from two deep (5000') geothermal wells drilled by Gulf Mineral Resources Co. in 1973. Additionally, 4 shallow thermal gradient holes were drilled by GeoProducts as part of the exploration program. The final site selection followed from this earlier work with some additional photogeologic interpretation and geologic mapping in the area northwest of the Wendel Hot Springs where a north striking structure dipping 60-70 west was indicated. This structure can be traced from a linear group of tufa mounds at the hot springs for at least 2.5 miles north through the western Skedaddle Mountains.

DRILLING SUMMARY

The well was spudded on August 1, 1981 and was completed in 51 days. Lost circulation zones were encountered at 209',1016' and 5304'. At the latter two depths casing strings were run, and at 5304' a sand plug was set to protect what was believed to be a production zone, prior to setting pipe. The well was drilled to a total depth of 5823' using a light gel-lignite to 5304' and water below. The producing interval has been completed open hole below 9 5/8" casing at 5068'.

LITHOLOGY

In WEN-1 volcanic rocks were penetrated from 300' to approximately 4550' variosly interspersed with sections of sediments largely coarse grained conglomerates. To a depth of 2600' andesite to basaltic flows and tuffs were abundant. In the section from approximately 2600' to 4550' there was little if any flow rock but was largly comprised of highly altered tuff, tuffaceous sediment, mudflow and conglomerate of uncertain origin(probably both epiclastic and volcanic). These latter lithologies were very clay rich and showed little or no permeability. From 4550' to the top of the granitic basement at 5050' a conglomerate of clearly distinct character, with somewhat less clay component was penetrated, however it also showed no permeability. From 5050' to total depth "Sierran" granodiorite was encountered with some evidence of veinning(qtz) and intrusion by more acidic dikes. A fault zone occurs at approximately 5300' and serves as the main production zone. The well lithologies are summarized in figure 2.

GEOPHYSICAL LOGGING

Geophysical logging runs, made at casing points and total depth, were performed by Gearhart Industries, as follows:

-1016': Dual Induction Laterlog(DIL) w/ Spontaneous Potential(SP), Temp.

-5304': DIL w/SP, Temp.(2), Neutron-Density(CD-CNL)w/ Gamma Ray(GR) and Caliper(CAL), Sonic(BHC)w/GR,Cal, and Siesmic Spectrum(SS)

-5823': DIL w/SP, BHC w/GR,Cal,and SS, Temp.(2)

Additional temperature logs along with a caliper and two spinner surveys were also run as part of the testing program. The log suites shown above represent a modification of the initial plans due to hole conditions and other factors considered at the time of logging.

The logging program was designed to serve two objectives: (1) evaluation of lithology and formation fluid, and (2) correlation with past and future wells in the area. Each of these objectives has been realized due to the overall excellent quality of the logs obtained. The evaluation of lithology was especially important due to the heterogeneous rock types which occur in close sequence. Statistical techniques were used to correlate observed

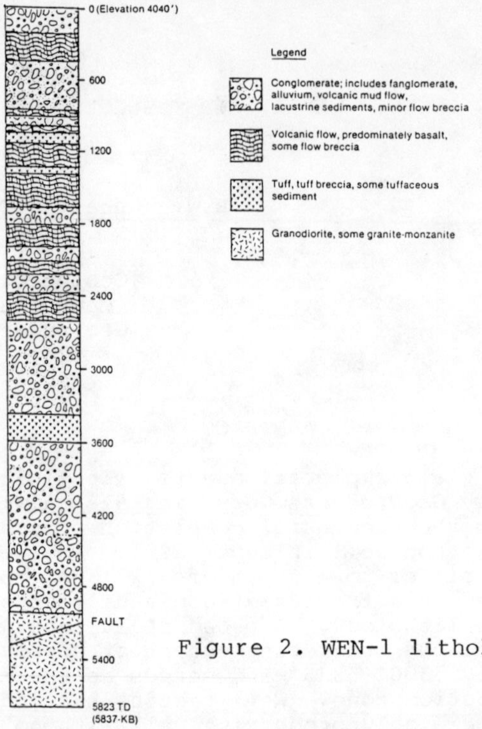

Figure 2. WEN-1 lithology.

lithologies with their geophysical signatures Original estimates of fluid resistivity also agreed very well with those determined later from produced samples. The correlation of the WEN-1 logs with those from the two previous deep thermal wells in the area has been of great importance in defining a basin model which appears to include a migration of volcanic centers across the region.

In addition to the above objectives the log suites were valuable in delineating fracture/potential production zones within the granodiorite section. A combination of criteria were used in this evaluation as outlined in Sanyal and others, 1980. Subsequent spinner surveys, although complicated somewhat by large hole size variations, generally confirmed the geophysical interpretation.

Another important aspect of the geophysical logging program was the prediction of bottom hole temperatures, particularly at the intermediate casing point. Temperatures were recorded regularly at the mud return line as well as downhole, using maximum reading thermometers(unjacketed) run in a sub on each directional survey. Little could be determined from these data. In order to determine the equilibrium bottom hole temperature before committing intermediate casing at 5300' a two day temperature build-up was observed and used for projection. The build-up was recorded using jacketed maximum reading thermometers on each of the geophysical tools run in the hole. Due to the long period of non-circulation for logging a simple Horner plot was used to obtain an estimate of the true equilibrium bottom hole temperature. The plot projected to a temperature of approximately 254 F which agrees well with the highest temperature recorded during the long term flow testing of 251.2 F.

FLOW TESTING

Inital flow testing of WEN-1 was conducted with the rig still on location and it indicated a "keeper". One month later a comprehensive test program was begun using the testing assembly shown in figure 3.

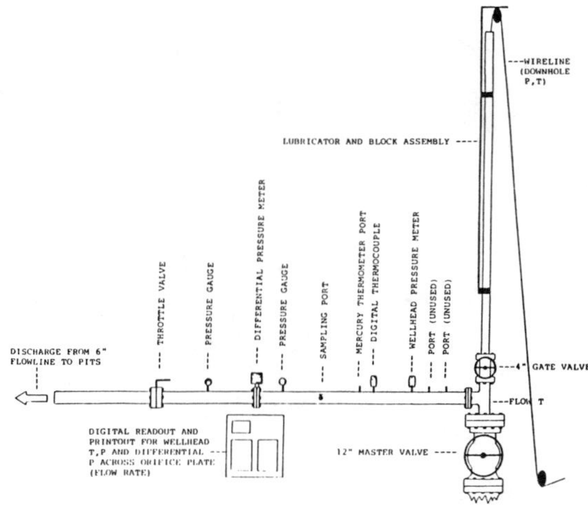

Figure 3. Well test assembly.

During the early phases of the testing a number of anomalies were noted in the pressure behavior. A subsequent spinner log showed a break in casing some 230' from the surface. Immediate steps were taken to rectify the situation and a packer was set below the break. Later a televiewer survey was run to assess the damage, which was then repaired by squeeze cementing and tying back with smaller casing. Final flow testing was begun in March of 1982.

The test format called for three short term pulse tests, followed by a longer constant flow/recovery test. During the tests wellhead and downhole temperatures and pressures were contiuously recorded and several water samples were collected. Complex hydrologic conditions indicated earlier by the geophysical logs were also reflected in the pressure transient data. Complications such as fracture flow in the well-reservoir system penetrated in WEN-1 makes the application of theoretical methods to calculate hydrologic parameters of uncertain value. However, the following generalizations can be made; WEN-1 is a highly productive well, although this productivity is compromised by large wellbore losses, the producing zone transmissivity is very high(approx. 100,000 gpd/ft), a constant head or recharge boundary may exist relatively nearby. A more complete analysis of the pressure data is in progress which should allow a more accurate modelling of the subsurface flow conditions

Water samples taken from the flowline during testing have been analysed for standard and trace elements as shown in table 1. The produced fluid is very similar to that found in the neighboring Wendel and Amedee Hot Springs. Figure 4 shows a comparison of geothermometer values for the WEN-1 fluid and those of the two springs.

ELEMENT	CONCENTRATION (PPM)
Na	349
K	20
Ca	38
Mg	0.5
Fe	LT0.25
SiO_2	130
Ba	LT0.63
Cr	LT0.05
Mn	LT0.25
Cu	LT0.06
Pb	LT0.25
Zn	LT0.13
Cd	LT0.06
As	LT0.63
Sb	LT0.75
B	7.7
SO_4	410
Cl	284
HCO_3	73
F	6.5
TDS	1320
pH	8.3

Table 1. Chemical analysis of fluid from WEN-1(LT= less than).

GEOTHERMOMETER		W-1	AHS	WHS
max. steam loss	qtz	268	253	257
no steam loss	qtz	275	257	262
	chal	226	207	214
Fournier	Na/K	333	261	268
Truesdell	Na/K	261	178	185
	NaKCa	276	257	262

W-1= WEN-1
AHS= Amedee Hot Springs
WHS= Wendel Hot Springs

Figure 4. Geothermometer values in F .

FUTURE PROJECT DIRECTIONS

Further data collection and study is in progress in and about the GeoProducts leasehold in preparation for the drilling of additional production and reinjection wells. This work includes the reinterpretation of the existing surface geophysics in the light of the deep drilling information, downhole elemental geochemistry on cuttings, surface soil mercury survey, regional water sampling program including deuterium and oxygen 18 analyses and extensive geologic mapping. The synthesis of all of the available data is aimed at refining the geothermal system model of the Wendel-Amedee KGRA for future numerical treatment.

ACKNOWLEDGEMENTS

The U.S. Dept. of Energy has provided a Geothermal Energy Research and Development Grant in partial support of this project in addition to cost sharing on the drilling operations under the User-Coupled Confirmation Drilling Program. Support from the University of Utah Research Institute, EGG Idaho and Lawerence Berkeley Laboratory is also gratefully acknowledged.

REFERENCES

Sanyal,S.K., Wells,L.E. and Bickham,R.E., Geothermal Well Log Interpretation State of the Art, LASL informal rpt. LA-8211-MS, 1980.

GEOCHEMISTRY OF THE PALIMPINON GEOTHERMAL FIELD, SOUTHERN NEGROS, PHILIPPINES

C. W. KLEIN*, J.R. McNITT*, S. K. SANYAL*
and
M. CHE*, B. S. TOLENTINO**, A. ALCARAZ**, and R. DATUIN***

*GeothermEx, Inc., California, U.S.A.
**Philippines National Oil Company, Philippines
***National Power Corporation, Philippines

ABSTRACT

Areal patterns of aquifer fluid chemistry at Palimpinon are consistent with patterns of gas chemistry in indicating that the principal source of deep fluid feeding the production zones of existing wells lies to the south of the central part of the reservoir. Enthalpy-chloride mixing relationships indicate a deep parent fluid at about 310°C with about 3,800 mg/l Cl. Boiling and mixing processes have produced variations of chloride within the reservoir from about 3,000 to 4,100 mg/l. Fluids at the western end of the reservoir appear to have evolved principally through mixing of the deep fluid with dilute recharge, whereas along the eastern fringe of the central area mixing has followed concentration by boiling. The least evolved fluids are tapped by wells which lie along a N-S trend which enters the central part of the reservoir from the south. This trend is fairly well defined by contours of the chemical geothermometers, Na/K, and Cl. It may be related to structures along which the principal upwelling into the reservoir, and/or feeding from the south, is occurring.

INTRODUCTION

As part of an assessment of the Palimpinon Field of South Negros, Philippines, we have studied approximately 600 chemical analyses of fluids obtained between 1976 and 1981 from 19 exploration and production wells.

Our objective was an overview of trends in the reservoir as a whole, with data analysis concentrated on major element chemistry, chemical geothermometry and concentrations of CO_2 and H_2S in the vapor phase. There is ample evidence that many wells in the field produce from more than one zone at depth, and there is geochemical evidence indicating small differences in fluid compositions between an upper and lower zone. Convincing quantitative calculation of the differences in upper and lower zone fluids compositions has not been possible, and all wells were treated in terms of average fluid composition at the wellhead. The data represented samples collected by field personnel from wells on horizontal discharge both at atmospheric pressure (cippoliti wier box) and higher pressure using a webre separator; sampling points were provided along both high and low pressure sides of the orifice plate in the discharge pipe. Various orifice plate diameters were used during the tests,but without significant effect on fluid chemistry. Discharge enthalpies included in the raw data had been calculated using James' method. For some wells all available data were processed, but samples collected during periods of initial well clean-out and unstable chemistry after drilling were discarded when longer-term, more stable flow data were available. The final data base comprised over 270 individual water samples and a similar number of gas samples.

For each analysis the enthalpy and pressure of separation were used to calculate steam fraction, and the analysis was then recalculated to concentrations prior to steam separation.

EXCESS STEAM

To monitor the samples for excess enthalpy (steam), we relied upon the recent revision of the quartz-adiabatic geothermometer (Fournier and Potter, 1981), which applies to 330°C at any specified temperature or pressure of separation. This was used to calculate the expected total fluid enthlapy and steam fraction for each sample, assuming no excess enthlapy. The value (steam fraction calculated from silica-steam fraction from James' method measurement) was then defined as "excess steam", and averaged for a representative set of samples from each well. Results for each well were remarkably consistent.

Of 15 wells for which suitable data were available, 11 showed "excess steam" between ± 3%, 2 between ±(3 to 5)%, and 2 between + 10 and 12%. We believe that the estimates between ± 5% represent small, systematic biases in sampling and analysis. Wells OK-2 and OK-5, in contrast, gave good evidence of an excess of 10 to 12% steam entering the wellbore (OK-5 has exhibited cyclical behavior, with and without excess steam). This finding is consistent with the fact that these two wells produce from the shallowest part of the central reservoir area, where a steam cap is most likely to exist. Subsequent calculations of aquifer liquid chemistry relied upon the adjustment of analytical data using measured steam fraction at the wellhead, except at OK-2 and OK-5 where corrections for the excess steam were applied.

RESERVOIR CHARACTERISTICS

The silica and NaKCa temperatures, ratio Na/K and reservoir chloride all provide evidence that the highest temperature fluids of deepest origin and highest salinity are being tapped by wells in the south-central part of the drilled area (OK-7, -9, -10, PN-13). Only the geothermometers are illustrated herein (figure 1), but contours on reservoir Na/K and Cl show similar patterns.

Differences between the quartz and cation temperatures can be explained as due to the relatively stronger effects of mixing, cooling and re-equilibration on the former. On figure 2, Line A-A' separates wells in the north of the field (lower right) from wells to the south and west (upper left), the northern wells showing greater adjustment of silica to falling temperature. QTZA is the quartz, adiabatic temperature, whereas QTZC is the conductive temperature following correction for measured steam fraction.

Figure 3 shows good evidence of mixing in the reservoir with all fluids sampled having evolved from a parent at 310°C with 3,800 mg/l chloride at point B, and a cool mixing end-member at 30 cal/gm and 10 mg/l Cl^-. Point B as shown is based upon enthalpies determined using the quartz, adiabatic geothermometer and temperature of wellhead steam-water separation. Its position shifts slightly, to B', if quartz-conductive temperatures following adjustment for measured steam fraction are employed. Wells in the west part of the field have evolved from fluid B by mixing along line B-E. Boiling of B produces the fluid of well OK-7, which mixes to produce OK-9 (along C-F) and further boiling followed by mixing produces fluids tapped by wells along the eastern edge of the field (line D-G). Note the coincidence of the relatively primitive character of OK-5, -7, and -9 with the central location of these wells and maxima of the chemical temperatures (figure 1).

GAS CHEMISTRY

The average production well in the reservoir has 1.6% (wt.) CO_2 in the steam fraction at 0.65 MPa-abs, with a range across the reservoir from 0.8% to 3.2%. In the total discharge CO_2 ranges from 0.2 to 1.0% (wt.) and H_2S plus residuals comprise about $4 \pm 1\%$ by weight of the total gases (figure 4). It appears that the western part of the reservoir has a more even gas content and lower levels of CO_2 than the central part, and CO_2 is highest in OK-7, which is believed the only well to produce an unmixed fluid (figure 3).

The ratio CO_2/H_2S (wt.) shows a somewhat distinct pattern from that of gas concentration, with the lowest values along the south side of the central area, somewhat to the south and east of the highest CO2 concentrations. CO_2/H_2S is lowest in wells OK-5, -9, -10 and -13, which aside from OK-7 are believed to tap the least mixed fluids in the reservoir (see figure 3).

CONCLUSION

The areal patterns in aquifer fluid geothermometry (figure 1) and gas chemistry (figure 4) are remarkably consistent with measured temperature contours in the reservoir (see contours and cross-section by McNitt, et al elsewhere in this volume). This is particularly true of the quartz temperature data. Contours on the cation temperatures have been strongly affectd by a relatively low value at OK-5 (256°C), which may not be representative. The well was tested early during its production history when its behavior was erratic, and new tests are needed. The chemical data predict that the principal source of deep fluid feeding the production zones of existing wells lies to the south of the central part of the reservoir, that is, south of wells OK-5, -9, -7 and PN-13.

ACKNOWLEDGEMENTS

We would like to thank PNOC and its personnel for providing the data which made this study possible, and for permission to publish it.

REFERENCES

Fournier, R. O. and Potter, R. W. II, 1981 (unpubl.), A revised and expanded silica (quartz) geothermometer.

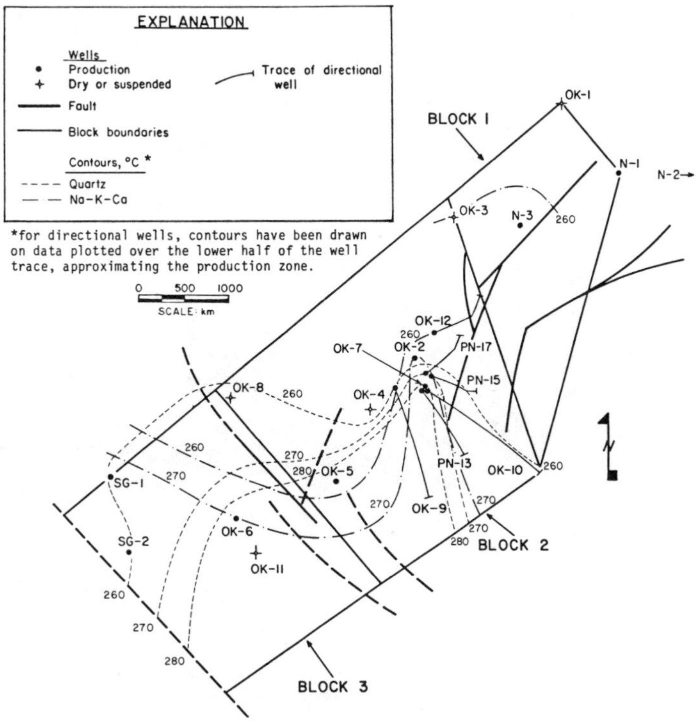

FIGURE 1. Temperature distribution based on chemical geothermeters.

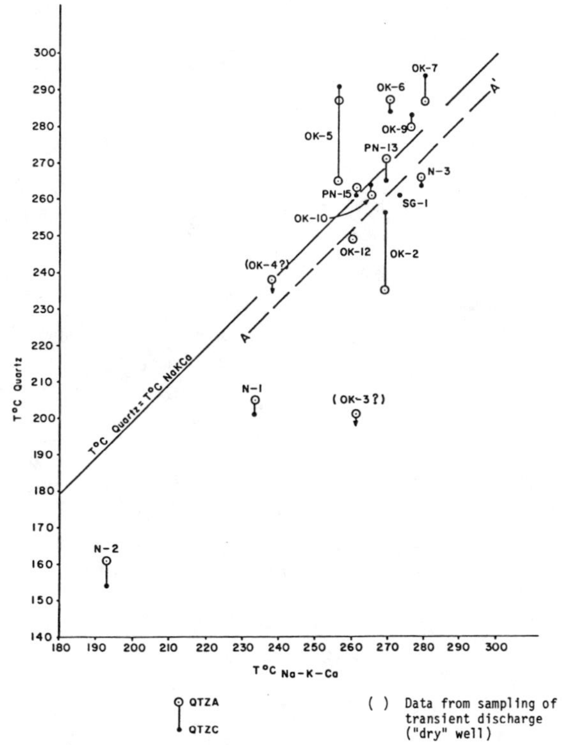

FIGURE 2. Diagram showing Quartz and Na-K-Ca temperature of Palimpinon wells.

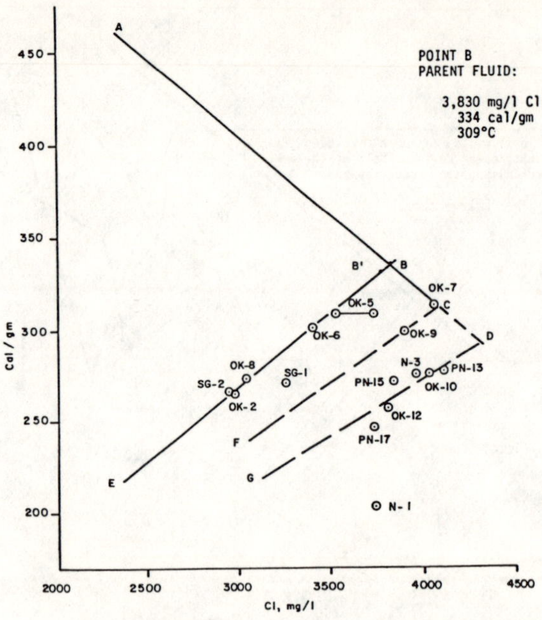

FIGURE 3. Enthalpy-Chloride diagram for Palimpinon well Fluids. Line A-D connects to enthalpy of steam (664 cal/gm) at 0 mg/l Cl.

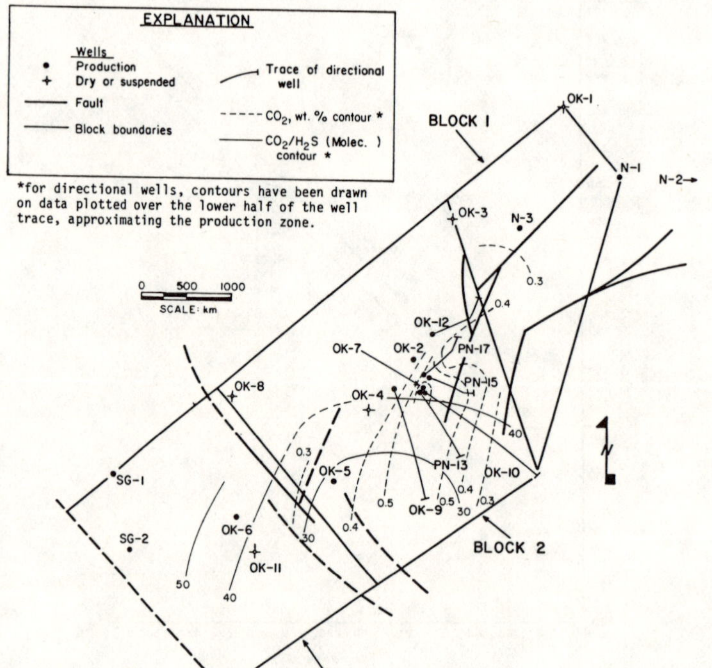

FIGURE 4. CO_2 wt.% and molecular ratio CO_2/H_2S in total discharge.

Geothermal Resources Council, TRANSACTIONS Vol. 6, October 1982

RESERVOIR ENGINEERING OF THE AHUACHAPAN GEOTHERMAL FIELD

P. Liguori*, F. Jarach*, M. Choussy**, T. Campos**, D. Escobar**

* ELC-Electroconsult, Milano, Italy
** CEL Comisión Ejecutiva Hidroeléctrica del Rio Lempa, San Salvador, El Salvador

ABSTRACT

Reservoir engineering in the Ahuachapán geothermal field, where there is an installed capacity of 2x30 MW + 1x35 MW, is at present mainly oriented towards field management, so as to predict production declines and to permit one to make the correct, timely intervention in order to guarantee the full rate of extraction.

From the conceptual model on which reservoir engineering is based, a mathematical model was derived, a modified version of the original by Faust and Mercer. The match of observed against simulated evolution for the first 12 years of exploitation provided the best matching parameters; these were then used to predict the future behaviour of the field under different production/reinjection rates and sites.

The main result to date has been the realization that two additional wells are needed if one is to sustain the full rate of extraction (95 MW) for the next twenty years.

INTRODUCTION

When, in 1968, the first exploratory well confirmed the existence of the Ahuachapán field, little was known on how to evaluate a water-dominated geothermal field. Geochemical, geological and geophysical data were used to form a preliminary conceptual model and to assess the field potential, according to volumetric approach and ground water resource techniques.

However, both the favourable potential assessment and the encouraging steam discovery hastened the construction of the power plant, whose first unit commenced operation in June 1975, followed by the second in June 1976 and the third in November 1980. The first two units rate 30 MW, single pressure; the third unit rates 35 MW, double pressure, utilizing water discharged by the first two units as well.

By the first year of exploitation it was clear that the preliminary conceptual model had to be modified, allowing for thermodynamic and chemical changes occurring in the reservoir.

An excellent set of data was logged since the very beginning of exploitation, mainly covering geochemistry and thermodynamics, permitting an accurate reconstruction of the field behaviour and leading to the development of an updated conceptual model of the field (C.E.L., 1978; C.E.L., 1981).

THE FIELD BEHAVIOUR

The values monitored during exploitation consist primarily of chemical constituents, pressure and temperature in the wells during shut-in, rate and enthalpy of production and pressure and temperature transients in the wells (CEL, 1981). Mapping and cross-checking of changes of several values made evident that the exploitation was influencing their distribution in the field.

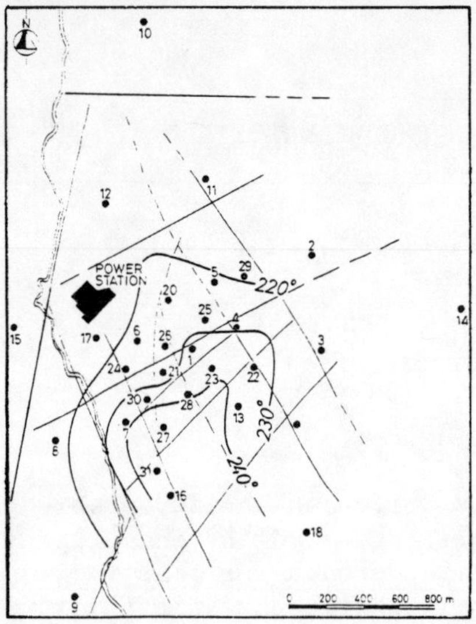

FIGURE 1a. SiO_2 temperature contours (1976)

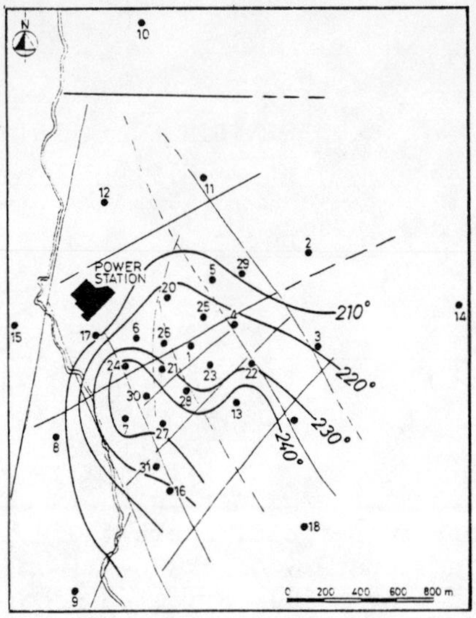

FIGURE 1b. SiO_2 temperature contours (1980)

Temperature contours calculated by SiO_2 geothermometry indicate (Fig.1) a progressive heating up of the inner southern part of the field and cooling down of the outer part; on the other hand, the changes in SiO_2 temperature at a fixed location, when compared with direct temperature measurements, indicate a progressive cooling down together with evaporation (Fig.2).

Pressure contours prior to and after exploitation (Fig.3) indicate an alignement of the piezometric lines along two main southerly directions; on the other hand, pressure measurements at a fixed location indicate a progressive depletion of the field together with an extreme sensitivity to mass extraction rate (Fig.4).

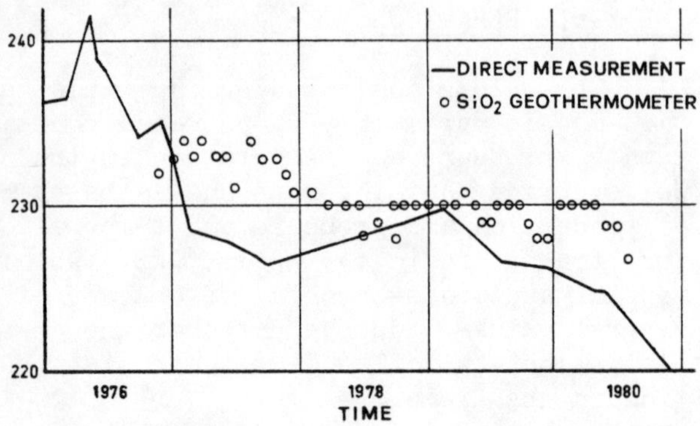

FIGURE 2. Temperature vs. time at a fixed borehole

Pressure and temperature declines are indirectly confirmed by declines in production of the wells (Fig.5); it has to be noted however that a proper relationship between both declines is made difficult by simultaneous changes in enthalpy and by the effects of the overall rate of production/reinjection.

Enthalpy increased in most of the wells, revealing the formation of a "dynamic" two-phase zone around the wells, which disappeared during shut-in.

THE CONCEPTUAL MODEL

The study of the field behaviour, using conventional ground water techniques, soon appeared impractical because of the evidence of important changes in temperature and even of a second phase (steam) in the field.

Thermodynamics, and related geochemistry, become the prime disciplines in understanding reservoir physics through the evolutionary behaviour of the field. Nevertheless the conceptual model based on thermodynamics cannot be separated from a full comprehension of fluid patterns and distribution of recharge areas; to fullfill this requirement a new geological investigation was started in 1980, aimed at clarifying this point (Aumento et al., 1982).

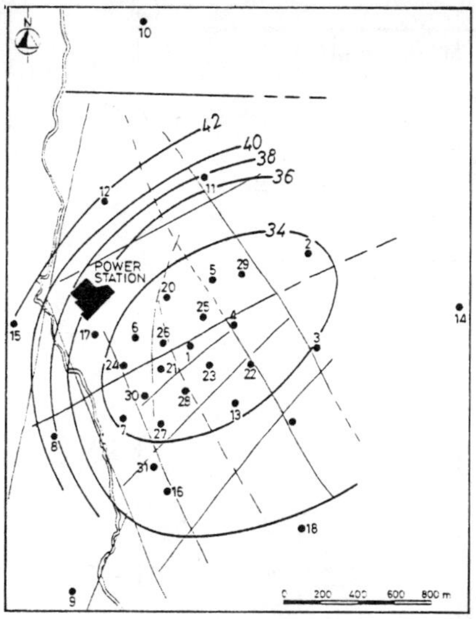

FIGURE 3a. Pressure contours (1975)

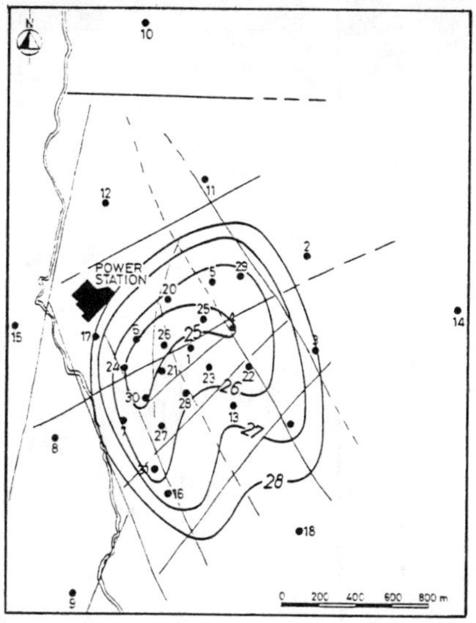

FIGURE 3b. Pressure contours (1980)

All the information led to form a conceptual model in which 1) the area drilled with its high fracture permeability, is fed from the south by two main fractures trending SSE-NNW 2) the area drilled, initially filled with liquid water, undergoes a general depletion facilitating local evaporation at points of minor pressure 3) these points are both structural highs and drawdown lines around the wells (the latter causing the "dynamic" two-phase zone) 4) evaporation implies a decline in rock temperature as well, caused by heat transfer to the guest liquid which will be boiling under ever decreasing pressures 5) reinjected water is partially re-cycled by nearby wells.

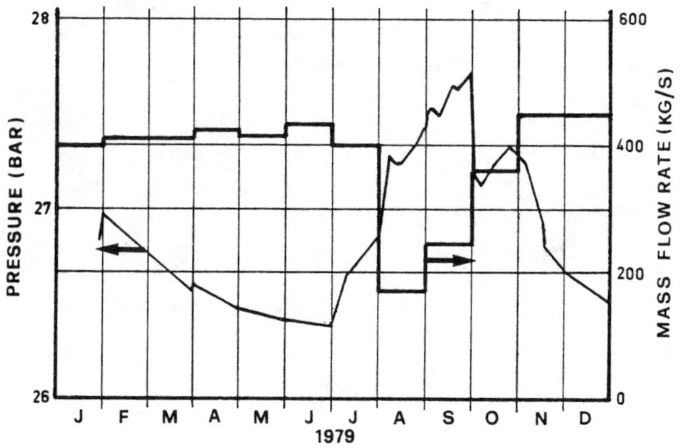

FIGURE 4. Pressure vs. time at a fixed borehole and net mass extraction rate of the whole field vs. time

THE MATHEMATICAL MODEL

The overall handling of several sets of data from many wells is impractical without the use of a frame able to store the data and correlate them with each other. A mathematical model was prepared to act as an intelligent frame. The model is a modified version of the one by Faust and Mercer; modifications do take into account the conceptual model, allowing for two-phase production, reinjection at variable temperatures, and heat transfer between rocks and fluids (Faust and Mercer, 1977).

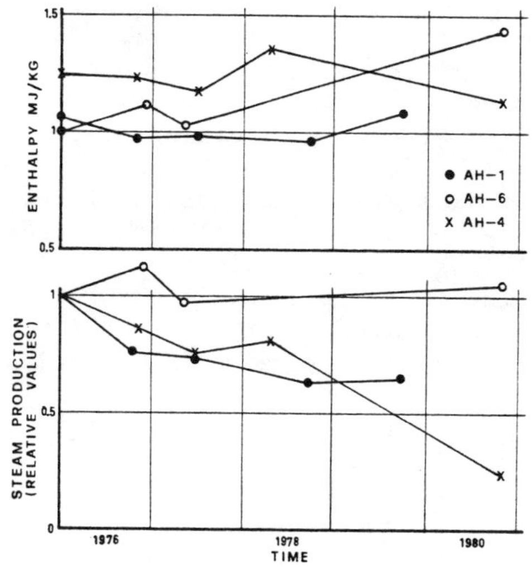

FIGURE 5. Steam production (relative to initial conditions) and enthalpy of some wells vs. time

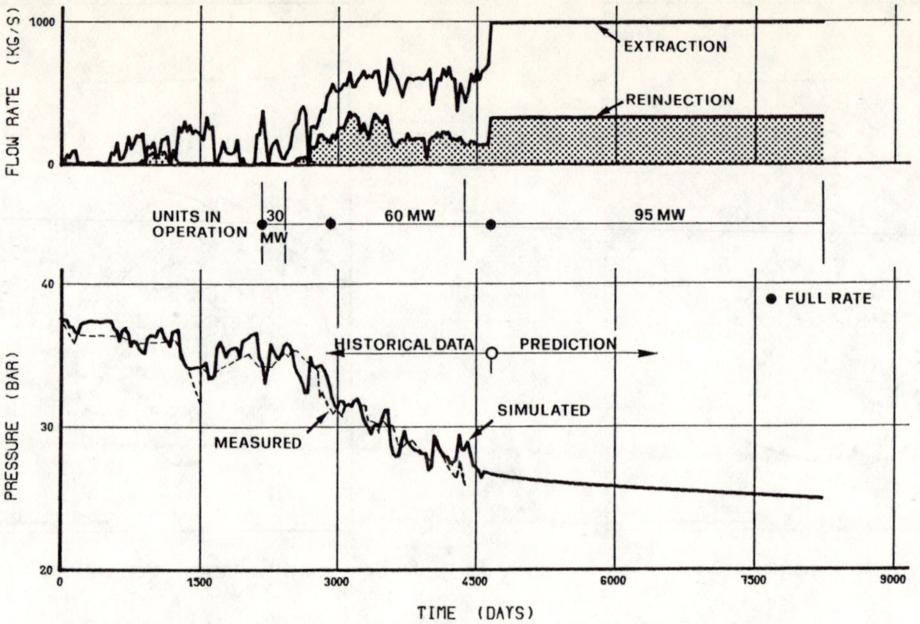

FIGURE 6. Pressure and mass flow rate (extracted and reinjected) vs. time. Match of data from 1968 to 1980 and prediction of future evolution.

SIMULATION BY MATHEMATICAL MODEL

A first calibration of the model was attempted by matching the observed evolution of the field, during its first 12 years of life. The first parameters considered were average reservoir pressure against mass and enthalpy extracted/reinjected; the match was attempted by varying parameters, such as drainage areas, preferencial fluid patterns and storages, heat transfer coefficients, and so forth. The match obtained is quite satisfactory, as can be appreciated in Fig.6. Using the best match configuration, prediction of future behaviour was made, varying rates and site of production/reinjection.

Studies are still under way, but preliminary results were sufficient to determine the need for two additional wells along the fault feeders, in order to sustain a full rate of extraction (95 MW) for the next twenty years (ELC, 1981).

The model is expected to undergo both numerical and conceptual modifications, as they become necessary according to the development of the studies; however, the model will be used in the future as a main instrument both in understanding the field evolution and in improving its exploitation.

REFERENCES

Aumento, F., Viale, P., Choussy, M., Santana, A., 1982, Alteration mineralogy of the Ahuachapán geothermal field, submitted to these transactions.

CEL, 1978, Consideraciones geoquímicas sobre el area geotérmica de Ahuachapán, CEL, Superintendencia de recursos geotérmicos, internal report, San Salvador.

CEL, 1981, Consideraciones geoquímicas sobre la prueba de máxima extracción campo de Ahuachapán, CEL, Subsección de investigaciones geotérmicas, internal report, San Salvador.

CEL, 1981, El campo géotermico de Ahuachapán durante los años 1979-1980, CEL, Superintendencia de recursos geotérmicos, internal report, San Salvador.

ELC, 1981, Need for additional drilling. Planta geotérmica de Ahuachapán, ELC, internal report, Milano.

Faust, C.R. and Mercer, J.W., 1977, Finite-difference model of two dimensional, sigle and two phase heat transport in a porous medium, U.S. Geol. Surv. Report 77-254.

Geothermal Resources Council, TRANSACTIONS Vol. 6, October 1982

TEMPERATURE AND PRESSURE DISTRIBUTION IN THE PALIMPINON GEOTHERMAL FIELD, SOUTHERN NEGROS, PHILIPPINES

J. R. McNITT*, S. K. SANYAL*, C. W. KLEIN*,
M. CHE*, B. S. TOLENTINO**, A. ALCARAZ**, and R. DATUIN***

*GeothermEx, Inc., California, U.S.A.
**Philippines National Oil Company, Philippines
***National Power Corporation, Philippines

ABSTRACT

This paper presents a case history of definition of the temperature and pressure distribution in a high temperature geothermal field in Southern Negros, the Philippines. This was achieved through a careful analysis of some 200 measured temperature and pressure profiles from 21 wells and combination of geological and drilling data. Based on the analysis the following flow model of the local natural convection system was derived: upflow near a young volcanic neck; lateral and upward migration to the highest point of the local block-faulted structure; partial downflow beneath the volcanic axis; upflow into the drilled depths in two levels along two different faults until convecting water reaches groundwater saturation zone and finally outflow until water reaches ground surface as springs.

INTRODUCTION

Two of the most important parameters determining the economic viability of a geothermal field are permeability and temperature. In a natural convection system these parameters are interrelated. In such a system, hot water will flow where permeable openings allow it to flow, consequently, highest temperatures are found in permeable rocks, or in rocks close to permeable channels in which hot water is circulating. A third parameter, pressure, is the result of temperature distribution in such a system. Water circulation is driven by the density differences caused by heating. The upflow area over the center of heating has higher flow potential than the outflow and downflow regimes located around the periphery of the system. This paper discusses the distribution of permeability, temperature and pressure in the Palimpinon geothermal field in Southern Negros, The Philippines, and how these distributions were deciphered from geological and drilling data, aerial and topographic maps, as well as some 200 measured temperature and pressure profiles from 21 wells.

Although the general direction of flow in a convecting hydrothermal system such as Palimpinon is upward above the center of heating and outward and downward around its periphery, in detail the location of actual flow paths is determined by the spatial distribution of permeability. The distribution of horizontal permeability at Palimpinon is determined by the stratigraphic sequence. The permeable stratigraphic units were located based on occurrence of lost circulation, interpretation of temperature and pressure logs and well test data. Vertical permeability in the Palimpinon Field is provided by a network of high-angle conjugate faults which strike NW and NE. These faults occur on a regional scale and their surface traces control the orientation of the reticulated drainage pattern throughout the region. Aerial and topographic maps as well as surface and downhole geology were used to define these faults. Being a conjugate set, these faults most probably have left- and right-lateral (shear) displacements related to an approximately N-S regional compressive stress field. The topographic expression of many of the faults, however, also shows significant vertical displacement: generally down to the SE on the NE-trending set and down to the NE on the NW-trending set.

The manner in which these faults in conjunction with the aquifers described earlier control water circulation, and therefore, temperature and pressure distribution, in the Palimpinon Field will be discussed in the following section.

TEMPERATURE DISTRIBUTION

The objectives of defining the initial temperature distribution that is, the distribution before it was disturbed by drilling and production, are the following: (1) The distribution helps to reveal the geometry of the convection cell and the location of permeable paths controlling flow through the cell, and therefore, provides important information for locating exploration and development wells; (2) it is necessary to know the temperature distribution in three dimensions to interpret well test data, calculate energy reserves and forecast reservoir performance.

The first step in defining the initial temperature distribution at Palimpinon was to process and plot the available temperature profiles for each well at a common scale with respect to elevation and temperature. The profiles measured during the first few days of heat-up after injection and the profiles with the longest time since last shut-in were used to define the initial temperature distribution.

divided for this study into 3 blocks (Figure 1) based on hydro-geology and temperature distribution. Wells in Blocks 1 and 2 encounter sedimentary rocks throughout most of the drilled depth, with interlayering of volcanic rocks such as andesite. The sedimentary and volcanic rocks encountered in these two blocks allow relatively rapid drilling compared to diorite, a hard intrusive igneous rock, encountered over half of the drilled depth in wells in Block 3. Permeable zones in Block 1 and 2 wells occur mainly in the volcanic layers or fracture zones associated with several steep faults that are present in the drilled area. Thus, there is a certain amount of regularity in the occurence of permeability in Blocks 1 and 2 because of its association with stratigraphic (volcanic layers) or structural (faults) features. On the other hand, in Block 3 (diorite) occurrence of permeability is entirely random, not being associated with any identifiable structural or stratigraphic feature. Also, contact metamorphism around the diorite intrusive has destroyed some of the permeability normally expected in the sedimentary section overlying diorite. Because of these reasons Block 3 wells are harder to drill and there is considerable chance of drilling dry holes. This point will be discussed further below.

APPLICATION OF LEARNING CURVE CONCEPT

In order to isolate the most sensitive parameters that affect drilling in this area, the data in Table 1 were analyzed for statistical correlation between certain parameters. The effective drilling rate (in meters per day) was chosen as the main criterion of drilling efficiency. Effective drilling rate was defined as the measured drilled depth divided by the interval between the spudding date and completion date of a well. Thus, this criterion includes the time of actual drilling, tripping, coring, logging, fishing, waiting for tools or services, completion, etc. The correlations between the effective drilling rate and various independent variables, such as drilled depth, rock types, year of drilling, etc. were investigated.

Figure 2 shows one aspect of this analysis. Here the effective drilling rates for wells have been plotted against the calendar time when they were drilled. It shows that the data points for wells N-1 and N-2, the two earliest wells, do not fall on the same trends as the others, presumably because these were shallow discovery holes and were not a part of the long range planning, which evolved later on. From N-3 to OK-5, six wells were drilled at progressively higher effective rates because of the classical learning curve situation. Thus, between November, 1977 and October, 1978, the effective drilling rate increased from 18.3 to 45.9 meters per day in a nearly linear fashion. All wells up to OK-5 were drilled in Blocks 1 and 2. The next well drilled, OK-6, showed a sharp drop in effective drilling rate because it was in Block 3 and encountered diorite. OK-7 again showed a higher effective rate because it was in Block 2. OK-8 showed a sharp drop in effective rate because of its proximity to Block 3 and due to its encountering diorite. OK 9D and 10D acheived high rates because of their location in Block 2. OK 11D drilled in Block 3 diorite showed a decline in rate again. This observation, that Block 1 and 2 wells show sharply higher effective drilling rates than those in Block 3, is evident for all wells except PN-13D and OK-12D, which had unusual fishing problems (several twist-offs).

On Figure 2, we have indicated a learning curve (Learning Curve 1) through the non-diorite (Block 1 and 2) wells and another (Learning Curve 2) through the diorite (Block 3) wells. Notice that the Learning Curve 1 has already reached a plateau (around 45 meters per day). Therefore, achievement of any significant improvement in drilling rate for the Block 1 and 2 wells is not possible unless different, higher efficiency equipment or substantially improved drilling techniques are applied. On the other hand, the Learning Curve 2 still shows signs of climbing. This indicates that the drilling rate in Block 3 will probably improve slowly with time until it reaches a plateau, perhaps a little above 30 meters per day. Therefore, drilling in Block 3 is never expected to be as efficient as in the other two blocks.

One interesting fact to note is that the drilling rates for vertical and deviated wells are very similar. Thus, while directional drilling requires drilling of more footage for the same vertical penetration, per meter of drilling it has about the same efficiency as drilling of vertical holes.

In order to check the effect of the length of drilling on rate, we plotted the effective drilling rate versus measured drilled depth (Figure 3). Block 3 wells form a cluster because these wells have similar depths and drilling rates. Block 2 wells show the apparently contradictory fact that the rate improved with increasing depth. This is not true; simply it reflects the effect of the Learning Curve 1 on drilling rate which masks the effect of depth. Block 1 wells and well N-2 (drilled east of Block 1 beyond the drilled area shown in Figure 1) show similar apparent contradiction for the same reason.

Based on our analysis it was concluded that drilling in Block 3 is going to remain at least 50% more expensive than in the other two blocks for the rest of the the development period of this field. This higher cost plus the lower success ratio of wells in Block 3 warrants and investigation into the possibility of deriving all further steam deliverability needed for the proposed plant from Blocks 1 and 2, or their extension.

the Okoy Sedimentary Formation. The high temperature waters flow from Cruernos Volcano to the NW, within the Okoy Sedimentary Formation and parallel to, but NE of, the Sogongon Fault, to feed the high temperature nose located between the Sogongon Fault and Lagunao Fault. The Southern Negros Formation appears to function as a cap, except where it is penetrated by the Ticala Fault which provides a vertical conduit for fluid moving from the deep, high temperature nose in the Okoy Sedimentary Formation to feed the tongue of shallow horizontal discharge flowing to the NE. The flow paths, as deduced from geologic structure and temperature distribution, are illustrated in the schematic block diagram of figure 3.

The following section will examine formation pressure in the Palimpinon Field to determine if this proposed flow pattern is also reflected in the pattern of pressure distribution.

PRESSURE DISTRIBUTION

Without the benefit of drillstem tests of each permeable zone in all wells, the determination of formation pressures in geothermal fields, like the determination of formation temperatures, is a matter of approximation. The typical long open-hole completions of geothermal wells, combined with the common occurrence of multi-aquifer systems at different flow potentials, allow interflow between aquifers. Consequently, the pressure profiles measured in wells reflect the average pressure of the open interval rather than the pressure of any one aquifer. If, however, variation of the vertical component of potential is significantly less than variation of the horizontal component existing over the area of investigation, then it is still possible to obtain an approximation of the piezometric surface of the reservoir from this average pressure data.

The variability of the pressure gradients measured in any one well is a function of the production/injection history of the well and the transient well temperature profile existing at the time the survey was made. Profiles measured shortly after cold water injection give relatively high gradients, while conversely, gradients measured shortly after production give relatively low gradients. If a well has been flow tested or injected into, considerable shut-in time may be required to restore the equilibrium pressure profile in the well. Consequently, the data chosen for compiling the piezometric map (figure 4) were taken from those profiles which were measured after the longest period of shut-in condition. The plotting datum for figure 4 was chosen at -1,000 msl.

The piezometric surface mapped on figure 4 shows a pressure high of over 12 MPa developed around well OK-5 with potential falling to the E, N and W indicating a component of horizontal flow toward those directions. This result is in good agreement with the conclusion derived from the field temperature distribution; that is, the deep high-temperature nose (located below OK-5) is receiving thermal fluids directly from the center of upflow located to the SE beneath Cuernos volcano, and the horizontal component of potential, therefore, drops from the nose outward to the E, N and W.

For the reason explained above, variations in vertical potential are difficult to assess from pressure profiles in completed wells. To overcome this problem partially, two methods have been used to qualitatively assess the direction in which the vertical component of potential is changing, that is, to determine if potential is increasing or decreasing with depth. Both methods involve interpretation of transient temperature profiles.

The first method is based on the permissive argument that temperature inversions should not occur in those parts of the convection cell which have a vertical component of upflow, but only in those areas which are dominated by outflow and/or downflow.

The second method originates from the observation that temperature profiles measured during well heat-up opposite zones that are relatively impermeable conform to segments of boiling-point curves having their points of origin well below known static water levels. This condition has been observed in many of the Palimpinon wells and is believed to be due to underpressuring of two-phase zones because of steam condensation during the cooling process. As the formation reheats, pore pressure increases by the formation of steam in the pore spaces, rather than by the outflow of wellbore water into the formation, because the relative impermeability of the rock retards water outflow. Whether the point of origin for the boiling point curve segments moves up or down relative to the origin of curves developed in higher and lower impermeable zones appears to be a sensitive indicator of whether formation potentials are becoming higher or lower with depth.

These two analytical methods have been applied to the temperature profiles of the Palimpinon wells and the results included in figure 4. Significant temperature reversals occur in wells OK-8, SG-2, PN-13, PN-15, OK-12, N-3, OK-1 and N-1, indicating that the parts of the reservoir intersected by these wells do not have an upward component of flow. Based on the second method of analysis, wells SG-2, OK-6, OK-7, PN-15, PN-13 and PN-17 all have indications of a component of downflow, while wells OK-11, OK-5, OK-4 and OK-9 all have indications of a component of upflow. The line labelled "upflow area" encloses the four wells show indications of upflow based on the boiling-point curve interpretation, and which do not have temperature inversions. As would be expected, this area coincides with the area of the field which also has the following characteristics: (a) the highest component of horizontal potential; (b) the highest temperatures at depth; and (c) the highest structural position of the Okoy Sedimentary Formation.

The correlation of these four features just discussed supports the flow model diagrammed in figure 3, that is, upflow near the

Cuernos volcanic neck; lateral and upward migration to the highest point of the local structure caused by block faulting; partial downflow in the eastern part of the area beneath the volcanic axis; upflow into the drilled depth along Lagunao Fault, or a fault close to it; partial upflow along the Ticala Fault to the top of the saturation zone; and finally, northeastward outflow until the water reaches the surface as springs.

ACKNOWLEDGEMENT

The authors are indebted to the staff of the Philippines National Oil Company for collecting the data used in this study.

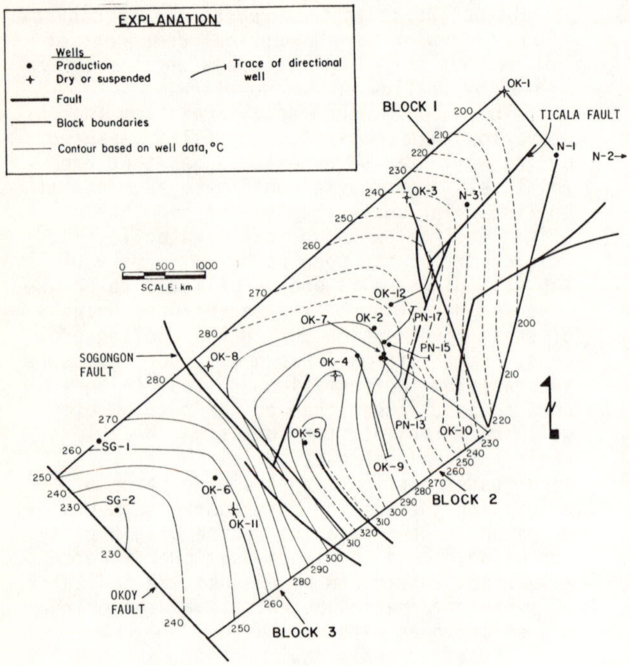

FIGURE 1. Temperature Contours at -1400 msl.

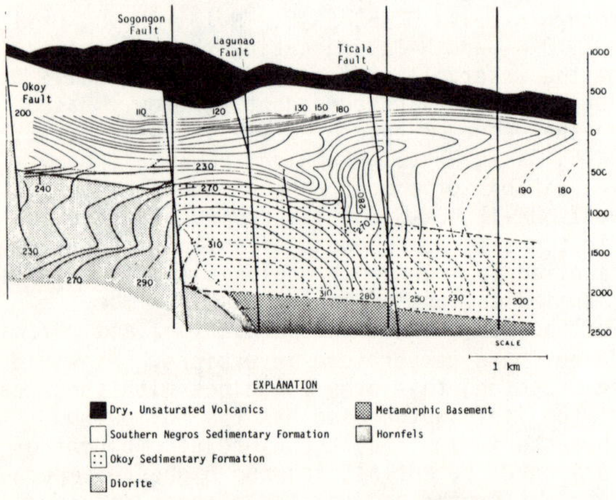

FIGURE 2. Southwest-Northeast Geologic and Temperature Cross-section.

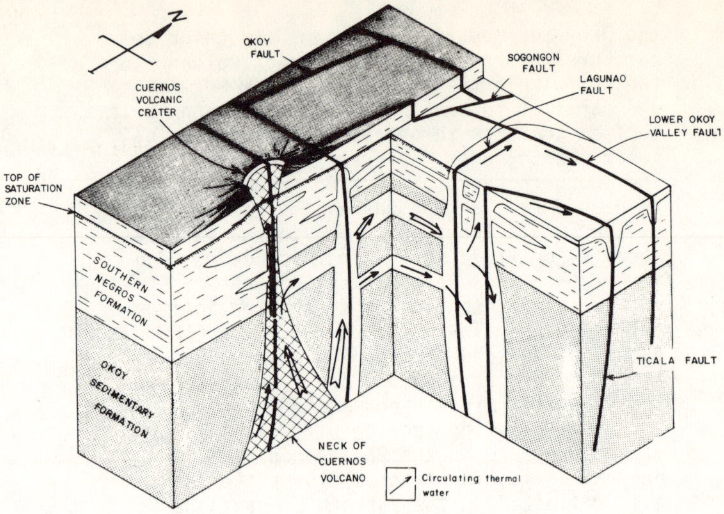

FIGURE 3. Schematic Block Diagram of Palimpinon Geothermal Field.

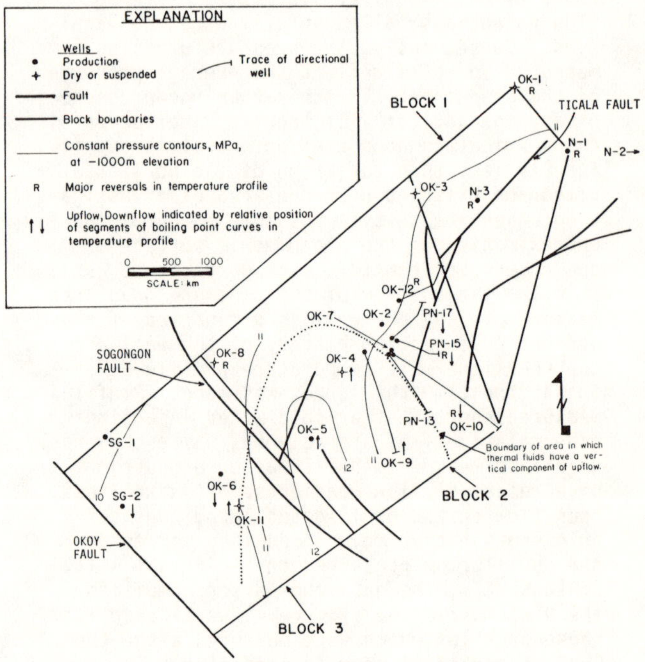

FIGURE 4. Piezometric Surface Map.

Geothermal Resources Council, TRANSACTIONS Vol. 6, October 1982

CHEMICAL EXPERIMENTS WITH FRESH, HOT, PARTLY-FLASHED HYPERSALINE BRINE

Donald E. Michels

Republic Geothermal, Inc.
11823 East Slauson Avenue
Santa Fe Springs, California 90670

ABSTRACT

Chemical behavior of single-phase liquid obtained at a wellhead has been quantified by novel test procedures that yield novel results. Chemical reagents were injected into the brine stream and the mixture passed through a series of chambers that stabilize at different temperatures between 400 and 200°F. Results show the brine's overall behavior is due to multiple reactions that respond unequally to added chemicals. Significant changes in the texture of deposited solids are elicited by plain water as well as by other chemicals. Methods to control or moderate scale deposition and formation of sludges can take advantage of these behaviors.

INTRODUCTION

Hypersaline geothermal brines from the Salton Sea KGRA have acquired a legendary reputation for depositing obnoxious hard, tenacious, multicomponent scales in wellbores and surface equipment. Considerable experimental work has been reported on them in regards to the Geothermal Loop Experimental Facility (GLEF), for example, Harrar, (1981), and Featherstone, et al, (1979).

Harrar describes experiments with soluble additives intended to prevent or retard the polymerization and deposition of silica. Relative effectiveness was measured by chemical analysis for silica remaining (stablilized) in the brine after a particular waiting protocol.

Featherstone, et al, describe a method to make silica deposition, as a solid, more efficient without using additives. They recycled part of the suspended solids into still-reactive brine so that new deposition was encouraged to occur on the suspendate, rather then on the surfaces of equipment.

This report shows that simple chemical additives can accelerate the deposition of silica and its cohorts and stimulate formation of soft sludgy scales that are easily dislodged from pipes and other equipment.

Republic Geothermal, Inc. has drilled two wells near Niland that tap the hypersaline resource. During a 31-day test-flowing of one of those wells (Britz No. 3) in March-April 1981, chemical experiments were made that provide the subject matter for this report.

It is important here to note that between the production zone at the bottom of the well and the wellhead the fluid temperature cools by 70 to 100 celsius degrees in 10 to 25 minutes, depending on the flow rate of the well. This kind of shock to a chemical system initially at equilibrium, has no counterpart in classical chemistry.

EXPERIMENTAL

Single-phase liquid was obtained as a sidestream from a tap at the wellhead (Figure 1). Depending on the flow status of the well, the temperature at the tap point could exceed

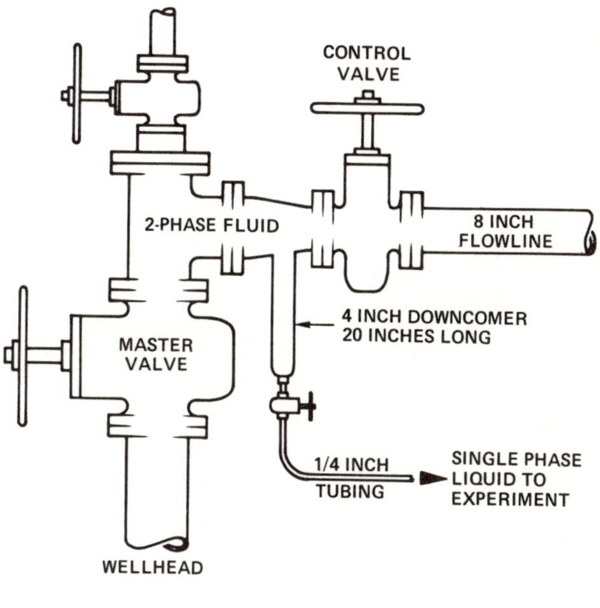

Figure 1. Tap to obtain single phase liquid from a 2—phase flowline.

400°F. From there, the liquid, driven by the wellhead pressure, was directed through a series of experimental chambers, (Figure 2). The chambers were connected with stainless steel tubing

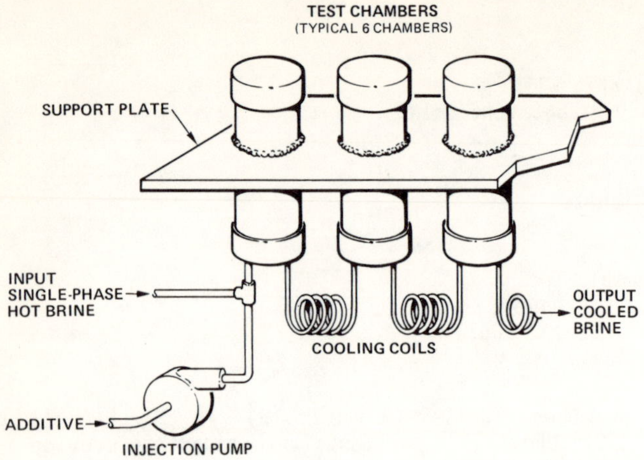

Figure 2. Schematic diagram of test chambers in a temperature cascade.

and the brine was cooled as it flowed between chambers, eventually discharging to the atmosphere, too cool to flash. The series of chambers establishes a cascade of experimental temperatures that involve the same brine in a continuous flow-through mode with a short residence time in each chamber. In addition, the input of the first chamber was connected to an injection pump so that the fresh, hot brine could be dosed with aqueous solutions.

Results of the brine's chemical behavior, with or without chemical additives, were quantified in several ways. Pre-weighed coupons placed in the chambers acquired scale deposits. These coupons were sized so as to be directly accomodated by X-ray diffractometers or scanning electron microscopes. This permits nondestructive tests to give data on amount and rate of deposition, mineralogy, gross chemical composition, localized (individual grain) composition, morphology, and texture for scales formed at a series of temperatures.

The (treated, reacted, cooled) brine which exited the apparatus could be sampled, filtered, and preserved for laboratory analysis and its pH and density measured in the field. Filterable solids are especially interesting because they can be subsequently weighed and analyzed to determine their yield and composition. A series of injection dosages yields a set of filtered solids whose compositional differences trace the brine's responses to the additive.

Dealing with fresh brine in this way permits one to view its chemical behavior from several perspectives. More importantly, the brine can be studied while still in a severely upset condition due to its recent plunge in temperature. This approach is highly relevant to solving problems of scale deposition because it partly imitates the circumstances for geothermal brine produced in an industrial setting.

RESULTS

One example of results yielded by this method is shown in Figure 3. These data concern the solids filtered from samples of brine that exited the series of test chambers. The set of samples represented additions of NaOH at seven different concentration levels.

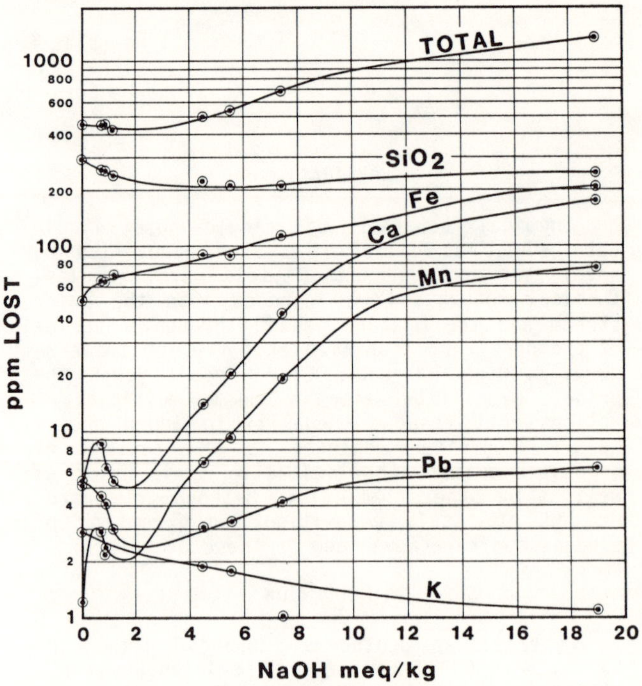

Figure 3. Effect of NaOH on losses from brine.

To focus better on how the brine responded to treatment, the composition of the solids has been recast into terms of ppm of components lost by the brine when the solids formed. Increasing dosages of NaOH caused increases in the total amount lost by the brine, but losses of some components were diminished with increased dosage whereas others exhibited maxima or minima. Few components behaved in covarient fashion although Ca and Mn show a striking parallelism. The mineral kutnahorite, a manganese-bearing analog of dolomite, has been identified in related experiments with Britz No. 3 brine.

At the lowest dosages of NaOH the final pH values pass through a minimum value condition (Figure 4). This reflects how non-intuitive the behavior of fresh brine can be. This curious response seems not to be an experimental artifact since a similar response was obtained in a separate experiment which used $NaHCO_3$ instead of NaOH.

The counterpart effects of injectates on solids deposited inside the chambers involve compositions and textures of the solids. For example, untreated brine deposits a hard tenacious black scale analogous to the scale which

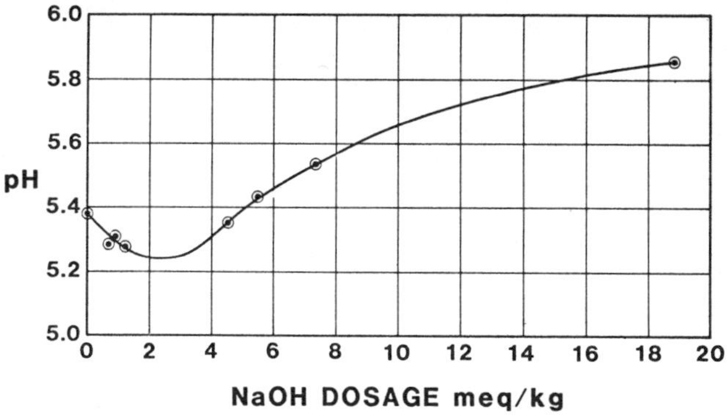

Figure 4. Final pH vs NaOH dosage.

forms in the piping and other equipment which carries the main flow from the geothermal well. However, treated brine decorates the test chamber surfaces and coupons with a pale gray, gelatinous deposit that has low strength and adheres only weakly.

On the relatively quiescent surfaces of the chamber the thickness built up 2 to 3 millimeters in about 2 hours, reflecting intrinsically high rates for polymerization reactions. However, the 1/4-inch tubings which connected the chambers were not clogged. The shear strength of the deposit is too small to withstand much hydrodynamic stress. Thus, one approach for controlling geothermal scale involves redirecting the scale-forming reactions to yield a low-strength sludge that flushes off pipe and valve surfaces. By itself, this approach may give less than total control of scale deposition, but it could be applied to improve the effectiveness of the process described by Featherstone, et al, for example. Additionally, it could be implemented upstream of steam separators, even in wellbores.

Blending of fluids in the wellbore is not limited to artificial injections. For example, the production zone of Britz No. 3 intersects two reservoirs with contrasting compositions. The lower is a typical hypersaline brine with total dissolved solids (TDS) content near 250,000 ppm and an acidic pH. The upper reservoir contains only about 55,000 ppm TDS and is inferred to be pH neutral or basic. They are produced in the proportions near 1:2, yielding a mix of 116,000 ppm TDS.

No deleterious consequences of this blending have been noted, but distinct advantages accrue compared to the alternatives of producing either fluid alone. For example, the dilute brine would, by itself yield a carbonate scale at the flash point; the more concentrated brine would, after flashing, be supersaturated in NaCl. Drilling and completion and workover procedures can be adapted to the choice of blending or not blending such brines, thereby constituting a method of scale control that is essentially chemical in character.

The rates of silica polymerization reactions appear greatly increased by the addition of even plain water, since similar textures of "scale" were obtained by injecting water with and without NaOH. The iron and silica contents in the solids are affected differently by H_2O compared to the effects of NaOH solutions. Figure 5 shows the SiO_2/Fe ratios plotted vs temperature for three kinds of solids: (1) hard, tenacious scales formed in the main flow line from unaltered brine; (2) gelatinous scale formed in test chambers when only H_2O was injected; and (3) gelatinous scale formed in test chambers when water with NaOH was injected.

The NaOH injection favors the deposition of silica, suggesting that the silica's polymerization reaction is kinetically more able to utilize the OH^- than can the mechanism for iron hydroxide formation. Alternatively, one can note that parallelism of the trend lines in Figure 5 suggests there is a reaction involving iron and silica which is perhaps unchanged. The higher overall silica/iron values for treated fluids would represent a separate kind of silica deposition which is enhanced and superimposed on the "ordinary" silica-iron reaction.

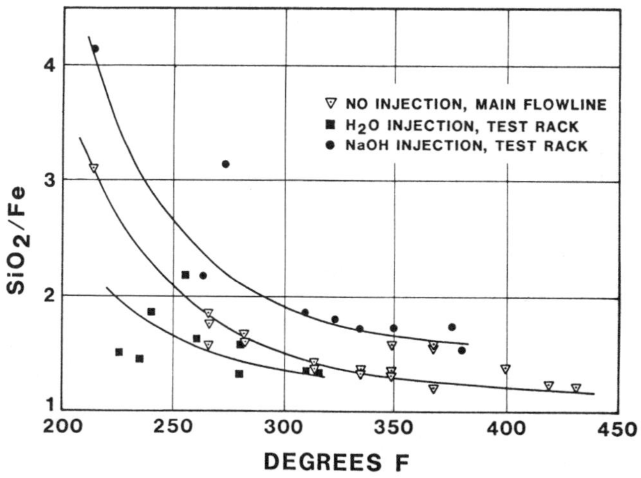

Figure 5. Silica/iron weight ratio in scales.

Reasons for the lowermost position of the trend line for plain water addition are speculative at this time. Perhaps the effect of plain water may be due to its having structural differences compared to the brine. If so, then the effect may be temporary, lasting only so long as required to establish a new microhomogeniety in the new mixture. Furthermore, until such homogeniety becomes established, an interface of sorts will exist between the end-members, brine and injectate. The gradients of structure and composition there contribute additional scope for transient and perhaps exotic reactivity.

CONCLUSIONS

This approach to investigating brine behavior allows the experimentor to disect the complicated mix of chemical reactions that occur, somewhat independently in the same space. One result shows that plain water is not merely a diluent. It must be considered also as a chemical reagent when added to brine since it changes the texture and composition of scales which deposit.

Additives injected with water also have their own effects. Those effects are in addition to the effects of water alone, a feature which complicates experimental work.

Additional insight is obtained by viewing the brine's response from the point of view of the brine, instead of the additive. The severe upset of temperature, that is concommitant with flashing flow, irrespective of additive injections, causes a host of chemical events to become energetically possible. A single additive triggers these potential reactions in different senses and degrees depending on its dosage. Some reactions will be inhibited. Single components of the brine that participate in multiple reactions have their distributions altered by the additive.

Existing chemical theory is inadequate to predict the consequences of the severe temperature upset the brine has experienced, especially in conjunction with the short residence times the brine has along its flow path. Therefore, an observational approach to these phenomena is required.

REFERENCES

Harrar, J. E., 1981, Final report on studies of brine chemistry and scaling at the Salton Sea geothermal field, 1977-1979: Lawrence Livermore Laboratory, Report UCID-18917, 42 pp.

Featherstone, J. L., Powell, D. R., and VanNote, R. H., 1979, Stabilization of highly saline geothermal brines: Soc. Pet. Engr. paper no. 8269, 54th Ann. Tech. Conf., Las Vegas, Sept. 23-26.

Geothermal Resources Council, TRANSACTIONS Vol. 6, October 1982

BACA GEOTHERMAL DEMONSTRATION PROJECT — RESERVOIR DEFINITION REVIEW

Martin W. Molloy

Geothermal Energy Division
U.S.Department of Energy
1333 Broadway
Oakland, CA 94612

ABSTRACT

In January 1982, the participants in the 50 MW Geothermal Demonstration Power Plant at Baca NM, signed an agreement of nonviability terminating the Baca project. This joint action by Union Geothermal Company of New Mexico (Union), Public Service Company of New Mexico (PNM), and the U.S. Department of Energy (DOE), was the result of technical difficulties encountered in drilling the steam supply wells. Since May 1981, a review team formed by DOE has conducted an independent assessment of reservoir definition activities. The conclusions of that team are summarized.

INTRODUCTION

The Baca Geothermal Demonstration Power Plant Project was the result of three preceeding activities. In the early 1960's, exploratory drilling recognized the geothermal potential of the surrounding Valles Caldera in northwestern New Mexico. In 1970, Union drilled the discovery well in Redondo Creek, and followed that success with 10 more wells by 1976. In 1977, DOE solicited cooperative participation from industry in a geothermal demonstration project. Union and PNM won this competitive solicitation.

When the DOE Demonstration Project started in 1978, about one-third of the power plant requirements were considered proven (320,000 lbs/hr of steam at line pressure). Review of exploration and well test data by DOE and its geothermal resource advisors, the US Geological Survey and Lawrence Berkeley Laboratory, concluded that there was very little risk that the reservoir could not support the plant requirements. Potential problems were foreseen with calcite scaling of production wellbores, and possible silica scaling of reinjection wells. Overall, the risk of resource failure was judged to be low.

DEVELOPMENTAL DRILLING RESULTS

Prior to the DOE Cooperative Agreement, five of the 11 Union wells in Redondo Creek flowed at commercial rates and pressures. The casing of one of these subsequently collapsed. Union estimated that 10 more successful wells were needed to supply the additional 600,000 lbs/hr of steam required for the proposed plant. Subsequent drilling yielded only two successful wells out of 13 attempts, including redrills. Proven steam production was increased to only 353,000 lbs/hr.

The problems encountered can be grouped in two categories: well targeting and mechanical difficulties. Targeting problems relate to the inability to unravel the key factors controlling the distribution of permeability in the volcanic rocks which form this fractured geothermal reservoir. A conceptual geologic model that reliably predicts drilling targets at depth for step-out and in-fill wells has yet to be developed. Zones of fracturing along the major faults, which form the down-dropped block of Redondo Creek, appear to be a necessary, but not sufficient condition, for successful wells. Additional factors which may hold the key to well targeting at Baca include cross faulting, fractures (including columnar jointing and fissure zones or linears), sandy layers within the sequence of volcanic rocks, and differences in cementation or hydrothermal alteration along the faults and fractures.

Mechanical problems are routinely encountered in drilling through hard, fractured volcanic rocks in an underpressured reservoir. While mud can be used as the drilling fluid until the production zone is reached, the weight of the fluid column forces mud into steam production fractures and permeable zones, where it impairs productivity. Standard procedure is to change over to drilling with air, air/water or foam, in order not to override the formation pressure in the production zone. These lower density drilling fluids add to the difficulty of drilling through hard, fractured rock.

Union experienced serious drilling and completion problems in 23 of 31 original holes, redrills and deepenings, or a total of 74% of the drilling operations. Nine wells (29%) were lost; only five wells (16%) are considered to be commercial producers. Failure of fishing operations, which generally resulted from twisted-off or stuck drill pipe, were the major cause of lost or damaged wells at Redondo Creek. Eight wells, or 26 percent of the total, were lost from this cause alone. Worn and collapsed casing affected two wells, resulting in the loss of another well. Other difficulties included extensive fishing (23%), bad sloughing (19%), extensive steam and water entries (10%), and wellbore scaling (10%). This high rate of mechanical failures resulted in significant loss of information on geologic structure and stratigraphy, which could have contributed to the conceptual model and to well targeting. In addition, drilling difficulties increased costs and lost steam production capacity.

As part of DOE's Geothermal Well Stimulation Program, hydraulic fracture stimulation was attempted on two wells, with the cooperation and support of Union. Production tests subsequently indicated that high conductivity fractures were created or reopened near the wellbores. However, the resulting steam flows were inadequate for commercial production. The conclusion drawn is that the hydrofrac procedures were successfully executed, but that major fluid producing features in the reservoir were not intersected by the created fractures (Morris, Hanold and Pearson, 1982).

A significant factor contributing to the economic failure of the project was the low permeability of the production zone, which was compounded by underpressure. Although bottom hole temperatures are substantial (646 °F max.), average transmissivity (6,000 md-ft) is considerably less than most commercial fields. This has contributed to substantial differences in opinion on the estimated producibility of the project area, in terms of actual generating capacity versus long-term capacity of the entire reservoir. The use of different reservoir models by Union (lumped parameter) and LBL (distributed parameter), with different assumptions and interpretations of reservoir data (e.g., internal discharge, Grant and Garg, 1981) have contributed to the variation in estimates.

When the seriousness of well targeting and drilling difficulties was recognized in May 1981, power plant activities were suspended. Deep exploratory drilling was undertaken in the search for steam production in the underlying andesite, limestone and basement granite. The experiments in hydraulic fracture stimulation were undertaken in parallel. These attempts were unsuccessful, and drilling operations were brought to a close in December 1981. Well tests were completed during Summer 1982 for evaluation of permeability directions and reservoir boundaries.

CONCLUSIONS

A review team, formed by DOE, conducted an independent analysis of Union's reservoir results, conclusions and plans (Goldstein, Molloy and Holman, 1982). Scarcity and unpredictability of production zones, low permeability and difficult drilling were identified as the reasons for the unsuccessful outcome. Complexity and limitations of the major production zones were not recognized until Union began infill drilling in 1979. While Union recognized that productivity was impaired in some of the earlier wells, it took extensive drilling throughout the project area to define the areal extent of the productivity impairment factors (Mohinder Gulati and Richard Dondanville, private communication). Failure of many infill wells caused Union to develop a structural geologic model for well targeting which emphasized fracture permeability along the major graben faults. Other potential avenues of fluid flow, such as cross-faults, columnar jointed units, and sandy volcanic units were not defined or integrated into the targeting of subsequent wells.

Reexamination of lithologic logs prepared during well drilling indicates that tuffaceous sandstones and sand-sized tuff beds within the Bandelier can be correlated between neighboring wells (Hulen, 1982). These permeable units appear to be important stratigraphic controls on fluid flow at Baca, coinciding with 12 of 23 major steam or hot water entries identified by Union.

Reconnaissance geophysics, including gravity and electromagnetic surveys, were conducted early in the exploration phase to determine the regional structure and basement depth, and to locate low resistivity anomalies that may be associated with geothermal reservoirs. These initial surveys were not followed up with more detailed investigations that might have delineated the extent of the reservoir. Magnetotelluric soundings reveal a deep, low-resistivity zone that correlates well with the production zone at Redondo Creek. Telluric profile data, which are useful in locating lateral changes in resistivity, define a 10 sq. km. area of anomalously low resistivity, at a depth of one to 10 km. All the successful deep wells at Baca lie within this zone (Wilt, Goldstein and Vonder Haar, 1982). Additional magnetotelluric and telluric stations would be necessary if the low resistivity zone is to be mapped in detail. Self potential surveying is also suggested as a method for locating fluid-filled fractures or fault zones. This information might define field limits and drilling targets.

Geochemical conclusions from the Baca wells were very limited. Further sampling and analysis can yield valuable information regarding conditions within the reservoir, as has been accomplished at many geothermal fields.

Careful analysis of Baca well test data shows that a two-phase zone was initially present in part of the reservoir, and that the extent of this zone is very sensitive to the carbon dioxide content of the reservoir (Garg and Riney, 1982). The initial conclusion that the DOE Project Area could support a 55 MW power plant was believed to be conservative. Subsequent studies by LBL using a distributed parameter model conclude that the lumped parameter model had grossly overestimated the generating capacity of this low permeability geo- thermal reservoir.

The extensive data resulting from Union's development of Redondo Creek has significant value to the industry for understanding Baca and similar geothermal reservoirs. All the well logs are being made available through Rocky Mountain Well Log. Exploration data and reports will be open filed at the University of Utah Research Institute, Earth Science Laboratory, 420 Chipeta Way, Salt Lake City, UT, 84108. Reservoir engineering information is available on a similar basis at Lawrence Berkeley Laboratory. Similar arrangements are being developed with Lawrence Livermore National Laboratory and the US Geological Survey.

ACKNOWLEDGMENTS

This summary represents a synthesis of the ideas and conclusions of the contributors to the Final Report of the DOE Baca Reservoir Definition Review Team: Dennis Nielson and Jeffrey Hulen, University of Utah Research Institute; Norman Goldstein, Michael Wilt, Joan Delany, Chin-Fu Tsang and Gudmundur Bodvarsson, Lawrence Berkeley Laboratory; Alfred Truesdell, U.S. Geological Survey; A. William Laughlin, Robert J. Hanold and C. F. Pearson, Los Alamos National Laboratory; Charles W. Morris, Republic Geothermal Inc.; Sabodh Garg and T. David Riney, Systems, Science and Software; and, William R. Holman and the author; DOE San Francisco Operations Office.

REFERENCES

Grant, M. A. and Garg, S. K., 1981, Baca Geothermal Demonstration Project: Interpretation of Downhole Data from the Baca Geothermal Field: Systems, Science and Software, La Jolla, California, DOE/ET/27163-12, 112 p.

Goldstein, N. E., Molloy, M. W., and Holman, W. R., 1982, Final Report of the Department of Energy Reservoir Definition Review Team for the Baca Geothermal Demonstration Project, Lawrence Berkeley Laboratory, (in press)

Hulen, J. B., 1982, Stratigraphy, Structure and Permeability in the Redondo Creek Project Area, in Goldstein et al., 1982

Morris, C. W., Hanold, R. J., and Pearson, C. F., 1982, Fracture Stimulation Experiments, in Goldstein et al., 1982

Wilt, M., Goldstein, N. E., and Vonder Haar, S., 1982, Geophysics, in Goldstein et al., 1982.

Geothermal Resources Council, TRANSACTIONS Vol. 6, October 1982

THERMAL DECOMPOSITION OF HYDROCARBONS IN THE CERRO PRIETO, MEXICO, GEOTHERMAL RESERVOIR

Nancy L. Nehring[1], David J. Des Marais[2], and Alfred H. Truesdell[1]

[1]U.S. Geological Survey, Menlo Park, CA 94025
[2]NASA--Ames Research Center, Moffett Field, CA 94035

ABSTRACT

The distribution of low-molecular-weight hydrocarbons (C1-7) in geothermal systems depends on hydrocarbon source and temperature/time regime. In the Cerro Prieto, Mexico, geothermal reservoir, thermal decomposition of coal to form methane is extensive at temperatures near 340° C. In high-temperature wells, only methane, ethane, propane, and benzene are present. At temperatures near 290° C, thermal decomposition is extensive, as evidenced by such intermediate products as n-butane, n-pentane, and C4 and C5 branched hydrocarbons. The wide variety of hydrocarbons and low concentrations of methane relative to other hydrocarbons in surface samples indicate only moderate thermal decomposition of the source carbon and intermediate hydrocarbons.

INTRODUCTION

Hydrocarbons in geothermal fluids were reported by Gunter (1978), who analyzed C1-4 hydrocarbons in steam from Yellowstone National Park. Nehring and Truesdell (1979) reported C1-6 hydrocarbons from various volcanic and geothermal localities, and noted similarities in the hydrocarbon distribution among systems in similar geologic environments. Isotopic studies of ^{13}C in the individual hydrocarbons from nonmagmatic geothermal systems have been undertaken by Des Marais et al. (1981) to determine the origin of these hydrocarbons. They have concluded that sedimentary or ground-water organic matter is the principal source for all the hydrocarbons. In particular, coal from a deep cold well (Prian 1) near the Cerro Prieto, Mexico, geothermal field was shown to be the source material for hydrocarbons occurring in the well fluids. A more extensive investigation of the hydrocarbons present in Cerro Prieto wells and surface features has revealed systematic differences in these hydrocarbons, as discussed below.

METHOD

Samples were collected in evacuated bottles containing NaOH, according to a procedure similar to that described by Giggenbach (1975). After analysis of residual gases, a 1-cm^3 aliquot of the headspace was taken in a gas-tight syringe and analyzed on a Hewlett-Packard 5700A gas chromatograph* equipped with a flame-ionization detector. Dual columns, 6 ft. by 1/8 in., were packed with 60- to 80-mesh Porapak Q, and helium carrier gas was used at 60-psi inlet pressure. Each analysis was temperature programmed as follows: -30° C for 2 min, 8°/min rise to 150° C, and 150° C for 32 min. Table 1 lists the peaks identified.

DISCUSSION

The distribution of hydrocarbon gases from wells and springs at Cerro Prieto, Mexico, falls into three groups that are related to the temperature/time regime: (1) wells about 3,000 m deep, with measured temperatures near 340° C (for example, well E-2); (2) wells about 1,000 m deep, with measured temperatures near 290° C (for example, well M-31); and (3) surface features with temperatures of 20° to 100° C (for example, N-31, a 100° C fumarole). Figures 1 through 3 plot the hydrocarbon distributions in these three samples; Table 1 identifies the hydrocarbons present by peak number and lists their semiquantitative concentrations. Although compositions intermediate to those presented here also were obtained, those listed are representative.

Samples near 340° C showed very few hydrocarbons, and the abundances of those present relative to methane were low (Figs. 1, 4). Two explanations are possible: (1) Coal is decomposing directly into small straight-chain hydrocarbons (C1-3) and benzene, or 2) intermediate compounds form, which are highly unstable near 340° C and decompose almost immediately to C1-3 hydrocarbons. The increased

*Any use of trade names and trademarks in this report is for descriptive purposes only and does not constitute endorsement by the U.S. Geological Survey.

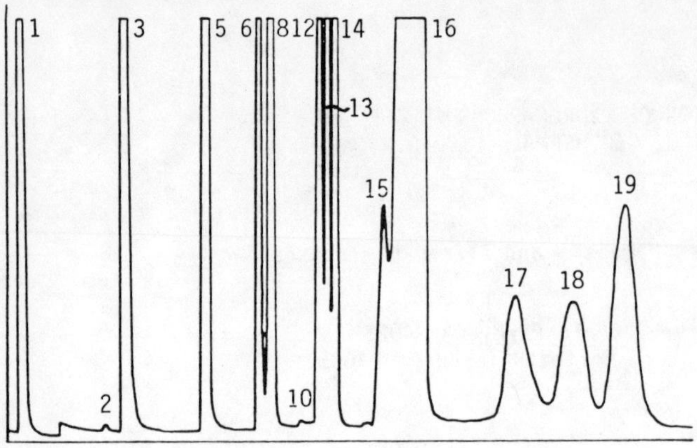

Figure 1. Chromatogram of C1-7 hydrocarbons from 2,000-m well (E-2) at 340° C.

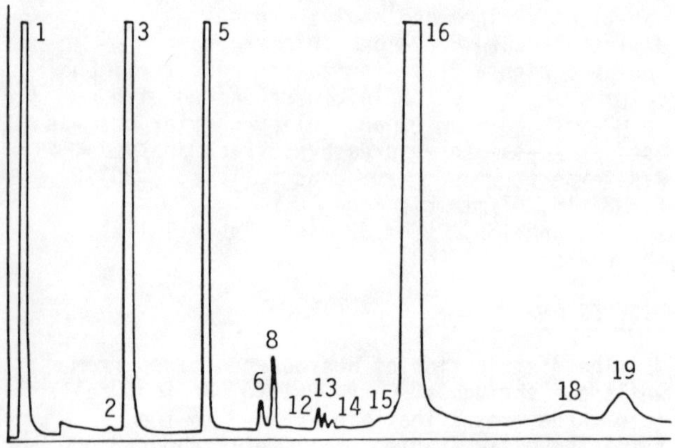

Figure 2. Chromatogram of cl-7 hydrocarbons from 2,000-m well (M-31) at 290° C.

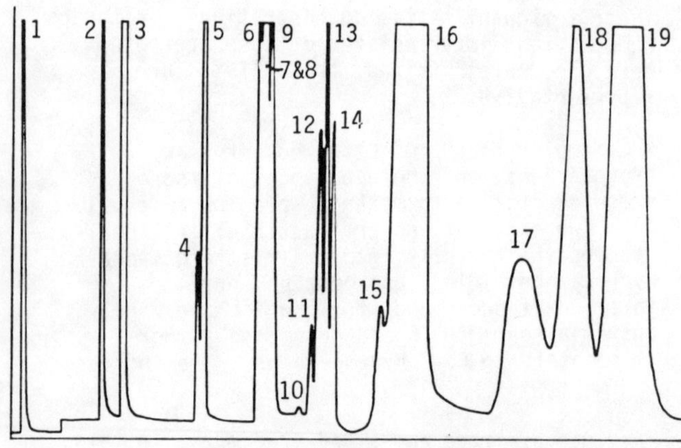

Figure 3. Chromatogram of C1-7 hydrocarbons from surface feature (fumarole N-31) at 100°C.

concentrations of higher molecular weight (C4-7) straight-chain and branched hydrocarbons at 290° C (Figs 2, 4) favor the second explanation. Many of these compounds would be considered intermediate between coal and methane. Lower temperatures would allow higher concentrations of these compounds to accumulate before decomposition occurred.

Slow decomposition of intermediate hydrocarbons is even more evident in the surface samples (Figs 3, 4). Methane concentration in these samples is substantially lower relative to other hydrocarbon concentrations than in samples from higher temperature regimes. Hexanes (C6) and heptanes (C7) are second only to methane in concentration. At 100° C or less, intermediates appear to break down very slowly, if at all.

CONCLUSION

Hydrocarbon distribution in geothermal systems depends on the temperature/time regime. High temperatures (about 350° C) favor low-molecular-weight straight-chain compounds (C1-3) plus benzene, whereas lower temperatures (about 100° C) favor higher molecular weight compounds (C6-7) probably owing to slower decomposition of compounds intermediate between the source organic carbon and methane at lower temperatures.

Table 1. Peak identification and semiquantitative concentrations (in parts per million) of hydrocarbons in representative wells.

Peak	Hydrocarbon	Well E-2	Well M-31	Well N-31
1	Methane	25,000	50,000	2,000
2	Ethene	0.05	2	9
3	Ethane	3,500	15,000	45
4	Propene	0	0	3
5	Propane	250	4,000	15
6	2-methylpropane	6	550	10
7	1-butene	0	0	15
8	n-butane	10	1,000	75
9	2-butene	0	0	7
10	2,2-dimethylpropane	0	2	0.01
11	1-pentene	0	0	1
12	2-methylbutane	2	250	4
13	n-pentane	2	300	7
14	Cyclopentane and 2-pentene	1	200	4
15	n-hexane	1	1,000	30
16	Benzene	2,000	6,000	900
17	Unidentified C7	0	300	25
18	Unidentified C7 (n-heptane?)	2	300	50
19	Toluene	8	1,000	250

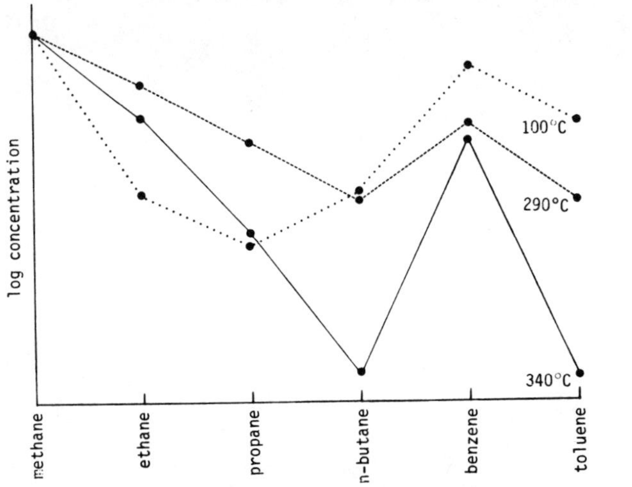

Figure 4. Concentrations of higher molecular weight hydrocarbons relative to methane.

REFERENCES

Des Marais, D. J., Donchin, J. H., Nehring, N. L., and Truesdell, A. H., 1981, Molecular carbon isotopic evidence for the origin of geothermal hydrocarbons, Nature, v. 292, p. 826-828.

Giggenbach, W. F., 1975, A simple method for the collection and analysis of volcanic gas samples, Bull. Volcanol., v. 39, (special issue), p. 132-145.

Gunter, B. D., 1978, C1-C4 hydrocarbons in hydrothermal gases, Geochim. Cosmochim. Acta, v. 42, p. 137-139.

Nehring, N. L., and Truesdell, A. H., 1979, Hydrocarbon gases in some volcanic and geothermal systems, Geotherm. Resour. Counc. Trans., v. 2, p. 483-486.

Geothermal Resources Council, TRANSACTIONS Vol. 6, October 1982

ANALYSIS OF PRODUCTION TEST DATA FROM SELECTED BACA WELLS

T. D. Riney and S. K. Garg

S-CUBED
P. O. Box 1620
La Jolla, California 92038

ABSTRACT

Results are presented of analyses of the downhole pressure buildup data for wells located in the Redondo Creek area of the Baca Geothermal Field. The downhole drilling information and pressure/temperature surveys are first interpreted to locate zones at which fluid enters the wellbore from the fractured formation and to estimate the initial reservoir temperature and pressure in these zones. Interpretation of the buildup data for each well considers wellbore effects, the CO_2 content of the fluid and differentiates between the single-phase and two-phase portions of the data. Estimates for the effective formation kh are made for selected wells.

INTRODUCTION

Analysis of the downhole data from wells in the Redondo Creek area of the Baca field indicates that the bulk of the formation permeability is in a fracture network (Grant and Garg [1981]). The performance of a well depends largely upon whether it intersects one or more fractures, how large each intersected fracture is and the degree to which it is connected to the rest of the network. The reservoir pressures identified from the downhole analysis define a vertical gradient of 0.35 psi/ft.

Chemical analyses of the discharge fluids from Baca wells indicate that the reservoir fluid is of low salinity ($\leq$ 8,000 ppm) with an incondensible gas content (principally CO_2) of about 0.4 to 1.5 percent by mass (α = 0.004 to 0.015). The probable extent of the initial two-phase region in the Baca reservoir and the extent to which the wells will induce flashing in the formation upon production are very sensitive to the CO_2 content of the fluid (Pritchett, et al. [1981]).

The S-CUBED reservoir simulator CHARGR was recently employed in a series of radial flow calculations to investigate the response of initially single-phase and initially two-phase reservoirs undergoing production or injection from a single well (Garg and Pritchett [1980]). Figure 1 shows the Horner plot for the simulated

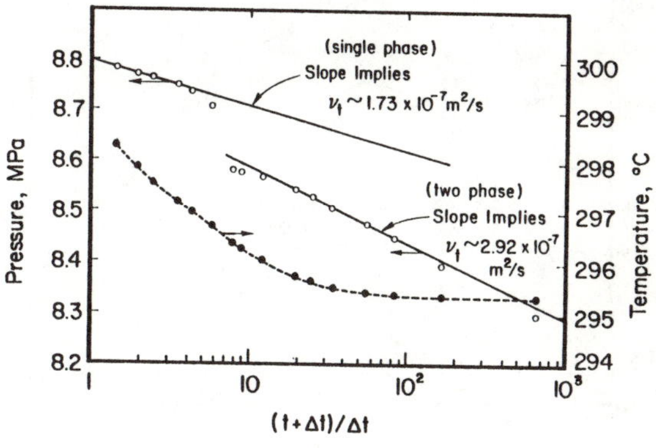

Figure 1. Pressure and temperature buildup data for Case B-1 simulated well test (Garg and Pritchett [1980]).

buildup history of an initially single-phase reservoir that undergoes flashing upon production. The initial buildup behavior (wellbore storage effects were not simulated) is governed by the two-phase region of the reservoir; the slope on the Horner plot of the pressure yields a value for the total kinematic viscosity ν_t which is characteristic of the two-phase region created during drawdown. The detailed calculations show that the pressure buildup is accompanied by the propagation of a condensation front, originating at the well, into the formation; the condensation front eventually engulfs the entire two-phase region after which the buildup behavior is essentially that of a single-phase fluid.

We (Riney and Garg [1982]) have used the results of the numerical simulations to provide guidance in our interpretation of the flow data from selected Baca wells (B-4, B-6, B-13, B-15 and B-20 through B-24). The interpretation considers the effects of the CO_2 content, the fracture permeability of the reservoir and the fact that the downhole pressure/temperature gages are usually located several hundred feet from the primary production zone. We will present the buildup analysis of data from well B-20 to illustrate the procedure.

WELL BACA NO. 20

The temperature surveys, in Figure 2, especially S5 and S10, indicate that relatively low temperature fluid enters the wellbore at approximately 4000 ft (5165 ASL). This implies that the primary production zone is located at 4000 ft and the fluid entering the wellbore at this depth during buildup subsequent to Flow Test No. 4 has been cooled by flashing within the formation. After being shutin for over six months the recorded temperature toward well-bottom is lower than it was after one month (compare S10 and S13). This implies minor downflow of colder water from the 4000 ft entry into a lower permeable zone.

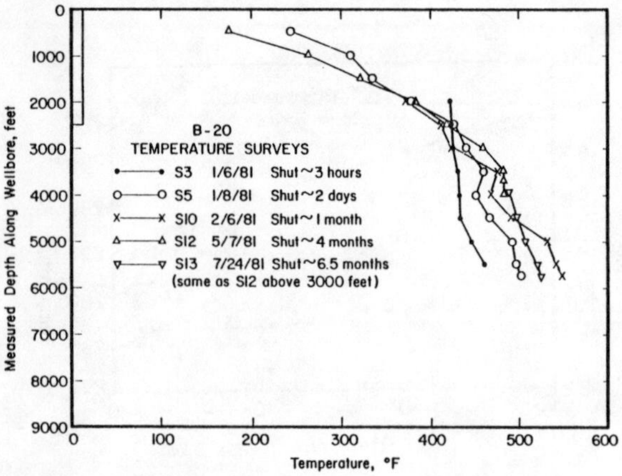

Figure 2. B-20 temperature survey following Flow Test No. 4.

The initial pressure at 4000 ft is estimated from S2 (not shown) and S13 (Figure 3) to be ~ 975 psig (p(a) = 67.9 bars)*. The initial temperature at this depth is estimated from S13 to be ~ 488°F (253°C). From S3 the temperature in the primary production zone at 4000 ft (5165 ASL) is estimated to be reduced from 488°F to ~ 432°F during Flow Test No. 4.

The values of the temperature and pressure recorded at 5750 ft (3567 ASL) during survey S10 are consideed representative of the initial conditions in the lower permeable zone. These values, 549°F (287°C) and 1443 psig (p(a) = 100.2 bars) were measured before significant cooling by the inferred downflow of colder water from the upper permeable zone.

Four flow tests have been performed on B-20 but only Flow Test No. 4 (9/24/80-1/6/81) was of sufficient duration (104 days) and with sufficient pressure transient measurements to warrant analysis. During the test the well flowed through a vertical separator. The wellhead pressure and total flow rates varied over the test period but stabilized over the last two months. For the purposes of analysis we use

$\dot{M}$ = 56,100 lbm/hr (averaged over last 10 days)

~ 7.06 kg/s

t = 2,653 hours (equivalent production time)

Time t is calculated by dividing the cumulative fluid mass produced during the 104-day Flow Test No. 4 by the total mass flow rate averaged over the last ten days of the production.

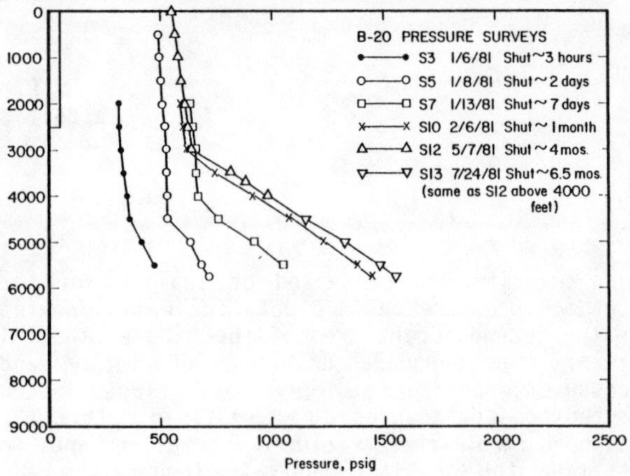

Figure 3. B-20 pressure surveys following Flow Test No. 4.

Figure 4 presents a semi-log plot of the pressure buildup at a depth of 4500 ft (4700 ASL) where most of the recordings were made. Since the primary production is at 4000 ft, it

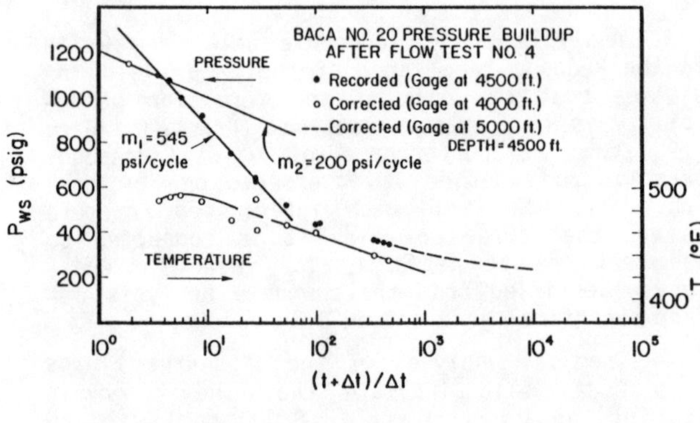

Figure 4. B-20 pressure and temperature buildup following Flow Test No. 4.

is necessary to account for the pressure difference in interpreting the data. The pressure profiles in Figure 3 show that the wellbore fluid is two-phase during the buildup period. The fluid below 4500 ft is two-phase at three hours (S3) and two days (S5) after shutin. After seven days (S7) the fluid below 4500 ft is single-phase liquid, the fluid above 4000 ft is single-phase gas and there appears to be a two-phase section at 4000-4500 ft. After one month there is liquid below 3000 ft with a gas cap above this depth.

* Here and in the sequel we add 10 psi (0.7 bars) to the gage pressure to correct for the atmospheric pressure at ~ 9000 ASL.

Since the wellbore pressure at the gage depth must be corrected by different amounts at different buildup times, we have replotted the Horner plot in Figure 5. The data points at 4000 ft (5165 ASL) recorded during pressure surveys are indicated as are the estimates made from surveys in which recordings were only made at 4500 ft (4700 ASL). In this case the estimates for the primary production depth (4000 ft) contain corrections which account for changes in the wellbore state with changes in buildup time. From the corrected plot (Figure 5) we see that the transition from slope m_1 to m_2 is actually completed at $p_{ws} > \sim 900$ psig rather than the much larger apparent value inferred from Figure 4.

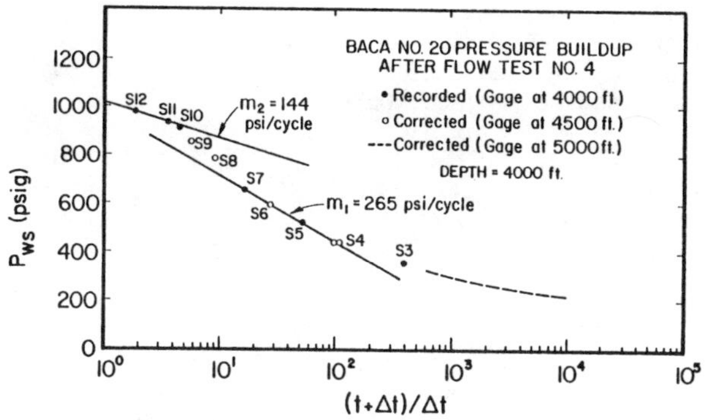

Figure 5. Corrected B-20 pressure buildup following Flow Test No. 4.

A CO_2 mass fraction of 1.47 percent has been measured in the discharge fluid from well B-20. This is higher than that measured in all but one other well and may be higher or lower than the in-situ value (Pritchett, et al. [1981]). If we assume that the CO_2 content in the primary production zone is in fact the same as measured in the discharge fluid (α_0=0.0147) and that the initial temperature at 4000 ft is T_0 = 253°C as estimated from the downhole data, then the corresponding pressure at the single-phase/two-phase boundary is P_{max} = 73.5 bars (> P_0 = 67.9 bars). This implies that the zone at 4000 ft is initially two-phase. According to the equation-of-state for CO_2/water mixtures, at 253°C the initial in situ gas saturation would be S_g = 0.08. In this case, the transition from slope m_1 to slope m_2 in Figure 5 would merely reflect the recovery of the production zone to its initial conditions.

On the other hand, if we assume the change in slope corresponds to a transition from two-phase to single-phase behavior during buildup, then the equation-of-state for CO_2/water mixtures may be used to determine the set of permissible points (T_0, α_0) for which the pressure at the single-phase/two-phase boundary is P_{max} = 62.8 bars (910 psia = 900 psig). A value of T_0 = 253°C would imply $\alpha_0 \sim 0.01$. Such a value is closer to the average CO_2 content of the discharge fluids from the Baca wells.

Because of the uncertainty in α_0 and T_0 we cannot establish if the primary production zone is initially single-phase liquid (e.g., P_0 = 67.9 bars, T_0 = 253°C, α_0 = 0.01) or initially two-phase (e.g., P_0 = 67.9 bars, T_0 = 253°C, α_0 = 0.0147). In either case, however, the slope m_1 in Figure 5 approximates the two-phase portion of the pressure build-up for a full log cycle and this portion of the data can be used to infer formation properties.

From the value of m_1 = 265 psi/cycle (18.3 x 10^5 Pa/cycle) we can calculate the corresponding value of the two-phase kinematic mobility-thickness product

$$\frac{kh}{\nu_t} = \frac{1.15\,\dot{M}}{2\pi m_1} = \frac{1.15\ (7.06)}{2\pi(18.3 \times 10^5)} = 7.06 \times 10^{-7}\ \mathrm{ms}$$

To estimate the formation kh product we first note that during the last ten days of Flow Test No. 4 the wellhead pressure and effective (total) enthalpy of the produced fluid were essentially constant: WHP = 116 psig; h_t = 794.3 BTU/lbm. We have used these values, and an isenthalpic model for two-phase flow in the wellbore that assumes no slippage between the liquid and gas components, to estimate the downhole flowing conditions during the final ten days of drawdown. The calculated pressure at 4000 ft (5165 ASL) is p_{wf} = 180 psig (p(a) = 13.1 bars) and the corresponding downhole flowing kinematic viscosities of the liquid and gas components and the mass ratio of the mixture entering the wellbore are as follows:

$$\nu_\ell = 1.59 \times 10^{-7}\ \mathrm{m^2/s}$$

$$\nu_g = 2.29 \times 10^{-6}\ \mathrm{m^2/s}$$

$$m_g/m_\ell = 1.12$$

The linearized equations in the Appendix can now be used to estimate ν_t. From Eq. (A-5) we compute

$$\frac{k_{rg}}{k_r} = 1.12\ \frac{22.9}{1.59} = 16.2$$

If we assume that the flow within the formation is primarily through a fracture network then from Eq. (A-6) we can calculate the individual relative permeabilities k_{rg} = 0.942 and $k_{r\ell}$ = 0.058. From Eq. (A-2) we estimate the total kinematic viscosity during the last ten days of drawdown.

$$\nu_t = \left(\frac{0.058}{1.59 \times 10^{-7}} + \frac{0.942}{2.29 \times 10^{-6}}\right)^{-1}$$

$$= 1.29 \times 10^{-6}\ \mathrm{m^2/s}$$

We assume that the kinematic viscosity during the two-phase portion of the buildup is approximated by this value. An estimate of the formation kh product from the two-phase portion of the buildup data can now be made.

$$kh = (1.29 \times 10^{-6})\ (7.06 \times 10^{-7})$$

$$= 90.8 \times 10^{-14}\ m^3 = 3030 \text{ md-ft}$$

DISCUSSION

The primary objective of this paper was to illustrate an analysis procedure which is based on the synthesis of field measurements and theoretical results in interpreting geothermal well flow data. The interpretation given here for well B-20 is based on the two-phase portion of the buildup data and a correction of the Horner plot to account for the fact that gage is located several hundred feet from the primary production zone. A conventional Horner analysis would not be applicable.

Table 1 summarizes the results of the analyses of the selected Baca wells considered by Riney and Garg [1982]. In cases where two permeable zones are listed, the higher productivity is generally from the first entry (upper permeable horizon). The two-phase portion of the pressure buildup data were used to estimate the kh product for wells B-15, B-20 and B-21. The single-phase portion of the buildup data were used for the other wells listed. Well B-15 has one of the largest kh values; it is also one of the most productive wells even though the two-phase nature of the productive zone retards the flow through relative permeability effects.

Table 1

DEPTH, TEMPERATURE, PRESSURE AND kh OF PRODUCTION ZONES FOR BACA WELLS

Well	Permeable Horizons	Temperature	Pressure	kh	Remarks
4	4800 ft (4546 ASL)	~500°F	1170 psig	5050 md-ft	Commercial
6	3700 ft (5058 ASL) 4750 ft (4033 ASL)	524°F ------	968 psig --------	6480 md-ft	Commercial
13*	4500 ft (4836 ASL) 6000 ft (3377 ASL)	491°F >500°F	1100 psig 1600 psig	2600 md-ft	Commercial
15	2525 ft (6610 ASL) 5505 ft (3755 ASL)	~500°F ------	450 psig 1405 psig	23800 md-ft	Commercial
20**	4000 ft (5165 ASL) 5750 ft (3567 ASL)	488°F 549°F	975 psig 1443 psig	3030 md-ft	Marginal
20***	5000 ft (4237 ASL)	507°F	1313 psig	---------	Sub-Commercial
21	2840 ft (6559 ASL)	436°F	519 psig	10250 md-ft	Sub-Commercial
22	---------------	--------	-------	---------	Sub-Commercial
23***	3250 ft (5526 ASL)	450°F	835 psig	4340 md-ft	Sub-Commercial
24	5502 ft (3273 ASL)	>500°F	1525 psig	15900 md-ft	Commercial

* Well cycles. kh values is suspect

** Prior to hydraulic fracturing.

*** After hydraulic fracture.

ACKNOWLEDGEMENTS

This work was performed under subcontract 4514510 with the Lawrence Berkeley Laboratory as part of the DOE evaluation of the Baca reservoir.

References

Garg, S. K. and J. W. Pritchett (1980), "Pressure Transient Analysis for Hot Water and Two-Phase Geothermal Wells: Some Numerical Results," Topical Report No. DOE/ET/27163-5, Prepared by S-CUBED Under Baca Geothermal Demonstration Project, October.

Grant, M. A. and S. K. Garg (1981), "Interpretation of Downhole Data from the Baca Geothermal Field," Topical Report No. DOE/ET/27163-12 Prepared by S-CUBED Under Baca Geothermal Demonstration Project, May.

Pritchett, J. W., M. H. Rice and T. D. Riney (1980), "Equation-of-State for Water-Carbon Dioxide Mixtures: Implications for Baca Reservoir," Topical Report No. DOE/ET/27163-8, Prepared by S-CUBED Under Baca Geothermal Demonstration Project, October.

Riney, T. D. and S. K. Garg (1982), "Pressure Tests for Individual Baca Wells in Redondo Creek Area," Topical Report Prepared by S-CUBED under subcontract No. 4514510, April.

APPENDIX

Several authors (e.g., Garg and Pritchett [1980]) have obtained approximate analytical solutions for a geothermal reservoir undergoing two-phase production at a constant rate. In these studies the balance equations for radial two-phase flow in porous media are linearized by assuming that the total and the component kinematic mobilities are related as follows:

$$\left(\frac{k}{\nu}\right)_t = \frac{k\,k_{r\ell}}{\nu_\ell} + \frac{k\,k_{rg}}{\nu_g} \qquad \text{(A-1)}$$

$$\frac{1}{\nu_t} = \frac{k_{r\ell}}{\nu_\ell} + \frac{k_{rg}}{\nu_g} \qquad \text{(A-2)}$$

Given the flowing enthalpy h_t of the two-phase mixture, it is also possible to obtain the separate kinematic mobilities.

$$\frac{k\,k_{r\ell}}{\nu_\ell} = \left(\frac{k}{\nu}\right)_t \left[\frac{h_g - h_t}{h_g - h_\ell}\right]; \quad \frac{k\,k_{rg}}{\nu_g} = \left(\frac{k}{\nu}\right)_t \left[\frac{h_t - h_\ell}{h_g - h_\ell}\right] \qquad \text{(A-3)}$$

The expression

$$h_t = m_g\, h_g + (1 - m_g)\, h_\ell \qquad \text{(A-4)}$$

relating the total and fluid component enthalpies may be used with Eqs (A-3) to write

$$\frac{k_{rg}}{k_{r\ell}} \frac{\nu_\ell}{\nu_g} = \frac{m_g}{m_\ell} = \frac{m_g}{1 - m_g} \qquad \text{(A-5)}$$

Here m_g and $1-m_g$ are the gas and liquid components of the fluid flow.

If we assume that the flow within the formation is primarily through a fracture network, then

$$k_{r\ell} + k_{rg} = 1$$

Geothermal Resources Council, TRANSACTIONS Vol. 6, October 1982

QUALITATIVE RESPONSE PATTERNS ON GEOPHYSICAL WELL LOGS FROM THE GEYSERS, CALIFORNIA.

S. K. SANYAL, M. CHE, RICHARD E. DUNLAP*
and MARY K. TWICHELL*

Geothermex, Inc., Berkley, California
*Aminoil USA, Santa Rosa, California

ABSTRACT

This paper presents a set of qualitative response patterns on the geophysical well logs from The Geysers geothermal field, California. The response patterns were diagnosed from the analog prints of gamma ray, density, neutron and caliper logs. These patterns were verified from core data, production test results and a detailed computer analysis performed later. The paper shows that much information can be gained by simply scanning the analog prints of well logs from The Geysers.

INTRODUCTION

Well logs can help drilling and completion decisions at The Geysers by allowing diagnosis of different lithologic units and steam entry zones. Various geologic decisions at The Geysers such as identification of lithology, subdivision of the massive graywacke sections, structural and/or stratigraphic correlation, diagnosis of rock alteration and fracture zones, etc. can be best accomplished by well log analysis. In reservoir engineering decisions, well log analysis is indispensible in location and correlation of steam entry zones and in estimation of porosity and steam quality.

Much information can be gained by simply scanning the analog prints of well logs. In the present study, the analog prints amounting to 160,000 curve-feet and corresponding mudlog data were compared carefully in order to arrive at some generalizations, if possible, on the qualitative response patterns on geophysical logs. Several conclusions were arrived at. These conclusions are:

1. In all apparent fracture zones, including steam-entry zones, gamma ray response increases and N decreases (ϕ_N increases) and both LS and SS increases (ρ_a decreases), while caliper log may or may not show any hole enlargement (figure 1). The increase in gamma ray is most probably due to deposition of uranium salts in the fractures by steam or hydrothermal fluids. Uranium, being highly mobile, is often dissolved from the rock matrix and deposited in fractures during the flow process. The decrease in N (increase in ϕ_N) is not always due to an obvious increase in porosity, because LS and SS do not necessarily increase (decrease in density). In most cases, the fractures probably are filled with a hydrated mineral enriched in uranium.

2. In all steam-entry zones, LS and SS show sharp increase (ρ_a decreased) indicating presence of porosity. Even the relative magnitude of this increase sometimes appears to be proportional to the steam-entry rate (and pressure kick). Figure 2 is an example of processed Densilog/neutron log section from a well. The zone from about 6,708 to 6,720 feet shows high GR, high ϕ_N and low ρ_a anomaly. The mudlog records that there was a pressure kick at this depth.

3. Hole enlargement is easily identified from rapid drilling rate, positive caliper anomaly, low GR, low N (high ϕ_N) and high LS/SS (low ρ_a). Hole enlargements are mostly caused by sloughing argillite and schists and correlate with ρ_a anomalies.

4. Greenstone can be distinguished by its lower gamma ray, higher N (lower ϕ_N) and lower LS/SS (higher ρ_a than for graywacke. Figure 3 shows a section of the processed Densilog/neutron log from a well showing a greenstone-graywacke contact at 4,230 feet.

5. Argillite appears to have higher GR, lower N (higher ϕ_N) and generally lower (but occasionally higher) ρ_a than graywacke. Argillite often causes hole enlargement and consequently high LS and SS (low ρ_a).

6. Biotite rhyolite is readily identified by its very high gamma ray and relatively high N (low ϕ_N). Figure 4 shows an example of biotite rhyolite response as identified on a section of GR-N log from a well. This type of response is expected, because biotite has potassium and rhyolite has less intrinsic porosity than graywacke.

7. Tourmaline-rich zones are identified by high GR and low N (high ϕ_N). Figure 5 is an example of the response of a tourmaline-rich zone (at 5,200 feet) in Densilog/Neutron log for a well. The high ϕ_N in tourmaline is due to its high thermalizing index and extremely high thermal neutron capture cross-section.

These conclusions were thoroughly verified by a later computer analysis and from core data and production test results.

NOMENCLATURE

- GR = Gamma ray intensity (API unit)
- N = Neutron log response (API unit or neutron porosity unit)
- LS = Long-spaced detector response (cps)
- SS = Short-spaced detector response (cps)
- ρ_a = Apparent bulk density (gms/cc)
- ϕ_N = Apparent neutron porosity

FIGURE 1: A Section of Gamma Ray-Neutron Log

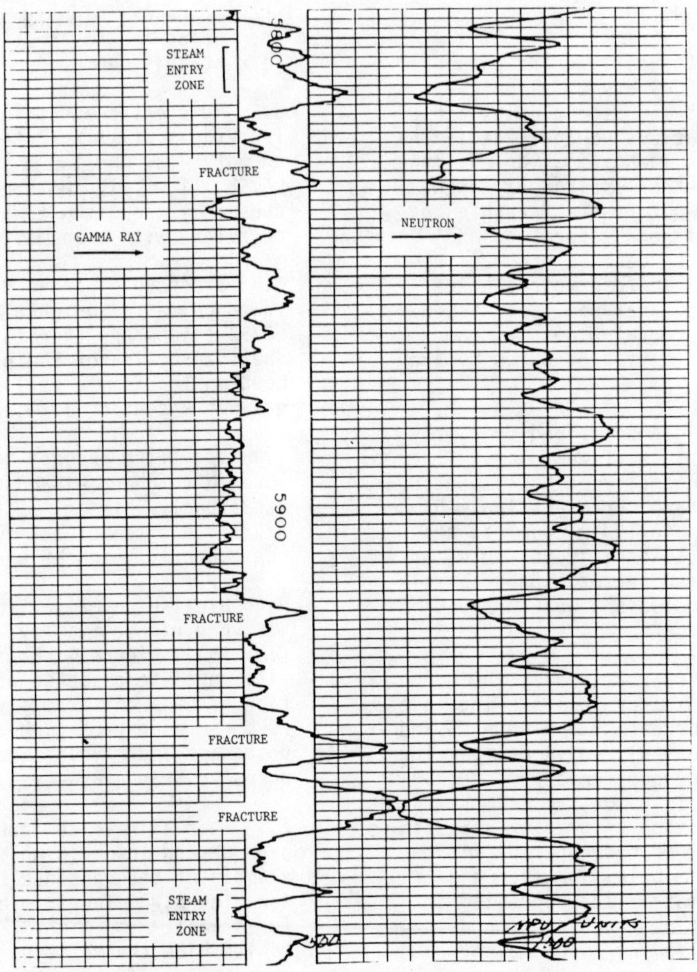

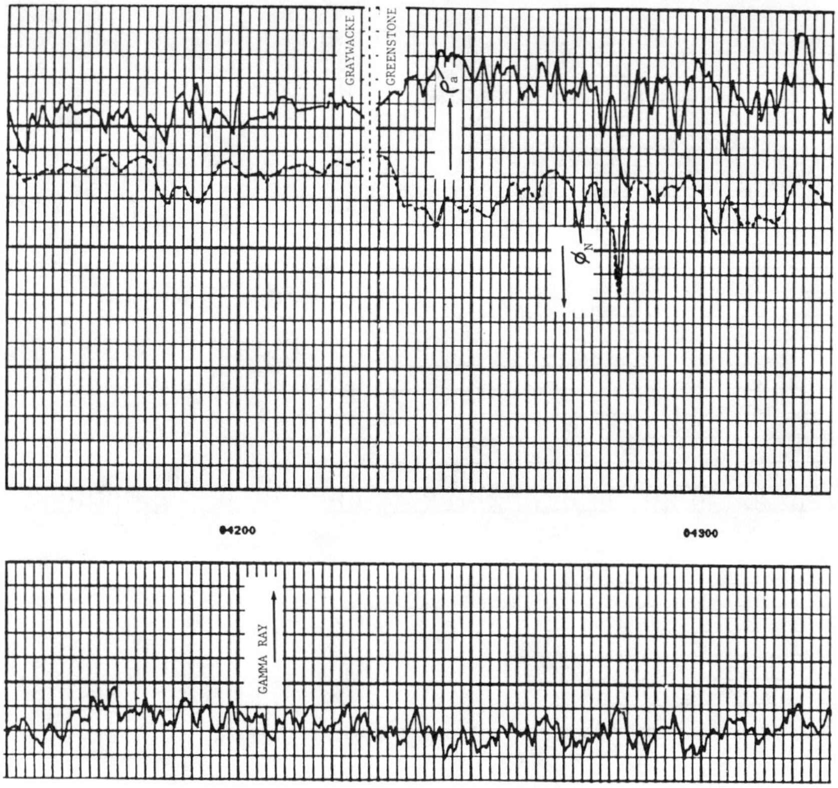

FIGURE 3: A Section of Densilog/Neutron Log

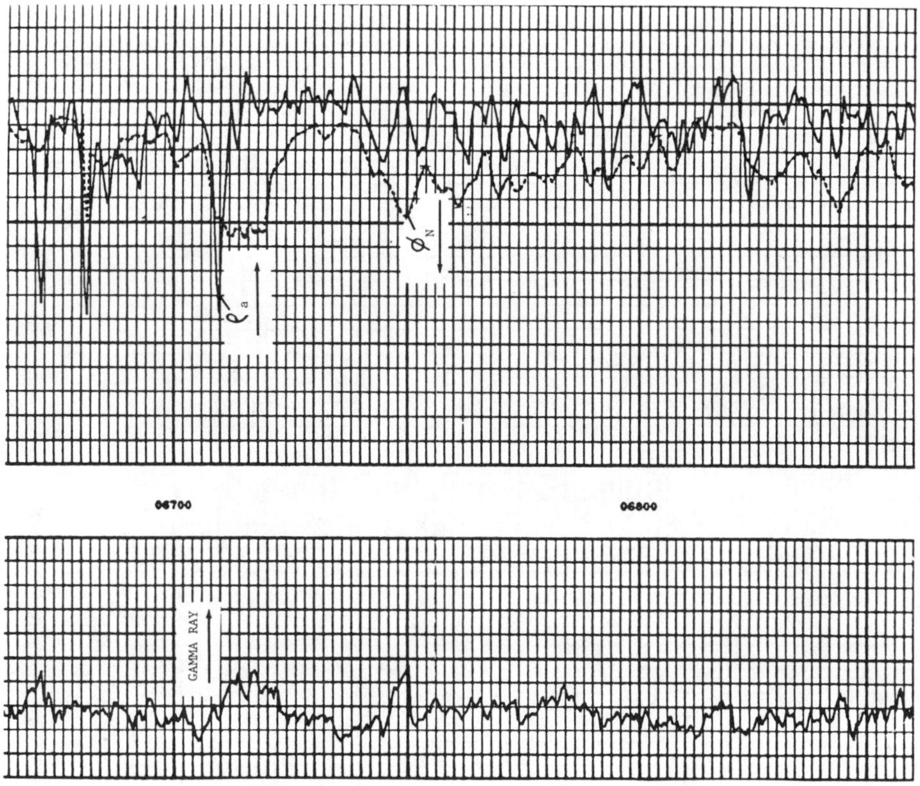

FIGURE 2: Densilog/Neutron Log Section

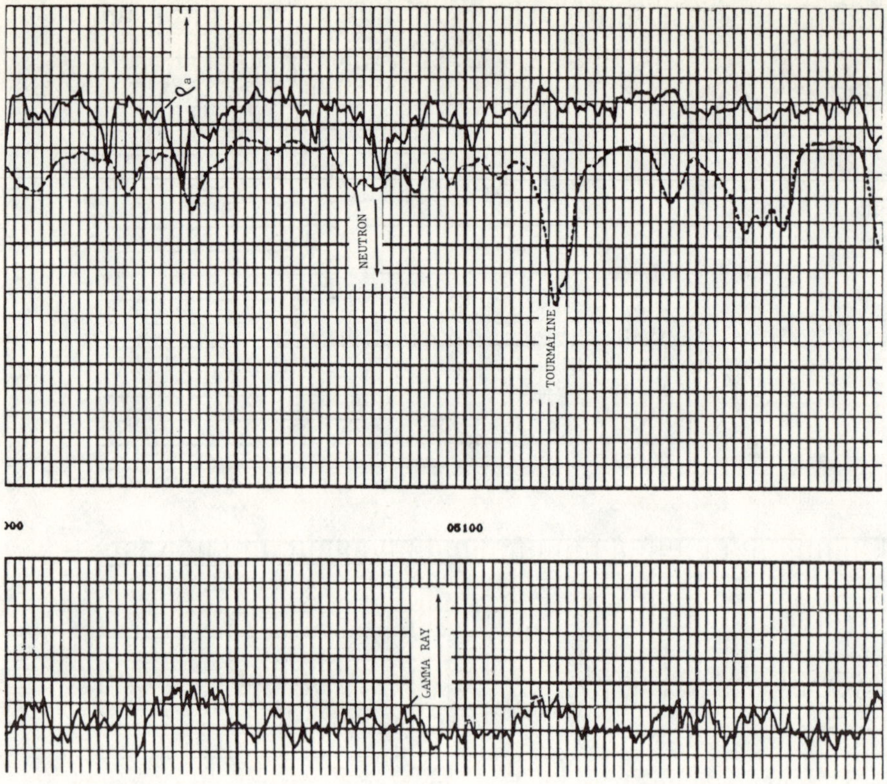

FIGURE 5: A Section of Densilog/Neutron Log

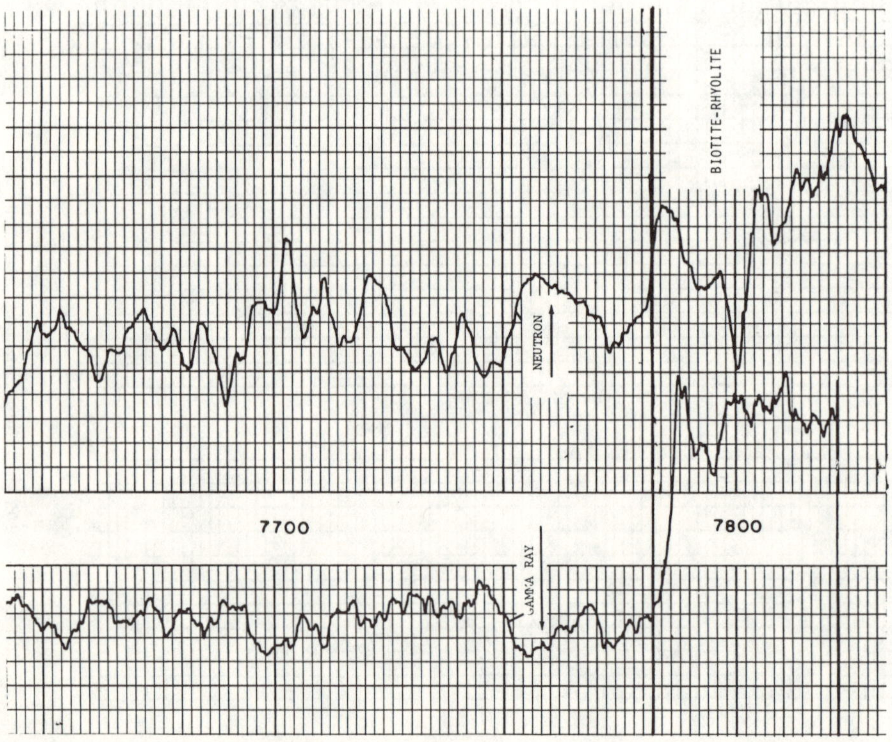

FIGURE 4: A Section of Gamma Ray/Neutron Log

Geothermal Resources Council, TRANSACTIONS Vol. 6, October 1982

AN INTERPRETATION OF PRECISION TEMPERATURE LOGS IN A DEEP GEOTHERMAL WELL NEAR DESERT PEAK, CHURCHILL COUNTY, NEVADA

Thomas C. Urban and William H. Diment

U.S. Geological Survey
345 Middlefield Road
Menlo Park, California 94025

ABSTRACT

Three precision (~ 0.001°C), continuous (0.6 m sensing interval) temperature logs (7/25/79, 7/27/79, and 10/19/79) were obtained in Phillips Petroleum Company's deep (2931 m) geothermal well B23-1 (39°45.6'N, 118°55.6'W, elevation 1393 m) a few months after completion of drilling (5/24/79). The measurements were made in a small diameter (6.2 cm I.D.) tubing capped at the bottom and extending from the surface to near total depth in the hole which ranged in diameter down to 20 cm. A 34 cm (O.D.) casing was cemented into place in the upper 901 m of the hole. The remainder of the hole was open, allowing the exchange of fluids among the many permeable zones encountered during drilling. Analysis of the temperature logs suggest the following: (1) The linearity of the temperature plot in the interval 200-900 m (after allowing for decay of the drilling disturbance) suggests the existence of a hydrothermal system of considerable antiquity (> 10^3 years). The high gradient (~180°C/Km) and high heat flow (> 5 microcal/cm^2 · sec) in this interval require that heat be transported convectively to the base of this interval. (2) The interval between about 900 and 1300 m exhibits anomalous warming since drilling. Although some of this warming represents decay of the thermal disturbance due to invasion of cool drilling fluids, a part of it may represent an invasion of hot water into the hole which then descends along the hole and exits into many permeable zones at greater depth. This could, of course, obscure the true nature of the thermal regime of the rocks surrounding the lower part of the hole.

INTRODUCTION

A considerable effort has been made in the exploration for geothermal resources in the area in western Nevada approximately 80 km east of Reno. Located in the Hot Springs Mountains, the Desert Peak geothermal system encompasses an area of roughly 250 sq. km. It was originally discovered by Phillips Petroleum Company while drilling temperature-gradient holes near Brady's Hot Springs, about 15 km to the northwest. Subsequent drilling in the Desert Peak area established it as a separate system. Evaluation of this thermal anomaly was partially funded by the Department of Energy and much of the information gathered at this site has been, or is being, released (e.g., Benoit, 1978; Benoit et al., 1980; Sethi and Fertl, 1979).

Although there is a considerable quantity of data that could be examined, our attention is restricted to the interpretation of three precision temperature logs that we obtained several months after drilling in the deepest of the exploratory wells (Phillips Petroleum Company B23-1). The results are of interest because they show that the fine structure of the temperature variation with depth can be reliably and repeatably obtained in a deep, high-temperature (~205°C) environment, and that this fine structure can yield useful information about conditions in and immediately surrounding the hole.

EXPERIMENTAL PROCEDURES AND UNCERTAINTIES

The logging system consists of a thermistor lowered at the end of a steel-armored, 4-conductor, teflon-insulated cable. The resistance of the thermistor is digitally recorded at the surface. When malfunctions occurred they were usually the result of leakage at the cable head which connects the probe containing the thermistor to the cable. In logging B23-1, the filler between the leads and the armor charred, thus making it difficult to rehead the cable. Our problem is reminiscent of that encountered by Mathews et al. (1981) in a more extreme environment. Uncertainties due to convection in the hole and other noise are discussed elsewhere (Diment and Urban, 1982).

ANALYSIS OF TEMPERATURE LOGS

The general characteristics of the temperature and temperature gradient profiles are illustrated in Figure 1. These profiles can be divided roughly into three parts: (1) a near-surface zone (0-900 m) of high but rather uniform gradient, which shall be referred to as the "conductive cap"; (2) a zone of high temperatures and variable gradients, here termed the "injection zone" (900-1300 m); and (3) a zone of high temperature, but of low or even negative gradients, here termed the "reservoir." In discussing these zones, we rely heavily on the summaries of Sethi and Fertl (1979) and Benoit et al. (1980).

The "conductive cap" appears conductive in the sense that the temperature profiles are rather

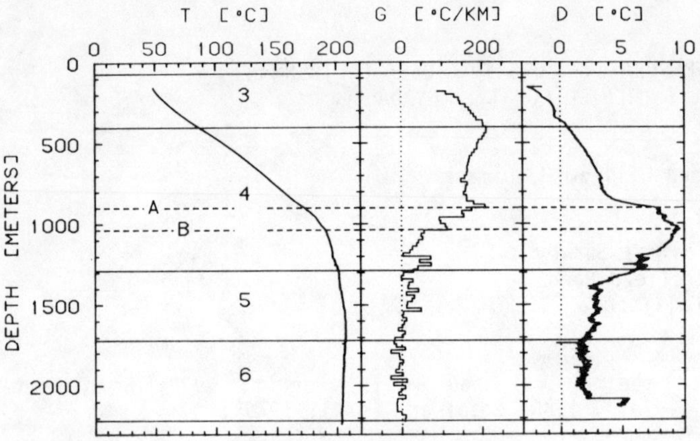

Figure 1. Temperature log (T) recorded at 0.6 m intervals in Desert Peak well B 23-1 on 10/19/79. Temperature gradients (G) were determined over 20 m intervals. Temperature differences (D) represent T (10/19/79) minus T (7/25/79). No attempt has been made to remove spikes due to electrical noise. Solid horizontal lines separate major geologic units which are designated by numbers (Table 1). Level A is the base of the casing (901 m) which is close to a zone of lost circulation (928 m). Level B (1036 m) is a major zone of porous rock with low formation pressure (Sethi and Fertl, 1979, p. 11). All depths are relative to ground level which is 8 m below the Kelley bushing.

linear (Fig. 1) and would be more so if a correction for the drilling disturbance were applied. Note that the cross over point in the cooling/heating history (Figs. 1 and 3) is near 350 m. Also note that this is the only interval in the hole in which a cemented casing (0-901 m) prevents exchange of fluids among aquifers. Silicified zones in this interval attest to past geothermal activity. Zones of hot-water flow at shallow depths nearby indicate present hydrothermal activity (Benoit, 1978). Judging from the linearity of the profile in B23-1, the temperature at the base of the cap has been rather uniform for some time. Given the calculations that we made in estimating the age of The Geysers Steam field (Urban et al., 1976, Fig. 4), the temperature of the base of the cap has maintained near its present level for at least 10^3 years and possibly much longer. The heat flux through the cap is so large ($\sim$180°C/km x $\sim$3 + millical/cm · sec. °C = $\sim$5 + micro cal/cm^2 · sec) and the gradient below it so low, that heat must be delivered to the base of the cap by the flow of hot water.

The behavior of the "injection zone" is complex. Given the considerable rise in temperature in the injection zone between 7/25/79 and 10/19/79 (Figs. 1 and 3), it is probable that the temperatures will increase further. How much more they will increase is difficult to estimate, because the effect of the invasion of drilling fluid into permeable formations can be large and long-lasting (e.g., Nathenson et al., 1980). Moreover, we are not sure that our temperature logs made in the tubing reflect the true range of temperatures in the outer regions of the hole. Temperature logs in some other injection zones, as sensed both in small diameter holes (e.g., Diment et al., 1980) and in large diameter holes (e.g., Diment, 1980, Fig. 2.3), exhibit large temperature variations (several degrees) over short intervals of depth (several meters).

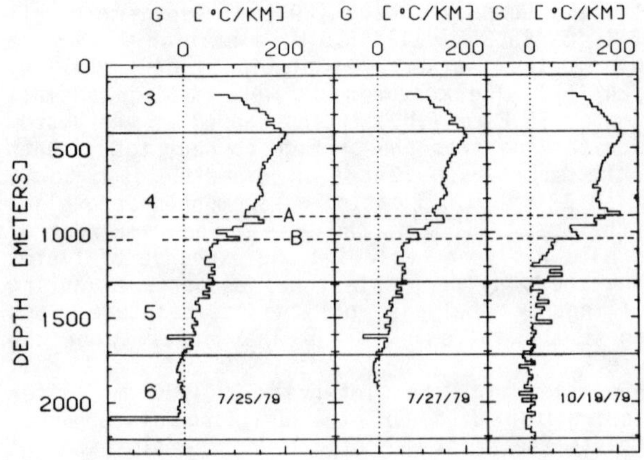

Figure 2. Temperature gradient logs (20 m interval) obtained in Desert Peak well B23-1 on the dates indicated. See Figure 1 for explanation of symbols.

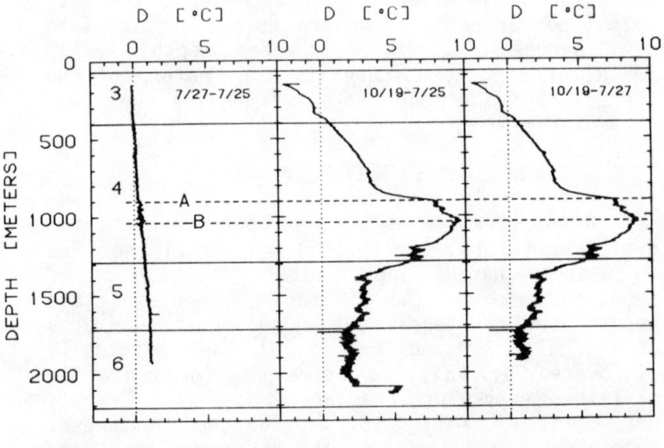

Figure 3. Temperature differences between logs obtained in Desert Peak well B23-1 on the dates indicated. See Figure 1 for explanation of symbols.

We suggest that hot water is cascading down the hole from various permeable intervals in the injection zone and that this has lead to the decreasing gradients with time in the region immediately below the injection zone (Fig. 2). It also appears that the maximum temperature in the upper part of the hole (<2500 m) may occur ultimately at level B, with an "equilibrium" temperature profile being nearly linear from 200 m to level B (1034 m) and thence declining.

The "reservoir" extends from near the base of unit 4 (1288 m) nearly to the bottom of the hole. The zone is marked by relatively high temperatures and low thermal gradients. The temperatures and the temperature gradients (Fig. 2) change with time in a manner that cannot be explained in terms of the conductive thermal-decay of the drilling disturbance. They are, however, explainable in terms of a hot fluid flowing downward through the hole and flowing out at various levels of low pressure and high permeability; many such intervals were documented in the process of drilling (Benoit et al., 1980).

The temperature profile may be rather like that observed by Blackwell and Morgan (1976) in a deep geothermal well (MGE No. 1) near Marysville, Montana. The form of the temperature-depth curve is similar to that in B23-1, although the amplitude is not. Moreover, the changes of the temperature curves with time after drilling are similar (Blackwell and Morgan, 1976, Fig. 5). Flow velocity logs obtained in MGE No. 1 indicate a large downward flow of water from fractures from 900 to 1100 m. The "reservoirs" in both holes are largely in granitic rocks, and both holes, mostly uncased, are sub-hydrostatic at the well-head.

The true rock temperatures in the "reservoir" are not known. They are less than those indicated by the temperature logs at least to a depth of about 2700 m where the measured gradient turns mildly positive (14°C/km, Benoit et al., 1980, p. 16). The question is: How much less?

An alternative speculation is that the "reservoir" is a weakly developed vapor-dominated system (e.g., White et al., 1971) whose identity has been masked by the drilling process, that involved drilling with mud to 927 m, with water to 2800 m, and with aerated water to the bottom (Benoit et al., 1980). In the well-known vapor-dominated system at The Geysers, the temperature and pressure in the main steam reservoir (> 1500 m) are 240°C and 40 bars (near the maximum enthalpy of steam).

CONCLUSIONS

With a little bit of luck and considerable attention to detail, precision temperature logs can be obtained in high temperature environments (~200°C) with modern, largely off-the-shelf components. Precision temperature logs, especially when repeated so that temperature difference logs can be constructed, may provide useful information about the motion of fluids within and surrounding a hole.

Table 1. Major geologic units encountered in well B 23-1 (abstracted from Phillips Petroleum Co., 1979, Fig. 9; and Benoit et al., 1980, Fig. 5)

INTERVAL	(M)	GEOLOGIC UNIT
1	0-12	Sand, gravel and boulders
2	12-69	Lacustrine sediments of Pliocene Trukee Formation occasional silicification.
3	69-404	Basalt, basaltic andesite, and andesite of Chloropagus Formation.
4	404-1288	Andesitic to rhyolitic ash-flow tuffs - Miocene. Basal 20 m contains gravel, talus(?), soil(?), and both phyllite and volcanic chips.
5	1288-1719	Metamorphosed Mesozoic sedimentary rocks, predominately phyllite.
6	1719-2216	Roughly equal amounts of granite and chlorite schist and chlorite schist and hornfels.
7	2216-2939	Mostly granite to T.D. (2939 m).

The hydrothermal system near Desert Peak, as revealed by temperature logs and other information obtained from well B23-1, appears similar to many others in the sense that: (1) the temperature field is dominated by the sub-horizontal flow of anomalously hot water along a confined aquifer of limited thickness that owes its source of hot water to an adjacent conduit which taps a deeper high-temperature reservoir and (2) the probable movement of water down the hole obscures the true nature of the thermal regime below the high temperature aquifer that provides the descending hot water.

ACKNOWLEDGEMENTS

We thank R. T. Forest of the Phillips Petroleum Company for arranging access and support at the site and for sharing his knowledge of the history of the project. The manuscript was read and improved by W. B. Joyner and A. H. Lachenbruch.

REFERENCES

Benoit, W. R., 1978, The use of shallow and deep temperature gradients in geothermal exploration in northwestern Nevada using the Desert Peak thermal anomaly as a model: Geothermal Resources Council, Transactions, v. 2, p. 45-56.

Benoit, W. R., Sethi, D. K., Fertl, W. H., and Mathews, M., 1980, Geothermal well log analysis at Desert Peak, Nevada: Society of Professional Well Log Analysts 21st Annual Symposium, July 8-11, 1980, Paper AA, 41 p.

Blackwell, D. D., and Morgan, P., 1976, Geological and geophysical exploration of the Marysville geothermal area, Montana, USA: Second United Nations Symposium on the Development and Use of Geothermal Resources, Washington, D.C., U.S. Government Printing Office, p. 895-902.

Diment, W. H., 1980, Geology and geophysics of geothermal areas, in Kestin, J., ed., A source book on the production of electricity from geothermal energy: Washington, D.C., U.S. Government Printing Office, p. 7-103.

Diment, W. H., Urban, T. C., and Nathenson, M., 1980, Notes on the shallow thermal regime of the Long Valley Caldera, Mono County, California: Geothermal Resources Council, Transactions, v. 4, p. 37-40.

Diment, W. H., and Urban, T. C., 1982, Temperature changes with time in the slotted interval of a deep, shut-in geothermal well near thermal equilibrium - East Mesa well 31-1, Imperial County, California, 1977-1982: Geothermal Resources Council, Transactions, v. 6.

Mathews, M., Pettit, R. A., and Miler, D. J., 1981, High temperature logging for basic development of HDR reservoirs: Geothermal Resources Council, Transactions, v. 5, p. 299-302.

Nathenson, M., Urban, T. C., Diment, W. H., and Nehring, N. L., 1980, Temperatures, heat flow, and water chemistry from drill holes in the Raft River geothermal system, Cassia County, Idaho: U. S. Geological Survey Open-File Report 80-2001, p. 29.

Phillips Petroleum Company, 1979, Geothermal resource assessment case study, northern Basin and Range Province, final report for the period 1 October 1978-30 September 1979, U.S. Department of Energy Report DOE/ET/27099-1: Springfield, Virginia, U.S. Department of Commerce, National Technical Information Service, 114 p.

Sethi, D. K., and Fertl, W. H., 1979, Geophysical well logging operations and log analysis in geothermal well Desert Peak No. B-23-1: Dresser Atlas Report for Los Alamos Scientific Laboratory, LA-8254-MS, 69 p.

Urban, T. C., Diment, W. H., Jamieson, I. M., and Sass, J. H., 1976, Heat flow at the Geysers, California, in Proceedings of the Second United Nations Symposium on the Development and Use of Geothermal Resources: Washington, D. C., U. S. Government Printing Office, v. 2, p. 1241-1245.

White, D. E., Muffler, L. J. P., and Truesdell, A. H., 1971, Vapor-dominated hydrothermal systems compared with hot-water systems: Economic Geology, v. 66, p. 75-97.

Geothermal Resources Council, TRANSACTIONS Vol. 6, October 1982

REVIEW OF DEVELOPMENT STRATEGIES FOR TWO GEOTHERMAL FIELDS IN THE PHILIPPINES

Graham H. Wheeler*, Pablo V. Malixi** and Norman W. Firth*

*Kingston Reynolds Thom & Allardice Limited (KRTA)
C.P.O. Box 4498, Auckland, New Zealand
**Vice-President, PNOC-Energy Development Corp,
Fort Bonifacio, Metro Manila, Philippines

ABSTRACT

This paper describes the policies and strategies adopted in the development of the Tongonan and Okoy geothermal fields in the Philippines. The general setting within which geothermal development in the country is taking place is first discussed. This leads to a presentation of general development and design policies. Finally, the differences of strategy between the two fields are contrasted. The conclusion reached is that development at Tongonan is dominated by wide regional influences, whilst that at Okoy is affected more by the local physical characteristics of the resource area.

1. INTRODUCTION

This paper describes the principal features of development strategy of two geothermal fields in the Philippines. These are the Tongonan Field (Northern Leyte) and the Okoy Field (Southern Negros) (refer Fig 1).

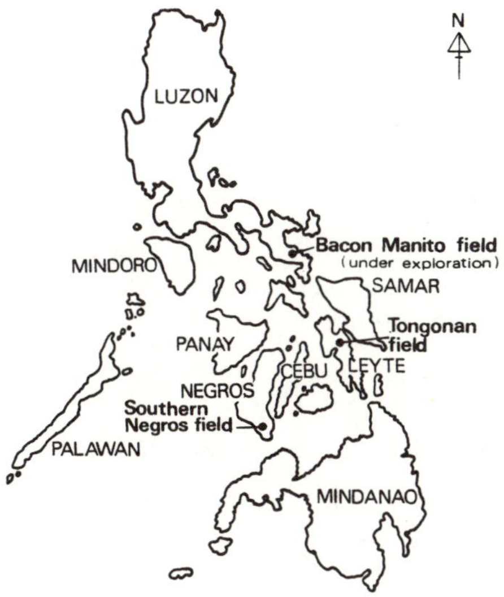

Fig 1.

Responsibility for the development of these fields lies with the Ministry of Energy of the Philippine Government through its agencies, the Philippines National Oil Company - Energy Development Corporation (PNOC-EDC) and the National Power Corporation (NPC). The PNOC-EDC area of interest covers exploration and exploitation of geothermal resources inclusive of installation and operation of wells, steam collection systems and wastewater disposal systems. The NPC responsibilities cover installation and operation of power generating plant and electricity distribution.

Kingston Reynolds Thom & Allardice Ltd., (KRTA) are consultants to NPC for the design of the Tongonan I power station. In conjunction with PNOC-EDC staff, KRTA are also engaged on the scientific exploration; on drilling management and resource assessment of each field; on the planning, design and construction management of each steam collection and disposal system; and on power development feasibility studies. Much of this work is now carried out through Geothermal Technology Corp (GTC), a PNOC/KRTA joint venture.

2. GENERAL SETTING OF GEOTHERMAL DEVELOPMENT

The geothermal resources of the Philippines are considerable, and their exploitation and development is proceeding at a pace which is unequalled anywhere else in the world. The impetus for this endeavour comes from the Government's determination to reduce the country's dependence upon imported oil, and achieve energy self-sufficiency by exploitation of indigenous energy resources.

To this end, the Ministry of Energy has laid down a five-year accelerated programme of energy resource development, covering the years 1981-86. In this programme, the goal is to match not only the expected increase in the total power generation energy requirement from 33.1 MMBOE in 1981 to 49.4 MMBOE in 1986, but also to reduce the proportion of imported oil in this total, from 59% to 3% in the same period (1).

The contribution of geothermal energy in achieving this objective requires the installation of an additional 995 MW(e) of geothermal generating capacity by 1986.

This is clearly a very ambitious programme. To achieve it will require the successful exploitation and development of not only the two fields described herein but also Bacon-Manito, Luzon, and some expansion of other existing geothermal generating facilities. Approximately 180-200 successful wells will need to be drilled, requiring the simultaneous working of some 10-12 rigs. In addition, continued exploration must maintain the impetus of geothermal resource development in these and other prospects throughout the country. Under this programme, the Philippines will become a world leader in the generation of electric power from geothermal resources. It is in this context of a vigorous pursuit of geothermal development that the following strategies were established.

3. COMMON DEVELOPMENT STRATEGIES

A number of policies have been adopted which have general application to those power developments with which KRTA is involved. A selection of the more significant of these includes:

. The undertaking of "prefeasbility" studies as soon as possible after initial drilling success has been achieved. The purpose of this is to assess early the prospects for development and to maintain the momentum of the National Energy Programme. Furthermore, it enhances the chances of being able to incorporate exploration and delineation wells into the final production system.

. The employment of two-phase fluid collection systems feeding centralised separator stations. This is appropriate for the liquid-dominated fields of the Philippines, and minimises the surface piping required for both steam collection and wastewater disposal, with consequent economic benefits.

. The adoption of reinjection as a general policy for the disposal of wastewaters, in order to reduce the environmental impact of development, and as an aspect of reservoir management. In general the system will be a closed one so designed as to ensure that injection takes place at as high a temperature and pressure as practicable. The problems of silica deposition are thereby minimised. Deep reinjection at the perimeter of the reservoir is usually sought, with the exception of power station condensate which must be injected at locations remote from the wastewater disposal zone.

. Well testing programmes, until the establishment of a reinjection system, require discharge of separated water to surface waterways, and are constrained so that at any one time the chemical load on the rivers does not exceed acceptable limits.

4. THE TONGONAN FIELD

4.1 General

Initial investigation in the Tongonan area commenced in 1973 as a New Zealand/Philippines Technical Co-operation Project. Following reconnaissance, surface exploration and some shallow drilling (up to 600 m), three deep (greater than 1500 m) exploration wells were drilled. Each was successful with temperatures in excess of 320°C.

The initial phase of exploration concentrated in the Mahiao and Sambaloran sectors of the field (refer Fig. 2), then later progressed south-eastwards in step-outs into the Malitbog, Mamban and Mahanagdong sectors. Some 40 wells have been drilled thus far. The wells have been consistently successful, with an average well discharge output equivalent to about 8 MW(e).

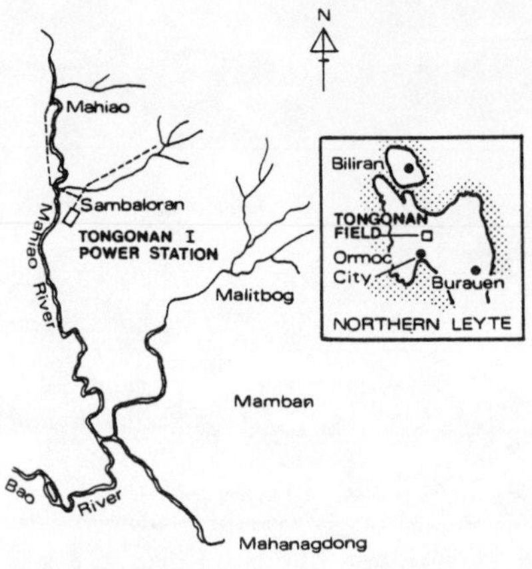

TONGONAN FIELD Fig 2.

At the present time (May 1982) the southern limit of the resource has yet to be delineated. The current prognosis is that two anomalies probably exist, one centred about the Sambaloran Valley and the other at Mahanagdong. Calculations of the volumetric stored heat within the anomalies based on the drilling to date give an assessed generating potential of at least 450 MW(e) in the Mahiao, Sambaloran, Malitbog and Mahanagdong sectors.

4.2 Tongonan I Power Station

Commitment to the development of the first stage of the Tongonan field was made by the Philippines Government when the first deep wells in the exploration programme proved to be successful. Continued drilling success resulted in a further decision to locate in the region a copper smelter and associated fertilizer plant, both of which would rely upon the geothermal resource for energy. Thus, the development of the Tongonan I Geothermal Power Station commenced, knowing that it was to be initially the single major power generating facility in an under-developed region, that a major portion of the station load would be provided by a single customer, and that the total load would not be continuous but fluctuating. Furthermore, a very short development timetable of 3 years was imposed to match the commissioning date of the copper smelter. (refer Fig.3).

Several important policies and design decisions arose out of these factors:

(a) Diversity of Steam Supply and Reliability of Plant Operation.

The nature of the operation carried out by the copper smelter, and the absence of any other generating facilities in the region, required that as reliable a supply of power as possible be available from the Tongonan station. This resulted in the following design features:

. Steam supply to the station was initially to be provided along three separate routes, each capable of supplying sufficient steam to power the smelter. This diversity of supply was felt necessary in view of the vulnerability of pipelines to damage in such a seismically active area, and in the existing steep terrain. In the event, a decision to continue immediately with further development stages of the Tongonan field reduced the need to provide such a degree of diversity, and the pipeline routes were subsequently reduced from three to two.

. The total initial station output of 75 MW is to be met with two 37.5 MW(e) turbo-generator sets, with a third 37.5 MW unit as standby.

. Each unit is to have its own unitised electrical system covering the ancillary equipment and extending to the main unit switchboards. Separate station services switchboards supply the buildings and provide the dc power.

. The reinjection system is to be backed up with emergency holding ponds so that essential maintenance and/or repairs can be carried out without interruption to the power supply.

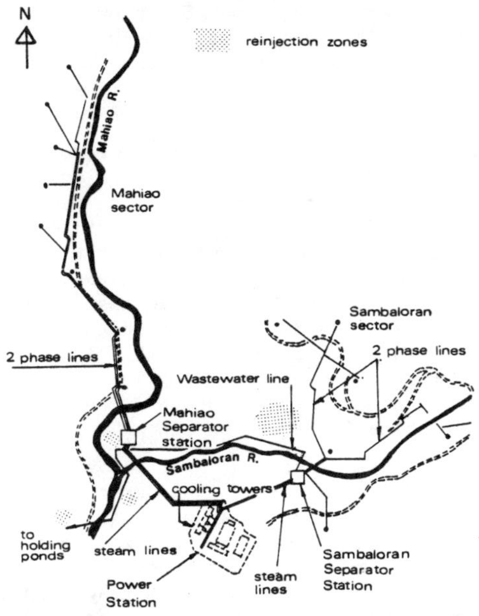

TONGONAN I POWER STATION DEVELOPMENT
Scale 0 200m
Fig 3.

(b) Fluctuating Station Load
The station is required to match the fluctuation of both the domestic and commercial demands of the region. As a result a series of unique control systems and operating procedures has had to be developed to maintain steam pressure to the turbines within specified tolerances and to regulate total well output under the variable load conditions. These systems include:

. Throttle valves in the two-phase pipelines to provide pressure adjustment at selected wells.

. Steam pressure control valves at the first stage separator vessels to provide fine adjustment.

. Automatic start-up/shut-down of selected wells, with control over the full range of load fluctuation being completed with manual start-up/shut-down.

. Blow-off valves to discharge excess steam to atmosphere in the event that total well output exceeds demand.

(c) Accelerated Development Programme
The three year timetable planned at the start of the work was unusually short for such a development. In order to achieve this a number of particular project management steps were taken:

. The feasibility study and preliminary design report were combined into a single stage.

. A series of multiple supply and construction contracts were drawn up so that work could be progressively released as required by the critical path to completion.

. Bid documents were so written as to provide for early tender call for plant while still allowing for revisions to detail design, in order to incorporate the latest data coming forward from the ongoing drilling of the steamfield.

4.2 Further Development at Tongonan and North-West Leyte

Beyond Tongonan I, a further potential generating capacity of at least 350 MW(e) appears to exist in the Tongonan field. In addition, two further prospects in the nearby region, situated at Biliran and Burauen (refer Fig. 2) show promise that the total potential capacity from the Northern Leyte region may exceed 650 MW(e). This is a considerable block of power which, even with the best will in the world, cannot be absorbed within the region for a considerable period. Consequently, the development of further geothermal generating facilities now hinges upon the concept of exporting the power out of the region and into the major power demand centres situated on the island of Luzon. The proposal is to transmit power from Leyte through Samar, across the San Bernadino Strait to Luzon by submarine cable, and thence by extra-high-voltage transmission lines to the major load centres around Manila.

5. THE OKOY FIELD (NEGROS)

5.1 General

In contrast to Leyte, the island of Negros possesses a power distribution grid, at least to a limited extent, and diesel generation facilities of about 50 MW already exist. Development of the Okoy geothermal field, however, is being pursued to match expected demand growth in Negros, and in response to the planned connection of the neighbouring islands of Cebu and Panay into a common grid. Cebu, particularly, has a considerable industrial power demand which must be serviced, so the opportunity to use geothermal resources for the provision of base-load energy is more immediate. The purpose of the proposed geothermal resource development is two-fold:- to stimulate industrial development in the region, and to enable early substitution to be made for the existing oil-fueled generating plant.

Exploration of the Okoy field has been continuing since 1975, and has resulted in the confirmation to date of at least 170 MW(e) of potential generating capacity. The field generally occupies the southern flanks and headwaters of the valley of the Okoy River. The topography is particularly vigorous and has necessitated developing the resource in two separate sectors, named Puhagan and

Nasuji/Sogongon (refer Fig 4).

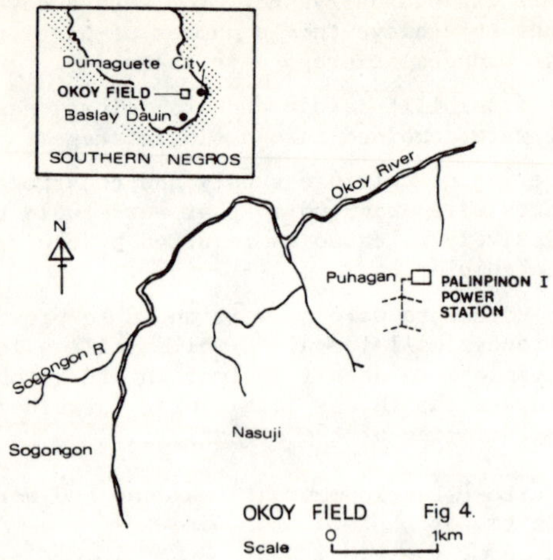

OKOY FIELD Fig 4.

5.2 Palinpinon I Power Station (Puhagan)

The first development is well under way in the Puhagan sector, with construction of the pipeline system about to commence, and the power house already half built. The turbines are similar to those for the Tongonan I station, and indeed, were selected on the basis of tenders received for Tongonan (i.e. three 37.5 MW(e) units).(refer Fig 5).

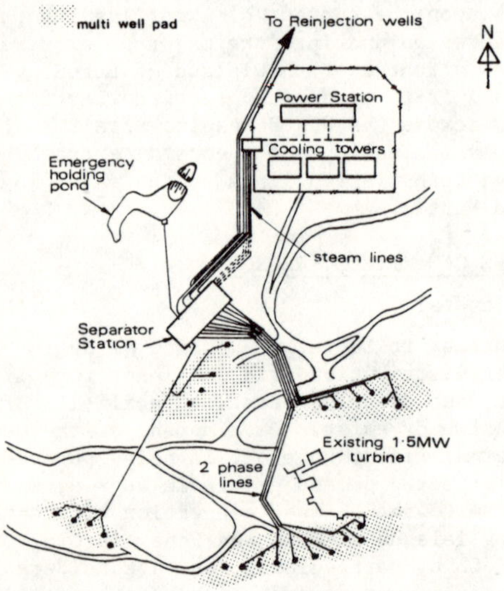

PALINPINON I POWER STATION DEVELOPMENT

Fig 5.

The topography and location of this field have imposed several important influences which distinguish the development from that at Tongonan:

. Since large areas of the resource zone are virtually inaccessible, a policy of drilling directionally from multi-well pads has been adopted. This applies to both the production and reinjection systems. As a result, an extremely compact fluid collection system, feeding two-phase fluids to a single centralised separator station, has been evolved.

. The field location is at high elevation (700-800 m), so that the depth to the reservoir is correspondingly greater. Significant differences in well casing design have resulted, and a deliberate decision taken to case off producing zones high in some of the wells in order to reduce the long-term likelihood of drawing in cold water following drawdown.

. The existing generating facilities in the region, which provide back-up, have reduced the degree to which diversity and reliability of power supply needs to be ensured. The compact steam collection system is considered acceptable in this situation. In addition, emergency provisions as back-up to the reinjection system are not as stringent, since it has been considered acceptable to shut the station down should repairs or maintenance to the reinjection system require it.

5.3 Further Development at Okoy (Palinpinon II)

The next stage in the Nasuji/Sogongon sector is about to enter the preliminary design phase. A recently completed feasibility study for the project has already pin-pointed several major constraints:

. Topography, again, imposes limitations on both power station sites and the corridors which are available for running pipelines. The high elevation of the area (800-1000m) poses a potential problem in respect of long-term drawdown in the wells, and it may be that a bias of wellhead locations towards the lower end of the sector could ensue. This has yet to be tested by drilling. Ironically, the best power station site lies towards the higher end of the sector, which has design implications for the location of separator stations and the limits to which two-phase lines can be run.

. Environmental constraints are of greater significance. As a result the development strategy for Palinpinon II is aimed at keeping site modification of all kinds (platforms, roading, well pads) to an absolute minimum. In order to cope with the chemical problem, it is also proposed that reinjection wells be amongst the first of those drilled in the development programme, and that all or part of the reinjection system be built early. This will enable well testing to proceed without hindrance whilst maintaining the chemical loading on the river at acceptable levels.

6. CONCLUSIONS

Whilst a number of common policies have been adopted in the development of the Tongonan and Okoy geothermal fields, significant differences exist. On the one hand, development strategy decisions at Tongonan have been dominated by the regional context and external influences of regional power demand upon the design. Development at Okoy, by contrast, is less constrained by such influences, and is instead significantly influenced by the physical characteristics of the resource area. This comparison demonstrates the site-specific nature of geothermal development.

REFERENCES

1. Ministry of Energy, 1981: The National Energy Program, 1981-86.
2. Various published project reports prepared by PNOC-EDC and KRTA staff.

NON-DARCY FLOW IN GEOTHERMAL RESERVOIRS

George Zyvoloski

Los Alamos National Laboratory
Earth and Space Science Division
Los Alamos, New Mexico 87545

ABSTRACT

The effects of non-Darcy flow laws are investigated for two geothermal reservoir types: multiphase and Hot Dry Rock (HDR). Long-term thermal behavior is emphasized as short-term pressure transient behavior is addressed in the oil field literature. Comparisons of Darcy and non-Darcy flow laws are made.

INTRODUCTION

The recent interest in geothermal energy has given rise to the development of numerical models that could be used to predict geothermal reservoir behavior. There have been a number of models developed that can simulate single and multiphase reservoirs. These models have provisions only for Darcy flow, but in many situations there is evidence of non-Darcy flow in geothermal reservoirs.[1-4] Geertsma[5] studied the results of many researchers, and obtained correlations for the non-Darcy coefficients. It is the purpose of this paper to investigate the effects of non-Darcy flow on long-term thermal behavior of geothermal reservoirs, and to assess the importance of non-Darcy flow on the different types of geothermal reservoir.

GOVERNING EQUATIONS

The detailed derivation of the governing equations has been presented by several investigators (see Brownell et al.[6] for example) and thus only a summary of the equations will be presented here.

The conservation of mass and energy for multiphase flow may be written

$$- \underline{\nabla} \cdot (D_m \underline{\nabla} P) + \frac{\partial}{\partial z} (D_{m\ell} \rho_\ell + D_{mv} \rho_v) + \frac{\partial A_m}{\partial t} = 0 \qquad (1)$$

and

$$- \underline{\nabla} \cdot (D_e \underline{\nabla} P) + \frac{\partial}{\partial z} (D_{m\ell} \rho_\ell h_\ell + D_{mv} \rho_v h_v) + \frac{\partial A_e}{\partial t} = 0 \qquad (2)$$

where P is the pressure and the subscripts m and e refer to mass and energy respectively. The accumulation terms A_m and A_e are given by

$$A_m = \phi(S_v \rho_v + S_\ell \rho_\ell) \qquad (3)$$

$$A_e = (1 - \phi)\, \rho_r u_r + \phi(S_v \rho_v u_v + S_\ell \rho_\ell u_\ell) \qquad (4)$$

Here ρ refers to the density, u to the internal energy, S to the phase saturation and ϕ to the porosity. The subscripts v and ℓ refer to the vapor and liquid phases respectively. The transmissibilities D_m and D_e are given by

$$D_m = D_{m\ell} + D_{mv} \qquad (5)$$

and

$$D_e = h_\ell D_{m\ell} + h_v D_{mv} \qquad (6)$$

with

$$D_{m\ell} = \frac{kR_\ell \rho_\ell}{m\ell} \delta_\ell \ , \quad D_{mv} = \frac{kR_v \rho_v}{\mu} \delta_v \qquad (7)$$

Here h refers to the enthalpy, R to the relative permeability, μ to the viscosity, and k to the intrinsic permeability. The term δ_ℓ, appearing in Eq. (7) may be expressed as:

$$\delta_\ell = 1/\left(1 + \frac{\beta_\ell \rho_\ell k \ |V_\ell|}{\mu_\ell}\right) \qquad (8)$$

where V_ℓ is the velocity of the liquid phase and β is the non-Darcy coefficient of the liquid phase. An analogous expression exists for δ_v. Geertsma [5] has correlated δ, k, and ϕ for a wide variety of materials and has presented the following relation:

$$\beta = 0.005 \frac{1}{\phi^{5.5} \ k^{0.5}} \qquad (9)$$

This empirical relation is valid for 100 percent liquid or gas saturation. For a mixture of phases a modified relation is presented:

$$\beta_v = \beta \cdot \frac{1}{S_v^{5.5} R_v^{0.5}} \tag{10}$$

Geertsma points out, however, that Eq. 10 may not be valid for high water saturations (greater than 40%) because Eq. 10 implies irreducible water saturations, and at these high saturations the water is likely to move and thus not be irreducible. There is no mention, however, of correlations for the nonlinear coefficient in multiphase flow for the liquid phase.

The dependent variables pressure (P) and mixture enthalpy (h) are used as dependent variables. The mixture enthalpy is related to the other varibles by the formula:

$$h = (\rho_v h_v S_v + \rho_\ell h_\ell S_\ell)/(\rho_v S_v + p_\ell S_\ell) \tag{11}$$

The functional dependence of the thermodynamic and transport quantities in Eqs. 1 and 2 on P and h are not given here. The formulae given by Faust and Mercer[7] were used and the reader can refer to that paper for details.

Equations 1 and 2 are highly nonlinear and are solved with a finite element based algorithm. Zyvoloski[8] describes the numerical procedures.

NUMERICAL EXAMPLES

The examples presented will represent multiphase and Hot Dry Rock geothermal reservoirs. It can be shown from Eqs. 8, 9, and 10 that non-Darcy flow terms are insignificant in most compressed liquid geothermal systems. Table 1 shows the data for the reservoir presented in Figure 1. This problem was originally investigated by Toronyi and Farouq Ali[9] with solutions by several researchers. The problem is one dominated by the steam phase and this serves to illustrate the effects of non-Darcy flow. There is some movement of the liquid phase since the saturation is 0.2 by liquid, and the residual liquid saturation is specified as 0.05, nevertheless the movement is slight and Eq. 11 is used to calculate the non-Darcy coefficient for the vapor phase. Figure 1 presents the results of this simulation with a comparison solution from Ref. 10. An initial time step of 8.3 days with 7 time steps of 10 days, resulted in a simulation time of 78.3 days. The results show that there is good agreement between the Darcy flow results presented here and those presented in Ref. 10. The results also show that the non-Darcy solution exhibits a greater drawdown than the Darcy solution. The pressure differences between the two flow types (not shown) for the multiphase system are less than the compressed water system because of boiling, however, the saturation difference between the Darcy and non-Darcy solutions are great. The relatively greater drying

Table 1. Parameters for Toronyi Example.

Parameter	Symbol	Value
Permeability	k	$9.869 \times 10^{-13} m^2$
Thermal Conductivity	K	1.730 W/(m°C)
Porosity	ϕ	0.05
Rock Density	P_r	2563 kg/m^3
Rock Specific Heat	C_r	1010 J/kg°C
Aquifer Length	--	1828 m
Aquifer Width	--	182.8 m
Initial Water Saturation	s^o_{ij}	0.2
Initial Pressure	p^o_{ij}	4.486 MPa
Residual Liquid Saturation	S_{RL}	0.05
Residual Vapor Saturation	S_{RL}	0.05
Discharge	q_m	0.082021 Kg/(sec m)

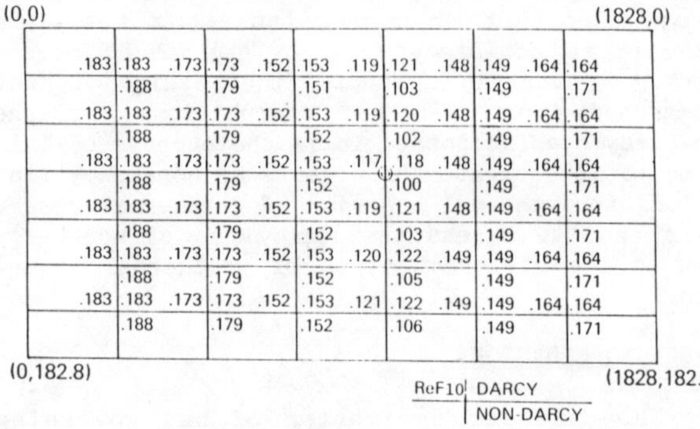

Figure 1. Flow field and results for Toronyi multiphase example.

out near the wellbore of the non-Darcy solution may enhance scale deposition which could clog the wellbore system.

The next example considered is a model Hot Dry Rock Fracture System. The model consists of a single two dimensional high permeability fracture embedded in a 3-dimensional block of granite. The reservoir data are presented in Table 2 and a model of the reservoir is given in Figure 2. Figure 3 shows a temperature contour map of the fracture face after 1000 days and shows graphically the non-Darcy effect of retarding the buoyancy drop of the incoming fluid, thereby decreasing the sweep area of the fracture. The additional resistance provided by the non-Darcy

Table 2 Parameters for Fracture Flow Example

Parameter	Symbol	Value
Permeability: Rock	k_R	$10^{-18} m^2$
Fracture	k_F	$10^{-8} m^2$
Thermal Conductivity	K	2.9 W/(m °C)
Porosity: Rock	ϕ_R	0.001
Fracture	ϕ_F	1.0
Rock Density	r	2700 kg/m^3
Rock Specific Heat	C_r	1000 J/kg° C
Reservoir Width	--	60 m
Fracture Horizontal Length	--	100 m
Fracture Vertical Length	--	400 m
Fracture Width	--	0.001 m
Initial Pressure	p^o_{ij}	hydrostatic
Initial Temperature	T^o_{ij}	140.0° C + .055(C/m)x2(m)
Discharge Production (Production Controlled by Pressure)	q_m	0.05 kg/sec
Injection		0.05 kg/sec at 118° C

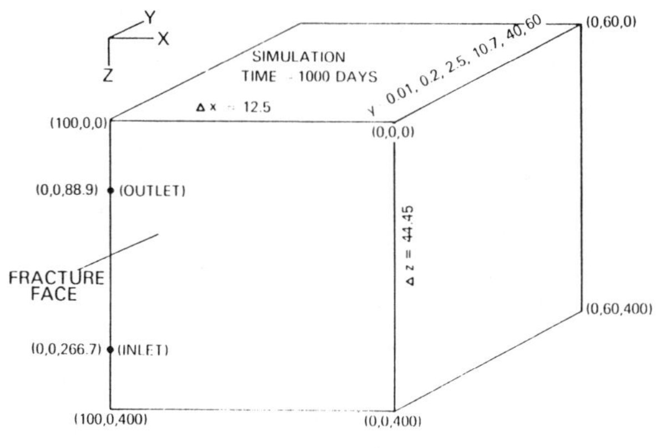

Figure 2. Model of Hot Dry Rock reservoir.

flow near the wellbore negates some of the bouyancy-induced spreading of the flow.

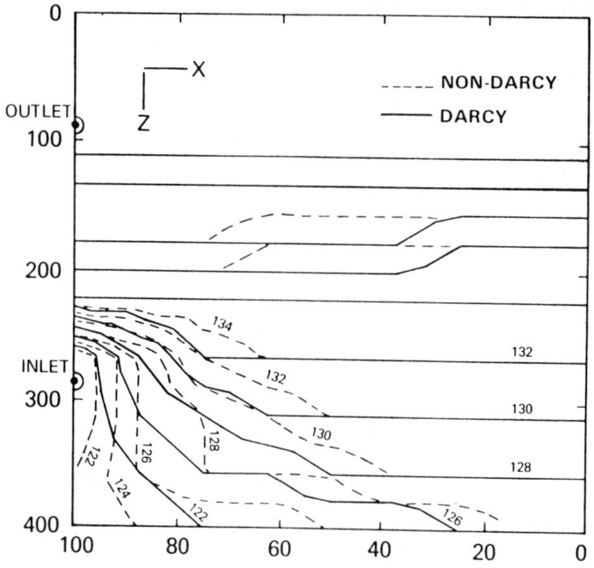

Figure 3. Temperature contour map for Hot Dry Rock mode.

DISCUSSION

The examples presented have indicated that non-Darcy flow can effect the performance of a geothermal reservoir system. The effects are presented for a multiphase system and for a Hot Dry Rock reservoir.

The results are dependent on Geertsma's correlation of flow data. Calculation of flow coefficients is experimentally possible if flow tests are run at several different flow rates. This procedure also helps to differentiate between effects of skin damage near the well and non-Darcy flow: skin damage effects vary linearly with flow rates whereas non-Darcy effects do not. At higher flow rates and hence higher pressures highly fractured systems may also experience fracture pressurization which tends to oppose the effects of non-Darcy flow. Work needs to be done in the area of detecting and characterizing non-Darcy flow in fracture dominated reservoirs.

CONCLUSIONS

1) Non-Darcy flow can be modelled in geothermal reservoirs.

2) Non-Darcy flow can be significant, and primarily effects multiphase and highly fractured reservoirs.

ACKNOWLEDGEMENT

The author would like to thank Rajagapal Raghavan for providing several of the references on non-Darcy flow. Financial support for this work was provided by the Department of Energy.

REFERENCES

1. K. H. Guppy, H. Cinco-Jay, and H. J. Ramey, Jr., "Effect of non-Darcy flow on the constant pressure production of fractured wells," Soc. Pet. Eng. J., 21, 390-400 (1981).

2. G. Rowan, and M. W. Clegg, "An approximate method for non-Darcy radial gas flow," Soc. Pet. Eng. J., 4, 99-114 (1964).

3. H. N. Fisher and J. W. Tester, "Pressure transient testing of a manmade fractured geothermal reservoir: an examination of fracture versus matrix dominated flow effects," Los Alamos National Labortory Report LA-8535-MS, (September 1980).

4. O. A. Plumb and J. C. Huenefeld, "Non-Darcy natural convection from heated surfaces in saturated porous media," Int. J. Heat Mass Transfer, 24, 765-768 (1981).

5. J. Geertsma, "Estimating the coefficient of inertial resistance in fluid flow through porous media," Soc. Pet. Eng. J., 14, 445-450 (1974).

6. D. H. Brownell, Jr., S. K. Garg, and J. W. Pritchett, "Computer simulation of geothermal reservoirs," paper SPE 5381 presented at 45th California Regional Meeting of the Soc. of Pet. Eng. of AIME, Ventura (1975).

7. C. R. Faust and J. W. Mercer, "Mathematical modeling of geothermal systems," <u>in</u> Proc. Second United Nations Symposium on the Development and Use of Geothermal Resources, San Francisco (1975).

8. G. A. Zyvoloski, "Finite element methods for geothermal reservoir simulation," Int. J. for Numerical and Analyticl Methods in Geomechanics, (in press).

9. R. M. Toronyi and S. M. Farouq Ali, "Two-phase two-dimensional simulation of a geothermal reservoir and wellbore system," Soc. Pet. Eng. J. 17, 171-183 (1977).

10. L. K. Thomas and R. G. Pierson, "Three dimensional reservoir simulation," Soc. Pet. Eng. J., 18, 151-161 (1978).

POWER GENERATION

Geothermal Resources Council, TRANSACTIONS Vol. 6, October 1982

GEOTHERMAL ELECTRIC POWER FORECAST IN THE UNITED STATES AND THE IMPACT OF R&D ADVANCEMENTS: 1982-2000

T.A.V. Cassel, G.T. Shimamoto, C.B. Amundsen and P.D. Blair

Technecon Analytic Research, Inc.
2400 Chestnut Street
Philadelphia, Pennsylvania 19103

ABSTRACT

This paper describes the use of modern techniques from the fields of decision science and econometrics to provide statistically-sound forecasts of geothermal electric power development in the United States. The computerized forecasting model first predicts the timing and quality of future geothermal resource discoveries from an underlying resource base, developed with the cooperation of several highly qualified geothermal geologists. The model then forecasts the rate and extent of utilization of these geothermal resource discoveries, based on the joint investment behavior of resource developers and electric utilities as statistically determined from extensive industry interviews. The paper concludes that, depending on the extent of R&D achievements in related areas of geosciences and technology, geothermal power development is expected to reach between 7700 and 17300 MWe by the year 2000. This represents between 8 and 18% of the expected electric energy demand (GWh) in western and northwestern states.

OVERVIEW OF GEOTHERMAL POWER FORECASTING

For the past 4 years we have been working with the geothermal industry and the United States Department of Energy (USDOE) to develop a sound approach for estimating the future growth of geothermal electric power in the United States. This work benefits from the cooperation of 20 geothermal resource companies, 13 western electric utilities and a number of financial institutions and governmental agencies. Additional expertise was provided by highly qualified geologists, reservoir engineers and power systems engineers. To this wealth of industry information, we added pertinent resource data from extensive USDOE and USGS computer files and then applied modern econometric and decision science techniques to develop a statistically-sound forecasting instrument. This paper provides an overview of this computerized instrument and presents current national geothermal power forecasts. Detailed descriptions of this work may be found in Cassel, et al. (1982) and Blair, et al. (1982).

The forecasting model sequentially addresses the following factors which directly influence the rate and extent of geothermal power development:

- o The level of wildcat well drilling activity in future years;
- o The number of resource confirmations in future years as a result of wildcat drilling activity;
- o The quality of future resources; and
- o The likelihood of resource developers and electric utilities undertaking well field and power plant projects at new, as well as at currently known, resources.

GEOTHERMAL WILDCAT DRILLING ACTIVITY

A careful review of more than 2000 records of geothermal well drilling activity reveals that, from 1958 to 1980, there were 110 wildcat wells drilled in the United States outside The Geysers resource area. Table 1 summarizes this drilling activity. Using statistical regression techniques we found that this wildcatting activity can be explained in terms of three parameters; namely: (a) the expected real, risk-adjusted rate of return on geothermal projects as estimated by detailed probabilistic cash flow analyses based on marginal competition with coal-fired power generation; (b) the risk-adjusted return for new oil discoveries adapted from short-term forecasts

Table 1. U.S. Geothermal Exploration Activity

Year	Wildcat Wells	Discoveries	Cumulative Drilling Success Ratio	Discovered Areas
1958	1	0	0	
1959	4	0	0	
1960	1	1	16.7	Mono-Long Valley
1961	2	0	12.5	
1962	2	0	10.0	
1963	3	0	7.7	
1964	9	3	18.2	Salton Sea; Brady-Hazen; Sulfur Bank
1965	2	0	16.7	
1966	0	0	16.7	
1967	2	0	15.4	
1968	1	1	18.5	Wilbur H.S.
1969	0	0	18.5	
1970	1	1	21.4	Baca Location #1
1971	0	0	21.4	
1972	10	2	21.1	East Mesa; Heber
1973	7	1	20.0	Power Ranches
1974	9	1	18.5	Soda Lake
1975	9	3	20.6	Brawley: Beowawe; Roosevelt H.S.
1976	5	1	20.6	Westmoreland
1977	7	2	21.3	Coso; Clear Lake
1978	10	1	20.0	Humboldt House
1979	15	3	20.0	Dixie Valley; Steamboat H.S.; Redondo
1980	10	1	19.1	S. Brawley

of new U.S. crude oil discoveries; and (c) the enactment of the Geothermal Steam Act in 1970. These three parameters combine in a linear equation (estimated by ordinary least squares techniques) to accurately fit historic wildcat data and to provide the basis for forecasts of future wildcatting activity.

The direct relationship that we find between annual wildcat drilling and the expected return on geothermal projects is readily understandable. Likewise, it is understandable that the enactment of the Geothermal Steam Act, which expanded leasing opportunities and improved the investment climate for geothermal resource development, caused a jump of about 10% in the average rate of drilling. However, our initial perception of geothermal projects "competing" with oil projects for drilling funds proved invalid. Instead, we find that geothermal wildcatting increases with increases in oil profitability. This is verified by oil industry executives who indicate that their geothermal budgets tend to reflect discretionary investment of oil-related income.

GEOTHERMAL WILDCAT SUCCESS RATIO

In the next step of the forecasting procedure we estimate the numerical probability that the wildcat wells will be successful in discovering a producible resource. As shown in Table 1, the success ratio for wildcat wells was 19.1% through 1980. However, due to depletion effects as more discoveries are made, we expect this success ratio to eventually decline. To estimate the probability of wildcat success, we use a technique known as "probit analysis" based upon data from geothermal well records. We derived a statistically-sound discovery function based upon three parameters; namely: (a) the geographic area of exploration; (b) cumulative number of successes to date; and (c) exploratory effort measured in cumulative well depth.

GEOTHERMAL RESOURCE QUALITY FORECAST

Future geothermal resource discoveries will be drawn from an underlying resource base. Early estimates of this geothermal potential appeared in USGS Circular 726, later updated in 1978 (USGS Circular 790). In response to requirements of the work discussed here, however, the Earth Science Lab of the University of Utah Research Institute,assisted by Eugene Ciancanelli of Cascadia Exploration Corp. and Dr. Subir Sanyal of Geotherm-Ex, Inc. under subcontract to Technecon, developed an updated and considerably refined characterization of the undiscovered resource base. A limited set of "generic" resource types was constructed, each of which describes a class of reservoirs having similar characteristics. These included temperature, well cost, well flow, salinity, areal extent and volume. Characterization of the undiscovered resource base then is completed by estimating the regional frequency of occurrence of each generic type.

As resources are discovered and withdrawn from the undiscovered resource base, increasing effort must be devoted to find additional reservoirs. Not only are reservoirs increasingly difficult to find, but the average quality of newly found discoveries will tend to decline. Following Barouch and Kaufman's analysis of oil supply (1976) and subsequent elaborations by Smith and Ward (1980), we view the discovery process as a sampling experiment in which a population of reservoirs is sequentially sampled without replacement according to a probability law that reflects the industry's search for high rather than medium or low temperature resources. Our rigorous translation of this oil industry methodology to the geothermal industry is described in detail in Cassel, et al (1982).

WELL FIELD/POWER PLANT INVESTMENT DECISIONS

The final forecasting step focuses on the investment behavior of resource developers and electric utilities. The probability of investment at each known and new resource is evaluated using a methodology which incorporates two modern techniques from the decision sciences, namely "multiattribute utility analysis" and "logit choice estimation" (see Blair, et al. 1982).

Results from detailed well field and power plant cash flow simulations (see Cassel, et al. 1981) are provided to decision models. This methodology compares time-wise geothermal investment opportunities to other investment opportunities available to developers and utilities. From the developers' point of view, comparisons are made in terms of rate of return, duration of investment exposure, project size, amount of capital at risk, and the probability of project failure. From the electric utilities' point of view, comparisons are made in terms of electric energy production cost (mills/kWh), plant availability, net plant output, and risk to corporate or municipal bond rating. A computerized multiobjective analysis then provides estimates, over time, of the joint likelihood of investments to develop well fields and to construct on-site power plants and transmission facilities. These models are statistically-strong in their ability to reproduce investment preferences as expressed in extensive interviews and demonstrated field practice.

NATIONAL GEOTHERMAL POWER FORECAST: 1982-2000

National geothermal electric power forecasts using the integrated approach outlined in this paper are provided for 2 cases:

Current Technology and Incentives (Case I): Assumes current state of geothermal technology with the introduction of a mature binary-type power plant in the mid-1990's. Tax incentives are assumed consistent with regulations in effect in early 1982.

Significant Technological Advances and Enhanced Incentives (Case II): Assumes appreciable R&D effort resulting in performance and cost improvements during the late 1980's in geosciences, well drilling, well stimulation, downhole pumps, piping materials and binary- and flash-type power plants. Also assumes a mature advanced binary-type power plant is introduced in the early-1990's for use at lower temperature resources. A 15% energy investment tax credit is assumed to be available for both well field and power plant investments through 1995.

Forecasted power development at both new discoveries and at the currently known resources listed in Table 1 is given in Table 2 for the two cases. Figures 1 and 2 illustrate the confidence intervals surrounding these two forecasts.

Table 2. Expected U.S. Geothermal Power (MWe net)

	Actual[a] 1981	Survey[b] 1985	Forecast[c] 1985	1990	1995	2000
THE GEYSERS ALONE	902	1753	1720	2380	2680	2890
LIQUID-DOMINATED ALONE						
Case I..............	15	391	280	1760	4050	4820
Case II.............	15	391	390	2770	6600	14420
TOTAL GEOTHERMAL						
Case I..............	917	2144	2000	4140	6730	7710
Case II.............	917	2144	2110	5150	9280	17310

Note: "Case I" assumes current technology and tax incentives
"Case II" assumes significant technological advance and enhanced incentives

[a]Adapted from Cassel (1982).
[b]Includes both "Planning Commitments" and "Speculative Additions" from 1982 survey of electric utilities and resource developers (ref. Cassel, 1982).
[c]Expected Values; confidence intervals are displayed in Figures 1 and 2.

Case I. Current Technology. Under Case I assumptions, wildcat drilling levels abruptly stabilize, ending a 20-year period which saw a compound annual growth rate of 9% through 1980. Stagnant drilling levels cause a decline in the resource discovery rate through the mid-1990's. With fewer discoveries to exploit, expected power capacity (including The Geysers) grows to only 7710 MW by the year 2000.

Case II. Significant Technological Advance. The assumptions of Case II have a notable, positive effect on every stage of geothermal development -- wildcat well drilling, discovery, and power plant construction. Wildcat well drilling increases at the historic rate of 9% per year through 1995, and discovery rates are significantly higher than those of Case I. These discoveries are more efficiently and economically utilized by the technological advances of this case. Major contributors are the assumed availability, in the late 1980's and early 1990's, of high-temperature (to 425F) downhole pumps, more cost-effective drilling technology and advanced binary-type power plants. Total expected capacity grows to 17,310 MW by the year 2000, more than doubling levels expected under Case I.

REGIONAL BENEFIT OF GEOTHERMAL POWER

The regional distributions of our geothermal power forecasts for the year 2000 are provided in Figure 3. Upwards of 95% of the geothermal power in the U.S. is expected to be developed within the California-Nevada Power Area and the Northwest Power Pool (excluding Canada) which includes

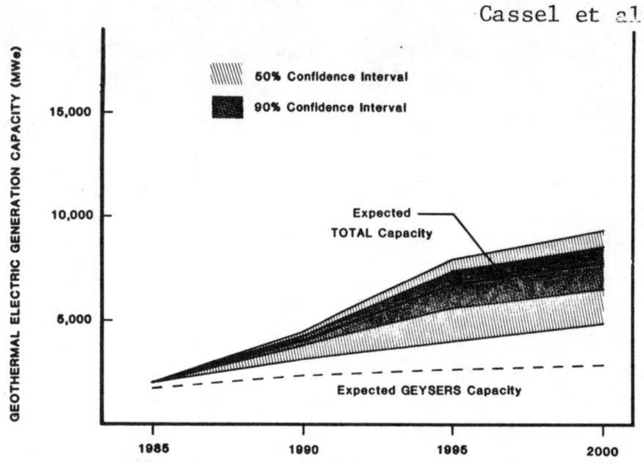

Figure 1. Geothermal Power Forecast With Current Technology (Case I)

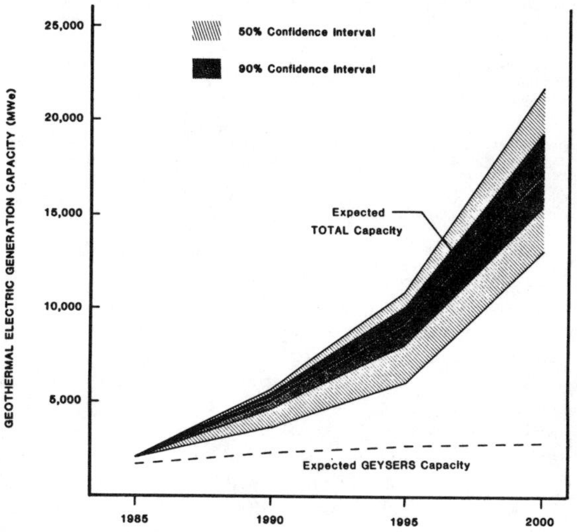

Figure 2. Geothermal Power Forecast With Technology Advancements (Case II)

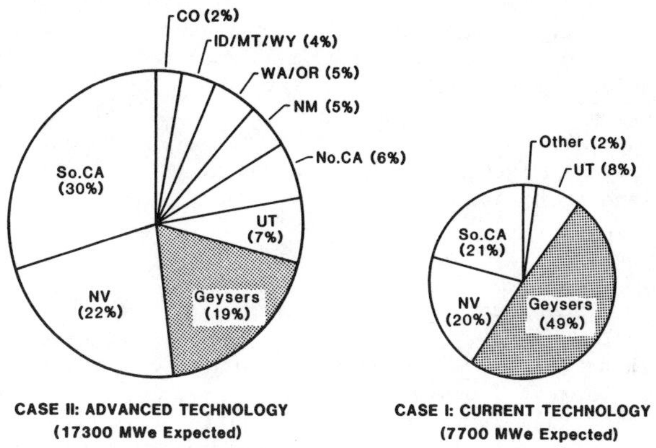

Figure 3. Regional Geothermal Power in Year 2000

Table 3. Regional[a] Geothermal Power Impact: Year 2000

	Geothermal Capacity Forecast (MWe)	Percent of Regional Baseload Capacity[b]	Percent of Baseload Additions 1981-2000[c]	Percent of Regional GWh Energy Production[d]
THE GEYSERS ALONE	2890	2.1%	3.2%	3.0%
LIQUID-DOMINATED ALONE				
Case I..............	4820	3.4%	7.6%	5.1%
Case II.............	14420	10.3%	22.9%	15.1%
TOTAL GEOTHERMAL				
Case I..............	7710	5.5%	10.8%	8.1%
Case II.............	17310	12.3%	26.1%	18.1%

Note: "Case I" assumes current technology and tax incentives
"Case II" assumes significant technological advance and enhanced incentives

[a]Region comprised of California-Nevada Power Area and Northwest Power Pool (excluding Canada).

[b]Regional baseload capacity in year 2000 estimated to be 140,440 MWe (net) based upon Western System Coordinating Council (WSCC) forecast of 104,280 MWe for 1990 and a compound annual growth rate of 3.0% (WSCC, 1981).

[c]Regional baseload additions from 1981 to 2000 estimated to be 62,900 MWe based upon installed baseload capacity of 77,540 MWe in 1980 and abovementioned estimate for 2000.

[d]Geothermal electric energy estimate based upon an 85% capacity factor. Regional net electric energy production in year 2000 estimated to be 709,600 GWh based upon WSCC forecast of 511,100 GWh for 1990 and a compound annual growth rate of 3.3% (WSCC, 1981).

California, Nevada, Utah, Oregon, Washington, Idaho and Montana. As illustrated in Table 3, the regional impact of geothermal power in this area is likely to be quite significant. Depending upon the degree of R&D achievements over the next two decades, geothermal power plants are expected to comprise between 10% and 26% of the new baseload generating capacity installed through the year 2000 in the California-Nevada and Northwest regions. By the year 2000, this geothermal capacity is expected to provide between 8% and 18% of the total regional electric energy demand. Based upon this forecast, geothermal capacity may be able to supplant a significant amount of new baseload additions currently planned as nuclear or coal-fired units (see WSCC, 1981). This may be desirable because, as noted in the WSCC reliability study, "nuclear and coal-fired baseload units are highly susceptible to delays...and substantial or sustained delays in these installations can be expected to have serious consequences on the adequacy of the region's bulk power supply". The relatively short construction lead times and minimal environmental impact of geothermal plants offer an attractive planning option.

CONCLUSION

The substantial promise of geothermal electric power in the United States is demonstrated in these forecasts of resource development. However, the future expansion of the geothermal industry will be quite sensitive to technological advances that reduce investment risk and/or increase investment return. Only a few industry participants (typically large oil companies) are sufficiently capitalized to undertake major, high-risk and long-payoff R&D ventures. Because of limited and scattered leaseholdings, the potential payoff at a particular resource may be insufficient to warrant significant R&D investment. While the aggregate benefit of geothermal power to the nation appears substantial, the benefits of R&D program investments to individual firms may, at best, be marginal. A central question, which remains to be addressed, is whether the fragmented geothermal industry is sufficiently motivated and capable of producing technological advances, through R&D activities, within a time frame that is consistent with national needs as they unfold over the next several decades.

REFERENCES

Barouch, E., and G.M. Kaufmann (1976), "Probabilistic Modeling of Oil and Gas Discovery," in Energy: Mathematics and Models, SIAM Institute for Mathematics and Society.

Blair, P.D., T.A.V. Cassel, R.H. Edelstein (1982), Geothermal Energy: Investment Decisions and Commercial Development, John Wiley and Sons, New York.

Cassel, T.A.V. (1982), "National Outlook for Geothermal Electric Power Development", Energy Technology IX - Energy Efficiency in the Eighties, Government Institutes, Inc., Rockville, MD, July.

Cassel, T.A.V., C.B. Amundsen, R.H. Edelstein, P.D. Blair (1981), Geothermal Well Field and Power Plant Investment Decision Analysis, Report No. DOE/ET/27242-T1, National Technical Information Service, Springfield VA.

Cassel, T.A.V., G.T. Shimamoto, C.B. Amundsen, P.D. Blair, W.F. Finan, M.R. Smith, R.H. Edelstein (1982), National Forecast for Geothermal Resource Exploration and Development, Report No. DOE/ET/27242-T2, National Technical Information Service, Springfield VA.

Smith, J.W. and G.W. Ward (1980), Maximum Likelihood Estimates of the Size Distribution of North Sea Oil Deposits, MIT Energy Lab Working Paper No. MIT-EL 80-027WP.

Western Systems Coordinating Council (1981), Coordinated Bulk Power Supply Program 1980-1990, Salt Lake City, April.

Geothermal Resources Council, TRANSACTIONS Vol. 6, October 1982

FINAL REPORT ON HGP-A WELLHEAD GENERATOR PROOF-OF-FEASIBILITY PROJECT

Bill H. Chen and Louis P. Lopez

University of Hawaii at Hilo

Research Corporation of University of Hawaii

ABSTRACT

The construction of the 3 MW geothermal power plant at HGP-A well was completed in June, 1981. The HGP-A started production on June 12, 1981. Numerous problems developed during the plant start-up and shakedown periods. However, the plant was formally turned over to Hawaii Electric Light Company (HELCO), the local utility company on March 1, 1982. Currently, all systems are functioning well and all major objectives have been met. The plan is to operate the power plant through 1990 where long term equipment and material information and reservoir characteristics can be obtained.

INTRODUCTION

The HGP-A Wellhead Generator Proof-of-Feasibility Project began in June, 1978 with funding from US Department of Energy (DOE), State of Hawaii, County of Hawaii and Hawaii Electric Light Company (HELCO) (Ref. 1). The initial objectives of the project were: (Ref. 2)

1. To design and construct a 3 MW condensing generator system with full environmental controls.
2. To operate the plant for two years with an environmental monitoring program to monitor the effectiveness of the environmental controls.
3. To collect the reservoir production data and to assess the reservoir potential.
4. To build a visitor center where the general public will have a firsthand look at an on-going operating geothermal power plant.

In addition the initial projected plant-on-line data was December 1, 1980. However, due to the unexpected development of well casing problems in 1979 (Ref. 3) a well workover program was carried out and the projected plant-on-line date was delayed to June 1981.

START-UP

The construction of the power plant was completed in June, 1981 and the HGP-A well was brought into production on June 12, 1981. After a month of shaking down, the plant was generating electricity successfully and the plant was dedicated on July 17, 1981.

Soon after the dedication, some serious problems developed. These were:

1. Control valve seats
2. Main lube oil pump
3. Turbine inter stage drain alarms and trips.

To remedy the above problems, the turbine was frequently shut down and steam piping pulled open. These frequent exposures of the steam piping to atmosphere have resulted in oxidation of the piping system and subsequently the oxides coated onto the turbine blades. On August 30, 1981, the turbine rotor experienced a vibration problem and resulted in slight damage to the rotor. Consequently, the well was shut down on September 4, 1981 so the rotor could be repaired.

After the turbine rotor was repaired and reassembled in November, 1981, minor changes were made in the generator package by the manufacturer. This change resulted in another delay due to the necessity of rebalancing the generator. After considerable difficulties in balancing the generator, the plant was formally on line to HELCO on March 1, 1982. Currently, HELCO is operating and maintaining the plant under contract with the project. Table I shows the production of the well fluids and the net KWH generated since June 12, 1981.

HYDROGEN SULFIDE ABATEMENT SYSTEM

The main H_2S abatement system employed at the power plant is the incinerator/absorber system. The non-condensible gas ejected from the condenser is fed into the incinerator where the H_2S is oxidized to SO_2. Subsequently, the SO_2 is absorbed in a column where 10% caustic solution is circulated. The system is performing very well and periodic checks revealed that 100% of the H_2S and 99.9% of the SO_2 were removed. The off-gas is then routed to the cooling tower and dispersed. With this incinerator/absorber system on line, the power plant is approximately 98% efficient in removing the H_2S.

During plant down time, the steam is vented through a rock muffler where 10% caustic solution is again injected to remove the H_2S. Three moles of sodium hydroxide to one mole of H_2S is used and we consistently remove 90% of the H_2S.

ENVIRONMENTAL MONITORING PROGRAM

Ambinet air quality was monitored by the Houston-Atlas continuous H_2S monitoring system and colortec cards. Table II shows the average

H_2S concentration monitored approximately 1 mile downwind from the plant. We plan to install two additional ambient air monitoring stations downwind in June, 1982 to expand our monitoring effort.

Five wells surrounding the HGP-A were periodically sampled to monitor effects of the surface percolation of geothermal spent fluid. So far, no effect has been observed. Water from water catchment systems around the area was also sampled and analyzed for its sulfide content. No increase in sulfide content is observed.

CONCLUSION

The HGP-A Wellhead Generator Proof-of-Feasibility Project has achieved all of its initial objectives in building a plant with full environmental controls and our monitoring program generally confirms our environmental control effort. We plan to operate this plant under contract to HELCO until 1990 where long term equipment and materials information and reservoir production characteristics can be obtained.

TABLE I

HGP-A PRODUCTION DATA*

Dates	Steam (Million Lbs)	Brine (M Lbs)	Total (M Lbs)	Net KWH Generated
June, 1981**	24.26	24.25	51.19	
July, 1981	40.92	40.92	81.14	
August, 1981	39.43	42.78	82.21	124,600
September, 1981***	4.59	5.13	9.72	
December, 1981***	25.30	42.85	68.15	
January, 1982	37.94	49.99	87.93	116,600
February, 1982	34.27	41.93	76.20	227.400
March, 1982	37.94	45.35	83.29	866,000
April, 1982	36.72	43.89	80.61	1,440,000
May, 1982	37.94	45.35	83.29	1,788,000

* Wellhead pressure is set at 160 psig.

** Well production commenced on June 12, 1981

*** Well was shut down between Sept. 4 and Dec. 11, 1981

TABLE II

Monthly Average of Ambient H_2S Concentration At Schroeder's Station (Approximately One Mile Downwind of HGP-A)

Month	Ave. (Peak) in Parts Per Billion
March, 1981	4 (27)
April, 1981	5 (28)
May, 1981	4 (21)
June, 1981	6 (27)
July, 1981	6 (28)
August, 1981	4 (45)
September, 1981	5 (21)
October, 1981	3 (13)
November, 1981	2 (9)
December, 1981	3 (41)
January, 1982	5 (40)
February, 1982	4 (42)
March, 1982	3 (48)

Note: HGP-A well was in production from 6/81 to 9/4/81 and from 12/11/81 to present.

ACKNOWLEDGEMENT

This work is performed under the Department of Energy Contract DE-AC03-78-ET28420, and fundings from the State of Hawaii, County of Hawaii and Hawaii Electric Light Company. The plant is designed by Rogers Engineering, Inc. of San Francisco and environmental monitoring work is done by Environmental Analysis Laboratory in Richmond, California.

The authors also wish to thank Anthony Adduci, Program Manager of DOE/SAN for his participation in the start-up.

REFERENCES

Chen, B.H. and Grabbe, E.M., "Planning for Geothermal Development in Hawaii", Geothermal Resources Council, Transactions, Vol. 2, July 1978, p. 95-98.

Chen, B.H., Lopez, L.P., Farrington, R.J., Kuwada, J.T. and Uemura, R.T., "HGP-A Wellhead Generator Feasibility Project", Geothermal Resources Council, Transactions, Vol. 3, September 1979, p. 103-106.

Chen, B.H., Lopez, L.P., Kuwada, J.T. and Farrington, R.J., "Progress Report on HGP-A Wellhead Generator Feasibility Project", Geothermal Resources Council, Transactions, Vol. 4, September, 1980. p. 491-494.

Geothermal Resources Council, TRANSACTIONS Vol. 6, October 1982

HEBER GEOTHERMAL PROJECT

Norman J. De Haven

Southern California Edison Company

ABSTRACT

The Heber Geothermal Project a joint venture between Southern California Edison Company and Chevron USA, Inc. consists of a dual flash 52 MW gross/47 MW net power plant. This plant was conceived in 1978 with establishment of the Brine Supply Contract in November, 1978. Approval of the California Public Utilities Commission was sought on March 10,1980 and was rejected by the Commission due to the project unfavorable financial burden upon the rate payer. Subsequently, a revised project was resubmitted to the CPUC and is presently awaiting an approval.

INTRODUCTION

Southern California Edison (SCE) and Chevron USA, Inc. (Chevron) have entered into contractual arrangements consisting of a geothermal energy contract and geothermal sales contract, for the development of a geothermal project with a power plant nameplate rating of 52 MW.

PLANT SITE

The 28 acre site for the proposed plant is located southeast of the community of Heber. (See Figure 1). The land is currently vacant and not in agricultural use, although land use around the site is predominantly agricultural.

PLANT

Edison is engineering and will construct and operate a 52 MW gross/47 MW net geothermal electric generating facility, with an expected annual capacity factor of 75 percent. The major equipment components to be owned and operated by Edison include the turbine generator, flash vessels, cooling tower, condenser, pumps, transformers and switchgear. The major components of the Chevron facility for delivery of geothermal fluid to Plant No. 1 include the production island, a group of production wells drilled into the geothermal reservoir, an injection pipeline system which includes fluid injection pumps and a pipeline; and an injection island, a group of injection wells to return spent fluid and condensate to the geothermal reservoir.

Edison will own and operate the generating power plant requirements. Chevron will own the production and injection islands, and the fluid pipeline system, including the pumps, between the islands and the power plant.

PROJECT LOCATION

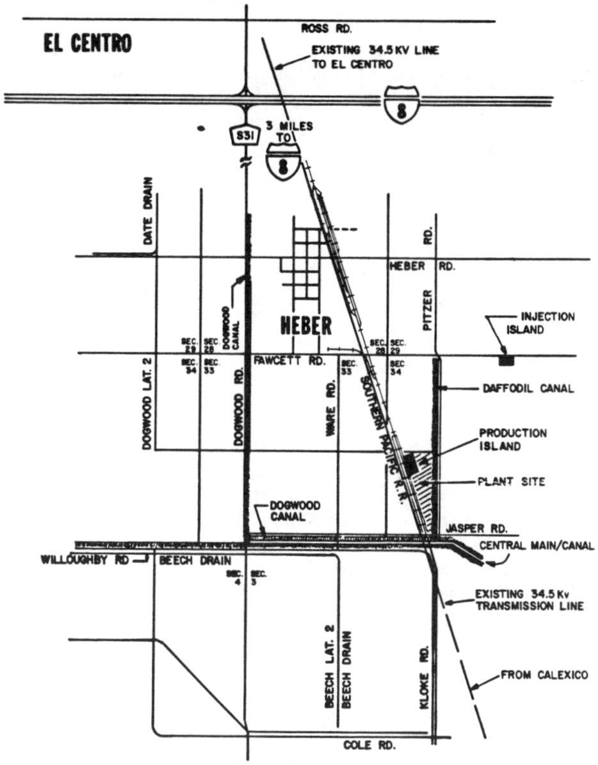

FIGURE 1

WATER

Operation of the power plant will require approximately 5000 acre feet of water annually for makeup use in the cooling tower. Arrangements have been made with Imperial Irrigation District (IID) to use water from a nearby canal for the operation of Plant No. 1 for five years. Discussions will continue with IID to obtain water up to the first ten years of operation. Edison will own and operate the required canal pump station, piping and its associated equipment. When canal water is no longer available, the cooling water will be obtained by using plant condensate. When this occurs and in order to maintain 100% reinjection of geothermal fluid into the reservoir, the amount of condensate lost for reinjection will be made up by treated water from the New River located nearby.

TRANSMISSION

Operation of Plant No. 1 is not anticipated to require the construction of new transmission facilities. An existing IID 34.5 kV transmission line (T/L) runs along the western boundary of the site and will be used to transmit power generated at the site. Due to the increased power flow associated with Plant No. 1's output, approximately 7.2 miles of that line will be reconductored to increase its power carrying capacity. Within the Project site, a 34.5 kV switchyard will be constructed and the existing 34.5 kV line will be looped into the Plant switchyard.

In the future, as additional geothermal plants are built, it may become necessary to build an additional 220 kV T/L from Devers Substation north of the Palm Spring area to the Heber area to collect and transmit the geothermal power from the Heber and other KGRA's. That transmission line would be sited in accordance with the "Transmission Corridor Element" of the Imperial County General Plan, adopted August 12, 1980.

PROJECT SCHEDULE

The schedule for Plant No.1 is shown in Figure 2. Engineering and construction have started and will be completed in mid 1984. The turbine roll is scheduled for April 1, 1984 with firm operation scheduled for August, 1984.

CAPITAL COSTS

Capital investment requirements associated with the Heber Geothermal Project are summarized in Table 1. Edison is responsible for the turbine generator plant and the associated flash system. These facilities constitute only about sixty percent of the total project investment in constant dollars. Chevron is responsible for the remaining capital costs: all expenditures and facilities associated with brine production and injection.

TABLE I

SOUTHERN CALIFORNIA EDISON COMPANY
HEBER GEOTHERMAL PROJECT
CAPITAL COST
(THOUSANDS OF DOLLARS, IN YEAR OF EXPENDITURE)

DESCRIPTION	COST
STRUCTURES AND IMPROVEMENTS	7,780
FUEL HOLDERS, PRODUCER, AND ACCESSORIES	1,280
PRIME MOVERS	20,860
TURBINE - GENERATOR	26,430
ACCESSORY ELECTRICAL EQUIPMENT	10,180
MISCELLANEOUS POWER PLANT EQUIPMENT	5,090
TRANSMISSION - STATION EQUIPMENT	190
COMMUNICATION EQUIPMENT	190
RIGHT OF WAY AND LAND	100
TRANSMISSION (IID)	500
TELECOMMUNICATION	500
TOTAL	73,100

GOVERNMENTAL AUTHORIZATIONS

State

On March 10, 1980, Southern California Edison submitted an application to the California Public Utilities Commission (CPUC) for an order certifying that the public convenience and necessity required Edison to construct and operate such power plant for service to Edison's customers. After conducting public hearings, and considering the evidence presented to it, the Commission concluded that the costs of the project were not competitive, and for that reason, among others, denied the application.

Edison and Chevron thus restructured the project so that it would be more competitive with the costs of power from other sources, and resubmitted an application to the CPUC on December 1, 1981. A decision on this amended application is still pending.

County

An application for a Conditional Use Permit (CUP) for the original project was granted by the County of Imperial on January 22, 1980. In compliance with the California Environmental Quality Act (CEQA) an Environmental Impact Report for the project entitled "Final Master Environmental Impact Report (EIR) was prepared by the County of Imperial as lead agency prior to the issuance of the CUP. The EIR describes the scope of Plant No. 1 in relation to potential full field development. On December 28, 1981, Edison filed an amendment to its Conditional Use Permit with Imperial County to reflect the changes in the project from that originally proposed. The amended application was granted by the County of Imperial on March 23,1982.

PLANT DESIGN CONSIDERATIONS

1. Plant Processes

Edison will utilize a double-flash system to separate steam from fluid, with the steam being used to generate electricity from a multi-pressure turbine-generator. Edison studied various geothermal cycles to determine which technology would be the most appropriate for the moderate temperature and low salinity fluid found at Heber. Dual-flash, four-flash, and binary technologies were compared considering commercial practicality and availability, economics, and environmental effects of each technology.

Because of the quality of the brine at the Heber Geothermal anomaly, it was determined that flash steam cycles would be practical. Based on the criteria of proven technology, lower capital cost and availability of commercial equipment, a dual-flash cycle was selected for brine found at Heber. Such dual-flash plants are currently in operation in several other countries. An essentially identical plant to that planned at Heber is operating successfully in Hatchobaru, Japan, under similar temperature and salinity conditions.

Geothermal fluid to be utilized in the proposed dual-flash power plant cycle will be produced by Chevron Resources Company. Chevron's production facilities will be located contiguous to the power plant site, making the production pipelines as short as possible. At full plant load, approximately 8,100,000 lbs/hr of geothermal fluid will enter the first stage flash (or separator) tanks where steam is separated and then flows to the throttle of a steam turbine-generator. Throttle steam conditions are expected to be 57 psia and 287°F. The anticipated heat rate at this pressure and temperature is about 28,300 Btu/kWh. The heat rate is high compared to conventional power plants because of the inherently low pressure and temperature of the geothermal resource. Fluid from the bottom of the first stage tanks flows to a set of second tanks where additional steam at 16 psia and 216°F is separated for use at a lower pressure region of the steam turbine.

Spent fluid from the second stage tank is redelivered to Chevron for return to the geothermal reservoir. Exhaust steam from the turbine will go to a steam condenser and the condensed steam (condensate) will also be returned to Chevron for return to the reservoir. Canal water will be used for cooling water and for miscellaneous power plant service water requirements.

The plant's heat rejection load is dissipated in a mechanical draft evaporative cooling tower. The cooled water passing down through the tower is collected in a concrete basin below the tower. Circulating water pumps convey the water from the basin through the steam condenser and back to the top of the cooling tower.

2. Plant Electrical System

Electrical power is expected to be generated at 13.8 kV and fed through the generator breaker to a 13.8/34.5 kV step-up transformer. The transformer will be connected to IID's transmission system through 34.5 kV switchgear located in the plant switchyard. The existing IID 34.5 kV transmission line is located within a few feet of the western boundary of the proposed plant site. Power supply to turbine-generator auxiliary equipment and other station load is provided from 4.16 kV and 480 V auxiliary power systems fed from the 13.8kV system.

Construction power will be provided from IID's existing distribution system. This power source will also serve as a permanent AC emergency supply when normal auxiliary power from the generating unit is not available.

Chevron will purchase power from IID for the operation of the production plant and injection facilities.

The electrical system also includes indoor and outdoor plant lighting, switchyard lighting, a plant and switchyard grounding system and a communication system for in-plant and system-wide needs.

SAFETY & RELIABILITY

1. Safety

The plant will be designed in accordance with all the applicable codes and standards to provide for the safety of the plant operators and the general public. Specific safety features will include insulation on fluid and steam pipes, noise control, warning signs, protection against overpressurization, and automatic equipment shutdown provisions in case of abnormal system conditions. In addition, operating procedure will be developed for the plant operators to provide a safe start-up, normal operation and shutdown of the plant.

2. Reliability

The plant is expected to operate reliably at 75 percent or greater capacity factor. This high reliability expectation is the result of consideration of the following:

(a) Minimization of component failure and expeditious repair based on Edison's experience in the operation of similar power plants.

(b) Provision of redundancy in the critical plant systems/components. For example, dual train is provided for the steam flashing system and a 100 percent back-up is provided in the brine injection booster pump capability.

(c) Plant design based on the use of prudent safety factors.

(d) Plant design based on a thorough investigation of the soil conditions, seismic history and climatic data of the Imperial Valley, and other applicable site-specific considerations.

SUMMARY

The Heber Project, when it goes on line in 1984, will be Edison's first commercial geothermal plant. Its annual estimated 308,000,000 kWh of energy will save the annual conservation of 500,000 barrels of oil.

PROJECT SCHEDULE
HEBER GEOTHERMAL PROJECT

INITIAL 52 MW DEVELOPMENT

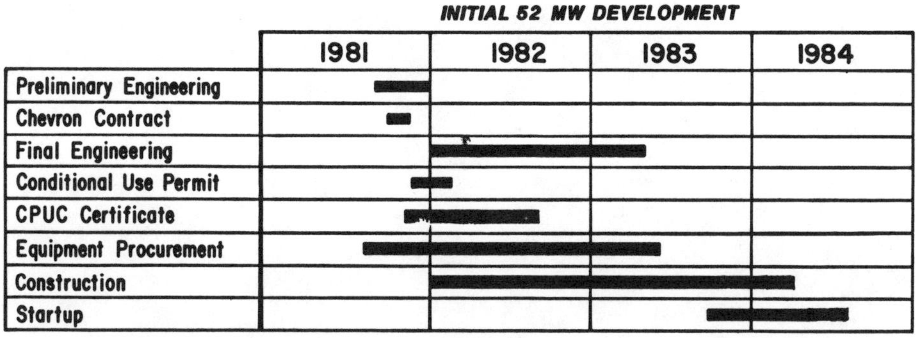

FIGURE 2

Geothermal Resources Council, TRANSACTIONS Vol. 6, October 1982

The Effect of Expansion-Ratio Limitations on Positive-Displacement, Total-Flow Geothermal Power Systems

by

Ronald DiPippo*

Mechanical Engineering Department
Southeastern Massachusetts University
North Dartmouth, Massachusetts 02747

ABSTRACT

Combined steam-turbine/positive-displacement engine (PDE) geothermal power systems are analyzed thermodynamically and compared with optimized reference flash-steam plants. Three different configurations of combined systems are considered. Treated separately are the cases of self-flowing and pumped wells. Two strategies are investigated that help overcome the inherent expansion-ratio limitation of PDE's: pre-flashing and pre-mixing. Parametrically-obtained results show the required minimum PDE efficiency for the combined system to match the reference plant for various sets of design conditions.

INTRODUCTION

Various types of "total-flow" expanders have been proposed for geothermal power applications, including an axial-flow impulse turbine, a rotary separating expander, a Lysholm or helical-screw expander, and a Wankel-type engine. For those of the positive-displacement type (Lysholm engines, for example), the most severe practical constraint is the expansion ratio - the ratio of the specific volume at the exhaust to that at the inlet. Expansion ratios of 8-15 have been estimated as practical for a helical-screw expander [1].

The value of a positive-displacement engine (PDE) in a geothermal plant stems from the replacement of a totally lossy process - a throttling process - with one that generates power, and at the same time also produces steam for use in a conventional turbine. The limitation on expansion ratio for the topping PDE means that plants having a PDE plus a steam turbine are not always thermodynamically superior to a simple, but optimized flash-steam plant.

The effect of the limitation is shown in Fig. 1 which gives a comparison between two expansion processes, one starting from a saturated liquid condition (upper curve) and one starting from a saturated vapor condition (lower curve). The contrast is striking. For a volume ratio of 4.5, the enthalpy drop for the latter case is about 750 times larger than the former case.

*Also, Division of Engineering, Brown University, Providence, Rhode Island 02912

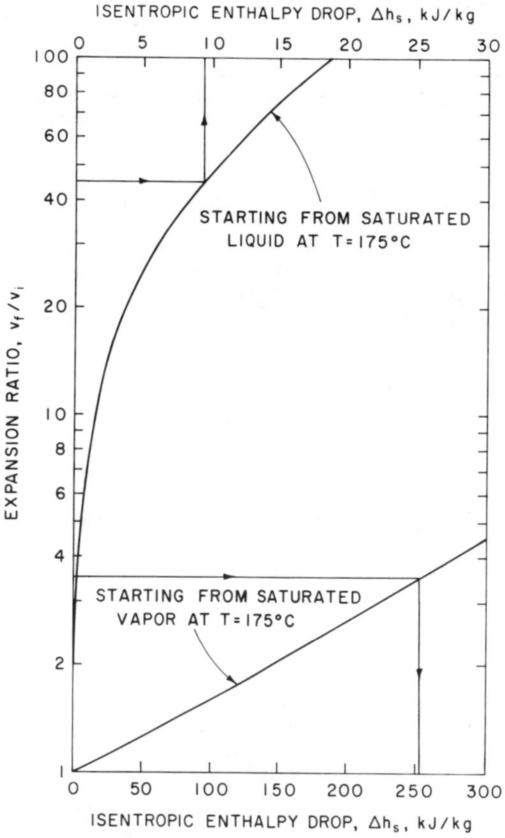

Fig. 1 Isentropic enthalpy drop versus expansion ratio.

REFERENCE FLASH-STEAM PLANTS

We will develop comparisons between plants incorporating a PDE together with a steam turbine, and basic flash-steam plants. We distinguish two cases: (1) the geothermal wells are self-flowing, giving a mixture of liquid and vapor at the wellhead, and (2) the wells are pumped, the product being a pressurized liquid. In the first case, we use a "double-flash" plant as the reference; in the second case, we use a "single-flash" plant. Each reference plant will be optimized to produce the maximum work output per unit mass of geofluid, yielding the highest efficiency for the given wellhead conditions. The combined PDE/steam turbine plants will then be compared to the thermodynamically optimum flash-steam plants. For various design options on the PDE we will seek the minimum PDE efficiency necessary to make the combined plant equal to the reference plant in overall efficiency.

BASIC PDE STRATEGY

To help overcome the volume-ratio limitation imposed on PDE's working near the saturated liquid line, the basic technique is to admit the working fluid to the PDE with an amount of vapor already present. This increases the inlet volume, and hence mitigates the volume-ratio limitation, allowing a "deeper" expansion into the two-phase region, i.e., a greater enthalpy drop. We examine two practical ways of achieving this: (1) a controlled pre-flash of saturated liquid prior to admission to the PDE, and (2) by pre-mixing some separated steam with the saturated liquid before admission. Obviously only the first option is available in the case of pumped wells.

COMBINED PDE/STEAM TURBINE CONFIGURATIONS

We will study three configurations: Case A: Self-Flowing Wells/Pre-Flashing Mode; Case B: Self-Flowing Wells/Pre-Mixing Mode; and Case C: Pumped Wells/Pre-Flashing Mode. Schematic diagrams for each are shown in Figs. 2-4.

From Figs. 2 and 4 we see that the reference plants are obtained by replacing the PDE with a by-pass. The analysis given here is unaffected by certain details such as whether a dual-admission turbine or two separate turbines is used. Since we do not account for well-pumping power (Case C), the performance figures cited here are to be taken as gross rather than net.

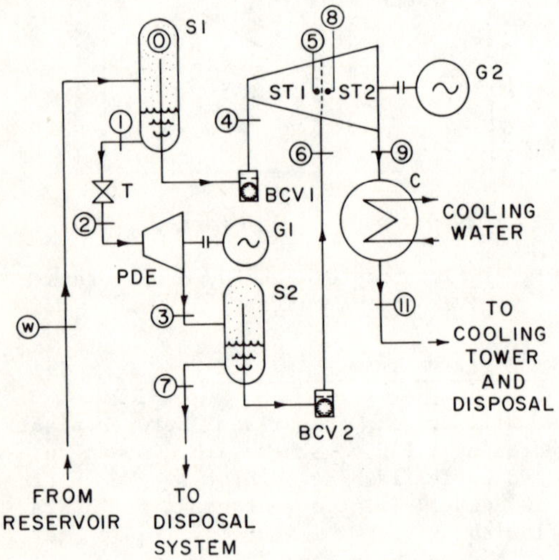

Fig. 2 Plant diagram for Case A: self-flowing wells, pre-flashing mode.

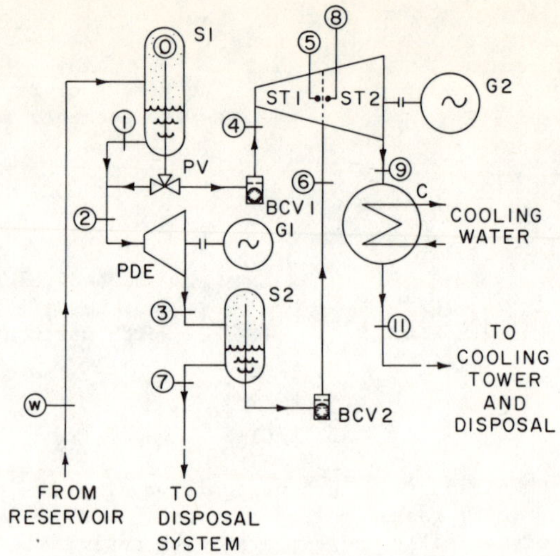

Fig. 3 Plant diagram for Case B: self-flowing wells, pre-mixing mode.

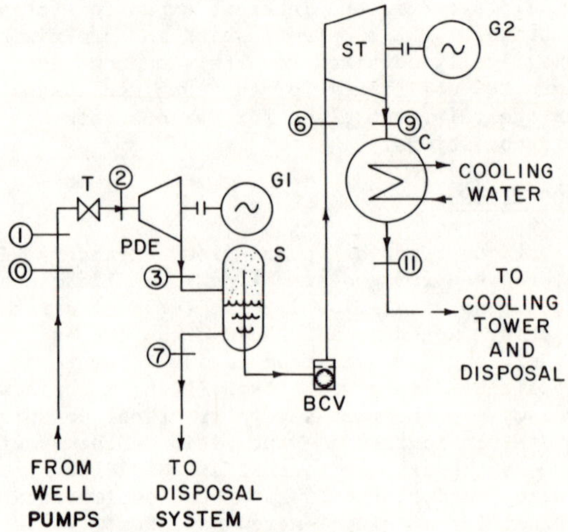

Fig. 4 Plant diagram for Case C: pumped wells, pre-flashing mode.

THERMODYNAMIC ANALYSIS AND CALCULATION METHODOLOGY

The thermodynamic analysis assumes the plants operate steadily, all components are perfectly insulated, and there are no pressure losses in pipes and heat exchangers. The working fluid is pure water with properties according to Keenan et al. [2]. The governing equations result from applying the First Law of thermodynamics for open systems in steady operation, and are given elsewhere [3].

The performance is measured by the utilization efficiency, η_u, which is based on the Second Law and defined as

$$\eta_u = \dot{w}/\dot{e}_o, \tag{1}$$

where $\dot{w}$ is the power output of the plant (per unit mass flow rate of geofluid) and $\dot{e}_o$ is the rate of exergy (or available work) associated with the flow of geofluid at wellhead conditions [4]. To calculate the exergy we chose a dead-state temperature of 323.15 K (50°C).

The PDE is described by its isentropic efficiency, η_{PDE}:

$$\eta_{PDE} = (h_2 - h_3)/(h_2 - h_{2s}). \qquad (2)$$

In the computations, η_{PDE} was varied from 0 (pure flashing process) to 1 (ideal isentropic expansion). Steam turbine efficiencies were found from a modified Baumann rule [5], i.e.,

$$\eta_{ST} = \eta_d(1 - \alpha y), \qquad (3)$$

where η_d is a dry turbine efficiency ($\eta_d = 0.85$), α is a wetness loss coefficient ($\alpha = 1.2$), and y is the wetness at the turbine exhaust.

Calculations were performed parametrically as follows. Three wellhead temperatures were selected: T_o = 150, 175 and 200°C. The condenser temperature was fixed at T_{11} = 50°C. For Cases A and B, three wellhead dryness fractions were chosen: x_o = 0.10, 0.15 and 0.20. Four volume ratios across the PDE were selected: v_3/v_2 = 5, 10, 15 and 20. For Case A, five pressure drops were selected for the pre-flash process 1-2: $(P_1 - P_2)/P_1$ = 0, 10, 20, 30 and 40%. For Case B, the dryness fraction x_2 was chosen as 0.05, 0.10, 0.15, up to 0.20, but not greater than x_o. The exhaust state from the PDE (state 3) fixed the intermediate temperature level T_7.

Performance calculations were made for optimum reference plants and combined plants, and from the results we obtained by interpolation the minimum PDE efficiency, η^*_{PDE}, needed for the combined plant to match the reference plant. In some case, η^*_{PDE} > 100%, meaning that the reference plant is always superior.

RESULTS

Owing to space limitations, only sample results can be given here; full results may be found elsewhere [3]. Table 1 lists the performance figures for the reference plants. Values given for x_o = 0% are for a single-flash plant; all others are for double-flash plants. Table 2 gives η^*_{PDE} for Case A and T_o = 200°C. Similar results hold for 150 and 175°C. In general we conclude that the combined system is not thermodynamically competitive, at least for volume ratios up to 20, without a pre-flash. For low volume ratios (say, VR = 5), rather high η_{PDE} are needed unless one is prepared to take very large pre-flash pressure drops of, say, 40%. Combined systems compete favorably under conditions of high volume ratio (VR > 15) together with large pressure drop prior to admission ($\Delta P/P_1$ > 20%). As long as the design is restricted to these limits, then combined systems will be superior to the reference double-flash plants if η_{PDE} > 27% (for the range of conditions studied here). A combined system with a PDE having an efficiency of 40%, a VR = 15, $\Delta P/P_1$ = 20%, and operating on a 200°C wellhead fluid with x_o = 15% would produce 120.65 kJ/kg, about 2% more than the optimum reference plant. Only 5.8% of the plant output would come from the PDE.

Table 1. Output and efficiency for reference flash plants

	Wellhead Temperature, °C					
	150		175		200	
x_o %	w kJ/kg	η_u %	w kJ/kg	η_u %	w kJ/kg	η_u %
0	22.7	25.9	34.3	28.0	48.0	39.6
10	61.5	41.0	77.8	40.9	94.8	40.5
15	81.0	44.7	99.6	44.4	118.3	43.8
20	100.0	47.3	121.4	47.0	141.7	46.4

Table 2. Minimum η_{PDE} needed for combined system for Case A to match optimum double-flash plant for wellhead temperature = 200°C

	$(P_1 - P_2)/P_1$, %				
VR	0	10	20	30	40
5	(*)	(*)	(*)	(*)	47.8
10	(*)	(*)	64.8	20.0	3.6
15	(*)	88.5	22.7	3.7	0.0
20	(*)	46.6	8.3	0.0	1.6
5	(*)	(*)	(*)	(*)	51.0
10	(*)	(*)	68.0	21.0	4.0
15	(*)	95.0	25.0	4.0	0.0
20	(*)	48.0	5.0	0.1	1.0
5	(*)	(*)	(*)	(*)	56.6
10	(*)	(*)	76.0	24.2	4.6
15	(*)	(*)	27.3	4.8	0.0
20	(*)	54.8	10.3	0.4	1.6

(*) Optimum reference plant always superior.

Table 3 gives a sample of results for Case B. The pre-mixing mode suffers from an inherent handicap; namely, geothermal steam that would normally be expanded in a high-efficiency steam turbine is used (via a mixing process) in a relatively low-efficiency PDE. The most competitive combined plants are those designed for high volume ratios and low (but not zero) inlet qualities. The case of x_{PDE} = 0% is the same as Case A with $\Delta P/P_1$ = 0%, and from Table 2 we know such plants are not competitive. A combined system with a PDE having 40% efficiency, a VR = 15, an admission quality of 5%, and working with a 200°C wellhead geofluid having 15% quality would produce 119.97 kJ/kg, 1.4% more than the optimum reference plant.

Table 4 summarizes the findings for Case C, pumped wells. Although some might argue that a flash plant is not appropriate as the reference in this case, it was selected because it is commercially available whereas binary plants are still developmental. The same trends appear in this case as with Case A. A combined system with $\eta_{PDE} > 20\%$ will be superior to the reference plant provided that VR > 15 and $\Delta P/P_1 > 20\%$. A combined system with a PDE having an efficiency of 40%, a VR = 15, $\Delta P/P_1 = 20\%$, and operating on a 175°C wellhead geofluid (saturated liquid) will generate 37.97 kJ/kg of output, 10.7% more than the optimum reference plant; the PDE contributes 16.4% of the plant output. It is interesting that combined systems in this case appear thermodynamically most competitive for resources of lower temperatures. The reason however is that the reference plant is relatively inefficient under these conditions. Binary plants, which can be designed, if not yet built, to show a higher efficiency, would present more of a challenge to combined PDE/steam turbine systems.

Table 3. Minimum η_{PDE} needed for combined system for Case B to match optimum double-flash plant for wellhead temperature = 200°C

	x_{PDE}, %			
VR	5	10	15	20
5	(*)	81.5	na	na
10	54.5	43.2	na	na
15	33.6	36.1	na	na
20	26.4	34.3	na	na
5	(*)	84.6	67.9	na
10	56.8	43.8	45.3	na
15	34.3	36.2	41.7	na
20	26.7	34.3	41.2	na
5	(*)	88.1	69.8	63.9
10	59.4	44.4	45.6	48.3
15	35.3	36.4	41.8	46.1
20	27.0	34.4	41.3	46.0

(*) - See note Table 2; na - Not applicable.

Table 4. Minimum η_{PDE} needed for combined system for Case C to match optimum single-flash plant.

Wellhead Temperature, T_o/°C		$(P_1 - P_2)/P_1$, %				
	VR	0	10	20	30	40
150	5	(*)	(*)	(*)	50.4	11.0
	10	(*)	(*)	27.5	2.5	0.6
	15	(*)	60.7	5.0	0.3	5.9
	20	(*)	26.7	0.3	3.2	11.5
175	5	(*)	(*)	(*)	60.5	26.8
	10	(*)	(*)	45.4	9.7	0.4
	15	(*)	75.6	13.1	0.6	1.2
	20	(*)	37.3	3.5	0.2	4.6
200	5	(*)	(*)	(*)	(*)	40.8
	10	(*)	(*)	55.9	16.8	2.9
	15	(*)	76.8	19.2	3.0	0.1
	20	(*)	39.9	6.8	0.1	1.6

(*) - See note Table 2

CONCLUDING REMARKS

The decision whether to select a combined PDE/steam turbine system over an alternate system will be based on economics. Since a combined system will likely have greater capital and maintenance costs relative to a conventional plant, it must gain its advantage in lower "fuel" costs, i.e., it must be significantly superior in energy conversion efficiency.

The need for small, simple, reliable, and inexpensive wellhead power units for geothermal applications is now recognized. In field tests of a helical-screw expander (used in a stand-alone mode) at Cerro Prieto [6], a peak efficiency of 68% was attained. The engine generated between 100-930 kWe, but malfunctioned 8 times during a 1000-hour endurance run. Full operating conditions under which the peak effiency was achieved were not reported. Through testing of this type, which is continuing elsewhere, positive-displacement expanders may eventually be developed to compete both thermodynamically and economically with conventional geothermal power plants.

ACKNOWLEDGEMENTS

The author expresses his thanks to the following people: S. Kampf and D. Noble, for help with the calculations; J. Kestin and P.F. Maeder, for valuable suggestions; R.J. Hanold and C.B. McFarland, for support and encouragement. This work was partly funded through Contract DE-AC02-79ET27225 of the U.S. Dept. of Energy, Div. of Geothermal Energy, with Brown University.

REFERENCES

[1] House, P.A., "Helical-Rotor Expander Applications for Geothermal Energy Conversion", Lawrence Livermore Laboratory, UCRL-52043, Livermore, CA, April 1976.

[2] Keenan, J.H., Keyes, F.G., Hill, P.G., and Moore, J.G., Steam Tables (International Ed.-Metric Units), John Wiley & Sons, New York, 1969.

[3] DiPippo, R. "The Effect of Expansion-Ratio Limitations on Positive-Displacement, Total-Flow Geothermal Power Systems", Brown University Rep. No. GEOFLO/15, Providence, RI, February 1982.

[4] DiPippo, R., Geothermal Energy as a Source of Electricity, U.S. Dept. of Energy, DOE/RA/28320-1, Washington, DC, 1980, pp. 9-18.

[5] Moore, M.J. and Sieverding, C.H., Two-Phase Steam Flow in Turbines and Separators, Hemisphere Publishing Corp./McGraw-Hill, Washington/New York, 1976.

[6] Gonzales Rubio, J.L. and Illescas, F., "Test of Total Flow Helical Rotor Screw Expander at Cerro Prieto, Mexico", Geothermal Resources Council, Transactions, Vol. 5, 1981, pp. 425-427.

Geothermal Resources Council, TRANSACTIONS Vol. 6, October 1982

A BENCH SCALE TEST FOR HYDROGEN SULFIDE OXIDATION

J. Ivester, S. Attiga, M. Castillo, and E. Goldman

Bechtel Group, Incorporated
Research and Engineering
San Francisco, California

ABSTRACT

The oxidation of hydrogen sulfide in geothermal steam condensate by hydrogen peroxide was tested in a specially designed, continuous flow bench scale treatment system. The effectiveness of the treatment was monitored using a silver/sulfide specific ion electrode in a technique developed specially for this study. The effects of pH, treatment ratio, and catalyst on the oxidation reaction were studied.

The results of the study were used to operate a pilot plant treatment system to determine optimal peroxide treatment conditions. The bench scale studies identified iron as the catalyst of choice and indicated that the pilot plant testing could concentrate on the pH expected in the full scale plant, pH 8. The bench scale studies further indicated that pilot plant testing would not need to include hydrogen peroxide to sulfide treatment ratios of greater than 4:1 and that lower ratios might be adequate.

INTRODUCTION

If not adequately abated, hydrogen sulfide in the geothermal steam used for electric power production in The Geysers area could cause a public nuisance. The California Department of Water Resources (DWR) is planning to construct a 55 MWe geothermal power plant in the Bottle Rock area of Lake County, California that would utilize steam containing 300 to 450 ppm_w H_2S. The primary abatement system for the Bottle Rock Power Plant is a Stretford system to remove virtually all the H_2S from the noncondensible gases withdrawn from the condenser. Operating experience by Pacific Gas and Electric Company at its Geysers geothermal electric generating Units 13 and 15 indicates that 65 to 80 percent of the H_2S in the geothermal steam is removed from the surface condenser in this way.

The remaining H_2S is dissolved in the condensate which is used as cooling tower makeup water. In the cooling tower, the H_2S could be air stripped and released to the atmosphere. If the condensate is not treated to remove H_2S before it is admitted to the cooling tower, the H_2S released to the atmosphere could cause a public nuisance. Quantitatively, DWR has committed to limiting H_2S emissions from the plant to 5 lb/hr (CEC, 8/80), and this limitation has been determined to be adequate for compliance with ambient air quality standards (CEC, 10/80). Since manufacturers of the Stretford equipment expect to limit H_2S emissions to less than 1 lb/hr for that equipment, the H_2S emitted from the cooling tower must be less than 4 lb/hr. This implies that the H_2S concentration in the condensate must be reduced to less than 4 ppm_w since the condensate flow rate will be approximately 1,000,000 lb/hr.

Hydrogen peroxide oxidation is currently employed for geothermal condensate hydrogen sulfide abatement with apparent success, but information regarding optimal treatment conditions is limited. A test program was recently conducted to provide data demonstrating the effect of treatment ratio, catalyst type, residence time, and pH on H_2S removal efficiency by monitoring the sulfide concentration in a continuous flow treatment system. Prior to operation of a pilot plant treatment system, a smaller bench scale system was used to develop adequate analytical methods and to provide guidance for the pilot plant tests. The concern of this paper is the development, operation and results of the bench scale test program and the analytical methods that the program employed.

TEST APPROACH

The H_2S oxidation test program was conducted in the field using steam from a representative well to achieve realistic test conditions for condensate composition. A 10 x 50 ft trailer was set in place at the Francisco leasehold well pad. The trailer was outfitted for the tests as a field laboratory and included equipment for chemical analysis.

The geothermal condensate for the tests was obtained by tapping into the steam line from the Francisco 3-5 well developed by MCR Geothermal. This well was producing a bleed flow (nominally 10,000 lb/hr) of steam during the tests. Approximately 2,000 lb/hr of this steam was routed to the test facility sampling system where a fraction of the 2,000 lb/hr flow was bled off and condensed. The remainder was returned to a vent facility. This flow arrangement assured that representative condensate for testing was continuously obtained from the well. H_2S oxidation tests were conducted as bench scale batch tests, as bench scale flow tests, and later as pilot plant flow tests. This paper describes the bench scale batch and flow tests.

Batch H_2S oxidation tests, carried out in a beaker using approximately 150 ml of condensate, served as a cost-effective first check to screen various treatment parameters such as pH and type of catalyst. A bench scale flow system designed to treat approximately 25 lb/hr of condensate in in a continuous flow process was used to establish the range of test parameter values for later tests using a larger scale pilot plant system. The bench scale tests were conducted from June 1981 to November 1981.

The steam for the bench scale test was condensed in 40 to 80 ft (depending on air temperature) of 1/4" O.D. stainless steel tubing at approximately atomospheric pressure. The condensate thus obtained had the following characteristics:

pH = 6.0
H_2S = 70 ppm_w
Alkalinity = 360 ppm_w as $CaCO_3$
NH_3 = 100 ppm_w
Conductivity = 735 umhos/cm
Iron = 0.05-0.25 ppm_w

THE ANALYTICAL METHOD

To carry out the oxidation tests, it was necessary to have an analytical method for the determination of total dissolved H_2S ($H_2S + HS^- + S^=$) in the presence of H_2O_2 and the various reaction products of the sulfide-peroxide redox reaction. It was also desirable to have an analytical method that gave quick results so that the effect of residence time on the reaction could be evaluated. Further requirements were that the analytical method work over the range of pH tested (6-10) and not interfere with the redox reaction by altering the reaction conditions. Ease and accuracy of analysis were also significant considerations. Because of interferences in established analytical methods, such as iodine titraation or methylene blue test, these requirements constituted a major problem for the oxidation test program.

An analytical techique was chosen for sulfide determination that employs a silver/sulfide electrode with a millivolt readout on a specific ion meter, a pH reading, and a programmable calculator. The method was expected to be free from hydrogen peroxide interference, to give quick results, and to be flexible enough to be used over a wide range of pH conditions. The use of electrodes under oxidizing conditions requires stringent analytical quality control efforts and frequent recalibration and replacement of the electrodes, but since they were to be used by experienced analytical chemists over a fairly short time period, the problem of electrode maintenance was considered acceptable.

The silver/sulfide electrode produces a potential proportional to the log of the concentration of the sulfide ($S^=$) ion in solution, according to the Nernst equation. By combining the Nernst equation with the acid-base equilibrium equations for sulfide, the electrode potential can be expressed as a function of the total dissolved hydrogen sulfide, pH, the electrode slope, and some thermodynamic constants.

In this study, the electrode slope was determined for each electrode pair used by a standardization technique in which a series of sulfide standards were prepared by standard dilution and measured by iodine titration. The millivolt difference corresponding to a ten-fold change in sulfide concentration was recorded as the electrode slope.

The analytical method for both batch and flow tests consisted of:

- Recording the millivolt reading from the silver/sulfide electrode immersed in untreated condensate
- Recording the pH of the untreated condensate stream
- Measuring the total sulfide in a sample of the untreated condensate by iodine titration
- Adding peroxide and catalyst to the condensate
- Recording the millivolt reading from the silver/sulfide electrode immersed in the treated condensate
- Recording the pH of the treated condensate
- Calculating the total sulfide in the treated condensate from the difference between the two millivolt readings (corrected for any pH change) and the previously established electrode slope

The calculation, based on the Nernst equation and the sulfide acid/base equilibrium equations, had the form:

$$(H_2S)_t = 10\left(\frac{E_u - E_t}{M} + (pH_t - pH_u) + \log\left(\frac{10^{-pH_t} + 10^{-7}}{10^{-pH_u} + 10^{-7}}\,(H_2S)_u\right)\right)$$

where: K = dissociation constant of hydrogen sulfide
E = millivolt reading
M = electrode slope
(H_2S) = total dissolved hydrogen sulfide,
u,t = the subscripts u and t refer to the treated or untreated condensate measurements

The calculation has a correction term for differences in pH between the treated and untreated condensates, but experience indicated that the correction was only valid for changes of less than 1/2 pH unit. At differences greater than 1/2 pH unit, the precision of the method decreased significantly.

Interference with the method could occur as a result of complexation of the sulfide ion by elemental sulfur produced by the oxidation reaction. These complexes and polysulfides produced directly by oxidation may release sulfide ions under certain circumstances but are not detected by the electrodes. In these tests, confirming analyses indicated that complexation, if occurring, was effectively removing H_2S from the condensate.

The development of adequate confirming tests for the instrumental method was partially successful All wet methods used to check the silver-sulfide electrode method had some problems. Confidence in the instrumental methods, however, was established by repeated confirmation with several somewhat imperfect wet chemical methods.

The overall measurement accuracy (95% confidence) was determined by standard and duplicated measurement to be about $\pm 10\%$ at 80 ppm_w, $\pm 50\%$ at 5 ppm, $\pm 60\%$ at 3 ppm_w and $\pm 80\%$ at 1 ppm_w.

THE BENCH TEST PROGRAM: BATCH TESTS

The first phase of the bench scale investigations consisted of a series of batch oxidation tests employing a broad spectrum of pH, catalyst, and hydrogen peroxide conditions. The purpose of these tests was to provide preliminary information that would allow subsequent flow tests to focus on the most promising conditions, especially with regard to catalyst selection. In addition, the batch tests afforded an opportunity to examine electrode behavior in an oxidizing medium without the complications of flow, pressure, or solids build-up. The batch tests also provided some information on the kinetics of the oxidation reaction.

The oxidation reaction in a beaker of condensate was monitored using a silver/sulfide electrode and specific ion meter manufactured by Orion Company, as shown in Figure 1. The condensate in the reaction beaker was stirred with a magnetic stirrer and the electrode readings were recorded as a function of time by a digital printer.

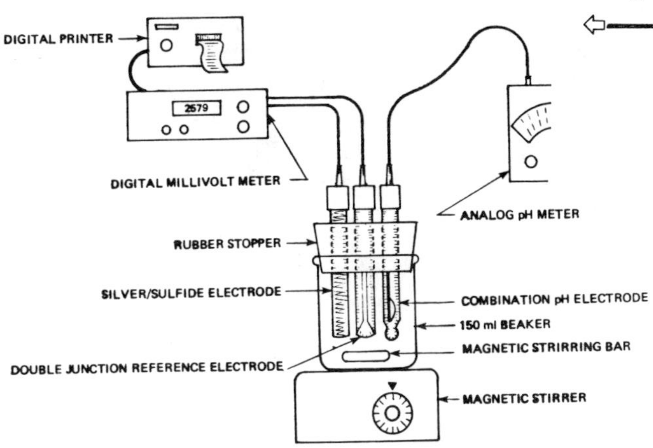

Figure 1. Batch test apparatus

The batch test procedures involved adjusting the pH of a sample condensate by buret addition of sodium hydroxide, and stabilizing the pH by pipet addition of a concentrated buffer solution. This was followed by pipet addition of catalyst solution, and syringe injection of hydrogen peroxide. Raw data in the form of millivolt and pH readings were converted to concentration and H_2S removal data.

The conditions studied in the batch tests included pH conditions from less than 6 to nearly 9. Hydrogen peroxide to sulfide treatment ratios of 1:1 to 4:1 were studied. A total of eight different catalysts were investigated including chelated and non-chelated types. Metals tested as catalysts were iron, manganese, cobalt, and nickel.

THE BENCH TEST PROGRAM: FLOW TESTS

The bench scale flow tests constituted a small scale, approximate model of possible treatment schemes using on-line pH control and oxidant injection so that H_2S oxidation occured under plug flow reactor conditions. (Plug flow refers to the condition in which there is no mixing between adjacent quantities of liquids as they flow down the line. Obviously, there is some back-mixing, but the plug flow assumption has proven to be reasonable for simple flowing systems.)

The bench scale flow apparatus had five major components: the condensate preparation, collection and transport system; the reagent injection system; the reaction loop; the monitoring and sampling systems; and the cooling tower. All of these components, except the initial condenser and the cooling tower, are shown in Figure 2.

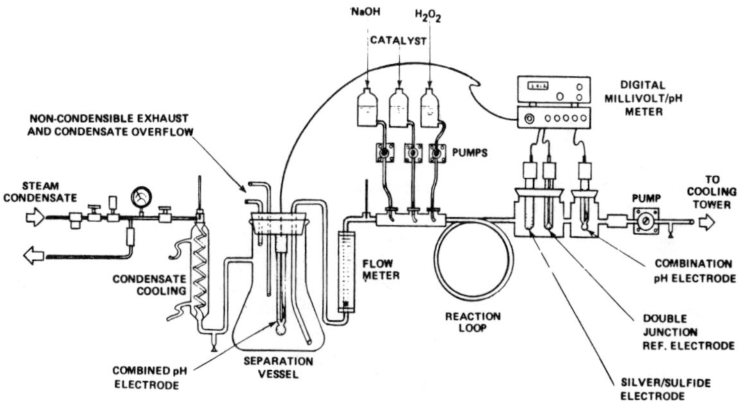

Figure 2. Bench scale flow apparatus

The condensate preparation, collection and transport system consisted of an air-cooled, stainless steel condenser, followed by a series of valves to control pressure, temperature, and flow rates of the condensate. This in turn was followed by a condensate cooling heat exchanger. The condensate then flowed into a phase separation vessel equipped with a noncondensible gas exhaust line and a condensate level maintenance control. From the phase separation vessel, condensate was drawn through the treatment system by means of a variable speed peristaltic pump located downstream of the reagent injection, reaction and monitoring sections of the treatment system.

The reagent injection system consisted of a three-port manifold on the condensate line into which caustic, catalyst, and peroxide could be added at rates determined by three separately controlled peristaltic pumps. Total reagent addition was normally about 5% of the condensate flow. The reaction loop component of the flow system began at the point of hydrogen peroxide addition with a bed of raschig rings to promote reagent mixing followed by a section of Tygon tubing that was varied in length to change the reaction time.

The silver/sulfide, reference, and pH electrodes were wired through a switching device to a digital readout specific ion meter.

The bench scale flow test was operated at atmospheric pressure and relatively low temperatures, 95-105°F. In general, run times were less than one hour. The tests were designed to further narrow the field of investigation for the pilot scale tests that followed and to further investigate apparent trends observed in the batch tests.

The bench scale flow tests covered a more limited variety of test conditions than the batch tests but provided more flexible and precise controls over many of the parameters, including temperature, treatment ratio, and pH. The flow system also produced a number of data points for each condition thus providing a high degree of data reliability. In addition, the flow test results were free from uncertainty due to handling errors or air contamination that could occur in the batch tests.

THE BENCH TEST PROGRAM: RESULTS

The following were the most significant bench scale results:

- The batch test data indicated that at pH 6, in the absence of a catalyst, increasing the treatment ratio had little effect on the H_2S concentration remaining, while at a higher pH (pH 8.5), an increase in the treatment ratio by a factor of two decreased the residual H_2S by a factor of ten. For the continuous flow system, however, a clear correlation between increased treatment ratio and decreased residual H_2S was found at all pH and catalyst conditions.
- The batch tests indicated that at 1:1 and 2:1 treatment ratios, maximum removal is obtained at pH 7; while at 4:1 treatment ratio, removal efficiency increases with increasing pH. The bench flow test data show increasing H_2S removal with increasing pH for all treatment ratios.
- In the batch tests the following effects of catalysts were noted:
  - --At pH 6, iron and nickel were effective catalysts; while manganese and cobalt had very little effect.
  - --At higher pH, iron was the most effective catalyst.
  - --With these results, and in view of the relative cost of the metals, iron was selected for the study in the bench flow tests and later in the later pilot plant tests.

CONCLUSIONS

The analytical method developed in the bench scale study successfully measures total dissolved hydrogen sulfide in hydrogen peroxide-treated geothermal condensate. The method is free from sulfite or thiosulfate interference and is accurate to within 2 ppm in the range 0-4 ppm_w. The method requires careful maintenance of equipment and stringent analytical quality control, and thus is not practical for sustained process monitoring applications at this time.

The bench scale studies identified iron as an adequate catalyst for the oxidation reaction. The studies indicated that pilot plant testing could concentrate on the pH expected in the full scale plant (pH 8). The studies also indicated that pilot plant investigations would not need to include testing of treatment ratios greater than 4:1, and that ratios of 2:1 or lower might be adequate.

ACKNOWLEDGMENT

This paper presents a summary of work funded by the California Department of Water Resources (under DWR Agreement B52957) with the DWR project management of Donald L. Smith, Supervising Mechanical Engineer. The Bechtel Project Manager was Frank Cochrane; Howard H. Hipkin was the Project Engineer. Randall Soo-Hoo was the Field Engineer.

REFERENCES

California Energy Commission, August 1980, Environmental Impact Report (revised draft), Bottle Rock Geothermal Power Plant.

California Energy Commission, October 1980, Environmental Impact Report (final), Bottle Rock Geothermal Power Plant.

Geothermal Resources Council, TRANSACTIONS Vol. 6, October 1982

BEOWAWE #1-A 10 MW GEOTHERMAL UNIT IN NORTHERN NEVADA

Lee Keilman

Sacramento Municipal Utility District
Sacramento, CA 95813

ABSTRACT

This paper describes a project to build and operate a nominal 10 mw electrical generating unit using the geothermal heat from the Beowawe, Nevada, geothermal reservoir to power an isobutane binary unit. This 10 mw unit would be fabricated on portable skids by equipment supplier for shipment to the site. The project will be owned and operated by the NORNEV Demonstration Geothermal Company which is made up of Pacific Power and Light, Eugene Water and Electric Board, Sierra Pacific Power Company, and Sacramento Municipal Utility District. The geothermal brine for powering the 10 mw binary WGU will be purchased from Chevron Resource Company. This first unit is a research and development unit and will, hopefully, lead to total development of the 300 mw plus Beowawe reservoir.

PROJECT HISTORY

The NORNEV project had its beginning in early 1978 when five utilities -- Eugene Water and Electric Board (EWEB), Pacific Power and Light (PP&L), Portland General Electric (PGE), Sacramento Municipal Utility District (SMUD), and Sierra Pacific Power Company (SSP Co.) began discussing the feasibility of constructing a 50 mw geothermal flash steam plant on a yet to be determined northern Nevada site. This region has long been recognized as an area of considerable geothermal potential. PGE has dropped out of the project.

After a year of informal but earnest activity, the utilities entered into a "Phase A" agreement in May 1979, to commit up to $100,000 each to finance a defined program that would look into the geothermal resource evaluation of various sites; the development of a conceptual generating plant design, together with cost estimates; a review of State/Federal statutes relating to construction and operation of commercial geothermal power plants; and investigation of funding sources including possible Federal assistance.

While the Phase A project effort continued, two new developments took place. First, it was becoming increasingly apparent that some of the resource producers were not sufficiently confident of their respective Nevada resources to embark upon large scale drilling operations (20 to 30 deep wells for a 50 mw plant) until they learned more about the geology of the strata they were working in. All agreed hydrothermal development in Nevada's relatively unknown "basement rock" formations was considerably different from the well defined dry steam reservoirs at The Geysers area in northern California.

The second development was to consider a new concept that would use a 10 mw binary plant that called for factory fabrication of transportable modules ready for site installation on concrete pads. This concept offered a number of advantages: first, its cycle efficiency was presented as being at least equivalent to that of larger steam "flash" plants; second, in the event a geothermal resource failed or declined, it was possible to disassemble the modules and relocate the plant; and third, the binary technique, coupled with air condensor cooling and reinjection of geothermal fluids, minimized air and ground water pollution. Since the geothermal fluids would be totally encased during their passage from the production wells through the plant, thence back into the ground through re-injection wells, neither the ambient air nor the groundwater geochemistry would be affected. Essentially, only the BTUs were extracted. Environmentally, everything else would stay the same.

Figure number 1 shows the reservoirs studied, namely, Steamboat Springs, Desert Peak, Dixie Valley, and Beowawe. Dixie Valley had the hottest bottom hole temperatures, 460°F plus, while Beowawe the lowest at 360°F. The Beowawe reservoir was selected because of the short electrical transmission line required and the receptiveness of Chevron to enter into a development-type brine supply contract.

Figure 1

PLANT DESCRIPTION

The 10 mw binary unit will be contained in a fenced facility area approximately 500' x 500'. Figure number 2 shows a simplified cycle diagram of the power conversion cycle. The cycle, as built, will have an economizer heat exchanger to improve cycle efficiency.

Figure 2

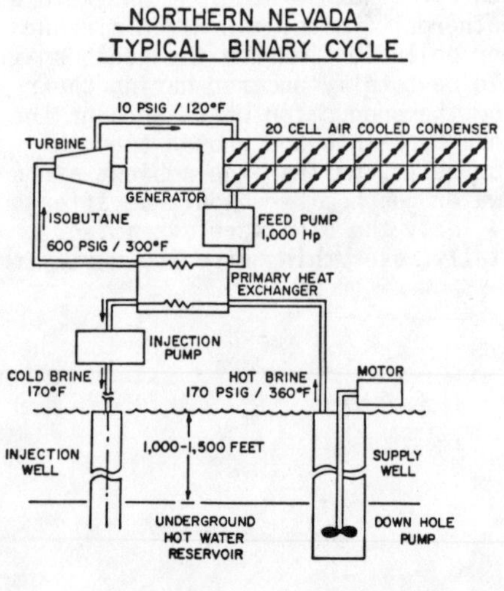

The plant is sized for a gross generating capacity of 13 megawatts. The plant auxiliary load will vary, but the plant is sized for 10 megawatts net. Power supplied for Chevron's onsite loads in exchange for geothermal heat reduces the net saleable energy to about 8 megawatts.

Individually packaged on four separate skids will be an air compressor section, fire protection equipment, an electrical switchgear and motor control center module, and a lube oil/seal oil section. One module will include a solids separator complete with valves, piping, and instrumentation for control of the geothermal fluid input coming from the producing wells. Two modules, each expected to contain seven shell and tube heat exchangers 44 feet long and 3 feet in diameter arranged for parallel flow, will be connected in series and contain all necessary piping, valves, and instrumentation. Each of these modules will measure 44 feet x 12 feet. Another module will include an economizer section of two shell and tube heat exchangers that will reclaim surplus heat exiting from the turbine-generator and use it to preheat the liquid isobutane passing from the air coolers to the main heat exchangers.

The turbine-generator module will include a turbo-expander with speed decreasing gear and a 1200 rpm TEFC 4160 V. generator rated at 13 megawatts. The electrical switchgear, synchronization panel, a solid state microprocessor control panel and other related control equipment will be packaged in a controls module. Another module will include a 24,000 gallon isobutane (IC_4) storage vessel and a 15,000-gallon, 250 lb. pressure, nitrogen vessel with piping, valves and controls. Additionally, there will be an air condensor yard area (110' x 180') for cooling the isobutane working fluid after it has passed through the turbine-generator. This area will include 34 pre fabricated aerial coolers complete with inlet and outlet header sections, piping, valves, blow-down system, and two-speed electric motor driven fans.

Fire Protection

A fire protection system for the switchgear/control module building and the turbine-generator skid will be provided.

Hydrogen Sulphide (H_2S) Concerns

Geochemical analysis of Beowawe's resource indicates H_2S present at this site is very low compared to The Geysers resource. The plant design has taken this into account. It is anticipated that the only H_2S released to the atmosphere will be from brine heat exchanger vents.

PROJECT PERMITTING CLEARANCES

Once the NORNEV Group had selected the Beowawe, a six-month baseline air monitoring program began. Concurrently, a study on what impact the project would have on the air quality (established as the norm by the monitoring program), soils, surface water, flora, threatened and endangered plants, fauna, and land use was undertaken.

The information was compiled in a special report in August 1981 to the Nevada Public Service Commission requesting issuance of a construction permit endorsement. Public notices were placed in newspapers when the application was filed, publicizing NORNEV's proposed plans for Beowawe. A similar application was filed with the Nevada Division of Environmental Protection.

There were no intervenors during the 45-day public comment period provided by law, and the Division of Environmental Protection officially endorsed the project. The Nevada PSC issued a construction permit on October 26, 1981.

PROJECT SCHEDULE

SMUD estimate of project schedule is as follows:

TABLE NO. 1 NORNEV 10 MW BINARY WGU

NORNEV members authorize all agreements	June 1982
Release detailed design and fabrication	June 1982
Start site construction of foundation and support facilities	August 1982
Start assembly of equipment	March 1983
Roll expander turbine-generator	July 1983
Commercial operation date	August 1983
Terminate project	August 1993

ESTIMATED PROJECT COSTS

Capital Loss

The following Table 2 is SMUD's order of magnitude estimate of the total capital cost to construct the 10 mw binary WGU:

TABLE NO. 2 CAPITAL COSTS IN ON-LINE 1983 DOLLARS

Purchased Equipment	$12,796,000
Site Constr. and Assembly	3,761,000
Overhead and Interest during Construction	3,620,250
Sub-Total	$20,177,250
Contingency	3,000,000
TOTAL EST. PROJECT COST	$23,000,000

A 1983 on-line cost of eight to ten cents/kwh has been used as an order of magnitude estimate, and it should be recognized that there are still many unknowns about the development of this first-of-a-kind unit.

PROJECT FINANCE PLAN

A unique financing plan will use the following approach to project financing:

- a conventional interim financing program providing construction funds for up to 18 months, and
- a lease agreement with a lessor who would purchase the unit after commercial operation has been demonstrated.
- A non-profit, taxpaying cooperative corporation will own the project. This cooperative is made up of the NORNEV members. The NORNEV Corporation will enter into a ten-year lease agreement with a lessor eligible to receive both the basic ten percent investment tax credit plus the additional 15 percent credit provided for geothermal equipment leases.

NEVADA RECAPTURE LAW

Developing power plants in Nevada involves power recapture provisions. Pursuant to Nevada law, each out-of-state utility will have an agreement that provides Sierra Pacific the right of first refusal for membership in geothermal units constructed in the out-of-state utility's service territory to the extent of the out-of-state utility's ownership share in subsequent (follow-on) plants constructed by NORNEV. Agreements between Sierra Pacific and each out-of-state NORNEV members will be executed separately. Sierra Pacific does not intend to recapture any of the 10 mw from the WGU during its ten-year operating life.

SUMMARY

Each of the utilities involved in the NORNEV group have had different ideas of how to develop a geothermal demonstration plant in Northern Nevada. The fact that the project has gone forward and is presently in the construction phase proves that with hard work and a compromising attitude, utilities can develop jointly owned facilities.

The group hopes that additional units will be built at Beowawe or at the other reservoirs studied. However, this will require that the brine suppliers and the utility work together to get the cost lowered so that hot water geothermal can compete with the cost of coal-fired power plants in the area.

Geothermal Resources Council, TRANSACTIONS Vol. 6, October 1982

UTILITY INDUSTRY ESTIMATES OF GEOTHERMAL ELECTRICITY

Paul Kruger* and Vasel Roberts

Electric Power Research Institute
Palo Alto, CA 94303

ABSTRACT

The 1982 annual EPRI survey of the nation's electric utilities shows a 14% decline in the forecast of possible geothermal electric generating capacity by the year 2000. The decline to 9350 MWe follows three consecutive years of constant forecasts of 10,800 MWe by 2000 and increases in announcements of new geothermal capacity. Factors accounting for the 14% drop in expectations most likely include: (1) the cancellation of the 50 MW flash demonstration plant in New Mexico; (2) the curtailment of Federal programs in geothermal research and development; (3) the general climate of licensing and regulations confronting the nation's utilities; and (4) the general depressed economy of reduced demand, scarce capital, and high interest rates.

Over the next 10 years generation additions in the West are expected to be fueled by 47% coal, 27% nuclear, 16% hydro, and 10% other. This trend is indicative of the fuels against which geothermal energy must compete. The price that utilities may be willing to pay for geothermal power will range from the equivalent of the most economic competing fuel to avoided cost, depending on unique market conditions in the subregions.

INTRODUCTION

The potential for electric power generating capacity from indigenous geothermal resources as estimated by the nation's electric utilities has been reported in each of the past five proceedings of the Geothermal Resources Council annual meeting. Data for these reports are compiled from two sources: (1) forecasts of future year generating capacity as detailed by the National Electric Reliability Council (NERC); and (2) informal estimates in response to the annual EPRI survey from 36 electric utilities interested in geothermal energy as a possible source of electric power.

Since most of the NERC forecast of geothermal capacity is from the western states comprising the Western Systems Coordinating Council (WSCC), these data are compiled separately. Data for the Gulf states region, where interest in the potential geopressured resources exist, are also compiled separately. A third type of geothermal resource, the petrothermal deposits that may be widely distributed has not yet been investigated sufficiently for commercial development by the electric utilities.

NERC DATA

A summary of the projections reported by NERC (1981) for the contiguous United States and Canadian parts of the U.S. network is given in Table 1. The data show the actual installed capacity in gigawatts by energy source for both 1980 and 1981 and the estimated values for the respective 10-year periods 1989 and 1990. The estimated growth rate over the two 10-year periods for total installed capacity is 2.8% per annum reflecting an anticipated decline compared to earlier 10-year estimates of 3.8%/a (Roberts and Kruger, 1981). The 10-year forecast shows an 8% decrease in oil/gas generating capacity compared to a 60% increase for coal and a 150% in nuclear generating capacity. Geothermal energy for the nation remains small, accounting for 0.4 percent of the total generating capacity in 1990, but planned growth continues at a rate of 13%/a over the 10-year period.

Data from the 10-year coordinated plan of WSCC (1981) are given in Table 2. Additions of 50.6 GWe generating capacity indicates a mean growth rate of 3.6%/a over the 10-year period. By major energy source, the data show essentially no decrease in the dependence on oil and gas, a modest growth of 1.7%/a in hydropower, and major growths of 7%/a in coal and 17%/a in nuclear power. Geothermal energy approaches a significant contributor at 2% of the total generation capacity, growing at a mean rate of 13%/a.

If this growth rate of geothermal capacity continues through the next 10 years, installed capacity would reach 10,500 MWe by the year 2000. From the point of view of the potential for growth in the utilization of geothermal energy, it is important to note the heavy emphasis given to coal and uranium generating plants, accounting for almost 75 percent of all the additional capacity over the next 10 years. If the installation of coal and nuclear power plants are delayed or cancelled, pressures to expand geothermal energy production would likely result.

*Stanford University, consultant to EPRI

EPRI SURVEY RESULTS

The results of the sixth annual EPRI survey of the geothermal plans of the nation's electric utilities most likely to include geothermal capacity are shown in Table 3. The results are grouped by regions of the United States, the Gulf states comprising the states having potential resources of geopressured thermal and natural gas deposits, and the contributions from the Canadian provinces' part of the WSCC are included with the northwest states region. The data in Table 3 show considerable distribution among the western states, indicating significant growth in utilization of liquid-dominated resources in addition to the steady growth of the ubiquitous vapor-dominated resources in northern California. The Pacific northwest states continue to show a significant probable growth to the year 2000, as do the Rocky Mountain states. The large gradients between announced, probable, and possible outside of California still represent the uncertainties in the electric utilities on the commercialization of hydrothermal resources. Successful demonstration of the binary cycle conversion system could be a stimulus in lowering the gradient.

The seven-year trend in forecasts given in Table 4 shows the utilities' responses to annual stimuli superimposed on an orderly growth in installed capacity over the 15-year period from 1985 to 2000. Expectations were relatively low in 1977, the year of the first survey and were relatively high in 1978 when geothermal development appeared to accelerate with Federal assistance and development of new fields outside of northern California. Since then, forecasts appear to be reaching asymptotic levels of about 1550 to 1600 MWe installed capacity by 1985 independent of external stimuli. However, for the year 2000, the announced to possible capacities vary by a factor of three. The asymptotic values appear to have changed this year. The possible capacity of 9350 MW for the year 2000 represents about a 15% decline from the average of 10820 MW for the prior three years. The decline may be attributed to several factors affecting the utility industry perception of the future of geothermal energy for electric power production. These include: (1) cancellation of the 50 MW flash demonstration plant at Baca, New Mexico; (2) curtailment of Federal programs in geothermal research and development; (3) difficulties in obtaining licenses and complying with diverse regulations from Federal, state, and local agencies; and (4) depressed economy with reduced demand, scarcity of capital, and high interest rates.

The results of the survey were not explicit on value vs price of geothermal energy, but all utilities responding in this area felt that geothermal must be economically competitive in one of two ways. Some indicated that geothermal energy must be equal to or less than system avoided cost. The avoided cost may be considered to be the highest allowed cost for acquiring power within the system. Others felt that it would have to be comparable to specific dominant fuels within their respective service areas. The WSCC data show that the generation additions in the West over the next 10 years will be comprised of 47% coal, 27% nuclear, 16% hydro and 10% other. These categories indicate the fuels against which geothermal may have to compete depending on unique conditions in each subregion.

REFERENCES

Kruger, P. and V. Roberts, Utility Industry Estimates of Geothermal Electricity Generating Capacity, Proceedings, GRC Transactions, Vol. 1, May, 1977.

Kruger, P. and V. Roberts, Utility Estimates of Geothermal Electricity Generating Capacity, Proceedings, Geothermal Resources Council Transactions, Vol. 2, July, 1978.

Kruger, P. and V. Roberts, Utility Industry Estimates of Geothermal Electricity, Proceedings, GRC Transactions, Vol. 4, September, 1980.

National Electric Reliability Council, "Electric Power Supply and Demand, 1981-1990", July, 1981.

Roberts, V. and P. Kruger, Utility Industry Estimates of Geothermal Electricity, Proceedings, Geothermal Resources Council Transactions, Vol. 3, September, 1979.

Roberts, V. and P. Kruger, Utility Industry Estimates of Geothermal Electricity, Proceedings, Geothermal Resources Council Transactions, Vol. 5, September, 1981.

Western Systems Coordinating Council Technical Staff, Ten-Year Coordinated Plan Summary 1981-1990, May 1981.

Table 1

ELECTRICITY GENERATION BY ENERGY SOURCE*

	NERC (U.S., GW)					
	Oil/Gas	Coal	Hydro	Nuclear	Geothermal	Total
1981 - Actual	198.80	243.18	61.72	54.34	0.91	570.77
(% Total)	(34.8)	(42.6)	(10.8)	(9.5)	(0.2)	
1980 - Actual	198.72	231.78	66.09	49.20	0.80	558.24
(% Total)	(35.6)	(41.5)	(11.8)	(8.8)	(0.1)	
Annual Change (%)	0.0	4.8	-6.8	9.9	12.9	2.2
1990 - Estimated	182.96	338.19	68.17	133.69	3.10	749.61
(% Total)	(24.4)	(45.1)	(9.1)	(17.8)	(0.4)	
1989 - Estimated	184.41	330.53	67.87	131.06	2.89	738.38
(% Total)	(25.0)	(44.8)	(9.2)	(17.7)	(0.4)	
Annual Change (%)	-0.8	2.3	0.4	2.0	7.0	1.5

*National Electric Reliability Council, "Electric Power Supply and Demand, 1981 - 1990", July, 1981.

Table 2

WSCC GENERATION BY ENERGY SOURCE*

	Capacity (GWe)					
	Oil/Gas	Coal	Hydro	Nuclear	Geothermal	Total
1981 - Actual	34.9	24.8	54.1	3.5	0.92	118.5
(% Total)	(29.5)	(20.9)	(45.7)	(3.0)	(0.8)	
'81-'90: Additions	-0.4	23.6	9.8	13.4	2.19	50.6
(% Total)	(-0.1)	(46.6)	(19.4)	(26.5)	(4.3)	
1990 - Estimated	34.5	48.4	63.9	16.9	3.11	169.1
(% Total)	(20.4)	(28.6)	(37.8)	(10.0)	(1.8)	

*from Western Systems Coordinating Council Technical Staff, Ten-Year Coordinated Plan Summary 1981 - 1990, May, 1981.

Table 3

1982 EPRI UTILITY GEOTHERMAL SURVEY

	Capacity (MWe) by Year				
	1981 Actual	1985 Est	1990 Est	1995 Est	2000 Est
AZ/NM/NV	0				
Announced		0	0	0	0
Probable		20	41	188	280
Possible		20	73	330	494
CA/HI	928				
Announced		1515	2267	2491	3091
Probable		1517	2922	3292	3912
Possible		1534	3239	4444	5364
OR/WA/BC/AL	0				
Announced		4	4	4	4
Probable		12	170	320	1120
Possible		16	342	1168	2186
CO/ID/MO/UT	0				
Announced		21	120	120	120
Probable		30	220	420	770
Possible		40	245	770	1270
Gulf States	0				
Announced		0	0	0	0
Probable		0	0	0	5
Possible		0	2	10	40
Totals	928				
Announced		1540	2391	2615	3215
Probable		1579	3353	4220	6087
Possible		1610	3910	6722	9354

Table 4

COMPARISON OF THE ANNUAL EPRI SURVEYS

	U.S. Geothermal Electric Power Capacity (MW)			
	1985	1990	1995	2000
Announced				
1977	1178	1378	1678	1828
1978	2019	3019	3619	3919
1979	2057	2242	2512	2832
1980	1574	2294	2599	3299
1981	1865	3399	3920	4481
1982	1540	2391	2615	3215
Probable				
1977	2528	3258	4358	5358
1978	2664	5414	7473	9023
1979	2564	4577	6108	7288
1980	1912	4216	5681	7416
1981	1887	3849	5440	7851
1982	1579	3353	4220	6087
Possible				
1977	2858	4268	6268	8868
1978	3374	7664	11323	14723
1979	2999	6443	9188	10888
1980	2177	5203	8106	10761
1981	1887	4448	7390	10811
1982	1610	3910	6722	9354

Geothermal Resources Council, TRANSACTIONS Vol. 6, October 1982

HEBER BINARY PROJECT
BINARY CYCLE GEOTHERMAL DEMONSTRATION POWER PLANT

Robert G. Lacy and Tiffany T. Nelson

San Diego Gas & Electric Company
Post Office Box 1831
San Diego, CA 92112

ABSTRACT

Recognizing the desirability of demonstrating the operation of a commercial scale, binary cycle, geothermal power plant, San Diego Gas & Electric Company (SDG&E) has formed a consortium consisting of three California utilities, the Electric Power Research Institute (EPRI), and the Department of Energy (DOE) to carry out the Heber Binary Project. The overall goal of the Project is to establish the economics and prove the operational viability of the binary cycle geothermal power plant. The Project has progressed through the conceptual and preliminary engineering phases, and completed such major milestones as receipt of major state and local regulatory approvals, execution of a contract for the purchase of geothermal heat, and award of the turbine-generator contract. System design has been finalized, and the assembly of the plant design into construction packages is underway. In addition to the progress on engineering, an estimate encompassing the engineering, equipment, construction, start-up, and Project management costs, as well as the cost for geothermal heat and plant O&M for the two-year plant demonstration period has been prepared for the DOE.

PROJECT BACKGROUND AND OBJECTIVES

The Heber Binary Project had its beginning in studies performed for the Electric Power Research Institute (EPRI), which identified the need for commercial scale (50 Mw or larger) demonstration of the binary cycle technology. In the late 1970's, SDG&E formed a consortium of other California utilities, along with EPRI to initiate preliminary engineering and environmental studies on a 65 Mw (gross) binary cycle power plant on the Heber geothermal reservoir, and to seek federal funding for a portion of the Project costs. In September 1980, San Diego Gas & Electric Company (SDG&E) and the Department of Energy (DOE) signed a Cooperative Agreement calling for DOE to share in 50% of the Project costs. Similarly, SDG&E signed Project participation agreements with EPRI, and the Imperial Irrigation District, California Department of Water Resources, and Southern California Edison Company.

The objectives of the Heber Binary Project are (1) to demonstrate the potential of moderate-temperature (below 410°F) geothermal energy to produce economic electric power with binary cycle conversion technology; (2) to allow the scaling-up and evaluation of the performance of binary cycle technology in geothermal service; (3) to establish schedule, cost and equipment performance, reservoir performance, and the environmental acceptability of such plants; and (4) to resolve uncertainties associated with the reservoir performance, plant operation, and economics.

The Project will be the first large-scale power generating facility in the United States utilizing the binary conversion process. It is expected that information resulting from this demonstration plant will be applicable to a wide range of moderate-temperature, low-salinity hydrothermal reservoirs, which account for 80% of geothermal reservoirs in the United States. With a moderate-temperature reservoir, the binary cycle has a significant advantage over the already proven dual flash process by being capable of converting a greater amount of the available geothermal heat into electric power.

To accomplish the plant engineering, design, and equipment procurement, SDG&E hired Fluor Power Services, Inc. (FPS) of Irvine, California. In parallel with the FPS engineering effort, SDG&E implemented an availability enhancement program using Pickard, Lowe and Garrick, Inc., as consultants. In late 1981, as the design became more complete, SDG&E contracted for construction management services with Dravo Utility Constructors, Inc. (DUCI) of New York. DUCI will cast the FPS design into construction packages, let the construction contracts, and oversee the construction in the field. Figure 1 shows the Project schedule and major milestones.

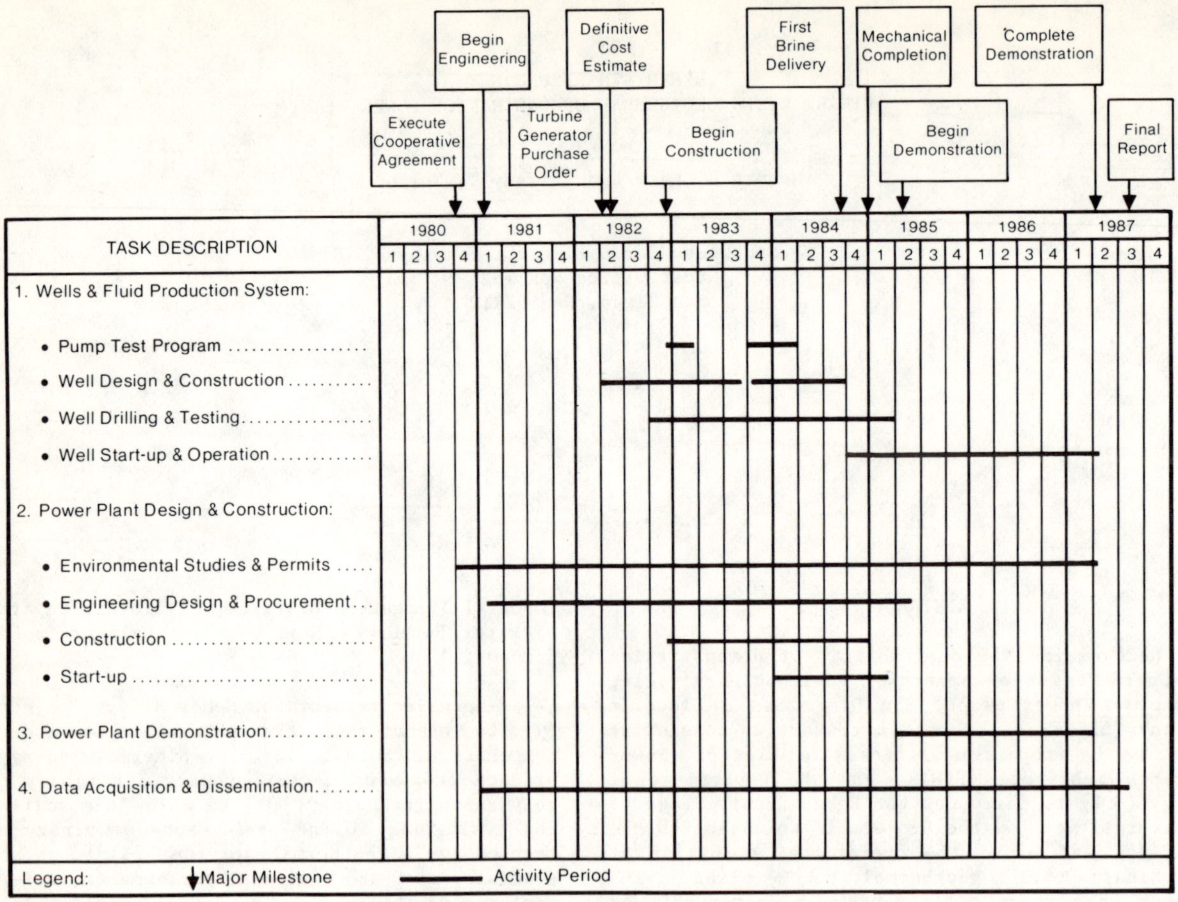

Figure 1. Heber Binary Project Schedule

POWER CYCLE DESCRIPTION

The major elements of the power cycle consist of a geothermal brine loop, a hydrocarbon binary loop, and a cooling water loop as shown in Figure 2. The geothermal brine is produced at a facility adjacent to the plant and is delivered to the plant under liquid phase (nonflashing) conditions from pumped wells at a temperature of approximately 360°F and a pressure of 250 psia. The brine loop contains a bank of eight shell and tube heat exchangers arranged in two trains of four heat exchangers each. The thermally spent brine is returned for injection to the geothermal reservoir at a temperature of about 160°F. The injection wells are located about 2¼ miles northwest of the plant site. The pressure for injection is provided by variable speed pumps, which can accommodate the wide range of injection pressures needed for both plugged and unplugged wells.

The binary loop contains the hydrocarbon working fluid and provides the medium for the transfer of geothermal energy from the brine to the hydrocarbon turbine. The hydrocarbon is pressurized and heated under supercritical conditions before entering the turbine throttle at 575 psia and 305°F. The working fluid is a mixture of 90 mole percent isobutane and 10 mole percent isopentane.

The Project will incorporate a floating cooling cycle, which will permit the optimization of yearly power output by taking advantage of variations in wet bulb temperatures. The turbine back pressure, as controlled by the condensing temperature, will be allowed to fluctuate with the ambient wet bulb temperature in order to generate the maximum amount of power. By incorporating the floating cooling concept, the plant average efficiency is improved (with a corresponding reduction in brine requirements) while maintaining an average net power output of 45 Mw.

TECHNICAL STATUS

The current plant design has evolved extensively from the conceptual designs of earlier studies. Design improvements have been made in all areas of the plant from the process loops to the electrical system. System design is now essentially complete and detailed engineering is underway. With the completion of system design, the engineering effort is shifting toward the goal of developing the individual construction packages that will be bid and managed by the Construction Manager.

The procurement of the major equipment items is well underway, kicked off by the award of the turbine-generator contract to the Elliott Company.

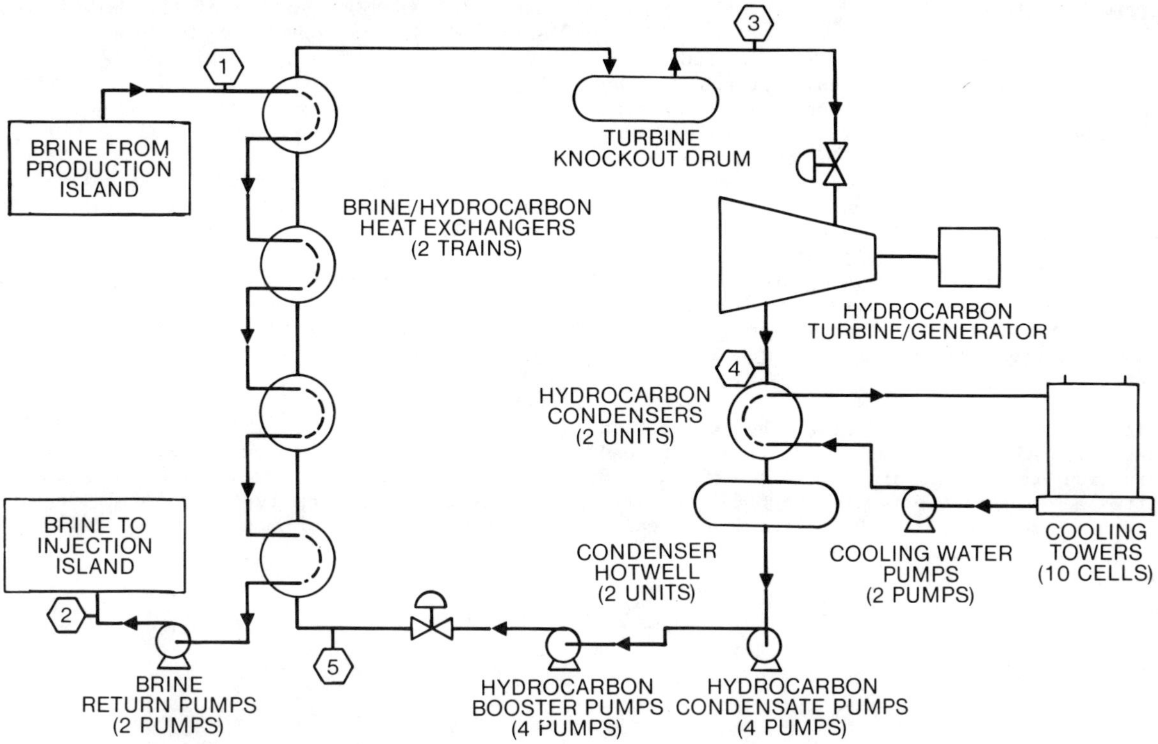

	1	2	3	4	5
PRESSURE - PSIA	250	320-595	575	73	630
TEMPERATURE - °F	360	162	305	152	101
FLOW - 10^6 LB/HR	7.9	7.9	8.3	8.3	8.3

Figure 2. Heber Binary Project Process Flow Diagram

The Elliott machine is a four stage, double flow, 3,600 RPM, axial turbine within a horizontally split shell contained in barrel enclosure. The turbine is mated to an Electric Machinery Company hydrogen cooled generator rated at 70 MWe. The turbine's barrel casing construction does not limit the location of the exhaust piping, allowing the elimination of a costly turbine pedestal.

Engineering studies recently completed have involved the optimization of the heat exchangers and condensers, the evaluation of a turbine exhaust recuperator, the selection of tube materials for the heat exchangers and condensers, and the verification of the plant's power cycle control scheme. The optimization study reduced the cost of the heat exchangers by 42%, settling in on a two train, four shells per train, two passes per shell design. The condenser design was essentially unchanged, but surface area optimization resulted in a 12% cost reduction. The recuperator study evaluated the economics of a recuperator for preheating the hydrocarbon working fluid before entry into the heat exchangers. The study revealed a balance between the capital costs and savings in operating costs. The recuperator was not incorporated to prevent schedule delays for re-engineering. The materials selection study recommends Allegheny Ludlum AL 29-4 as the tube material for the heat exchangers and condensers based on cost and expected life. AL 29-4C was considered an acceptable alternate, but is still awaiting ASME code approval. Additional heat exchanger and condenser savings of about 25% can be achieved with the use of AL 29-4C. Trent Tube's Sea Cure, which is comparable in cost to AL 29-4C, was found acceptable for the condensers, but it too is not ASME-approved. Finally, the power cycle control study verified the operation of the "process-following-the-turbine" control scheme. The control logic will protect the turbine from working fluid expansion through the "wet region" of the vapor dome, and limit turbine load changes to no more than heat exchanger response time.

DEFINITIVE ESTIMATE

The Cooperative Agreement between SDG&E and the DOE provided for a phased approach to the Project to allow better definition of the estimated total Project cost of $122 million at a time when the design engineering would be essentially complete. This estimate has been completed, and it was formally delivered to DOE in June 1982. The estimate covered the total cost of the Project, including engineering, equipment, construction, start-up, and Project management, as well as heat and O&M costs during a two-year demonstration period. A credit for the energy generated during the demonstration period was also applied to the estimate.

The primary effort during the definitive estimate was FPS's estimate of equipment and construction cost. To aid in this effort, FPS submitted a "benchmark estimate" to SDG&E in early January 1982 for review by DUCI, the Construction Manager. This "benchmark estimate" served as the basis for the FPS portion of the definitive estimate, which was submitted to SDG&E on April 1, 1982. While FPS's engineering effort has been broken down by plant systems, the construction and equipment estimate was broken down into the nine construction packages, which will be bid by DUCI.

Total Project costs are still estimated to be approximately $122 million, although a certain amount of "softness" is still inherent in numbers, particularly in the areas of heat costs and value of the energy generated. One of the goals of the Project is to establish these Project costs, and as equipment contracts are let the estimate will be updated.

ACKNOWLEDGEMENTS

We would like to acknowledge the Fluor Power Services' engineers whose work is summarized in the Technical Status section of this paper. They are Paul Linam, Paul Meiran, Surjit Pabby, and Ken Svoboda.

GRAVITY-HEAD GEOTHERMAL ENERGY CONVERSION SYSTEM

Hugh B. Matthews
Warren D. McBee

Sperry Research Center
100 North Road
Sudbury, MA 01776

ABSTRACT

A novel electric power generating system for low-temperature hot water geothermal resources is described. This so-called "gravity-head" system is a binary plant with the primary heat exchanger in the production well to a depth of 2200 feet. This configuration results in a novel heat-engine cycle which eliminates the "pinch effect" in the primary heat exchanger and the working fluid feed pump and provides a major decrease in brine pumping power. The net effect is a significant increase in net electric output power over a "conventional" binary plant under equivalent conditions and costs.

INTRODUCTION

Hot-water geothermal resources in the 250°F to 400°F temperature range have been little-used for electric power generation. A major reason for this is the low efficiency of existing utilization technology, which leads to high plant and well costs per kilowatt and large brine consumption per kilowatt-hour. This is so even for binary plants, which offer the best efficiency.

Sperry Research Center has been engaged for 10 years under Government and Sperry sponsorship in developing a concept known as the "gravity-head system", which is a well head binary cycle differing from a Rankine cycle and promising a 30% improvement in efficiency over a conventional well head binary plant and perhaps an additional 5% to]0% advantage over a binary gathering plant, primarily because of more effective condensing means useful to a small plant but difficult to adapt to a large one.

In this paper we first describe a "conventional" binary plant and discuss some of the limitations on its output power. Then we will describe the gravity-head system and its features which increase net output power.

THE "CONVENTIONAL" BINARY PLANT

For purposes of discussion, assume a hot water geothermal resource capable of providing 1000 GPM of brine flow rate from one production well with a down-hole pump at about 2000 ft. depth adding 700 psi pressure at a brine temperature just below the pump of 350°F. Also assume a wet-bulb temperature of 50°F.

A schematic diagram for this plant is shown in Fig. 1.

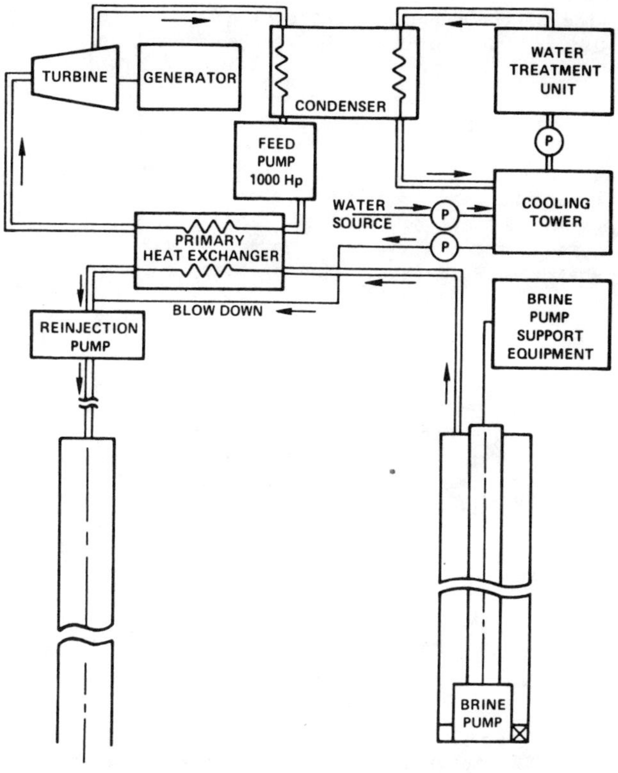

Fig. 1 Conventional organic Rankine cycle

There is a brine pump in the production well and support equipment for it on the surface. The purpose of the pump is to add pressure to the brine to prevent its flashing into steam or evolving CO_2 as it loses its pressure in approaching the surface.

The brine then passes through the primary heat exchanger where its heat is transferred into an organic working fluid. The cooled brine is then reinjected into the ground in another well, after additional pressurization from a surface reinjection pump.

The principal components of the organic fluid cycle are, as shown, the feed pump, one side of the primary heat exchanger, the turbine, and the condenser.

The feed pump increases organic pressure from 30 psi at condenser-outlet to 900 psi into the heat exchanger. It typically consumes 20% of the gross output power of the system.

The turbine, because of the high energy density of the organic fluid, will be a fraction of the size of an equivalent steam turbine operating at the same low temperatures.

The condenser is a conventional shell and tube design with the organic condensing on the outside of the tubes and the cooling water inside the tubes. The cooling tower is also conventional.

Fig. 2 is a pressure-enthalpy diagram of R-114 fluid ($C_2Cl_2F_4$) with a supercritical Rankine cycle superimposed. R-114 is used here, since it has qualities desirable for the gravity-head system to be described later, and we wish to be able to make direct comparisons. Isobutane was analyzed under the same circumstances, and was somewhat less efficient. The operating parameters assumed are as follows:

Well Flow	1300 GPM
Brine Temperature	350°F
Wet-Bulb Temperature (This would be an average throughout the West, not the worst case which would be over 80°F).	50°F

The processes are as follows:

1-2	Feed pump work
2-3	Heat added at constant pressure
3-4	Expansion in turbine
4-1	Heat rejected to atmosphere

The turbine, of course, drives the alternator which, in turn, furnishes the energy for the auxiliaries including the feed pump, cooling pumps and fans , and the brine pump.

Fig. 3 is a heat transfer diagram of the Fig. 1 system. It has the normal design goals of 20° mean effective temperature difference in the primary heat exchanger; 20° average temperature difference in the condenser; a 10° approach of the exit cooling tower water to the wet-bulb temperature; and 20° ΔT between the brine entering the heat exchanger and the working fluid leaving at the peak of its cycle.

The 10° ΔT which results from the feed pump work represents a loss in overall efficiency.

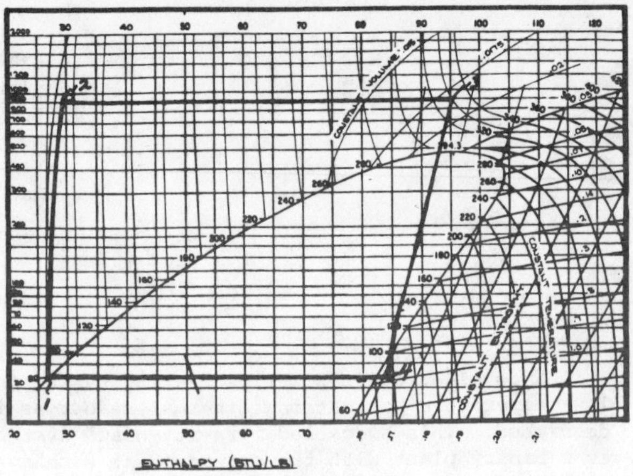

Fig. 2 Cycle diagram for conventional Rankine cycle-R114 fluid

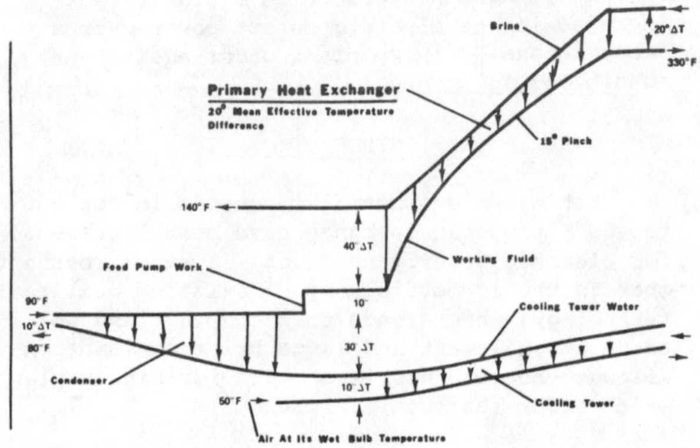

Fig. 3 Heat transfer diagram Conventional binary system

The pinch effect in the primary heat exchanger is caused by the heat capacity variation of the R-114 at constant pressure.

The problems in the binary system which are alleviated by the gravity-head system are as follows:

- The large portion of the available power used by the brine pumps and feed pump in a binary system.
- The "pinch effect" - a loss in available energy in transferring heat from water to an organic fluid at constant pressure.
- The cost of the feed pump and heat exchanger shell.
- The lost temperature rise in working fluid produced by feed pumping.

THE GRAVITY-HEAD SYSTEM

Fig. 4 is a simplified schematic of the gravity-head cycle. In this system, heat transfer between the brine and the organic takes place entirely in the production well from the brine flowing upward in the outside annulus to organic fluid flowing downward in the second annulus. Thus, the brine arrives at the surface cold, requiring no pressurization to prevent it from flashing, but merely sufficient pressure to reinject it into another well. The organic arrives at the bottom of its downward path hot. The organic is then routed through a small turbine driving a brine pump, where a small portion of its thermal power is extracted. It then returns to the surface in an insulated riser, drives the power generating turbine on the surface, condenses in an evaporative condenser, and then returns to the primary heat exchanger input at well-head.

Fig. 5 shows the gravity-head cycle superimposed on the same R-114 pressure-enthalpy diagram of Fig. 2. The assumed operating parameters are the same. An important difference between the two cycles is obvious. In the Rankine cycle, the heat is added at constant pressure, whereas in the gravity-head cycle, the heat is added at a constantly increasing pressure. In other words, heat is added to the fluid as it flows down the 2000-foot vertical heat exchanger with a constantly increasing pressure due to the gravity head.

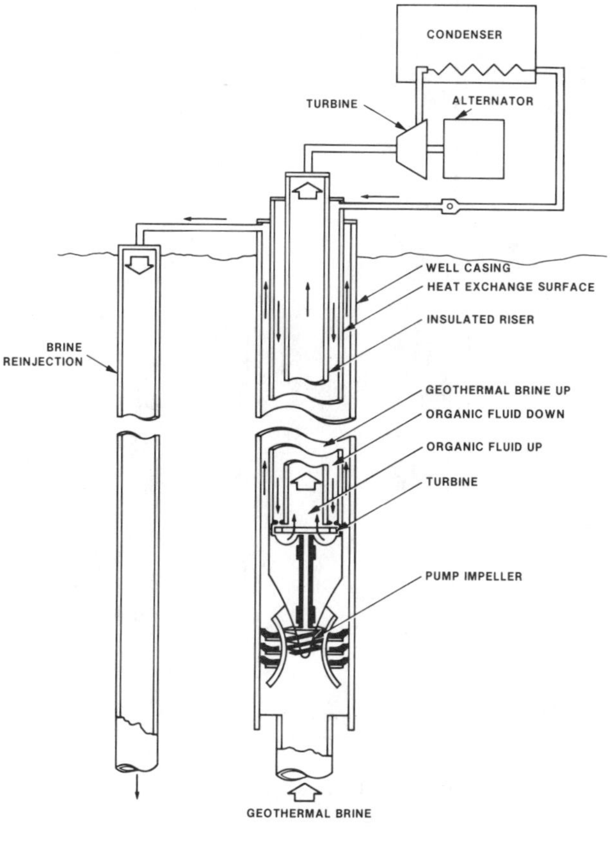

Fig. 4 Gravity-head system schematic

The processes are as follows:

1-2 Heat input to working fluid from brine. Also, there is an added enthalpy (heat energy) contribution due to the potential energy of the fluid (energy due to height) being transferred into heat energy as the fluid flows down. This is the Z/J term in the general energy equation.

2-3 Down-hole brine-pump turbine work.

3-4 Lift work. This is the inverse of the process of 1-2 in which enthalpy is being transferred back into potential energy.

4-5 Expansion in surface turbine to drive alternator.

5-1 Heat rejected to atmosphere in an evaporative condenser.

The path of process 1-2 can be altered within limits by varying the amount of heat transfer surface per foot of depth. The one shown has a nearly constant heat capacity for the R-114. This means that the pinch effect with its attendant loss in available energy is eliminated.

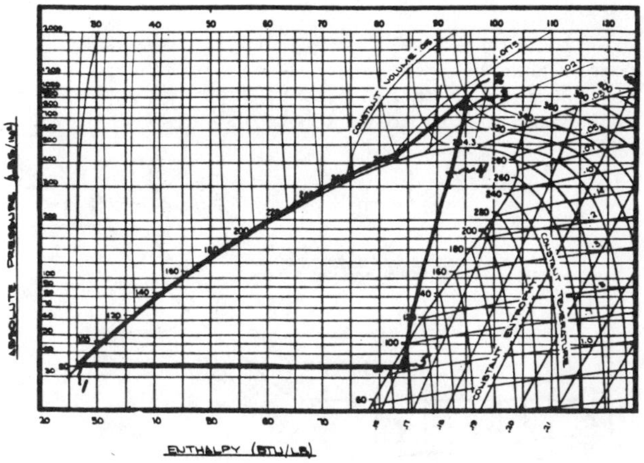

Fig. 5 Cycle diagram for Gravity-head system - R114 fluid

Fig. 6 is the heat transfer diagram of the gravity-head system, showing the elimination of the pinch effect and the 10° ΔT during feed pumping and a total 30° ΔT between wet bulb and condensing temperature rather than the 40° ΔT, which is the reasonable minimum of a heat rejection system composed of a separate cooling tower and condenser. The net effect is that the brine is rejected at $100^{\circ}F$, compared with $140^{\circ}F$ for the conventional binary system.

The down-well heat exchanger design may be of interest. The system parameters on which the design is based are as follows:

Well Flow	1300 GPM
Well Temperature	350°F
Wet-Bulb Temperature	50°F
Working Fluid (Dichlorotetrafluoroethane)	R-114
Working Fluid Flow	637 lbs/sec

As mentioned previously, the desired curve on the heat transfer diagram is obtained by varying the heat transfer surface per foot of depth. Fig. 7 is a table listing this variation. The design fouling factor was .0015 on the brine side and .0005 on the organic side. Fluid velocities were chosen as a compromise between low flow losses, high heat transfer coefficients, low fouling rates and small well casing size; they vary from 1.75 ft./sec. at the top to 5 ft./sec. at the bottom for the brine and approximate 5 ft./sec. in all sections for the organic. Heat transfer coefficients approximate 150 BTU/HR-FT^2-°F in all sections.

Fig. 7 is the design of one of the tube bundle sections. Extensive engineering test work has been done with this design. It is planned to install turbular heat exchanger sections to the 1000 ft. depth and an annular heat exchanger from this to the 2180 ft. depth.

The planned well casing profile to accommodate this heat exchanger design is 32" o.d. to 1000 feet, then 24" o.d. to 2200 feet, then conventional size to well bottom.

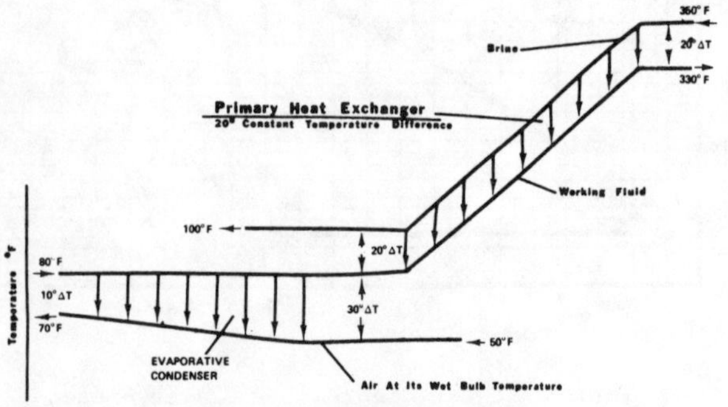

Fig. 6 Heat transfer diagram Gravity-head system

SYSTEMS COMPARISONS

The performance comparison between the gravity-head system and the conventional binary system is shown in Fig. 8 . Note the 37% larger net electric output power from the gravity-head system. This is almost entirely due to four advantages:

- Elimination of pinch effect
- More efficient brine pumping
- Elimination of mechanical feed pump
- Use of evaporative condenser for heat rejection.

Cost comparisons are inexact. However, the increased cost of the larger production well required for the gravity-head system is offset by elimination of the mechanical feed pump, the primary heat exchanger shell, and the brine pump support equipment. We think the combined cost of plant and wells will be less for the gravity-head plant per unit of brine flow rate, in which case the $/KW advantage would be greater than the efficiency advantage. Another way of looking at the comparison in cost-performance between the gravity-head plant and the conventional binary plant is that if one assumes equal cost per unit brine flow rate and equal brine flow rate, the gravity-head plant would produce with a brine temperature of 350°F the same net electric output power as a conventional binary plant with a brine temperature of 400°F.

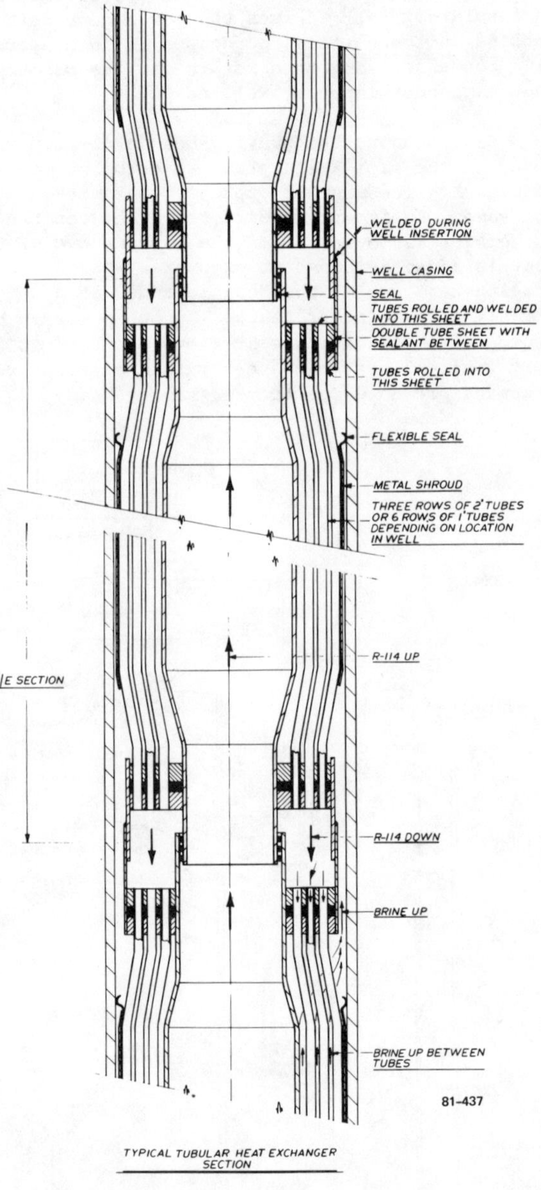

Fig. 7 Gravity-head heat exchanger schematic of one tube bundle section

	Binary Gathering Plant	Gravity-Head Well Head Plant
Well Flow	1300 GPM	1300 GPM
Mean Effective Temperature - Difference in Heat Exchangers	20°F	20°F
Working Fluid	R-114	R-114
Wet-Bulb Temperature	50°F	50°F
Condensing Means	Cooling Tower & Shell & Tubes	Evaporative Condenser
Condensing Temperature	90°F	80°F
Brine Pump - Mechanical Efficiency	.49	.56
Surface Turbine Efficiency	.96	.96
Condensing Parasitics	313 kW	375 kW
Working Fluid Flow	555 lbs per sec	637 lbs per sec
Net Electrical Output	3670 kW	5045 kW

Fig. 8 Performance Comparisons of One Well of a Binary Gathering Plant vs. a Gravity-Head Well Head Plant

Geothermal Resources Council, TRANSACTIONS Vol. 6, October 1982

A NEW BINARY-FLASHING PLANT FOR THE PRODUCTION OF ELECTRICITY

Efstathios E. Michaelides and Farid Fakhre-Shafaie

Department of Mechanical and Aerospace Engineering
University of Delaware
Newark, Delaware 19711

ABSTRACT

The next expansion of geothermal power intallations will cover the resources with temperatures less than 150°C. Binary plants are the only suitable for the utilization of these resources. A Modification to the conventional binary plant design is proposed here aiming at the production of more power. The new plant uses a flashing system and partial evaporation of the secondary fluid in the heat exchanger. The thermodynamic analysis shows that the proposed installation would provide approximately 25% more power than the conventional installation, a feature that makes it attractive for further technical and economic analyses.

I. INTRODUCTION

Most of the geothermal resources utilized at the present for the production of electricity have temperatures above 150°C. The utilization is mainly achieved by direct-steam or flashing installations, which require high resource temperatures. It seems that the low temperature resources have been ignored for the production of electricity and are only contemplated for supplying direct heat. These low temperature resources though are the most abundant on earth (Kunze, 1979). Actually, there are geothermal reservoirs with temperatures close to 100°C in many regions of the U.S.A.. These resources are considered of great importance to the local economies, and if utilized would become valuable assets to the region.

It is believed that the next step in the utilization of geothermal resources for the production of electricity will be the development of low temperature areas (below 150°C). The successful development of the other resources, the vast knowledge and experience accumulated in the past twenty years and the uncertainty of the prices and of the environmental impact of the conventional energy sources will prompt the utilization of resources with temperatures in the range of 80° to 140°C. The current rate of growth of geothermal power plants (19% in the last five years) and their economic success would ascertain to the expected continuous development of lower temperature geothermal resources.

Binary geothermal plants are touted to be the best type of plants for low temperature applications. This, because of the low vapor yield of any flashing system and the constraint to keep flashing pressures above 1 atm (otherwise non-condensable air inleakage cannot be avoided). One of the disadvantages of the binary plants with complete evaporation of the secondary fluid is that the brine exit temperatures are usually high and, hence, they waste a great deal of the available work. This project examines an alternative binary-flashing system, which would enable the secondary fluid to absorb more heat from the brine, and to produce more work by rejecting the brine at lower temperatures.

2. ASSUMPTIONS

Throughout this work it will be assumed that the geothermal well produces liquid water free of impurities. The properties of this substance are taken from tables (Keenan et al.). This simplifying assumption is justified in view of the fact that most of the wells (with the exception of the ones in the Niland area) produce low salinity water. Furthermore, the study will concentrate only on the comparision of two types of binary geothermal plants. The same fluid will be assumed to be the input in both the installations and all the quantities for comparison will be given per kg of brine supplied to the plant. This assumption bypasses practical questions such as corrosion or deposition in the main heat exchanger, which can be avoided by the appropriate treatment of the brine.

The available power will be calculated for both the installations (Kestin, 1980) according to the formula:

$$\dot{W}^{\circ} = \dot{m}_i [h-h_o - T_o (s-s_o)], \qquad (1)$$

where $\dot{m}_i$ is the mass flow rate, h the enthalpy and s the entropy of the secondary fluid as it enters the turbine. The subscript o refers to ambient conditions.

This presumes ideal operation of the turbines of the two installations. The power actually produced by each plant is about 70% of the above quantity depending on the turbine efficiency. Bearing in mind that the two installations under comparison would have the same turbine efficiency, the comparison of the available powers is justified completely.

For the calculations T_o is taken as 30°C and a minimum of 3°C temperature differences was assumed between the two fluids in the heat exchanger. The secondary fluid in all cases is taken to be isobutane (C_4H_{10}) and its properties are taken from the Goodwin tables (1978).

3. THE EXISTING BINARY INSTALLATION

A schematic diagram of existing binary installations is shown in Fig. 1a. The secondary fluid (usually a mixture of isobutane and pentane) enters the counter flow heat exchanger (HX) and evaporates completely by cooling the brine. It then enters the turbine and produces work. A condenser, a pump (P) and a reinjection well complete the

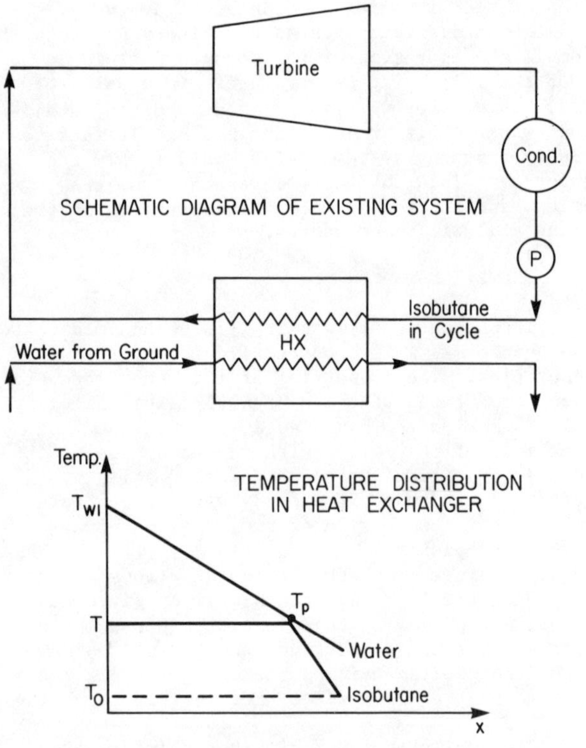

Figure 1a. Schematic diagram of the existing installations.
1b. Temperature diagram for heat exchanger of the existing installations.

closed loop for the secondary fluid and the open loop for the brine. The temperature diagram of the heat exchanger is shown in Fig. 1b. Brine cools from T_{wl} to T_{out} while the secondary fluid (isobutane) absorbs heat and evaporates at temperature T.

The available work that can be produced by such an installation may be calculated for any evaporating temperature T. This is depicted in Fig. 2, where T is regarded as a parameter. If $T = T_o$ then $h-h_o - T_o(s - s_o) \equiv 0$ and if $T = T_{wl}$, $\dot{m}_i \equiv 0$. Hence, the available power is zero at these two points and exhibits a maximum for

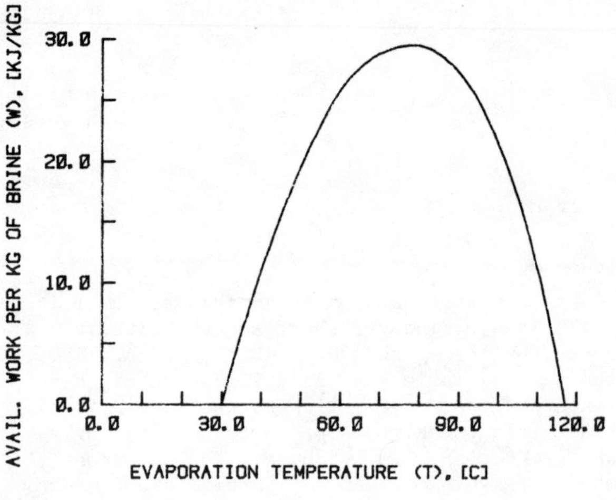

Figure 2. Work produced for T_{wl} = 120°C

$T_o<T<T_w$. This maximum is obtained graphically in Fig. 2. It occurs at 78°C and the plant delivers 29 kW of power per kg/s of brine input at T_{wl} = 120°C.

It is observed that even at this optimum operation (T = 78°C) of the binary installation the brine must be reinjected at T_{out} = 58°C at the expense of more available power. The suggested installation whose description follows aims at extracting part of this dissipated power by utilizing more of the heat from the brine.

4. THE PROPOSED BINARY-FLASHING SYSTEM

The binary-flashing system proposed here is designed to extract more heat from the geothermal brine than the conventional system and thus deliver a higher amount of available work. A schematic diagram of this installation is shown in Fig. 3a. Here the secondary fluid comes out of the heat exchanger (HX) partly evaporated (0<x<1) and passes through a separator (S) where the vapor is removed and directed to the high pressure side of the turbine. The liquid from the separator passes through a flasher-separator (F,S) where it flashes at temperature T_1 and produces more vapor. This vapor is directed to the low pressure side of the turbine and produces an additional amount of work. At the exit of the turbine the fluid is directed to a condenser at temperature T_o and through a pump (P) to the heat exchanger. The second stream of liquid at temperature T_1 from the flashing chamber is directed to an auxiliary pump (P) and then to the same heat exchanger.

Thus, the duty of the heat exchanger is to

heat up the first stream from T_o to T and the second from T_1 to T. The evaporation in the heat exchanger is partial and the quality of the heat exchanger effluent x may be used as an optimization parameter for the system. The operation of the heat exchanger is shown in Fig. 3b and the equations describing the heat flow in the two parts are as follows:

$$\dot{m}_w[h(T_{w1}) - h(T_p)]_w = \dot{m}_i\, x\, h_{fg}(T)\ , \qquad (2)$$

$$\dot{m}_w[h(T_p) - h(T_2)]_w = \dot{m}_i\, [h_f(T) - h(T_1)]_i\ , \qquad (3)$$

$$\dot{m}_w[h(T_2) - h(T_{out})]_w = \dot{m}_i[x+x_1][h(T_1) - h(T_o)]_i . \qquad (4)$$

These equations may be solved simultaneously to yield the variables $\dot{m}_i$, T_{out} and T_2 for a given x and for $\dot{m}_w$ = 1 kg/s. The quality x_1 in the flashing chamber is obtained from the constant-enthalpy equation:

$$h_f(T) = h_f(T_1) + x_1 h_{fg}(T_1). \qquad (5)$$

Here h is the enthalpy, obtained at the temperature shown in the parenthesis, and the subscripts denote the following: w = water, i = isobutane, f = saturated liquid, fg = latent heat, p = pinch point.

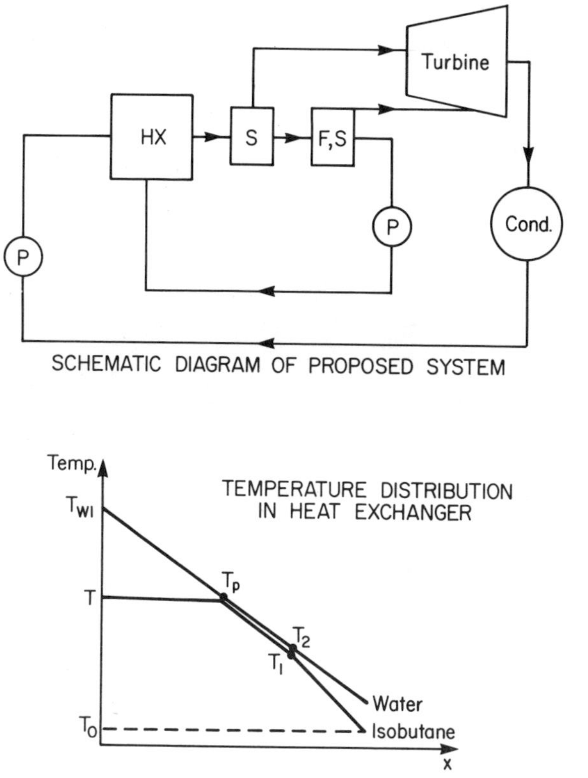

Figure 3a. Schematic diagram of binary-flashing system.
3b. Temperature diagram for heat exchanger in the binary-flashing system.

Equations (3), (4), (5), and (6) are solved, subject to the constraints $T_2 \geq T_1 + 3°C$, and $T_p \geq T + 3°C$. The temperature T_1 is obtained as the mean value of T and T_o in order to optimize the performance of the flashing part of the installation. This practice is commonly used in single and dual flashing systems.

It is obvious that the design of the proposed binary-flashing installation has two parameters for optimization namely the evaporating temperature T and the amount of vapor produced in the heat exchanger x. In order to optimize both of them the following procedure is followed: x is fixed at a certain value and T varies between T_o and T_{w1}. The amount of available work is plotted against T and the highest value of it $\dot{W}°_{max}$ is selected together with the corresponding T. This is shown in Fig. 4 for x = 0.6, T_{w1} = 120°C and

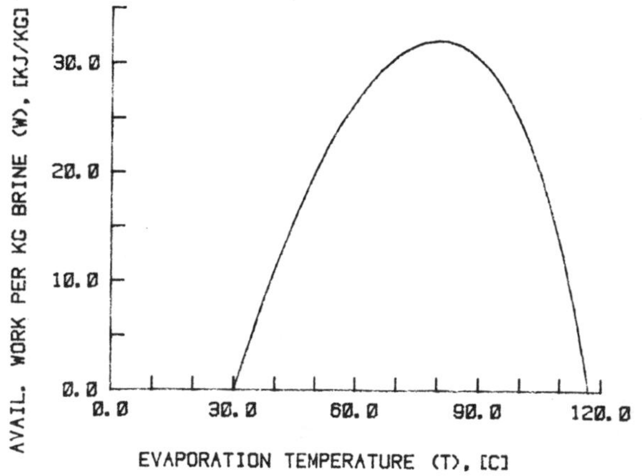

Figure 4. Work produced for T_{w1} = 120°C in binary-flashing system with x = 0.6.

T_o = 30°C. It is observed that the maximum available power is 32.1 for T = 81°C. This procedure is repeated for other values of x and the values $\dot{W}°_{max}$ thus obtained are plotted against x; this curve has a maximum point in the range of variation of x (0<x<1). The maximum point yields the optimum amount of power $\dot{W}°_{opt}$ and the conditions for this optimum (x_{opt} and T_{opt}) to be achieved. The curve for $\dot{W}°_{opt}$ is shown in Fig. 5 where x_{opt} = 30% for T_{w1} = 120°C and T_o = 30°C. This optimum point also corresponds to T_{opt} = 95°C. The geothermal installation therefore, must be designed to operate at the above optimum conditions in order to yield the maximum possible amount of work. This amount of work is higher than the amount produced with the existing binary installations, as estimated in Fig. 2. The augmentation of work is due to the fact that the proposed

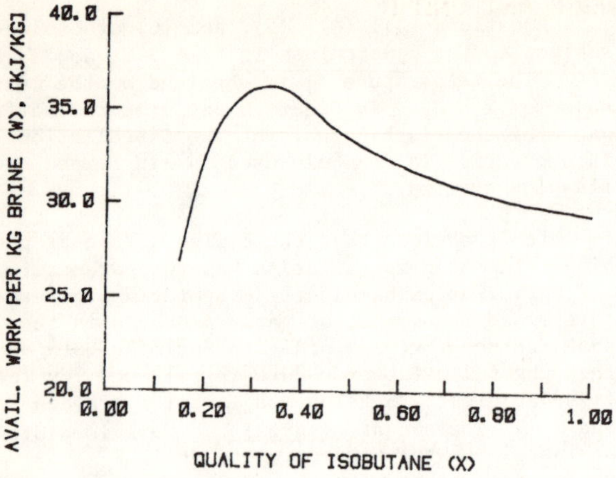

Figure 5. Work produced with variable x for T_{w1} = 120°C.

system has two parameters to be optimized (T and x) instead of one (T).

A measure of the comparison for the two binary installations (proposed and existing) is the ratio of the maximum possible power obtained by both. This ratio would yield the percentage of power the proposed system delivers in excess of the existing system. For that matter the maximum power $\dot{W}^{\circ}_{max}(E)$ is calculated for resource temperatures T_{w1} = 80, 100, 120 and 140°C for the existing system (total evaporation of isobutane). The same quantity $\dot{W}^{\circ}_{max}(P)$ is calculated for the above resource temperatures, but with variable x. The ratio R of these quantities $\dot{W}^{\circ}_{max}(P)/\dot{W}^{\circ}_{max}(E)$ and plotted as a function of x in Fig. 6. It is observed that the proposed system is always superior than the existing one and yields up to

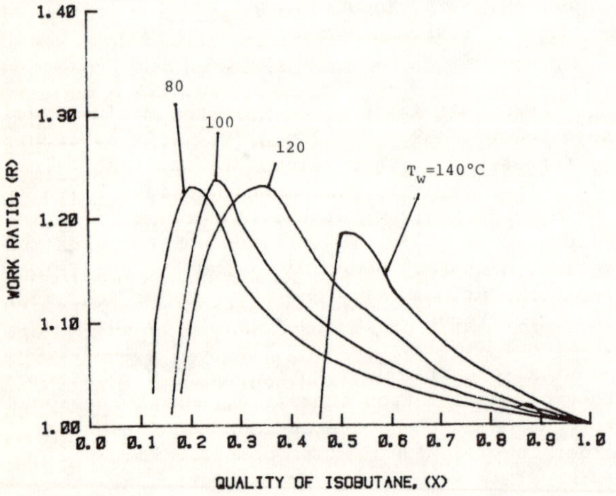

Figure 6. Work ratio for different resource temperatures.

20% more power for T_{w1} = 140°C or up to 25% more power for T_{w1} = 80°C.

It is realized that the proposed system is more complex in the design and economically more costly. Although it produces more power, the excess cost associated with the extra components could offset the additional revenue of more power. A detailed economic study is needed to answer this question and this is not the scope of the present brief study. It seems, however, that at the current level of prices for energy there is economic justification for this system. If the economic feasibility of this system is demonstrated, one may even add more flashing stages.

5. CONCLUSIONS

A new binary geothermal installation is proposed here, which makes use of a flashing stage. This installation could produce about 25% more work than the existing installations when designed to operate at its optimum conditions. It is believed that the proposed installation is ideal for the utilization of geothermal resources with temperatures less than 150°C. These resources are more abundant than the high temperature ones and their utilization is forthcoming.

6. REFERENCES

Goodwin, R. D., Isobutane: Provisional Thermodynamic Functions from 114 to 700 K at pressures to 700 bar, U.S. Dpt. of Com. no. NBSIR 79-1612, July 1979.

Keenan, F. G., Keyes, F. G., Hill, P. G. and Moore, J. G., Steam tables (International edition - metric units) Wiley, 1969.

Kestin, J., Available work in Geothermal Energy, ch. 3 in Sourcebook on the production of Electricity from Geothermal Energy, ed. J. Kestin, U.S. Dpt. of Energy, Washington 1980.

Kunze, J. F., Utilizing Geothermal Resources Below 150°C, Trans. ASME J. of En. Res. Tech., 101, p. 124, 1979.

THE SALTON SEA 10 MWe POWER PLANT, UNIT 1

W. E. Moss
O. D. Whitescarver
R. N. Yamasaki

Southern California Edison Company
Union Oil Company of California
WESTEC Services, Incorporated

ABSTRACT

The Southern California Edison Company's Salton Sea Geothermal Electric Project is the second of two flash-steam projects located in the Imperial Valley of California to successfully demonstrate the feasibility of utilizing steam from highly saline geothermal fluids for electric power generation.

The objective of Edison's Power Plant Unit 1 program at the Salton Sea KGRA is to develop design, operating, and economic criteria for commercial geothermal developments in the Imperial Valley of California. The Edison plant is designed specifically for utilization of geothermal steam and employs principles found in conventional fossil-fueled electric generating plants. This plant serves as a model of a full scale commercial plant, using systems and components which likely will be utilized in large scale follow-on units.

INTRODUCTION

The Salton Sea Geothermal Demonstration Unit is one of three power plants Edison has planned for the Imperial Valley of California. The first unit to go on line was the Brawley 10MWe Unit 1 on June 21, 1980. The Salton Sea 10MWe Unit 1 went on line on July 9, 1982. The third unit, a 47MWe plant at Heber will be the first commercial scale power plant to use heat from the Heber KGRA. All three projects represent another step in Edison's commitment to develop alternate and renewable resources during the 1980's.

The Salton Sea Unit 1 power plant is an experimental effort by Southern California Edison and its preceding companies which have long been interested in developing geothermal electric generating plants. In 1973, Edison began looking strongly in California's Imperial Valley area for geothermal energy development. Original field research and development plans at the current Salton Sea demonstration site were inaugurated in October 1978. Actual construction began in February 1981. The unit was commissioned on July 9, 1982.

Joining Edison in the Salton Sea Project are the Union Oil Company of California, primarily responsible for the steam production facility, the Southern Pacific Land Company and the Mono Power Company, an Edison subsidiary.

Total project cost, including the power plant and Edison's share of the field development costs, is approximately $30 million.

GEOTHERMAL RESERVOIR

The Salton Sea Geothermal Reservoir is located on the largest of several geothermal anomalies within the Salton Trough in the Imperial Valley. It is defined by a large region of anomalously high near-surface heat flow. As early as 1925 a plan was completed for a proposed geothermal steam plant to be built on Mullet Island, located about 5 miles northeast of the Salton Sea demonstration site. The plan called for up to 40 wells to be drilled to supply the steam for six turbine generator units. The plan, however, was abandoned with the decrease in the price of oil. The discovery well (Sinclair No. 1) for the Salton Sea Geothermal Field was drilled in 1957 by the Kent Imperial Corporation. In June 1976, start-up operations were initiated for the San Diego Gas & Electric/DOE Geothermal Loop Experimental Facility (GLEF) located in Niland. The major objective of the GLEF was to determine the technical and economic feasibility of generating electricity from high salinity brines. This 7½ year program provided important information for the Salton Sea 10MWe Power Plant, Unit 1.

The geothermal reservoir portion associated with the Salton Sea 10MWe Demonstration Plant is located at the southern end of the Salton Sea, about a mile southwest of Obsidian Butte. The reservoir is a large hot-brine resource with commercial temperatures in excess of 500F (260C). Reservoir brines are quite saline, up to 250,000 ppm (mg/l) Total Dissolved Solids. Production testing (simulating actual power generating conditions) revealed that the hot brine caused fouling and scaling throughout the surface equipment. Sufficient testing was conducted, however, to design and fabricate a surface brine handling system to continuously supply high quality steam to the power generating facility.

The Salton Sea 10MWe geothermal demonstration facility site consists of some 20 acres of unincorporated vacant agricultural land in the County of Imperial, California. The major facility site is located approximately 8 miles southwest of the town of Niland. The site is bounded on the west and north by the Salton Sea, which is separated from the project by levees. The site is approximately 5 feet below the current level of the Salton Sea.

STEAM GATHERING SYSTEM

The geothermal steam production system at the Salton Sea facility is owned and operated by the Union Oil Company of California. The flashed-steam system utilizes up to four production wells centrally located within the reservoir supplying up to 1,416,000 lbs/hr (642,298 kg/hr) of hot effluent to the Steam Gathering System. The effluent from each well is a mixture of vapor and brine. The flowrate is controlled by adjusting the wellhead pressure. Three production wells have sufficient capacity to supply the total brine required to the system, should one of the wells require shutting-in for maintenance.

The effluent from each production well flows through an insulated pipeline from the wellhead to a dedicated wellhead separator (flash vessel) located at the Steam Gathering Facility. The first-stage flashed vapor and brine are separated in this vessel. The vapor from each separator flows into the main steam line to the turbine. Meanwhile, the brine from each separator flows to the high pressure crystallizer. Steam flashed from the brine in the crystallizer also flows into the main steam line to the turbine.

The brine from the high pressure crystallizer flows to a low pressure crystallizer where the pressure is further reduced to a pressure slightly above atmospheric. The remaining steam and brine from the low pressure crystallizer flows to an atmospheric separator for final vapor/brine separation.

As steam is evolved in the separation processes, the brine is cooled and the concentration of dissolved material increases. In this supersaturated condition, various compounds tend to come out of solution and form solid particles suspended in the brine. The crystallizers are designed and operated so that the particles coming out of solution tend to grow to a size which can be readily separated from the brine. This particle growth process reduces the tendency of the very small particles to adhere to the vessel walls and downstream piping. Maintaining balanced brine flow and stabilized vessel conditions was initially forecasted as difficult. For this reason, two parallel trains of high and low pressure crystallizers and atmospheric separators are utilized. Each train is capable of handling 100% of the plant flow. One train can be shutdown for servicing without interrupting the plant operations.

From the atmospheric separator the brine flows into the clarifier tank. Here the suspended particles grow larger and fall to the bottom of the tank where they are removed. The clarified brine rises to the top of the tank and is removed for final filtering. The filtered brine is pumped through insulated pipelines to the four field injection wells. The solids collected from the clarifier and filters are hauled to a disposal site.

POWER GENERATION SYSTEM

The Southern California Edison Company's Salton Sea 10MWe Power Plant is located on a 4½ acre parcel adjacent to the Union Oil steam production facility.

Steam is delivered to the power plant at a design rate of 187,629 lbs/hr (85,108 kg/hr) at a single pressure, 105 psia (7.38 kg/cm^2) at an approximate saturation temperature of 338F (170C).

At the power generation facility, steam first enters two 50% capacity demisters (known as the moisture separator at the Brawley Unit 1). The demisters are designed to remove 97.2 percent of the moisture contained in the main steam.

Steam enters the turbine at a design rate of 170,655 lbs/hr (77,409 kg/hr). The turbine is a 10,000 kW, 3600 rpm, single flow, single cylinder unit with six reaction stages and two low pressure stages. Steam enters the turbine at an approximate pressure of 103 psia (7.24 kg/cm^2) and at an approximate temperature of 337F (169C).

Steam from the turbine is exhausted to the main condenser at a back pressure of 4 inches Mercury absolute (102 mm of HgA). Unlike the Brawley Unit 1, which uses a surface condenser located above the turbine, the Salton Sea main condenser is located directly below the turbine deck. The condenser has a design heat rejection rate of 156,274,760 Btu/hr (39,381,240 kCal/hr). Corrosion resistant stainless steels are used because of the oxidation potential of air, oxygen, and hydrogen sulfide present in the system.

All condensate is returned to Union Oil for field use and reinjection via carbon steel pipe. The Brawley Unit 1 utilizes 304 stainless steel condensate return pipe. It was expected that the pH of the condensate would be below 7.0. In fact it has been proven at Brawley that the steam condensate pH is in the range of 8-9. Thus carbon steel pipe was chosen for Salton Sea at a significant cost savings.

Noncondensible gases are removed from the main condenser by four 25% steam-jet air ejectors. Each ejector requires 4110 lbs/hr (1864 kg/hr) of 90 psia (6.33 kg/cm^2) motive steam to remove a design capacity of 1635 lbs/hr (742 kg/hr) each of air and noncondensible gases from the main condenser.

Noncondensible gases from the main condenser and motive steam to the steam-jet air ejectors flow to the air ejector condenser with a back pressure of 15.75 inches of Mercury absolute (400 mm of HgA). This shell-and-tube condenser of stainless steel construction has a design heat rejection rate of 20,168,500 Btu/hr (5,082,462 kCal/hr). Condensate from the air ejector condenser flows to the main condenser from where the condensate is returned to Union Oil.

Noncondensible gases are extracted from the air ejector condenser by an induced flow caused by two vacuum pumps. Unlike the Brawley Unit 1, the Salton Sea plant uses no second stage ejectors, but relies solely on the electrical vacuum pumps. This has shown to be an economical as well as a more reliable design based upon Brawley plant experience.

Discharge of the noncondensible gases from the power plant is vented to atmosphere through the cooling tower fan plenums. It is expected that by mixing with the cooling tower plume, the concentration of gases, especially H_2S, will be dispersed to the point that ambient air quality standards will not be exceeded.

A conventional four cell, induced draft, counterflow, wet cooling tower with a rated heat load of 1.98 million Btu/hr (498,960 kCal/hr) provides 18,000 gpm (1135 l/sec) of cooling water to the main condenser and plant auxiliary heat exchangers. The tower is designed to cool water from 113F (45C) to 91F (33C). Makeup water to the tower is provided by a single irrigation canal located to the east of the plant.

The Salton Sea Plant is designed to produce 10,000 kilowatts of gross electric power. The load to run the plant itself is approximately 900 kilowatts resulting in a net plant output of 9.1 MW. Electric power is generated at 13.8 kV and stepped up to 34.5 kV for the Imperial Irrigation District (IID) electrical system. Power from the plant is presently transmitted into the IID's service territory for consumption in the Imperial Valley. The amount of energy generated by this plant is sufficient to meet the needs of approximately 9100 residential consumers and can save approximately 100,000 barrels of oil per year.

OPERATING EXPERIENCES

As of July 15, 1982 the Salton Sea unit has been on line for 67 hours and has generated 540,000 kilowatt hours of electricity. Geothermal steam was first admitted to the turbine on the 4th of July. On July 7, 1982 the turbine was brought up to 3600 rpm for the first time. The unit was placed on line for the first time at 1215 hours on July 9, 1982 with one production well on line. Three production wells were in operation by July 12th when the unit attained 10MWe for the first time.

Start-up problems have been few. On July 6th a flashover occurred on the 34.5kV transmission line which cut off power to the plant. The arc-over was caused by dirty insulators and possible moisture from geothermal steam vapor. Flashovers were a significant problem at the Brawley Plant, but have since been remedied by installing new larger insulators.

Initial steam quality has been within the SCE-Union Oil contract limits. Table 1 shows the steam chemistry for the Salton Sea Plant. Brawley Plant chemistry is also listed for comparison.

REFERENCES

County of Imperial Planning Department, "Salton Sea Ten Megawatt Geothermal Demonstration Facility, Final Environmental Impact Report", December 1980.

San Diego Gas & Electric, "Geothermal Loop Experimental Facility, Final Report", prepared for U.S. Department of Energy under contract no. DE-AC03-76ET28443, April 1980.

County of Imperial Planning Department, "North Brawley Ten Megawatt Geothermal Demonstration Facility, Final Environmental Impact Report", April 1979.

Siegfried, H.N., "Geothermal Exploration in the First Quarter Century", for Southern Sierra Power Company, 1925.

SALTON SEA GEOTHERMAL PROJECT

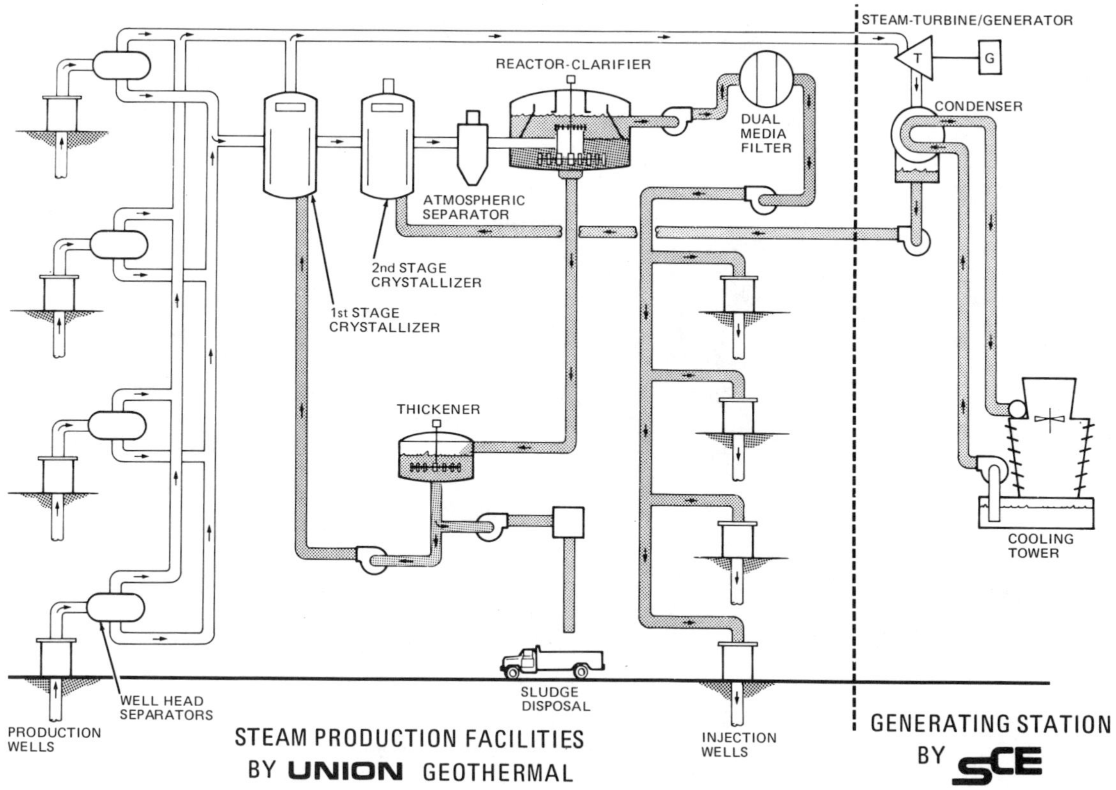

TABLE 1
MAIN STEAM CHEMISTRY AT INLET TO POWER PLANT
FOR WEEK OF July 5-11, 1982

	Salton Sea Unit 1	Brawley Unit 1
Pressure (psig)	100	101
Temperature (F)	338	336
Moisture (% by wt.)	*	0.58
pH	6.1	6.0
Conductivity (μmho/cm)	1700	1450
Chlorides (mg/l)	1.9	1.5
Total Dissolved Solids (mg/l)	14.7	15
Noncondensible gases (% by wt.)	*	0.26

*sampling set up yet to be completed.

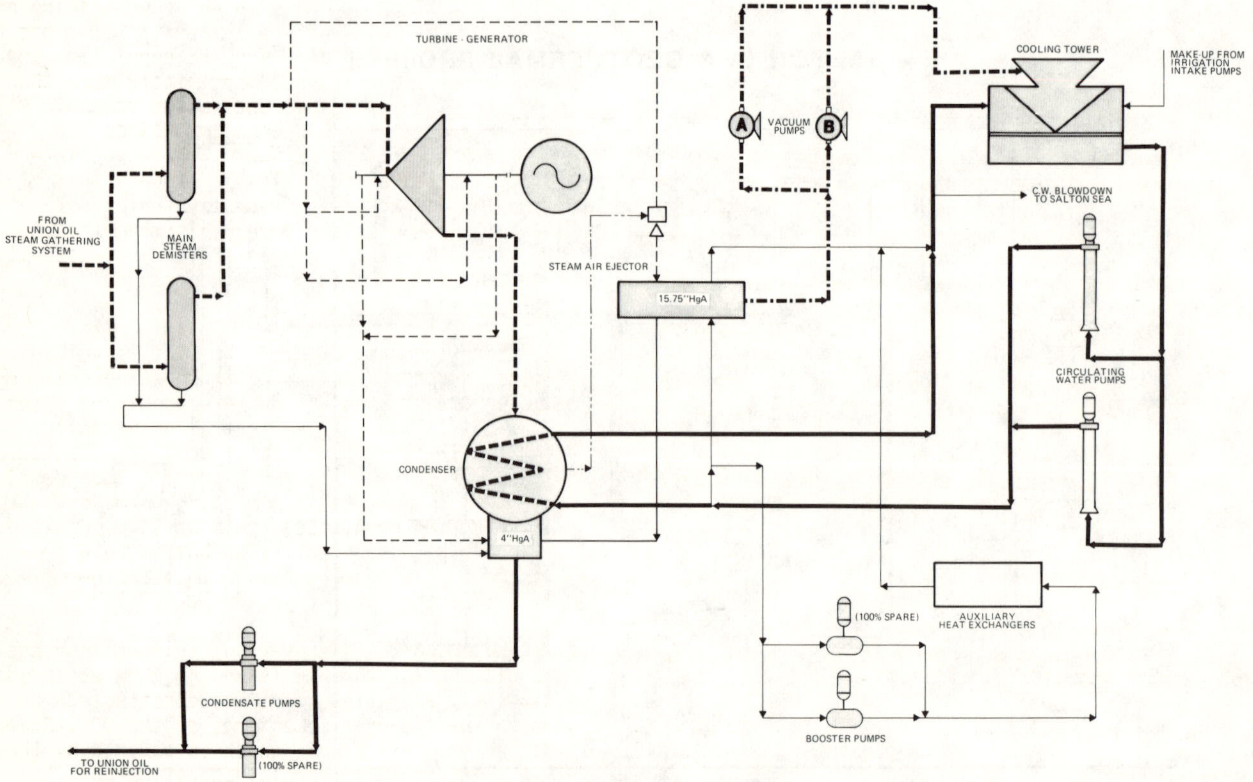

SALTON SEA GEOTHERMAL PROJECT, UNIT 1
POWER GENERATION SYSTEM DIAGRAM

HAWAII'S GEOTHERMAL FUTURE AND THE DEEP WATER CABLE

G. N. Okura — Hawaiian Electric Co.

R. T. Uemura — Hawaiian Electric Co.

G. A. Chapman — The Ralph M. Parsons Co.

ABSTRACT

The State of Hawaii presently utilizes costly imported fuel-oil for over 90 percent of its electrical power generation. With the uncertainties associated with the price and supply of this oil, the State has set, as one of its highest priorities, a goal of electrical energy self-sufficiency. Geothermal power, because of its proven technology and apparent abundance, appears to be the major hope for achieving this goal within the immediate future. A problem exists, however, in transporting the electricity generated from a potentially large resource on the Island of Hawaii to the market on the Island of Oahu. This paper discusses geothermal activities in Hawaii and the relationship of a deep water power cable demonstration program to those activities.

INTRODUCTION

The State of Hawaii is blessed with, among many other assets, abundant indigenous renewable alternate energy resources (DPED and LBL, 1981). Continuing research, development and demonstration (RD & D) programs have shown the viability of harvesting the energy of the earth's heat, winds, waters and plants to produce electricity. Some of these sources have, in fact, been developed to the extent of making a significant contribution to offsetting the amount of costly imported fuel-oil required for conventional electrical power generation.

Table 1 lists the present methods of electrical power generation on the various islands of the State.

Table 1

Electrical Power Generation Methods

Island	Generating Capacity (MW)	Generating Method (percent) Oil-Fired	Biomass	Hydro
Kauai	105	55.2	32.0	12.8
Oahu	1209	98.3	1.7	-0-
Molokai	5	100.0	-0-	-0-
Maui	112	79.4	19.1	1.5
Hawaii	124	60.9	33.6	5.5
TOTALS	1555	91.0	7.6	1.4

Although significant inroads have been made on individual islands (primarily Hawaii and Kauai) to reduce the dependency on oil, statewide use of fuel-oil for electrical power generation still remains at a greater than 90 percent level. It is evident, therefore, that work still remains to achieve the State of Hawaii's goal of electrical energy self-sufficiency.

The development of geothermal energy and of an interisland underwater cable system are basic necessities to this achievement (DPED and LBL, 1981).

HAWAII'S GEOTHERMAL PROGRAM

Appreciation of the geothermal energy potential in the State begins with an understanding of the volcanic origins of the islands. The eight major islands (seven of which are populated) and the numerous minor islands, are little more than the peaks of an extensive underwater volcanic mountain range. The eight major islands are shown in Figure 1 (Kahoolawe is the unpopulated island). Based on the volcanic process that formed the island chain, Kauai is the oldest island while the island of Hawaii (known as the "Big Island") is the youngest. Volcanic activity on the Big Island has occurred as recently as April/May, 1982.

Using the knowledge of the State's volcanic origins, geothermal resource exploration drilling began in 1961 at one of the more promising sites on the Big Island. After several unsuccessful attempts, the first successful well was flashed on July 2, 1976. Known as HGP-A (Hawaii Geothermal Project - "A" for Agatin Abbott, chairman of the site selection committee), this well was drilled to a depth of 6,450 feet. The bottom hole temperature of 676 degrees F made this geothermal well one of the hottest in the world.

Following the discovery of this well in the Puna District of the Big Island, a 3 MW wellhead geothermal generating plant was constructed through the cooperative efforts of the DOE, State and County. This plant, which is powered by the HGP-A well, began commercial operation on February 12, 1982. Construction costs for this 3 MW plant were $8.5 million ($2,833/kw). Drilling costs for the HGP-A well were $1,609,000.

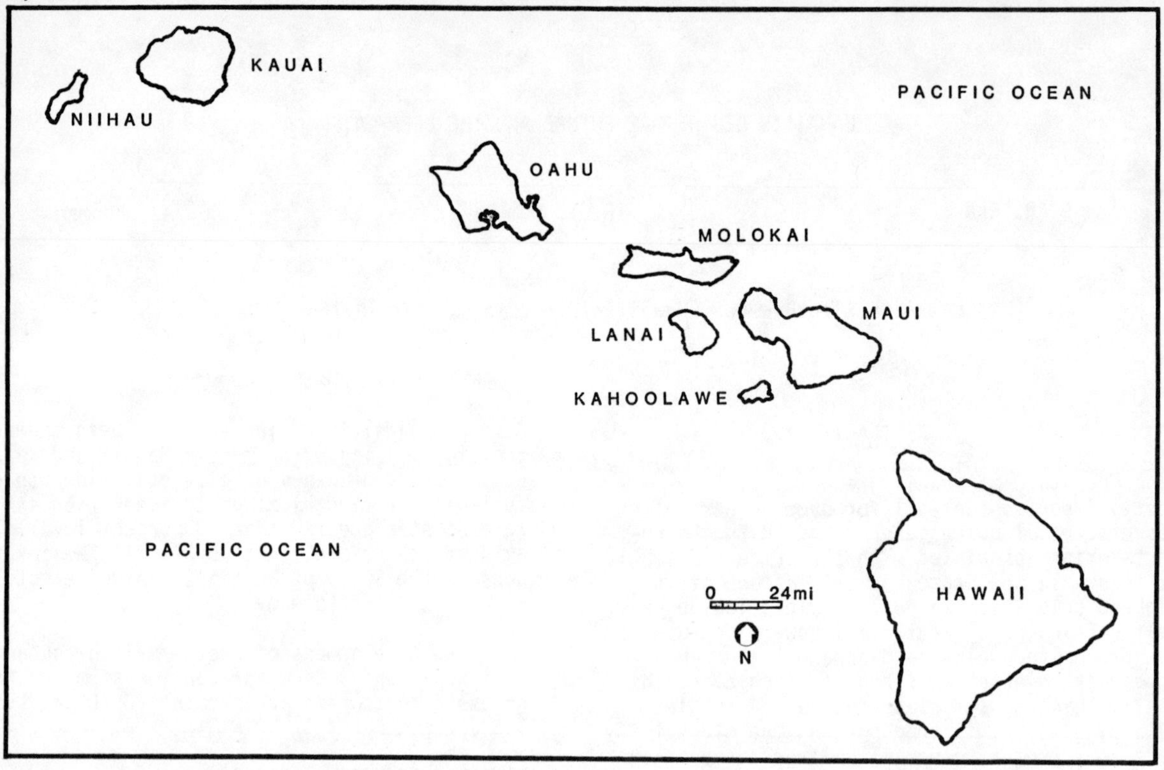

Figure 1. Major Hawaiian Islands

This successful demonstration of a geothermal resource generating electric power provided an added incentive to three private geothermal resource development organizations who responded to the Hawaii Electric Light Company (the Big Island's utility) Request-for-Proposal (RFP) for a 25 to 50 MW geothermal electric power development project. The RFP covers the areas of 1) exploration and development of geothermal well fields, 2) design and construction of geothermal power plants, and 3) sale of the electrical energy produced to the Hawaii Electric Light Company.

The three companies are presently performing resource confirmation and development work to satisfy not only the RFP, but, more importantly, to confirm the existence of a presently estimated geothermal resource of 1000 MW of electrical power. It is interesting to note that these companies have taken two diverse approaches to the question of geothermal development.

Two organizations are actively drilling exploratory wells to confirm the resource and to test the longevity and quality of the resource. Subsequent to this activity, these two organizations will plan the development of their respective resources and secure additional permits to develop well fields and construct power plants. The third organization has chosen to acquire all necessary permits/approvals for the full development of 250 MW of geothermal resources prior to the initiation of any drilling activities. Since the potential geothermal resources of this latter organization are situated on landholdings zoned for conservation district purposes, acquisition of the permits/approvals is a major constraint in their plans.

Present forecasts indicate that the Big Island cannot accept more than 40 to 50 MW over the next 30 to 40 years. Unless another viable market is found, the full potential of the Big Island's geothermal resource will not be realized. As implied by Table 1, the largest load center is on the Island of Oahu where the additional generation can be accepted into its electrical system. It is anticipated that, if a means of transporting the electrical power to Oahu can be found, a significant block of power could be accepted for use immediately (Okura and Chapman, 1982a). Out of the need to transport the potentially huge block of power grew the Hawaii Deep Water Electrical Transmission Cable Demonstration (HDWC) Program.

THE HDWC PROGRAM

Under most circumstances, the electrical interconnection of the islands in the State via an underwater cable system would be an achievable task. For example, the electrical linking of the islands of Molokai, Maui and Lanai using presently available underwater cables and cable-laying vessels has been deemed to be feasible (Hwang and Young, 1979). Similar type cables have been utilized to intertie islands in Japan and to interconnect Vancouver Island with mainland Canada. High voltage direct-current (HVDC) underwater cables have been installed in Europe to link many areas including Norway and Denmark. The

Skagerrak Cables, as this latter system is commonly called, are deployed to a depth of approximately 1,800 feet and over a distance of about 78 miles. They are the deepest and longest HVDC cables deployed to date (Hauge, et al., 1978).

As noted earlier, however, in Hawaii the major potential source of alternate energy electrical power, specifically geothermal, is on the Big Island while the major marketplace is Oahu. Separating these two locales are the Alenuihaha Channel (between the Big Island and Maui) which is approximately 7,000 feet deep and the Kaiwi Channel (between Molokai and Oahu) which is approximately 2,200 feet deep. Further, depending on the routing of the cable system, lengths ranging from 130 to 260 miles would be encountered. Because the depths and distances noted will be encountered in the establishment of a Hawaiian interisland cable system, cables capable of withstanding deployment tensions of over 125,000 pounds and water pressures up to 3,000 psi must be designed, tested and proven reliable and economically feasible (Chapman, 1981).

The HDWC Program was therefore designed with three major goals. First, to determine the technical and economic feasibility of establishing an interisland electrical grid system. Second, to determine the ocean engineering problems and solutions of deploying, retrieving and repairing a deep water cable. And third, to develop a deep water cable criteria document that can be used for the design, installation and maintenance of deepwater electrical transmission cable systems (Okura and Chapman, 1982b).

The HDWC Program was initiated in 1981 with the release of $300,000 of State of Hawaii funds for the Phase I, Preliminary Definition Study, work efforts. This phase of work was completed in late April, 1982, when the results of the five major tasks were submitted to the State of Hawaii's Department of Planning and Economic Development.

The five major tasks consisted of two administrative-type tasks and three technically-oriented tasks. The administrative tasks resulted in the preparation of program planning/management documents, schedules and costs which will be used to guide the program in future phases of work. The technical tasks were preliminary studies covering 1) cable routes, 2) cable designs and 3) cable vessel and handling equipment designs (Okura and Chapman, 1982b).

The preliminary cable route survey has resulted in the identification and analyses of over 40 separate routings or route segments for an electrical intertie between the Big Island and Oahu. The selection of a preferred alignment will depend on additional economic, system planning, environmental and at-sea survey studies to be conducted in future phases. For the cable design task, detailed analyses of 16 candidate designs were performed and five were selected for further consideration. Additional mechanical/electrical analyses and cost-benefit studies to be conducted in future phases will result in a preferred cable design. This design will be the basis for manufacture of a test length to be deployed in the Alenuihaha Channel. Finally, the cable vessel and handling equipment task has resulted in a thorough literature search of presently available vessels and equipment and a preliminary conceptual design of the vessel and equipment to be used on the HDWC Program. Future phases of work on this task will include additional design and cost tradeoff studies to determine not only the HDWC requirements, but also the ultimate cable vessel and handling equipment needs.

Federal and State funding support have been requested for the presently estimated $17 million HDWC Program. This multi-phase program, scheduled for completion in December, 1984, has brought together a multi-discipline team composed of a public utility, a program management specialist, a cable manufacturer, a cable vessel/handling equipment specialist, and various State organizations. It is signficant to note that, similiar to the initial geothermal development activity, government funding support is required since none of the above program participants could, individually, support the high risk, high technology, long-term investment nature of deep water cable design and manufacture.

CONCLUSION

This paper is a brief summation of geothermal power development activity in the State of Hawaii and the important role to be played by the HDWC Program in the successful development of the large potential geothermal resource. It is clear that, without the HDWC Program, geothermal power could not be developed to the maximum extent practicable. Conversely, without the emphasis placed on geothermal development, the HDWC Program would not be required. Hence, both programs are proceeding concurrently and it is acknowledged that, if one fails to meet its goals, activities on the other could substantially decrease or stop altogether.

If, however, both programs should successfully meet its goals, a significant step will have been taken towards achieving the State of Hawaii's goal of electrical energy self-sufficiency.

REFERENCES

Chapman, G. A.; 1981; Submarine Transmission Cables for Hawaii; Hawaii Integrated Energy Assessment; Volume II Alternate Energy Technologies for Hawaii; pp. 37-50.

Department of Planning and Economic Development (DPED), State of Hawaii and Lawrence Berkeley Laboratory (LBL), University of California; June, 1981; Geothermal Energy for Hawaii; Hawaii Integrated Energy Assessment; Executive Summary and Volume II Alternate Energy Technologies for Hawaii.

Hauge, O., A. Berg, J. N. Johnson, G. Wettre and K. Bjorlow-Larsen; 1978; The Skagerrak HVDC Cables; International Conference on Large High Voltage Electric Systems; CIGRE Paper No. 21-05.

Hwang, H. H. and B. Young; 1979; A study of the Feasibility of Linking the Islands of Maui, Molokai and Lanai with Submarine Electrical Power Cables; Hawaii Natural Energy Institute Report No. 79-02.

Okura, G. N. and Chapman, G. A.; 1982a; Hawaii Deep Water Electrical Transmission Cable Demonstration Program Phase I Program Review.

Okura, G. N. and Chapman, G. A.; 1982b; Hawaii Deep Water Electrical Transmission Cable Demonstration Program Phase I Executive Summary.

Geothermal Resources Council, TRANSACTIONS Vol. 6, October 1982

PERFORMANCE TESTING OF GEOTHERMAL STEAM TURBINES

W. O. Printup and S. Sigurdson

General Electric Company
Medium Steam Turbine Department
Lynn, Massachusetts 01910

ABSTRACT

At the present time the ASME Performance Test Codes do not address the testing of geothermal turbines that contain non-condensable gases in the inlet steam. This paper presents a method for conducting performance tests for this class of turbine. The determination of the separate quantities of steam and non-condensable gases as well as their respective enthalpies is necessary to the calculation for determination of compliance with the guarantees.

TESTING GEOTHERMAL STEAM TURBINES

INTRODUCTION

Steam turbine-generator efficiency is more important than ever as a bid evaluation factor in the purchase of modern goethermal power plants. It is in the best interests of the user to perform an accurate performance test to verify the performance guarantees used in the evaluations and award of contract.

The performance testing of geothermal steam turbines is not covered by the ASME Performance Test Codes. Although the Performance Test Code for Steam Turbines PTC6-1976,[1)] covers the testing of nuclear turbines which are quite similar to geothermal turbines, there is one major difference not covered by the Code. This is the possible addition of non-condensable gases of a significant amount to the inlet steam. A proposed method of measurement of flow and enthalpy necessitated by the addition of non-condensable gases in geothermal installations is presented in this paper.

INSTRUMENT DIAGRAMS

Figure 1 shows a typical instrument diagram for a geothermal turbine exhausting into a surface condenser and Figure 2 shows an installation with a direct contact condenser. The recommended procedures for installing flow measuring devices and making pressure, temperature and generator power measurements as indicated on Figures 1 and 2 are set forth in PTC6-1976[1)]. In this paper we will discuss recommendations for making throttle flow and enthalpy determinations.

THROTTLE FLOW DETERMINATION

The determination of throttle flow containing a mixture of steam and non-condensable gases should be performed somewhat differently for an installation with a surface condenser than one with a direct contact condenser. PTC6-1976[1)] recommends measuring condensate flow with a flow nozzle in a calibrated test flow section and then accounting for any extraneous flows entering or leaving the system in order to determine throttle flow. We recommend using this method for an installation with a surface condenser in order to keep the testing uncertainties to a minimum.

For an installation with a direct contact condenser, since the condensate and cooling water are mixed, this method will not work. In this case we recommend making a direct measurement of the throttle flow mixture of steam and non-condensable gases in the inlet approach pipe to the turbine. An orifice flow section should be considered for this measurement since the orifice as a primary element in an uncalibrated flow section will provide the best means of predicting the flow coefficients. The orifice should also be less susceptible to changes in flow coefficients due to incrustation in the throat (such as might occur in flow nozzles) from impurities carried over in the steam and non-condensable mixture.

SURFACE CONDENSER INSTALLATION

In order to determine the throttle flow after measuring the condensate flow, it is necessary to measure the steam flow to the air ejector since this condensed steam is included in the condensate flow measurement. This flow is not chargeable to the turbine and should be subtracted from the condensate flow in determining the throttle flow. It is also necessary to measure the gas flow in the air ejector gas removal pipe as this flow is chargeable to the turbine and must be added to the condensate flow to determine the total throttle flow. In order to minimize the pressure drop in the gas removal pipe, it is recommended that a commercially available flow tube be used.

DIRECT CONTACT CONDENSER INSTALLATION

It is expected that the flow section for the throttle flow measurement will be designed as part of the station piping and will also be used for station indication of throttle flow. Some design considerations for reducing the uncertainty of this measurement in an uncalibrated flow section are as follows:

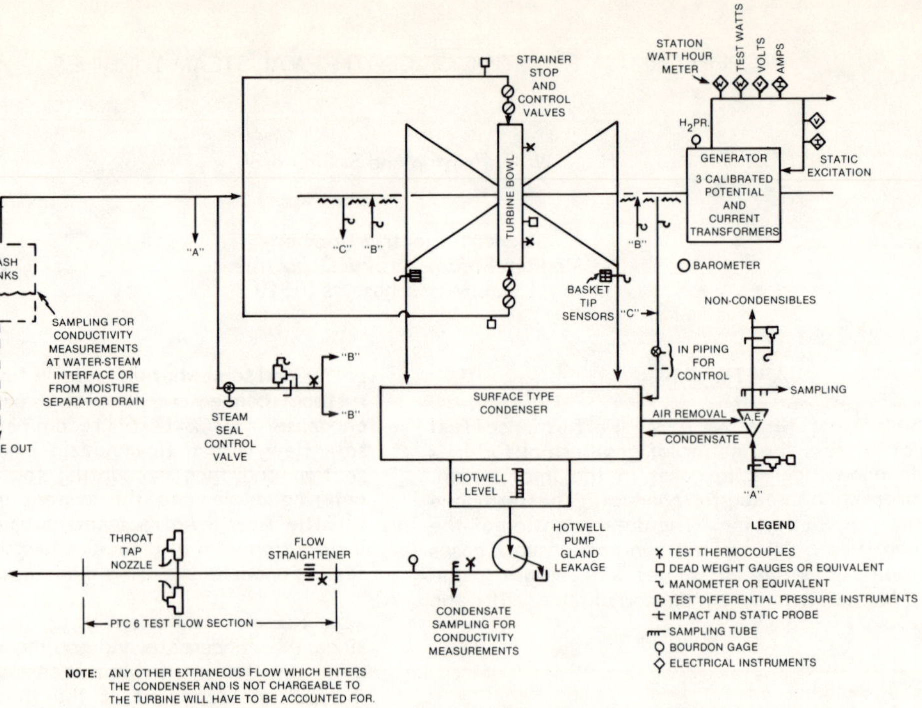

Figure 1. Instrumentation Diagram,
Surface Condenser

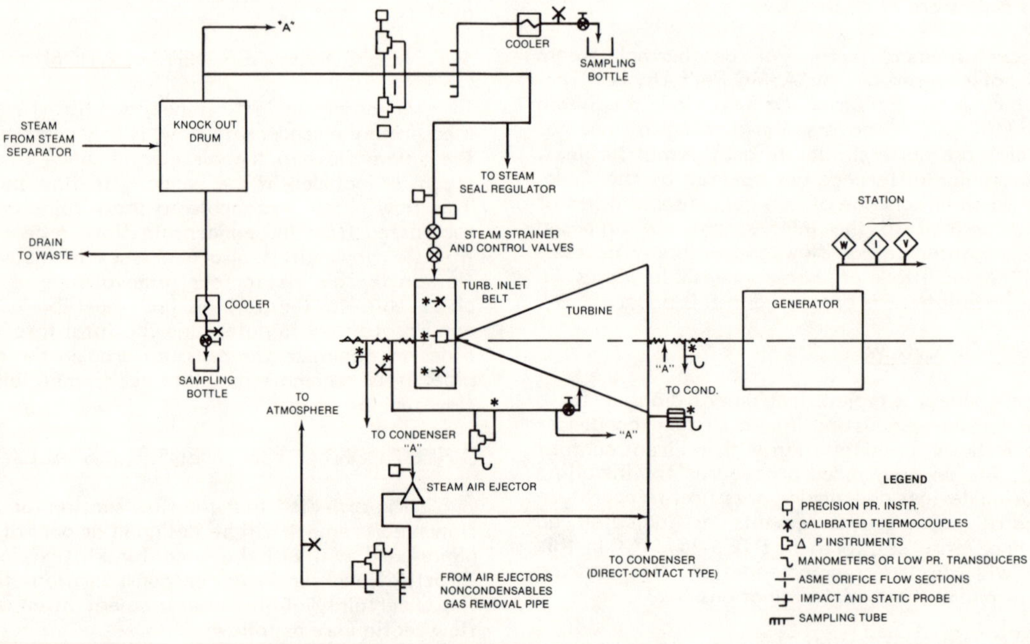

Figure 2. Instrumentation Diagram,
Direct Contact Condenser

1. The flow section lengths and orifice configuration should be as described in the ASME publication "Application Part II of Fluid Meters Sixth Edition 1971".[2)]

2. The beta ratio should not be greater than 0.6 as recommended by PTC6-1976.[1)] To achieve this beta ratio, the flow section portion of the inlet piping may have to be made from pipe which is larger in diameter than is being used for the rest of the inlet piping.

3. In order to minimize the pressure loss across the orifice and errors in flow measurements due to the steam being saturated or slightly wet, the orifice bore should be sized to produce a relatively low pressure differential across the orifice. High accuracy electronic differential transducers with low full scale range should then be used to measure differential pressure. Duplicate measurements requiring two sets of taps located on the horizontal centerline of the flow section are recommended.

4. The orifice plate primary element should be inspected before and after a performance test is run. To avoid having to remove the orifice plate from the pipe to perform this inspection, it is recommended that an inspection port as described in the Bornstein-Cotton ASME paper 80-JPGC/ Pwr-15[3)] be provided in the flow section.

Measurements of the non-condensable gas flow for an installation using a direct contact condenser can be accomplished by the same method as used for an installation using a surface condenser. An additional problem, however, becomes apparent for the direct contact condenser installation. This problem is the addition of air to the mixture in the air ejector gas removal pipe which is due to the entrained air in the circulating water. One method of isolating this air from the non-condensables is to provide a sampling tube in the gas removal pipe and use an oxygen analyzer to determine the amount of air. The gas flow mixed in with the air ejector motive steam must also be accounted for. This can be done by accurately measuring the air ejector inlet steam pressure and calculating the steam-gas mixture flow to the air ejector using the air ejector nozzle diameters and a predictable flow coefficient. The gas portion of this flow and the entrained air flow then is subtracted from the total non-condensable flow measurement for establishing the non-condensable flow in the turbine throttle steam.

Both Figures 1 and 2 call for a pressure measurement in the turbine outer shaft seal cells. During the test these cells should be run with a slight positive pressure to prevent air leakages at these points from entering the condenser and appearing as additional air flow in the non-condensable flow measurements.

THROTTLE ENTHALPY

As is the case for nuclear units, the measurement of steam quality of the throttle steam is required when testing a geothermal unit and it should be possible to use the tracer method for this purpose. However, instead of using radioactive tracers as discussed in ASME PTC6-1976, it should be possible to use the impurities in the steam supply as a tracer. Conductivity or sodium measurements of the water removed from the moisture separators in the turbine throttle steam line are used along with the same measurements taken from a sample of water drawn off using a sampling tube and sampling cooler in the throttle steam line or from the condensate line from the condenser. The moisture in the steam leaving the steam source is the result of water carryover. Thus, if there is any moisture in the steam after the moisture separator, the impurities in the steam source will also be found in the throttle steam. The amount of impurities in the throttle steam are diluted by the condensed steam. Therefore, if it is assumed that the droplets of moisture carried over contain the same amount of impurities as in the steam source, the steam wetness fraction or moisture content in the throttle steam can be established by the following relationship:

> The concentration of impurities is related by the balance $C_W W = C_c W_c$ where C_W is the concentration of impurities in steam source, W is the amount of water in the wet steam, C_c is the concentration of impurities in the condensed throttle steam, and W_c is the amount of condensate from wet throttle steam.

With the impurities concentration known from test measurements, the steam wetness fraction is determined from the ratio $M = C_c/C_W$. For further discussions on the use of tracers for steam quality measurements, reference may be made to PTC6-1976[1)] and Deming-Feldman ASME paper 71-WA/PTC-2.[4)]

For single flow steam turbines, a check of throttle steam quality measurements can be obtained by accurate pressure and temperature measurements in the steam seal piping from the high pressure end of the turbine or, for a double flow turbine, in the steam seal supply line after the steam seal regulator. This steam should be more than 25°F superheated after being throttled down to the steam seal level of pressure. If it is assumed that this steam represents a homogeneous mixture of the steam entering the turbine first stage steam nozzles with negligible heat loss, the enthalpy of this steam at the temperature and pressure measurement point in the steam seal line should equal the enthalpy of the throttle steam. The amount of moisture in the throttle steam can then be determined from this relationship and should agree reasonably well with the impurities measurements method.

The test energy input to the turbine includes the energy in the throttle flow non-condensables as well as the energy in the steam. To arrive at a corrected turbine performance, these energies must also be accounted for when making comparisons with guarantees or past performance of the turbine. Flow measurements of these non-condensables have been previously discussed in this paper. Since the non-condensables in geothermal installations are mostly carbon dioxide, the properties of carbon dioxide probably can be used with negligible error in the results for establishing the total enthalpy of the throttle flow and the effect of non-condensables on the turbine available energy.

TESTING UNCERTAINTIES

If guidelines presented in this paper are followed, it should be possible to test performance of geothermal steam turbines with reasonable accuracy. For a turbine installation utilizing a surface condenser, the two sigma uncertainty will be approximately 0.5%. This compares favorably with the expected uncertainty of a PTC-6 test on nuclear units. For an installation utilizing a direct contact condenser the uncertainty increases to about 2.5% and is mainly due to the use of an uncalibrated flow section for measuring the throttle flow. This uncertainty, however, can be reduced to approximately 1.2% by calibrating the flow section before installation. In this case, most of the 1.2% uncertainty is due to the use of extrapolated calibration data derived from calibration data derived from calibrations using water and the uncertainties in establishing the specific volume of the throttle steam - non-condensables mixture for use in the test flow calculations.

Normal station instruments can also be used to run performance tests, but with considerable loss in accuracy. If these instruments are of a type which produce good repeatability and the instrument systems are calibrated prior to a test, it is expected that the uncertainty will be about 5.0%.

COST OF TESTING

It is estimated that the cost of running an accurate performance test on a geothermal steam turbine installation will be between $50,000 and $100,000. These costs include the incremented cost of upgrading the station primary flow section, additional instrument taps which would be required in the station, and the labor required for test preparations, running the test, and analysis of the test data.

CONCLUSIONS

A reasonable method is available for testing geothermal turbines to a sufficient degree of accuracy to warrant conducting performance checks to ascertain compliance with performance guarantees. Accurate in-service performance data will provide useful base line information for operators and information essential to future purchasers of geothermal steam turbine units for evaluation of manufacturer's quoted performance guarantees.

REFERENCES

1) ANSI/ASME PTC6-1976 Steam Turbines

2) ASME PTC19.5 Application Part II of Fluid Meters Sixth Edition 1971.

3) Bornstein, B. and Cotton, K.C., A Simplified ASME Acceptance Test Procedure for Steam Turbines, ASME Paper 80-JPGC/Pwr-15.

4) Deming, N.R. and Feldman, R.W., Non-radioactive Tracer for Performance Tests of Steam Turbines in PWR Systems, ASME Paper 71-WA/PTC-2.

Geothermal Resources Council, TRANSACTIONS Vol. 6, October 1982

ECONOMIC AND ENGINEERING CONSIDERATIONS FOR GEOTHERMAL DEVELOPMENT IN THE MAKUSHIN VOLCANO REGION OF UNALASKA ISLAND, ALASKA

John W. Reeder [1], Michael J. Economides [2], and Donald R. Markle [3]

[1]Alaska Div. of Geological & Geophysical Surveys, 3001 Porcupine Drive, Anchorage, AK 99501
[2]Petroleum Engineering Department, University of Alaska, Fairbanks, AK 99701
[3]Alaska Div. of Energy and Power Development, 338 Denali Street, Anchorage, AK 99501

ABSTRACT

Large vapor-dominated hydrothermal reservoirs are suspected to exist in the region marked by fumarole fields on the southeast flank of Makushin Volcano on Unalaska Island, Alaska. In this paper, economic and engineering considerations with respect to potential hydrothermal development in the Makushin Volcano region are presented.

INTRODUCTION

Unalaska Island is part of the Aleutian Island arc. The Makushin Volcano on Unalaska Island is one of at least 36 volcanoes on this island arc which have been reported active since 1760 (Coats, 1950). Such Volcanic regions, with shallow magma bodies and deep tectonic fracture systems, represent a favorable setting for the existence of large hydrothermal reservoirs.

During volcano investigations in the Aleutian Islands, Drewes et al. (1961) observed fumaroles and hot springs on the top and on the south flank of Makushin Volcano. Later, Miller and Smith (1977) suggested that a high-level magma chamber exists under the summit caldera of this volcano. During more recent geologic investigations of Unalaska Island, Reeder (1981 & 1982) discovered active fumarole and hot-spring fields on the flanks of the volcano. Many of these fumarole fields are probably the surface expressions of vapor-dominated hydrothermal systems (Reeder, 1981 & 1982; and Motyka et al., 1981), i.e., the hydrothermal reservoir fluids consist mainly of vapor instead of liquid.

The Unalaska community presently serves the largest American fishing fleet for the Bering and North Pacific region and it could play a major role in the development of a bottomfish industry. In addition there is nearby potential for outer continental shelf oil, gas, and mineral production. Present peak electric utility demands for Unalaska is 15 MW, including both the publicly owned diesel generators and those operated by the private fish processors.

PLAN

A northwest oriented zone roughly marked by fumarole fields nos. 1, 2, 3, and 4 (Figure 1) has been identified as possibly containing large hydrothermal reservoirs (Reeder, 1982). Although further exploration may better define the nature of such reservoirs in the Makushin Volcano region, deep exploratory drilling along with appropriate well testing such as described by Economides et al. (1982) will be required to make any estimates of the nature and the development potential of such geothermal reservoirs. Republic Geothermal, Inc. has been awarded a contract by the State of Alaska Power Authority to undertake deep drilling in the Makushin Volcano region with the capability of reaching depths of 2 kms.

With respect to the northeast oriented target zone, there are only three potential road approaches: up Glacier Valley, up Makushin Valley, and from Driftwood Bay (Fig. 1). Of these, the Glacier Valley and Makushin Valley approaches would require, especially in the upper reaches of Makushin Valley, the routing of the road across large rivers and through fairly deep canyons. In contrast, a good part of the 16 km Driftwood route as shown in Fig. 1 avoids canyons and major drainages. In this route, one bridge would be required at point A and extensive bedrock blasting would be required at point B (Fig.1). An existing road does connect Driftwood Bay with Sugarloaf Cone. However, this road is washed out at numerous places and considerable repairs would be necessary to make it passable. An abandoned airstrip, 100 m by 35 m, is located at Driftwood Bay as shown in Fig. 1. It is expected that this airfield would serve as the logistical base for any planned drilling. The surface of the airstrip is currently in poor condition but it can be upgraded to receive heavy traffic in a short period of time. The cost for building the proposed road to Site C and for rebuilding the airstrip at Driftwood Bay should be under 2 million dollars, Table 1.

Once entering the northwest oriented target zone at site C, any further road construction would require extensive rock removal and major bridge constructions. Further surface exploration could be conducted to determine drilling targets, but logitical constraints may limit the drilling target to site C. Directional drilling from this site could greatly extend the target region.

Helicopter support would allow the siting of the initial deep well almost anywhere in the upper

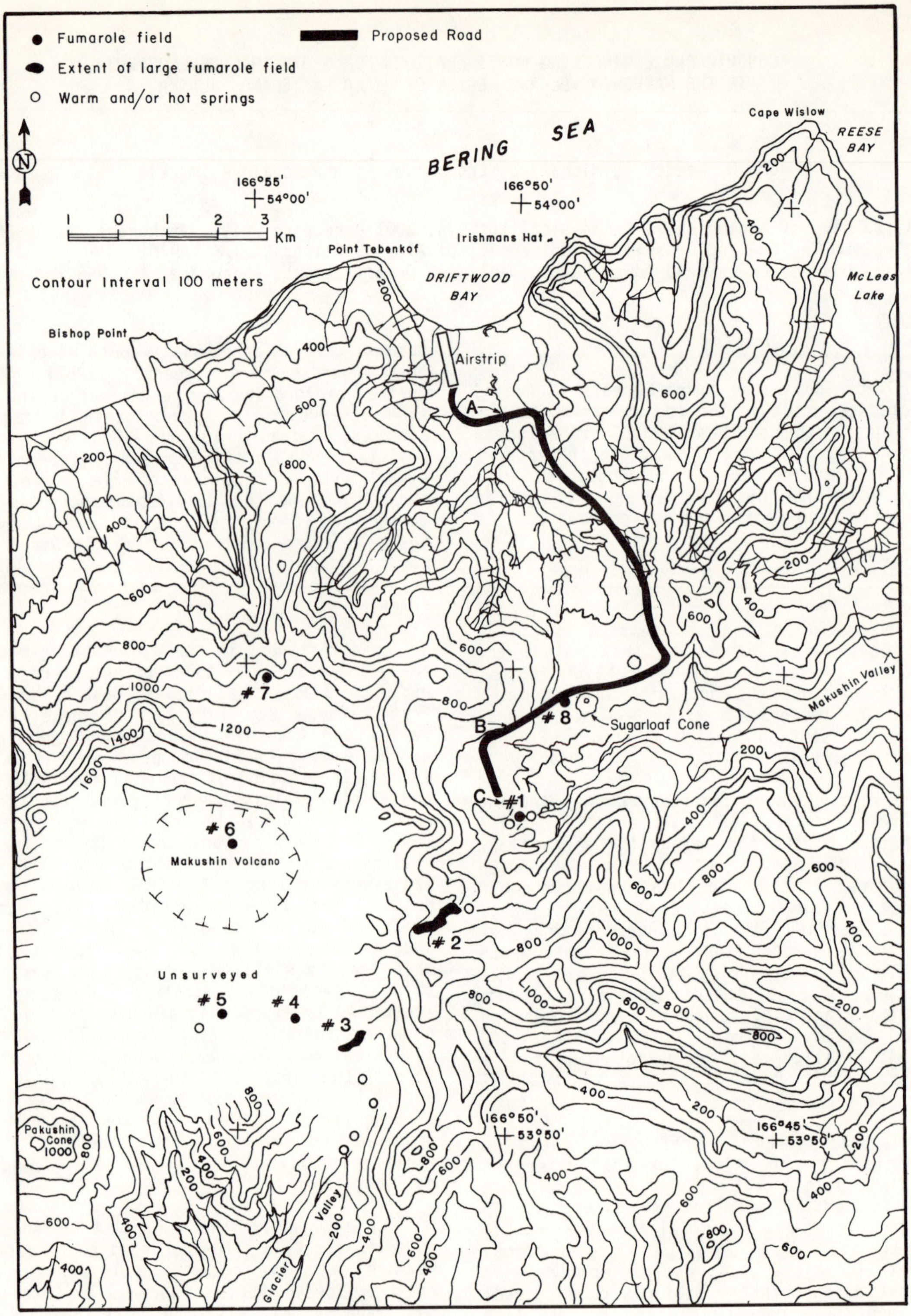

Figure 1. Topographic map of the Makushin Volcano region showing the location of fumarole fields, hot and/or warm springs, a proposed deep exploratory drilling site (Site C), and an access road to this proposed drilling site which will require a bridge at Site A and extensive bedrock excavation at Site B.

regions of Makushin Valley and Glacier Valley. Unfortunately, access to such regions would be marginal if not impossible a good part of the time due to weather conditions. But, such an approach should not be ruled out.

Emphasis on deep drilling should not rule out other exploratory surveys conducted prior to and/or during deep exploratory drilling (Reeder et al., 1980). For example, if shallow temperature drill holes (i.e., less than 500 meters in depth) could be drilled by helicopter support, they would be of value in obtaining temperature gradients and even rock permeability values for the metavolcanic and the plutonic bodies found in the region southeast of Makushin Volcano. Low temperature gradients and high permeability values would be expected for most of the unaltered volcanics on Unalaska Island, thus shallow exploratory drilling in these bodies would not be recommended. These unaltered volcanics, because of their physical nature, could also be difficult to drill in.

In addition, self potential and gravity surveys along with more detailed geologic mapping would be recommended prior to drilling. These surveys would be geared toward defining the extent of the metavolcanic and plutonic bodies underneath the unaltered volcanics, and toward better defining the fracture systems. Passive seismic surveys which should be conducted over a long period of time, i.e., at least one year, could be very helpful in recognizing any magma bodies and/or reservoirs. During the drilling operation, a portable seismograph station should be operated near the drilling operation because of volcanic hazards.

Development Economics

Geothermal energy is "location intensive". Unlike fossil fuels, there is a need for the benefited market to be in close proximity with the resource. Unalaska Island, with its relatively sizeable population and a large fishing and processing industry presents an attractive target for geothermal development. A report by the U.S. Bureau of Land Management (1981) projects an electric utility demand of 50 MW by the year 2000 under a base case. Anticipated offshore petroleum exploration in the Bering Sea will increase the demand. The same BLM report projects a demand of 56 MW if moderate oil and gas leasing in the outer continental shelf proves successful. Their projections appear on Fig. 2.

The present economy of the island is dependent on its harbor and the associated fishing and processing operations. More than 15 plants are located on the island, processing a variety of seafoods, dominated by king crab and salmon. In 1978 Dutch Harbor (Unalaska) was rated as the number one port by the National Marine Fisheries Service on the basis of the value of the seafood caught.

The 1980 employment on Unalaska was 1600 "average annual full time jobs". This figure is the result of an extremely non-homogeneous employment picture with a peak of 6000 laborers during the fishing season. The BLM report projects a total employment of 9000 by the year 2000 with moderate oil and gas development.

Electric power at this time is generated by the city and by each of the private processors. However, consolidation is likely if an attractive single power source (such as geothermal) becomes available.

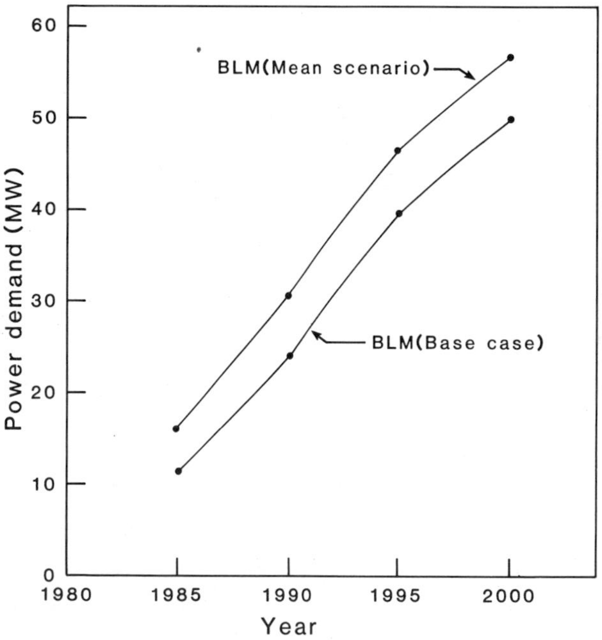

Figure 2. Electric utility demand for Unalaska Island from a U.S. BLM report (1981).

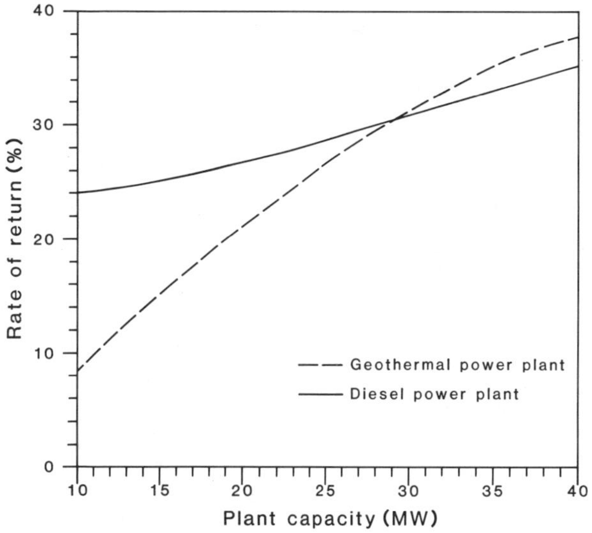

Figure 3. Comparative economics for geothermal and diesel power plants on Unalaska Island (Economides et al., 1981).

Table 1. Capital Investment for a 30 MW Geothermal Power Plant, Unalaska Island

ITEM	NUMBER	DESCRIPTION	COST
Well	6	2,500 m, 20 cm diameter (assumed 50% dry wells)	$12,000,000
Piping	-	1 km, 20 cm diameter pipe, installed	250,000
Road from Driftwood		16 km of service road, 5.5 m wide, gravel $125,000/km	2,000,000
Generator	1	55 MW maximum capacity generator, installed	20,000,000
Transformer Station	1	55 MW at $30/kW, installed	1,375,000
Transmission Line		18 km of line overland (helicopter installed), 8 km underwater, $62,000/km	1,612,000
Road along Transmission Line		26 km of 5.5 m wide gravel road, $125,00/km	3,250,000
SUBTOTAL			$40,487,000
Contingency		10% of Capital	$ 4,048,700
Total to be depreciated			$44,535,700

Table 1 presents a best estimate capital investment scenario for a 30 MW geothermal power plant on Unalaska Island. Such a geothemal power development would require a 36 km service road from Unalaska and possibly 16 more kms of road connecting the site with the abandoned military airstrip at Driftwood Bay. A similar calculation was done by Economides et al. (1981) for a variety of plant sizes in increments of 10 MW.

Because any new energy venture needs to be measured against existing or other possible options, a comparison between geothermal and diesel power plants of same maximum output sizes is presented in Fig. 3. While geothermal power plants of less than 30 MW appear less attractive when compared with diesel generators, they rapidly become desirable at larger plant sizes. Taking into account the BLM estimate of a 50 MW demand by year 2000, geothermal development on Unalaska appears to be feasible. The picture becomes even brighter when one contemplates future prices for petroleum fuels. The analysis presented by Economides et al. (1981) presumed a price for diesel that would remain constant in relation to 1981 dollars. This is, however, highly optimistic, a fact that makes geothermal energy development more attractive.

Conclusion

Any geothermal resources under the southeast flanks of Makushin Volcano on Unalaska Island may be of the sizes and types that could be developed for electrical energy production. Such energy development appears to be economically attractive if the electric needs of the island exceed 30 MW.

The design of any output power plant must follow the resource evaluation and the projection of future needs. The former point addresses the need for deep exploratory drilling within the geothermal resource area in order to define the potential of the reservoirs. The latter point touches on significant social and economic considerations that need to be addressed by the local residents, and by the local and state governments.

References

Coats, R.R., 1950, Volcanic activity in the Aleutian arc: U.S. Geological Survey Bull. 974-B, pp. 35-49.

Drewes, H., Fraser, G.D., Snyder, G.L., and Barnett, H.F., Jr., 1961, Geology of Unalaska Island and adjacent insular shelf, Aleutian Islands, Alaska: U.S. Geological Survey Bull. 1028-S, pp. 583-676

Economides, M.J., Reeder, J.W., and Markle, D., 1981, Unalaska geothermal development: The New Zealand geothermal workshop, Proceedings, University of Auckland and the New Zealand Ministry of Works, pp. 7-12.

Economides, M.J., Ogbe, D.O., Miller, F.G., and Ramey, H.J., Jr., 1982, Geothermal steam well testing: state of the art: Journal of Petroleum Technology, pp. 976-988.

Miller, T.P., and Smith, R.L., 1977, Geothermal potential of high-level magma chambers in Alaska, in The relation of plate tectonics to Alaska geology and resources, Program and Abstracts: Alaska Geological Society, p. 56.

Motyka, R.J., Moorman, M.A., Liss, S.A., 1981, Assessment of thermal spring sites, Aleutian arc, Atka Island to Becherof Lake - Preliminary results and evaluation: Alaska Div. of Geological and Geophysical Surveys, Open-File Report 144, 173 p.

Reeder, J.W., Motyka, R.J., and Wiltse, M.A., 1980, The State of Alaska geothermal program: Geothermal Res. Council, Trans., v. 4, pp. 823-826.

Reeder, J.W., 1981, Vapor-dominated hydrothermal manifestations on Unalaska Island, and their geologic and tectonic setting: 1981 IAVCEI Symposium-Arc Volcanism, The Volcanological Soc. of Japan and the Int'l Assoc. of Volcanology and Chemistry of the Earth's Interior, pp. 297-298.

Reeder, J.W., 1982, Hydrothermal resources of the northern part of Unalaska Island, Alaska: Alaska Division of Geological & Geophysical Surveys, Alaska Open-File Report, in press.

United State Bureau of Land Management, 1981, St. George basin petroleum development scenarios local socioeconomic systems analysis: OCS Technical Report no. 59.

Geothermal Resources Council, TRANSACTIONS Vol. 6, October 1982

HYDROGEN SULFIDE OXIDATION TEST PROGRAM

Randall Soo-Hoo, Howard Hipkin, and Frank Cochrane

Bechtel Group, Incorporated
Research and Engineering
San Francisco, California

ABSTRACT

This paper summarizes the results of a project carried out for the California Department of Water Resources (DWR) by Bechtel at The Geysers area to field test the oxidation of hydrogen sulfide (H_2S) in geothermal steam condensate using hydrogen peroxide. The objective was to find the oxidant-catalyst combination that will remove sufficient sulfide to allow power plant operation without violating air pollution H_2S emission standards.

Batch screening tests, bench scale flow tests, and pilot plant flow tests were made on condensed steam from one of the wells that will supply the DWR Bottle Rock power plant. The effects of treatment ratio, pH, catalyst, residence time, and temperature were tested. Based on the test results, a conceptual design was prepared for a recommended abatement system for the power plant.

INTRODUCTION

If not adequately abated, hydrogen sulfide in the geothermal steam used for electric power production in The Geysers area could cause a public nuisance. DWR is planning to construct a 55 MWe geothermal power plant in the Bottle Rock area of Lake County, California that would utilize steam containing 300 to 450 ppm_w H_2S. The primary abatement system for the Bottle Rock Power Plant is a Stretford system to remove virtually all the H_2S from the noncondensible gases withdrawn from the condenser. Operating experience by PG&E at Units 13 and 15 indicates that 65 to 80 per cent of the H_2S in the geothermal steam is removed from the surface condenser in this way.

The remaining H_2S is dissolved in the condensate which is used as cooling tower makeup water. In the cooling tower, the H_2S would be air stripped and released to the atmosphere. If the condensate is not treated to remove H_2S before it is admitted to the cooling tower, the H_2S released to the atmosphere could cause a public nuisance. Quantitatively, DWR has committed to limiting H_2S emissions from the plant to 5 lb/hr (CEC, 8/80), and that limitation has been determined to be adequate compliance with ambient air quality standards (CEC, 10/80). Since manufacturers of the Stretford equipment expect to limit H_2S emissions to less than 1 lb/hr for that equipment, the H_2S emitted from the cooling tower must be less than 4 lb/hr. That implies that the H_2S concentration in the condensate must be reduced to less than 4 ppm_w since the condensate flow rate will be approximately 1,000,000 lb/hr.

PROGRAM OBJECTIVE

The purpose of this test program was to study oxidation of sulfide from H_2S present in steam condensate from a geothermal well at the DWR Bottle Rock Plant site. The goal of the test program was to find the oxidant-catalyst combination that will remove sufficient sulfide to allow power plant operation without violating air pollution H_2S emission standards.

The H_2S oxidation test program was conducted in the field using steam from a representative well to achieve realistic test conditions for condensate composition. A 10 x 50 ft trailer was set in place at the Francisco leasehold well pad along with a pilot plant unit. The trailer was outfitted for the tests as a field laboratory and included equipment for chemical analysis. H_2S oxidation tests were conducted as batch tests, as bench scale flow tests, and as pilot plant flow tests.

The geothermal condensate for the tests was obtained by tapping into the steam line from the Francisco 3-5 well developed by MCR Geothermal Corporation. This well was producing a bleed flow (nominally 10,000 lb/hr) of steam during the tests. Approximately 2,000 lb/hr of this steam was routed to the test facility sampling system where a fraction of the 2,000 lb/hr flow was bled off and condensed. The remainder was returned to a vent facility. This flow arrangement assured that representative condensate for testing was continuously obtained from the well.

The batch H_2S oxidation tests, carried out in a beaker using approximately 150 ml of condensate, were used to screen various treatment parameters such as pH and type of catalyst. The batch tests were quick and inexpensive compared to testing with continuous flow systems. Thus, this phase of testing served as a cost effective first check and screening method. These tests were conducted from June 1981 to August 1981.

The bench scale flow oxidation tests were used to

establish the range of test parameters for tests using the larger scale pilot plant system. The bench scale system, designed for approximately 25 lb/hr of condensate, allowed treating condensate in a continuous flow process as is planned for the full-size power plant. The bench scale flow tests were conducted from July 1981 to November 1981. During the bench scale test program, the analytical methods were defined and refined as needed. The successfully demonstrated on-line application of sulfide-selective ion probes on treated condensate was a significant milestone. In addition, the following supporting tests were performed:

- turbidity
- iron content of condensate
- chelation of catalyst
- pH control
- electrode effects
- effect of vacuum on pH
- total steam analysis
- removal profiles
- other oxidants

PILOT PLANT TEST PROGRAM

The objective of the pilot plant testing was to run a series of tests, with a range of potentially successful operating conditions based on promising results from the bench test program. The pilot plant tests were conducted with a system designed for a flow rate of 1,000 lb/hr of geothermal steam. This size was chosen as a reasonable compromise between the bench scale flow system and the commercial plant. The equipment was small enough to be easily modified and maintained while being large enough to provide valid information for the commercial operation.

The pilot plant not only scaled up the size of the H_2S oxidation treatment system from the bench scale flow apparatus but in general offered more uniform condensate flow, more accurate pH control, a wider range of residence time, and condensate temperature control.

The pilot plant system was designed as a single transportable unit with a three-stage condenser scheme to provide geothermal steam condensate at three H_2S concentrations, varying from 6 ppm_w to 170 ppm_w. It contained treatment facilities and monitoring instrumentation for hydrogen sulfide oxidation at various pH conditions, reaction residence times, hydrogen peroxide/hydrogen sulfide ratios, and catalyst addition rates.

The design of the pilot plant did not attempt to duplicate the conditions of condensing steam at the exit of a power plant geothermal steam turbine. Power plant condensers operate under vacuum, typically 3.5 in Hg at 120 F. The pilot plant condensers were designed to condense the incoming geothermal steam at 75 psig at 308 F. The cost of larger and additional equipment, in addition to generally more complex operation, were key factors in electing not to condense steam under vacuum conditions. The difference in operating conditions affects the partitioning the H_2S from the incoming steam. The objective of the pilot plant was not to model the power plant condenser, but to provide a condensate stream for testing that would contain at least as much H_2S as the power plant condensate.

The pilot plant unit basically consisted of the following categories of major equipment:

- steam condensers and subcoolers
- oxidation test loop
- sulfide monitoring system
- chemical feed pumps and storage tanks
- temperature, pressure, and flow instrumentation
- condensate surge tank
- automatic and manual control valves
- electrical control system
- cooling tower

These components, with the exception of the electrical control system, are shown in the pilot plant process flow diagram, Figure 1.

In the pilot plant test program, the oxidation of H_2S was performed with the following variables:

- treatment ratio
- condensate pH
- chelated metal catalyst
- condensate temperature
- reaction residence time

The pilot plant runs were typically 45 minutes to 180 minutes long. In addition, data were obtained from a 24-hour pilot run to demonstrate the consistency of the H_2S removal and the stability in the operation of the pilot unit.

TEST PROGRAM RESULTS

The results from the bench scale tests and pilot plant runs can be summarized as follows:

- Effectiveness - Hydrogen peroxide is capable of reducing hydrogen sulfide dissolved in geothermal condensate from 70 - 90 ppm_w to less than 4 ppm_w.

- Treatment ratio - The required ratio of hydrogen peroxide to hydrogen sulfide is 2.5:1 (molar or weight) or higher for condensate at 100 to 125 F. For condensate at 60 to 70 F, a ratio of 1.5:1 or higher is adequate to reduce the hydrogen sulfide below 4 ppm_w.

- Catalyst - Of the soluble metallic catalysts tested, ferrous iron is the best and the least expensive. The iron salt needs to be chelated to prevent precipitation of iron sulfide or ferric hydroxide. The chelating agent used in this work was the sodium salt of ethylene diaminetetraacetic acid (EDTA). A ratio of EDTA to ferrous iron of 3:1 was adequate.

- Residence Time/Catalyst Tradeoff - Catalyst is not necessary if the residence time is 75 seconds or longer. At 120 F

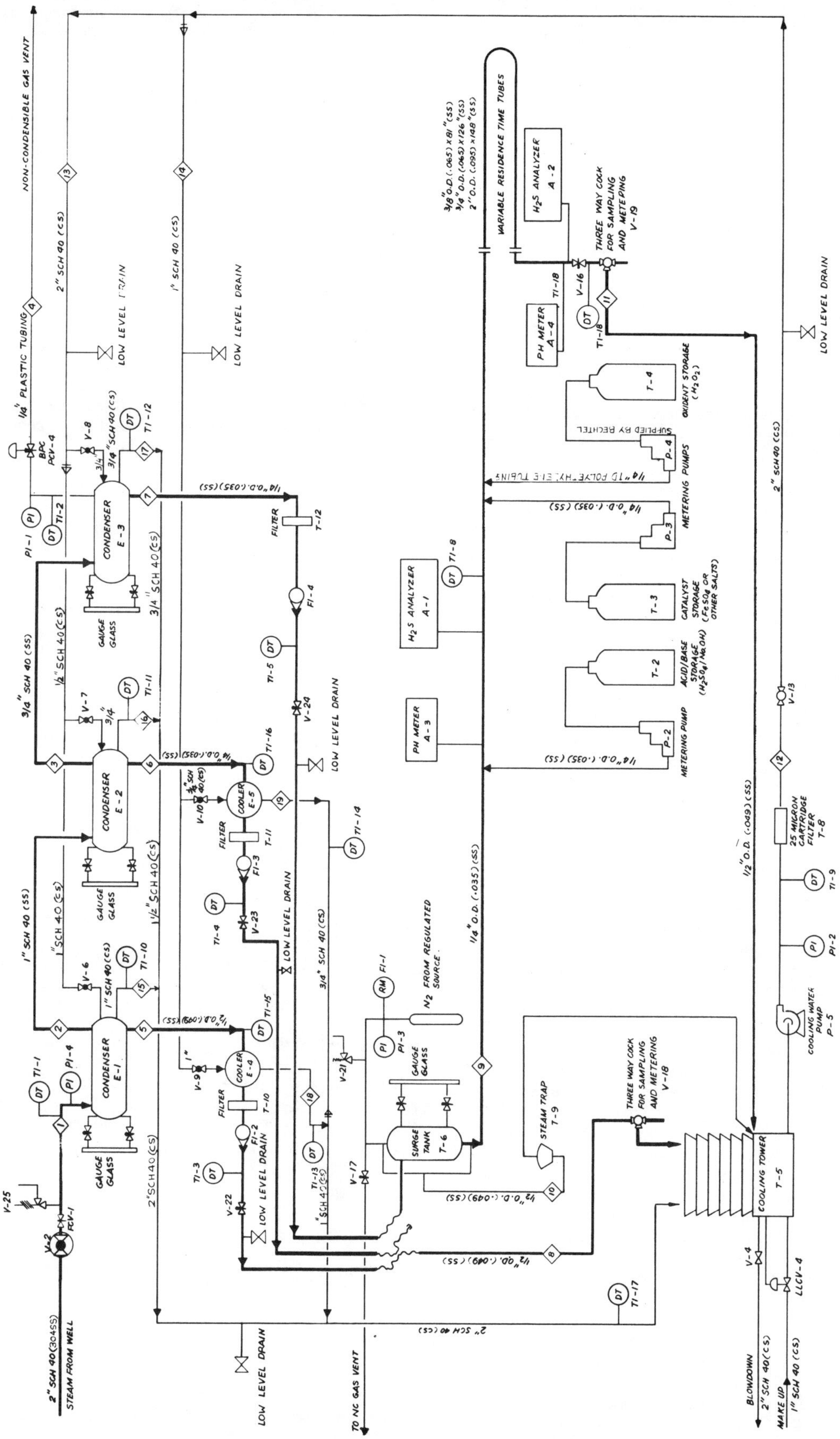

Figure 1. Pilot plant process flow diagram

and 75 seconds, addition of catalyst did not increase removal, presumably because the hydrogen peroxide was completely consumed in less than 75 seconds. In lower temperature tests, iron catalyst was shown to accelerate the oxidation reaction.

- Residence Time - These tests showed that a residence time of 50 to 75 seconds is needed to reduce the H_2S below 4 ppm_w. The bench scale tests covered a range of residence times from 10 seconds to 150 seconds. The pilot plant tests were at three residence times, 10, 75, and 715 seconds, but not at all values of pH and temperature. Residence time of 10 seconds was found to be marginal. The 715 second tests showed that, while removal is slightly better than at 75 seconds, the difference is small.

- pH - The pH of the condensate is important. Most of the data were obtained at pH 8 because the dissolved ammonia in the condensate is expected to yield a power plant condensate of about pH 8. The tests showed that pH above 7 gives better removal, other conditions being equal. Moreover, the turbidity of the treated condensate is lower at higher pH both because high pH favors formation of soluble sulfate, and because any sulfur, formed by the sulfide oxidation reaction, is partly soluble in alkaline solutions. However, if the condensate is treated at pH 9, which gives excellent removal and low turbidity, the treated condensate would require acidification to prevent cooling tower damage due to delignification of wooden components.

- Temperature - At 120 F (the power plant condensate design temperature), the reaction mechanism is different from the the mechanism at 60 - 70 F. At the higher temperature, the treated condensate has very little suspended solids (low turbidity), a high treatment ratio is required, and the reaction appears to be faster. At the lower temperature, the treated condensate turbidity is high indicating that the reaction is making sulfur to a large extent. A hypothesis for these phenomena is that the reaction shifts at the higher temperature to favor formation of sulfate and thiosulfate rather than sulfur. Sulfate and thiosulfate are soluble and also require more oxidant than the reaction to sulfur.

- Suspended Solids - Turbidity (suspended sulfur in the treated condensate) was high at low temperature (60-70F) and at low pH. The turbidity was considerably reduced at pH 9, and was essentially eliminated at pH 10, at low temperature. Sulfur, formed by the sulfide oxidation reaction, is partly soluble in alkaline solutions and is completely soluble in concentrated sodium hydroxide. It is reasonable that, at pH 10, the relatively small amount of sulfur produced in the condensate would be dissolved. When low temperature condensate was treated at pH 10 with treatment ratios of 3.0:1 or less, acidification of the treated condensate produced turbidity. However, at a treatment ratio of 4.0:1 and pH 10, acidification did not produce turbidity, presumably indicating oxidation to sulfate.

  At pH 8 and 60 - 70 F, the turbidity was quite high, varying from 200 to 1300 Jackson Turbidity Units (JTU). High treatment ratio and the presence of catalyst tended to increase the turbidity. At pH 8 and 120 F, the turbidity was essentially eliminated (JTU readings of 0 to 20), regardless of residence time, treatment ratio or catalyst. Moreover, when the treated condensate cooled, sulfur did not appear.

  Therefore, turbidity, with possible consequent condenser fouling and cooling tower sludge, will not be a problem if the condensate is treated at 120 F with 2.5:1 or greater treatment ratio, 75 sec or greater residence time, and no catalyst. This was the treatment method recommended for the full-size power plant H_2S treatment system.

CONCLUSION

Based on the bench and pilot flow test results, the H_2S oxidation test program was successful in determining oxidant-catalyst combinations that should allow operation of the DWR Bottle Rock power plant within legal H_2S air pollution limits. As in all pilot plant testing, the true test of the accuracy in the modelling will be determined when the full-scale H_2S treatment section of the power plant commences operation.

ACKNOWLEDGEMENT

This paper presents a summary of work performed for the California Department of Water Resources (DWR Agreement B52957) under the DWR project management of Donald L. Smith, Supervising Mechanical Engineer. The Bechtel designed mobile analytical facility was operated by Eugene Goldman, Mario Castillo, Jonathan Ivester, and Salem Attiga of Bechtel.

REFERENCES

California Energy Commission, August 1980, Environmental Impact Report (revised draft), Bottle Rock Geothermal Power Plant.

California Energy Commission, October 1980, Environmental Impact Report (final), Bottle Rock Geothermal Power Plant.

Geothermal Resources Council, TRANSACTIONS Vol. 6, October 1982

AN ANALYSIS OF GEOTHERMAL RESOURCE DEVELOPMENT ON UNALASKA ISLAND, ALASKA

S. G. Spencer, G. A. Long and W. Chapman-Riggsbee

Morrison-Knudsen Company, Inc.
P. O. Box 7808
Boise, Idaho 83729

ABSTRACT

A rapid expansion in the seafood industry and projected oil, gas and mining developments have resulted in a shortage of power on Unalaska Island. Currently, all power is supplied by small diesel generators at a cost of 340 mills/kwh for the local utility system. Available data indicate the potential for a significant high temperature geothermal resource on Makushin Volcano, west of the town of Unalaska. A summary of the considerations affecting the development of the Makushin resource to supply power to Unalaska is presented. A preliminary economic analysis of various resource and development assumptions indicated that geothermal power can be competitive with diesel power even though capital investment is high.

INTRODUCTION

Unalaska Island is located in the eastern Aleutian Islands, 1300 kilometers by air southwest of Anchorage. General concerns about rising energy costs and uncertain fuel supplies are magnified on the Aleutians, where transportation and logistics contribute to an expensive and vulnerable energy picture.

The situation is complicated by boom-town growth as a result of expanding seafood industries, oil and gas exploration and seafloor mining ventures. Villages in the Aleutians are faced with outdated energy production and transmission facilities, energy demands which exceed capacities and few, if any, alternative energy resources. In many locations, improvements in living and economic conditions are severely constrained by insufficient energy supplies.

Most of Unalaska Island is uninhabited. The town of Unalaska on the northeast coast consists of the traditional Aleut village of Unalaska and the settlement of Dutch Harbor (Figure 1). Unalaska is one of the few deep water ports in the Aleutians and serves as a transportation link to the western Alaskan mainland, the North Slope and the western Aleutian Islands.

Unalaska has recently experienced a major population growth to become the largest non-military settlement in the Aleutians. The expansion of the seafood industry in the last decade resulted in a population increase of 290%. The 1980 census indicated a permanent resident population of 1322, which increases to an estimated 5000 during the peak fishing season. Recent socioeconomic assessments project a population in the year 2000 of 13,000 to 17,000 with moderate oil and gas and bottom fishing development (BLM, 1981).

ENERGY USE

There is no central system which supplies the total energy needs of the city of Unalaska. The city utility has two 600 kw and two 300 kw diesel generators which provide electricity to residential and small commercial users on the main island. The estimated average demand for the system is 300 kw, with a current peak demand of 450 kw. Individual diesel generators supply power to fish processors and dock facilities on Amaknak Island. In 1979, the processors had a combined total installed capacity of 12,250 kw, with a noncoincidential peak load of 7460 kw. Experience in other Alaskan communities has indicated that processors will utilize central power generating systems if they are available and economical.

Only a rough estimate of the total diesel fuel use on the island can be determined. At the 1980 level of seafood processing, an estimated 11,000,000 liters are consumed annually for processors' energy requirements. An estimated additional 4.9 million liters are used annually for residential space heating, small commercial needs and the city utility.

Power demand projections for Unalaska are varied. A 1979 electrification study projected a peak demand of 21 MW in 1995 and a total demand of 105,000 MWH for residential, small commercial and large industrial users (Retherford, 1979). These projections are based on an energy growth rate of 10% per year through 1985 and 7% per year thereafter. The BLM leasing report for outer continental shelf oil and gas development projected total capacity requirements of 50 MW in the year 2000 under base case development, and 56 MW with moderate oil and gas development (BLM, 1981). These projections are summarized in Figure 2.

Unalaska is faced with major changes in its energy system, due to (1) the need for repair and improvement of the existing generating and transmission facilities, (2) the lack of sufficient capacity to supply the present needs of large industrial users or meet projected future residential, commercial or industrial requirements, and (3) the community's total dependency on oil-based fuels. Several alternatives for meeting Unalaska's future energy needs are presently being considered, including large diesel generators, gas turbines, hydroelectric and geothermal.

Current utility rates for electricity in Unalaska are 340 mills/kwh, of which 135.9 mills are presently subsidized by the State of Alaska. These rates were increased in 1981 from 170 mills/kwh to provide a balanced utility budget. Estimates of total cycle costs for the fish processors range from 130 to 200 mills/kwh. With future system improvements, including new generating equipment, the city utility hopes to bring rates down to about 290 mills/kwh.

GEOLOGY AND GEOTHERMAL RESOURCE

Surface manifestations of geothermal resources have been identified at two locations on Unalaska Island: Summer Bay and Makushin Volcano. The Summer Bay warm spring occurs near the base of the exposed Unalaska Formation in a boggy swamp 6 km east of the town of Unalaska. The spring discharges an estimated 4 lmp at a surface temperature of 35°C. The preliminary quartz (adiabatic) geothermometer for this spring indicates a resource temperature of 60°C.

The Makushin Volcano resource shows greater potential for development than the low temperature Summer Bay resource. The Makushin Volcano is a thick pile of unaltered and little-deformed lava and pyroclastic rocks which overlie the volcanic and sedimentary rocks of the Unalaska Formation. The broad volcanic dome is more than 1800 meters high and 16 kilometers wide. These rocks, which consist of basalt and subordinate andesite, are designated the Makushin volcanics. Locally, they are unconformably capped by basaltic flows and pyroclastic rocks that retain their constructional forms. The thickness of the Makushin volcanics is highly varied but probably does not exceed a few thousand meters. The formation is of Pliocene (?) and Pleistocene age; the bulk of the rocks are probably late Pleistocene (Drewes, et al, 1961).

Eight fumarole fields have been identified on Makushin Volcano (Figure 3) at elevations ranging from 430 meters above sea level to the summit of the volcano. Several faults striking between N 40° W to N 70° W have been identified in the vicinity of the fumarole fields. Chemical analyses of fumarole samples indicate low TDS fluids with relatively high concentrations of silica and sulfates and low chloride concentrations (Motyka, et al, 1982).

Due to the limited nature of the existing data, an assessment of the resource potential on Makushin is difficult at this time. Based on the locations of the fumarole fields and geologic field mapping conducted by the State of Alaska, it is reasonable to project a fault/fracture controlled resource with temperatures at depth in the range of 200° to 230°C.

There is a reasonable probability of a deeper regional thermal anomaly underlying the Makushin volcanics. This is evident by the existence of extensive fracture systems with southeast trends and identification of younger intrusive bodies underlying the Makushin volcanics. Exploration funded by the Alaska Power Authority and being conducted by Republic Geothermal, Inc. this year will provide much more definitive information about the extent and nature of the Makushin geothermal resource.

LOGISTICS

One of the primary considerations in the development of geothermal resources for power generation on Makushin is logistics. Not only is Unalaska Island a remote location with difficult access, Makushin Volcano is one of the remotest areas on Unalaska Island. The only maintained roads on the island are in and around the town of Unalaska. There are remnants of a road partway up Makushin Valley on the east flank of the volcano and another road in marginal condition runs from Driftwood Bay on the north side of the volcano to within two kilometers of the lower fumarole fields. There is an unused air strip at Driftwood Bay which could be utilized to mobilize equipment and crews for resource development. Service to the commercial airport at Dutch Harbor (vintage WWII) is currently limited to small propellor aircraft.

INSTITUTIONAL CONSIDERATIONS

Lands on the island of Unalaska are controlled either by the Bureau of Land Management or by the Ounalashka (native village) Corporation. The Ounalashka Corporation was entitled under the Alaska Native Claims Settlement Act (1971) to select a total of 46,600 hectares, including Makushin Valley. Although the village corporation's stock cannot be sold for 25 years, the land conveyed to the corporation is surface real estate that can be sold with no restrictions. The Ounalashka Corporation has established a policy of leasing rather than selling their land, and became the first village corporation in Alaska to pay a dividend to its stockholders.

The Aleut (regional) Corporation, in general, owns the subsurface rights under land held by village corporations in the Aleutians. According to Alaska state law, geothermal resources are defined as "the natural heat of the earth at temperatures greater than 120°C, measured at the point where the highest temperature resources encounter, enter or contact a well or other resource extraction device...(Reeder, et al, 1980)." Under this definition, rights to geothermal resources with temperatures above 120°C on Native lands are controlled by the regional corporation. The Aleut Corporation has not yet established a policy for geothermal leasing or development.

The State of Alaska is actively involved in supporting power developments. In 1981, the State put $500 million into energy projects on a grant basis. The State also supports power developments with loans at low interest rates, providing the opportunity for creative financing.

ECONOMIC ANALYSIS

To determine if geothermal power could be competitive with other energy alternatives on Unalaska, a preliminary economic analysis was performed. Because of the current uncertainty regarding the resource potential on Makushin, several power development options were considered. The results of the economic analysis provide an initial assessment of the sensitivity of the cost of geothermal power on Unalaska to resource temperature, well production, plant design and location. Fourteen development options were analyzed (Table 1).

The economic analysis developed the required mill rate per kilowatt-hour of electricity sold to recover operating costs, loan principal payments and interest payments. This rate was compared to the cost of supplying an equivalent amount of power with diesel generation. Rather than escalate the cost of diesel power at an assumed (and probably erroneous) rate, a "breakeven" escalation rate for diesel generation was determined. This escalation rate ranged from a low of 1.8% for case 13 to a high of 6.7% for case 1. If diesel costs escalate at a higher rate, the geothermal power plant is more economical.

The analysis of the geothermal options was based on the following assumptions:

- an annual discount rate of 10%
- financing by a tax-exempt entity at 10% interest
- seven years for well field development and plant construction (on-line in 1990)
- investment and operating costs escalated at 10.5% in 1982 and declining to 7.5% by 1988
- a 10 MW plant producing a total of 65,7000,000 kwh annually
- interest on debt rolled over during construction
- debt is repayed over a 20-year term beginning in 1990
- an initial diesel power price of 300 mills/kwh in 1990, escalated at the breakeven rate.

The total geothermal investment varied significantly with the options considered. A breakeven of the investment for the case 8 is shown in Table 2. Capital costs for the power plant were based on similar plants in the lower 48 states and included a construction cost factor of 2.5 for the remote Makushin location. Costs for roads, transmission lines and wells were based on recent estimates at comparable locations by the Alaska Division of Geological and Geophysical Surveys and Morrison-Knudsen. Average (escalated) costs for geothermal power for each option are shown in Table 1. It should be emphasized that the variation in the rates is more interesting than the absolute value for this preliminary analysis.

CONCLUSIONS

Available data on the geothermal resource potential on Unalaska Island indicates that there is a promising potential for economic power development. The economic analysis of the fourteen development options indicates that while investment costs for a 10 MW power plant near Makushin Volcano look very high, geothermal can be competitive with diesel generation.

The geothermal power costs are particularly sensitive to the number of wells required to produce 10 MW and to the power plant location. At individual well flow rates of 115,000 kg/h and 225,000 kg/h, the costs are less sensitive to the resource temperature and steam fraction than at lower production rates.

ACKNOWLEDGEMENT

A portion of this study was funded by the U.S. Department of Energy through the Alaska Division of Energy and Power Development.

REFERENCES

Bureau of Land Management, 1981. St. George Basin Petroleum Development Scenarios Local Socioeconomic Systems Analysis, OCS Technical Report No: 59.

Drewes, H., G. D. Fraser, G. L. Snyder and H. F. Barrett, 1961. Geology of Unalaska Island and Adjacent Insular Shelf, Aleutian Islands, Alaska, U.S. Geological Survey Bulletin 1028-5, p. 1439-1453.

Motyka, R., M. Moorman and S. Liss, 1982. Assessment of Thermal Spring Sites, Aleutian Arc, Atka Island to Becharof Lake, AOF-144.

Reeder, J. W., P. L. Coonrod, N. J. Bragg and D. R. Markle, 1980. Alaska Geothermal Implementation Plan, U.S. Department of Energy.

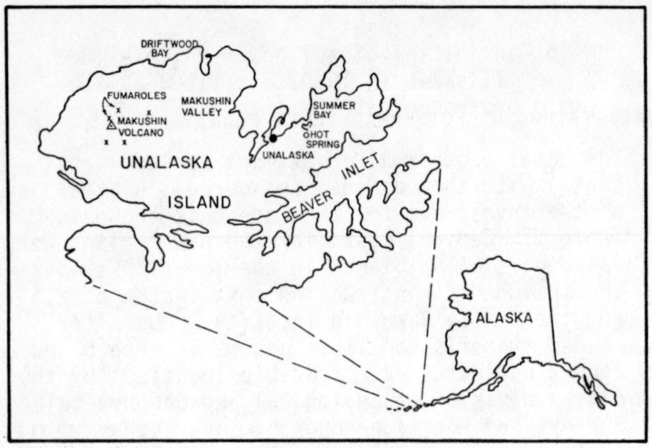

FIGURE 1. LOCATION MAP

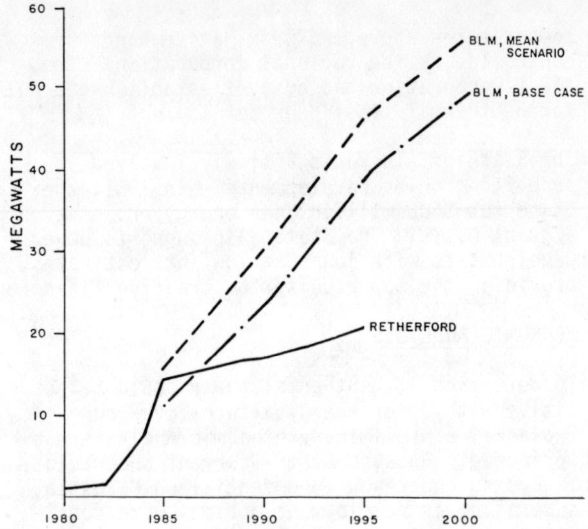

FIGURE 2. ELECTRIC POWER CAPACITY PROJECTIONS

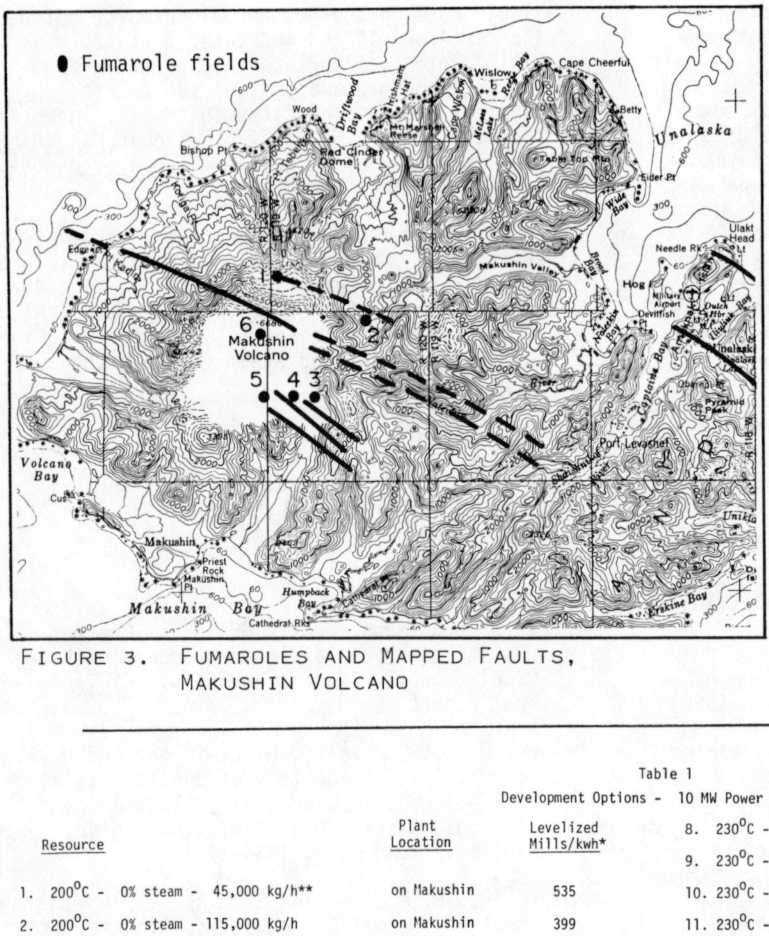

FIGURE 3. FUMAROLES AND MAPPED FAULTS, MAKUSHIN VOLCANO

Table 2
Investment - Case 8
(1982$)

Wells (5)	$ 20,500,000
Plant and Engineering	$ 20,340,000
Brine Handling and Treatment	$ 6,600,000
Permitting and Land	$ 600,000
Transmission	$ 6,200,000
Roads and Bridges	$ 5,600,000
Total	$ 59,840,000

Table 1
Development Options - 10 MW Power Plant

Resource	Plant Location	Levelized Mills/kwh*
1. 200°C - 0% steam - 45,000 kg/h**	on Makushin	535
2. 200°C - 0% steam - 115,000 kg/h	on Makushin	399
3. 200°C - 0% steam - 225,000 kg/h	on Makushin	357
4. 200°C - 10% steam - 45,000 kg/h	on Makushin	450
5. 200°C - 10% steam - 115,000 kg/h	on Makushin	370
6. 200°C - 10% steam - 225,000 kg/h	on Makushin	345
7. 230°C - 0% steam - 45,000 kg/h	on Makushin	493
8. 230°C - 0% steam - 115,000 kg/h	on Makushin	385
9. 230°C - 0% steam - 225,000 kg/h	on Makushin	358
10. 230°C - 10% steam - 45,000 kg/h	on Makushin	440
11. 230°C - 10% steam - 115,000 kg/h	on Makushin	370
12. 230°C - 10% steam - 225,000 kg/h	on Makushin	335
13. 230°C - 0% steam - 115,000 kg/h (no brine treatment)	Makushin Valley	345
14. 230°C - 0% steam - 115,000 kg/h (significant brine treatment)	on Makushin	441

* 10% discount rate
**total production per well

Geothermal Resources Council, TRANSACTIONS Vol. 6, October 1982

OPTIMIZATION OF NON-CONDENSABLE GAS REMOVAL SYSTEM IN GEOTHERMAL POWER PLANT

Shigeru Tajima and Mitsuo Nomura

Toshiba Corporation
Tokyo, Japan

ABSTRACT

Optimization of non-condensable gas (hereinafter called N.C.G.) removal system in geothermal power station, in a special case that the geothermal steam contains large amount of non-condensable gas, is subject to be discussed. Four different alternative N.C.G. removal systems are studied, which are 1) Steam jet gas ejectors, 2) Centrifugal gas compressors, 3) Combined systems of steam ejectors and centrifugal compressors and 4) Back pressure turbine-without N.C.G removal system. This report summarizes the results of a study, and give recommendations as to the most suitable gas removal system, and also as to optimum condenser pressure, in case of large quantity N.C.G. content in geothermal steam.

INTRODUCTION

Geothermal steam contains non-condensable gases in large amount compared with that of conventional thermal power plant. For that reason, geothermal power plant requires large capacity non-condensable gas removal system and it plays very important role in geothermal power generation, and also occupies large portion in its total plant cost and total auxiliary power consumption. So, it becomes major concern to be studied at planning and basic design stage of geothermal power station to select the type of N.C.G. removal system, together with determination of rated turbine exhaust pressure (=condenser pressure).

Many factors affects on optimization of N.C.G. removal system, and the most important one is N.C.G. content in geothermal steam. Generally, N.C.G. content in geothermal steam lies between 0.03 and 2% by volume in most geothermal resource area, but in the case treated in this report, it reaches around 8% by volume. The main objective of this report is to select the optimum N.C.G. removal system in such large N.C.G. content of geothermal steam. The optimization is made based on economic evaluation (Comparison of electrical generation cost for each alternatives).

BASIC CONDITION FOR OPTIMIZATION

Following conditions are assumed as a base for optimization of N.C.G removal system in this case study.

Rated output at generator terminal	50,000 kW
Turbine inlet steam pressure	4.5 kg/cm^2 abs
Turbine inlet steam temperature	147 °C
NCG content	7.25% by volume
NCG composition	CO_2 : 98.1 %
	H_2S : 1.2 %
	H_2 : 0.02 %
	N_2 : 0.42 %
	CH_4 : 0.26 %
Condenser pressure	75 mmHg abs(0.102 kg/cm^2 abs), 100 mmHg abs(0.136 kg/cm^2 abs), 125 mmHg abs, ((0.170 kg/cm^2 abs), 0.3 kg/cm^2 abs, 0.5 kg/cm^2 abs (varied as parameter)
Cooling water inlet temperature	22 °C
Steam price	0.6 yen/kg
Capacity factor	90 %
Fixed charge rate	18 %

Based on above mentioned conditions four(4) alternative N.C.G extraction systems are evaluated. (Fig. 1)

1. Steam jet gas ejectors (single stage and two stage),
2. Centrifugal gas compressors.
3. Combined system of steam ejectors and centrifugal compressors.
4. Back pressure turbine.

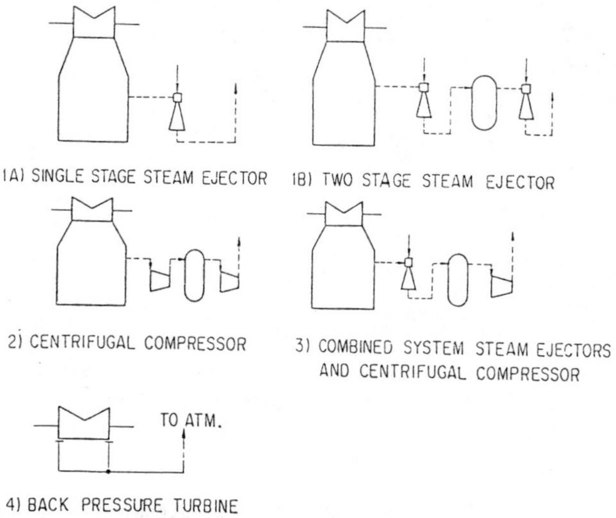

Figure 1. Alternative N.C.G. removal system

ALTERNATIVE 1: STEAM EJECTORS

Fig. 2 shows the turbine inlet steam flow, N.C.G. flow extracted from condenser, driving steam flow of steam ejector v.s. condenser pressure, in case of steam ejector system. As condenser pressure becomes high, turbine requires more steam flow to produce rated power output because of decreased available energy, and N.C.G. flow extracted from condenser increases together with turbine exhaust flow. While, driving steam flow of steam ejector decreases as condenser pressure becomes high because of decreased pressure ratio.

Two kinds of steam ejector system, which are single stage steam ejector system and two stage steam ejector system, are studied. In case that the condenser pressure is relatively low, two stage steam ejector system is indispensable, with inter condenser to reduce the capacity of second stage ejector. In case that the condenser pressure is relatively high, it can be possible to extract N.C.G. from condenser by single stage steam ejector, and no inter condenser and cooling water is required. The study shows that single stage steam ejector is enough to extract N.C.G. from condenser for condenser pressure above 0.22 kg/cm^2 abs, and below 0.22 kg/cm^2 abs of condenser pressure two stage ejector is needed. Driving steam flow of two stage steam ejector system decreases up to 0.25 kg/cm^2 abs of condenser pressure when condenser pressure increases. Above condenser pressure of 0.25 kg/cm^2 abs, driving steam flow increases again, because of the design limit of flow ratio (=extracted N.C.G. flow/driving steam flow).

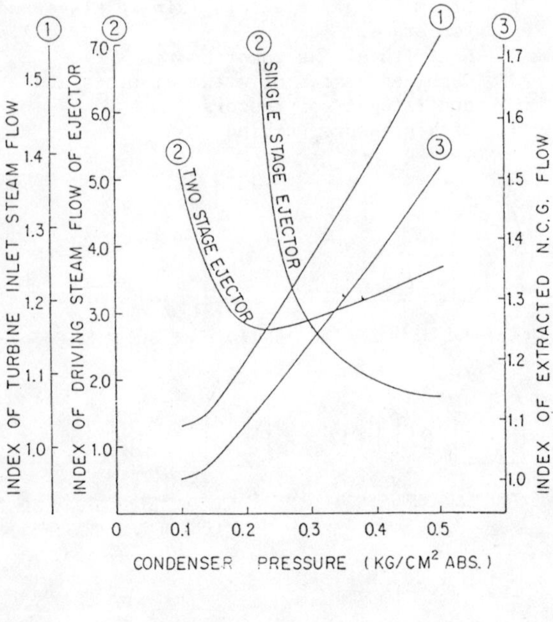

Figure 2. Effect of condenser pressure on steam consumption of steam ejector system

For optimization of steam ejector system, it is important to minimize the driving steam flow. In this view-point, single stage gas ejector system is better choice than two stage gas ejector system when the condenser pressure is higher than 0.3 kg/cm^2 abs. And two stage steam ejector system is prefered for condenser pressure less than 0.3 kg/cm^2 abs.

ALTERNATIVE 2: CENTRIFUGAL GAS COMPRESSOR

Steam ejector mentioned above has many advantages such as simple construction, easy operation and maintenance and low capital cost, and is commonly used as N.C.G. removal system in geothermal power plant. But when N.C.G. content in geothermal steam is large, as in the case treated in this paper, driving steam consumption of ejector becomes considerably large amount, and centrifugal compressor comes into spotlight as an effective alternative because of its relatively high efficiency.

Fig. 3 shows N.C.G quantity extracted from condenser and required power for driving gas compressor. In case that the condenser pressure is low, turbine inlet steam flow is small and N.C.G. flow extracted is also small accordingly. While size or number of gas compressor installed becomes large because of large specific volume of N.C.G. in such a low suction pressure condition and capital cost also becomes high. As condenser pressure increases, required power for gas compressor decreases because the pressure ratio decreases.

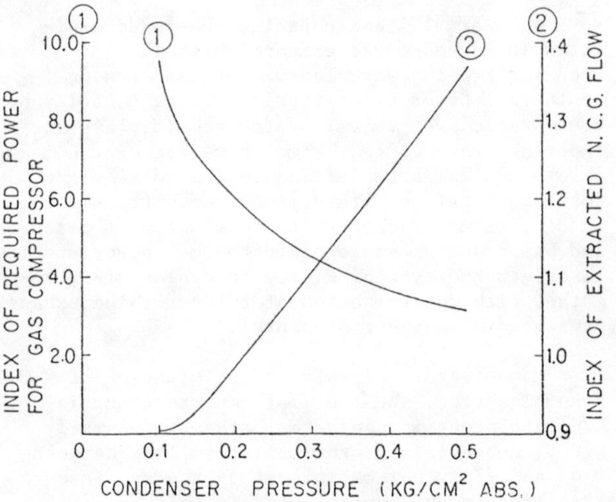

Figure 3. Effect of condenser pressure on required power for gas compressor

ALTERNATIVE 3: COMBINED SYSTEM OF STEAM EJECTOR AND CENTRIFUGAL COMPRESSOR

To improve the characteristics of steam ejector system (low capital cost and easy maintenance, but require large driving steam consumption) and centrifugal compressor system(relatively high efficiency but high capital cost), combined system of steam ejector and centrifugal compressor is considered. The system comprises steam ejector as first stage, intercondenser and centrifugal compressor as second stage, as shown in Fig. 1.

Fig. 4 shows N.C.G extracted, electrical power required, driving steam of gas ejector and turbine inlet steam flow. Driving steam flow of steam ejector and required power for gas compressor decreases as condenser pressure increases.

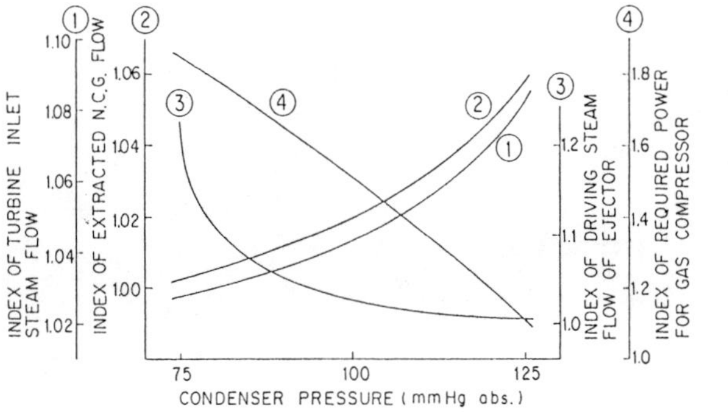

Figure 4. Effect of condenser pressure on driving steam flow of steam ejector and required power of gas compressor

ALTERNATIVE 4: BACK PRESSURE TURBINE

Back pressure turbine is considered to be another effective alternative in the case of large N.C.G. content in geothermal steam. Major specifications of back pressure turbine studied here are as follows.

Type	Tandem compound, four flow, two casing, back pressure turbine
Rated output	50,000 kW
Exhaust pressure	1.11 kg/cm^2 abs
Number of stages	2 stages x 4 flow (2 casing)

The capital cost of back pressure turbine system is the lowest among other alternatives because no N.C.G. removal system is necessary. Furthermore auxiliary steam and electrical power is not needed. These advantages are emphasized when N.C.G. content in geothermal steam is considerably large. On the other hand, steam consumption of back pressure turbine is the largest among other alternatives. So effectiveness of back pressure turbine system depends on steam price as well as N.C.G. content in geothermal steam.

EVALUATION

Evaluation is made by comparison of electrical power generating cost for above mentioned four alternative N.C.G. removal systems. Electrical generating cost is calculated taking equipment cost, steam consumption of turbine, driving steam flow of ejector and electrical power for compressor into consideration.

(Electrical Generation Cost) = (Steam Cost)* + (Capital cost)

*Steam cost means cost of geothermal steam required to produce unit electrical power (yen/kwh)

The result is shown in Fig. 4, and the conclusion is summerized below.

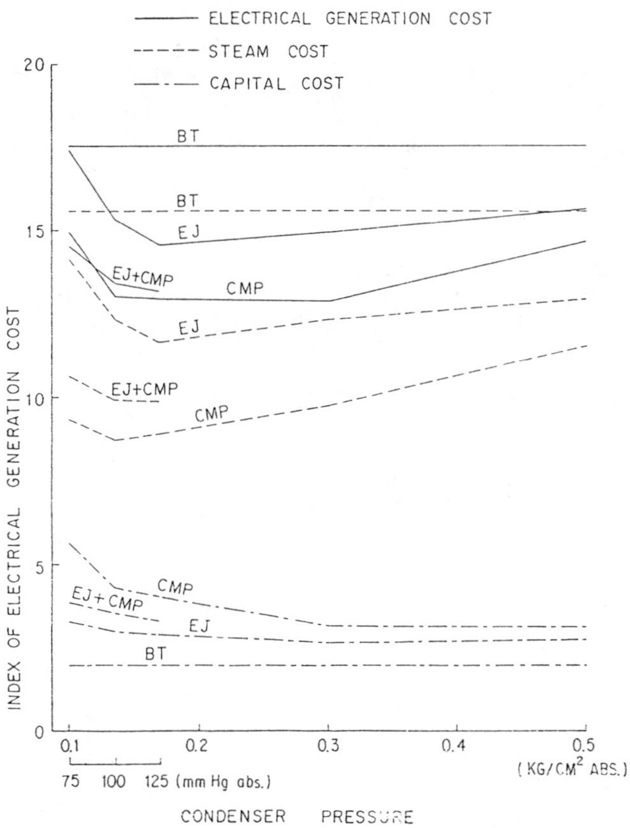

EJ	: Steam ejector
CMP	: Centrifugal compressor
EJ+CMP	: Combied system of steam ejector and compressor
BT	: Back pressure turbine

Figure 5. Effect of type of N.C.G removal system and condenser pressure on electrical generation cost

CONCLUSIONS

In case N.C.G content is 7.25% by volume, and under the given condition as mentioned above:

1. Centrifugal compressor system, and condenser pressure between 125 mmHg abs(0.170 kg/cm^2 abs)and 0.3 kg/cm^2 abs, is the optimum selection among all alternatives.

2. Centrifugal compressor is the most economical selection when condenser vacuum is higher than 0.12 kg/cm^2 abs. Below 0.12 kg/cm^2 abs, combined system of steam ejector and centrifugal compressor becomes the most economical selection.

3. Centrifugal compressor gives the highest capital cost and the lowest steam cost. Back pressure turbine gives the lowest capital cost and highest steam cost.

4. Optimum condenser vacuum that gives the lowest capital cost lies around 0.3 kg/cm^2 abs for all alternatives except for back pressure turbine.

5. Back pressure turbine is not recommended in the case of relatively large capacity unit such as 50MW, even in the case the N.C.G. content is large.

6. Optimum vacuum for steam ejector system lies around 125 mmHg abs(0.170 kg/cm^2 abs)

Geothermal Resources Council, TRANSACTIONS Vol. 6, October 1982

PROCESS CHEMISTRY MONITORING AT THE HGP-A POWER PLANT: ANALYTICAL RESULTS, PROCESS PROBLEMS AND MODIFICATIONS

Donald M. Thomas

Hawaii Institute of Geophysics
2525 Correa Road, Honolulu, Hawaii 96822

ABSTRACT

The HGP-A generator plant began operations on June 12, 1981 and came on-line on a continuous basis on March 1, 1982. During this period process problems were identified and, in most cases, plant modifications have eliminated the difficulties. Silica in the brine was stable at a pH 7.5, however, at a pH above 9.5 deposition of silica was triggered in a brine disposal system and required abandonment of the hydrogen sulfide abatement process originally proposed for the brine system. The steam phase sulfide abatement system for standby conditions was 90% effective, although superheat in the treatment system reduced abatement efficiency. Brine carryover through the separator was very low; however, scale deposition on the turbine blades resulted in substantial damage to the turbine. Non-condensable gases in the condenser were weakly partitioned into the liquid phase, and about 99% were carried into the off-gas treatment system which was found to be approximately 99% effective.

INTRODUCTION

Start-up operations at the HGP-A generator facility on the island of Hawaii were initiated on June 12, 1981; the power plant began providing power to the utility grid on a continuous basis on March 1, 1982. During this shakedown period and continuing to the present, a number of mechanical and process problems have arisen from the geothermal fluid characteristics or from proposed operational procedures that were incompatible with the geothermal fluids. Most of the initial problems have been resolved and modifications are continuing on the plant facility to correct those that remain.

BRINE PROCESS MONITORING

The HGP-A well, which powers the generator facility, has an output capacity of approximately 23,000 kg/hr (50,000 lb/hr) of brine and 23,000 kg/hr (50,000 lb/hr) steam. The total fluid flow from the well is directed into a centrifugal separator/demistor vessel (Fig. 1) where the steam

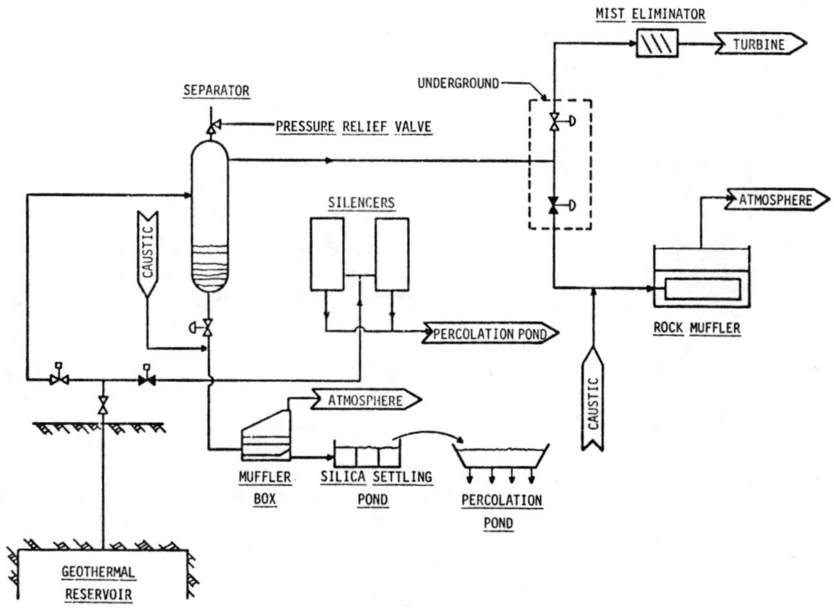

Fig. 1 HGP-A Geothermal Generator Project steam supply process.

Hawaii Institute of Geophysics Contribution 1268

and brine flows are separated at approximately 1,200 kPaa (175 psia). The observed major element chemical composition from the separator is summarized in Table I. Although a detailed interpretation of the fluid chemistry has not been completed, several general observations can be made. The total dissolved solids content has increased substantially since flow was initiated in June 1981. This increase may be the result of increased seawater intrusion into the production aquifers or, alternatively, it may reflect an increasing fraction of steam formation from the single phase reservoir fluid as the flash front migrates away from the wellbore. It is also noteworthy that silica concentrations in the brine have decreased, rather than increased, since the early production flow. The decrease is thought to result from rapid reequilibration of the reservoir silica concentrations to lower temperatures encountered in the flash zone and in the wellbore. If reequilibration is controlling the silica concentrations, it is probable that substantial silica precipitation is occurring both within the flash zone and in the wellbore; this situation may, in the long term, reduce the output capacity of the HGP-A well.

TABLE I. MAJOR ELEMENT CHEMISTRY OF HGP-A FLUIDS (mg/kg)

	Cl^-	Na	K	Mg	Ca	$SO_4^=$	SiO_2
6/17/81	1593	806	148	.061	15.9	50	1233
9/4/81	3700	2190	250	0.1	72.	63	795
Well Shut In	--	--	--	--	--	--	--
12/22/81	3261	1745	281	0.038	58.3	ND	1004
2/5/82	4226	2253	364	0.054	84.4	ND	850
4/5/82	5119	--	--	--	--	--	--

From the separator, the brine is passed through a pressure reduction valve into an atmospheric flash tank. The remaining liquid fraction is passed through a silica settling tank and then into a percolation pond. The steam phase formed is presently being emitted to the atmosphere. The brine phase fed into the low-pressure flash tank contains approximately 30 ppm hydrogen sulfide and approximately 60 ppm carbon dioxide; about 90% to 98% of these gases are partitioned into the steam phase formed in the flash tank, resulting in an appreciable pH change in the brine (from about 7.4 to 8.5). The steam discharge, as a result of the gas loss from the brine, becomes a small but significant source of hydrogen sulfide emissions from the plant. The increase in brine pH occasioned by the gas loss also has significant impact on the silica present in the liquid phase. Whereas the dissolved silica in cooled (non-flashed) samples of brine appeared to be stable indefinitely, the increased pH of the flashed brine triggered silica polymerization resulting in the formation of a moderately stable colloidal suspension of silica polymers. Under these conditions, silica precipitation in the silica settling tank is very slow; after nearly five months of flow, the silica cake on the walls of the pond is less than 2 cm thick.

The hydrogen sulfide emissions in the steam phase from the flash tank were considered sufficiently large to justify the application of abatement procedures. The proposed process was an adjustment of the brine pH to approximately 10 pH units immediately downstream of the pressure reduction valve. This procedure was effective in reducing the sulfide emissions from the brine. However, it also destabilized the colloidal silica formed at pH 8.5; the colloidal suspension rapidly flocculated to form a moderately viscous sludge having a water content of approximately 95% to 99%. The sludge formed was sufficient to fill the silica settling tank in approximately 48 to 72 hours and, if passed through the settling tank, caused an immediate and drastic reduction in the performance of the percolation ponds. Although a number of procedures for handling the silica sludge were investigated, none proved to be economically workable, and thus direct caustic injection into the brine as hydrogen sulfide abatement procedure was abandoned. Work is currently under way to construct a hood and treatment stack for the hydrogen sulfide emissions from the brine flash tank.

STEAM PROCESS MONITORING

The steam fraction from the separator, during plant stand-by conditions, is diverted through a pressure reduction valve and into a rock muffler. The non-condensable gas concentrations present in the steam phase are summarized in Table II. The sulfide production rate from this well amounts to approximately ½ a metric ton per day in the steam phase; thus it was necessary to apply hydrogen sulfide abatement procedures to the geothermal steam emitted to the atmosphere. The abatement technique applied was to inject a 10% solution of sodium hydroxide into the low-pressure steam line upstream of a static mixer spool. This procedure was found to be about 90% effective at a mole ratio of 3 sodium hydroxide to 1 hydrogen sulfide. Although this may not be the peak efficiency for this system, we did find that substantially increased caustic injection rates provided only a minimal improvement in the abatement efficiency.

TABLE II. GAS CHEMISTRY IN HGP-A STEAM

	CO_2	H_2S	N_2	H_2	CO
6/14/81	670	609	130.5	9.	∿1
2/8/82	964	1297	125.	12.9	∿1

The degree of superheat in the steam was found to be a very important factor in the efficiency of the hydrogen sulfide abatement procedure. During the initial start-up period, insufficient water was added to the steam line downstream of the pressure reduction valve and as a result, caustic was being injected into a superheated steam. Under these conditions, the efficiency of the caustic abatement procedures was reduced by 30% to 50% and substantial back pressures were created (due to solids deposition) throughout the low-pressure system. The latter effect may have precipitated the failure of a temporary port in the rock muffler box. Further work with the system indicated that sulfide abatement and system back pressures were optimized with a slight excess (10%-20%) of desuperheat water being injected. Additional testing of this abatement system is planned.

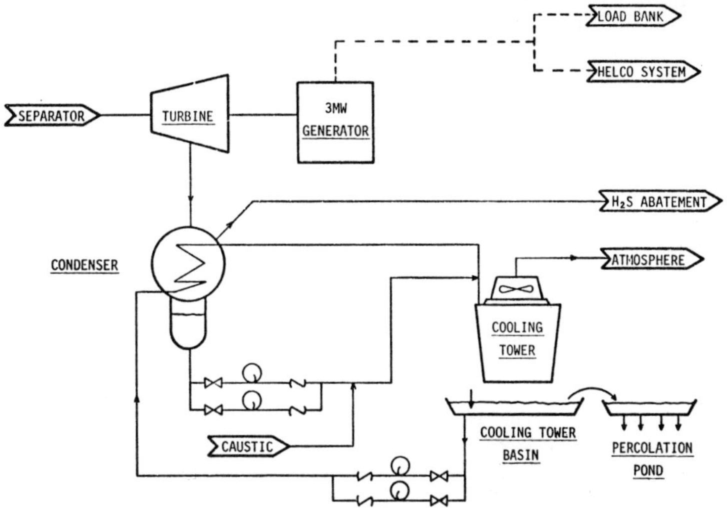

Fig. 2 HGP-A Geothermal Generator Project output and heat rejection.

The steam phase from the separator, under normal conditions, passes through a secondary demistor vessel and into the turbine through trip and control valve manifolds (Fig. 2). To determine the fraction of brine carried past the separator and demistor, the chloride concentration in the steam phase was tested immediately downstream of the plant separator and downstream of the demistor. Silica concentrations were tested there as well. The chloride levels observed in the steam samples were generally between 0.3 ppm and 0.7 ppm; sample blank levels (using distilled water) for the analytical procedures were generally about 0.1 ppm to 0.2 ppm. These chloride levels corresponded to about 0.01% to 0.02% brine carryover in the steam phase.

In spite of the low brine carryover observed in the steam phase, scale material was discovered on the turbine blades during the initial start up period of the facility. The turbine scale material was discovered subsequent to a generator start-up procedure that resulted in excessive turbine vibration that inflicted substantial damage to the turbine rotor and shaft assembly. Removal of the turbine casing revealed that significant amounts of scale material had been deposited on the turbine blades and on the internal components of the turbine housing. Analysis of the scale material indicated that it was primarily a mixed iron oxide with smaller amounts of iron sulfides. An evaluation of the plant chemistry and operating procedures up to the excessive vibration incident suggests that the source of the scale was the piping and manifold section between the turbine inlet and the block valve upstream of the demistor vessel. During the start up, this section of pipe, which is constructed of mild steel, had been exposed alternately to both sulfide-bearing steam and to the atmosphere a number of times due to required modifications on the control valve assembly. This treatment of the pipe had allowed substantial quantities of iron oxide scale to form, which was subsequently carried into the turbine and mechanically deposited on the blades. To alleviate this problem during future turbine starts, the plant piping was modified to allow the entire steam manifold, down to the turbine inlet, to be blown clear after any extended period of turbine shutdown. The effectiveness of this modification has not been determined; turbine vibration at critical speed has not increased substantially during subsequent turbine start-up procedures.

The spent steam from the turbine is passed through the exhaust duct into a stainless steel surface condenser maintained at a pressure of about 0.8 kPaa (6 in. Hg abs.). Analysis of the steam condensate formed indicated that the liquid had a pH of about 4.5 and that approximately 0.6% of the hydrogen sulfide present in the steam was carried into the liquid phase. This latter value contrasts sharply with steam plant performance at Geysers where as much as 30% of the hydrogen sulfide remains in the steam condensate. The virtual absence of ammonia in the steam from HGP-A probably accounts for the observed difference in hydrogen sulfide solubilities.

The condensate liquid is pH-adjusted to approximately 9 and then pumped into the cooling water return line as make-up water for the cooling tower. The pH of the cooling water was found to be extremely important in minimizing corrosion of the mild steel piping used in the cooling water lines and to inhibiting emission of hydrogen sulfide from the cooling tower.

The non-condensable gases from the condenser are removed through a two-stage ejector assembly and are injected into an incinerator backed by a packed column scrubbing tower. The hydrogen sulfide fraction of the gases is incinerated to sulfur dioxide, which is subsequently removed by sodium hydroxide treatment in the packed column. During the initial start-up of this abatement system, insufficient air flow into the incinerator resulted in an incomplete reaction of the hydrogen sulfide to sulfur dioxide, leading to the following reaction: $2H_2S + SO_2 \rightarrow 3S^\circ + 2H_2O$. Elemental sulfur formed very rapidly, reduced the scrubber column flow capacity, and required a shut down of the system for removal and replacement of the tower packing. Modification of the air blower system to provide an excess mole ratio of oxygen in the incinerator apparently has eliminated this problem.

Although the abatement efficiency of the incinerator scrubber was quite high (>99%), the gaseous emissions from the column were found to carry a significant quantity of sulfur particulates. Because this was considered to be a source of visual pollution, piping was installed to transmit the stack gases to the air intake of the cooling tower. Passage of the gas through the cooling water cascade removed a substantial fraction of the sulfur particulate and diluted the remainder to the point that it was not visible in the cooling tower plume. No adverse effects on the cooling tower water from this procedure have been observed.

PRESENT OPERATION

At the present time the power plant is operating at a capacity of about 2.8 MWE with an overall plant hydrogen sulfide abatement efficiency of approximately 98%. Work is currently under way to reduce the condenser back pressure to 0.5 kPaa (4 in. Hg abs.) or less in order to bring the plant capacity up to 3 MWE. Modifications are also under way on the brine steam emission system that will attempt to improve the total hydrogen sulfide abatement efficiency of the facility to better than 99%.

ACKNOWLEDGMENTS

Sampling and analytical assistance of David Mills, Chuck Fraley, Robert Mitiguy, and Robert Kochy, and assistance in the preparation of the manuscript by Leslie Kajiwara and Jan Danek are gratefully acknowledged. This work was supported by Department of Energy Grant DE-AC03-78ET 28420.

Geothermal Resources Council, TRANSACTIONS Vol. 6, October 1982

SAMPLING AND ANALYTICAL METHODS DEVELOPMENT AT THE HGP-A GENERATOR FACILITY

Donald M. Thomas

Hawaii Institute of Geophysics
2525 Correa Road, Honolulu, Hawaii 96822

ABSTRACT

During shakedown operations for the HGP-A generator plant sampling and analytical problems were encountered during the process chemistry monitoring effort. Acid-preservation of brine for cation analysis required the use of nitrous oxide-acetylene flame for accurate A-A analysis of calcium. Analysis of gases for carbonate and sulfide was by specific ion electrode and alkalinity titration, respectively. Sulfide caused substantial interferences with the alkalinity method and corrections for sulfide were required. Sulfide also interfered with chloride analyses in the steam phase requiring removal of the sulfide by boiling. Analysis of dissolved silica in the brine was complicated by the presence of colloidal silica which produced erratic analytical results. An accurate evaluation of the hydrogen sulfide abatement system was possible only when the hydrogen sulfide concentrations in the treated and untreated steam were compared with a second component in the steam phase that was unaffected by caustic injection.

INTRODUCTION

Construction of the HGP-A turbine generator facility on the island of Hawaii was completed in June of 1981 and operations for start up of the plant were initiated on June 12, 1981. Beginning then, and continuing to the present, the chemical compositions of a number of process streams throughout the plant have been monitored.

PLANT PROCESS

The HGP-A generator facility (Figs. 1 and 2) is fed by a single geothermal well with an output capacity of approximately 23,000 kg/hr (50,000 lb/hr) of steam and 23,000 kg/hr (50,000 lb/hr) of water. Mixed fluid from the well is piped to a centrifugal separator/demistor vessel where the brine fraction is removed from the steam phase at 1,200 kPaa (175 psia). The brine stream is passed through a pressure reduction valve to atmospheric pressure and, after passing through a muffler box and silica settling pond, it is allowed to percolate into the ground.

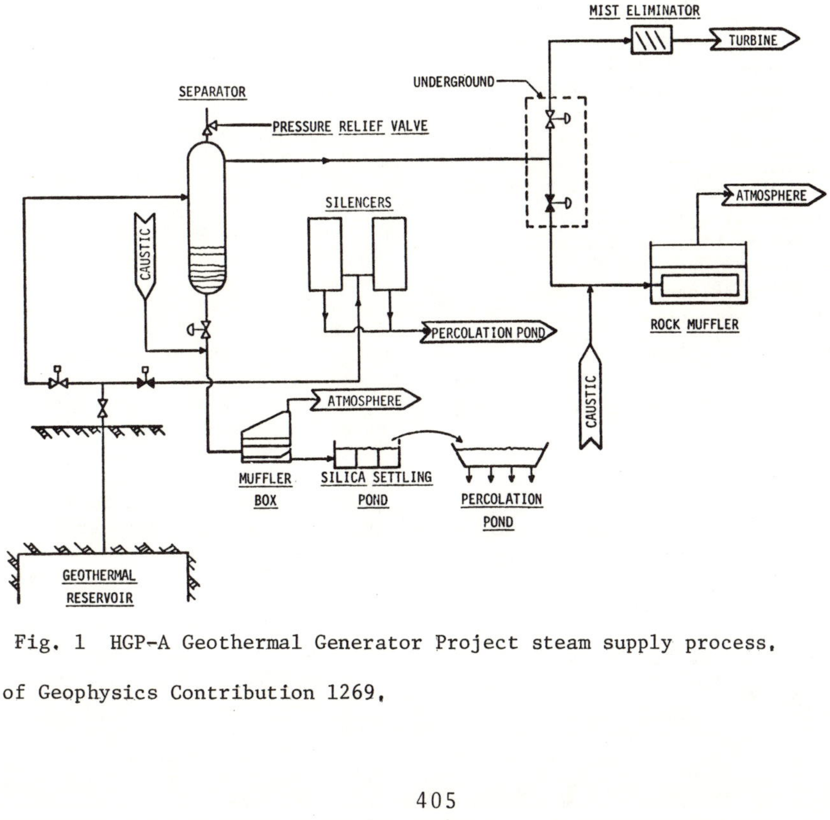

Fig. 1 HGP-A Geothermal Generator Project steam supply process.

Hawaii Institute of Geophysics Contribution 1269.

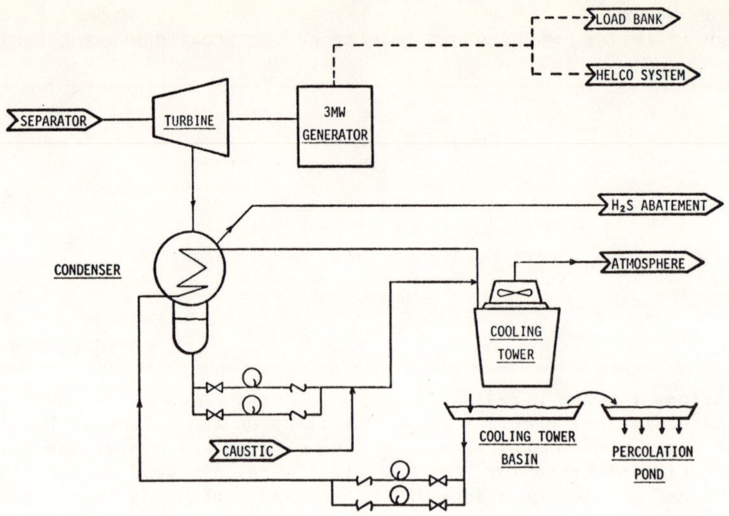

Fig. 2 HGP-A Geothermal Generator Project output and heat rejection.

The steam phase from the separator, under normal operating conditions, passes through a second mist eliminator and the power plant trip and control valve manifolds to the turbine. The turbine exhaust is piped to a stainless steel surface condenser and the steam condensate, after pH adjustment, is pumped into the cooling water return line as make-up water for the forced draft cooling tower. The non-condensable gases are ejected from the condensate into a thermal oxidizer where the hydrogen sulfide fraction is incinerated to form sulfur dioxide that is subsequently scrubbed with caustic soda in a packed column. The tail gas from the scrubber column is piped to the air intake of the cooling tower where, as a final treatment, the cooling water cascade removes sulfur particulates from the gas stream.

When the power plant turbine is not operating at full capacity, part or all of the steam from the plant separator is diverted to a pressure reduction valve where it is passed to the atmosphere through a caustic abatement system and rock muffler.

PROCESS MONITORING

The locations at which the process chemistry was monitored, and the species for which analyses were made, are as follows.

(1) The high pressure brine discharge line immediately downstream of the plant separator was monitored for major element compositions including SiO_2, Cl^-, H_2S, Na, K, Ca, Mg, SO_4. Silica was monitored to indicate the changes in reservoir temperatures; chloride, to determine the stability of the production aquifers; and H_2S to determine sulfide emission rates from the brine discharge point. The other species were analyzed as a check on scale formation within the production zone of the well.

(2) The high-pressure steam line immediately downstream of the plant separator was monitored for non-condensable gases, chloride, and silica. The N-C gases were used to evaluate the gas load and air requirements of the incineration system, and chloride was used to determine the fraction of brine particulate that was being passed by the plant separator/demistor vessel.

(3) The condensate discharge line from the plant condenser was monitored for pH and N-C gas content and for chloride concentrations. The former data were used to evaluate partioning of the N-C gases between the condensate and gas phase, and the latter to check on the chloride measurements upstream of the turbine.

(4) The incinerator column gas discharge line was monitored for hydrogen sulfide and sulfur dioxide to evaluate the abatement efficiency and compliance with emission requirements for the plant.

(5) The silica precipitation tank was monitored for silica concentration to check the effectiveness of the pond in removing dissolved silica.

(6) The rock muffler discharge was monitored for hydrogen sulfide to check the effectiveness of the abatement system and compliance with emission regulations, and for pH to ensure that the steam discharge was not emitting significant quantities of the caustic soda injected into the steam line for hydrogen sulfide abatement.

SAMPLING AND ANALYTICAL PROCEDURES

The sampling procedure for the pressurized brine was essentially that described by Kindle and Woodruff (1981). Although their recommended on-site analytical procedures for chloride, silica and dissolved gases were satisfactory, later analysis of the major element chemistry did encounter one significant problem. The sample preservation and analytical method recommended for cation samples was acidification of raw and filtered samples and later analysis by atomic absorption spectroscopy. Using this method, we found that the calcium concentrations in the acidified samples were indicated to be approximately 30% of those in the non-acidified samples for both filtered and unfiltered splits. Subsequent analysis indicated that a nitrous oxide/acetylene flame was required to obtain accurate concentration values for calcium ion. This situation suggests that silica polymerization or complexing is in some way binding the dissolved calcium ions. It is not presently known whether a similar effect is occuring for the trace elements in the brine; however, this possibility is being investigated.

Initially, steam samples were obtained from a ½-inch insulated line attached to a port on the side of the main steam line. This sampling procedure was found to be unacceptable because of condensation in the sampling line. Subsequent experimentation found that, lacking sample probe access directly into the steam line, an acceptable alternative was a steam jacketed 1/8-inch probe (Fig. 3). The steam samples obtained were condensed into a cooling water-jacketed evacuated bottle containing 4 N sodium hydroxide.

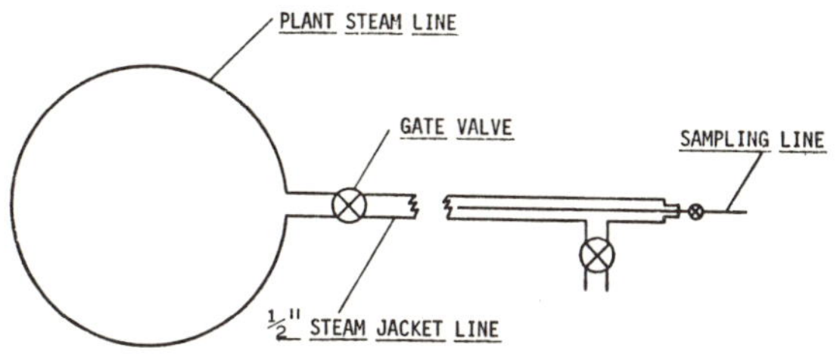

Fig. 3 Steam Jacket Sample Line.

Analysis of the samples for non-condensable gases was by gas chromatography for the non-absorbable gases (N_2, H_2, CO) and by specific ion electrode and alkalinity titration for H_2S and CO_2, respectively. Sulfide concentrations in the samples were high enough to result in substantial interferences from HS^- ion in the carbonate determinations. Precipitation and oxidation of the sulfide were attempted, but the results were inconsistent; the only reliable method found for overcoming the sulfide interferences was to perform the alkalinity titration over a pH range of 11.5 to 3.5 and then correct the acid volume used by that required for a similar titration (over the same pH range) of a carbonate blank having a sulfide concentration identical to that of the sample.

Chloride analyses of the steam condensate also encountered interferences from dissolved sulfide ion even for raw condensate taken without caustic. Precipitation of the sulfide by nickel nitrate or lead nitrate did not reduce the sulfide concentration sufficiently to eliminate the interference. The only method effective for eliminating the interference was to acidify and boil the sample down to one-tenth its original volume to drive off the sulfide and concentrate the chloride in the remaining liquid. Another method presently being investigated is pre-concentration of the chloride and removal of sulfide through the application of ion exchange resins.

The condensate discharge sampling procedures for non-condensable gases were straightforward provided that the liquid was introduced directly into a caustic medium (Kindle and Woodruff, 1981). Without this precaution, dissolved H_2S and CO_2 were rapidly lost to the atmosphere. The analytical procures were identical to those described above for carbonate, sulfide and chloride ions.

The incinerator packed-column gas discharge was monitored with a lead acetate tape continuous monitoring system. Under normal operating conditions this method was successful; however, under flame-out of the incinerator, the sulfide levels in the column gases were sufficiently high to affect the entire tape spool in the instrument and render it useless. Sulfur oxides from the column effluent were initially sampled intermittently by disposal gas detection tubes; however, absorption of a measured volume of effluent into a caustic impinger is presently believed to yield more reliable results for total sulfur emission rates.

Analysis of silica concentration in the silica precipitation pond proved to be an intractable problem. Whereas the silica present in the high-pressure brines seemed stable enough to allow dilution before polymerization occurred, the change in pH of the liquid phase from 7.5 to 8.5, brought about by the atmospheric flashing of the hot pressurized brine, induced the rapid formation of colloidal silica in the brine phase. Filtered and diluted (10:1) brine samples yielded inconsistent results with the molybdate blue method even after several days of reequilibration. Later analysis of these samples by nitrous oxide/acetylene atomic absorption provided better results, although the variations in the data suggest that there still may be analytical problems to be resolved.

A precise evaluation of the sulfide concentration present in the steam emissions from the rock muffler was a relatively straightforward procedure. Steam condensate and non-condensable gases were drawn into a caustic solution through an evacuated flask. The major requirement with this procedure is the exclusion of air from the sample solution; the introduction of atmospheric oxygen into the sample allows oxidation of sulfide to sulfate to occur in the basic medium.

Although the above procedure yields an accurate representation of the sulfide present in the atmospheric steam emissions, an evaluation of the efficiency of the abatement system, in terms of percent removal of the original sulfide present, required a further step. As the hydrogen sulfide-to-water vapor ratio in the steam phase is altered by the addition of de-superheat water to the steam line downstream of the pressure reduction valve, it is necessary to ratio the H_2S concentration in the emitted steam to another component in the steam phase that is unaffected by caustic treatment. In the present case, the hydrogen concentration in the steam emissions was used as the second component. Comparison of the $H_2S:H_2$ ratios in the steam before and after caustic treatment yielded the required abatement efficiency values.

SUMMARY OF CURRENT DATA

A summary and discussion of the plant process chemistry data acquired during this monitoring program is presented elsewhere in this volume (Thomas, 1982). However, in general terms the following observations have been made during the first nine months of operations of HGP-A:

1. The HGP-A well fluid chemistry is following a slow but predictable trend of increasing chloride concentration.
2. The non-condensable gases make up about 0.2% of the steam phase.
3. Brine carryover through the separator/demistor system is about 0.03%.
4. Less than 1% of the non-condensable gases are partitioned into the exhaust steam condensate from the exhaust steam condenser.
5. The turbine off-gas hydrogen sulfide abatement system is operating at better than 99% efficiency.
6. The abatement efficiency of hydrogen sulfide in the turbine standby (rock-muffler) system is about 90% at a mole ratio of 3 NaOH to 1 H_2S.
7. Silica precipitation in the silica settling pond is occurring at a very slow rate.

ACKNOWLEDGMENTS

Sampling and analytical assistance of David Mills, Chuck Fraley, Robert Mitiguy, and Robert Kochy, and assistance in the preparation of this manuscript by Leslie Kajiwara and Jan Danek are gratefully acknowledged. This work was supported by Department of Energy Grant DE-AC03-78ET 28420.

REFERENCES

Kindle, C.H. and Woodruff, E.M., 1981, Techniques for geothermal liquid sampling and analysis: Pacific Northwest Laboratory Rept. PNL-3801/UC-66d.

Thomas, D.M., 1982, Process chemistry monitoring at the HGP-A power plant: analytical results, process problems and modifications (this vol.).

Geothermal Resources Council, TRANSACTIONS Vol. 6, October 1982

EQUIPMENT CONSIDERATIONS FOR A BINARY CYCLE GEOTHERMAL POWER PLANT

William C. Thorleifson and Alex P. Ibe

Fluor Power Services
3333 Michelson Drive
Irvine, California 92730

ABSTRACT

The binary cycle geothermal power plant incorporates existing hydrocarbon handling technology proven in use by the petrochemical industry. Equipment sizing and hydrocarbon cycle control on the commercial plant scale, however, introduce some unknowns. This report discusses the various technical factors considered in the design, selection, and sizing of the major equipment for use in the Heber Binary Cycle Geothermal Demonstration Power Plant.

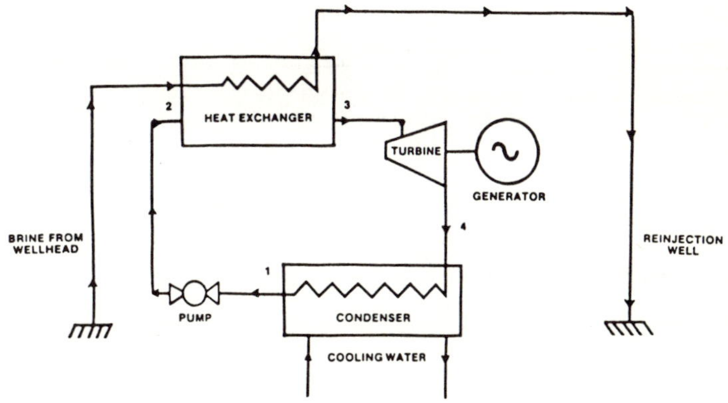

Figure 1: Simplified Binary Cycle

(Reference 1)

INTRODUCTION

BACKGROUND

The recovery and productive use of heat energy from the liquid dominated, moderate temperature KGRA's, or known geothermal resource areas, has not been actively pursued in the past because of the marginal efficiencies and high cost of the major equipment in the commonly used "flashed cycle" energy conversion plant. Several new state-of-the-art conversion methods are now being evaluated. Considered to be the most promising of these in technical viability and near term availability is the binary cycle geothermal energy conversion method.

The Heber anomaly, located in Southern California's Imperial Valley, is typical of the abundant liquid dominated, moderate temperature KGRA's found throughout the continental U.S.A. This site has been selected for the construction and operational demonstration of the first commercial sized binary cycle geothermal energy power plant.

This paper discusses the technical factors considered in the design, sizing, and selection of the major equipment for the Heber Binary Cycle Geothermal Demonstration Power Plant.

THE HEBER POWER GENERATING PLANT

The Heber demonstration facility is composed of the two fluid loops as shown in Figure 1, the hot geothermal brine and the hydrocarbon working fluid.

The hot geothermal brine is delivered by use of downhole pumps to a production island where desanding occurs. The brine, still in the liquid state at 360°F and 250 psi, is transported to the plant heat exchangers where energy is transferred to the shellside hydrocarbon. The spent brine is then transported via variable speed pumps to a reinjection island where, still in the liquid state, it is returned to the peripheral areas of the reservoir.

On the working fluid side the hydrocarbon is pressurized to the supercritical condition and on passing through the plant heat exchangers experiences a gradual phase change to the dense vapor state prior to introduction into the turbine. Figure 2 shows the simple heat operated Rankine cycle where energy is absorbed above the hydrocarbon saturation dome and expansion through the turbine remains clear of the phase boundary. Turbine exhaust pressure is approximately 75 psi, floating with the ambient wetbulb temperature, where the water cooled condenser

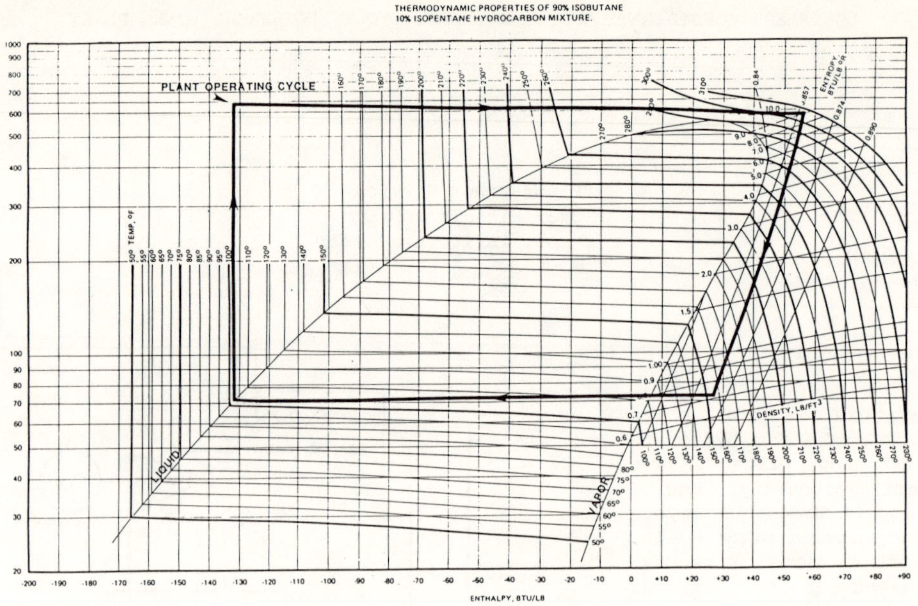

Figure 2 (Reference 2)

returns the hydrocarbon to the liquid state. The hydrocarbon is again pressurized and the cycle repeats.

The Heber Power Generating Plant operating parameters for the rated operating power point at start of run and for the plant ambient at 55°F wetbulb temperature are listed in Figure 3. Our discussion will address the various technical factors considered in the design, sizing and selection of the plant heat exchangers, hydrocarbon turbine and hydrocarbon condensers.

ESTIMATED
HEBER POWER GENERATING PLANT
OPERATING PARAMETERS
START OF RUN
(55°F WETBULB AMBIENT TEMP)

Plant Rating	70 MWe
Brine Flow Rate	7.953×10^6 lb/hr
HT Exchanger Inlet	360 °F
HT Exchanger Exit	160 °F
Reinjection Temp (Min)	150 °F
Hydrocarbon Flow Rate	8.34×10^6 lb/hr
HT Exchanger Exit Temp	305 °F
Turbine Inlet	305 °F
	575 psia
Turbine Exit	73 psia
Condenser Inlet Temp	159 °F
Condenser Exit Temp	101 °F

Figure 3

DISCUSSION

BRINE/HYDROCARBON HEAT EXCHANGERS

The usual concerns of geothermal brine corrosiveness and fouling are, for the binary cycle energy conversion method, applicable only to the brine/ hydrocarbon heat exchangers. The remainder of the plant equipment sees only the clean and inert hydrocarbon working fluid. Additionally, the tubeside brine flow remains near delivery pressure with a relatively high heat exchanger discharge temperature, as noted in Figure 3. These tubeside flow conditions serve several important purposes:

- the tubeside flow path for the geothermal brine reduces eddy flow and stagnation areas that could result in excessive solids precipitation and dissolution of noncondensable gases. Also, tube cleaning is easier than cleaning the shell cavity.
- preserving the brine pressure, except for minor flow losses, through the heat exchangers also discourages the dissolution of the highly corrosive non-condensable gases.
- the relatively high heat exchanger brine discharge temperature serves to optimize energy transfer per unit area of heat exchange surface and, importantly, could minimize scale deposition to the extent that batch chemical cleaning will be effective.

- o maintaining the hot brine in the liquid state through the heat exchanger significantly enhances the heat transfer efficiency.

- o maintaining minimum pressure and temperature values will preserve the geothermal brine identity so that reinjection of the spent brine into the peripheral areas of the reservoir will not adversely affect the bed operation.

On the shellside, the hydrocarbon liquid flow at supercritical pressure will gradually transition into a dense vapor flow for transport to the turbine. Operating shell side pressures of 575 psia are required but transient excursion plus margin result in a design pressure of 850 psig. Since the working fluid system can tolerate no oxygen, both from a safety and corrosion aspect, the shellside must be designed for full vacuum for purification.

In order to reduce field fabrication the heat exchangers are to be factory assembled in components with sizes and weights that will permit rail or truck shipment. This results in a multiple set of exchangers, two sets of four units each, with shell and tubeside flows as indicated in Figure 3. The envelope dimensions of each of the heat exchangers is 76½ feet overall length and a shell outside diameter of 7 feet. Empty weight exceeds 380,000 lbs.

HYDROCARBON TURBINE

The Heber turbine, although a first-time design for this magnitude of power generation using a hydrocarbon working fluid, is soundly based on existing technology and fabrication techniques. The nature of the working fluid, a potentially hazardous material in the proper mixture with oxygen, requires that all leakages be kept to a minimum. Leakage that does occur must be captured and routed to flare equipment or returned under pressure to the working loop. To further enhance the safety aspect, the Heber major equipment is located in open areas with, at most, overhead protection from the elements.

The Elliot Company turbine to be applied in the Heber power plant is a single nozzle, outer casing midpoint entry, double flow with dual discharge ports, axial type with four stages per side for expansion. Although this turbine is standard technology, several areas of turbine operations are of specific interest in this first binary cycle geothermal demonstration power plant these are:

- o turbine controls for the hydrocarbon working fluid will be closely investigated. Frequency stability on isolated grid at low or house load power levels will be demonstrated. Load stability during intertie operations will be established. Load transient capabilities will require thorough definition.

- o referring back to Figure 2 it is evident that the expansion leg of the simple Rankine cycle runs close to the hydrocarbon mixture analytical phase boundary line. As can be seen at the turbine inlet station even minor perturbations in either fluid pressure or temperature may result in hydrocarbon phase change in the turbine. This operating arena will be closely scrutinized during the demonstration interval of the Heber power station operations.

- o the turbine seals for the hydrocarbon working fluid can be greatly simplified from the elaborate and power consuming typical steam turbine seals. This results because the hydrocarbon turbine discharge, as noted in Figure 3, is always substantially above ambient pressures in the operating mode. Statically, the system is subject to full vacuum for purification purposes. The Elliot turbine is provided with the proprietary "iso-carbon seal" configuration proved in millions of operating hours of rigorous hydrocarbon compressor service.

HYDROCARBON CONDENSERS

Two parallel operating, water cooled hydrocarbon condensers are planned for the Heber Station. The hydrocarbon turbine discharge pressure, ranging upwards from 70 psia depending on the ambient wetbulb temperature, exceeds the tubeside cooling water pressure. Hydrocarbon leak dectectors are required in the cooling tower arrangement.

The hydrocarbon condensers are designed with the integral capacity for desuperheating as well as condensing the turbine discharge vapor and to accommodate full bypass of the turbine during plant startup and shutdown. The condensers will be shop fabricated with envelope dimensions of 80 ft by 11 ft shell diameter. Empty weight for each unit is nearly 640,000 lbs.

CONCLUSION

The major equipment for the Heber Binary Cycle Geothermal Demonstration Power Plant is produced and installed using existing technology and manufacturing techniques. The universal concerns associated with geothermal brine handling are applicable only to the brine/ hydrocarbon heat exchangers. And even the heat exchanger exposure is considered minimal since the brine is maintained in its liquid state throughout the energy transfer process.

REFERENCES

1. Geary, C. H., Oct. 1977, "Binary Cycle" published in Geothermal '77 Seminar Proceedings, Bulletin H-38.

2. Fluor Power Services, Inc., May 1982.

USE OF AVAILABLE GEOTHERMAL MODERATE TEMPERATURE RESOURCES FOR POWER GENERATION RE: THE HEBER BINARY CYCLE DEMONSTRATION PLANT

William C. Thorleifson and Alex P. Ibe

Fluor Power Services, Inc.
3333 Michelson Drive
Irvine, California 92730

ABSTRACT

The Heber Binary Cycle Demonstration Plant is undertaken to provide, on a commercially viable scale, an energy conversion technology utilizing our liquid dominated, moderate temperature geothermal resources. Use of these abundant resources, available in many areas of the U.S.A., would be denied the nation without an effective and economical means to convert the energy of the hot geothermal brine into economical and available electric power. This paper discusses the technical considerations supporting the plant design criteria and power cycle definition for this first commercially sized binary cycle geothermal power plant.

INTRODUCTION

BACKGROUND

The use of geothermal energy for the production of electrical power is a maturing technology that is being implemented on a commercial scale in many areas of the world. The U.S. geothermal energy conversion technology, for liquid dominated resources, with this nation's inheritance of economical and readily available fossil fuels for power production, has however, remained relatively undeveloped. The U.S. geothermal scene today may be likened to that of the petroleum industry at the turn of the century.

The recent threat to our energy independence has emphasized the desirability, if not the essential nature, of a diverse power base. Leading in promise and near-term availability in this struggle for diversification is the relatively untapped national resource of liquid dominated geothermal energy.

KGRA

The known geothermal resource areas (KGRA's) span the continental U.S.A. with the major high temperature resources located predominantly in the West and Southwest. The KGRA's are primarily liquid dominated, with some notable exceptions such as the Geysers in northern California and may be broadly categorized by temperature regime, as shown in Figure 1. This simple curve, showing the cumulative geothermal heat content versus the resource temperature, indicates that approximately one third of the known energy is considered "high temperature", while the remainder is below the 410°F (210°C) high-to-moderate temperature boundary. This boundary approximates the practical economic limits for the direct flash conversion technology generally in use for commercial power production.

GEOTHERMAL ENERGY CONVERSION METHODS

Of the several methods available for converting geothermal energy from liquid dominated KGRA's into useful electrical power, the most common process is that of the direct flash cycle. Here the hot geothermal fluid, typically a brine solution, is allowed to "flash" to its vapor state which is then, in its turn, expanded across a conventional turbine driving a generator. This conversion method exhibits inherent disadvantages and penalties, particularly if the

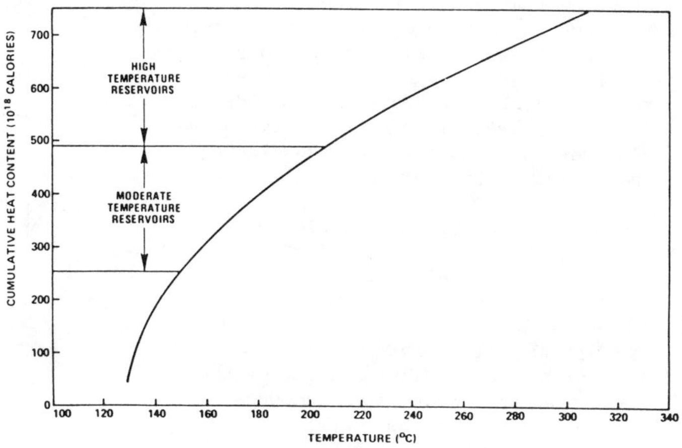

Figure 1: Cumulative Geothermal Energy Content of Known Hydrothermal Systems as a Function of Temperature. (Reference 1)

geothermal resource temperature is at or below the 410°F (210°C) boundary. The impact on the environment, from concentrated brine and precipitate disposal and from noncondensable gas emissions, can be severe. Equipment size and cost escalate as the temperature differential across the turbine narrows. Corrosion resistant materials and anti-fouling measures are expensive.

Another geothermal energy conversion method exists, however, that eliminates or effectively reduces the disadvantages and penalties associated with the direct flash cycle, and one that is more compatible with the moderate temperature KGRA. This method has come to be known as the "binary cycle" conversion method.

THE BINARY CYCLE

The binary cycle, from a thermodynamic viewpoint, is simply a heat-operated Rankine cycle using geothermal fluid as the heat source. The term "binary" does not necessarily imply a two-component working fluid, but is used here to distinguish this system from those other systems in which the geothermal fluid powers the turbine directly.

The binary cycle typically consists of two loops, as shown in Figure 2; the brine side, or primary fluid, and the working side, or secondary fluid. The brine side may pass through the heat exchanger in the liquid state or in the vapor state. The Heber system will maintain the hot brine in the liquid state since heat transfer efficiency is greater for the liquid-to-liquid exchange than for the vapor-to-liquid exchange. The binary cycle working fluid is generally a low boiling point refrigerant. The Heber plant will use a hydrocarbon mixture of 90% isobutane and 10% isopentane.

Figure 3 provides a visual comparison of the binary cycle versus flashed steam cycle performance while operating in KGRA's with resource temperatures of 390°F (200°C) and below. Considering that 80% of the identified fields fall in this temperature classification, and also considering the attractive environmental package offered by the binary cycle plant that reinjects the geothermal fluid, then the binary cycle geothermal energy conversion method is an excellent candidate for commercial demonstration.

THE HEBER POWER GENERATING STATION

The Heber station is proposed as a binary cycle geothermal demonstration plant. From the beginning of its anticipated 30 year operating life it will provide data on various aspects of the plant operating modes from resource bed behavior to hydrocarbon working fluid behavior. Although the binary cycle technology comes to us from the petroleum industry, this technology has never before been scaled to the commercial power generating station proportions. Also, this will be the first combination, again on the commercial scale, of the binary cycle technology and the geothermal energy conversion technology.

DISCUSSION

THE HEBER BRINE LOOP

The Heber KGRA is located in Southern California's Imperial Valley. It is nearly 3 km deep and approximately 3 km in diameter. Although this KGRA has been thoroughly tested and evaluated, several areas of reservoir performance remain to be demonstrated under plant operating conditions. Wells will be drilled, by use of directional drilling techniques, to different locations in the bed from a common production island. Each well, with its own downhole pump, will deliver brine to the production island where it is desanded and transported, still in the liquid state, to the power plant heat exchangers. Brine temperature and pressure at the heat exchanger inlet are 360°F (182°C) and approximately 250 psi. The tube-side brine flow through the heat exchanger, minimizing eddy flow and stagnation areas, will reduce, if not preclude, solids precipitation and deposition fouling. The minimum brine exit temperature of 160°F (71°C) serves to optimize energy transfer per unit area of the heat exchanger surface. It will also preserve the brine fluid identity and should reduce scale deposition to the extent that batch chemical cleaning will be effective.

The spent brine will be reinjected into the KGRA peripheral regions. This reinjection process has been initiated, in addition to the obvious environmental advantages, to reduce, perhaps preclude, surface distortion and water table decline.

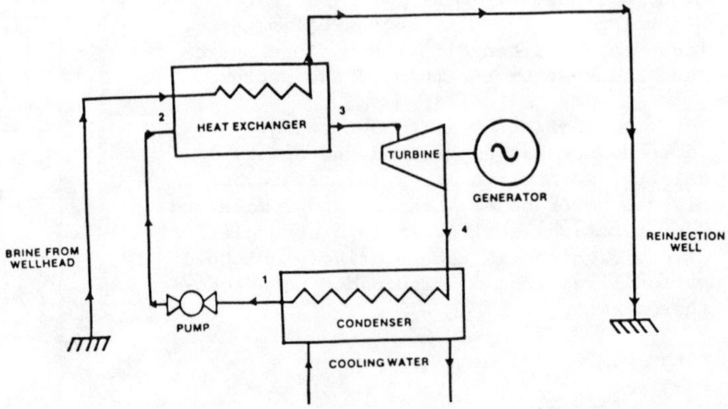

Figure 2: Simplified Binary Cycle

(Reference 2)

BRINE SIDE INVESTIGATIONS

Several areas of the brine side operation will be subject to investigations of varying duration and intensity. The downhole pump operations and reliability in the geothermal bed environment is of particular interest and, if single phase brine flow is to be maintained, of essential import. Brine fluid behavior and identity under conditions of extraction, transport, energy exchange and reinjection, all in the liquid state, will be closely monitored. The heat exchanger itself, aside from the corrosion, erosion, fouling and thermal transient considerations, will be the subject of detailed study because of the supercritical cycle for the hydrocarbon. Perhaps the most intriguing element to be evaluated in the brine side operation will be the Heber KGRA reservoir behavior under conditions of maximum plant energy demand and the predictable event of power black-out. The operational survivability of the downhole pumps, well casings and brine transport equipment under impact of both hydraulic and thermal shock, resulting from total loss of pumping capability or other brine flow anomaly, will be the subject of intensive study and investigation. The brine side operation alone is sufficient cause for the "demonstration" characterization of the Heber Power Generating Station.

THE HEBER HYDROCARBON LOOP

The Heber plant working fluid is a hydrocarbon mixture composed of 90% isobutane and 10% isopentane. This mixture has been carefully selected for its compatibility with the brine exchange temperature regime and with the Heber ambient wet bulb temperature. The hydrocarbon mixture condensation temperature is 105°F (40°C). This excellent "fit" of the primary and working fluid characteristics is further enhanced by a design point minimum brine/hydrocarbon heat exchanger approach, or pinch, temperature of 10°F (5.5°C).

To futher augment the working side performance the hydrocarbon pressure leg will operate in the supercritical cycle. Figure 4 provides a stylized PH curve showing the sub and supercritical Rankine cycles. For our working mixture, the supercritical pressure in the heat exchanger is 575 psia which satisfactorily clears the saturation dome. With a hydrocarbon heat exchanger discharge temperature of 305°F (151°C) we are ready for vapor introduction into, and expansion across, the turbine. From the supercritical cycle curve it is evident that the expansion leg, from point 3 to point 4, remains clear of the phase change boundary until turbine discharge.

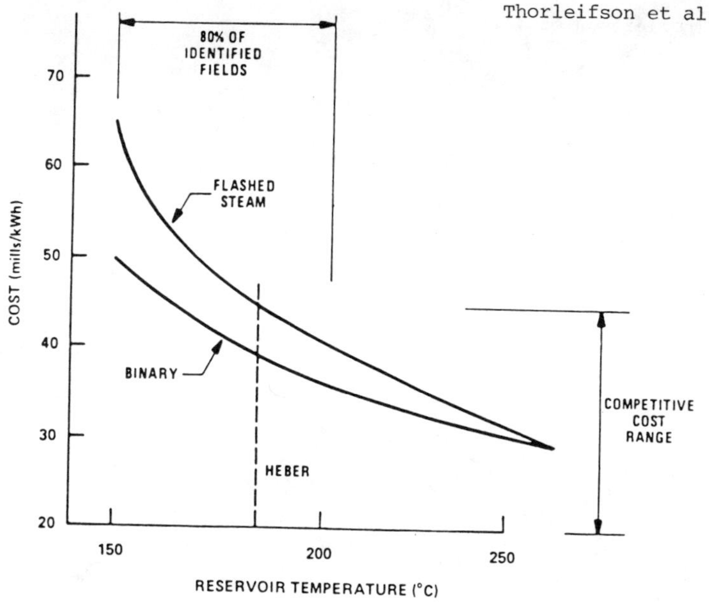

Figure 3. Power Cost vs Reservoir Temperature (Reference 1)

The hydrocarbon supercritical cycle provides for a smooth and continuous phase change to effect the transition from the liquid state to the vapor state with no discontinuities in the heating curve. Figure 5 provides an optimized power curve for the Heber cycle. The two brine temperature profiles noted on the curve reflect start-of-run parameters and, 30 years later, end-of-run brine temeratures.

The spent working fluid is condensed and then pressurized for reintroduction into the heat exchanger and the cycle repeats.

The hydrocarbon handling techniques defined for the Heber process were developed by the petroleum industry. There are, however, some interesting challenges in the Heber process.

HYDROCARBON SIDE INVESTIGATIONS

As previously noted, the binary cycle technology comes to us in well developed form. The magnitude of the hydrocarbon handling equipment required for commercial power production is, however, altogether new. The sizing and configuration of the hydrocarbon turbine and the working loop control system will be a first for manufacturers as well as designers and operators. Discussions on Heber equipment sizing and selection, and on control system definition are provided in other papers.

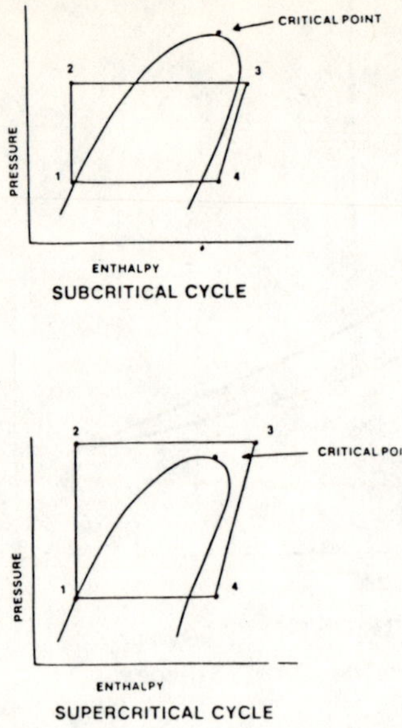

Figure 4: Sub- & Supercritical Cycles

(Reference 2)

Figure 5

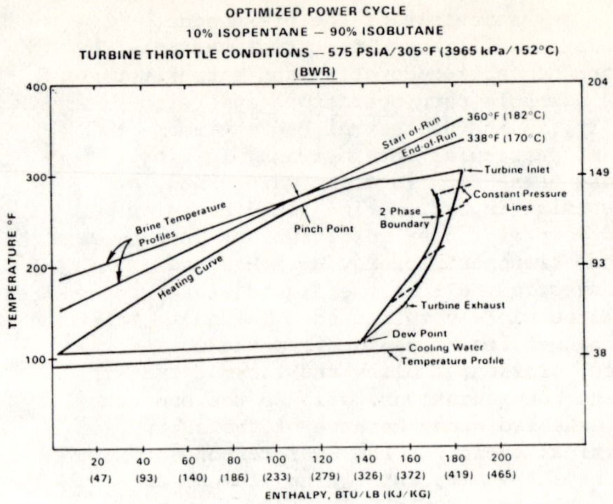

(Reference 3)

Figure 6 shows the supercritical Rankine cycle in more detail. It is evident from this curve that any reduction in hydrocarbon vapor temperature at the turbine inlet, for constant vapor pressure, will result in probable infringement on the phase boundary during the expansion leg. Likewise, an increase in vapor pressure at turbine inlet, for a constant temperature, will have similar results. This operating arena will be the subject of keen interest during the "demonstration" run of the Heber Power Generating Station.

HEBER EXPERIENCE BENEFITS

The preceding discussions, describing both the brine side and hydrocarbon side areas delineated for evaluation and specific subjects for test and investigation, imply a degree of risk in the Heber demonstration power plant undertaking. As with all worthwhile enterprises, some risk is involved. In the case of successfully demonstrating the binary cycle geothermal energy conversion technique, the rewards are exceptional:

- °the binary cycle will make available moderate temperature geothermal energy resources previously considered to be unrecoverable or at best, marginally exploitable. The Heber KGRA alone is conservatively estimated to be capable of producing 500 MWe for a 30 year period.

- °prospective owners/operators of geothermal based power producing facilities who will be facing major industry decisions in the early and mid 1980's will derive direct benefit from the Heber operating experience.

- °the Heber binary cycle process provides an effective alternate approach to environmental preservation for power producers.

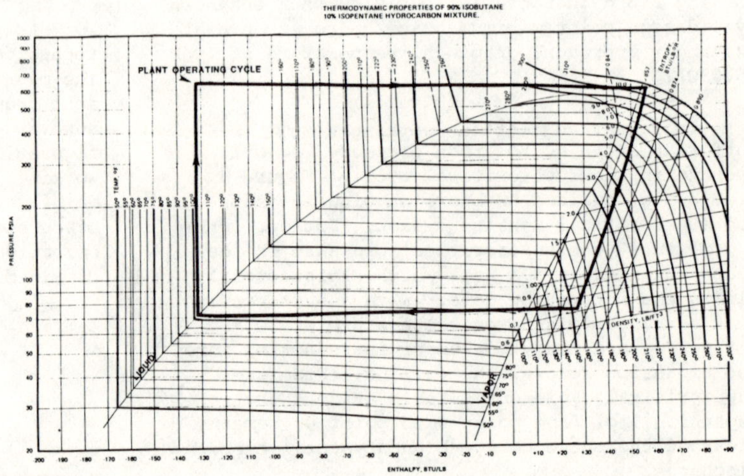

Figure 6

(Reference 4)

- the data base describing geothermal reservoir management techniques will be significantly expanded.

- binary cycle plant operations, on a commercial scale, will be defined.

CONCLUSION

Some minor risk exists in the scaling-up activity of the binary cycle equipment and some unknowns are recognized in the management and control of the Heber reservoir and geothermal fluid. The Heber demonstration power plant, however, is soundly based on existing technology and exhibits excellent potential for successful exploitation of previously marginal or unuseable geothermal resources.

REFERENCES

1. Roberts, V., Oct. 1978, "Rational for the Heber Geothermal Demonstration Power Plant and Need for Broad Utility Participation"

2. Geary, C. H., Oct. 1977, "Binary Cycle" published in Geothermal '77 Seminar Procedings, Bulletin H-38.

3. Unitt, S. G., June 1979, "Heber Geothermal Demonstration Power Plant", EPRI ER-1099 Project 580-2.

4. Fluor Power Services, Inc., May 1982.

Geothermal Resources Council, TRANSACTIONS Vol. 6, October 1982

GEOTHERMAL TURBINES - MATERIALS EVALUATION AND DESIGN ASPECTS

S. Toney, O. Tuncel, and M. Cohen

General Electric Company
Medium Steam Turbine Department
Lynn, Massachusetts 01910

ABSTRACT

A considerable number of on-site material test programs have been conducted at different geothermal sites to provide a data base which permits characterization of materials used, or considered potentially useful, for geothermal turbine applications. These programs have included the exposure of test specimens of different materials to geothermal steam from a steam resource, as at the Geysers, or in flashed steam from a hydrothermal source, such as at Baca, New Mexico (1) and Cerro Prieto, Mexico (2). The results of an on-site test program, performed at the Geysers for more than three years, are presented. Some aspects of geothermal turbine design which take into account materials behavior in geothermal steam are also discussed.

INTRODUCTION

Geothermal steam, whether from a steam-dominated source such as the Geysers and Lardarello, or produced from a steam flashing or separation process from a liquid-dominated geothermal system as in Wairakei and Cerro Prieto, has a markedly greater level of corrosive contaminants than would be advisable in a conventional steam turbine power plant. On the basis of laboratory materials test data and field experience with conventional steam turbine-generator power plants, geothermal steam contaminants are of concern with regard to corrosion and related phenomena such as erosion-corrosion, corrosion fatigue and stress corrosion cracking. Fracture toughness is also an important consideration for geothermal turbines which must operate at relatively low temperatures. Materials selection and the turbine design criteria must take these aspects of materials behavior into account, particularly with regard to allowable stress levels for critical turbine components.

It is recognized that on-site test programs represent a useful contribution towards the development of a better understanding of material suitability for a given environment. However, conditions within test chambers do not fully represent the service environment of a steam turbine; in addition, the tests generally are not designed to simulate concentration of steam contaminants at interfaces between stressed components. Consequently, results so obtained are used for screening purposes and to identify those materials which should be avoided because of the adverse effects of geothermal steam contaminants.

It cannot be over emphasized that turbine design requirements and experience with turbine materials in contaminated steam environments are important factors in the optimal selection of turbine materials for geothermal applications.

Design calculations for geothermal turbine components utilize the data obtained from these field tests and experience with other turbines exposed to contaminated steam. Necessary modifications are made for turbine operations in a geothermal steam environment.

TEST EQUIPMENT

The materials evaluation test, MET, chamber used at the Geysers and other geothermal sites is shown in Figure 1 (1,2). The overall dimensions of the carbon steel test chamber are 610 mm (24 in.) Ø (diameter) x 914 mm (36 in.). Steam at the temperature and pressure of interest flows into the test chamber from the top, where a baffle plate prevents direct impingement of the specimens, to the outlet on the bottom; steam is generally exhausted at atmospheric pressure.

In tests performed at the Geysers, geothermal steam was admitted to the test chamber at temperature and pressure conditions of 340/345°F and 105/110 psig; the steam flow rate was approximately 2700 lbs./hour. The nominal composition of geothermal steam at the Geysers is given in Table I (2).

TABLE I

NOMINAL COMPOSITION OF GEOTHERMAL STEAM, PPM[(a)]

Field	CO_2	H_2S	NH_3	H_2	pH(Condensate)
Geysers	3260	222	194	56	7.7-8.1
New Mexico	44460	445	2	4	4.7
Cerro Prieto	14100	1500	110	NA[(b)]	7-8

NOTE: (a) By Weight
(b) Data not available

TESTS PERFORMED AND MATERIALS

Four basic types of specimens, corrosion discs, U-bends, elastically stressed bent-beams and fatigue, were tested. Since some investigators have reported Ni-

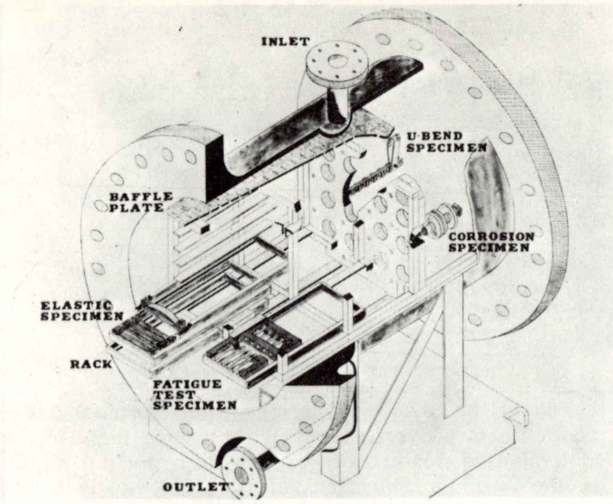

Figure 1. Materials evaluation test chamber used for exposure of turbine alloys to geothermal steam.

containing steels tend to be undesirable for sulfide-contaminated geothermal steam (3), particular attention was given to a determination of the tendency of such steels towards preferential corrosion attack along grain boundaries.

The nominal composition and typical mechanical properties of the turbine materials tested are given in Tables II and III. These alloys are among those used, or considered potentially useful, for steam turbine components such as buckets, diaphragm rings and partitions, packing rings and springs and rotors. Details of the tests performed have previously been reported (1,2). These tests were designed to obtain corrosion rate data, stress corrosion cracking susceptibility and determination of the fatigue strength for pre-pitted specimens. The data are compared with those obtained in test programs performed at New Mexico and Cerro Prieto (1,2).

TABLE II

NOMINAL CHEMICAL COMPOSITION OF TURBINE ALLOYS

Alloy	C	Mn	Cr	Ni	Mo	V	Others
12Cr	0.10	0.55	12.00	0.20	0.15		
12CrMoWv	0.22	0.75	12.00	0.75	1.00	0.25	1.00W
12Cr 4Ni	0.06	0.85	11.75	3.85	0.40		
M152	0.14	0.85	12.00	2.75	1.75	0.30	0.03N
A286	0.06	1.50	14.75	25.50	1.25	0.30	0.30Al,2.1Ti
Inconel X	0.06	0.80	15.50	72.00	—	—	0.70Al,2.5Ti
Monel	0.10	1.00	—	67.00	—	—	30Cu,1.5Fe
Leaded Nickel Brass	—	—	—	12.00	—	—	65Cu,15Zn, 6Pb
Ductile Nickel Cast Iron	2.70	1.90	—	22.50	—	—	2.50Si
NiMoV	0.25	0.55	0.20	2.75	0.50	0.08	
NiCrMoV	0.25	0.40	1.60	3.60	0.45	0.10	
CrMoV	0.30	0.75	1.05	0.20	1.25	0.25	
CrMo	0.15	0.60	2.25	—	1.00	—	
C-Steel	0.60	0.75	—	—	—	—	

TABLE III

HARDNESS AND NOMINAL MECHANICAL PROPERTIES OF TURBINE ALLOYS

Alloy	Hardness Rockwell-A	1000 PSI TS	1000 PSI 0.02% YS	% El	% RA
12Cr	61	105	75	22	70
12CrMoWV	62	120	80	22	55
12CrMoWV	68	145	115	15	45
12Cr-4Ni	68	150	120	18	69
M152	70	170	140	20	68
A286	67	155	105	20	40
Inconel X	69	190	140*	23	42
Monel	43	80	30*	50	65
Ductile Ni Cast Iron	—	60	30*	20	—
NiMoV	60	105	85	22	65
NiCrMoV	60	110	95	24	70
CrMoV	60	115	90	20	60
CrMo	57	95	80	22	62
C-Steel	69	—	—	—	—

* Yield strength at 0.2% offset

RESULTS

Corrosion rate data for the alloys tested are given in Table IV together with results obtained at Cerro Prieto and New Mexico geothermal fields (1,2). These data are based on weight losses after 39 months exposure of the corrosion discs to geothermal steam. The highest corrosion rates were obtained for the Cu-containing alloys, Monel and CuZnNiPb. Corrosion rates for the remaining alloys and the AISI 1060 carbon steel were low, less than 0.007 mm/year. In general, the corrosion rates for alloys tested at the Geysers were consistently below corresponding data obtained at Cerro Prieto and New Mexico, possibly because of the higher condensate pH and the absence of chloride in the Geysers steam.

TABLE IV

CORROSION RATE DATA FOR TURBINE ALLOYS EXPOSED TO GEOTHERMAL STEAM

Alloy	Corrosion Rate, mm/Yr Cerro Prieto	Corrosion Rate, mm/Yr New Mexico	Corrosion Rate, mm/Yr Geysers
12Cr	0.004	0.004	0.002
12Cr-Mo-W-V	0.006	0.003	0.001
12Cr-4Ni	0.004	0.005	0.002
M152	0.004	0.003	0.001
12Cr-Al	0.004	--	0.002
A286	Nil	Nil	0.002
Inconel-X	Nil	--	Nil
Monel	0.099	--	0.049
Ductile Ni Cast Iron	0.016	0.018	0.004
Cu-Zn-Ni-Pb	0.040	0.016	0.010
Ni-Mo-V	0.012	0.018	0.003
Ni-Cr-Mo-V	0.008	0.011	0.001
Cr-Mo-V	0.020	0.014	0.002
2-1/4Cr-1Mo	0.014	0.018	0.002
C-Steel	0.033	0.032	0.007

Metallographic examination of polished-and-etched cross-sections through stress corrosion test specimens containing nickel as a major alloy element did not reveal any evidence to suggest preferential paths of corrosion attack after 42 months exposure. The 12Cr-4Ni martensitic stainless steel and the bainitic NiCrMoV and NiMoV alloys did not exhibit any tendency towards intergranular corrosion.

Of particular interest relative to stress corrosion cracking susceptibility were the turbine rotor alloys. Of the three rotor alloys tested as bent-beams at the Geysers, only the CrMoV alloy was found to contain intergranular stress corrosion cracks. The appearance of such cracks in a bent-beam specimen of CrMoV stressed to 80% of its yield strength and exposed to geothermal steam for 42 months is shown in Figure 2. Shallow intergranular cracking also occurred in CrMoV specimens at lower stress levels.

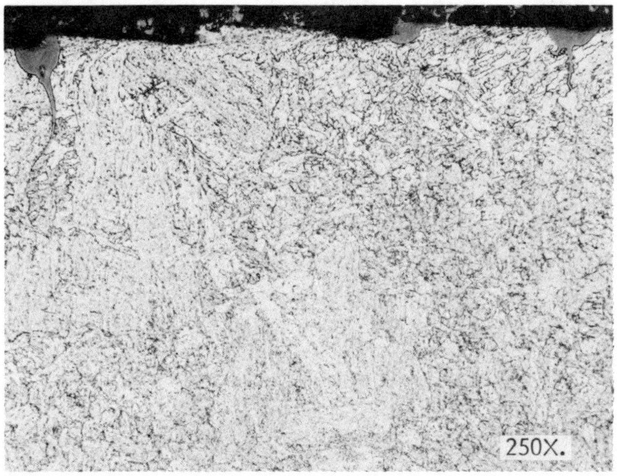

Figure 2. Intergranular stress corrosion cracks in the CrMoV rotor alloy after 42 months exposure at 80% of its yield strength.

While the 12Cr martensitic stainless steel, tempered to hardness levels of the order of 62 Rockwell-A, is resistant to stress corrosion cracking, it becomes highly susceptible when left in the untempered condition. The microscopic appearance of an intergranular stress corrosion crack in a 12Cr heat-affected zone, HAZ, caused by over-critical heating during simulation of an improper erosion-shield braze procedure is shown in Figure 3.

Fatigue endurance limit data for the 12Cr and Modified-12Cr alloys tested after 40 months exposure to geothermal steam are shown in Figure 4 together with fatigue data obtained for specimens pre-pitted in geothermal steam at Cerro Prieto and New Mexico (1,2). The 12Cr martensitic stainless steel, tempered to a hardness of 61 Rockwell-A, exhibited the smallest reduction in the fatigue endurance limit. Compared with the 12Cr alloy, M152 exhibited a larger reduction in its fatigue endurance limit; however, because of its initially high fatigue endurance limit, it remained the stronger alloy under high cycle fatigue test conditions.

DESIGN ASPECTS

In any design, understanding material behavior within the environment in which the component will

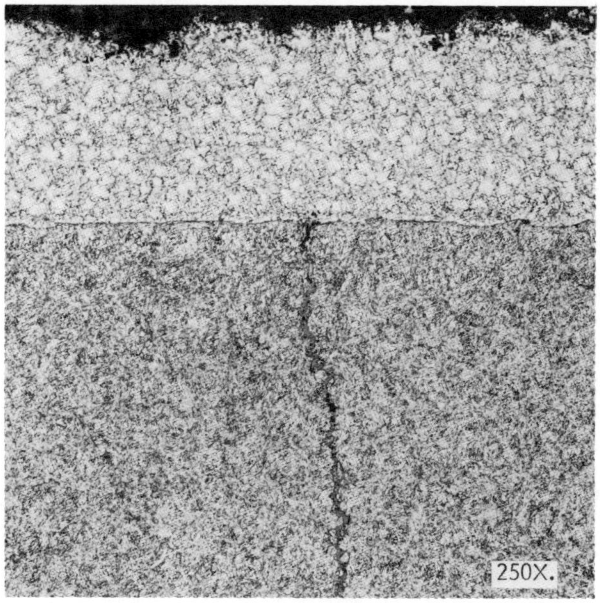

Figure 3. Intergranular stress corrosion crack in 12Cr bucket alloy HAZ after 42 months exposure at 80% of unaffected base metal yield strength.

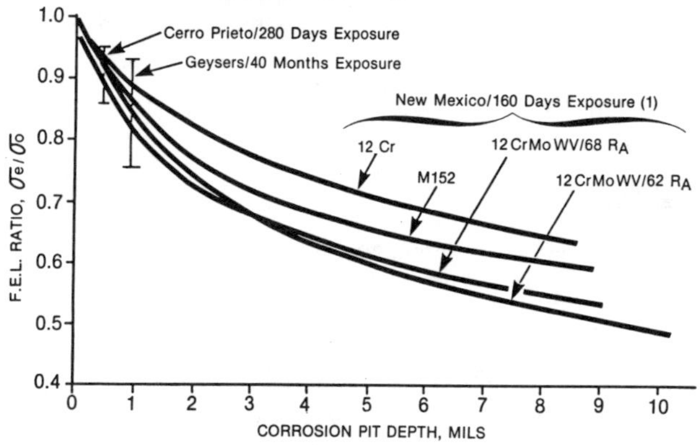

Figure 4. Fatigue Endurance Limit data for specimens tested after exposure to geothermal steam.

operate is essential. Design requirements are related to important material properties such as strength, toughness and fatigue strength. The critical components must be designed to minimize potentially harmful environmental effects identified in the materials test program.

In the wet stages of the steam turbine, which comprises all the stages of geothermal turbines, both erosion and corrosion resistance are emphasized. Chemical attack caused by impurities in the steam may result in pitting or stress corrosion cracking; in addition, the corrosive geothermal steam environment may result in preferential paths of corrosion attack in some materials.

Erosion is a problem in the wet stages of the turbine where the water droplets would impact the leading edges of the bucket at very high relative velocities, such as 1000 ft/sec. In these applications bucket protection is provided either by flame hardening or brazing of 59Co-30Cr-4-1/4W-2Ni-1.2C shields.

Corrosion pitting attack will adversely affect fatigue behavior of turbine alloys and may result in failure of critical components. Most of the common turbine alloys are susceptible to this type of attack. Under these conditions, fatigue strength is reduced considerably depending upon the chemistry of steam and increased corrosion pit depth. This reduction in fatigue strength, as shown in Figure 4, can be as much as fifty percent or even greater. This challenges the design engineers to create a design which can compensate for environmental attacks on the material by improved design capabilities. Such a fatigue strength reduction for geothermal turbine applications necessarily entails rugged steam path designs. For instance, in bucket designs this is done by keeping the net loading on the buckets to a minimum, i.e., low steam bending stress. As always, attention is given to secure operation of buckets free from any vibratory conditions which could otherwise amplify the vibratory stress above the already reduced fatigue strength. Again a rugged steam path design together with the appropriate material selection minimizes the effects of combined corrosion and fatigue.

In addition to material selection that would meet the design aspects of the components, many special features are introduced to give added margin. Some examples of these special design features are, large radius wheel fillets, hand polished wheel and dovetail fillets, shot peening of balance grooves and wheel dovetails, L-shaped bands to protect tenons from erosion, and shot peening of tenon fillets. Reliable service of geothermal steam turbines is achieved primarily through the combination of proper materials and designs which give better operational capability.

DISCUSSION

As previously noted, the selection of materials for low temperature geothermal turbine applications involves the consideration of not only such factors as corrosion rates and stress corrosion cracking susceptibility but also fracture toughness, particularly for the turbine rotor. Since a review of fracture mechanics concepts, vital to turbine design, is beyond the scope of this paper, it will only be noted that the fracture mechanics approach permits an evaluation of crack-like flaws in terms of the stress intensity factor, K, the size and geometry of the flaw and the surrounding stress field. For an internal flaw, catastrophic failure will occur when a critical stress intensity factor, K_{Ic}, for plain-strain conditions, is reached. This critical stress intensity factor is a material property commonly refered to as fracture toughness.

Alloy steels which possess higher fracture toughness are able to sustain larger flaws prior to catastrophic failure. Newhouse and DeForest have summarized K_{Ic} results for turbine and generator rotor steels (4). It is apparent from their data summary that the NiCrMoV rotor alloy possesses the greater fracture toughness. While it is possible to heat treat the CrMoV rotor alloy to improve its fracture toughness, it is not anticipated that a fracture toughness level comparable with that of NiCrMoV will be attainable. The chemical composition of the CrMoV alloy has been designed to achieve a high level of creep-rupture strength for high temperature rotor applications rather than low-temperature toughness. On the basis of available K_{Ic} data, the NiCrMoV alloy is considered to be the preferred turbine rotor alloy for those geothermal steam environments which do not promote the occurrence of preferential paths of corrosion attack. On the basis of results obtained in this investigation, the NiCrMoV alloy did not exhibit preferential paths of corrosion attack and for the given test conditions it was more resistant to stress corrosion cracking than the CrMoV high-temperature rotor alloy.

CONCLUSIONS

1. Reliable service of geothermal steam turbines is achieved primarily through the combination of proper materials and designs which give better operational capability.

2. Corrosion rates for the alloys which do not contain copper are low; in the presence of hydrogen sulfide, copper-containing alloys do not possess the desired corrosion behavior and should be avoided where long-term service is a requirement.

3. Ni-containing alloy steels, exposed up to 42 months in geothermal steam, are not susceptible to preferential intergranular corrosion attack.

4. The 1Cr-1-1/4Mo-1/4V high-temperature turbine rotor alloy was more susceptible to stress corrosion cracking than the 3.6Ni-1.5Cr-0.3Mo-0.1V low-temperature rotor alloy.

5. The most commonly used turbine bucket alloy, 12Cr/Type 403 stainless steel, is resistant to stress corrosion cracking when tempered to optimize the combination of strength and toughness; bucket designs which reduce cyclic stresses are essential.

6. Although fatigue testing of pre-pitted specimens of the 12Cr and Modified-12Cr alloys resulted in fatigue endurance limit reductions of up to about 25%, the design engineer recognizes that this reduction does not fully take into account the corrosion-fatigue behavior of these alloys in wet geothermal steam.

REFERENCES

1. Toney, S., Cohen, M., and Cron, C.J., Sept. 1977, "Metallurgical Evaluation of Materials for Geothermal Power Plant Applications", Geothermal Magazine, Vol. 5, p. 8.

2. Toney, S., Cohen M., Mercade, S., and Manon, A., March 6, 1978, "Observations of Materials Exposed to Geothermal Steam at Cerro Prieto", National Association of Corrosion Engineers, Symposium on Corrosion in Power Generation Equipment, Houston, Texas.

3. Manon, A., May 16, 1975, "Corrosion Problems at the Cerro Prieto Geothermal Project", Workshop on Materials Problems Associated with the Development of Geothermal Energy Systems, El Centro, California.

4. Newhouse, D.L., and DeForest, D.R., May 6-9, 1970, "Meeting Requirements for Larger Generator Rotors...A Metallurgical Challenge", International Forgemasters' Meeting, Terni, Italy.

Geothermal Resources Council, TRANSACTIONS Vol. 6, October 1982

OCCIDENTAL GEOTHERMAL NO. 1
AN INNOVATIVE PLANT DESIGN

Robert E. Tucker
Stone & Webster Engineering Corp., Denver, Colorado

R. F. Ward
Occidental Geothermal, Inc., Bakersfield, California

ABSTRACT

Occidental Geothermal, Inc. (Oxy) is currently in the design phase of a geothermal power plant in The Geysers area of Northern California. Oxy plans to own and operate both the well field and the power plant. This is the first time in The Geysers that a resource developer has elected to produce power from a geothermal resource to be sold under PURPA.

This paper discusses the unique features which make this unit different from others now operating in The Geysers. These features include a twin turbine-generator arrangement incorporating partial-arc admission and sliding pressure operation, a series condenser arrangement, and a condenser bypass.

INTRODUCTION

The Geysers-Calistoga Known Geothermal Resource Area (KGRA) located in Northern California is the world's largest geothermal field, with power generation development approaching 2,000 megawatts (MW) of electricity.

Stone & Webster Engineering Corporation is currently engineering an 80 MW geothermal power plant for Occidental Geothermal, Inc. (Oxy) in The Geysers. Oxy is a resource developer, planning to own and operate the power plant as a small power producer as defined by the Public Utilities Regulatory Policy Act (PURPA) of 1978.

With control of both the well-field steam gathering system and the power plant several innovative features have been incorporated which make the design and operation of this plant different from those currently in operation in The Geysers.

SMALL POWER PRODUCER

The steam production capability of the Oxy lease will support an approximately 80 MW plant. The Oxy Geothermal Plant No. 1 has been designed to produce 80 MW of net (sellable) power, and thus also meets the PURPA fuel use criteria. Under PURPA, utilities are required to interconnect small power production facilities and to purchase generated electricity at avoided (or marginal) cost rates.

Occidental will own the power plant as well as the steam wells, well-field piping, water reinjection system, and access roads to the plant and well pads. For this reason, Oxy will be the first in The Geysers to control both the operation of the steam field and power plant, lending flexibility to both the design and operation of the entire system.

TURBINE CHARACTERISTICS

The Oxy Geothermal Plant No. 1 is designed to use 115 psia saturated steam during normal operation. The turbine operates at 3600 rpm in order to eliminate the "shrink-fit" parts used in the construction of 1800 rpm turbines. It is felt that such construction is not suitable for use in geothermal applications where steam is so corrosive.

The corrosion, erosion, and scale deposition experienced in the operation of turbines in The Geysers may be jointly termed "turbine fouling". Turbine fouling contributes to a loss in generating capability (perhaps as high as 10 percent) and a consequential increase in heat dissipated in the condenser. The increase in condenser duty raises the condenser pressure - a fact that must be considered in the design and optimization of the circulating water system.

The performance of four feasible geothermal turbines is illustrated in Figure 1. They are shown in order of increasing exhaust annulus area. Annulus area may be increased by either longer last stage blades (LSB) or increasing the number of ends. The larger exhaust areas reduce what is known as "exhaust losses" while increasing the capital cost of the turbine and turbine pedestal. In Oxy's case, the value of added performance far outweighed added equipment cost, resulting in the selection of the most efficient turbine available that can be accomodated on the site. Turbines with

GENERATION vs BACKPRESSURE

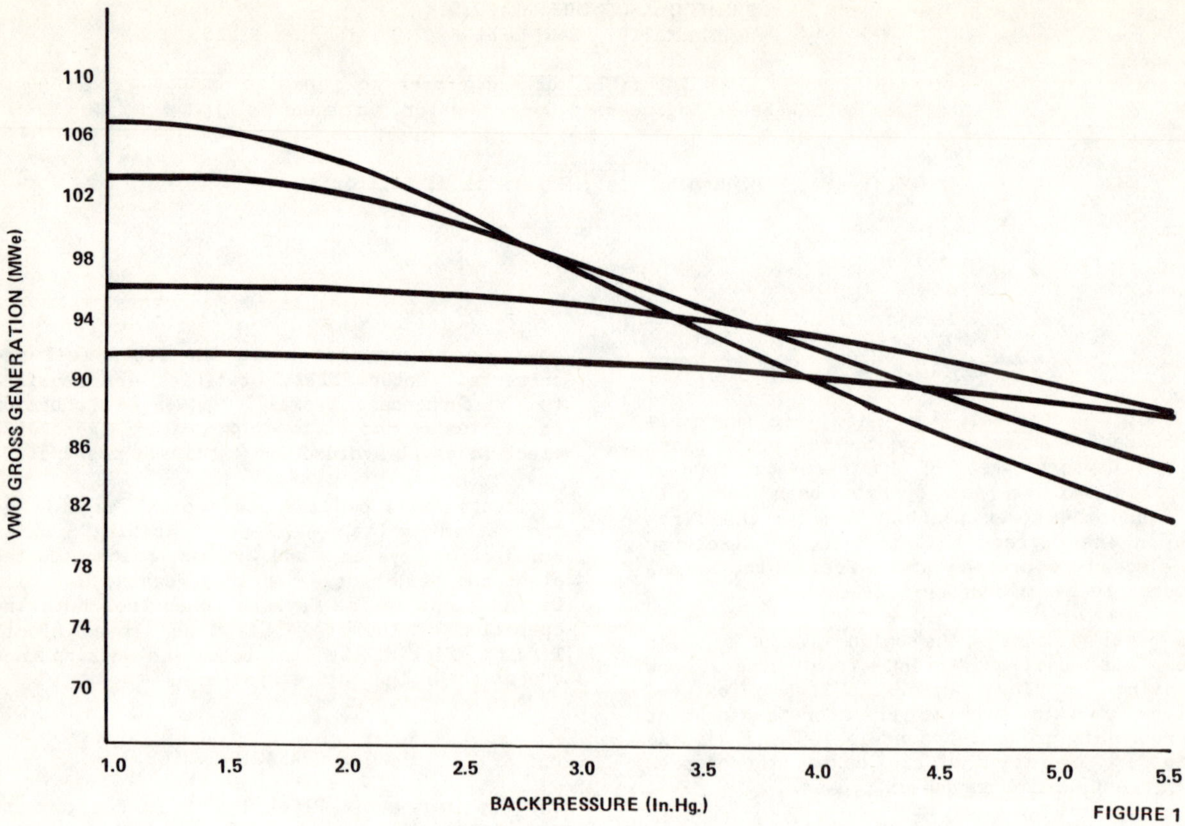

FIGURE 1

larger LSB are supplied for fossil power plant applications; however, it is felt that the higher stress corrosion and erosion due to high tip speeds and high moisture content associated with LSB greater than 25 or 26 inches is unacceptable in geothermal applications. This blade length has an equivalent annulus area of approximately 40 square feet per exhaust end.

Instead of using a four-flow single turbine arrangement (Figure 2) the two low pressure turbine casings have been separated for the Oxy plant, providing two separate turbine-generators. The two turbine arrangement allows approximately 90% generation during one turbine operation, as turbine throttle pressure is allowed to buildup to as high as 165 psia. The higher pressure provides both added flow to the operative turbine and higher available energy per pound of steam, resulting in considerably more power production than that produced by an individual turbine when in the normal two-turbine operation mode. This type of steam pressure flexibility is possible because Oxy supplies its own steam. This increase in sellable power during one turbine operation, brought about by both forced and scheduled outages, more than pays for the increased cost of two separate generators. The twin turbines are arranged along a common axis with generators at opposing ends.

The turbines purchased from Oxy Geothermal No. 1 are the first geothermal turbines to employ nozzle governing/partial arc admission. Each turbine is equipped with two valves, which control steam flow to the nozzle governing stage. The larger valve controls flow to approximately 83 percent of the nozzles, while the smaller controls flow to the remaining 17 percent. During normal (two turbine) operation the larger valves are wide open and the smaller are shut, thereby minimizing the need for throttling and the associated losses. When only one turbine is in operation due to either forced or scheduled outage; the second, smaller valved is opened in order to pass additional flow. The single operating turbine is capable of producing significantly more generation in this mode due to the higher inlet pressure and the increased valve orifice area.

The combination of increasing well-field pressure and the use of partial arc admission maximize the use of the geothermal reservoir during all modes of plant operation.

CONDENSER BYPASS

In order to obtain both the maximum output from a single turbine during one-turbine operation, and provide the maximum use of the steam during two-turbine operation, the Oxy plant incorporates a multi-pressure condenser arrangement. The condenser bypass piping (shown in Figure 3) allows each of Oxy's condensers to be operated while the other is undergoing maintenance. The full capacity of the cooling tower is therefore used to reject the higher single condenser duty required by the increase steam flow from the operating turbine. This configuration provides sufficient cooling in summer operation to keep from exceeding allowable turbine backpressure and to maximize generator output during winter conditions.

During two-turbine operation, the series flow condenser arrangement provides two different operating pressures, providing performance superior to single-pressure operation. Multi-pressure condenser arrangements are standard practice in fossil and nuclear applications when turbines with two or more low pressure casings are used.

TURBINE BYPASS

In an effort to minimize unabated atmospheric emissions of hydrogen sulfide from the plant, the Oxy plant has been designed to include a turbine bypass system. This system will be used on plant startup and shutdown as well as in the case of a full load rejection or turbine trip.

Oxy's system will allow full single turbine flow to be bypassed into one condenser even if the other condenser is shut down, without lifting the safety valves on the main steam line with the resultant unabated stacking. The plant will have a sufficiently large cooling tower basin to allow for a negative water balance which could occur during the turbine bypass mode of operation.

SINGLE FOUR FLOW TURBINE ARRANGEMENT

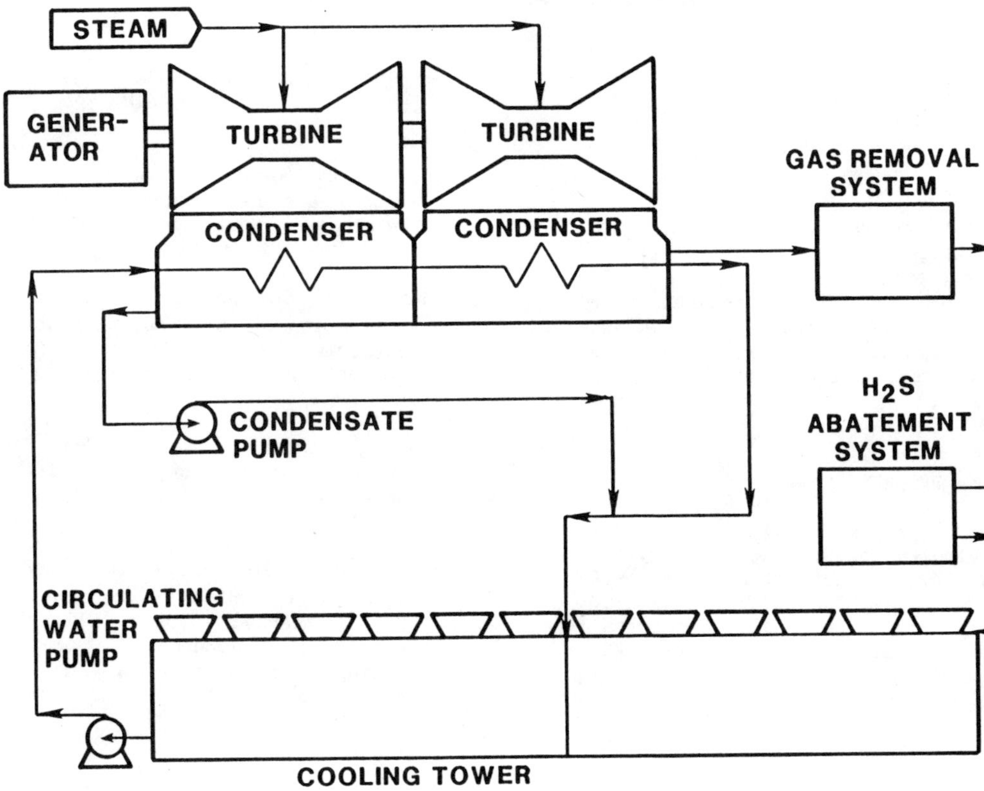

FIGURE 2

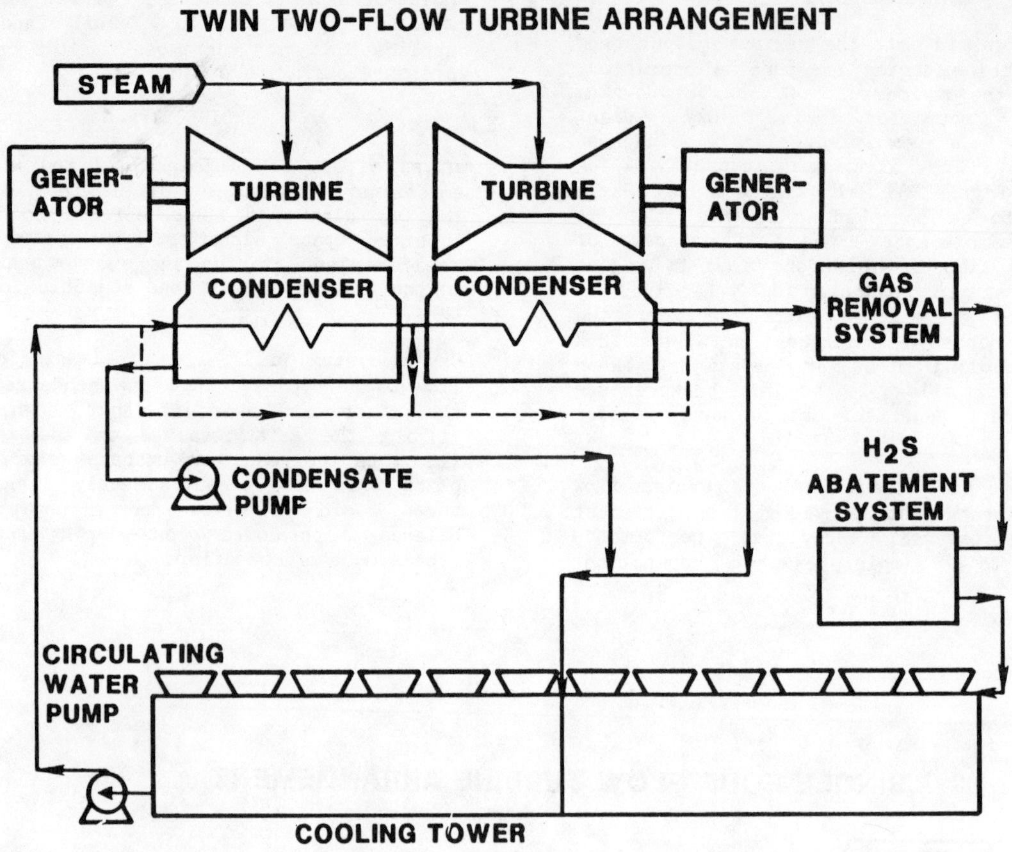

FIGURE 3

SUMMARY

As a small power producer under PURPA, with control of both well-field and plant operation, Occidental's Geothermal Plant No. 1 will incorporate the use of two turbine-generator units, variable turbine throttle pressure and partial arc admission, a multi-pressure condenser arrangement, and condenser and turbine-bypasses.

REFERENCES

Dan, F.J., Herbam, D.E., Kho, S.K., & Kramland, 1975. Development of a Typical Generating Unit at The Geysers Geothermal Project - A Case Study: Second UN Symposium on the Development and Use of Geothermal Resources, San Francisco. Proceedings, Lawrence Berkely Lab., University of California.

Forster, L.L. and Pietruzkiewicz, J., 1978. History and Development of Condensers at The Geysers Geothermal Power Plant, ASME Paper 78-JPGC-Pwr-18.

Occidental Geothermal, Inc., 1981. Application for Certification, Oxy Geothermal Plant No. 1, Bakersfield, California.

Palmer, W.E. and Miller, E.H., 1965. Why Multi-Pressure Surface Condenser Turbine Operation?, American Power Conference, April 27-29, 1965, Chicago, Illnois.

Tucker, R.E., Kleinhans, P.V., and Keilman, L.R., 1980. SMUDGEO #1 - Economic Impacts on Geothermal Power Plant Design, Geothermal Resources Council Annual Meeting, September 9-11, 1980, Salt Lake City, Utah.

DIRECT USE

Geothermal Resources Council, TRANSACTIONS Vol. 6, October 1982

DISTRICT SPACE HEATING FROM A SINGLE GEOTHERMAL WELL, WARREN ESTATES, RENO, NEVADA

David E. Carlson

William E. Nork, Inc.
1026 West First Street
Reno, NV 89503

ABSTRACT

WILLIAM E. NORK, INC. has designed and tested a geothermal well which is capable of meeting the heating requirements for 60 homes located at Warren Estates Subdivision, Reno, Nevada. A 833 feet deep well was completed April 1982. A temperature log of the well showed a bottom-hole temperature in excess of 211°F (99.5°C). A pumping test conducted in May 1982 indicated that the well is capable of yielding 150 gpm indefinitely. In the near future more tests will be conducted involving output potential from a custom-designed downhole heat exchanger.

REGIONAL GEOLOGY AND HISTORY

The Moana (KGRA) Geothermal Resource Area is the geothermal heat source for approximately 150 residences and businesses in southwest Reno (Figure 1). The water tempertures vary from less than 100°F to greater than 200°F, depending upon location.

Subsurface geology in the Moana Area is not terribly complex. It consists of basement rock (heat source) which is made up of Tertiary Age volcanic units that are overlain by a series of diatomaceous siltstones and unconsolidated sands. The diatomaceous siltstone is commonly referred to as "blue clay" by well drillers and "hot well" owners of the Moana Area. Overlying the siltstone and sand layers is alluvium. The thickness of the different units varies depending upon location. Along Lakeside Drive and Manzanita Lane the average well depth to the heat source is between 200 and 300 feet. As one moves westward and to the northwest, well depths increase to as much as 1400 feet.

There are indications, through surface features and surface seismic analysis, that many faults exist throughout the Moana Area. The faults are parallel and trend generally north-south which would be expected of an echelon (step) type faulting, given the basin and range geology which exists in this area. These faults probably convey the hot geothermal water upward to a level where the water meets an impermeable boundary, which may be another fault or the blue clay. Once the boundary or contact has been reached, the water moves laterally and hydraulically downgradient unless tapped by a well.

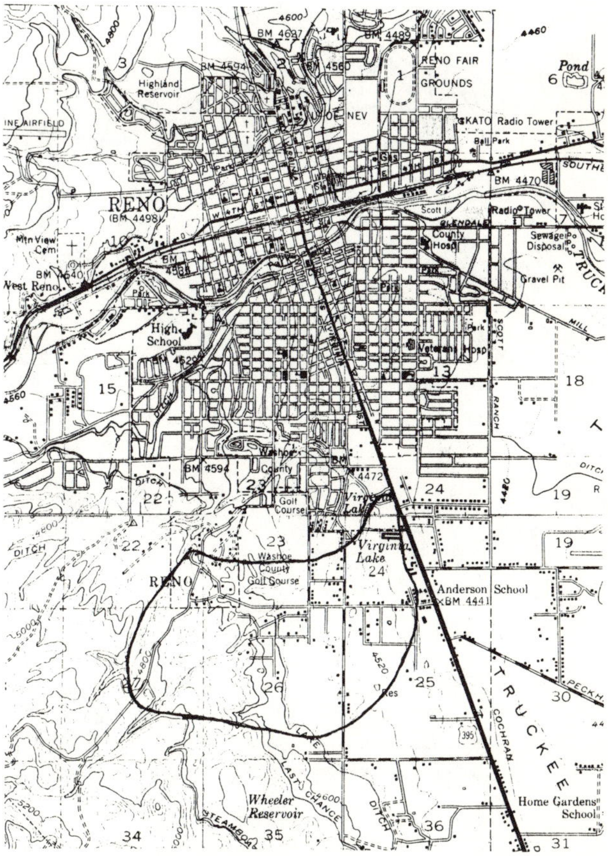

Figure 1. Approximate boundary of Moana geothermal resource.

The total extent of the geothermal resource area is unknown, however, its general limits and temperature yielding capability are somewhat known.

Since the early 1900's the Moana Geothermal Waters have been used for space heating in homes through the use of springs and wells. The majority of the presently used wells are for single family residences and are equipped with downhole heat exchanger loops installed in wells. In addition to the heat exchanger, a small pump is usually installed in each well which is operated periodically to remove the cooler water which may accumulate in the upper portion of the well. In the last few years, single family residences, quasi-municipal (more than one home), and businesses have developed a keen interest in the geothermal potential of this area. The number of wells tapping the geothermal resource is likely to increase in the future as more homes and other users take advantage of this inexpensive heat source.

SITE SELECTION

On February 19, 1980, drilling began on an exploration hole located in the south-central portion of the Warren Estates property (Figure 2).

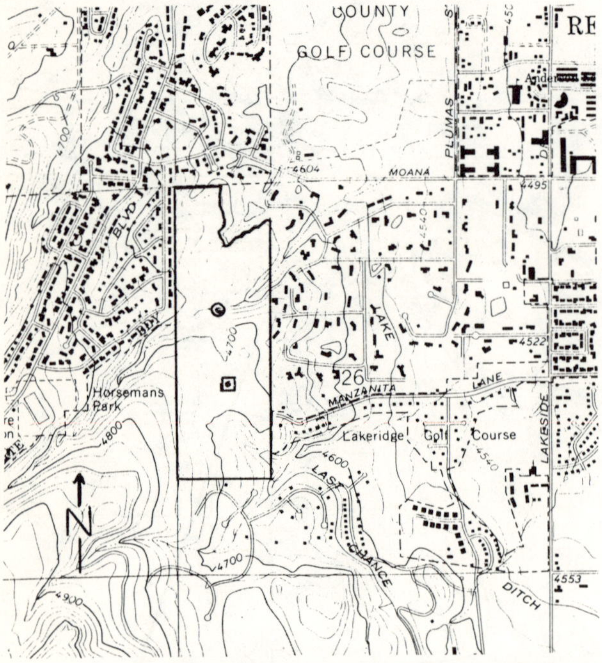

Figure 2. Warren Estates Project boundary with well locations.

⊙ Production well site
⊡ Exploration well site

Site selection was based on surface geology, interpretation of geophysical data, and accessibility by drilling equipment. On April 1, 1980 a 696 feet deep exploration hole was ready for testing. Temperature log of the borehole showed temperatures of up to 204°F. During the pumping test the initial discharge temperature was 204°F, dropping to 176°F midway in the test and rose to 183°F at conclusion of testing. The upper cold water zones were not cemented off, which probably resulted in cold water leakage around the casing after the initial start up. At the conclusion of 17 hours of step-drawdown testing, the total drawdown was 17.39 feet with a final pumpage rate of 91 gpm. Specific capacity was determined to be 5.25 gpm/ft. The pumping test indicated that a properly constructed production well at this site could yield as much as 500 gpm of geothermal water at a temperature of about 200°F, or more.

As the housing market slumped and interest rates rose, a decision was made to develope the subdivision in phases. The initial phase would consist of 60 units located within the northern portion of the property. A production well site was then chosen within the area of the first phase (Figure 2). As in the siting of the exploration hole surface geology and geophysical data were used. In addition a newly developed water temperature contour map (by the author) contributed to the production well site location.

DRILLING AND COMPLETION

The drilling of Warren Estates Production Well No. 1 was commenced on February 19, 1982 by Paul Williams and Sons, Sparks, Nevada. A 12-inch diameter pilot hole was drilled to a depth of 810 feet using the mud-rotary method. A temperature and electric log profile was run to evaluate the capability of the well site to produce the expected and desired geothermal resource. Logging data indicated a geothermal resource potential which would meet the required subdivision needs. The pilot hole was reamed to 20 inches diameter to a depth of 733 feet. 16-inch O.D. blank casing was installed and cemented into place from 733 feet to the ground surface. Through use of the cement, the cold water zones were effectively sealed off from the hot water resource located below 733 feet depth. After the cement had set-up properly, a nominal 15-inch borehole was drilled from 733 feet to 833 feet depth using the air-rotary method. Hot water was encountered at a depth of 791 feet with a flow of approximately 200 gpm. Upon completion of the drilling, a combination 12-inch diameter blank casing and 0.250 slot bird cage mild steel screen was installed in the bottom 100 feet of the hole.

TESTING AND RESULTS

A two-step drawdown test was conducted May 12, 1982, 11:00 a.m. to 2:00 p.m. on Production Well No. 1. The prepumping depth to water was 91.55 feet from the top of the well casing. The initial water temperature as the pump was activated was 150°F. The pumping rate of the first step was 100 gpm; this rate was maintained for one hour. At the end of the first step the drawdown was 22.18 feet, water temperature had risen to 190°F and the specific capacity was 4.51 gpm/ft drawdown. The depth to water was rising slightly during the last 30 minutes of the step. This rise was probably due to thermal expansion of the water as its temperature rose from 150°F to 190°F. At the completion of the first step, the pumping rate was increased to 150 gpm. The second step was two hours in duration, the drawdown at the end of the step was 51.67 feet, water temperature had risen to 205°F, and the specific capacity was 2.90 gpm/ft drawdown. Transmissivity, the ability of the aquifer to transmit water, was determined to be 47,000 gpd/ft.

The exploration hole, located 1,100 feet south of the production well was monitored during the pumping and recovery sequence. A drawdown of approximately one-half foot was observed at the end of the three hours of pumping. Transmissivity of the aquifer at this location was calculated to be 48,000 gpd/ft. A Storage Coefficient of 2.9×10^{-4} was calculated from the data gathered.

A third step was originally planned, in addition to longer pumping duration in each step interval. However, as the pumping water temperature approached 206°F or 207°F, a decision was made to terminate the test once sufficient data had been collected. If a third step of 200 gpm had been attempted, flashing within the well undoubtedly would have occurred.

When the decision was made to stop the test, recovery readings were taken at both the pumping well and exploration hole. At the pumping well, within ten minutes, the water level was five feet higher than the prepumping level. This water rise is attributed in large part to thermal expansion.

Based on the information gathered to-date, including extrapolation of test data, an estimate can be made that Production Well No. 1 could produce 150 gpm for an indefinite period of time with no adverse effect on the existing geothermal resource.

PROJECT POTENTIAL

The present design for the production well is to install a downhole heat exchanger. This heat exchanger will circulate city water initially fed into the closed loop system, throughout the entire subdivision piping system. A review of this design may be necessary after testing of the installed heat exchanger.

Other options which are available include a pumping/injection system. This system would involve either pumping the geothermal water from the production well and putting the same water through the subdivision loop, or installing a surface heat exchanger using geothermal water as the heat source with a separate water supply going through the subdivision loop. In both systems the spent geothermal water will be injected via a separate well. If the former mentioned system were to be used, it is estimated that the entire 60 homes within the first phase of development could be heated. The water chemistry of the production well waters is such that special consideration for possible silica and calcite precipitation must be addressed. The potential silica precipitation could be avoided by maintaining a minimum temperature within the loop system. The potential calcite precipitation may be handled by lowering the pH of the geothermal water in the loop. It is estimated that the use of a surface heat exchanger with a separate subdivision loop could meet the heating requirements of about 30 homes.

A complete well potential may be known once the remaining tests are completed. These tests will be conducted in the coming months and will permit determination of final design for meeting the subdivision heating requirement.

In addition to meeting the winter heating requirement, another possible application of the geothermal heat source would be in conjunction with a Lithium Bromide air conditioning unit. In the summer months when daytime temperature in Reno is around 100°F, the geothermal water could be used to cool the homes in a most efficient manner. Further study is required on the air conditioning aspect before possible adaption into the subdivision system.

Geothermal Resources Council, TRANSACTIONS Vol. 6, October 1982

INDUSTRIAL MARKET FOR GEOTHERMAL ENERGY IN COLORADO

Barbara A. Coe
Richard H. Pearl
Kevin McCarthy

Colorado Geological Survey
1313 Sherman Street, Suite 715
Denver, Colorado 80203

ABSTRACT

The industrial potential of geothermal energy in Colorado has been little discussed in the past but industrial use is an attactive prospect for both consumers and producers/distributors of geothermal energy. This investigation was conducted to help identify the industrial market demand for geothermal energy. The 58 known low- to moderate- temperature geothermal sites in Colorado afford opportunities for industrial applications. The agriculture, manufacturing, travel, and to some extent, the mining industry, have energy requirements that could be easily supplied by geothermal energy.

INTRODUCTION

The industrial potential of geothermal energy in Colorado has been little discussed in the past, probably because Colorado's major industrial area is along the Front Range, whereas the known geothermal sites dot the mountains. Industrial use is, however, an attractive prospect for both consumers and producers/distributors of geothermal energy. The heat load of the industrial consumer is usually relatively high and stable year-round. Rising energy bills stimulate the search for energy alternatives. The energy producer/distributor can more economically supply energy to one or a few major geothermal energy consumers than to numerous small users. Furthermore, he stands to receive a greater return in a shorter time because industrial consumption is relatively steady, compared to more seasonal commercial and residential consumption.

To help identify what the industrial market demand for geothermal energy in Colorado might be, this investigation was conducted as part of the geothermal marketing effort by the Colorado Geological Survey, under their contract with the U.S. Department of Energy. This paper describes the results of that investigation.

The investigation focused upon two major questions: 1. what existing industrial facilities which are co-located with geothermal energy are a potential market for the energy and; 2. what industrial facilities not currently co-located with geothermal energy might reasonably be expected to consider establishing new facilities in areas that have geothermal energy? To answer these questions, processes with temperature requirements of 400°F or under and existing facilities which could use the energy were identified. These included manufacturers, mines, and tourist facilities, namely, hotels/motels and restaurants. Estimates of the energy requirements for the facilities were obtained or calculated. To understand the potential demand for geothermal energy among new industrial users, location criteria, such as raw materials, labor or a local market, transportation facilities or high-level research institutes were investigated. Secondly, assuming that industries currently located in a significant number (i.e., 10) of counties outside major metropolitan areas, are likely to continue locating in such areas, the total energy consumption and average plant consumption by industrial category by county were analyzed. Those with the highest thermal energy requirements and the highest percentage of the statewide thermal energy requirement in non-metro areas were considered to be the best candidates for use of geothermal energy in Colorado.

POTENTIAL INDUSTRIAL/GEOTHERMAL SITES

The geothermal energy in Colorado is sufficient to be seriously considered as an energy resource. Although probably not the entire number, fifty-eight geothermal sites have been positively identified and inventoried. Furthermore, the energy is widely distributed throughout the mountains; 23 of the 63 counties in Colorado have some geothermal surface expressions. Seventeen of these geothermal sites are within 5 miles of communities (Pearl, 1979).

Some research has been completed for all of the inventoried geothermal sites in Colorado. This includes geochemical techniques to estimate subsurface water temperatures. At several locations in Colorado, preliminary exploration has been done, and at some locations wells have been drilled and tested. Colorado's geothermal resources are apparently low to moderate temperature, with good water quality (Barrett and Pearl,

1978). At some of these sites, geothermal energy now provides heat for the agriculture and travel industries but has barely been tapped. Many of those counties in Colorado with geothermal energy are growing rapidly because of energy exploration and development, because of new or existing ski areas, or because of natural beauty, clean air and small populations offering a desirable quality of life. Other counties are economically distressed and seeking economic growth.

POTENTIAL INDUSTRIAL GEOTHERMAL DEMAND

In Colorado, the significant major industrial sectors requiring thermal energy that could be supplied directly by geothermal energy include agriculture, manufacturing, mining and travel. The first three sectors in 1979 were responsible for $586,055,000, $3,341,825,000 and $890,000,000 in 1979 or 2, 13 and 3 percent, respectively, of the total personal income derived in the state. Travel is estimated to have brought in $890,000,000 in 1979 (Colorado Department of Labor and Employment, June, 1980).

Agriculture. Within the agricultural sector, production of crops, especially for heating and cooling, greenhouses, and also for soil warming to extend growing seasons can be assisted with geothermal energy. With wholesale sales of flowers and plants alone in Colorado bringing in $31,000,000, in 1980, this industry represents a significant market opportunity for geothermal energy (Colorado Department of Agriculture, 1981). Agricultural production of livestock is another potential use, for warming feedlots and livestock water and feed, for warming barns, lambing and calving pens and pig farrowing pens, for nurseries and for poultry houses. Geothermal fluid is also ideally suited to clean-up of dairy barns and equipment and animal shelters. All of the counties known to have geothermal energy are a potential market for this application of geothermal energy, with 17 percent of the cattle, 56 percent of the sheep and 10 percent of the hogs and pigs raised statewide being produced in those counties (Colorado Department of Agriculture, 1981). Poultry houses can also be heated to enhance the production of eggs and chicks.

Hatcheries, raceways and ponds can also use geothermal fluid to increase the growth of such aquatic species as catfish, eels, the freshwater shellfish, Malaysian prawns, and even trout. The results of an informal survey of fish markets in the Denver area indicated that Malaysian prawns are much in demand but difficult to obtain, since they are generally shipped from Southeast Asia.

Several sites in Colorado currently use geothermal energy to enhance agricultural production including flower growing and swine and cattle raising (Coe, 1980). Owners of other facilities are interested. In the San Luis Valley, Colorado, a geothermal well was planned for use in mushroom growing. The heat would be used for compost preparation, spawning and production. As noted by Coury and Associates, although the mushroom industry has experienced rapid growth, currently most of the mushrooms consumed in the Denver region of Colorado are imported from Utah or California (1980). The operators of the Colorado Fish Hatchery near Salida, Colorado, indicated interest in using the heat to enhance trout growth (Coe, et al, 1982). As the demand for fish as a source of protein continues to grow, the opportunities for using geothermal fluid in aquaculture will expand as well.

Manufacturing. The manufacturing industry could use geothermal energy for process heat. As indicated by the Solar Energy Research Institute, 47 percent of the 1972 national industrial energy consumption was in the manufacturing sector. Although many manufacturing processes require temperatures higher than those expected from geothermal energy in Colorado, some manufacturing types in every manufacturing category have maximum temperature requirements within the range of low to moderate temperature geothermal energy. Some processes now requiring higher process heat temperatures might be candidates as well, because processes can be changed to use lower temperatures if it is economical. Manufacturing plants co-located in counties with geothermal energy include food processing plants, lumber mills, clothing manufacturers, chemical plants, concrete producers, and printers. As shown on Table 1, the manufacturing plants that could use geothermal energy in those counties (in 1977) were estimated to require over 5,000 billion Btu's of thermal energy (NMEI, 1982).

TABLE 1
Estimated Thermal Energy Requirements of Manufacturers Requiring Low to Moderate Temperatures* in Counties with Geothermal Energy (Billion Btu's)

County	Energy Requirements	County	Energy Requirements
Alamosa	5.05	Grand	11.17
Archuleta	62.31	Gunnison	9.12
Boulder	2,882.26	Jackson	21.69
Chaffee	10.81	La Plata	70.29
Clear Creek	.24	Moffat	8.78
		Ouray	.08
Conejos	7.55	Park	1.46
Delta	126.82	Pitkin	6.32
Dolores	29.77	Pueblo	1,472.02
Eagle	17.45	Routt	8.89
Fremont	285.91	Saguache	1.19
Garfield	17.30	San Miguel	1.78
		Total	5,058.58

*400°F maximum

Source of data: New Mexico Energy Institute, 1982

Additional manufacturing plants seem likely to locate in the future where geothermal energy is available. The analysis of current plant geographic distribution, energy requirements and site location criteria, indicates that of the manufacturing processes most likely to locate near geothermal sites, those that seem likely to have the greatest demand for geothermal energy are:

1. Food and kindred products
2. Lumber and lumber products
3. Chemicals
4. Clay, concrete, gypsum and asbestos

Some of the processes are drying, peeling, canning, washing, freezing, chilling, warming, scalding, blanching, clean-up, cooking, sterilization, pasteurizing and evaporating for food processing; drying and curing for lumber products; heating, drying, digesting, and washing for chemical manufacturing; and curing, heating water, and drying of clay, concrete, and stone products. In 1977, these industries required about 29,110 billion Btu's of thermal energy in Colorado, based on New Mexico Energy Institute estimates (1982). With 61 percent of the food products, 59 percent of the lumber products, 58 percent of the chemicals and 57 percent of the stone, clay and concrete products being produced in non-metropolitan counties, the potential market demand for geothermal energy in these industries is significant indeed.

TABLE 3
Coal Fields and Coal Mines
Co-Located with Geothermal Energy

Counties with Coal Fields and Geothermal Sites	Permitted Coal Mines
Archuleta	X
Boulder	
Delta	X
Dolores	
Fremont	X
Garfield	X
Grand	
Gunnison	X
Jackson	X
La Plata	X
Moffat	X
Ouray	
Park	
Pitkin	X
Routt	X
San Miguel	

Source: Kelso, Ladwig and Sitowitz, 1981.

TABLE 2
Estimated Thermal Energy Requirements
Selected Industries
Colorado and Non-Metro Counties*, 1977
(Billion Btu's)

	Food	Lumber	Chemicals	Stone, Clay, Concrete	Total
Colorado	8,835.6	583.0	7,377.6	12,312.34	29,109.0
Non-Metro Counties*	5,363.6	343.6	4,281.3	7,071.4	17,059.9
Percent Required in Non-Metro Counties	61	59	58	57	

*Excluding Denver Region and El Paso County

Mining. Several mining processes that use temperatures within the specified range were investigated as potential users of geothermal energy, including processing of iron ore, copper concentrate, bituminous coal, potash, phosphate rock and sulpher. Of these processes, one, bituminous coal drying, seemed to be a possible use of geothermal energy in the foreseeable future. Currently, two facilities in Colorado are known to be drying coal, the Dorchester Coal Company near Canon City and Mid-Continent Coke and Coal Company near Redstone. About 30 billion Btu's of thermal energy are required at the Dorchester facility for a heating season (DeLucio, pers. comm.).

In the future, additional coal could be dried in Colorado, either to dewater it to reduce shipping costs or following washing that is done to reduce pollutants. As shown on Table 3, 16 counties with identified geothermal energy have coal fields and 10 have permitted coal mines (Kelso, Ladwig and Sitowitz, 1981).

Travel and Tourism. Travel and tourism is a big business in Colorado. According to the Colorado Office of Tourism, expenditures by tourists totaled $1.6 billion in 1980, $1.8 billion in 1981, and are forecast to be $2.2 billion in 1982. Total travel expenditures in 1980 were more than double the tourism figure or $3.3 billion. The total travel expenditures in all of those counties with known geothermal energy were $839 million in 1980.

Numerous hotels, motels and restaurants to accommodate travelers in Colorado consume large amounts of hot water for cleaning, cooking and laundry as well as energy for space heat. As shown on Table 4, nearly 15,000 hotel and motel rooms and 2807 restaurants are estimated to be located in those counties that have geothermal energy (University of Colorado and City Chambers of Commerce and Ferd Dircks, pers. comm.). Those hotels and motels are estimated to require about 700 million Btu's of thermal energy, or 23 percent of the state-wide total. Restaurants in those counties were estimated to require, in 1977, about 796 billion Btu's or 26 percent of the statewide total (adapted from CERI data, 1979). Since many of the resort areas in Colorado are located near geothermal sites (in fact, many were originally hot spring resorts) they could easily use the energy to satisfy heating requirements.

TABLE 4
Number of Hotel and Motel Rooms and Restaurants in Counties with Identified Geothermal Energy, 1981.

County	Number of Restaurants	Number of Hotel/Motel Rooms
Alamosa	61	400
Archuleta	26	200
Boulder	662	1,000
Chaffee	78	1,055
Clear Creek	79	180
Conejos	31	39
Delta	10	153
Dolores	6	49
Eagle	160	10
Fremont	125	677
Garfield	123	717
Grand	110	8
Gunnison	77	2,093
Jackson	14	42
La Plata	159	1,523
Moffat	57	325
Ouray	25	188
Park	36	N.A.
Pitkin	146	2,744
Pueblo	643	1,386
Routt	116	1,406
Saguache	33	15
San Miguel	30	386
	2,807	14,596
Total Est. Thermal Energy Req. (Billion Btu's)	796	700

Sources: Business Research Division, University of Colorado, 1975, Ferd Dircks, pers. comm., Colorado Energy Research Institute, 1979.

CONCLUSION

Clearly, there is an industrial energy demand in Colorado that could be satisfied by geothermal energy. Particularly in the four sectors identified: agriculture, manufacturing, mining and travel, energy requirements are within the temperature, the geographic, and the technological ranges of geothermal energy. In the agriculture sector, greenhouses, livestock production and aquaculture could benefit from geothermal energy use. In the manufacturing sector, processing of food, lumber products and to some extent chemicals and clay, concrete and stone products are potential markets. In the mining industry, a demand may be found in the coal-drying process. In the travel sector, heating of hot water and space for hotels, motels and restaurants is clearly a strong potential market, given the co-location of tourism and geothermal energy.

ACKNOWLEDGEMENTS

The contribution of information and knowledge by numerous individuals is much appreciated. Most particularly, Evan Metcalf and Jack Olson, of the Colorado Division of Commerce and Development, helped us to better understand industrial location concerns.

Barbara A. Coe, a private Consultant, conducted the research for this paper for the Colorado Geological Survey under their contract with the U.S. Department of Energy.

REFERENCES

Barrett, James K. and Pearl, Richard Howard, An Appraisal of Colorado's Geothermal Resources, Colorado Geological Survey, 1978.

Business Research Division, Graduate School of Business Administration, University of Colorado, Colorado Business Review, February, 1982.

Business Research Division, University of Colorado, Colorado Regional Profiles, 1975.

Coe, Barbara A., Geothermal Energy Potential in the San Luis Valley, Colorado, Colorado Geological Survey, 1980.

Coe, Barbara A.; Dick, Jay D.; Galloway, Michael; Gross, James T.; Meyer, Richard T.; Raskin, Rob; and Zochall, John R.; Study of the Geothermal Potential for Commercial and Industrial Direct Heat Applications in Salida, Colorado, in process.

Colorado Department of Agriculture, Colorado Agricultural Statistics, July, 1981.

Colorado Department of Labor and Employment, Manpower Review, April, 1981.

Colorado Energy Research Institute, Colorado Energy Consumption Patterns, 1979.

Coury and Associates, Inc., Geothermal Agricultural Park Site Demonstration Project, 2nd Quarterly Report, March 20, 1980.

Dircks, Ferd, Colorado Department of Health, personal communication.

DeLucio, Ken, Dorchester Coal Company, personal communication.

Kelso, Bruce S., Ludwig, L. R. and Sitowitz, Linda, Map of Permitted Coal Mines in Colorado as of August 1, 1981, Colorado Geological Survey.

New Mexico Energy Institute, Physical Science Lab, Computer Results, Las Cruces, New Mexico, 1982.

Pearl, Richard H., Colorado's Hydrothermal Resource Base - An Assessment, Colorado Geological Survey, 1979.

Solar Energy Research Institute, End-Use Matching for Solar Industrial Process Heat, January, 1980.

Geothermal Resources Council, TRANSACTIONS Vol. 6, October 1982

EXPERIENCE WITH MINERALS RECOVERY FROM GEOTHERMAL AND OTHER BRINES

Christopher H. Crane

Engineering and Economics Research, Inc.
1951 Kidwell Drive
Vienna Virginia 22180

ABSTRACT

The commercial extraction of minerals from geothermal fluids is not a novelty, although there appear to be no current commercial activities. Products which have been commercially recovered from geothermal fluids include NaCl, $CaCl_2$, H_3BO_3, and NH_3. Sites of significant commercial activity include Larderello, Italy and the Imperial Valley in California. Furthermore, a few pilot-plant systems have been tested for this purpose. Commercial recovery from natural non-geothermal brines has a long history, which may have implications for geothermal energy applications. The technical feasibility and economic benefits of large-scale minerals production in conjunction with geothermal energy exploitation remain speculative. In any case, the uncertainties can be resolved only on a site- and product-specific basis.

INTRODUCTION

The concept of by-product recovery from geothermal fluids used in energy conversion arouses interest due to the large quantities of valuable materials which will pass through geothermal power plants. Although intuitively attractive, the technical and economic feasibility of such combined projects are fraught with uncertainties and will be highly dependent on site-specific variables. It is likely that the availability of large volumes of highly-saline brines from expected geothermal energy development in the Imperial Valley of California will continue to stimulate investigation into by-product recovery. This paper examines the status of geothermal and non-geothermal brine utilization for minerals production, especially as it pertains to geothermal energy systems.

MINERALS FROM GEOTHERMAL FLUIDS

Exhibit 1 identifies significant references, found in the open literature and through personal contacts, to actual or possible recovery operations using geothermal fluids. The most notable commercial operations have been at Larderello, Italy and California's Imperial Valley. As many as 8 compounds may have been recovered from steam at Larderello over the past 80 years, including H_3BO_3, NH_3, NH_4 compounds, and CO_2. In this country, considerable attention has been focussed on the Imperial Valley geothermal resource area as a source of brine minerals since at least the 1930's due to the very high salinity of many brines there. Products from the Imperial Valley have been CO_2 and chlorides of Na, Ca, and K.

Most, if not all, geothermal operations in the U.S. appear to have been terminated by the end of the 1960's due to technical, economic, and market developments. However, the prospect for serious energy development of geothermal resources in the Imperial Valley and consequent availability of tremendous volumes of "waste" brine has rekindled interest in by-product recovery.

Important current R&D by the Bureau of Mines is focussing on recovery of certain heavy metals (Pb, Zn, Ag, and Li) from Imperial Valley brines. BOM has sponsored R&D for two approaches to heavy metals extraction: recovery as sulfides and recovery as hydroxides. Preliminary results of the sulfide precipitation process (investigated by SRI International) indicate that, generally, 80-90% of the Pb and Zn, 30-40% of the Fe, and 20% of the Mn can be recovered as sulfide. This process has, so far, been investigated only in the laboratory.

Lime treatment to precipitate hydroxides (investigated by Hazen Research, Inc.) generally recovered 90-95% of the Pb, Zn, Li, Fe, and Mn. A pilot-plant (now dismantled) was constructed (by WESTEC) near Niland for the purpose of producing these hydroxide precipitates. It was found that further processing of the obtained precipitates to separate mineral values was more difficult than expected: only about 50% of the precipitated Zn could be usefully recovered, for reasons as yet unknown.

Due to the fact that higher Zn extraction was demonstrated with the sulfidation process and a better product for Zn reduction plants was obtained, BOM has not pursued the problem of low Zn recovery in the hydroxide process. BOM plans to concentrate further R&D on the sulfide recovery process, which needs more work to ascertain its technical viability. This effort will finish current Li extraction work and investigate methods to improve sulfide precipitate grain size and to produce a Zn product suitable for industrial Zn reduction.

EXHIBIT 1

STATUS OF MINERALS RECOVERY FROM GEOTHERMAL FLUIDS

SITE	RECOVERED/ RECOVERABLE PRODUCTS	STATUS OF SYSTEM	COMMENTS (REFERENCES)
Salton Sea, California (Magmamax and Woolsley Wells)	Pb Zn Ag Fe Mn	Bench-scale laboratory	• Precipitation as sulfides (4) • U.S. Bureau of Mines/ SRI International
Reykjanes, Iceland	NaCl KCl $CaCl_2$	• Economic feasibility study	(5)
		• Planned process plant	(8)
East Mesa, California	Water (desalinated)	Pilot plants for 3 systems: multistage flash evaporator, vertical tube evaporator, and electrodialysis unit.	• Desalination systems operated by U.S. Bureau of Reclamation (6,10) • $CaCO_3$, $BaSO_4$, and SiO_2 scales produced
Tuscan Pools,?	H_3BO_3	Commercial ?	Recovered since 1700's (2)
Imperial Valley (Niland), California	CO_2	Commercial	From high gas-content wells (2)
Kenya	CO_2	Commercial	For making dry ice (1)
Larderello, Italy	H_3BO_3 NH_3 CO_2 H_2S	Commercial	During 1930's (1,2,3)
	H_3BO_3 Borax Dry ice (CO_2) NH_3 (liquid) $(NH_4)_2CO_3$ NH_4Cl Sodium perborate Manganese borate	Commercial	• 1940, 8,000 tons/yr of boric acid and 4,500 tons/yr borax (8) • Borax production uneconomic by 1971.
Wairakei, New Zealand	As	Pilot Plant	• Iron floc process (2) • Could produce 200 T/yr.
Wairakei and Broadlands, New Zealand	SiO_2	Two techniques at pilot plant	• $CaSiO_3$ precipitation (2) • Fluidized bed SiO_2 precipitation
Cerro Prieto, Mexico	KCl	Pilot plant	• KCl-NaCl salts from evaporation ponds separated by flotation (2)
Iceland	Heavy water	"Investigation"	• From steam (2) • Development unlikely due to world oversupply
Danakil Depression, Ethiopia	Mg	Feasibility study	• Mg concentration 100X seawater (1) • Believed to be profitable
El Tatio, Chile	Li Cs Rb	Feasibility study	(1)
	H_3BO_3 KOH	Theorized	
Salton Sea, California	CO_2	Presume commercial	• Used for refrigeration from 1932-1954 (7)
	Other minerals including KCl		• Serious but unsuccessful effort in early 1960's (8)
Imperial Valley, California	NaCl $CaCl_2$ KCl	Apparently commercial	• Morton International 1964 - 67 (8) • Basically solar pond evaporation • 95% of product shipped was KCl
Imperial Valley, California	NaCl KCl $CaCl_2$	Pilot plant and process design	• Union Oil research (8) • KCl primary product • Sequential precipitation (NaCl, KCl, $CaCl_2$) using evaporators and crystallizers
Wairakei, New Zealand	NaCl KCl LiCl		• Ion exchange process (8) • Multiple-effect evaporation process
	NaCl KCl Li_2CO_3 Borax, Rb, Cs } possible	Laboratory and field tests	• Several design problems judged resolvable, but market prices made process uneconomic (8)
Imperial Valley, California	Pb Zn Li Fe Mn Al?	Pilot plant (recently dismantled) at Niland	• Precipitation as hydroxides (9) • U.S. Bureau of Mines/ Hazen Research/WESTEC

The Bureau of Reclamation has pilot-plant tested 3 desalination technologies for geothermal brine at East Mesa for producing fresh water to augment the Colorado River flow. All 3 technologies (multi-stage flash evaporator, vertical tube evaporator, and a high-temperature electrodialysis unit) were successful, although the multi-stage flash system may be preferred due to its relative simplicity and fewer plugging problems. This field work (conducted by Bechtel) was terminated in 1977-78 apparently because it was concluded that even commercial-scale operations could not significantly improve the local fresh water supply. The implications of desalination for minerals recovery from residuals has not been examined by BOR.

MINERALS FROM NON-GEOTHERMAL BRINES

Exhibit 2 shows that commercial recovery of minerals from non-geothermal brines (surface and subsurface inland brines and seawater) is well established, especially around Searles Lake, CA and Great Salt Lake, UT. Multi-product recovery processes (particularly simple chlorides) characterize operations in these areas. The lack of reported brine characteristics, process technology descriptions, comparative economic parameters, and other factors in many instances hampers evaluation of potential technology transfer to geothermal applications. Evaporation (virtually always solar) is a common element of many of these processes, but evaporation-based recovery techniques will be restricted in most, if not all, geothermal power projects due to the need to inject spent brine.

CONCLUSIONS

The feasibility of minerals recovery in tandem with geothermal energy production has not been demonstrated in the U.S. However, commercially-successful recovery from geothermal fluids, seawater, and other natural brines has occurred in the U.S. and abroad for many years. Recovery of chlorides and perhaps some other compounds is unlikely to be practicable due to injection-related constraints on brine evaporation. R&D in progress at BOM may lead to one or possibly two chemical precipitation methods for recovering Pb, Zn, Ag, and Li from geobrines. The Imperial Valley appears to be a good prospect for minerals recovery. However, the costs of recovery cannot be meaningfully calculated at this time in order to determine the net economic benefit, if any, to specific geothermal energy projects.

REFERENCES

(1) Barnea, J., Nov. 26 - Dec. 5, 1978, "Geothermal Minerals - The Neglected Minerals", The Future of Small Scale Mining, First International Conference, UNITAR/SEPAFIN, Jurica, Qro., Mexico.

(2) Kingston Reynolds Thom and Allardice, Ltd. (New Zealand), Oct. 1981, Preliminary Survey of Geothermal By-Products, P2430.

(3) Collins, A.G., Nov. 26 - Dec. 5, 1978, "Mineral Brines and Their Utilization Through Small-Scale Mining in the USA", The Future of Small Scale Mining, First International Conference, UNITAR/SEPAFIN, Jurica, Qro., Mexico.

(4) Farley, E.P., et.al., Dec. 1980, Recovery of Heavy Metals From High Salinity Geothermal Brine, U.S. Bureau of Mines, OFR 91-81, NTIS Report No. PB81-222218.

(5) Bjornsson, S., et.al., Nov. 13, 1982, "Economic Evaluation of Reykjanes Thermal Brine Area, Iceland", J. Am. Assoc. Pet. Geol. Bull., Vol. 56, No. 12.

(6) Lindemuth, T.E., et.al., June 27-28, 1977, "Experience in Scale Control With East Mesa Geothermal Brine", 1977 SPE-AIME International Symposium on Oilfield and Geothermal Chemistry, LaJolla, California, (Paper No. SPE 6605).

(7) Howard, J.H., and Towse, D., May 1977, "History of Exploitation at Salton Sea Geothermal Field, California", Am. Assoc. Pet. Geol. Bull., Vol. 61, No. 5, p. 796.

(8) Blake, R.L., Dec. 1974, Extracting Minerals From Geothermal Brines: A Literature Study, U.S. Bureau of Mines Information Circular 8638, NTIS Report No. PB-240681. (This report references 67 other sources of information).

(9) Schultze, Lawrence, April 1982, U.S. Bureau of Mines, Reno Research Center, personal communication.

(10) Fernelius, Wayne, April 1982, U.S. Bureau of Reclamation, Policy Planning Staff, Washington, D.C., personal communication.

(11) Berthold, C.E., et.al., Feb. 1975, Process Technology for Recovering Geothermal Brine Minerals, U.S. Bureau of Mines Report No. OFR 35-75, NTIS Report No. PB-241 867.

EXHIBIT 2

STATUS OF MINERALS RECOVERY FROM NON-GEOTHERMAL BRINES

SITE	RECOVERED/ RECOVERABLE PRODUCTS	STATUS	COMMENTS (REFERENCES)
Woodward, OK	I_2	Commercial 1	$2x10^6$ lb/yr (3)
U.S.			• Oil well brines (11)
Michigan			• Underground brines (11)
El Dorado, AK	Br_2	Commercial 2	• Brine from Smackover Formation (3)
U.S.		Commercial till '69 3	• Seawater (11)
Israel		4	• Dead Sea Bromine Co. (11) • In conjunction with $MgCl_2$
Clayton Valley, NV	Li	Commercial 5	(3)
Searles Lake, CA		6	• Salt brine lakes (2)
Smackover Formation, TX/AK		Research 7	• Oil field brines (3)
Trona, CA		Commercial 8	• Li_2CO_3 as co-product (11)
Silver Peak, NV		Commercial 9	• Li_2CO_3 from evaporation and chemical treatment (11)
Great Salt Lake, UT		Planned 10	(11)
• Ogden, UT • Ludington, MI • Freeport, TX • Port St. Joe, FL	Mg Others	Commercial 11	• Lake brines, well brines, and seawater (3)
California	NaCl	Commercial 12	• Number of companies (11) • Solar evaporation of seawater • 35-40 tons/acre
Great Salt Lake, UT			• Solar evaporation (11) • 300-350 tons/acre
Wendover, UT			• Kaisar Chemical (11) • By-product of KCl • 4% of 1.2m tons/yr. sold
Carlsbad, NM			• By-product of KCl (11) • Most discarded because market too distant
Latin America, Australia, Asian & African countries			• Solar evaporation of seawater and brines is major NaCl source (11)
Europe			• By-product of KCl (11)
Trona, CA	KCl	Commercial 13	• Kerr-McGee Chemical (11) • Brine under Searles Lake • Evaporation, crystallization, solid-liquid separation • Other products recovered (see items #8 & 18)
Wendover, UT		Commercial 14	• Kaisar Aluminum & Chemical (11) • Brine under Great Salt Lake Desert • Evaporation, crystallization, flotation
Great Salt Lake, UT	KCl	Commercial ? 15	• 2 operations (Great Salt Lake Minerals & Chemical; NL Industries) - KCl not major product (11) • Evaporation-cystallization
Regina, Saskatchewan		Commercial 16	• Kalium Chemical (11) • Solution mining, crystallization
Sodom, Israel (Dead Sea)		Commercial 17	• Evaporation, solids processing by decomposition/ crystallization or flotation (11) • By-products Br_2, MgO, HCl
Searles Lake, CA	Na_2CO_3	Commercial 18	2 processes (Kerr-McGee): (11) • Brine carbonation with boiler flue gas, then drying • By-product of KCl (see item #13)
		Commercial 19	• West End Chemical (11) • Similar to item #18
		Pilot Plant 20	• Occidental Chemical (11) • Evaporation
U.S.	$MgCl_2$ (Produced primarily as inter-mediate in production of Mg or MgO)	Commercial 21	• Dow Chemical (11) • Seawater
Norway		Commercial 22	• Norsk-Hydro Electrisk (11) • Seawater • Process similar to item #21
San Diego, CA		Commercial 23	• FMC Corporation (11) • Artificial evaporation of salt-work bitterns
Great Salt Lake, UT		Commercial ? 24	• NL Industries (11) • From residual brine from KCl production (see item #15)
Midland, MI		Commercial 25	• Dow Chemical (11) • Evaporation and other processing of underground brine to produce $MgCl_2$, $CaCl_2$, and Br_2
Bristol Lake, CA	$CaCl_2$	Commercial 26	• From brine seeping into halite mines (11) • Evaporation and other processes
Searles Lake, CA	B (as borax)	Commercial 27	• Kerr-McGee (11) • Co-product with Na_2CO_3 (see item #18) • Crystallization, residual solvent extraction
		28	• Stauffer Chemical (West End Chemical) (11) • From filtrate from $NaHCO_3$ cake in Na_2CO_3 production (see item #19) • Seeding and solid-liquid separation

Geothermal Resources Council, TRANSACTIONS Vol. 6, October 1982

WORLD USES OF LOW-TEMPERATURE (<150°C) GEOTHERMAL RESOURCES IN 1980

J. S. Gudmundsson* and G. Palmason

Geothermal Division, Orkustofnun
(Iceland Energy Authority), Reykjavik

ABSTRACT

The results of a recent world survey of low-temperature geothermal energy utilization are reported. It was found that 44 countries have low-temperature geothermal resources, 24 countries are involved in exploration, 12 countries are utilizing low-temperature geothermal energy and 9 countries have carried out assessment studies. The survey showed that about 8700 MW-thermal (above 15°C reference temperature) were installed at the end of 1980.

INTRODUCTION

In recent years the main goal of geothermal exploration and development projects in the world has been the generation of electric power. With few exceptions these projects have been undertaken in areas of known surface manifestations typical for high-temperature resources. At the same time an increased interest has been shown in the less noticeable low-temperature resources, particularly in countries where space heating is necessary for a good part of each year. In a few countries low-temperature resources have been utilized for decades on a commercial scale while elsewhere geothermal waters have traditionally been considered little more than a novelty. Recently, however, several countries have come to appreciate the great energy potential of low-temperature geothermal resources and are translating their interest into exploration and field developments for direct use applications.

In August 1981 the United Nations convened a Conference on New and Renewable Sources of Energy in Nairobi, Kenya. To prepare for this conference its preparatory committee formed several panels of experts to address issues concerning the principal renewable energy sources, including geothermal energy. The Geothermal Division of Orkustofnun (Iceland Energy Authority) was asked by this panel to carry out a World Survey of Low-Temperature Geothermal Energy Utilization (Gudmundsson & Palmason 1981). The survey was carried out in 1980 and in early 1981 the detailed results were sent for comment to all who contributed information. At the same time they were asked to update the information to the status of geothermal utilization in their country at the end of 1980. The main purpose of this paper is to report the findings of the world survey and the updated information.

*Presently, Petroleum Engineering Department, Stanford University, Stanford, California 94305.

A survey similar to the one reported here was carried out for high-temperature geothermal resources. This was done by the Ministry of Works and Development in New Zealand (Bolton 1981). For the purpose of the two surveys the technical panel decided to use 180°C as the temperature limit dividing high- and low-temperature geothermal fields. The utilization reported in the low-temperature survey, however, was mainly in fields with much lower reservoir temperatures, usually below 100°C. The reservoir temperature limit dividing high- and low-temperature geothermal fields in Iceland has traditionally been 150°C where most high-temperature fields are above 200°C and low-temperature fields below 100°C. There are important generic differences between the two types, but these will not be discussed here. In the United States the temperature limit of 150°C is now used for differentiating between high- and intermediate-temperature geothermal fields. In view of the above it seems appropriate to use 150°C as the arbitrary reservoir temperature defining the scope of the survey reported in this paper. It should be noted that the utilization reported here does not include any direct or nonelectrical uses in high-temperature fields. The temperature limit of 150°C represents the resource temperature but not the temperature of utilization.

PREVIOUS REVIEWS

A widely quoted review reference on non-electrical applications within various countries is that of Howard (1975a). It reported that in 1975 about 400 MW-thermal were used for residential and commercial applications, 5500 MW-thermal in agricultural applications and 200 MW-thermal in industrial applications, mainly from high temperature fields. These values were based on a 35°C discharge temperature. About 5000 MW-thermal for agricultural applications was reported as being associated with a large acreage of greenhouses in the Soviet Union. However, in the survey reported here (Gudmundsson & Palmason 1981) it was shown that the above 5000 MW-thermal was grossly overestimated. At the same time all the low-temperature waters used in Japan (estimated as 4475 MW-thermal above 15°C in the present survey) were omitted by Howard (1975a). A more detailed presentation of the same study is that of Howard (1975b). Barbier & Fanelli (1975) updated and expanded the above study and showed that the total capacity associated with nonelectrical application was about 6200 MW-thermal. By assuming that nonelectrical applications have an efficiency of 85% this value was raised to 7300 MW-thermal. Again, the Barbier & Fanelli (1977) study was faulted because of the anomalous high value for the Soviet Union.

The most recent available review of geothermal energy is that of Fanelli & Taffi (1980). Although they give no new information on the amount of geothermal energy used in direct use or nonelectrical applications, they review the status of exploration and exploitation in the world. Other relevant material includes an extensive list of thermal springs of the world given by Waring et al. (1965).

METHOD OF SURVEY

The survey was conducted by sending a questionnaire to all the countries known or thought to have low-temperature geothermal resources. The information requested was the following:

"1. Present utilization. A list showing uses of low-temperature geothermal energy in various fields. Specify flowrates and temperatures (entering and leaving installation) of the geothermal waters used and give information about the type of installation (district heating, greenhouses, baths, etc.) utilizing the waters. Information on the available (but not used today) geothermal waters should also be included.

"2. Past and future utilization. A list showing uses of low-temperature waters in recent years and scheduled expansion (where financing is available or being arranged) of present and future uses in the next few years. Specify flowrates and temperatures (entering and leaving installation) of the geothermal waters and give information about the type of installation utilizing the waters.

"3. Exploration projects. A list showing national and local programs and activities in the exploration of low-temperature fields which are still in the predrilling stage. Indicate present status of exploration.

"4. Regional assessment. A list showing resource base or total potential of low-temperature fields in your country if such assessments have been made."

The questionnaire was sent to 41 countries and replies were received from 29. Information was gathered from 7 other countries through meetings. The literature was consulted for data on 10 additional countries and where no reply had been received. The total number of countries was 58. The data gathering exercise proved on the whole to be successful although almost no information was received on past and future utilization. The replies reported mainly on present utilization and exploration.

In the survey the main countries utilizing low-temperature geothermal energy were dealt with in more detail than some others. And because new extensive data were being presented for several of these countries it was decided to write special notes on China (People's Republic), France, Hungary, Iceland (Gudmundsson 1982), Japan and the U.S.S.R. These include sections on geography, utilization, exploration and assessment. A short bibliography was also included. Information on other countries than the six mentioned above was written as individual sheets in the same manner but more concisely. Over 100 references were consulted in the survey (Gudmundsson & Palmason 1981). After the survey was completed, updated information was received from China (Xin Kuide), Hungary (Tibor Boldizsar), U.S.A. (Marshall Reed) and Yugoslavia (Neven Miosic). At the same time the results of the survey were confirmed by many of the other countries. The information presented here should show the status at the end of 1980.

RESULTS OF SURVEY

An overall view of the results of the survey is presented in Table 1. It shows in alphabetical order the 58 countries covered by the survey. The table shows what countries are known to have low-temperature geothermal resources and indicates the presence or absence of utilization, exploration and assessment within the following restrictions or definitions:

Utilization. This term is used if low-temperature geothermal waters are used for other purposes than simple bathing and washing at natural hot springs. This restriction is necessary to demonstrate which countries have put effort into drilling and distribution systems to serve a thermal market. Of the 44 countries known to have low-temperature resources, only 12 are reported to utilize them.

Exploration. This term applies if other measurements than heat flow have been carried out in prospective geothermal fields. The contention here is that only explicit geothermal prospecting or drilling constitutes exploration. Table 1 shows that 24 countries are involved in exploration work for low-temperature geothermal resources.

Assessment. This term is more difficult to define than utilization and exploration. For the purpose of the survey an assessment was said to have been carried out in a country if the hot water resources of a major geothermal region had been estimated. The table shows that 9 countries had carried out such assessment studies.

In the questionnaire, countries were primarily asked to report the flowrate and inlet/outlet temperatures of geothermal water presently being used. These data were to give the useful thermal power associated with all known geothermal installations and in turn the thermal energy used, if the load

TABLE 1.
An overview of the countries included in the world survey of low-temperature geothermal energy utilization (Gudmundsson and Palmason, 1981).

	Countries	Resources	Utilization	Exploration	Assessment
1	Algeria	+	-	-	-
2	Argentina	+	-	+	.
3	Australia	+	-	-	-
4	Austria	+	+	+	-
5	Bolivia	+	-	.	.
6	Brazil	+	-	.	.
7	Canada	+	-	-	-
8	China	+	+	+	.
9	Colombia	+	-	+	-
10	Costa Rica	.	.	.	.
11	Czechoslovakia	+	+	+	+
12	Denmark	-	-	-	-
13	Djibouti	.	.	.	.
14	Ecuador	+	-	.	.
15	Egypt	+	-	.	.
16	Eire	+	-	+	.
17	El Salvador	.	.	.	.
18	Fiji	+	-	+	-
19	France	+	+	+	+
20	Germany, East	.	-	-	-
21	Germany, West	+	.	+	.
22	Greece	+	-	-	-
23	Guatemala	.	.	.	.
24	Haiti	.	.	.	.
25	Honduras	.	.	.	.
26	Hungary	+	+	+	+
27	Iceland	+	+	+	+
28	India	+	-	+	.
29	Indonesia	+	-	-	-
30	Israel	+	-	.	.
31	Italy	+	+	+	+
32	Jamaica	.	.	.	.
33	Japan	+	+	+	+
34	Kenya	.	.	.	.
35	Korea, North	+	.	+	-
36	Korea, South	+	-	.	.
37	Mexico	+	-	-	-
38	Netherlands	+	-	+	-
39	New Zealand	+	-	-	-
40	Nicaragua	+	-	-	.
41	Nigeria	.	.	.	.
42	Panama	.	.	.	.
43	Peru	+	-	.	.
44	Philippines	.	.	.	.
45	Poland	+	-	+	-
46	Romania	+	+	+	+
47	Solomon Islands	+	-	-	-
48	Sweden	+	-	-	-
49	Switzerland	+	-	+	.
50	Tanzania	+	-	.	.
51	Thailand	+	-	+	.
52	Turkey	+	-	-	-
53	Uganda	+	.	.	.
54	USSR	+	+	+	+
55	UK	+	-	+	-
56	USA	+	+	+	+
57	Venezuela	.	.	.	.
58	Yugoslavia	+	+	+	.
	Total	44 +	12 +	24 +	9 +

+ = Known (Present)   - = Unknown (Absent)   . = No information

factor was known. It would then take a simple calculation to arrive at the tonnes oil/coal equivalent. When the replies were received, however, it became clear that this kind of information was not available except in rare instances. It could therefore not form a common basis for a worldwide survey.

Typical data was the flowrate and temperature from one well or a field but not what was being used in installations. It was therefore found convenient to report the data in terms of thermal power available for utilization. And because most of the thermal waters are reported used, it also represents the installed thermal capacity associated with low-temperature geothermal resource utilization.

Of great interest in the survey was the total installed thermal power of the countries utilizing geothermal energy in the world. Table 2 (column A) shows the total installed capacity of the 12 countries listed in Table 1 as having utilization other than hot springs used only for bathing and washing. Japan is clearly the largest user of low-temperature waters with 4475 MW-thermal installed above 15°C. Table 2 (column A) shows also that Hungary is the second largest user with 1540 MW-thermal and Iceland third with 1127 MW-thermal. China, Italy, the United States and the Soviet Union form a group of countries using hundreds of MW-thermal while Czechoslavakia, France, Romania and Yugoslavia use tens of MW-thermal. A surprising result of the survey was the large amount of geothermal water used for bathing in Japan. Information from several sources confirmed that the bulk of low-termperature waters produced there are used for bathing. It was estimated that 81 MW-thermal are used in space and and water heating, greenhouses and animal husbandry. Also, an examination of the data from Hungary showed that 38% of the installed thermal power was used for bathing purposes. It was therefore decided to estimate the installed geothermal power in the world not associated with bathing. The result is shown in Table 2 (column B) and amounts to 3468 MW-thermal. About 60% of the total world uses of low-temperature geothermal resources is therefore used for bathing, balneology and swimming.

It is opportune at this point to explain why the temperature limit of 15°C was used in the survey. When geothermal water cools down at ambient conditions it gradually approaches the temperature of its surroundings. It follows that the maximum amount of energy that can be extracted from geothermal water corresponds to its cooling to the average ambient temperature. The average **annual** temperature at the surface of the earth is reported to be close 15°C. This temperature was therefore used when considering geothermal energy on a worldwide basis. The average annual temperature in each country or region would be a better reference temperature in future and more detailed surveys. A remaining problem is to estimate the useful thermal power and energy associated with geothermal applications in the absence of actual measurements. From the data used in the survey it appears that a reference temperature of 35°C would be appropriate for most low-temperature applications.

TABLE 2
World uses of low-temperature geothermal resources in 1980, expressed as thermal power above 15°C.

Country	Thermal power (MW)			
	A	(%)	B	(%)
Austria	5	(0.1)	5	(0.1)
China	346	(4.0)	329	(9.5)
Czechoslovakia	43	(0.5)	43	(1.2)
France	56	(0.6)	56	(1.6)
Hungary	1540	(17.7)	959	(27.7)
Iceland	1127	(13.0)	1096	(31.7)
Italy	265	(3.1)	73	(2.1)
Japan	4475	(51.4)	81	(2.3)
Romania	36	(0.4)	36	(1.0)
U.S.A.	225	(2.6)	221	(6.4)
U.S.S.R.	555	(6.4)	555	(16.0)
Yugoslavia	14	(0.2)	14	(0.4)
Total	8687	(100.0)	3468	(100.0)

A: All utilization as defined in text.
B: Utilization excluding bathing.

CONCLUDING REMARKS

The worldwide survey reported here is probably the first of its kind. While several surveys have been done on geothermal electric power generation, nonelectrical low-temperature uses have not received much attention. It is, however, interesting to note that about the same number (12) of countries were involved in low-temperature direct uses as in high-temperature electric power generation in 1980 (Bolton 1981). As of May 1980 the installed electric generating capacity of the world was 2462 MW-electrical.

It was found here that about 8700 MW-thermal above a 15°C reference temperature are installed in the 12 main low-temperature countries. The low-temperature geothermal fluids used for other purposes than bathing and similar uses amounted to 3500 MW-thermal (above a 15°C reference temperature) at the end of 1980. The survey has shown new data on the type of use to which low-temperature resources are used. Instead of 5000 MW-thermal used for agricultural applications in the Soviet Union, as previously reported, it reports on nearly 4400 MW-thermal associated with bathing applications in Japan. This example perhaps demonstrates that surprising aspects on the utilization of low-temperature geothermal energy may yet come to light. An important contribution of this survey is the data from the People's Republic of China. It was found that about 350 MW-thermal are installed in several locations being used in traditional applications of thermal waters.

An important result of the survey was information on the worldwide availability of geothermal resources. It was reported that 44 countries have known low-temperature resources. This is a conservative number by the nature of the survey--it appears that geothermal resources are to be found in most countries of the world. In this respect it should be noted that a large number of countries have potential geothermal resources in sedimentary basins not considered important in past years. The examples of Hungary and France are now being followed worldwide in an effort to exploit geothermal energy in such environments.

ACKNOWLEDGMENT

The world survey reported here would of course not have been possible without the cooperation of numerous friends and colleagues in the various geothermal countries. We hereby thank all those who contributed data and information to the survey.

REFERENCES

Barbier, E., and Fanelli, M., 1977: Non-Electrical Uses of Geothermal Energy. Prog. Energy Combust. Sci., 3, 73-103.

Bolton, R. S., 1981: Notes on a Survey of High Temperature Geothermal Fields. New Zealand Ministry of Works and Development, 15 p.

Fanelli, M., and Taffi, L., 1980: Status of Geothermal Research and Development in the World. 26th Int. Geol. Cong. (Published in: Revue de l'Institut Francais du Pétrole, XXXV, 429-448.)

Gudmundsson, J. S., 1982: Low-Temperature Geothermal Energy Use in Iceland. Geothermics, 11, 59-68.

Gudmundsson, J. S., and Palmason, G., 1981: World Survey of Low-Temperature Geothermal Energy Utilization. Orkustofnun (Iceland Energy Authority), Report OS-81-005/JHD-02, 148 p.

Howard, J. H., 1975a: Principal Conclusions of the Committee on the Challenges of Modern Society Non-electrical Applications Project. Proceedings, Second U.N. Symposium Development Use Geothermal Resources, San Francisco, 2127-2139.

Howard, J. H. (editor), 1975b: Present Status and Future Prospects for Nonelectrical Uses of Geothermal Resources. Lawrence Livermore Laboratory, UCRL-51926 and NATO-CCMS Report No. 40, 162 p.

Waring, G. A., Blankenship, R. R., and Bentall, R., 1965: Thermal Springs of the United States and Other Countries of the World - A Summary. U.S. Geological Survey Professional Paper 492, 383 p.

Geothermal Resources Council, TRANSACTIONS Vol. 6, October 1982

HEATING THREE BUILDING COMPLEXES IN SCOTTSBLUFF, NEBRASKA, WITH GEOTHERMAL ENERGY - DESIGN OPTIONS AND ECONOMICS

Raymond A. Guenther

American Geothermal Energy Associates, Inc. (AGEA)
5603 Jackson Str., Omaha, Ne., 68106
and
University of Nebraska at Omaha

ABSTRACT

The topic of this paper is the feasibilty of heating three building complexes with geothermal energy. The complexes are a college, a hospital and the high schools of Scottsbluff, Nebraska. The resource is assumed to be 180°F water from aquifers 4700 to 5200 ft below the surface; the static water level is expected to be about 2000 ft below the surface. Reinjection will be necessary. We discuss some of the design probelms as well as the economics of utilizing this resorce. Various scenarios for heating the complexes are explored. The payback periods in terms of current energy prices range from 9 to 32 years. Using DOE projected fuel prices the payback periods range from 6 to 11 years. The most cost effective project would be to heat the College and part of the Hospital with geothermal energy.

INTRODUCTION

During the last year our organization has been working towards the commercialization of the Nebraska geothermal resource (Gosnold and Ingersoll, 1982). In the Spring of 1981 we were commissioned to study the heating/cooling alternatives of Nebraska Western College (NWC) in Scottsbluff Nebraska (AGEA, 1981). The geologist's report indicated that flow rates of 150 gpm and temperatures of 180° F are quite probable. The wells would have to to be drilled to a depth of 5200 ft and the artesian head is expected to bring the water to within 2000 ft of the surface. The water is expected to contain 5 to 15 ppm dissolved solids and reinjection of the used geothermal fluids is anticipated. The economic analysis indicated that an additional user would decrease the payback period.

Early in 1982 a second study was commissioned by a user group representing the College, the West Nebraska General Hospital (WNGH) and the Board of Education. This second study was completed in April of 1982 and the results were presented to the governing bodies of those institutions (AGEA, 1982). A report on the retrofitting of the Hospital was available from Kirkham, Michael & Associates (KM&A, 1982). In the second AGEA report this information was utilized together with studies on the College and the High Schools to form district heating schemes and for the purpose of cost-benefit analysis. In addition this study contains a fairly detailed discussion of the relevant geology in the Scottsbluff region .

DESCRIPTION OF THE PROJECT

Figure 1 shows the relative locations of the buildings. The letters A,B,C,D,E label the individual sites. At

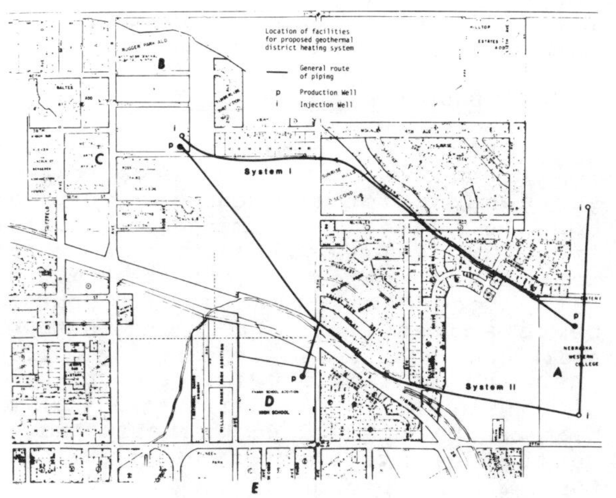

Figure 1 . A portion of Scottsbluff showing the locations of the three complexes.

the upper left is the hospital complex (labelled B and C); at the lower left is the School complex consisting of the Senior High School with associated Splash-Arena (labelled D) and the Junior High School (labelled E); at the lower right is the College (labelled A). The maximum distances between complexes are somewhat over one mile.

All of the buildings are heated with natural gas. The College (NWC) consists of three buildings which are mostly on one level. The bulk of the heat is delivered by circulating hot water.

The schools consist of the Junior and Senior high schools (JHS and SHS). Both buildings are heated with low pressure steam. The SHS has a modern steam heating system ; the JHS is an older building with cast iron steam radiators.

The hospital has a north and a south building. Both utilize low pressure steam in their mechanical systems and the main heating load is the ventilating air. The north building is a large multistory structure; the south building is a smaller and older structure.

To gain an insight into the economics of geothermal heating we first assumed that each complex is heated from its own associated well system. Next we explored the possibilities of supplying more than one user from one or more geothermal well.

There are many possibilities for district heating schemes. We choose to develop two of them and call them System I and System II. In System I we supply heat to the College and Hospital Complexes from one well located near the College; in System II we attempt to heat all three complexes from two production wells.

GENERAL DESIGN CONSIDERATIONS

Standard engineering methods were used to arrive at heating loads from the gas consumption and degree day data and from the ratings of the mechanical systems in place. Generally the designer's objectives were to heat the buildings with geothermal energy for outside temperatures above 20° F and to supply additional heat from the existing boilers when the ambient temperature drops below 20° F. In this way better than 90% of the heat can be supplied geothermally and the geothermal system operates at full load for a greater period of time.

Electricity costs for lifting and reinjecting the geothermal waters are a major expenses. Our calculations will assume that the pump is sited about 150 feet below the piezometric surface to allow for drawdown . We will also make allowances for a reinjection plant. The design should be such that the flow rate can be regulated. One way would be to throttle the flow with a valve, another way would be to install variable speed pumps and a third way would be to utilize a surge tank from which the geothermnal fluids can be drawn on demand. Our calculations will be based on the last option.

COSTS

The well costs were particularly difficult to pin down. For scenarios requiring flow rates of 150 gpm we used \$321,600 for the production well and \$143,000 for the reinjection well (P. Roberts, 1982) with an additional 10% for design and other considerations. For the College we used a slightly smaller figure, \$269,000, for the production well. We used \$400/hp (EG&G, 1981) to compute the price of pumps. The computed pump prices compare well with the actual prices. Maintainance expenses were taken as 3% of capital costs excluding retrofit costs.

Results

Table 1 (see next page) summarizes the relevant parameters for the various situations. There has been some controversy about the siting of the pump. Some persons believe that the pump has to go near the bottom of the well. Because we could not completely resolve this issue it was decided to show the effects of lowering the pump for System I. System IA assumes that the pump is sited near 2000 ft and System IB is for the pump placed at 5200 ft.

The next table (see Table 2) displays the payback periods. Column 2 lists the payback periods in terms of current energy prices. There is a wide range of payback periods, and if one assumed that the price of gas were to remain constant then only System I would be deemed attractive. If, however, one were to postulate rising gas prices then some of the other scenarios would also be acceptable.

ECONOMIC PROJECTIONS

The U.S. Department of Energy has been making projections of fuel prices which are published in its' report to Congress (U.S. Department of Energy, 1981). In our calculations we used the figures representing the midrange of those projections for commercial consumers. For current rates we used local 1982 prices. The projected prices are in 1980 Dollars and make no allowance for inflation. These numbers are displayed in Table 3.

Table 1. Summary of system parameters.

System	Flow Rate gpm	Total Head ft	Prod. & Inj. hp	Electricity Usage: Well Pumps	Electricity Usage: Other (10^5kWh)	Electricity Usage: Total	Oper. Electr. 1982	Exp. Maint. 10^3 $	Gas Savings 10^3 CCF	Gas Savings 10^3 $	Cap. Inv. 10^3 $
College	100	2150	70 30 100	3.77	---	3.77	13.9	15.3	185.5	53.8	697.
Schools	200	2300	160 70 230	8.66	.56	9.21	32.0	18.8	184	82.4	1,080
Hospital	170	2200	130 60 190	7.16	.16	7.32	27.1	17.3	373.6	108	1,238
System-IA	150	2250	115 50 165	6.22	.143	6.36	23.5	20.3	472.5	137	1,285
System-IB	150	5500	275 50 325	12.2	.143	12.4	45.9	22.7	472.5	137	1,342
				(System IB is the same as System IA except that the pump is at 5200 ft.)							
System II	235	2500	200 80 280	10.6	1.32	22.4	82.9	48.8	845	244.5	2,894
	x 2 470		x 2 560	x 2 21.2							

Table 2. Payback periods in years.

System	Current Energy Prices	Projected Energy Prices	Projected Energy Prices & 5% Infl.
College	28	9	8
Schools	32	11	9
Hospitals	19	8	7
System IA	9	6	5.5
System IB	10	-	-
System II	26	9	8

Table 3. Energy prices.

	1982	1985	1990	1995
Nat. Gas. $/CCF	0.29	0.519	0.726	0.857
Electr. $/kWh	.037	.0591	.0609	.0635

Using these prices we arrive at paybacks summarized in Table 2, column 3 and if one assumes an annual inflation rate of 5% one arrives at the figures in column 4.

We choose to include a more complete analysis for System I because it has the shortest payback period. Figure 2 shows the payback using projected energy prices and the effects of various interest rates; in Figure 3 an inflation rate of 5% was included. Figure 4 shows the effect of lowering the pump to 5200 ft and also the effect of decreasing electricity costs by 30%.

CONCLUSIONS

Our main concerns are:

1. The uncertainties in the placement of the pump.

2. Problems associated with the reinjection of the fluids.

3. Possible difficulties in reach-flow rates greater than 150 gpm.

4. Corrosion and precipitation of solids because of the chemistry of the geothermal fluids.

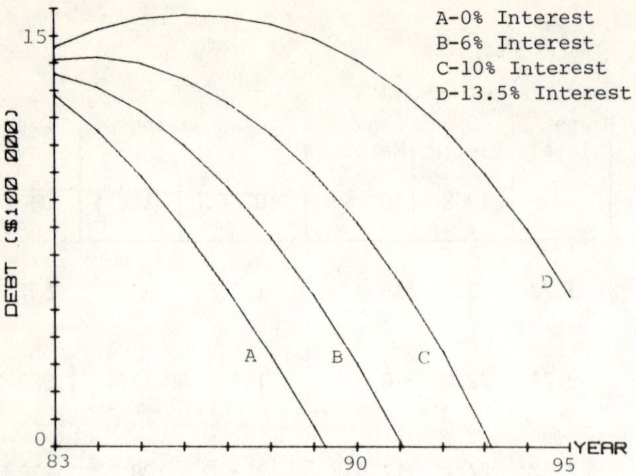

Figure 2. Indebtedness as a function of time with projected energy prices, zero inflation and showing the effects of various interest rates.

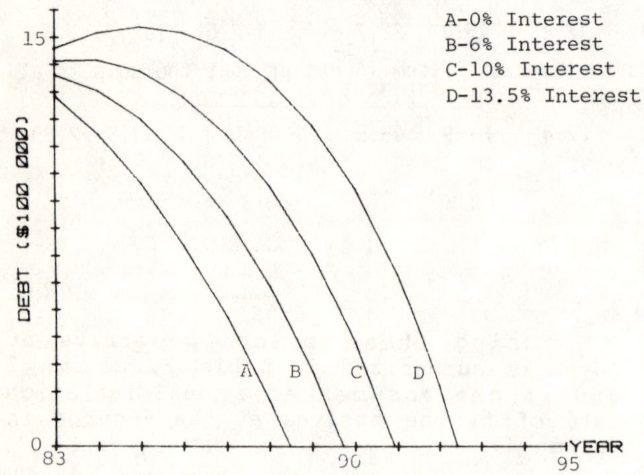

Figure 3. Indebtedness as a function of time with projected energy prices , 5% inflation and various interest rates.

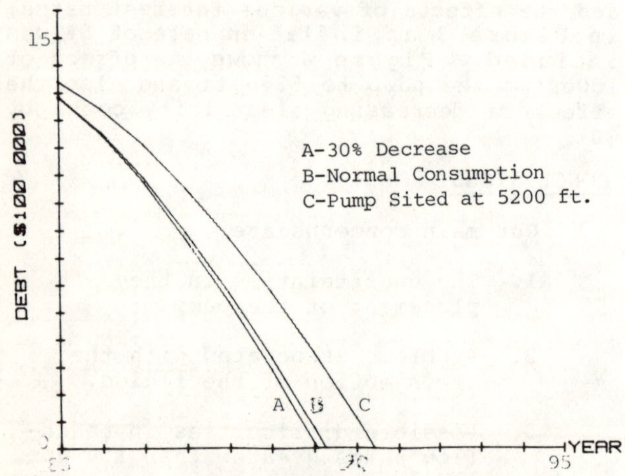

Figure 4. Indebtedness as a function of time showing the effects of varying electricity consumption.

These problems are commonly encountered in geothermal projects and they can be circumvented. One certainly should avoid exposure of the geothermal fluids to the atmosphere. However, relyable information on the extent of these difficulties will not be available till the first well is drilled and tested.

To minimize the risks we recommend that the pilot project be a system with a short payback period and one which is relatively insensitive to variations in electricity costs. System I would be the best option.

System I also has the virtue of being relatively simple from an engineering viewpoint. Because the heating system of the College is designed to operate at an average temperature of 160°F the retrofitting there is minimal if one uses the 180° F water to heat the College first. The water leaving that complex will still be above 140°F which is warm enough to heat a considerable portion of the hospital.

REFERENCES

Gosnold, W. D. and Eversoll, D. A., 1982, Geothermal resources of Nebraska. Map scale 1:500,000, Nebraska Conservation and Survey Division and Division of Geothermal Energy, U.S. Dept. of Energy.

American Geothermal Energy Associates (AGEA), 1981, Preliminary feasibility study of heating and cooling alternatives for Nebraska Western College, U.S. Dept. of Energy GTA Rept. No. 14, EG&G Rept. No 2147, 61 p.

American Geothermal Energy Associates (AGEA), 1982, Feasibility study of geothermal district heating for a portion of Scottsbluff, Nebraska, 112 p.

Kirkham, Michael & Associates, 1982, Feasibility Study for Geothermal Retrofit - West Nebraska General Hospital.

Roberts, P.- Dir., Nebraska Oil and Gas Conservation Div., 1982, letter, included in Appendix III of 1982 AGEA report.

EG&G Idaho, 1981, Rules of thumb for geothermal direct heat applications.

U.S Dept. of Energy, Energy Information Admin. , 1981, Annual report to Congress, v. 3, p. 154, 155.

CASE HISTORIES OF FOUR AGRICULTURAL AND MUNICIPAL DIRECT USE PROJECTS IN THE WESTERN U.S.

Alfred B. Longyear

Lahontan, Inc.

ABSTRACT

Engineering and economic studies have been conducted on four agricultural and municipal direct energy projects. Two of the projects located in Northern California are being constructed in 1981 and will be operating in 1981-82. Two others in Northern California and Southern Idaho have not progressed beyond the paper studies.

This paper will summarize the Susanville, Mountain Home, Kelley Hot Spring and Litchfield projects in terms of engineering and economics, institutional, permitting issues and financing and will draw conclusions from the evolution and maturing of geothermal direct use projects in the U.S. in the 1975-81 time period. The institutional, financial and management factors delaying two of the projects are addressed. The characteristics of successful geothermal direct use projects in the U.S. are discussed.

Three of the projects have been funded by the U.S. Dept. of Energy. One of the projects is a private-local government-state funded effort that will displace over 500,000 gallons ($2000m^3$) of oil per year.

1.0 INTRODUCTION

The major portion of the geothermal resources in the U.S. are most suitable for direct utilization for space heating and moderate-to-low temperature processes. Significant development of these resources has been impeded, initially, by regulatory and other institutional barriers and, more recently, by lack of public awareness and the immaturity of the resource as an investment.

This paper addresses the historical evolution of four direct use projects, occuring between 1974 and 1981. These are the Susanville Geothermal Energy Project--the prototype modern district heating system in the U.S.; the Mountain Home and Kelley Hot Spring Geothermal Projects--integrated agricultural applications; and the Litchfield Geothermal Project--a large scale cascaded system developed through a private developer/local government/state agency joint effort. The degree of success in terms of implementation for these projects and those factors affecting their individual degrees of success are discussed. And finally, those elements and factors found necessary for success in the future are described.

Since technology has not been a pacing factor in direct use development in the U.S., the technical aspects will be only minimally addressed in the paper. The technical aspects are addressed in depth in the published references.

2.0 OVERVIEW OF DIRECT USE GEOTHERMAL PROJECTS IN THE U.S., 1974-1981

Prior to the U.S. Department of Energy (DOE) incentive programs, geothermal, on a commercial scale, was concentrated at the Geysers, north of San Francisco, for power production. Direct use projects were limited to spas and a few small tomato greenhouse operations - most of which were put together by people inexperienced in the greenhouse vegetable raising business and, consequently, most subsequently suffered one or more significant reorganizations and a number of complete business failures. DOE initiated their engineering and economic studies (called PRDAs) in 1975 and their joint funded field experiments (PONs) in 1976. One of the more successful of these programs has been the Susanville Geothermal Heating District. This project started as a community-industry joint effort, research and development funding was made available by DOE, the Bureau of Reclamation and other government entities. (The DOE Geothermal Loan Guarantee Program has been initially inappropriate and subsequently found to be too expensive in front-end costs and calendar time for most direct use projects.) In the 1974-1979 time period, direct use projects were extensively studied by commercial banks, but the immaturity of the resource precluded financing without a federal guaranty. Direct use projects came of age recently when reservoir insurance was made available and, more importantly, when sophisticated private investors found that considerable tax advantage could be obtained through investing in geothermal as an alternative energy, pollution-mitigating natural resource. Hence, in 1980, geothermal direct use came of commercial age with the potential for a broad base of development. The Litchfield Project is a prototype for this type of development.

3.0 FOUR GEOTHERMAL DIRECT USE PROJECTS

The projects are described in chronological order of their initiation. All of the projects were initiated with private industry participation--at least in the project development phase. Federal participation was required for the prototype financing of the first three and was indirectly involved in the "seed" effort that justified private financing of the fourth. Local government was heavily involved in the first project and is an active participant in the fourth project.

3.1 SUSANVILLE GOETHERMAL ENERGY PROJECT

The City of Susanville, California is at 4200 feet (1280m) altitude on the east slope of the High Sierra mountains. Its 7,000 people have an agricultural and forest product-based economy in the rural northeastern part of the state. The core of this project is a fourteen building heating district started in early 1974 as a joint industry-city effort. Being a prototype, the resource was explored by the U.S. Bureau of Reclamation and the principal design and development funds came from the U.S. Department of Energy. A 126-home addition is being funded by the Farmers Home Administration and the Department of Housing and Urban Development. A local non-profit corporation is being established by the community to own, operate and expand the systems. Private developers are being encouraged by the City to develop the commercial and agribusiness uses of energy.

Like many of the hydrothermal resources in the western U.S., the geothermal fluids exist at typically moderate-to-shallow depth 3200 feet (1000m), moderate-to-low temperatures 206°F (370°K) and are suitable for agricultural or potable uses (total dissolved solids of 1000mg/1). In Susanville, the supply well, Susan No. 1, was drilled to 900 feet (274m) with a temperature of 170°F (352°K) at 800gpm (50.0 1/s). The Heating District, at this stage, requires a nominal geothermal flow of 550gpm (35.0 1/s). The transmission lines--supply and return--are principally asbestos-cement pipe. The supply line is insulated. Above ground, for bridge crossings, etc., they are insulated steel. One branch line is polybutylene donated by the Shell Chemical Co. as a demonstration item. Being all retrofits at this stage, peaking and emergency standby energy will be supplied by the existing heating systems in the buildings.

The HUD 126 home addition will utilize an existing 150°F (339°K) hot water well to supply a peak demand of 500 gpm (32.0 1/s). This is in construction at this time, with transmission line options of steel, asbestos-cement or polybutylene, with insulated supply and uninsulated return. The system will be interconnected to a nearby park of commerce by a Farmers Home Administration funded pipeline.

The pipeline routing and principal facilities involved in the primary Heating District are shown in Figure 1.

3.2 MOUNTAIN HOME GEOTHERMAL PROJECT

Mountain Home is a city in south central Idaho, located on the Snake River Basin hydrothermal resource area. In 1973, a fossil fuel exploratory well, Bostic 1A, was drilled to a depth of 9680 feet (2950m). Artesian flow of 1000gpm (63.0 1/s) occurred for a 2 week period. A bottom hole temperature of 370°F (462°K) was measured. This led EMMA, Ltd., a San Francisco investment firm, to seek development of a direct use project. Through a competitive application, DOE funded an engineering and economic study of a large, vertically integrated agricultural complex.

A system of feed production, swine raising, slaughter, potato processing and waste management was selected, based upon market trends, regional practices, available agricultural technology, use of commercial hardware, geothermal resource characteristics, thermal cascade and mass flow considerations, and input from regional agricultural advisors. The complex covers 160 acres (65 ha), (Figure 2); utilizes a peak energy demand of 1.14×10^8 Btu/hr. (1.2×10^{11}J/hr.) from geothermal heat between 300°F (422°K) and 70°F (294°K), (Figure 3); has an installed capital of $34.4 million; produces 150,000 hogs per year, 136 million lbs. (62 million kg) of processed potatoes per year, and on the order of 7.5×10^5W of continuous power from methane. The methane is produced from livestock and process wastes in a geothermal heat-augmented, anaerobic digestion system. The total effluent from the facility is 100 gpm (6.3 1/s) of water of irrigation quality and 3.5×10^4 lb./yr. (1.6×10^4 kg/yr) of saleable fertilizer. The entire facility has a peak demand of 1000gpm (63.3 1/s) of 300°F (422°K) geothermal fluid.

To reiterate, this was an engineering and economic analysis of an agricultural complex that was to utilize proven technology and commercial hardware.

3.3 KELLEY HOT SPRING AGRICULTURAL CENTER

Kelley Hot Spring is the second largest boiling hot spring in the U.S., flowing at 300 gpm (20 1/s). It is located at 4360 feet (1329m) altitude in south central Modoc County, on State Route 299 in northeastern California. The resource has been extensively explored and there have been two exploration wells drilled to 3200 feet (975m) and 3395 feet (1035m) with bottom hole temperatures on the order of 240°F (390°K). The fluids are of agricultural quality and are used for watering livestock. The Kelley Hot Spring Agricultural Center (KHSAC) complex would require on the order of 1% of the estimated reservoir capacity in thirty years of operation.

The KHSAC complex was conceived in 1977 as a direct use application of the geothermal resources under lease to Geothermal Power Corporation of California. The Project incorporated the results of the Mountain Home Geothermal Project and between

1977 and 1979, field demonstration funds were sought from the DOE. Alturas, the local county seat, was involved only for purposes of permitting.

The core activity is a nominal 1360 sow swine raising complex. A conceptual design was completed and extensive trade studies conducted to select equipment and design features optimized for swine raising in the project area. Subsequently, a preliminary design, up to, but not including, construction drawings, was completed. At the same time, a very extensive environmental assessment was conducted. This included an archeological field survey that resulted in definition of extensive archeological middens in the area. These would have to be avoided in final design and construction.

The swine raising was planned to be a totally confined operation for producing premium pork, in controlled environment facilities that utilize geothermal direct use energy. The complex would include a feed mill for producing the various feed formulas required for the animals from breeding through gestation, farrowing, nursery, growing and finishing. A barley sprout raising facility was incorporated to produce a green grass constituent for use in the breeding, gestation and lactating feed formulae. The market animals were to be shipped live by truck to slaughter to Modesto, California. A complete waste management facility would include manure collection from all animal raising areas, transport via an enclosed water flush system to a methane anaerobic generator, solids separation, settling ponds and disposition of the surplus agricultural quality water.

The facility would produce over 29,000 marketable hogs/yr at an average weight of 225 lbs. (103kg) each. Methane would be produced at over 1×10^5 ft^3/day (3×10^6 l/day), which would produce about 400KW of electricity. Effluent agricultural quality water would amount to 5gpm (0.34 l/s).

The geothermal system schematic for the complex is shown in the plot plan in Figure 4. Building piping details for space heating are shown in Figure 5.

3.4 LITCHFIELD GEOTHERMAL PROJECT

Litchfield in northern California is located about 16 km east of Susanville and is the site of a medium security, 1200 inmate, State Correctional Center. The Center consumes 750,000 gal. (2840m^3) of oil/yr. for space heating and consumptive hot water. As part of the Susanville project, a geothermal hot water resource was discovered within 2 miles (3 km) of the Center. The City of Susanville acquired the leases for the resource and, as a joint effort with Carson Development Co., initiated project development activities in 1980. The Project includes an energy supply system to the prison with cascading to an adjacent park of commerce and the retrofit of the prison.

This is the first modern geothermal project in the U.S. to be developed as a utility with private financing. Private funds are used to develop the resource; the energy is marketed by a local government entity to a state entity and subsequently to private industry users. Unique institutional agreements had to be reached for this multi-agency-private industry venture. The private developer had to have an agreement with the City that, in turn, acts as the distributor utility and simultaneously that owns the resource rights. The City had to have a sales agreement with a State agency that in turn required close participation of four supporting state agencies. Private financing was arranged through a limited partnership and industrial revenue bonds authorized by the State for a "pollution control" project. State facility financing for the prison retrofit was raised through the legislature in one year instead of the normal three year process. Surprisingly, the Project was developed, financed, had construction design started and a successful well drilled and tested in one year from project start. Other than for some political delays incurred in a tight budget year, this project is a model for future alternative energy projects in California.

The supply well has been drilled to 1400 feet (425m), flow tested to 1500gpm (95 l/s) at 172°F (351°K). It is expected that this resource can be economically used to displace 60+% of the prison fossil fuel consumption. Peak geothermal fluid design flow is nominally 1100gpm (70 l/s) at 170°F (350°K) for the Center.

4.0 STATUS AND UNIQUE FEATURES OF PROJECTS

The author was the lead initiator and either technical director or advisor on the four projects, as well as a participant in similar projects in the west and southern states. Hence, these observations on unique features are relative to a spectrum of direct use projects in the U.S. The unique features are highlighted to identify characteristics that affect the degree and rate of success or lack of success in such projects.

4.1 STATUS

First, the status of the projects as of November, 1981, is as follows:

Susanville: The Heating District is completing construction and subsystem checkout. This project, defined in 1977, is one of few that will complete on schedule and within original allocated funds. The Heating District will go on line in early 1982.

Mountain Home: This engineering and economic study was completed on schedule and within funds. However, as a project, it did not progress beyond the study effort, which was completed February, 1979.

Kelley Hot Spring: This field demonstration completed Phase 1 - Preliminary Design - within funds with the Phase 1 report issued August, 1980. Further work was suspended pending determination of the historical value of the archeological finds in the area.

Litchfield: This project started in 1980 and is entering the hardware stage with a successful supply well and the recent completion of Preliminary Design of the retrofit. The project's final outcome is tied, in part, with the political process of bringing a state facility into use of an alternative form of energy with private investment and local government participation. To date, the Project has progressed extremely well--in fact, better than originally expected by the State agencies involved. Delays have occurred primarily in the political process during a tight budget year.

4.2 PROJECT FEATURES

The unique features of these projects include:

Susanville: The City is a small rural community located in a cold, semi-arid climate. It has a relatively high government employment base and a depressed agricultural and forest product industrial base. However, a spirit of independence and self reliance prevails in the community. The decision for participation in the project was made simply and quickly by the City Council, resulting in a one-page letter of intent in 1974. Since then, the subsequent councils have maintained a very consistent and firm support of the project and its industry team members. This has permitted the industry project developers to consistently commit and follow through on their support of the projects. This local environment, coupled with the dedication and integrity of key city staff persons, has established a strong national and state base of political support. Project leadership has been staffed with persons with persistence, initiative and creativity in management and problem solution, and a strong sense of schedule and accomplishment of cost effective results. Without this total team capacity, the project could have ceased to exist on several occasions from 1974 through 1978.

Mountain Home: This project accomplished what was contractually required. In fact, is was considered as a good effort. However, the Project, at Mountain Home, did not continue. (The results were picked up in other subsequent projects). The project lead entity was a new, small alternative energy financing company which subsequently has been acquired by a petroleum development company. While having experience in financing, the firm had no experience in agricultural project development and management. Developing such talent or participation from entities experienced in the multiple, very high technology disciplines involved in the \$34 million complex became an insurmountable task. The geothermal expertise available in the lead entity applied only to a very small part of the overall project development.

Kelley Hot Spring: The Kelley H.S. project built upon the results of the Mountain Home project. And the Kelley effort started while the Mountain Home project was just completing. Hence, it could not perceive the magnitude of effort re-required to bring the experienced, high technology talent to bear on a new energy resource and its application to livestock operations. Again, the lead entity was a small firm in the development of geothermal resources. The CEO was an engineer with experience in utilities, but no experience in livestock operations. The original investors committed to a project in 1977 that took until 1979 to go on contract and required another year to produce the preliminary design. Needless to say, the investment environment and the investor targets changed considerably over that time.

Further, the archeological finds became at least a time barrier for further federal or state participation in terms of demonstration project money or financing guaranties.

Litchfield: This project built upon prior work and the existing institutional and political structures for geothermal developments in the City of Susanville. Carson Development Company was selected as a medium-sized, commercial-industrial development firm with a track record of turnkey projects and with a private financing capability for projects of this size. The availability of a slim test hole (by the Bureau of Reclamation), availability of the new reservoir insurance program, the definition and recognition of geothermal direct use as a depletable, alternative energy, non-polluting resource and the attendant tax provisions, the new state-sponsored pollution control industrial revenue bonds, and the state's commitment to cogeneration/alternative energy all contributed to the timely initiation of this project. The meshing of private financing, reservoir insurance and a city-state agreement for long-term alternative energy sales with long-term financing through I.R. bonds is a precedence-setting project that indicates that this resource is a commercially viable opportunity.

Characteristics of the four projects are summarized in Table I. One must recognize that the initiation calendar times and the corresponding stature and maturity of this industry, as well as the con-committent cost of financing, materials and services, tend to preclude absolute comparisons. It is also difficult to compare municipal heating districts with a private business complex. However, in spite of the foregoing, one can observe the potential of geothermal as a competitive resource.

In an overall summary of features of these projects, one can observe that the federal government procurement procedures that had to be used for these supposed seed projects precluded a normal business approach to the programs. The time delays associated with these procurements were completely in discord with normal project investment schedules. This is belatedly being recognized. The success of project completion on time and within funds, once underway, and the relatively accelerated schedule in the Litchfield project, is a result of the unique capability of very talented staff in the federal and state energy offices that permitted a true team approach. This is the exception rather than the rule. The features of the Litchfield project model are obviously a more effective means of accelerating the development and use of the extensive geothermal direct energy resource in the U.S.

5.0 CHARACTERISTICS OF A SUCCESSFUL DIRECT USE PROJECT IN THE U.S.

☐ First, the project must have a committed (preferably finacially committed) competent "initiator" or "spear carrier" - at least until such projects become more commercially routine.

☐ The resource and its use must continue to be treated by federal and state taxing agencies on a competitive basis with other natural resources in order to attract the necessary broad base of private investment.

☐ Direct use resources have their broadest application in the agribusiness and food process sectors of industry. The utilization, in turn, is optimized in the confined, intensive growing or controlled environment applications. And these, in turn, are high technology businesses, requiring staffing with experienced professionals. This is an absolute requirement.

☐ District heating, especially in the lower population-density western U.S., are economically marginal systems, thus requiring a significant commercial or industrial direct use load to make them attractive for private and commercial financing.

☐ Geothermal direct use systems and their commercial/industrial applications are economic development projects requiring all of the professional disciplines and expertise required to promote and carry forward a conventional commercial economic development project plus the professional expertise to successfully develop an economical geothermal hot water energy system. This expertise requires an experienced understanding of cost-effective geothermal resource development and cost-effective engineering for geothermal direct use energy distribution systems.

☐ The lower energy density in a hot water system, relative to other forms of energy transmitted through an equivalent system, precludes the tendency to "over-engineer" pipelines and the other elements of energy supply and distribution systems. Reasonable maintenance and repair must be traded off against the high initial cost of a maintenance-free design.

6.0 REFERENCES

Longyear, A. B., "Survey of Selected Industries for Potential Branch Plants - Susanville Geothermal Project," Fred Longyear Co., April, 1978.

Longyear, A. B. etal, "Mountain Home Geothermal Project," EMMA/Lahontan, Inc., DOE/ET/28442-1, February, 1979.

Longyear, A. B., Editor, "Kelley Hot Spring Geothermal Project," Geothermal Power Corporation/ Lahontan, Inc., DOE/ET/27041-1, August, 1980.

Longyear, A. B. and J.C. Jeskey, Editors, "Susanville Geothermal Energy Project-Workshop Proceedings," City of Susanville, SAN-1077-4, July, 1976.

TABLE I. FOUR GEOTHERMAL PROJECTS SUMMARY

CHARACTERISTICS	SUSANVILLE HEATING DISTRICT	MOUNTAIN HOME	KELLEY HOT SPRING	LITCHFIELD
Initiation Period	1974 - 1977	1977 - 1978	1977 - 1979	1980
Project Size	4.3×10^{10} Btu/yr. (4.5×10^{13} J/yr.)	6.3×10^{11} Btu/yr. (6.6×10^{14} J/yr.)	9.1×10^{10} Btu/yr. (9.6×10^{13} J/yr.)	6.6×10^{10} Btu/yr. (7.0×10^{13} J/yr.)
Energy Supply System Cost	$1,800,000[(1)]	$1,450,000[(2)]	$406,000[(2)]	$900,000
Application Cost				
Retrofit	$600,000[(3)]	---	---	$1,520,000
New Facilities	---	$33,000,000	$5,000,000	---
Payback Period		Owner:	Owner:	Energy System:
	7 - 10 yr	3+ yr	3+ yr	9 yr
		Debt:	Debt:	Retrofit:
		12 yr	20 yr	7 yr

(1) Commercial costs without prototype costs

(2) Rework existing well plus backup and reinjection wells

(3) Includes $300,000 revolving fund for current and future retrofits

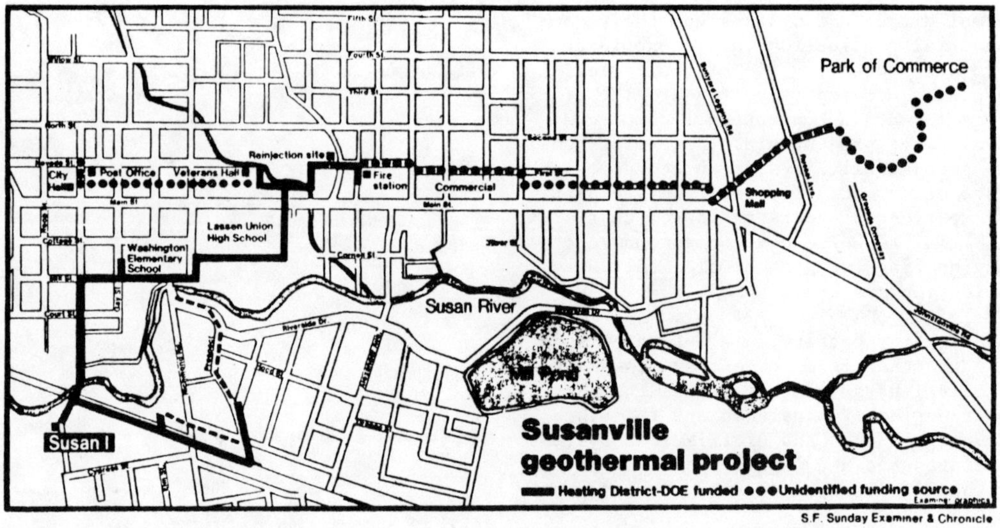

S.F. Sunday Examiner & Chronicle

Figure 1 - Susanville Geothermal Project Pipeline Route

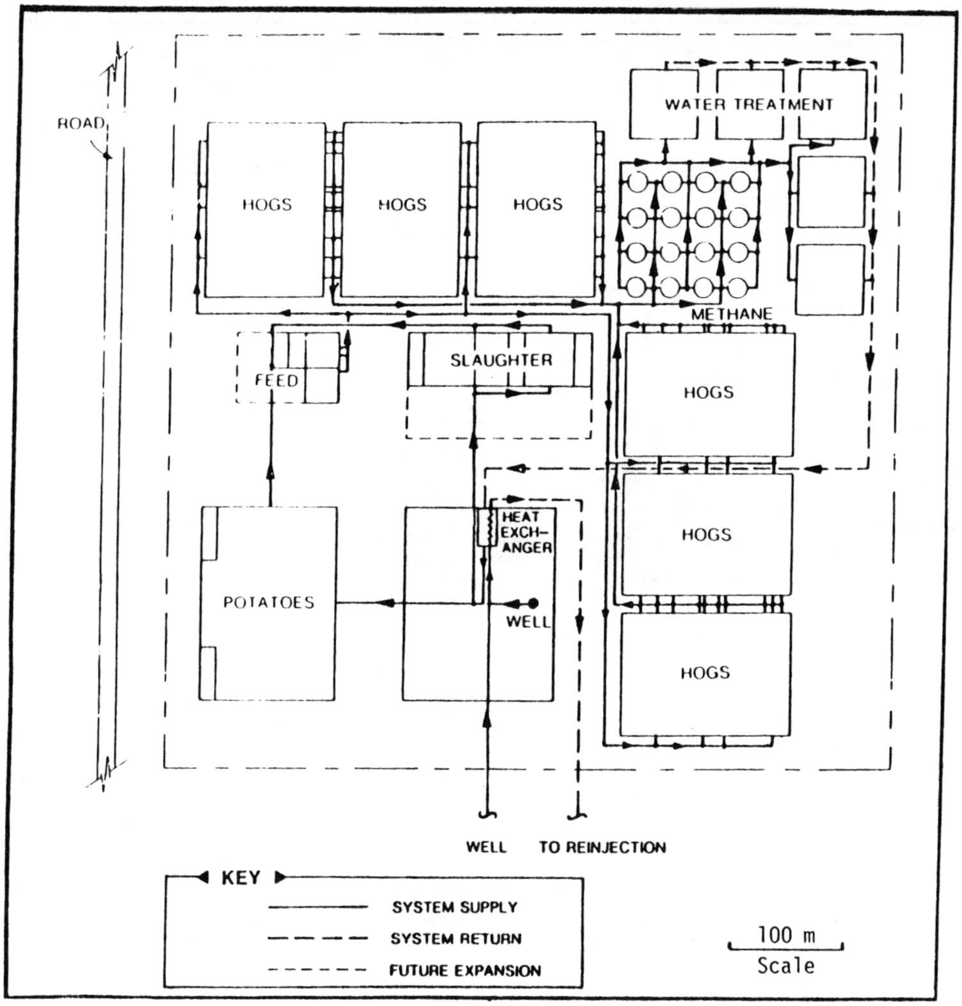

Figure 2 - Mountain Home Geothermal Project Facility and Geothermal System

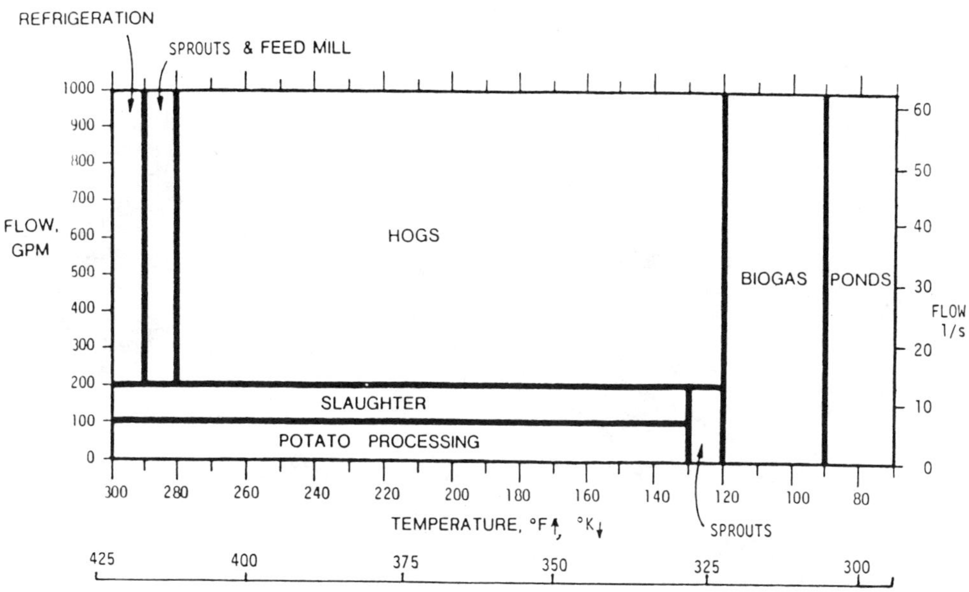

Figure 3 - Mountain Home Geothermal Project Utilization of Geothermal Energy.

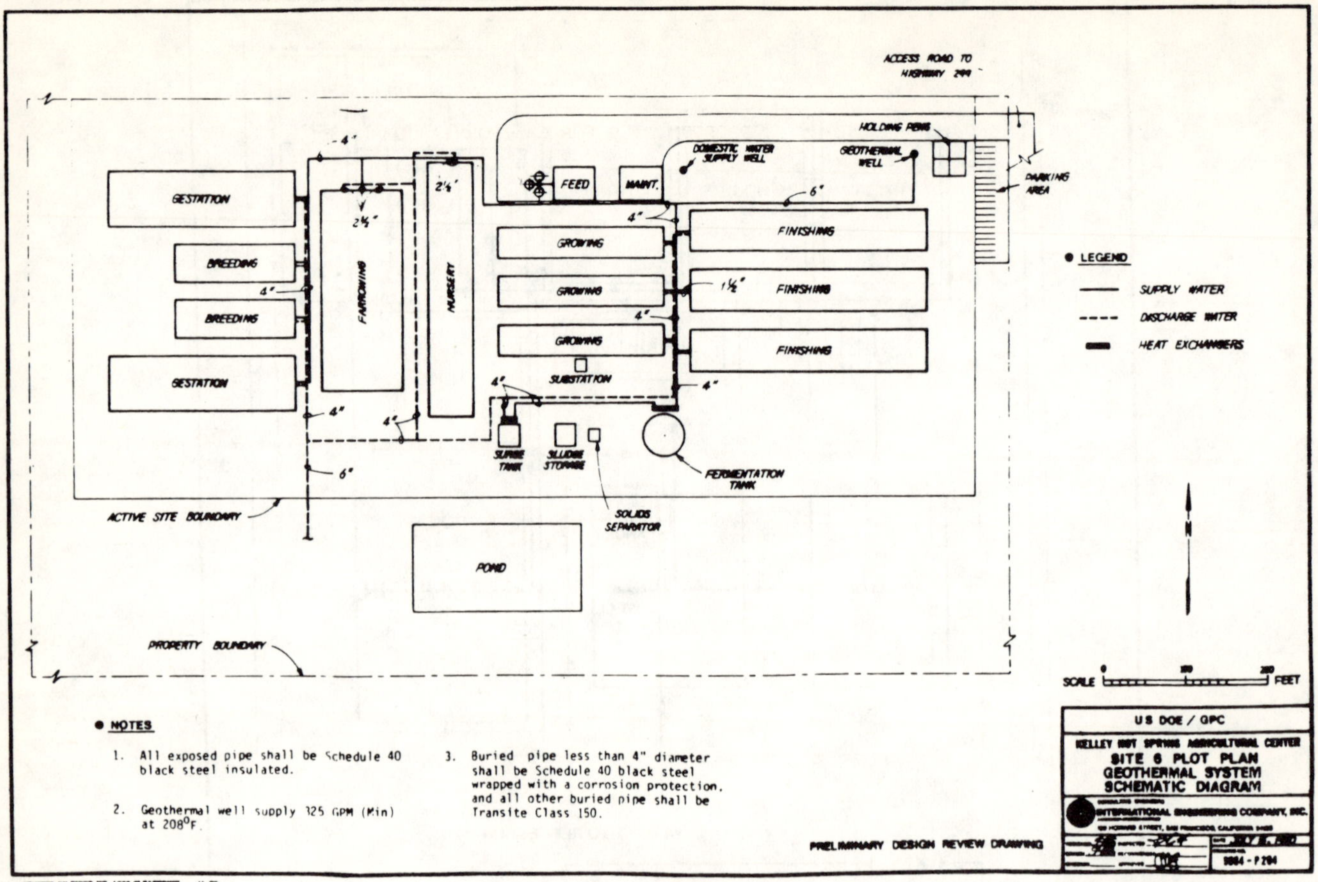

Figure 4 - KHSAC Geothermal System Layout

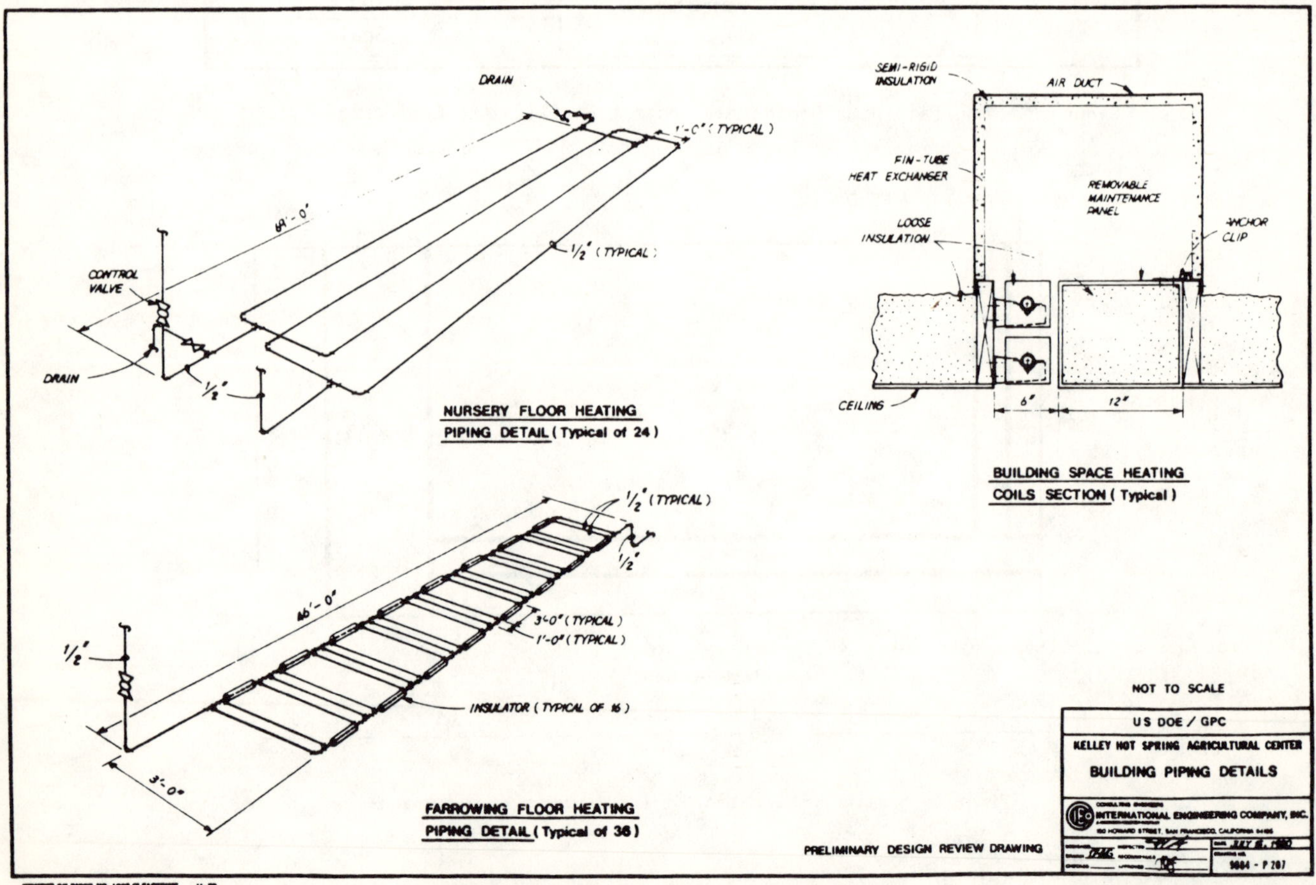

Figure 5 - KHSAC Building Piping Details

Geothermal Resources Council, TRANSACTIONS Vol. 6, October 1982

GEOTHERMAL ENERGY: OPPORTUNITIES FOR CALIFORNIA COMMERCE

Alfred B. Longyear and Justin N. Tierney

Lahontan, Inc., Alternative Energy Systems, Sacramento, CA
Geothermal Program, Development Division, California Energy Commission

ABSTRACT

To accelerate the development of abundant low- to moderate-temperature geothermal resources, the California Energy Commission (CEC) has identified those resource areas and direct use applications with the greatest potential for successful commercial near-term (2-5 years) development. An economic development approach was used to identify these areas and end-uses. This approach focuses on enhancing a community's local economy by developing its geothermal resources to attract industries which need a low-cost reliable energy source.

Twenty-seven areas of the state met strict geologic and economic criteria and are considered priority areas for immediate development activities. Twenty end-use applications, such as district heating and agricultural processes (e.g. greenhouse heating), met economic and energy use criteria and can successfully convert to using geothermal energy.

The CEC has initiated geothermal market development activities to overcome the major impediments to accelerated development of small-scale electric and direct use applications. These activities include infield technical and economic assessment and financial assistance to private developers and local governments. These programs are explained in the context of the evolving geothermal market development effort.

INTRODUCTION

The Geysers' vapor-dominated resource, which supplies the largest geothermal electrical generation complex in the world, often overshadows the fact that a major portion of California's resource is water-dominated and best suited for direct use. This low- to moderate-temperature (50°-150°C or 120°-300°F) resource can be found in at least 46 of California's 58 counties.

Direct use applications have several significant advantages to electric applications including: a far greater and more readily available resource; a higher conversion efficiency (usually 80-90%); the use of simple, off-the-shelf, conversion technology; shorter development time; and, most importantly, less expensive exploration and development costs. With continually increasing energy costs, many developers and investors are taking a serious look at the benefits of direct use applications. Unfortunately, many others are unaware of these benefits or lack the experience to economically develop a resource.

To guide the California Energy Commission's effort to accelerate direct use development, an intensive, methodical and assertive effort has been initiated to identify those resource areas and end-use applications with the greatest potential for successful near-term (2-5 years) commercial development. Building on results from previous market studies, the effort is being directed at target communities that have exploitable resources, and at venture capitalists and industries which could benefit substantially from converting to geothermal energy.

AN ECONOMIC DEVELOPMENT APPROACH

Before presenting the methodology and results of the marketing effort, it is important to understand the basic philosophical approach to geothermal commercialization - economic development.

Developers of geothermal resources for electrical generation pay little attention to how the energy is ultimately used once its been transmitted to the grid. However, when the heat from a resource is to be used directly, the cost of pipelines requires that an end-user be located within an economic distance from the hot spring or well. For marketing and developing the low- to moderate-temperature resource for direct use applications, attracting end-users goes hand in hand with exploring and developing the resource itself. This, in turn, requires the community to offer conventional economic development incentives (e.g., labor base, community services, etc.) normally used to attract industry to a given site.

Further, to be cost effective, several end-users are required in order to reduce resource development and distribution costs. The hot water can be cascaded through multiple applications, either in a large single complex or in a group of individual end-users who are united in a "Geothermal Park of Commerce." For example, water from one well can supply heat for large buildings in a downtown area, then be cascaded to food processors, then greenhouses and finally sold as irrigation water. A Geothermal Park of Commerce is similar to a conventional industrial park with one major exception: the offer of a lower-cost, reliable energy source. This lower cost energy is an important "carrot" for attracting industry. However, the benefit of lower energy costs usually has to be traded-off (at least in part) against increased transportation and/or other difficiencies (e.g., reduced labor base) that are associated with the rural sites where most geothermal resources are found. Hence, a full scale, local marketing effort is required to accomplish the successful development and utilization of a geothermal direct-use resource. Only through such an approach can the economic benefits accrue to the local community.

This strategy promotes geothermal development as a means to an end: economic development for a community or industry. If communities near a geothermal resource area can demonstrate the cost-effectiveness and reliability of the energy alternative and can offer conventional economic development incentives, they can attract industrial end-users (e.g., Susanville Park of Commerce). Industrial development leads to tax revenue, indirect benefits from employees and families moving to the community and decreased unemployment: these, in turn, stimulate the local economy. Successful businesses which use geothermal energy can achieve reduced energy costs and lead to adoption of the alternative by others in the industry and, therefore, improve the economic condition of the industry.

While many of the results of this effort are California-specific, this economic development approach and the methodology to determine and develop target resource areas and end-use applications can be replicated in other geothermal areas of the U.S. and elsewhere in the world.

PRIORITY GEOTHERMAL RESOURCE AREAS

California was divided into 5 broad geothermal regions according to general resource characteristics and proximity to major population and trade centers. Within these regions thirty-eight areas met or exceeded the minimum resource factors listed in Table 1. Three of these factors - temperature, well depth and amount of total dissolved solids in the geothermal fluids - determine at a very basic level of analysis whether a project will be economically feasible. For example, two independent studies evaluated the feasibility of constructing and operating a greenhouse complex using heat from geothermal wells (Lahontan, 1982). The economic cost effectiveness between a greenhouse complex at Wendel, near Susanville, using 200°F fluids with a TDS of 300 ppm from a 600 foot well and a greenhouse operation in Imperial County using 155°F fluids with a TDS of 25,000 ppm from a 2000 foot well is extreme. Considering the heating system costs alone, the Wendel project has an excellent return on investment compared with the unacceptably low return from the Imperial Valley project in Imperial Valley. Therefore, with few exceptions, the five resource criteria should be used as the absolute minimum standards necessary for most cost effective direct use projects.

Following this screening against resource criteria, the thirty-eight resource areas were then evaluated according to economic development factors (Table 1). These factors address the suitablity of the resource area for project development from a business perspective. As shown in Figure 1: Priority Resources for Direct Use Development, twenty-seven areas are rated Priority of I, II or III which qualified them for further marketing and development efforts. These areas, listed in Table 2 according to priority, should be used as a guide for government and industry in evaluating potential project sites, since some areas rated Priority IV (e.g. Lake County) may become more attractive if one or two major impediments are removed. Additonally, other areas not yet identified in this study could offer a good opportunity for future direct use project development.

PRIORITY END-USE APPLICATIONS

To identify potential applications, relevant market studies were reviewed and numerous state agencies, trade associations and industrial leaders were consulted. Twenty-nine general applications were determined to have potential for using low- to moderate-temperature geothermal fluids directly. These applications were then assessed against the economic and energy use criteria listed in Table 3. The applications having the highest potential for successful near-term commercial development, fell into three general categories: district heating, agricultural processes and small-scale electric systems.

Geothermal district heating systems have been successfully installed throughout the world. In the U.S., Boise (Idaho), Klamath Falls (Oregon), and most recently Susanville (California), space heat large sections of their communities with fluids ranging from 60-90°C (140-195°F). Bridgeport (California), south of Lake Tahoe, may be the next community to install a heating system. Included in the category of district heating, are "Parks of Commerce" where several end-users share the heated fluids for a variety of processes.

Agricutural is the most important industrial sector that can convert to direct use. Most geothermal resources have temperatures (50°-150°C or 120°-300°F) that match requirements of agri-industrial processes. Specifically, intensive, confined growing in greenhouses was determined to be the best application which could immediately use geothermal for space heating. Facing high interest rates and fierce foreign competition, the greenhouse industry, with over 3,000 acres under glass in California alone, is extremely sensitive to further increases in heating costs which have escalated more than 400 percent in the last five years. Several operators have already moved their operations to take advantage of geothermal energy. In addition to greenhouse operations, food processing and methane generation were also identified as having significant potential for successful near-term development.

Small-scale electric systems, 'well-head generators' (ranging in size from 40 kW to 10 MW) are beginning to be sold throughout the western U.S., despite uncertainty regarding operational performance and system reliability. These systems can dramatically improve the cost-effectiveness of developing many moderate temperature (90-150°C or 200-300°F) resources in the state. By cascading the heat from the well-head generator to direct use applications, several district heating and agri-business complexes become economically feasible. However, lacking a commercial track record, these small-scale electric systems may require unique approaches to financing, design, warranty agreements and serious attention to reliable operation and maintenance.

IMPEDIMENTS TO DEVELOPMENT

Following the assessment and ranking of resource areas and end-use applications, the next step was to identify the major impediments preventing the widespread adoption of geothermal energy in these areas and by these end-users. CEC geothermal staff, in preparing for the fourth Biennial Report to the Legislature, identified four major impediments: resource risk, capital availability, lack of technical expertise and regulatory disincentives.

1. While up to 50 percent of the total costs of a direct use project can be spent for resource development, there is no guarantee the well will produce the necessary temperature, flow rate or water quality for the application. Well costs increase exponentially with depth, making resource confirmation and development the major expense and greatest risk of a geothermal venture.

2. Given high interest rates and generally poor economic conditions, the lack of capital availability (at reasonable cost) for direct use projects are significantly slowing adoption of this technology. Direct use geothermal projects require large amounts of capital to develop the resource and delivery systems. The cost of financing must be added to the conventional project construction costs, and is the single largest annual expense.

3. Although geothermal resources occur in 46 of California's 58 counties, and are readily available, the following factors contribute to retarding rapid market development: a) the general public has little knowledge regarding possible low-temperature geothermal applications and the attendant problems are often greatly exaggerated; b) developments will not be seriously considered until they are proven feasible from both an engineering and an economic basis; and c) legal, institutional and environmental constraints peculiar to direct use development are beyond the experience of many communities and potential users.

4. Excessive regulations significantly impede development as most regulatory agencies continue to incorrectly perceive direct use development as a high risk venture with the same impacts as large-scale electric development at the Geysers or Imperial Valley. To achieve a maximum reasonable increase in use, regulatory agencies should recognize that, in most cases, direct use projects employ water well technology to drill to shallow depths and result in an environmentally benign development.

MARKET DEVELOPMENT

Successful commercial development of priority resource areas and end-use applications requires an intensive, cooperative effort between business and government that is aimed directly at mitigating the four major impediments.

Candidate projects at priority areas which employ priority applications, have been defined and analyzed as a continuation of the market development contract. Potential agricultural complexes have been identified at Wendel and East Mesa. The former includes small-scale electric and greenhouse operations and the latter includes animal feed production, swine raising and food processing. Heating districts cascaded from a small-scale electric system have been examined for Calipatria and Bridgeport. This market assessment effort confirmed that responsible businessmen are seriously considering this form of alternative energy. Their participation is expected to be paced only by the general economy and high cost of money. These businessmen are financially committed to their own businesses. Most expect that a community, private developer or cooperative district energy group will develop and supply the geothermal energy in a manner similar to other utilities (e.g., water and sewer).

To develop a market infrastructure and stimulate new projects, the CEC staff presents the results of the market study to communities near high priority resource areas. Emphasis is on explaining the geothermal development as a method of attracting industry and encouraging the area's economic growth. For industries rated as high priority end-use applications, contacts have been established with leaders in the industry, and information packets addressing the benefits of direct use for the specific industry can be furnished. Presentation at trade association meetings and articles for trade journals are also being implemented to communicate the benefits to the entire industry.

The CEC offers technical assistance for geothermal direct use and small-scale electric projects, through a contract with the Oregon Institute of Technology (OIT). In-field technical support services are offered to resource owners, developers and potential end-users. Such services include on-site investigations and consultation, and preliminary assessment of a project's engineering and economic feasibility. The assistance determines initial feasibility and can assist the developer in obtaining financial support by reducing investment risk.

Federal lease revenues dispursed to certain counties and to the CEC, provide funds for geothermal planning and development. With its portion of the revenues, the CEC has established the Local Government Grant Program. These grants can be used by local government entities for resource confirmation, as capital for demonstration projects or to plan for future development.

Direct use and small-scale electric demonstration projects can be partially funded by the CEC. Such projects will provide actual operating experience, produce data on performance and costs, and serve as models for duplication elsewhere by the private sector.

BIBLIOGRAPHY

"Guide to Financial Assistance for Geothermal Energy," March 1982, California Energy Commission (Publication No. P500-81-030).

Larson, Tod, May 1980, "Direct Use of Geothermal Energy in California: A Market Survey". Prepared by Science Applications, Inc. for the California Energy Commission (Publication No. P500-81-027).

Longyear, A.B. and J.C. Jeskey, Editors, July 1976. "Susanville Geothermal Energy Project - Worklshop Proceedings" City of Susanville and Department of Energy (SAN-1077-4).

Longyear, A.B., April 1978. "Survey of Selected Industries for Potential Branch Plants - Susanvile Geothermal Energy Project," Fred Longyear Co.

Longyear, A.B. et al., January 1982. "Geothermal Energy: Opportunities for California Commerce (Phase I)." Prepared by Lahontan, Inc. for the California Energy Commission (Publication No. P500-82-008).

Longyear, A.B. et al., June 1982, "Geothermal Energy Opportunities for California Commerce (Final Report)." Prepared by Lahanton, Inc. for the California Energy Commission.

"Map of Geothermal Resources of California," 1980. Division of Mines and Geology, California Department of Conservation.

TABLE 1

PRIORITY AREA EVALUATION CRITERIA

A. RESOURCE DEVELOPMENT

• TEMPERATURE	Above 50°C (122°F)
• WELL DEPTH	Less than 760 m (2,500 ft) Prefer 460 m (1,500 ft.) or less
• TOTAL DISSOLVED SOLIDS	Less than 5,000 ppm Prefer 1,000 ppm or less
• DISTANCE TO POTENTIAL APPLICATION SITE	Less than 1.6 km (1 mi)
• LOCAL PERMITTING FOR TOTAL PROJECT	Less than 9 months

B. ECONOMIC DEVELOPMENT FACTORS

- ADEQUATE LABOR BASE
- COMMUNITY SERVICES
- CONCENTRATED HEAT LOAD
- LAND COSTS
- ATTRACTIVE LIVING CONDITIONS
- OVERRIDING FACTORS
- ADEQUATE TRANSPORTATION TO MAJOR MARKETS
- RAW MATERIALS
- REASONABLE UTILITIY RATES
- FAVORABLE BUSINESS CLIMATE
- LOCAL FINANCIAL INSTITUTIONS

TABLE 2

TARGET COMMUNITIES IN GEOTHERMAL RESOURCE AREAS

Priority I	Priority II		Priority III
• Susanville	• Calistoga	• Sonoma Valley	• Napa Valley
• Litchfield	• Kelly Hot Springs	• Wendel - Amedee	• Surprise Valley
• Salton Sea Field	• Sierra Valley	• Bridgeport	• Likely
• Ontario Hot Springs	• Mammoth Lakes	• Lake Isabella	• Bassett-Kellog Springs
• Lake Elsinore	• Trona	• Desert Hot Springs	• Twenty-Nine Palms
	• Mecca	• East Mesa Field	• Winchester Area
	• Pasa Robles	• L.A. Basin	• Warner Hot Springs
	• San Bernardino		

FIGURE 1

PRIORITY GEOTHERMAL RESOURCE AREAS

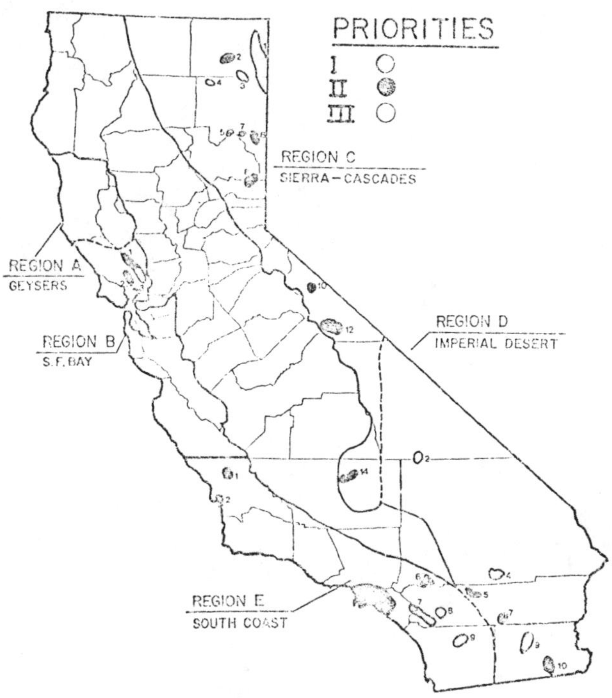

TABLE 3

PRIORITY APPLICATION CRITERIA

- ECONOMIC OUTLOOK
- ENERGY SENSITIVITY
- REPLICATION
- GEOTHERMAL COMPATIBILTIY
- ECONOMIC DEVELOPMENT FACTORS
- HISTORICAL USE (WORLDWIDE)
- BRANCH PLANTS
- OVERRIDING FACTORS

Geothermal Resources Council, TRANSACTIONS Vol. 6, October 1982

PRICING AND MARKETING GEOTHERMAL ENERGIES FOR SPACE HEATING IN BOISE, IDAHO

C. Mike Merz, CPA
Professor of Accounting
Boise State University
Boise, ID 83725

and

Phil Hanson, Director
Boise Geothermal
Boise, ID 83702

ABSTRACT

The Boise Geothermal System will begin delivering geothermal water for space heating in the fall of 1982. Boise Geothermal represents one of the first district heating systems to become operational from the ERDA and DOE programs to stimulate geothermal development. This paper discusses the economic aspects of the project. The pricing policy and operating cost structure are compared, along with the role of a drilling partnership from the private sector. Also described are Boise Geothermal's marketing activities to date and the obstacles that have been encountered in trying to convince building owners to convert their heating system to use geothermal water. The paper presents actual financial data and describes actual marketing experiences. Although the literature contains many feasibility studies, Boise Geothermal is the first of the new heating districts for which actual data are available.

INTRODUCTION

In October 1975, when the initial grant application was made to the Energy Research and Development Agency (ERDA), the Boise Geothermal System represented a "pie-in-the-sky" system concept; by fall 1982 it will have evolved into a fully operational district heating system which supplies geothermal energy for space heating to downtown Boise. Early feasibility studies indicated that the project made economic sense, i.e., that geothermal energy could be sold at a price that would cover operating costs but that was below the prices of alternative energy sources. All costs are now either known or can be estimated with considerable accuracy. A pricing policy has been established. Contracts are being negotiated with end-use customers. The actual financial data becoming available indicates that Boise Geothermal should achieve its goal of being financially self-supporting if the current marketing obstacles can be overcome.

SYSTEM HEATING CAPACITY

When it is completed in the fall of 1982, the Boise Geothermal System will consist of four production wells and distribution and disposal pipelines. Pump draw down tests which were completed in May, 1982 indicate a peak capacity in May, 1982 indicate a peak capacity of about 4,000 gallons per minute (GPM) at a temperature slightly above 170 degrees F. The distribution pipeline is 9,750 feet long and varies in diameter from 14 inches at the well heads down to 8 inches at end of the pipeline in downtown Boise. The pipeline has a nominal design capacity of 4000 GPM.

With 5,833 average degree days per year in Boise, and a minimum design temperature of 0 degrees F, the system will operate with a 25 percent average load factor. The number of degree-days occurring in a local area indicates the demand that the outside weather temperature places upon a building's heating system. When the outside air temperature drops below 65 degrees F, customers turn on their heating systems. If the outside air temperature stayed at 0 degrees F for a 24 hour period, 1 day, then a heat load of 65 degrees days would result. The theoretical maximum heat load would be 23,725 degree-days which would occur if the outside air temperature remained at 0 degrees F all year. With 5,833 average degree days, Boise's average annual load factor will be 25 percent (5,833 degree-days divided by 23,725 degree-days).

The load factor means that only 25 percent of the available energy will be sold each year. Customers will be added to the system until the aggregate customer demand reaches 4000 GPM when the outside air temperature drops to 0 degrees F. Below 0 degrees F, customers must depend on their backup heating system. In a typical year, however, the outside air temperature in Boise remains at or below 0 degrees F for only several hours. Therefore, the system's peak delivery rate will rarely reached.

The amount of heat extracted from geothermal water depends upon the temperature drop achieved by the building's heating system. All of Boise Geothermal's cost and pricing analyses assume a 50 temperature drop, or "delta T."

Boise Geothermal's heating capacity, then, depends upon three system parameters:

1) A peak flow rate of 4000 GPM.

2) An overall system load factor of 25 percent per year.

3) A temperature drop of 50 degrees F in

building's heating system which then extracts 417 BTUs from each gallon of geothermal water.

From these system parameters, the heating capacity of the Boise Geothermal System was calculated at 2.2 million therms per year. Pricing analyses assumed that this amount of heat would be sold each year.

PRICING POLICY

Early economic analyses indicated that Boise Geothermal should be able to deliver geothermal energy at a price which could be below alternative energy sources but which could still cover all costs. The Project Board of Control recognized that a significant price incentive would have to be present to motivate building owners to retrofit their existing heating systems for geothermal use. The Board also desired to operate the system at a profit which could be used to finance future growth. After considering preliminary cost estimates and comparing these cost estimates with the price of natural gas, the Board somewhat arbitrarily established the policy that Boise Geothermal would supply energy at least 30 percent below the price of natural gas.

In the Boise area, the Intermountain Gas Company (IGC) a privately owned utility, supplies natural gas. The average price per therm varies with: 1) the type of customer, 2) the quantity of gas consumed and 3) the season of the year, either heating or non-heating. The average prices in effect at June 1982 and expected to remain in effect until November 1982 are:

Customer Type	Average Price Per Therm
Residential	$0.64
Commercial	$0.56

Since a significant percent of the therms of natural gas which a customer purchases escape in the combustion process, the number of useful therms is less than the purchased therms. Therefore, the price of natural gas was adjusted for the relative efficiencies of natural gas and geothermal heating before applying the 30 percent price reduction for geothermal energy. Average heating efficiencies expressed as the percent of useful heating therms to total therms purchased were estimated at 75 percent for natural gas and 100 percent for geothermal. Adjusting the purchase prices of natural gas for these efficiencies and applying the 30 percent price reduction set as Boise Geothermal's policy results in the following maximum prices which Boise Geothermal may charge:

MAXIMUM PRICES FOR GEOTHERMAL ENERGY SET AT 70 PERCENT OF NATURAL GAS PRICES BASED ON INTERMOUNTAIN GAS COMPANY'S RATE SCHEDULES EFFECTIVE AT JUNE 1982

Type of Customer	Price Per Therm
Residential	$.60
Commercial	$.52

The minimum prices which Boise Geothermal might charge depend upon the costs to produce and distribute geothermal water. Boise Geothermal's costs are heavily influenced by an agreement with a drilling partnership formed to produce and sell water to Boise Geothermal.

ENTER THE PRIVATE SECTOR

When it became apparent that the Department of Energy (DOE) grants to Boise Geothermal would not be large enough to construct the desired geothermal system, participation was sought from the private sector. After almost two years of discussion and negotiations with a number of entrepreneurs and companies in the private sector, Boise formed an agreement in August 1980 with a limited partnership which agreed to drill the production wells. Boise's agreement with the partnership (subsequently referred to as the "drilling partnership") contained the following main provisions.

MAIN PROVISIONS WITH DRILLING PARTNERSHIP

1) Boise granted the drilling partnership a 30 year lease to the mineral rights to extract geothermal water from the property believed to encompass the geothermal resource.

2) The drilling partnership agreed to drill sufficient wells to produce 2000 GPM. The first well had to be drilled and tested within one year.

3) Boise agreed to obtain contracts for the use of 2000 GPM of geothermal water and up to 4000 GPM.

4) Boise agreed to purchase the water from the drilling partnership at a 46 percent discount from the price of natural gas in effect when the water was purchased. This provision tied the price of geothermal energy to natural gas prices.

The terms of this agreement were negotiated and agreed to with the specific intent of compensating the drilling partnership for incurring the risks inherent in drilling wells to search for an unproven resource.

By June 1982, the drilling partnership had drilled four wells with a combined peak production capacity of about 4,000 GPM. Since it is a private entity, the drilling partnership's investment in the project is not public information. The drilling partnership's total investment has been estimated, however, at about $2.1 million.

BOISE GEOTHERMAL'S COST STRUCTURE

Boise Geothermal will purchase water from the drilling partnership, distribute and sell the water to end-user customers and collect and dispose of the spent water. The total costs incurred by Boise Geothermal will include four cost elements.

1) Purchase From Drilling Partnership - At the price of natural gas in effect at June 1982

and expected to be in effect through November 1982, Boise Geothermal will pay the drilling partnership $.30 per therm equal to 54 percent of the natural gas price to commercial customers of $.56 per therm.

2) Depreciation - The total investment in the distribution and collection pipelines of slightly over $2 million will be amortized over 50 years with straight line depreciation of $40,000 per year.

3) Boise Geothermal's expenses for billing and administration are estimated at $35,000 per year comprised of $25,000 labor and $10,000 materials.

Cost items 2, 3 and 4 are considered fixed costs in that the total dollar amount of these costs will not vary with the amount of energy delivered by the system. The per therm cost of these items will depend upon the total therms delivered. The total per therm cost of geothermal energy is summarized in the table below for peak flow rates of 2000 GPM and 4000 GPM.

BOISE GEOTHERMAL'S COSTS
FOR PEAK FLOWS OF 2000 AND 4000 GPM
DOLLARS PER THERM

	2000 GPM Peak	4000 GPM Peak
To Drilling Partnership	$.302	$.302
Depreciation	.036	.018
Billing & Administration	.014	.007
Maintenance	.032	.016
Totals	$.384	$.343

As discussed above, the $.041 per therm difference results from spreading the fixed costs of depreciation, billing, administration and maintenance over a different number of total therms delivered.

PRICE - COST RELATIONSHIP

The upper and lower bounds of the price that Boise Geothermal will charge its customers for geothermal have been established. In order to motivate building owners to connect to geothermal energy, the Project Board of Control established a policy that geothermal energy should be at least 30 percent below the price of natural gas. This sets an upper limit on Boise Geothermal's prices at $.52 and $.60 respectively for commercial and residential users.

Boise Geothermal also established a policy that the geothermal system should be self-supporting, i.e., revenues from the sale of energy should cover all costs. The lower limits on prices that must be charged are the costs that Boise Geothermal will incur to purchase water from the drilling partnership and distribute it to end-users. This cost was estimated above at $.38 or $.34 per therm depending upon the volume of water sold. The following table summarizes the range within which Boise Geothermal will set its final price.

PRICE RANGE WITHIN WHICH
BOISE GEOTHERMAL WILL SET
FINAL PRICES TO CUSTOMERS

	Per Therm
MAXIMUM PRICE - set at 30 percent below natural gas prices........	$.52
MINIMUM PRICE - to cover all costs to purchase and distribute water..........................	$.34

Comparison of these maximum and minimum prices indicates that Boise Geothermal should not have much difficulty in setting prices which will cover all costs and which will still offer a significant incentive for building owners to convert to geothermal energy.

MARKETING STRATEGY

Concurrently with the technical and economic analyses, Boise Geothermal initiated a comprehensive marketing program. The purpose of this effort was to ensure that end user contracts to purchase geothermal energy would be finalized by the 1982-83 heating season when the system was ready to begin delivering hot water. This marketing program included the following activities.

1) Publicity - In order to make the general public aware of Boise's geothermal program, a large number of press releases, radio and television presentations, and speeches to various local groups were made.

2) Building Surveys - Each building in the downtown core area was carefully surveyed to determine if conversion to geothermal energy appeared practicable. The owners of about 150 buildings for which retrofit appeared feasible received a letter which asked if the owner wished to be included in the plans for a geothermal heating district. Almost all of these owners expressed an interest by signing and returning the letter. To a great extent, the pipeline route was designed to serve these buildings.

3) Letter-Of-Intent - A letter-of-intent was signed with interested building owners. Boise Geothermal guaranteed to deliver water at 170 degrees F at a price at least 30 percent below the price of natural gas. The building owner indicated his or her intent to convert to geothermal energy, if conversion turned out to be technically and economically feasibile. At June 1982, 45 downtown buildings have signed letters-of-intent out of a total of 142 buildings for which conversion may be feasible. The letters-of-intent which do not constitute a legal obligation, promise that Boise Geothermal will notify building owners six months before completion of the pipeline.

4) Free Consultation - A few hours of free engineering consulting were provided to some building owners to determine if conversion was technically feasible.

5) Site Specific Study - For a few, very large buildings, a comprehensive technical and economic feasibility study was completed. In addition to a preliminary design, this study showed: (1) the investment required to convert, (2) the annual savings in the building's heating bill, and (3) the rate of return on the investment under various financing alternatives.

By June 1982, many discussions had been held with the owners of many different types of buildings. With respect to potential use of geothermal energy, buildings seem to fall into several different categories.

BUILDING CATEGORIES

Each building presents a unique combination of characteristics which influences the potential for retrofit. Four dominant categories emerged from the many discussions and analyses.

1) Conversion not technically feasible - many buildings have heating systems, such as electric baseboard heaters, which cannot be retrofitted for geothermal energy.

2) Good existing system - some buildings have heating systems which could be converted, but their owners are not interested because the existing systems operate satisfactorily.

3) Already improved efficiency - some buildings have achieved significant improvements in heating efficiency over the last several years and the additional savings possible with geothermal energy does not warrant the required investment. For example, one large office building in Boise with over 200,000 square feet of floor space has achieved a 50 percent reduction in the number of therms of natural gas used per degree day between 1978 and 1981. Also, many of the public schools in Boise have achieved dramatic improvements in heating efficiency by making a large number of minor operating improvements. It is difficult to convince a building owner to invest $100,000 to save 30 percent on his or her heating bill after the same owner has just reduced the heating bill by 50 percent with practically no investment.

4) Conversion or replacement mandatory - some older buildings with inefficient heating systems have experienced such large increases in their heating bills over the past several years that the owners feel compelled to convert to geothermal energy or completely replace the existing system. Conversion to geothermal energy also appears desirable for many of the commercial and residential buildings which are being renovated near Boise's downtown core area.

Having several of the large, downtown commercial buildings convert to geothermal was a high priority for Boise Geothermal. The large buildings would be very cost-effective customers because they would use a relatively high volume of water compared to the required investment to bring the pipeline to the building. In addition, high public visibility would result from having a large, prominent building convert. Discussions with owners of large buildings, however, revealed several obstacles to conversion to geothermal.

MARKETING OBSTACLES

The most serious marketing obstacle which must be overcome appears to be the perceived risk in converting to geothermal energy. Building owners believe that electricity and natural gas will be available into the foreseeable future. The building owners are not certain that geothermal energy will be available forever. How long will the wells continue to produce? How do we know that the temperature won't drop over the years? Who will pay for and drill new wells if the existing wells fail? What about the bad odor from the hot water? Won't the water corrode our heating system? Will the city have enough money to keep the pipeline repaired? These are the types of questions asked by building owners. Some of the perceived risks appear mostly emotional and cannot be effectively addressed with purely rational answers. An important technical risk is whether a geothermal heating system in a large building will achieve the predicted operating efficiency. If the heating system extracts only 40 degrees F from the geothermal water rather than the assumed 50 degrees F, then 20 percent of the purchased therms will remain in the discharge water. No actual experience with geothermal conversions in large buildings appears available to prove that the projected efficiencies will occur.

Financial risk is also an important obstacle to conversion to geothermal energy by the large buildings. With the current high interest rates, achieving a satisfactory Internal Rate of Return (IRR) and a fast enough payback becomes difficult. For example, and IRR of about 18 percent and a payback period of five years were predicted for one large building. These were after the Geothermal Investment Tax Credit and the Economic Recovery Tax Act of 1981 provisions for accelerated depreciation. Compared to other investment alternatives, such financial performance is hardly spectacular and the building owners ask questions like, "what if your operating costs are higher or what guarantee do I have that the price won't increase faster than expected?" These marketing obstacles pertain mostly to large, commercial buildings. A number of residences in the Warm Springs Water District have converted to geothermal energy in the last few years. Most of the owners of these residences have experienced favorable results and they are quite enthusiastic about use of geothermal energy.

SUMMARY

At June 1982, Boise Geothermal is about to begin construction of a pipeline to distribute geothermal water to downtown Boise. Pricing and cost analysis indicate that geothermal water can be sold to customers for considerably less than natural gas. For a number of reasons, however, many building owners are reluctant to convert their existing space heating systems to use geothermal energy.

Geothermal Resources Council, TRANSACTIONS Vol. 6, October 1982

GEOTHERMAL RESOURCE, ENGINEERING AND ECONOMIC EVALUATION FOR THE CITY OF OURAY, COLORADO

Richard T. Meyer
Harold Prostka
Rob Raskin
John R. Zocholl

Western Energy Planners, Ltd.
11100 East Mississippi Avenue, Suite 1-102
Aurora, Colorado 80012

ABSTRACT

A geothermal energy feasibility study has been performed for the City of Ouray, Colorado, to determine the potential economic development opportunities to the City. The resource assessment indicates the resource to be associated with the Ouray fault zone, the Leadville limestone formation, the high thermal gradient in the area of the San Juan mountains, and the recharge from precipitation in the adjacent mountains. Four engineering designs of alternative sizes, costs, applications, and years of start-up have been defined to offer the City a range of development scales. Life cycle cost analyses have been conducted for cases of both public and private ownership. All systems are found to be feasible on both economic and technical grounds.

INTRODUCTION

The City of Ouray, Colorado, with partial financing from a DOE Appropriate Energy Technology Small Grant Award, has sponsored a geothermal resource, engineering and economic feasibility study. The purpose of the study project has been "to explore new ways for the City to use its geothermal energy for economic development and diversification." The project objectives have been (1) to create new tourism and commercial business opportunities, (2) to generate additional revenues for the City and its businesses, (3) to reduce energy costs for city and county buildings, for commercial businesses, and for residences, (4) to achieve year-round optimum use of geothermal energy, and (5) to preserve the rights and uses of geothermal springs and waters by parties already engaged in or entitled to use of geothermal energy.

RESOURCE ASSESSMENT

Geologic Setting: Ouray is located in a region of high heat flow, perhaps four times that of the world average (Zacharakis, 1981). Active faulting and a moderate to high level of seismic activity also characterize the region. Faults less than 2 million years old on the Uncompahgre Plateau are only 22 miles northwest of and on strike with Ouray (Howard and others, 1978). A seismic study of the Ridgway fault zone 12 miles north of Ouray shows it to be seismically active (Sullivan and Martin, 1981). Ouray lies at the intersection of several major faults. The Ouray fault that is associated with the hot springs at Box Canyon is one of several faults of WNW to ENE trend that make up the complex, 3-mile wide Ouray Fault Zone (O.F.Z.) that is the boundary between the Ouray Graben and Sneffels Horst.

The Ouray Geothermal System: The thermal springs at Ouray are probably the result of deep circulation of meteoric water in fractures and permeable sedimentary rocks of the Ouray Graben. The graben contains a maximum of 8800 feet of sedimentary rocks; about 3300 feet are aquifer zones. Within the graben, groundwater flow is probably largely confined to the two main aquifer zones. But in the basin margin fault zones, fracture permeability permits downward recharge of the aquifers and upward flow of deep thermal waters, as at Ouray and Ridgway. Recharge is mostly from the San Juan Mountains along the south flank of the graben, through steep intersecting fractures of the Ouray and other fault zones. The geothermal gradient is sufficiently high to heat the deeply circulating groundwater. The flow path, from recharge to discharge area, is much shorter for the Ouray thermal water than for the water of the Ridgway area. The resulting difference in subsurface residence time of the waters of the two areas is reflected in consistent differences of water chemistry and temperature. Orvis thermal water is more mineralized (2300 vs. 1650 ppm TDS) and has a higher geochemical reservoir temperature (78°C vs. 70°C) than Ouray thermal water (Barrett and Pearl, 1976, 1978).

Temperature within the graben can be estimated from the depth of fill and the geothermal gradient. The total sedimentary-volcanic section above PE basement is a maximum of 8800 feet thick. The geothermal gradient is 40°C/km, which amounts to a temperature difference of 117°C between the top and bottom of the section. Adding the average annual surface temperature of 7°C (Benci and McKee, 1977) gives 124°C as the maximum expectable temperature at the bottom of the graben. The "average" temperature within the graben would be about half that or 66°C, which checks closely with the 69°C temperature of the Box Canyon Pool Spring at Ouray. The chalcedony-silica geothermometer yields a subsurface temperature of only 69°C to 71°C (Barrett and Pearl, 1978). This is very close to the surface temperature of Pool Spring and may mean that the thermal waters are ascending rapidly with little cooling or high-level dilution.

The rate of recharge can be estimated from the average annual precipitation, which is about 40 inches per year (U.S. Weather Bureau Map). This computes to 93 million cubic feet of water per year per square mile, or 177 cubic feet per minute per square mile. Since 1 cubic foot = 7.48 gallons, the precipitation rate is 1323.5 gallons per minute per square mile. Only part of this water actually recharges the Ouray geothermal system. We assume that only one-eighth of the recharge, or about 3300 gallons per minute of hot water, is ascending beneath Ouray. The measured natural discharge from the thermal springs at Ouray ranges from 60 to 200 gpm. The greater part of the geothermal resource is apparently leaking out of the system elsewhere. A considerable amount of hot water may be lost up bedrock fractures into shallow surficial deposits, becoming diluted with cold groundwater.

It isn't possible to predict what effect drilling and pumping would have on the natural springs at Ouray. The permeability of the LL and fracture zones is unknown, hence the potential productivity of the system cannot be calculated. However, because the recharge rate is high, and the temperature and chemistry of the waters seem to indicate rapid ascent, the indications are favorable that the system may have the capacity to sustain a ten-fold increase in discharge without adversely affecting the existing springs.

ENGINEERING DESIGNS

Four engineering designs have been evolved to provide a range of development and/or application options for the City of Ouray. The City may choose to initiate geothermal development with the smallest scale option and proceed sequentially to enlarged geothermal systems until a complete city-wide district heating system is achieved. Or the City may choose to start out with an intermediate size system and progress from that design. However, the City may also choose to proceed with design and construction of a full-scale city-wide geothermal system. The four engineering designs presented herein give the City the opportunity to examine at this time the engineering design factors, the capital costs, and the construction aspects of four alternative size systems. The four engineering designs are briefly described here.

Case A: Space and Hot Water Heating of Public Buildings: The existing pipeline and geothermal waters from the Box Canyon Pool Spring to the Municipal Pool would be the source of hot water; no geothermal wells would be drilled. The existing pipeline would be tapped into at the base of Sixth Avenue and a portion of the Pool Spring water would be pumped up to the public buildings through a newly-installed insulated pipeline. The new pipeline system would serve the City Hall, the County Courthouse, the Museum, the School Gymnasium, and the proposed Emergency Services Building. The discharge water would be returned through a parallel insulated pipeline to the Pool Spring Pipeline at the base of Sixth Avenue and added to the flow to the Municipal Pool. The water in the existing pipeline is estimated to be 155°F and flow at 200 gpm. Under the peak demand situation, 124 gpm at 155°F would be required to provide thermal energy to the five buildings. This 155°F geothermal water would be lowered in temperature to 120°F and returned to the pool pipeline at 115°F. The resultant water flowing to the pool would be 130.2°F at 200 gpm, which represents the "worst case" condition.

Case B: Same as Case A, Except with the Addition of Space and Hot Water Heating for the Municipal Bathhouse and a Prospective Year-Round Recreation Center: Case B is basically the same as Case A, except that it provides for year-round use of the Municipal Pool, which requires space heating and hot water for the existing Bathhouse; it also allocates water from the Box Canyon Pool Spring pipeline to provide space and hot water heating for a prospective new year-round Recreation Center, which would be located in the immediate vicinity of the Municipal Pool and Bathhouse. Capital costs for Case B include retrofit costs for the Bathhouse but not building or heating system costs for the Recreation Center; only the energy load of the Recreation Center is incorporated into the design.

Case C: Main Street Commercial Strip, Including Public Buildings, Municipal Bathhouse and Recreation Center: Case C is designed to be a completely new geothermal system of geothermal wells and pipeline system. The design provides for two geothermal wells to be drilled at the northern edge of the City, to a well depth of 800 to 900 feet into the Leadville Limestone Formation. The two wells would be connected to a pipeline system which comes up Main Street and which has side-street branches in order to service most of the existing commercial businesses and motels. The Main Street pipeline would also connect with a major branch line to service the public buildings which are defined in Case A. Estimating the geothermal resource to provide 175°F and 400 gpm per well, two wells would be required to meet the thermal needs for the above locations. The total capital costs for Case C, including retrofit of the five buildings in Case A and all engineering design and contingencies, is estimated to be $720,000 in 1982 dollars.

Case D: Year 2000 City-Wide District Heating System: Case D is an engineering design incorporating six geothermal wells at the north end of town and a pipeline array which would supply most of the residences, commercial establishments, and light industry in the immediate area of the City in the Year 2000. The Year 2000 energy demand estimate is approximately 200 billion BTU per year. Peak commercial and residential demands are estimated equally at 30 million BTUH each. Estimating the geothermal resource to be available at 175°F and 400 gpm per well, six wells would be required to meet the city's thermal energy requirements.

The total estimated cost to develop the wells, for the well, circulating and booster pumps, including engineering design costs and contingencies, in 1982 dollars is $875,000. The total estimated cost for the well collection system and the transmission line, including engineering design, installation, and contingencies, is estimated to be $1,875,000 in 1982 dollars. The total capital costs for Case D is then computed to total $2,750,000.

ECONOMIC EVALUATIONS

Summary of Costs: Table 1 summarizes the estimated exploration, capital equipment, working capital, and annual operating and maintenance costs for the four engineering designs. Also listed in Table 1 are the total BTU equivalents of fossil fuel to be displaced by the geothermal systems, the total investment cost per MMBTU displaced, and the annual O&M cost per MMBTU displaced. The latter two sets of numbers suggest that significant economies of scale can be achieved with the larger geothermal systems.

Results of Evaluations: Table 2 summarizes the results. In each privately-financed case, the desired rate-of-return is specified and the first-year MMBTU-price to the Ouray consumer is calculated. For the publicly-financed case, the break-even price, which meets all operating, maintenance, and bond indebtedness costs, is determined. For comparison, Table 3 presents the expected costs of fuels currently in use in Ouray. It is seen that under all options, geothermal energy is highly competitive with propane and fuel oil. The highest first year geothermal price is Case A - Private at $5.39/MMBTU in 1983. Propane fuel oil prices are expected to be $10.10/MMBTU in that year. The previously projected effect of the economy of scale for Case D is confirmed here. At $2.70/MMBTU in 1986, Case D is also competitive with coal, which is projected to cost $3.63/MMBTU in that year. Year by year cash flows for the five options are available and have been provided to the City of Ouray.

Table 1

Cost Summary (1982 Dollars)

Cost Element	Case A	Case B	Case C	Case D
EXPLORATION COSTS	-	-	$30,000	$50,000
CAPITAL COSTS				
Well and/or Pump Costs	$10,000	$10,000	$279,700	$875,000
(No. of Wells)	(0)	(0)	(2)	(6)
Transmission Line Costs	$80,400	$80,400	$287,500	$1,875,000
Building Retrofit Costs	$21,800	$32,800	$32,800	-
Total Capital Costs	$112,200	$123,200	$600,000	$2,750,000
WORKING CAPITAL	$16,800	$18,500	$90,000	$412,500
TOTAL INVESTMENT	$129,000	$141,700	$720,000	$3,212,500
ANNUAL O&M COSTS	$11,900	$12,200	$28,600	$121,000
MMBTU DISPLACED	4,900	6,600	31,000	233,000
TOTAL INVESTMENT PER MMBTU DISPLACED	$26	$21	$23	$14
ANNUAL O&M COST PER MMBTU DISPLACED	$2.42	$1.85	$0.92	$0.52

Table 2

Economic Evaluation Results Summary

Parameter	Case A (Public)	Case A (Privately owned & operated franchise)	Case B	Case C	Case D
1st Year of Operation	1983	1983	1983	1985	1986
1st Year $/MMBTU	$3.50	$5.39	$4.22	$4.10	$2.70
DCFROR	---	15%	15%	20%	20%

Table 3

Projected Conventional Fuel Prices (Dollars per MMBTU)

	Propane and Fuel Oil	Coal
1982	$ 9.10	$2.50
1983	10.10	2.75
1984	11.21	3.02
1985	12.45	3.33
1986	13.81	3.63

REFERENCES

Barrett, J. K. and Pearl, R. H., 1976, Hydrogeological data of thermal springs and wells in Colorado: Colorado Geological Survey Information Series 6, p. 124.

Barrett, J. K. and Pearl, R. H. 1978, An appraisal of Colorado's geothermal resources: Colorado Geological Survey Bulletin 39, p. 229.

Benci, J. F. and McKee, T. B., 1977, Colorado monthly temperature and precipitation summary: Colorado.

Climatology Office, Climatology Report No. 77-1, Colorado State University.

Howard, K. A., and others, 1978, Preliminary map of young faults in the United States as a guide to possible fault activity: U.S. Geological Survey Map MF-916.

Sullivan, J. T. and Martin, R. A., 1981, Seismic hazard studies for Ridgway damsite, Colorado: Colorado tectonics, seismicity, and earthquake hazards: Colorado Geological Survey Special Publication 19, p. 30-31.

Zacharkis, T. G., 1981, Revised heat flow map of Colorado: Colorado Geological Survey Map Series 18, scale 1:1,000,000, DOE/ET/28365-12.

Geothermal Resources Council, TRANSACTIONS Vol. 6, October 1982

GEOTHERMAL DOMESTIC WATER HEATING AT WARM SPRINGS STATE HOSPITAL IN MONTANA

Allan D. Miller

MultiTech, Inc.
P.O. Box 4078
Butte, Montana 59702

ABSTRACT

We are presenting the background and necessary steps that were followed in developing a demonstration geothermal project at the Warm Springs State Hospital in Warm Springs, Montana. This Department of Energy sponsored program, under Cooperative Agreement No. DE-FC07-78ET-27055, was initiated to encourage the development of similar geothermal resources and to utilize the available geothermal energy to replace the natural gas utility requirements of the facility.

INTRODUCTION

Initial work involved obtaining and analyzing the components of the energy distribution system at Warm Springs State Hospital to gather pertinent data on the energy demand of the facility. Degree days and ambient temperature calculations were made to determine building heating requirements. Analysis of the central system plant loads were made to confirm the building heating needs and to substantiate the domestic hot water demand cycles during the nominal 12 hour day.

Recommendations from the preliminary work were to supply the domestic hot water requirement of 30,000 gallons per day (885,000 BTUH) and peak space heating requirements for the Warren Building, main hospital building, (456,000 BTUH) and the Food Service Building (1,943,000 BTUH) with the geothermal resources. A proposal for a production well, reinjection well, and the necessary conversion components was submitted to the Department of Energy Program Opportunity Notice for Field Experiments for Direct Uses of Geothermal Energy. That proposal was approved by the DOE and the project was underway.

ENVIRONMENTAL REPORT AND LEGAL/STATUTORY REVIEW

An environmental report was prepared to determine any adverse environmental or socioeconomic impacts resulting from the demonstration project. The report indicated that very little negative environmental impact would result from the development and demonstration of the geothermal resource. There were also no long-term effects in the social or economic environments anticipated.

A review of the permit requirements substantiate that well drilling permits and production well permits were required. Initial design and environmental considerations provided an option to discharge the spent geothermal water to a wetlands area for game habitat. No permit was required for discharge of spent geothermal water into an artificial impoundment, however, a permit is being obtained from the Montana Department of Fish, Wildlife, and Parks in the remote possibility that an accidental overflow into State waters could occur.

RESOURCE EVALUATION

Geophysical investigations were accomplished with a Resource Evaluation conducted by the Montana College of Mineral Science and Technology in conjunction with the Montana Bureau of Mines. Their charge was to determine a suitable well site, hopefully on the Warm Springs State Hospital property.

This evaluation consisted of a gravity and resistivity survey in the immediate area of the existing hot springs mound; evaluation of existing reports, maps, surveys, and studies of the immediate Warm Springs area; and review of two independent studies of the Deer Lodge Valley. Three probable faults were mapped running northeast to southwest and north to south. The final drilling location was chosen at a site northeast of the existing hot springs mound within 100 feet of the steam plant and a well rig was moved onto the site on October 11, 1979.

WELL DRILLING

Drilling proceeded through December 5, 1979 to a total well depth of 1,498 feet. A cemented surface casing of 12-3/4" O.D. was run to 896 feet and 8-5/8" O.D. production liner was run from 849 feet to a total depth of 1,498 feet. A slotted production liner with 16 slots per foot (3/16" x 2-3/4") was used from a depth of 1,037 feet to 1,371 feet. Schlumberger logged temperature, dual induction - laterlog, compensated neutron formation density and caliper of the hole for further evaluation and records of the well.

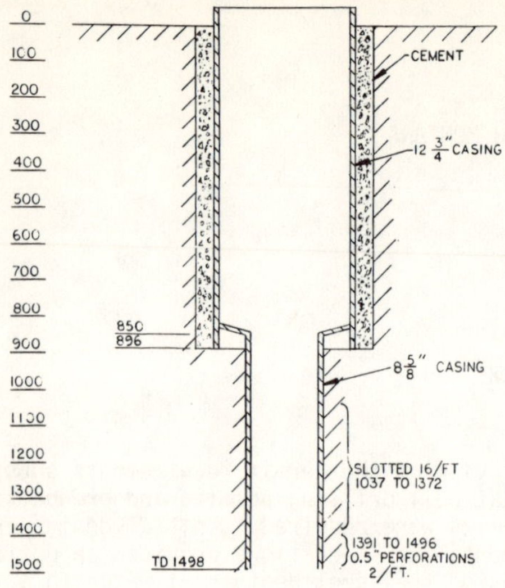

WARM SPRINGS PRODUCTION WELL

Initial preliminary pumping tests indicated that the well was producing considerably less than might be expected from the well log-formation evaluation. A decision was made to perform a hydrochloric matrix acid treatment on the well. A vertical line shaft turbine pump was then installed and pump tests were resumed. Vibration problems plagued the test and a total running of only 52 hours was recorded before removing the pump from the well. Subsequent examination revealed that the eslatomer spiders had failed, the bubbler tube had sheared off and the pump intake screen had been lost downhole.

WELL DEVELOPMENT

An evaluation was conducted on the well to present significant data and to evaluate the well in terms of the project requirements. The two most promising zones of geothermal productivity were identified at levels of 1,110 feet to 1,180 feet and 1,250 feet to 1,340 feet. The well was identified as having low values of permeability (2.84-5.66 millidarcies) and pump tests seemed to indicate wide variations in the production characteristics of the well. With this information it was recommended that the well be reworked and deepened an additional 50 feet to penetrate what is believed to be another zone of hot geothermal water.

This recommendation was followed and the bottom 220 feet of the well was bailed of drilling mud and rock chips. A solid metal item was encountered at the bottom of the well and could not be retrieved thus preventing increasing the hole depth an additional 50 feet. A decision was made to perforate the production liner from 1,391 feet to 1,496 feet with two 1/2" perforations per foot. A television camera was utilized to examine the production liner although the temperature prevented inspection to the total well depth. A proposal for screening was also abandoned as the well had been so plugged with drilling mud that screening would have served no useful purpose.

A submersible pump was intalled and initial testing indicated that the pump sizing was in excess of the producing capacity of the well. Another submersible pump was installed on May 19, 1981 at a setting depth of 987 feet. The well was then tested at rates of 50, 75, 95, and 110 gpm for 100 hours each and nine hours step drawdown tests were run at 25, 50, and 75 gpm. The gpm at 168°F can be maintained.

DESIGN AND CONSTRUCTION

The envisioned concept of utilizing the geothermal resource for both space heating and domestic hot water heating was modified to accommodate only the domestic hot water heating due to the reduced total geothermal resource available. The spent geothermal water is discharged to a wet-lands area being developed for waterfowl and wildlife instead of reinjected as previously considered. The designs of the heat transfer system proceeded with the incorporation of a 316 stainless steel plate type heat exchanger with capabilities for full geothermal pump discharge. A vent and vacuum valve, solids separator, flow indicator, pressure gauges, temperature gauges, and the necessary valves, etc. are used on the geothermal side of the system.

All domestic cold water makeup is routed through the heat exchanger to supply the demands of the Warm Springs facility. A Btu flow meter will compute the total amount of the geothermal resource utilized. Pressure gauges, temperature gauges, and safety devices are incorporated to provide safeguards for the system. A domestic water geothermal recirculation pump is used to provide flow from the existing domestic water recirculation system during times of low water demand.

SUMMARY

The energy extracted from the geothermal resource has been calculated based on a 60% utilization factor of the 70 gpm flow with a temperature reduction from 160°F to 90°F. This energy calculation of 1.47 x 10^6 BTUH can be further interpreted to a value of 15.52 x 10^6 cubic feet of natural gas (829.5 BTU/ft^3) replaced per year. The dollar value of this natural gas is $68,000 per year.

The installation was completed during the summer of 1982 and has proven to operate satisfactorily with only minor startup problems. The system has not been operational long enough to track the usage of water as calculated by the Btu flow meter. Further considerations for the use of spent geothermal water have been discussed although no positive action has developed from those discussions.

The development of this demonstration project has provided a show piece for direct use of a geothermal resource. The initial phases of the project provided the impetus to complete the project and present the geothermal energy resource concept to the general public and for their incorporation into similar projects.

BIBLIOGRAPHY

Barclay, Karen, L., January 1979, Geothermal Heat Survey for Warm Springs State Hospital, Montana Energy and MHD Research and Development Institute.

Carlson, Jon, R., April 1980, Drilling Report Warm Springs State Hospital Production Well No. 1, Montana Tech Foundation Mineral Research Center and Montana Energy and MHD Research and Development Institute.

Gould, R. W. and Kunze, J. F., January 1981, Evaluation of Data and Recommendations for the Warm Springs Montana Geothermal Well, Energy Services, Inc., and Montana Energy and MHD Research and Development Institute.

McLeod, Eugene, M., August 1979, Technical Progress Report on Geothermal Heating of Warm Springs State Hospital, Warm Springs, Montana, Montana Energy and MHD Research and Development Institute.

Stoker, Roger, C., June 1979, Geothermal Resource Evaluation of the Warm Springs, Montana Area, Energy Services, Inc. and Montana Energy and MHD Research and Development Institute.

Stoker, Roger, C., November 1980, Preliminary Resource Evaluation of the Warm Springs, Montana Geothermal Well, Energy Services, Inc. and Montana Energy and MHD Research and Development Institute.

A COMPARISON OF GEOTHERMAL DIRECT-USE PRICING TERMS IN SEVEN WESTERN STATES

Alex Sifford & Eliot Allen

Eliot Allen & Associates, Inc.
5006 Commercial Street, SE
Salem, OR 97306

ABSTRACT

A comparative survey of pricing terms in twenty-one direct-use projects in seven western states is summarized. The survey was undertaken to determine the character of pricing strategies which are currently in place or proposed for direct-use projects. A majority of projects are shown to base their geothermal price on a discounted natural gas rate. The average discounted rate for non-profit projects is calculated to be 56% of current gas rates; for profit projects the average is 66% of gas rates. The average delivered geothermal price for all projects basing rates on gas and oil was \$3.80/MMBtu as of March, 1982.

INTRODUCTION

Within the seven western states of California, Colorado, Idaho, Nevada, New Mexico, Oregon, and Wyoming, many geothermal direct-use projects involve contractual agreements between the resource supplier and user. Twenty-one such projects were surveyed for the prices and terms which are currently in place or proposed. A majority of the projects (43%) are still under construction; 33% are operating; and 24% are still in a proposal stage.

Factors which are examined in the projects include resource and load data, contractual terms, and delivered energy prices (exclusive of system efficiency). The survey is summarized in Table 1; and detailed by state and project in Table 2.

HEAT MEASUREMENT

The geothermal heat to be priced is measured in some manner, e.g. gallons per minute at a constant temperature. The majority of projects price the heat at the point of use, i.e. the temperature and flow of the resource as it enters the point of use and as it departs. In nearly all projects resources are not actually consumed, only heat is extracted.

Some projects price the resource at the wellhead, again based upon flow rate and temperature drop. A minority of projects, typically older operations, include the price of the heat in a larger sum, e.g. a building rental or land lease payment.

Table 1

SURVEY SUMMARY

Project Status (as of 3-82)	Number of Projects	% of Survey Sample
Operating	7	33
Construction	9	43
Proposed	5	24
	21	
Type of Wellhead Developer		
Municipal/Non-profit	10	48
Private/Profit	11	52
	21	
Projects with partial federal funding	11	52
Basis of Geothermal Price		
Based on percentage of oil cost	3	14
Based on percentage of gas cost	12	57
Based otherwise, e.g. flat rate	6	29
	21	

SYSTEM EFFICIENCY

In measuring the heat to be sold, system efficiencies are generally not considered in the price; thus prices shown in Table 2 are for usable heat. To clarify, fossil fuel systems usually have a maximum efficiency of 80% in converting delivered heat into usable heat. In contrast, direct-use geothermal systems meter heat used across the customer's system, so efficiency is generally not a factor. Therefore, a new geothermal customer may only use 80 MMBtu of geothermal heat for every 100 MMBtu of fossil fuel which was historically consumed. The result of neglecting to consider this difference in efficiency in computing geothermal heat prices is a geothermal revenue stream significantly lower than anticipated.

ALTERNATE FUELS

As in other areas of economics, the prices of alternatives directly influence the price of the commodity in question. In this case the alternatives are fossil fuels such as oil, natural gas, and coal. The majority of projects have pricing formulas tied to a percentage of natural gas costs, with a sufficient discount to make conversion to geothermal energy financially attractive. Natural gas has been selected in most projects because it is generally the least costly alternative fuel in an area; and because of its similarity to geothermal production and distribution practices and costs.

DISCOUNT RATES

Discount rates ranged from 0 to 50 percent of current alternate fuel prices. Some projects offered a flat fee (per MMBtu) at frozen 1981 fossil fuel rates. Others start with a base price equivalent to some percentage of current fossil fuel rates, but with an annual escalation factor which is not connected to fossil fuel prices. Still others offer the geothermal heat at a floating rate with a maximum price ceiling or limit.

It should be noted that some risk exists in indexing geothermal heat to alternate fossil fuels. Should these fuels drop in price, as oil did recently, revenue problems may develop for geothermal producers. Conversely, the rate of increase in fossil fuel prices is hard to predict, and thus, could possibly exceed allowable increases given in the escalation clauses of price agreements.

ECONOMIC DEVELOPMENT

In several of the surveyed projects, economic growth and diversification for the community, rather than energy savings, was the driving force behind the project. In these cases incentives to businesses which locate in "geothermal industrial parks" may include free heat for a certain period, or heat at even greater discounts than offered elsewhere in the community.

SUMMARY

The results of the survey are summarized as follows:

- A majority of the projects (57%) base their geothermal price on some discounted rate of natural gas, because it is usually the least costly alternative fossil fuel, and its similarity to geothermal production and distribution operations.

- For those projects basing geothermal prices on discounted natural gas rates, the average geothermal price for non-profit projects is 56% of current gas rates, and for profit projects the average is 66% of gas rates.

- The average price of delivered geothermal energy for those projects with prices based on gas and oil rates is $3.80/MMBtu as of March, 1982.

- A majority of the projects escalate their geothermal price annually with corresponding increases in fossil fuel prices; a small minority of the projects utilize other escalation factors, e.g. an arbitrary rate not indexed to fossil fuels.

- Only one project includes a fixed demand charge in addition to the price rate for heat consumed.

- The prices in place or proposed for 52% of these projects have been affected to an undetermined extent by various federal subsidies, e.g. DOE, EDA, HUD.

- Nearly all of the private developer projects are structured such that the developer "wholesales" the geothermal heat to another distribution entity, usually a local governmental unit, so as to allow the private developer to avoid public utility regulation; indicating that PUC jurisdiction is still a significant impediment to large-scale, i.e. district heating, geothermal commercialization.

ACKNOWLEDGEMENTS

Grateful acknowledgement is given to Charles Higbee and Diana King for their critical remarks.

REFERENCES

Atkinson, D.J., Marketing Geothermal Energy for Space Conditioning, G.R.C. Transactions Vol. 4, pp. 545-547, 1980.

Derrah, H., City of Klamath Falls, personal communications, February, 1982.

Dick, J., Chafee Geothermal Ltd., personal communications, February, 1982.

Goering, K., Coury & Associates, Inc., personal communications, February, 1982.

Geothermal Resources Council Bulletin, Vol. 10, No. 11, p. 13, December, 1981.

Longyear, A., Lahontan, Inc., personal communications, February, 1982.

Pinto, R.G., Esq., personal communications, February, 1982.

Rackley, I.S., Chilton Engineering, personal communications, February, 1982.

U.S. Department of Energy, Geothermal Direct Heat Applications Program Summary, September, 1981.

Table 2

PRICING & TERMS BY STATE AND PROJECT

Project (Status[1])	Resource	End-Use Load	Contract Terms	Price/MMBtu[2]
		CALIFORNIA		
A. (Construction)	160-170°F; 600 GPM; 2 wells; 900 ft.	14 public bldgs; 40% load factor initially	City to sell to customers at base price of 67% of current oil price; annual escalation is limited to a maximum of 5% of base price, not indexed to oil increases.	7.14
B. (Construction)	150°F	126 homes & greenhouse park	Same as above	7.14
C. (Construction)	180°F; 2500 GPM; 2 wells	State prison and greenhouse complex	Same as above except price drops to 50% of oil after equipment is amortized.	7.14
D. (Operating)	174-205°F; 6 wells	30 greenhouses	Resource developed by single user; landowner receives royalty of flat rate per greenhouse.	NA
E. (Proposed)	Power plant effluent	Unknown	Lessee/developer to sell effluent @ 50% of gas rate; also possibility of Federal royalty of 10% of coal price because of land-ownership.	2.57
		COLORADO		
F. (Operating)	116°F; 700-1000 GPM artesian; 3080 ft.	36,000 ft. shopping mall	60% of gas rate.	2.53
G. (Construction)	131-148°F; 1800 GPM 2 wells; 300 ft.	10 public bldgs; 54 businesses; 63 residences; initially 29 x 10^9 Btu/yr	40% of gas rate.	3.35
H. (Proposed)	150°F; 110 GPM springs; also 240°F @ 1600 GPM from 4 wells	Businesses, public bldgs; residences	60-90% of gas rate.	2.04-3.06
I. (Proposed)	122°F; 1200 GPM; 155 ft.	40,000 sq. ft., 4-story office bldg.	None developed, but likely to be % of gas rate.	NA
		IDAHO		
J. (Operating)	170°F; 900 GPM artesian; 2 wells, 400 ft. ea.	250 residences.	Water district sells both heat & domestic water based on actual comsumption according to pipe diameter.	NA
K. (Construction)	160-170°F; 700 GPM artesian; 6 local and state wells; 400-2000 ft.	500+ residences & businesses.	Wellhead developer sells to City @ 54% of gas rate; City then distributes and sells to customers @ 70% of gas rate.	2.58 to City; 3.35 to customers
L. (Operating)	159°F artesian springs; estimated 1000 GPM.	Approximately 60 homes & businesses	Only written agreements between resource/owner & 2 customers: motel and pool purchase resource temperature & flow under same terms; both flat fees; other customers purchasing by oral agreements which vary widely.	$200/yr flat rate.
		NEVADA		
M. (Construction)	180°F; 250 GPM; 1000 ft.	150-unit condo complex; total 15.6 x 10^9 Btu/yr	Developer to sell to individual users at 85-100% of gas rate.	3.40-4.00
N. (Construction)	180°F; 485 GPM artesian well	Bank, laundry, & motel	Utility/developer to sell heat at 50% of gas rate; escalation allows only 50% of gas escalation.	1.90

Table 2 continued

Project (Status[1])	Resource	End-use Load	Contract Terms	Price/MMBtu[2]
		NEW MEXICO		
O. (Construction)	NA	Industrial users	Developer negotiating with user for 70-100% of gas rate, with possible flat rate during first two years as incentive.	NA
P. (Operating)	140° F; 550-600 GPM; 2 wells	11 college campus bldgs.	System O&M cost given for comparative purposes; no sale occurs.	2.50
		OREGON		
Q. (Operating)	140°F; 70 GPM; effluent	2 bldgs. currently; 23 max. potential	First one or two years free if employment is created in industrial park, then approximately 33% of gas, plus demand charge.	1.85
R. (Construction)	219°F; 2 wells; 728 GPM	14 public bldgs; total 36 x 10^9 Btu/yr.	Initial rate and demand charge unknown, but roughly equal to 50% of gas; possible future reduction to 30-40% of gas, plus undetermined demand charge.	3.00
S. (Operating)	87°F; 100 GPM; springs.	Waste water to greenhouses.	Flat rate for land and resource combined; resource portion estimated to be $150/mo.	NA
T. (Proposed)	200°F; 2-8 wells; 870 GPM each; 2000 ft.; Base case, 8 wells	1170 residences; 120 commercial; 1 fuel plant; total; 183 x 10^9 Btu/yr.	Price estimated to cover developer's O&M and PUC-approved investment return; also, 16.25% royalty on gross revenues goes to underlying resource holder, and 3% to City as franchise fee.	8.20
	Mini-dist., 2 wells	Mini-District Case: 50 commercial & public bldgs; 1 fuel plant; total 17 x 10^9 Btu/yr.	Different service area, same terms, except no 3% City fee because of exemption.	10.30
		WYOMING		
U. (Proposed)	130-160°F; 10K-25K GPM; 13-33 wells; 500-900 ft.	Case 1: 1700 residences; 230 businesses; 1 hospital; total 603 x 10^9 Btu/yr. Case 2: 100 residences; 100 businesses, 1 hospital; 3 schools; total 153 x 10^9 Btu/yr.	70-100% of gas rate.	2.66-3.50

[1] As of March, 1982.

[2] Prices are for delivered heat as of March, 1982; system efficiency is not considered in calculating the prices shown.

Geothermal Resources Council, TRANSACTIONS Vol. 6, October 1982

ASSESSMENT OF LOW-TEMPERATURE GEOTHERMAL RESOURCES IN THE UNITED STATES

Michael L. Sorey, Marshall J. Reed, Robert H. Mariner, and Manuel Nathenson

U.S. Geological Survey, Menlo Park, CA

ABSTRACT

A nationwide assessment of geothermal resources < 90°C has recently been completed by the U.S. Geological Survey. Estimates were obtained of accessible resource base, resource, and beneficial heat for 1145 identified low-temperature reservoirs occurring in hydrothermal convection systems and conduction-dominated systems. For identified systems, the nationwide total of thermal energy recoverable at the wellhead (resource) is estimated to be 82 x 10^{18} J, from which a total beneficial heat of 39,000 MW_t for 30 years could be obtained.

INTRODUCTION

Geothermal resource assessment is the estimation of the amount of thermal energy that might be extracted from the Earth and used at costs competitive with other forms of energy at a foreseeable time under reasonable assumptions of technological improvement. The first systematic effort to estimate the geothermal resources of the United States was published in 1975 as U.S. Geological Survey Circular 726 (White and Williams, 1975). This assessment and a followup assessment published in 1979 as Circular 790 (Muffler, 1979) focussed on the quantities of geothermal energy stored in regional conductive environments, igneous-related geothermal systems, hydrothermal convection systems, and geopressured-geothermal systems. Estimates were given of the thermal energy recoverable from hydrothermal convection systems at temperatures of 90°C and above. In addition, the 1979 assessment included a compilation of data on the occurrence of low-temperature geothermal water less than 90°C (Sammel, 1979), but no attempt was made to estimate the associated thermal energy.

Low-temperature geothermal resources occur in two types of geohydrologic environments. These are hydrothermal convection systems, which commonly involve upward flow of thermal water along faults in areas of above-normal heat flow, and conduction-dominated systems in sedimentary basins and coastal embayments, where areally-extensive aquifers occur beneath a thick insulating blanket of rocks having low thermal conductivity. Low-temperature geothermal resources in each of these environments are currently being utilized in space heating, food processing, and other non-electric applications.

To provide estimates of the quantities of thermal energy at temperatures less than 90°C recoverable from geothermal reservoirs in the United States, the U.S. Geological Survey (USGS) recently completed an assessment based on an expanded inventory of low-temperature geothermal occurrences and using a new methodology for estimating recoverable energy. Many state agencies and several private contractors with the support of the State Coupled Geothermal Program of the Department of Energy's Division of Geothermal Energy have provided data for this assessment. Information on water chemistry, temperature, flow rate, and other parameters measured at over 1,000 low-temperature sites has been stored in the computer-based GEOTHERM information system maintained by the USGS. Results of the assessment will be published in a USGS circular in 1982. Additional information for each resource area contained in the GEOTHERM system is available on request.

METHODS FOR ASSESSING LOW-TEMPERATURE GEOTHERMAL RESOURCES

Assessment of geothermal energy resources involves determinations of the location, extent, and geohydrologic characteristics of each resource area and estimation of the accessible resource base (amount of thermal energy stored in each reservoir) and the resource (the amount of thermal energy that can be produced at the land surface). Identified resource areas are considered to meet the criteria that reservoir permeability is sufficient to supply long-term production and that reservoir temperatures exceed a minimum temperature-depth relation. We used a lower-temperature limit which is 10°C above the local mean annual temperature at the land surface and a 25°C/km temperature gradient with depth. This eliminates assessment of an enormous quantity of shallow groundwater having "normal" temperatures but includes water in reservoirs associated with hydrothermal convection systems and in regional aquifers in many deep sedimentary basins and coastal embayments where temperatures increase relatively rapidly with depth.

The accessible resource base for each low-temperature area identified in this assessment is calculated as:

$$q_R = \rho c \cdot a \cdot d \cdot (t - t_{ref}) \quad (1)$$

where: q_R = accessible resource base (J)
ρc = volumetric specific heat of rock plus water (2.6 J/cm^3°C)
a = reservoir area (km^2)
d = reservoir thickness (km)
t = reservoir temperature (°C)
t_{ref} = reference temperature (15°C)

The reference temperature of 15°C is an assumed average mean annual air temperature, and the volumetric specific heat of 2.6 J/cm^3°C is a weighted average for rock types found in low-temperature geothermal areas. Statistical methods outlined in Circular 790 were used to quantify the uncertainty in these calculations and to provide probability distributions for the estimates of total accessible resource base, resource, and beneficial heat. For each reservoir, estimates were made of the minimum, maximum, and most likely values for parameters area, thickness, and temperature; these values were then used to form triangular probability densities from which the mean and standard deviation for each parameter and for the various energy quantities were calculated.

Many identified low-temperature areas associated with hydrothermal convection systems are identified only on the basis of a single thermal spring or well because little or no data on subsurface conditions exist. For these areas, a standard reservoir volume of 1.0 km^3 was assumed from consideration of possible models of hydrothermal circulation patterns.

The approach used for resource determinations in previous assessments was to assume a recovery factor of 0.25 of the accessible resource base. This factor is based on a heat-sweep process involving injection of cold water into the reservoir to replace hot water withdrawn during production (Nathenson, 1975). In this assessment, resource determination involves estimation of the number of wells each reservoir can support over a development period of 30 years, based on the assumption that cold water will not be injected into the reservoir. Although injection of produced fluids following surface utilization may be legally required in certain areas, lower reservoir temperatures and larger reservoir areas make costly injection schemes for energy recovery less likely to be used in low-temperature geothermal areas than in intermediate- and high-temperature geothermal areas. The method we used to make resource estimates allows for induced recharge of water from permeable regions surrounding each reservoir as reservoir pressures decline. As a result, recovery factors approach 0.25 over 30 years for small-area reservoirs, but are significantly less than 0.25 for large-area reservoirs.

Resource estimates are calculated as:

$$q_{WH} = (\rho c)_f \cdot N \cdot Q \cdot (30 \text{ years}) \cdot (t - t_{ref}) \quad (2)$$

where: q_{WH} = resource (J)
$(\rho c)_f$ = volumetric specific heat of fluid (4.1 J/cm^3°C)
N = number of production wells
Q = volumetric discharge rate of each well (L/s)
t = fluid temperature at the wellhead (assumed equal to reservoir temperature) (°C)

A detailed analysis of well-field design for each resource area is not warranted for the purposes of this assessment. Instead, determinations of the optimum number of wells and their spacing were made for an idealized production plan consisting of regularly-spaced wells discharging at 31.5 L/s (500 gpm) for 30 years with a total drawdown at the center of the reservoir of 150 m (500 ft). Based on this production plan, the number of wells which together would produce the specified drawdown is determined by the ratio of reservoir area, a, to the area per well, a_w. The area per well is the square of the spacing between wells and is derived from curves presented in the circular that relate well spacing to reservoir area and transmissivity and to leakage properties of confining beds. Methods used to quantify the uncertainty in resource estimates follow those for accessible resource base estimates.

The amount of the resource that can be applied to a specific non-electric utilization is termed beneficial heat. Estimates of beneficial heat from geothermal water can be compared with thermal energy obtainable from other fuels. Selection of appropriate uses for low-temperature geothermal water depends partly on the reservoir temperature, and different applications involve different temperature drops during the heat-exchange process. We have used the following equation, which is based on data for five direct-use applications, to calculate the temperature drop, Δt, as a function of wellhead temperature:

$$\Delta t = 0.6\,(t - 25°C) \quad (3)$$

Beneficial heat estimates are obtained by substituting Δt into equation 2 in place of $(t - t_{ref})$. Units for reporting beneficial heat are megawatts thermal (MW_t) for 30 years. For comparison with estimates of accessible resource base and resource, one MW_t for 30 years = 0.945×10^{15} J. For values of $t \leq 25°C$ the useful Δt and the beneficial heat are zero. This limit results from defining low-temperature resources as being at least 10°C above the mean annual land surface temperature, which averages about 15°C across the country.

ESTIMATES OF RESOURCE BASE, RESOURCE, AND BENEFICIAL HEAT

A total of 1145 low-temperature reservoirs have been identified, exclusive of national parks (table 1). Most of these areas are associated with hydrothermal convection systems in the western United States. In the central United States, resource areas occur primarily in deep sedimentary basins in the Great Plains where regional aquifers contain waters meeting our temperature-depth criterion. In the eastern United States, 25 low-temperature areas in the Appalachian Highlands and Atlantic Coastal Plain have been identified.

The distribution of 1108 low-temperature hydrothermal convection systems as a function of reservoir temperature fits the exponential relation defined in Circular 790 of cumulative frequency versus reservoir temperature for intermediate and high-temperature systems (Brook and others, 1979, figure 11). This relation shows that the number of geothermal reservoirs in hydrothermal convection systems increases exponentially as reservoir temperature decreases.

The total accessible resource base (table 1) is dominated by the thermal energy stored in sedimentary basins in the central United States. However, the methodology for resource calculation yields relatively low recovery factors for this type of resource area. Hence, the total resource estimate, 82 x 10^{18} J, is more evenly distributed between the western and central United States. Accessible resource base and resource totals for the eastern United States are relatively small, reflecting in part a lack of exploration activity and in part the widespread existence of low heat flow in rocks of average thermal conductivity.

In identified systems, total beneficial heat is 39,000 MW_t for 30 years, which is equivalent to 37 x 10^{18} J. This represents a significant quantity of thermal energy when compared with equivalent requirements for other energy sources (for example, 39,000 MW_t of beneficial heat from a geothermal system is comparable to 39,000 MW_e if that electricity were used to heat a home).

Estimates of the undiscovered low-temperature resources (table 1) in systems whose locations are as yet unknown and in identified systems due to possible upward revisions of reservoir size and temperature are less than corresponding estimates for identified systems. This reflects the degree to which reservoirs meeting our temperature-depth criterion can be delineated at present.

ACKNOWLEDGEMENTS

Significant contributions to the preparation of this assessment were made by James D. Bliss, Charles A. Brook, Duncan Foley, Marianne Guffanti, Randolph Lieb, Robert J. Monroe, Amy Rapport, Joel L. Renner, John H. Sass, and Christian Smith. The authors would also like to thank Edward A. Sammel, Wendell Duffield, and L.J. Patrick Muffler for critical discussions and reviews of this work.

Table 1. Mean values of geothermal energy from identified and undiscovered reservoirs < 90°C in the United States, exclusive of national parks. (Numbers are rounded off to two significant figures or to three significant figures when the first digit is 1.)

Type of Geothermal System Region	Number of Reservoirs	Accessible Resource Base (10^{18} J)	Resource (10^{18} J)	Beneficial Heat (MW_t for 30 yr)
Hydrothermal Cenvection Systems				
Western U.S.	1048	196	28	12300
Central and Eastern U.S.	60	3.2	0.81	290
Conduction-dominated Systems				
Western U.S.	9	99	1.15	470
Central and Eastern U.S.	28	24000	52	26000
Identified Resources	1145	24000	82	39000
Undiscovered Resources	--	7200	65	30000

REFERENCES

Brook, C.A., Mariner, R.H., Mabey, D.R., Swanson, J.R., Guffanti, M., and Muffler, L.J.P., 1979, Hydrothermal convection systems with reservoir temperatures of ≥ 90°C, *in* Muffler, L.J.P., ed., Assessment of geothermal resources of the United States -- 1978: U.S. Geological Survey Circular 790, p. 18-85.

Muffler, L.J.P., ed., 1979, Assessment of geothermal resources of the United States -- 1978: U.S. Geological Survey Circular 790, 163 p.

Nathenson, Manuel, 1975, Physical factors determining the fraction of stored energy recoverable from hydrothermal convection systems and conduction-dominated areas: U.S. Geological Survey Open-File Report 75-525, 35 p.

Sammel, E.A., 1979, Occurrence of low-temperature geothermal waters in the United States, *in* Muffler, L.J.P., ed., Assessment of geothermal resources of the United States -- 1978: U.S. Geological Survey Circular 790, p. 86-131.

White, D.E., and Williams, D.L., eds., 1975, Assessment of geothermal resources of the United States -- 1975: U.S. Geological Survey Circular 726, 155 p.

POTENTIAL FOR RECOVERY OF BY-PRODUCTS FROM SPENT GEOTHERMAL FLUIDS

Michael S. Wei

Engineering and Economics Research, Inc.
1951 Kidwell Drive
Vienna, Virginia 22180

ABSTRACT

This paper presents the results of a preliminary evaluation of the potential for recovering by-products from geothermal fluids. Over 30 chemical species found in geothermal fluids were systemmatically evaluated. A hypothetical 50 MWe power plant, using high salinity fluids, was used as a base for our estimate of mineral values. Our study indicates that Li, B, CO_2, NH_3, Br_2, I_2, Sr and heavy metals appear to be the most promising minerals to be recovered from geothermal fluids. However, improvement in process technology are still required in order to make the minerals competitive with conventional methods of production.

INTRODUCTION

Geothermal fluids are chemically complex with total dissolved solid (TDS) levels ranging from a few hundred to a few hundred thousand parts per million. With the high flow rates needed for power production, a substantial amount of chemicals inevitability passes through the geothermal power plant. While these chemicals may cause operational problems such as scaling and corrosion, there is a potential to recover some of these chemicals as by-products. This paper investigates the feasibility of recovering chemicals from spent geothermal fluids taking into consideration the mineral values, market demand, technology readiness, resource uncertainties and competition from other supply sources.

MINERAL VALUE

A first consideration for recovering minerals from geothermal fluid is to determine the amount of minerals that can be recovered from geothermal fluid and their monetary values so that decisions can be made regarding the worthiness of such an investment. Obviously, fluids of high salinity will potentially yield more recoverable products than low salinity fluids. In order to establish a baseline for comparison, a hypothetical "best-case" 50 MWe geothermal power plant was established. The assumptions used for this 50 MWe plant are:

- Brine flow rate = 100,000 lbs per hour per well
- 14 production wells
- Plant operates 7,000 hours per year (0.8 capacity factor)
- High salinity fluids with dissolved solids contents comparable to those of Salton Sea

Based on these assumptions, the total amounts of minerals recoverable and their mineral values are shown in Table 1. These figures were arrived at assuming that the minerals present in the fluid can be totally recovered and are saleable at current market prices. Because of the differences in mineral values, chemical species present in the highest concentrations do not necessarily yield the highest monetary return. Looking at the most likely recoverable forms of the listed brine species, 14 products exceed $1,000,000 in value. They are RbCl, Au, Li_2CO_3, CsCl, KCl, $Na_2B_4O_7$, MoO, $Mg(OH)_2$, $ZnSO_4$, NaCl, $SrCO_3$, $CaSO_4$, I_2, and Ag.

Table 1. Annual Amounts and Values of Recoverable Materials in a Best-Case 50 MWe Plant.

Chemical Species	Brine Concentration* (ppm)	Recoverable Materials	Amount of Recoverable Materials (tons)	Mineral Values**
Na	74,000	NaCl	919,000	$4,135,000
Ca	40,000	$CaSO_4$	666,000	$3,430,000
K	25,000	KCl	233,000	$10,600,000
Mg	3,000	$Mg(OH)_2$	24,400	$4,950,000
Fe	3,000	$Fe(OH)_3$	28,100	$793,000
CO_2	1,000	CO_2	20,200	$304,000
Si	1,000	$CaSiO_3$	4,900	$206,000
Sr	750	$SrCO_3$	6,210	$3,600,000
B	740	$Na_2B_4O_7$	32,000	$5,340,000
Zn	740	$ZnSO_4$	8,980	$4,850,000
Ba	570	$BaSO_4$	4,750	$264,000
Al	450	Al_2O_3	4,170	$792,000
Li	300	Li_2CO_3	7,770	$20,200,000
NH_3	300	NH_3	1,470	$176,000
Rb	169	RbCl	828	$129,000,000
Pb	155	Pb	760	$654,000
Br_2	146	Br_2	715	$315,000
Mo	110	MoO	730	$5,110,000
I_2	40	I_2	196	$2,665,000
Cs	20	CsCl	124	$16,400,000
Cu	16	Cu	78	$172,000
Ni	6	Ni	32	$162,000
Au	2	Au	9.8	$83,000,000
Ag	1.4	Ag	7	$1,175,000

* Reeber, 1980 ** Calculated from BOM, 1980

In addition, 10 products have values falling between \$100,000 and \$1,000,000 per year. They are: $Fe(OH)_3$, Al_2O_3, Pb, Br_2, CO_2, $BaSO_4$, $CaSiO_3$, NH_3, Cu, and Ni.

MARKET DEMAND ASSESSMENT

Several factors were taken into consideration in order to identify a preliminary list of promising by-products recovered from geothermal power plant operations. These factors are:

- Quantity and form of materials producible from a "best-case" 50 MWe geothermal power plant
- Monetary values of by-products
- U.S. demand for commodity (ignoring regional demands)
- Import dependency
- Strategic consideration
- Existing sources of supply (competition)
- Technology readiness of recovery process
- Resource uncertainty

Table 2 lists in decreasing order the mineral monetary values obtained in the fluid of a "best-case" 50 MWe geothermal plant and some of the more readily quantifiable factors discussed above. Based on these considerations, the 24 minerals with values exceeding \$100,000 per year were classified as promising, marginally promising, speculative, and not promising. The rationale for this grouping is presented on a product-by-product basis in the following sections.

PROMISING MATERIALS

The following materials are considered to be promising for recovery from geothermal fluids:

- Lithium has a high mineral value and the process used to recover lithium, (aluminum oxide addition to produce lithium aluminate) appears to be promising. Lithium is currently produced from brine extraction and from the ore spodumene.

Table 2. Factors Considered in Determining the Marketability of Geothermal By-Products

Chemical Species	Brine Concentration (ppm)	Annual Mineral Value from 50MWe Plant	% of U.S. Demand from 50MWe Plant	% of U.S. Demand Met By Import*	Strategic Consideration*	Current Supply Sources*	Potential for Recovery from Geothermal Fluids**
Rb	169	\$129,000,000	72,000	0	-	Lepidolite	Not promising
Au	2	\$83,000,000	5	45	Stockpiled	Mines	Speculative
Li	300	\$20,200,000	46	0	-	Spodumene	Promising
Cs	20	\$16,400,000	700	100	-	Pollucite	Not promising
K	25,000	\$10,600,000	3.3	68	-	Seawater, Brine	Not promising
B	740	\$5,340,000	2.9	0	-	Brine	Promising
Mo	110	\$5,110,000	1.5	0	Strategic	Molybdenite	Promising
Mg	300	\$4,950,000	1.3	0	-	Dolomite, Brine	Marginal
Zn	740	\$4,850,000	0.4	51	-	Zinc Ore	Promising
Na	74,000	\$4,135,000	1.8	9.5	-	Seawater	Not promising
Sr	750	\$3,600,000	16.7	100	-	Celestite	Promising
Ca	40,000	\$3,430,000	1.2	2	-	Limestone	Not promising
I_2	40	\$2,665,000	4.5	72	Stockpiled	Brine	Promising
Ag	1.4	\$1,175,000	0.1	30	-	Ore	Speculative
Fe	3,000	\$793,000	0.008	33	-	Iron Ore	Promising
Al	450	\$792,000	0.003	93	Controlled	Bauxite	Marginal
Pb	155	\$654,000	0.05	15	Controlled	Ores with Zinc	Promising
Br_2	146	\$315,000	0.4	0	-	Brine	Promising
CO_2	1,000	\$304,000	0.01	0	-	Natural wells	Promising
Ba	570	\$264,000	0.16	33	-	Barite	Marginal
Si	1,000	\$206,000	0.7	13	Stockpiled	Sandstone	Not promising
NH_3	300	\$176,000	0.008	5	-	Natural gas	Promising
Cu	16	\$172,000	0.003	13	Strategic	Cu_2S Ore	Promising
Ni	6	\$162,000	0.015	65	Stockpiled	Sulfide Ore	Promising

* BOM, 1980

** See text for rationale of classification

- Boron has a relatively high mineral value and can be recovered as borax by the addition of sodium compounds. This process is similar to the current method of commercial production and thus, recovery technology is demonstrated. However, marketing boron from geothermal fluids will face stiff competition from alternate sources of production.
- Carbon dioxide can be recovered using technology similar to conventional commercial methods. Despite its marginal mineral value, CO_2 can penetrate local markets and can also be used on site in the drilling process to control the acidity of drilling muds,
- Ammonia can be recovered through the air stripping process. It has the potential of penetrating small local markets.
- Bromine can be recovered from geothermal using the same process as the commercial production process, i.e. chlorine addition. Br_2 from geothermal resources will face stiff competition from conventional sources.
- Iodine can be recovered in a similar method as bromine. However, the 40 ppm of iodine in brine is probably optimistic.
- Heavy metals (Ni, Cu, Pb, Fe, Zn, and Mo) can be collectively precipitated from geothermal fluids as hydroxides and/or sulfides. These metals are either strategic or imported to a large degree.
- Strontium can be recovered as $SrCO_3$ through the addition of carbonates. Its relatively high mineral value and the high import reliance make the recovery from geothermal fluids attractive.

MARGINALLY PROMISING MATERIALS

These minerals are considered as marginally promising for recovery from geothermal fluids:

- Magnesium is currently recovered from seawater which has a magnesium level of 1,350 ppm. Except for Salton Sea, most geothermal resources have Mg contents substantially less than seawater.
- Aluminum can be extracted as hydroxide through the addition of lime. Recovery of aluminum warrants consideration because of the large import dependencey of this critical material.
- Barium can be recovered as $BaSO_4$, which can be used as a weighting material in drilling mud. Thus some of the recovered products may be used on-site.

SPECULATIVE MATERIALS

Recovery of gold and silver from geothermal fluids is at best speculative primarily due to uncertainty in resource characteristics.

- Gold has a mineral content of $83,000,000 per year based on our best-case calculation. However, this value is derived from one data point. The uncertainty of Au content in geothermal fluids makes the recovery of this valuable mineral a highly speculative endeavor.
- Silver may appear to be a good compromise for gold. However, its mineral value is relatively small due to the low level of Ag in geothermal fluids.

NOT PROMISING MATERIALS

This group of minerals is considered to be not promising in terms of marketability in the U.S. market. Situations may be different in other countries to warrant their recovery.

- Rubidium from geothermal fluids would not be marketable because of the low domestic demand. One best-case 50 MWe geothermal power plant will produce enough Rb to supply 720 times the U.S. requirement.
- Cesium is not a good mineral to market because of its low market demand.
- Sodium, as NaCl (salt), is unattractive because the only viable technology for recovery is solar evaporation. In addition there are ample supplies of salt from seawater.
- Potassium is not an attractive mineral to recover because solar evaporation is the only viable technology. It will also face competition from other sources such as marine evaporates, brines, and seawater.
- Calcium will most likely be precipitated as $CaCO_3$ scales in the energy conversion plant. There is also ample supply of $CaCO_3$ as limestone.
- Silicon will most likely be precipitated as silica scales in the plant. It is also doubtful that high quality silicon compounds can be recovered from geothermal fluids.

CONCLUSIONS

Recovery of by-products from geothermal fluids appears to be an attractive option and may be feasible for certain chemical species. Li, B, CO_2, NH_3, Br_2, I_2, Sr and heavy metals are identified as most promising materials in this report. However, there are problems yet to be overcome even in this most promising category. There is a need to improve process efficiency so as to reduce cost for materials recovery. The uncertainty in resource characteristics is another major problem. This of course can only be resolved through extensive site specific monitoring.

REFERENCES

(1) Rebeer, R.R., 1980, "Coating in Geothermal Energy Production". Thin Solid Films, Vol. 72.

(2) Bureau of Mines, 1980, "Mineral Facts and Problems". United States Department of Interior, BOM Bulletin 671. U.S. Government Printing Office, Washington, D.C.

REGULATORY/ ENVIRONMENTAL

Geothermal Resources Council, TRANSACTIONS Vol. 6, October 1982

ECONOMIC ASSESSMENT OF GEOTHERMAL PLANT EFFICIENCY IMPROVEMENTS IN THE PG&E SYSTEM

Ezra Amir & Lon W. House

Technology Assessments Office
Assessments Division
California Energy Commission

ABSTRACT

This study examines the merits of investment in geothermal efficiency improvements in the PG&E system. The study compares the costs, composed of the equipment investment cost and the increased payment to steam suppliers, with the benefits, composed primarily of displacing existing oil-fired generation and cost savings due to geothermal fuel repricing. Two alternative oil cost real escalation rates were examined: 4.7 percent through 1987 and 3.0 percent thereafter, and a zero escalation rate throughout.

The analysis found for the higher oil price escalation case that the present value of future savings was about $17,000/kW of improved efficiency at a cost of $6,000/kW. Under the lower oil price escalation case, the present value of future savings was less than the investment. Whether or not improving the efficiency of geothermal units is a profitable investment is almost totally determined by future oil prices, given the resource plan used in the analysis.

BACKGROUND

The efficiency of PG&E's geothermal units has typically been on the order of 22,000 Btu/kWh electricity. SMUD has proposed to construct a plant which has a heat rate of about 19,000 Btu/kWh. This is 14 percent more efficient than PG&E's units.

This development raises two principal questions:

1. Is the major motivation for SMUD's more efficient design due to the different steam pricing methods used by SMUD and PG&E? PG&E's steam pricing formula is based on a steam energy charge per kWh of electricity produced. The steam pricing formula depends on the relative electricity production from fossil and nuclear sources. The formula is presented in Appendix A. In comparison, SMUD's geothermal steam cost is determined on the basis of steam energy supplied to the plant.

2. Is it possible that a geothermal steam efficiency improvement in the PG&E system would yield savings via displacement of more expensive fuel oil, natural gas, and other effects? Are they sufficient to warrant investing in steam efficiency improvement?

A CEC-sponsored study by Professor David Gallo (Ref. 1) showed that such investments are substantially cost-effective. Gallo's initial findings have prompted this more in-depth analysis. This relies on computer simulation of PG&E's system to determine the specific effects of improved geothermal power plant efficiency.

INTRODUCTION

The main economic question is whether the necessary investment to increase geothermal efficiency of plants coming on line in the PG&E system between 1986 and 1992 would yield sufficient savings to cover the investment costs at a reasonable rate of return. We will answer this question by capitalizing the future savings, that is, expressing them in terms of their present value as of the year 1992, and comparing them with the estimated investment necessary, again expressing it in terms of the same year, 1992, for consistency. If the value of the saving exceeds the investment, the investment may be considered economically sound.

The increment in geothermal power plant efficiency has two cost-related impacts:

- The efficiency improvements produce about 20 percent more electric energy from the same amount of geothermal steam. The increased supply of geothermal-derived electricity may be presumed to decrease the use of more expensive types of power plants, typically oil-fired units. However, under the existing steam contracts, the electricity producer will have to pay about 20 percent more to the steam supplier, "all other things being equal."

- All other things, though, are not equal. The increased geothermal-based electricity due to the efficiency improvement will primarily replace some of the oil-based generation, thereby changing the source of generation and, in turn, changing the cost of geothermal steam per kWh generated. Under the steam cost formula, if fossil generation decreases relative to nuclear, steam costs per kWh will decrease due to the cheaper nuclear fuel if the existing pricing formula is continued.

RESULTS OF THE SYSTEM SIMULATION

The system simulation produced cost estimates of running the PG&E generation system, for selected years (1992, 1994, 1998, and 2002) under two scenarios.

1. PG&E system baseline case. This case is based on 1980 PG&E supply plan filing with CEC as modified to be consistent with the CEC demand forecast.

2. Base case 1 with 20 percent increase of efficiency in geothermal plants coming on line between 1986 and 1992 with geothermal steam repricing due to the 20 percent efficiency increment and the subsequent change in the fuel mix.

Table 1 shows that the total savings due to fuel displacement and steam repricing are as follows for the years 1992, 1994, 1998, and 2002 ($ million): 98, 123, 173 and 263, respectively. (Computer runs were terminated in 2002 because utility resource plans are not available beyond that date.) Table 1 also reveals that, as expected, due to the relatively high fuel costs of oil-fuel generation, the primary displaced energy is oil-based generation.

Table 1
Difference: Base Case Versus
Increased Efficiency Case

	Cost, $(10)^6$ \$			
	1992	1994	1998	2002
Resid. oil and Cogen. gas	-150	-163	-222	-338
Coal	--	-3	-9	-8
Geothermal	+52	+43	+58	+83
Total System	-98	-123	-173	-263

NOTE: Totals do not add up because only the supply sources which are mostly affected by geothermal steam efficiency are recorded in this table. For this reason, nuclear and other supply sources are not presented in this table.

These saving estimates can form the basis for projecting future savings for the intervening years and the years after 2002 up to 2021.

OVERALL ECONOMIC ASSESSMENT

This section will discuss the cost effectiveness for two alternative cases relating to oil costs: (a) real escalation of 4.7% a year through 1987 and 3.0 thereafter (the Baseline case), and (b) zero real escalation rate.

BASELINE CASE

Table 2 derives the estimated yearly saving due to steam repricing alone, for the years 1992, 1994, and 1998, and 2002. Table 3 derives the adjusted total savings for these years. About 5/6 of these savings are attributed to fuel displacement and 1/6 to steam repricing.

We can now use the estimates on lines 1 and 2 of Table 3 to project yearly savings for the period 1992-2021, the assumed life of the geothermal facilities. The rate of increase in the saving from 1992 to 1998 is about 10 percent a year for both saving components (fuel displacement and steam repricing). Assuming that this 10 percent rate will continue and that the discount rate will be about 12 percent for the 1992-2021 period, a present factor of 22.4 may be calculated. Thus, the present value of the fuel displacement saving for 30 years is 82 (1992 saving) x 22.4=\$1,837 million for 132 MW (capacity increment resulting from 20 percent geothermal efficiency increase) and the present value of the steam repricing saving is 16 (1992 saving) x 22.4 = \$358 million for 121 MW (132 MW, less 11 MW which are not subject to the steam pricing formula). The corresponding 1992 present value of the saving per kilowatt of geothermal efficiency improvement is \$13,900 due to fuel displacement and \$3,000 due to steam repricing, a total of \$16,900 per kilowatt increment.

We may consider the 10 percent increase in future savings as the upper bound of future savings due to the reasonable expectation that as oil is progressively phased out from PG&E system, relatively less costly generation types will be displaced by the geothermal generation increment. As a result, the value of the savings after 1998 is expected to diminish.

Table 2
Saving due to Steam Repricing

	kWh 10^6 (1)	Mill./kWh (2)	Total 10^6\$ (3)
1992	14,454	1.120	16
1994	14,432	1.553	23
1998	14,457	1.960	29
2002	14,560	3.205	47

Source: Columns (1) and (2) are derived from Ref. No. 2 and supporting computer printouts.

Column 2 is the cost per kWh geothermal generation before the efficiency improvements less such cost after the efficiency improvements.

Table 3
Saving due to Fuel Displacement
and Steam Repricing
CEC Baseline Oil Price Escalation
(million $)

	1992	1994	1998	2002
1. Saving due to fuel displacement (line 3 less line 2)	82	100	144	216
2. Saving due to steam repricing (Table 2)	16	23	29	47
3. Total saving (Table 1)	98	123	173	263

The estimated saving of $16,900 per kilowatt compares favorably with the estimated cost of $5,000-$7,000 per kilowatt capacity increase. The range of $5,000-$7,000 was derived from a CEC staff estimate of $2,700/kW in 1986, and engineering consultants' (Ref. 5) estimate of about $3,800/kW in 1986, both escalated at 10 percent a year to 1992.

ZERO REAL OIL ESCALATION CASE

Table 4 presents this case. The main purpose of this table is to modify the savings due to fuel displacement under the baseline case (Table 3, line 1) as a result of the alternative assumption of a zero real escalation rate in oil price. The modified savings due to fuel displacement are derived in lines 1 through 5 in Table 4. The present value is based on a 6.8 percent yearly increase in the value of savings 1992-1998 and a 12 percent discount rate. These produce a present value factor of 15.6 and 1992 savings of $390 million for 132 MW capacity increment due to efficiency improvements, or $3,000/kW. The total saving is $3,600 per kilowatt, assuming the same ratio of repricing to fuel displacement savings obtained in the baseline case. This is less than the required investment per kilowatt efficiency improvement.

Table 4
Savings Due to Fuel Displacement and Steam Repricing
Oil Price: Zero Real Escalation

	1992	1994	1998	2002
1. Total savings in oil/gas costs, CEC baseline escalation rates, million $[1]	150	163	222	338
2. % reduction in oil/gas cost, assuming zero real escalation[2]	38	41	48	53
3. Reduction in oil/gas costs saving (line 1 x line 2), million $	57	67	107	179
4. Savings due to fuel displacement, CEC baseline escalation case (Table 3), million $	82	100	144	216
5. Savings due to fuel displacement, zero oil/gas escalation, million $ (line 4 less line 3)	25	33	37	37

Footnotes:

1. From Table 1.

2. The estimates on line (2) are = 1 less the ratio of oil cost under zero real escalation rate versus baseline escalation rates. The ratios are .62, .59, .52 and .47, respectively for the four milestone years.

CONCLUSIONS

1. CEC baseline oil/gas cost escalation assumptions indicate that the investment is substantially cost-effective ($5,000-$7,000 per kilowatt versus $16,900 saving per kilowatt). On the other hand, the assumption of a zero real cost escalation for oil/gas yields dismal results (again, $5,000-$7,000 investment per kilowatt versus $3,600 saving).

2. The investment necessary for the efficiency improvement is substantial ($660-$920 million in 1992 dollars, or $240-$340 million in 1982 dollars). The potential for saving, if the more favorable assumptions prevail, are also substantial. We, therefore, recommend that the merits of this potential investment be assessed on the basis of the reasonableness of the assumptions and, in particular, the reasonableness of the assumed oil escalation rates. It is also recommended that the investment merits be reassessed as new geothermal plants are built, in light of the most likely assumptions at that time. Lastly, the assessment should be based on comparison of all electricity generation and nongeneration technologies.

ACKNOWLEDGEMENTS

The authors would like to thank Dan Nix for his very valuable comments and Jerri Phillips and Bev Johnson for typing this manuscript.

REFERENCES

1. Gallo, David, 1982, The Cost Effectiveness of Geothermal Plant Efficiency Improvements, submitted to California Energy Commission, Sacramento, Ca.

2. House, Lon, 1982, The System Impacts of Geothermal Steam Plant Efficiency Improvements, California Energy Commission, Sacramento, Ca.

3. Cauchois, Scott, 1982, draft report Preliminary BR IV Electricity Price Forecast, California Energy Commission, Sacramento, Ca.

4. Teknekron Research, Inc., Cost Estimates and Cost Forecasting Methodologies for Selected Nonconventional Electrical Generation Technologies, CEC-sponsored study expected to be published summer 1982.

5. Gibbs and Hill, Inc., 1982, Turbine Back-Pressure Optimization Study, Northern Calif. Power Agency Geothermal Project No. 3.

Appendix A

Geothermal Steam Pricing Formula

CEC simulations for PG&E system, as is documented in Reference No. 2, utilized a production cost model (POWRSYM) and a geothermal steam pricing model (STEAMRAT). The latter is the subject of this Appendix.

The program calculates the geothermal steam cost (in mill/kWh and ¢/10^6 Btu) based on the costs, heat rates, and generated energy for the previous year by fossil-fuel and nuclear generating stations.

The formula for calculating the steam cost for the year Y is:

$$\frac{2.11 \times \dfrac{\text{FOS Cost}_Y}{\text{Oil Cost}_B} \times \dfrac{\text{Heat Rate}_Y}{\text{Heat Rate}_B} \times \text{FOSEN}_Y + [\text{NuCEN}_Y \times \text{NuCOST}_Y]}{\text{FOSEN}_Y + \text{NuCEN}_Y}$$

Where:

2.11 : A constant, mills per kWh;

FOS Cost_Y : Average fuel cost of fossil-fired generation (¢/10^6 Btu) at the end of the previous year Y;

Oil Cost_B : Average fuel cost of fossil-fired generation (¢/10^6 Btu) at the end of the base year (1968);

Heat Rate_Y : Lowest operating heat rate (Btu/kWh) of fossil-fired units at the end of the previous year Y;

Heat Rate_B : Lowest operating heat rate (Btu/kWh) of oil-fired units at the end of the base year (1968).

FOSEN_Y : Energy (GWh) generated by fossil-fired systems during the previous year Y.

NuCEN_Y : Energy (GWh) generated by nuclear systems during the previous year Y.

NuCOST_Y : Average cost (mills/kWh) of nuclear fuel at the end of the previous year.

NOTE: The analysis reported here reflects solely the conclusions of the authors. It does not necessarily represent the views of the California Energy Commission, other Commission employees or the State of California.

Geothermal Resources Council, TRANSACTIONS Vol. 6, October 1982

DECISIONS OF THE CALIFORNIA PUBLIC UTILITIES COMMISSION CONCERNING UTILITY PURCHASES OF GEOTHERMAL POWER

Dr. Priscilla C. Grew, Commissioner

California Public Utilities Commission
350 McAllister Street
San Francisco, California 94102

ABSTRACT

Pursuant to an order by the California Public Utilities Commission, California utilities have filed standard offers for purchase of as-available and firm electricity generated by small non-utility power producers. This paper summarizes recent rulemaking actions by the California Public Utilities Commission in establishing standards for these utility power purchases. Since 1979, the Commission has endorsed payment of full avoided costs to cogenerators and small power producers on the grounds that such a policy would maximize independent power production from diverse sources. The Commission believes that this production will benefit California ratepayers by increasing diversity of supply and developing in-state resources.

The purpose of this paper is to provide a status report on actions by the California Public Utilities Commission (CPUC) establishing standards governing the prices, terms, and conditions of electric utility purchases of electric power from small power producers, including geothermal facilities. In the oral presentation at the Annual Meeting, the author will review CPUC actions in the intervening months since preparation of this written text in May 1982.

In recent years, the CPUC has gone on record supporting the encouragement of small power production by non-utility business entities. The basis for this support has been the CPUC's conclusion that small power production allows a utility to diversify its resource plan, which increases the reliability of its system and minimizes the risk that results from excessive reliance on a single technology. Ratepayers do not have to pay the costs of unscheduled outages of non-utility small power facilities. Reduced dependence on oil and gas is yet another important advantage of opening up the electric generation business to non-utility developers.

For these reasons, California has for some time promoted small power production, in decisions by the Commission, starting in 1979, and in actions by the State Legislature.1/ In Decision 91109, issued on December 19, 1979, the PUC adopted "avoided cost" as the reasonable basis for payment by a utility to purchase power from cogeneration facilities. "Avoided cost" means the added cost the utility avoids by not having to produce an additional unit of electricity at its existing facilities or from a new powerplant. The price thus reflects the value of a producer's power to the utility system. The CPUC chose avoided cost because this payment would encourage full economic development of new cogeneration and small power resources.

On January 21, 1982, the CPUC issued Decision 82-01-103 concerning its Order Instituting Rulemaking No. 2 (OIR-2). The key action in the OIR-2 decision is the Commission's requirement that a utility purchase electricity from private non-utility suppliers at a rate equal to the utility's full avoided cost. The Commission established several different methods to be used in calculating these avoided costs, each of which is to be used by the utilities in the formulation of a so-called "standard offer" to be made to all small power producers. OIR-2 sets up standards to be used in determining how much of the utility's marginal costs are actually avoided under different kinds of contractual arrangements. It establishes requirements for allocation of costs for interconnection of the producers with the utility. The decision also states the Commission's willingness to review special nonstandard contracts between qualifying facilities and utilities that do not meet the terms of one of the standard offers.

The Commission's purpose in requiring certain "standard" offers is to enable most of the contractual arrangements between private producers and utilities to be executed without Commission review, in order to avoid governmental approval delays. The decision requires the development of four different standard offers designed to reflect variations in firmness of supply and to provide some flexibility in financial arrangements. The decision directs that each utility offer a choice of contract terms at the seller's option. These choices are intended to be economically equivalent over the life of the contract.

The first of the standard offers is the offer to buy electricity on an "as-available" basis, that is, electricity provided without a long-term contract and/or a commitment to meet certain performance standards. Such an offer might be appropriate for a wind facility or a cogenerator with widely fluctuating heat loads. Under this "as available" offer, a small power producer will receive an "as available" energy payment in ¢/kWh equal to the sum of the utility's avoided fuel cost at the time of delivery, plus an "as-available" capacity payment based on a percentage of the utility's annual capacity shortage costs. Because the utility's fuel costs are higher during peak periods, payments to private producers will be greater accordingly. For example, electricity produced during summer afternoons is worth the most. In other words, payments will be "time differentiated": the value of the electricity delivered to the utility will vary with the time of day and season.

If a producer can agree to certain performance standards, that is, the producer can guarantee delivery of electricity a set number of hours and during specified periods of the day or year, he or she is eligible for the "firm capacity standard offer." Under this contract, the small power producer must meet certain standards relating to availability at system peak, including dispatchability, reliability, scheduling of outages, and availability during emergencies. In return for adherence to these more stringent performance standards, the firm capacity offer allows the producer to receive a payment in ¢/kWh equal to the utility's avoided fuel costs plus a capacity payment reflecting the annual capacity cost of a combustion turbine.

The third and fourth standard offers required by the decision are more innovative. One of these simply adds a provision that can be used in conjunction with either of the offers described previously; namely, it allows the utility and a small power producer to fix an energy payment in advance for a five-year contract period. This "long-term energy price offer" allows both the utility and the private producer the certainty of a known price, based on a forecast of the utility's marginal fuel costs.

The fourth standard offer is to be based on the utility's long-run marginal costs. The other three standard offers I have just described are essentially based on the marginal costs of running the utility's existing system or its short-run marginal costs. Since it has been suggested that the utility's long-run marginal costs, namely the cost of building and running new power plants, would be different from the current short-run avoided costs (where a peaking gas turbine is used as a proxy for a short-run shortage cost), we have ordered the utilities to develop a long-run cost methodology on which long-term contract prices may be based.

In addition to requiring the utilities to set fair prices for purchases of power, the OIR-2 decision establishes rules regarding several other issues that can affect the viability of small power production. These include rates for sales of electricity to private producers, allocation of interconnection costs, requirements for availability during emergencies, and utility curtailment of purchases.

For example, in order to allow a small power producer to reap the full benefits of avoided cost pricing, our decision allows the private producer to buy electricity for its own needs at the going average rate, while selling all of the power it produces to the utility at what may be a higher marginal cost rate. This practice is called simultaneous purchase and sale.

An issue that has been in dispute in the past between utilities and private power producers is the question of who should pay for interconnection costs. The OIR-2 decision resolves this dispute by stating that the small power producer is responsible only for those costs of interconnection that exceed the cost the utility would otherwise incur to connect the producer as an ordinary customer.

With regard to availability during system emergencies, the OIR-2 decision requires small power producers to be available during system emergencies only if they are receiving the more attractive firm capacity payment. Under the firm capacity offer, the private producer will be expected to operate at maximum capacity on notice to meet utility needs during peak-load periods and emergencies.

After the decision was issued in January, applications for rehearing of the decision were filed by three California utilities. Responses to the applications for rehearing were filed by a number of parties, including Natomas Company, Thermal Power Company, and Union Oil Company. On April 12, 1982, the Commission issued Decision 82-04-071 modifying its original decision and denying rehearing.

The three utilities each questioned the use of a gas turbine proxy as the basis for calculating short-term capacity costs which is used for the as-available capacity payment and the firm capacity payment. The Commission acknowledged that the gas turbine is "merely a proxy for shortage costs," and does not suggest that it is the "desired incremental capacity choice. However its use is consistent with an incremental fuel cost that will for some time be based on oil or gas."2/ The Commission stated that it would entertain revisions in the as-available capacity methodology in utilities' general rate cases. The decision also stated that "as utilities reduce their dependency on oil, and other fuels fix their marginal cost, this fact should be taken into account in the calculations."3/

Several parties challenged the authority of the Public Utilities Commission to issue and enforce its January decision. The day after the decision, the District of Columbia Circuit Court of Appeals announced its decision in <u>American Electric Power et al. v. Federal Energy Regulatory Commission (FERC)</u>. The Court remanded to FERC two of the four aspects of FERC regulations under PURPA that were challenged by utility interests. The regulations in question deal with the rates that an electric utility must pay to qualified facilities under PURPA and the utility's obligation to connect these facilities to its grid. The Court found that FERC did not adequately consider whether payment below avoided cost would sufficiently stimulate cogeneration and small power production at a lower cost to the ratepayers and found that PURPA did not supersede those provisions of the Federal Power Act which require an evidentiary proceeding prior to the authorization of an interconnection. Southern California Edison and San Diego Gas & Electric have argued that the Court action means that the full avoided cost rates and blanket authority for qualifying facilities to interconnect are no longer in effect. The FERC petitioned the Court of Appeals for reconsideration and was unsuccessful. As of May 1982, it was anticipated that FERC would appeal the decision to the Supreme Court. The California PUC considers, however, that it is empowered to implement the OIR-2 decision. In its April decision, the PUC stated that "irrespective of the status of litigation on the FERC rules, Sections 2801 through 2804 of the [California] Public Utilities Code establish a comprehensive scheme 'to encourage private energy producers to competitively develop independent sources of natural gas and electric energy.'"4/

The three utilities and other parties took exception to the portion of the OIR-2 decision establishing periods during which purchases from qualifying facilities are not required. Their arguments focused on the potential situation in which the spill of utility-owned hydroelectric and/or curtailment of utility-owned geothermal plants would be required to permit mandated purchases from qualifying facilities. The Commission, considering these arguments, recognized the need for refinement of the standard price offer to avoid waste during these circumstances. As established in the OIR-2 decision, a utility can refuse to purchase electricity from qualifying facilities during any period during which, due to operational circumstances, purchases from qualifying facilities will result in system costs greater than those which the utility would incur if it did not make such purchases, but instead generate an equivalent amount of energy itself. The Commission uses the term "negative" avoided costs to define such periods. The Commission cites an example as the case when a base load or large oil-fired intermediate load plant is shut down at night due to an excess of QF electricity, but then cannot be restarted and brought up to its rated output for the next day's peak-load, thus necessitating instead the startup of a plant with very high generating costs (a gas turbine peaker) or an expensive energy purchase of capacity. The Commission was concerned that in the case where hydro is spilled a significant waste may occur if qualifying facilities (some of which burn nonrenewable fuels) are paid average avoided costs while water is wasted. The Commission decided that it would not permit a utility to refuse to purchase from qualifying facilities but would allow it to offer "hydro savings" prices to QFs during periods of potential hydro spill conditions on its own system, upon notice to the QFs. This would be a situation in which, if the utility were to accept full QF power, the utility would have to spill its own hydro resources. The Commission also stated that, since hydro savings pricing would create uncertainty and administrative burdens for QFs, utilities must consider all feasible alternatives before paying this price.

The OIR-2 decision also considered the issue of nonstandard contracts, those not filed pursuant to the "standard offers" filed by the utilities. The decision provides for utilities to file for review of nonstandard contracts with the Commission for a period of two years after the effective date of the decision. On April 21, the Commission issued Decision 82-04-087, approving certain provisions of a power sales agreement between U.S. Windpower, Inc. and Pacific Gas & Electric Company.5/ This case represents the first review of a nonstandard offer since the issuance of OIR-2. Since the procedure in this case is a good example of the "nonstandard review process" contemplated in the OIR-2 decision, it is an important case for small power producers to be aware of whether or not they are involved in wind power generation specifically.

PG&E and U.S. Windpower negotiated a power sales agreement with nonstandard pricing provisions calling for levelized payments above avoided costs. The payments above avoided costs would be offset by expected discounts to avoided costs in later years of the project, as well as interest payments. Under the agreement, PG&E will purchase all energy delivered from U.S. Windpower's 300MW wind generation facilities located in the Altamont Pass area of eastern Alameda County for a period of 30 years. PG&E and U.S. Windpower used PG&E's standard offer for wind facilities over 100kW as the starting point for negotiations. PG&E will pay U.S. Windpower a fixed price of 9¢ for each kWh actually delivered. The difference between the fixed price of 9¢ and 97% of the standard offer price at the time energy is delivered will be entered in a "Payment Tracking Account." In the early phase of the project, the fixed price of 9¢ will exceed 97% of the standard offer price. Once 97% of the standard offer price exceeds the fixed price, the balance in the Payment Tracking Account will be reduced as energy is delivered to PG&E. In the second phase, after the retirement of the Tracking Account balance, PG&E would pay U.S. Windpower 95% of the standard offer price for energy delivered. In the succeeding phase, Windpower will be paid 90%. This project was the subject of rigorous staff scrutiny, and the PUC staff analysis indicated that the allocation of project risk to the ratepayers under the specific terms of the contract was more than compensated by expected ratepayer benefits derived from the facility. The amount of ratepayer risk that is involved is limited to prudent levels by the provisions of the contract. In summary, the Commission's decision states that the agreement appears to offer the ratepayers high potential rewards at little risk; the fixed price is only slightly above current avoided costs, and there are safeguards written into the contract should the project experience early failure.

California utilities have filed standard offers effective May 12, 1982, for the following: (1) as-available power; (2) firm power--long-term capacity prices based on a gas turbine; (3) simplified offers for qualifying facilities under 100kW; and (4) related tariffs. Standard offers for 5-year forecasted energy prices and for energy and capacity prices based on long-run avoided costs have been filed in proposal form by three utilities. Hearings are scheduled to begin July 12, 1982 on the as-available, firm, and below 100kW offers to determine their compliance with the OIR-2 decision.

FOOTNOTES

1. State of California, 1980, Public Utilities Code, Ch. 7, Private Energy Producers, Sections 2801-2824.

2. California Public Utilities Commission, 1982, Decision 82-04-071, Order Modifying Decision (D.) 82-01-103 and Denying Rehearing and Stay Thereof, at 2.

3. Ibid, at 3.

4. Ibid.

5. California Public Utilities Commission, 1982, Decision 82-04-087, Opinion.

GEOTHERMAL PLANNING: A TALE OF TWO COUNTIES

Alex T. Hinds

Senior Planner
Lake County Planning Department
Lakeport, CA 95453

ABSTRACT

This paper examines the diverse settings, community concerns and planning procedures pertaining to geothermal development in Imperial and Lake Counties. The Imperial County planning process has predictably emphasized protection of its agricultural base. With fewer environmental problems, more comprehensive advance planning and fewer agencies, geothermal development has generally proceeded smoothly except when agricultural or recreational interests are threatened. Lake County development reflects the area's greater environmental sensitivity, conflicting land uses and fast growth. However, increased revenues and advanced environmental mitigation practices have accompanied development.

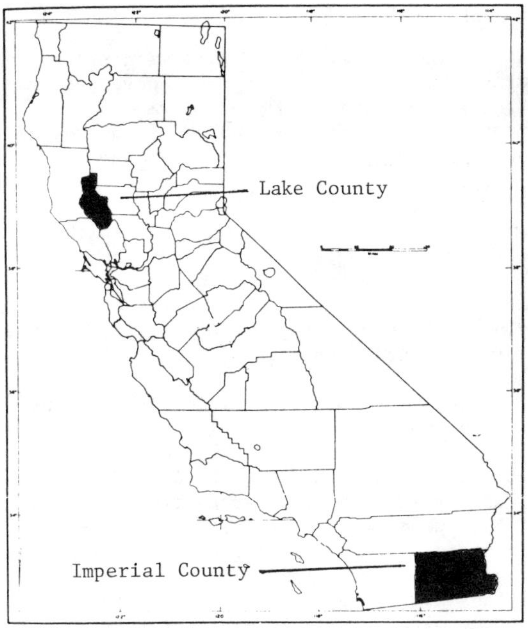

INTRODUCTION

Geothermal development activities span the state of California from level farmlands and parched deserts near Mexico to northern mountains, forests and meadows. In recent years most development has occurred in the "Geysers" region to the north and the Imperial Valley to the south. This paper examines the diverse settings, major community concerns and planning procedures pertaining to geothermal development in Imperial and Lake Counties.

LOCAL SETTINGS

The Imperial Valley is located approximately 80 miles east of San Diego, extending northward from the Mexican border to the Salton Sea. The valley is best known for its 475,000 acres of irrigated farm lands and warm climate. The valley usually ranks among the top five counties in the nation in agricultural production. The county has 95,000 inhabitants. Although agribusiness produced revenues exceeding $757 million in 1980, 1981 joint tax returns indicated the second lowest per capita income in the state. In addition, unemployment reached 26% in the spring of 1982. Between 1970 and 1980 the population growth rate was 2.4% per year.

The Imperial Valley also contains nearly one third of the nation's known hot water geothermal resources. Six Known Geothermal Resource Areas (KGRAs) are now considered to be economically viable and may be capable of producing 4,000 MW of electrical power sometime around the year 2010. Approximately 220 deep exploratory/production wells have been permitted, of which 120 have been drilled with 35 wells currently active. Mud drilling is used during exploration. Reservoir temperatures range from around 350°F to 644°F. High salinity levels, ranging up to 325,000 ppm, have in many of the hottest resource areas required the careful development and testing of elaborate control measures to prevent the "plugging" of geothermal equipment. At present, two 10 MW demonstration power plants are operational, with a third plant almost completed. The City of El Centro has drilled a geothermal well for direct use in the heating and cooling of the Community Center as well as other potential applications. Many other direct use studies involving agricultural processing have been and are being conducted.

Lake County is located 80 miles northeast of San Francisco and is best known for its recreational opportunities, large retirement population and walnut and pear orchards. The topography and vegetation are diverse, with coastal mountains, small farms, agricultural valleys, and lakes and creeks hosting a variety of forest, chaparral and riparian plant communities. The region's many lakes and mountain resorts are popular tourist attractions, resulting in a summer population roughly three times as large as its year round population of 40,000. The county is currently one of the fastest growing areas in the state. Between 1970 and 1980 the annual average growth rate was in excess of 8%. Most of this growth is attributable to in-migration, many coming to retire or resettle away from more urbanized areas. However, local economic conditions which featured 16% unemployment in Spring 1982, and the state's lowest 1981 joint per capita income are expected to, according to the Lake County General Plan, reduce the area's explosive growth rate to a high five to six percent per year.

A large portion of Lake County lies within one of four KGRAs. However, most geothermal development activities are occurring in the vapor-dominated, 350°-475°F "Geysers" field, located in the Mayacama Mountains near the Sonoma County boundary. Both mud and air drilling are used during exploration. At present, Lake County hosts one operational geothermal power plant--PG&E's Unit 13--which at 135 MW is the world's largest. In addition, Occidental Petroleum Company and the State Department of Water Resources are currently constructing power plants whose combined output will add another 135 MW of electrical generation.

The entire Geysers geothermal field currently produces around 906 MW of electrical power per year--with most of this production occurring in nearby Sonoma County. Projections for the Geysers area indicate roughly 2000 MW of power on line by 1990. Most of this projected development will occur in or adjacent to Lake County. Approximately 150 deep geothermal wells have active permits. Direct use applications are currently limited to a few privately owner warm water spas and individual water wells, although other options such as space and pool heating and agricultural applications are being examined.

COMMUNITY CONCERNS

Protection of the agricultural economy remains the paramount interest in Imperial County. Consequently, water consumption, subsidence prevention and the placement of transmission lines are among the most controversial issues facing geothermal developers. Subsidence prevention is a critical concern because of the region's extensive gravity flow irrigation systems. Other important local concerns include minimizing loss of agricultural lands, air quality impacts including potential damage to crops, protection of surface water quality and increased seismicity.

In the Salton Sea area, substantial opposition to geothermal development has been voiced when the possibility of off-shore drilling has raised local concerns regarding a potential loss of recreational amenities, increased water pollution and a proliferation of transmission lines. Diversion of surface waters for use as geothermal make-up water may further contribute to the sea's already high salinity levels which may soon rise to levels (40,000 ppm) capable of interfering with fish reproduction. Additionally, geothermal expansion may impact the extensive waterfowl associated with the area's wildlife refuges, primarily through habitat loss, noise and collisions with transmission lines. As the Salton Sea rises, threatening further inundation of surrounding agricultural and residential areas, local apprehension regarding geothermal induced subsidence also continues to rise.

However, county wide support for geothermal development is widespread as demonstrated in a 1977 University of California survey where 90% of the county respondents were in favor of it. The expectations that development would increase jobs, attract business and provide less costly power were cited as major reasons for favorable responses. Even in the Salton Sea area, a planning department questionnaire distributed at a 1981 Niland town meeting indicated that 90% were in favor of geothermal development in the Imperial Valley (versus only 48% approval of off-shore drilling in the Sea itself). Another factor contributing to favorable local attitudes is the extensive agricultural character of the valley, where the presence of large tractors, crop dusters and feedlot odors are normal daily occurrences. In general, the agricultural areas are flat, heavily impacted, sparsely populated and have a developed road system. Consequently, many environmental issues such as increased noise, dust, odor and traffic have not yet emerged as major concerns. Geothermal leases provide a source of supplemental income to the farmer and have thus far not seriously interfered with agricultural operations.

In Lake County, geothermal development frequently occurs in pristine, scenic areas in relatively close proximity to resort communities, recreation facilities and mountain retreats. Well pad and access road construction require considerable grading and removal of natural vegetation. As in the Imperial Valley, water is a major issue. Landslides, erosion and sedimentation are widespread concerns with potential impacts to the region's many creeks, which supply drinking water and are used for fishing and swimming. Diesel,sump fluids and other materials have accidentally entered area creeks on several occasions. Development near the community of Anderson Springs, which currently receives 70% of its domestic water from Gunning Creek, is of particular concern. Fast growing Lake County is largely dependent upon creeks and ground water for its water needs. In the long run, competition for water may prove to be the major issue. Many creeks in surrounding areas are continuously "harvested" for the sand and gravel needed to accomodate geothermal expansion.

Noise, dust and heavy truck traffic associated with geothermal development activities can be a source of considerable irritation to residents accustomed to the sounds of birds, running water and rustling leaves. Bleeding wells can damage surrounding vegetation and contribute to the distinctive aroma associated with H_2S. Some agricultural interests are concerned that boron emissions may someday damage local crops. All of the aforementioned issues are exacerbated when development occurs on privately owned surface lands with federally (or otherwise) reserved mineral rights. Rare, native plants and a legacy of archaeological resources must be taken into account. Growth inducement, and commensurately expanding roads, water, sewage, hazardous waste and classroom facilities are raising important fiscal considerations.

While local controversies continue regarding some of the environmental consequences of geothermal development, it is important to note that only one well has ever been denied. An opinion poll conducted by the University of California in 1975 indicated that 68% of the respondents felt the county should "go ahead" with geothermal development. In today's economy, the development of local geothermal is often viewed as the brightest star in a largely dismal sky. County government is receiving increased revenues from geothermal facilities. For example, according to the Lake County Assessor's Office, 1971-72 geothermal property taxes represented only .03% of the total county property tax base. By 1981-82, geothermal property values accounted for $2.7 million in taxes, more than 19% of the county's property tax base. In addition, over $700,000 in "AB 1905" geothermal lease funds have been returned to Lake County during the past two years. Continued geothermal expansion is a locally accepted fact of life. Today's environmental controversies generally center around geothermal site planning, performance and enforcement standards.

PLANNING PROCEDURES

Geothermal activities in the Imperial Valley are governed by a variety of planning documents including the county's "Terms, Conditions, Standards and Application Procedures for Initial Geothermal Development", Geothermal Element, Transmission Corridor Element and a variety of studies and application procedures. Except for temperature observation holes, all geothermal development activities require a conditional use permit. A distinction is made, however, between exploration activities which can occur in any zone and production activities which require a geothermal overlay zone. This practice reduces paperwork at the initial exploration stage while recognizing that production activities represent a long term land use commitment which requires additional review. In addition, the Geothermal Element adopted in 1977 promotes the use of Master Environmental Impact Reports (MEIRs) for each anomaly. The MEIR addresses the cumulative effects of multiple installations and usually allows for a streamlined permit process requiring only site specific analysis in subsequent applications. The MEIR is typically written (at the developer's expense) when a major power plant or geothermal overlay zone expansion is proposed. The MEIR identifies environmentally sensitive areas and recommends mitigation for each anomaly.

The Geothermal Element provides overall policy directives which are then specifically addressed in individual use permit conditions. For example, the element provides for the protection of agricultural water supplies and gravity flow irrigation systems by promoting the use of irrigation water only for demonstration power plants not exceeding a total of 75 MW in each anomaly for the first five years of operation. Subsidence monitoring networks are also promoted and participation in the network is generally required. The plan also promotes 100% reinjection of fluids back into the geothermal reservoir to protect against potential subsidence. In practice, reinjection requirements have varied from 80-98%. In accordance with general plan policy, wells are usually located in the corners of agricultural fields in order to minimize impacts to farming operations. The clustering of several wells per pad further reduces the loss of agricultural lands. Brine transmission pipelines are typically placed around the perimeter of agricultural fields. The Transmission Corridor Element addresses the location of electrical transmission corridors and substations in an effort to minimize interference with agricultural operations. Direct use applications are also briefly encouraged in the Geothermal Element.

Other county procedures also enable less comprehensive environmental documentation and faster permit processing if proposed exploration projects are located .5 miles from sensitive receptors such as residences, schools and wildlife refuges. However, the county's 1971 "Terms, Conditions and Performance Standards..." stipulates that wells only need to be a minimum of 300 feet from a residence, 50 feet from a public waterway, and 1,320 feet from a school. This document also requires a $50,000 indemnity bond for each well or a $150,000 blanket bond for all wells to indemnify the county for potential site repair or restoration costs. Other performance standards are also addressed.

In accordance with state law, Imperial County has been delegated "lead agency" status for processing geothermal exploration projects. This responsibility may be requested by local jurisdictions with comprehensive, adopted geothermal elements and equivalent processing procedures. While the county has not requested authorization to process geothermal power plants in excess of 50 MW from the California Energy Commission, the county is, in accordance with state law, processing all power plant applications under 50 MW. Coincidentally, no geothermal power plant application over 49 MW (net) generating capacity has been submitted in Imperial County. The county currently employs a full time geothermal planner as well as a coordinator located in the Public Works Department. Federal and state grant money has been widely used to finance a variety of geothermal planning studies, administration and staff.

The Lake County Geothermal Element was adopted in November, 1981. Three major policies are included in the document pertaining to:

1. the promotion of appropriate geothermal projects as a means of providing an alternative energy supply.
2. the protection of the natural environment

through proper mitigation of geothermal impacts.
3. the promotion of direct use applications including space heating and cooling, spas, frost protection, aquaculture, food and resource processing.

The Geothermal Element is a concise document which promotes the revision of local regulations and performance standards if necessary to ensure adequate protection of public health and safety. However, a geothermal overlay zone is not mandated and geothermal activities may be conditionally permitted in any zone subject to use permit requirements. The element does promote buffer zones around streams, water quality monitoring and use of the best available control technologies. The element also addresses the need to review existing insurance and bond requirements, revegetation and road improvement provisions and other issues. The element calls for the maintenance of an up to date public information map of all geothermal activities and sufficient staffing to plan for the orderly expansion of geothermal development.

Other regulations are contained in the county's "Conditions, Procedures and Performance Standards for Geothermal Regulation", the County Code, and individual use permits. County "Conditions..." stipulates that well sites shall be at least 500 feet from a residence, 2640 feet from a school and in no case use more than five acres of arable land. Bonding requirements are currently $5,000 for each well or $25,000 for a blanket bond for up to five wells. In addition, no geothermal well can be permitted within one-half mile of a populated area (defined as ten or more dwelling units within a one-quarter mile area) or a recorded subdivision, without the written consent of 75% of the property owners. Site specific regulations, however, are largely the result of individual use permit conditions. These conditions have become relatively standardized and generally require well siting at least 500 feet from a perennial stream, noise emissions not exceeding Ldn 55dBA, submittal of erosion control, drainage and spill protection plans and a variety of other conditions depending on the scale and location of the project. Noise control technologies include lead vinyl noise barriers, enclosing diesel equipment, rock mufflers and other techniques. Air pollution control measures to limit the release of H_2S include abatement of bleeding wells, treatment with hydrogen peroxide, Stretford units, turbine bypasses and other technologies. At the field development stage, school fees based on classroom construction costs and an annual survey are becoming standard conditions.

The Lake County Geothermal Element was not written to qualify the county for increased regulatory responsibilities. Consequently, the State Division of Oil and Gas is the lead agency for environmental review purposes on exploratory wells, and the California Energy Commission is responsible for power plant siting. It should be noted, however, that all exploration and steam field development projects occurring on privately owned land surfaces do require county use permits. While county procedures do not require MEIRs, the California Energy Commission is conducting "cumulative impact studies" with current hearings addressing impacts to schools, roads, domestic water systems, power plant efficiency and transmission system planning. Until being disbanded in late 1981, the GRIPS Commission, a four county joint powers agency, attempted to coordinate geothermal development studies in the Geysers area.

CONCLUSIONS

Vastly different topography, reservoir and land use features characterize geothermal development issues in Imperial and Lake Counties. Predictably, Imperial County planning procedures have emphasized the protection of its agricultural base. With fewer environmentally sensitive areas, more comprehensive advance planning, fewer regulatory agencies, most agricultural concerns addressed and only small demonstration plants currently on line, geothermal development activities have generally proceeded smoothly. However, this process has been severely stressed in the Salton Sea area and in instances where transmission lines and competition for water threaten agricultural interests.

On the other hand, Lake County development has reflected the area's pronounced environmental sensitivity, conflicting land uses and retirement/recreational character. These concerns have been exacerbated by limited water supplies, adverse impacts but not revenues from geothermal activities in Sonoma County, separate ownership of surface and mineral rights, less comprehensive advance planning, wider political fluctuations and fragmented regulatory responsibilities. Nevertheless, development has been accompanied by increased revenues and use of advanced environmental mitigation measures.

REFERENCES

Imperial, County of, 1971, Terms, Conditions, Standards, and Application Procedures for Initial Geothermal Development, Department of Public Works, El Centro, California.

Imperial, County of, 1977, "Geothermal Element", Imperial County General Plan, El Centro, California.

Imperial, County of, 1980, "Transmission Corridor Element", Imperial County General Plan, El Centro, California.

Irwin, Daniel, May 18, 1982, Imperial County Assessor's Office, Senior Appraiser, Personal Interview, Lakeport, California.

Lake, County of, 1972, Conditions, Procedures and Performance Standards for Geothermal Regulation, Lakeport, California.

Lake, County of, 1981, Lake County General Plan, Lakeport, California.

Vollintine, L.R. et al, 1977, The Lake County Economy: Potential Socio-Economic Impacts of Geothermal Development, University of California, Berkeley, California.

WESTEC Services, Inc., 1981, Final MEIR Salton Sea Anomaly, San Diego, California.

Geothermal Resources Council, TRANSACTIONS Vol. 6, October 1982

HAWAIIAN ATTITUDES TOWARD GEOTHERMAL DEVELOPMENT

Jerry Johnson

University of Hawaii at Hilo

Puna Hui Ohana

Puna, Hawaii

ABSTRACT

The results of a survey of the attitudes of a Hawaiian Community toward geothermal development are described. The survey was conducted by a grassroots community organization, and high response rate (85%) provides a good representation of community sentiment. The findings in general suggest a substantial polarization of attitudes toward the possible effects of geothermal development and toward potential uses of the geothermal resource. There is evidence of a balancing of perceived economic advantages of development against a long list of perceived negative effects on the physical, social and cultural characteristics of the community. Implications of the findings for assessing the social and cultural impact of geothermal development on the Puna Hawaiian Community are discussed.

INTRODUCTION

The Puna Hui Ohana is an organization of the Hawaiian Community of the Puna district of the Island of Hawaii. Its function is to address the needs of the youth, young adults, parents and elders of the community. The Hui's interest in geothermal development grew out of the public's concern over recent discoveries of geothermal potential in Puna, and the subsequent land-lease negotiations for geothermal rights. The activities of land speculators and pressure from agents representing several interested oil companies prompted 44 Native-Hawaiian land owners to appeal to the Puna Hui Ohana for advice, guidance, and an investigation of the overall implications of geothermal development in Puna.

Native Hawaiian rights groups were also expressing serious interest in the question of ownership of the geothermal resource. In addition they were concerned about traditional Hawaiian beliefs regarding the uses of the geothermal steam. For example, it was suggested that Madame Pele, the Hawaiian fire goddess, would be offended by geothermal drilling with potentially disastrous consequences for the Puna community.

The concerns described above were voiced by a number of Hawaiians, including representatives of the Puna Hui Ohana, at the Geothermal Resources Council Annual Meeting held in Hilo in July, 1978. The Department of Energy staff attending the conference indicated a willingness to consider funding a proposal from the Hawaiian Community to conduct a study to address the social and cultural implications of geothermal development for Aboriginal Hawaiians. A proposal submitted by the Puna Hui Ohana was funded by the Department of Energy; and the present paper summarizes a portion of the results of that study.

The present study describes the responses of the members of the Puna Hawaiian community to a set of attitude survey questions about the perceived impact of geothermal development, potential uses for and ownership of the geothermal resource, quality of life in Puna, and cultural characteristics of the Hawaiian community which might be affected by geothermal development. The aims of the survey were to collect data which would accurately describe the attitudes of the Hawaiian Community toward geothermal development, and to gather predevelopment baseline data about the Community.

METHOD

Sampling

The target area for the survey was defined by the Lower Puna census tract (Pahoa-Kalapana). A house-to-house census of the area was completed by the members of the Hui to identify all Hawaiians living in Lower Puna. Questionnaires were administered to all adult (18 years of age or older) Hawaiian and Part-Hawaiian residents of the area who could be located and who were willing to complete the questionnaire.

Data Collection

For organizational purposes Lower Puna was divided into six geographic areas with a team leader coordinating the survey administration in each area. The questionnaires were individually delivered and collected by a member of the Hui. In most cases, the survey team member knew the person completing the questionnaire. Information was provided by the respondent anonymously on the survey form, and the survey was returned in a sealed envelope.

Training for the members of the survey team consisted of three meetings in which questionnaire content, administration procedures, and potential problems were discussed.

Questionnaire Construction

Questionnaire items used a closed question format following a modification of the Consequence Analysis procedure (Sanford & Fawcett, in press) for community impact analysis. This procedure asks the respondent to indicate both the <u>magnitude</u> of the perceived effects of development (i.e., large or small), and the <u>value</u> of such effects (i.e., good or bad). Magnitude of impact is judged on a five-point scale and value of impact is judged on a seven-point scale. For questions about the potential uses of the geothermal resource only the seven-point value scale was used.

The final questionnaire was a refinement of an earlier instrument used to assess the impact on a group of 15 Hawaiian leaders of a visit to geothermal fields in New Zealand. The survey instrument was piloted three times with the Puna Hui Ohana Board of Directors and Project Advisory Board to clarify wording and item structure, and to be sure that all relevant topics were included.

The first section of the questionnaire addressed attitudes toward geothermal development. The second section addressed lifestyle, cultural practices and values of the Hawaiian Community.

The content of the items on the questionnaire was determined by a review of the literature about numerous discussions by the Hui Board of Directors about Community concerns, and from the information acquired from the earlier survey.

RESULTS

The census of the Community identified a total of 413 Hawaiian or Part-Hawaiian adults in Lower Puna. An attempt was made to contact each adult Hawaiian personally to explain the rationale for and nature of the survey, and to solicit their cooperation in completing the questionnaire. This procedure led to a return-rate of 85% (351 questionnaires). Missing data is due to an inability to contact the respondent, the respondent's refusal to complete the questionnaire, or to a blank questionnaire being returned to the survey team member.

The 1980 Federal Census identified 1712 households and 4696 individuals in the Lower Puna census tract. The adult respondents who completed the questionnaire represent 255 households in which 928 people reside. The survey thus includes data from 14.9% of the households and 19.8% of the population of Lower Puna.

Sources of Information about Geothermal Development

Respondents were asked about their level of knowledge about geothermal development. The majority felt that they had a small (25%) or moderate (30%) amount of information, with relatively few perceiving themselves as having large (8.5%) or very large (4.5%) amounts of information. Almost 30% of the Community reported having either a very small amount or no information about geothermal development.

The newspaper was clearly the most common source of information about geothermal development for the Lower Puna Hawaiian Community. Two-thirds of the respondents indicated the newspaper as a source of information, with radio (47%), friends (42%) and television (37%) also frequent sources. Thirty percent of the respondents reported receiving information directly from the Hui, while only 13% had attended geothermal workshops or conferences.

While the sources described above provide secondary information about geothermal development, it is also possible to gain information firsthand by visiting the HGP-S wellsite in Puna or by visiting some other geothermal field. Slightly less than half (49%) of the members of the Lower Puna Hawaiian Community reported having visited the HGP-S wellsite, while even fewer (7.3%) had visited some other geothermal field. It seems clear that media reports and other sources of indirect experience have provided most of the information to the Community to date, and that direct experience has played a relatively minor role in the formation of attitudes toward geothermal development.

Perceived Impact of Geothermal Development

The questionnaire asked each respondent to rate both the magnitude and the favorability of a number of possible effects of geothermal development in Puna. Table 1 summarizes the attitudes expressed toward these possible impacts.

Table 1

PERCEIVED IMPACT OF GEOTHERMAL DEVELOPMENT

GOOD*	NEITHER GOOD NOR BAD	BAD*
Economy	Social Conditions	Hawaiian Culture
	Community Closeness	Historical Sites
	Employment	Traditional religion
	Overall Impact	Hunting, Fishing Gathering
		Traffic
		Agricultural Land
		Land Taxes
		Physical Environment
		Quakes/Eruptions
		Plants/Animals

*All impact categories reported show nonchance ($p<.05$) frequencies in the indicated direction using a binomial test of significance.

The economic impact of geothermal development was perceived as positive, but all other effects were perceived as either negative or neutral. It is particularly interesting that the item asking about the overall effect of geothermal development falls in the neutral category, given this ten to one ratio of perceived negative to positive effects. This apparent contradiction is clarified somewhat by the information in Table 2, which shows that only 18.5% of the sample actually perceived the overall impact to be "neither good nor bad."

TABLE 2

DISTRIBUTION OF RESPONSES TO "NEUTRAL" IMPACT ITEMS

IMPACT CATEGORY	RESPONSES			
	GOOD %	NEITHER %	BAD %	NO RESPONSE %
Overall	(32.5)	(18.5)	(40.2)	(8.83)
Social Conditions	(35.6)	(21.9)	(34.5)	(7.98)
Community Closeness	(31.3)	(31.9)	(27.4)	(9.40)
Employment	(38.5)	(19.9)	(30.5)	(11.11)

Forty percent of the Community perceive an impact on the "bad" side of the continuum and 32.5% perceive an impact on the "good" side of the continuum. While the average of these values falls in the "Neither good nor bad" category, this position does not reflect the views of three quarters of the sample. A similar conclusion can be drawn from the distribution of responses to each of the other three apparently "neutral" items. This polarization of attitudes is not limited to the four "neutral" impact categories. Both age and number of years of residence in Puna were investigated as variables which might account for the variability in the data; however neither of these variables was systematically related to attitudes toward development.

In addition to questions about the favorability of the possible impacts of geothermal development respondents were asked to rate the expected magnitude of the impacts. Responses were consistently near the "large" end of the continuum for all categories, regardless of whether the value of the impact was perceived to be good or bad.

Uses of Geothermal Energy

The six uses of geothermal energy which have been most frequently proposed for Hawaii were presented to respondents for their evaluation on a seven-point scale from good to bad. Table 3 presents the results of this evaluation.

TABLE 3

ATTITUDES TOWARD POTENTIAL USES OF GEOTHERMAL ENERGY

GOOD*	NEITHER GOOD NOR BAD	BAD*
Agriculture/ Aquaculture	Electric Power for Oahu	Large Industries
Small Industries	Hotels/Spas	
Electric Power for Big Island		

*All impact categories reported show nonchance ($p<.05$) frequencies in the indicated direction using a binomial test of significance.

The favored uses imply either support of existing Island needs and activities (i.e., agriculture, Bit Island electric power) or relatively small scale industrial activities, rather than large scale development.

Ownership of the Geothermal Resource

In response to the question of who should own geothermal energy and receive the income from it, the majority (52.4%) of the respondents indicated Native Hawaiians. In order, the other choices were the surface land owner (23.8%), the owner of the mineral rights for the land (12.5%) and the State government (11.38%).

Quality of Life in Puna

In order to assess the degree of satisfaction among members of the Hawaiian Community with their present lifestyle, respondents were asked to indicate on a seven-point scale how happy or unhappy they were with the quality of life in Puna. Responses to this question produced greater concensus than did those to any other question in the survey. The great majority (81.9%) responded that they were happy with the present quality of life in Puna, while only 9.5% were unhappy, and 8.6% were neither happy nor unhappy.

Attitudes of Subgroups Within the Community

Information available about the background of the Community members makes it possible to investigate potential differences in attitudes for different subgroups of the Community. The variables of particular interest are 1) whether either HGP-A or other geothermal wells have been visited, 2) the amount of information people believe they have about geothermal development and 3) the geographical area of Lower Puna in which people reside.

Approximately one-half of the sample had visited the HGP-A wellsite; however the responses to the survey questions for those who have visited HGP-A did not differ from those who had not.

Twenty-five people indicated that they had visited a geothermal well other than HGP-A. At least 17 of these people would have seen the geothermal fields in New Zealand. Those who had visited other sites rated large industrial uses as more negative than those who had not, and they also perceived a more negative impact on Hawaiian culture.

A relationship was found between the amount of information that respondents felt they had about geothermal development and their overall attitudes toward it; but the relationship is not a simple one. Those who reported having either "very small" or "large" amounts of information had more negative overall preceptions than did those who reported "very large," moderate," "small" or "no information." In addition, those feeling they had "large" amounts of information perceived large industrial use of the geothermal resource as more negative than did the other groups.

The final variable, geographical area of residence in Puna, did not reflect differences in attitudes toward either the expected impacts or the uses of the geothermal resource.

DISCUSSION

The picture of the Lower Puna Hawaiian Community which emerges from background information provided by the survey is one of a community with many young families, and one which has grown substantially during the last 10 years. Occupational status varies considerably, formal education is typecally completed with high school graduation, and approximately one fourth of the adult Community receives social services from government or private agencies. What is the relationship of these characteristics to the possible effects of geothermal development in Puna?

One of the most stable of the findings of the survey was that the Hawaiians of Lower Puna are quite satisfied with the present quality of life in their community. The second major point of agreement among the respondents was that the impact of geothermal development would be large in scale. However, a consensus about the desirability of these potentially large impacts was not so readily apparent. There seems to be a balancing of the potential economic benefits of geothermal development with the environmental and social costs of development; and there is considerable polarization of attitudes. This situation is not unique to the Puna Hawaiian Community, and has also been described among the residents of Lake County in the Geysers geothermal field in California (Vollintine & Weres, 1976). A large number of impacts were perceived as negative by the respondents; and only one, economic impact, was reported to be clearly positive. It is not clear from the survey exactly what the respondents see the economic gains to be, especially since they were fairly evenly divided on the impact of development on jobs for Hawaiians (43% positive, 22% neutral, 34% negative). While only 8.5% of those answering a question about employment indicated that they were unemployed, 16% did not answer the question, fully 48% of those responding were not engaged in income-generating work. There may be a sizeable need for employment opportunities among the members of the Hawaiian Community, but there is considerable disagreement about whether geothermal development would meet this need. It is interesting that this issue generated one of the highest frequencies of write-in comments. Most of these comments can be summarized by one respondent's statement that the jobs would be "not for Hawaiians" but "for Haole and Japanese only". It would seem that the high school educated residents of this rural Community do not necessarily see the highly technological geothermal industry as an answer to their employment needs.

Of particular interest in assessing the cultural impact of geothermal development is the extent to which the Community members engage in traditional subsistence activities which could be in conflict with geothermal use of the land. There is reported a high frequency of such activities with a majority of the sample fishing (66%), shoreline collecting (62%) and food gathering (59%). The practices of gathering medicinal plants (48%), gathering maile (38%) and hunting (38%) are also quite common.

The impact of geothermal development on the traditional culture of the Puna Hawaiian Community is likely to focus on conflict over the use of the land in culturally congruent ways, and the potential interference with the application of certain Hawaiian cultural values. The importance of the land to the modern Hawaiian resident of Puna comes through very clearly both in the questionnaire responses and in the numerous write-in comments about the use and meaning of the land for the Hawaiian. Many traditional cultural activities require access to fairly large areas of land that is "underveloped" in the Western sense, but highly productive of things necessary for the practice of traditional Hawaiian culture.

CONCLUSION

The study which has been described was a Community-based approach to the assessment of attitudes toward geothermal development. It was sponsored by a grassroots Community organization, and the survey data was collected by its members. The survey instrument itself was created in continuous interaction with the Puna Hui Ohana Board of Directors to insure that it not only reflected their concerns about possible impacts of development, but also would be as clear as possible in wording and format to the people responding to it. Contrasting the 85% return-rate of the present study with the 31% return-rate for a similar study in Lake County, California (Vollintine & Weres, 1976) illustrates one clear advantage of taking the time to actively and meaningfully involve the Community in such undertakings. Information about community attitudes is necessary in order to plan geothermal development in a way that is responsive to community concerns and minimizes the social, cultural and economic costs to the affected community. It is recommended that such community involvement be a routine part of any new or expanded geothermal development.

ACKNOWLEDGMENTS

The study described in this paper was supported by the U.S. Department of Energy through a grant to the Puna Hui Ohana (Peter Hauanio, President and Principal Investigator; Everett Kinney, Project Director; Jerry Johnson, Project Consultant); Contract No. DE-FC03-79ET 27133.

REFERENCES

Sanford, Fonda L. & Fawett, Stephen B., 1980, Consequence Analysis: Its Effects on Verbal Statements About an Environmental Project. Journal of Applied Behavior Analysis, 13, p. 57-64.

Vollintine, L. & Weres, O., March, 1976, Public Opinion Concerning Geothermal Development in Lake County, California. University of California Lawrence Berkeley Laboratories, March, (LBL-4447).

Geothermal Resources Council, TRANSACTIONS Vol. 6, October 1982

GEOTHERMAL ELECTRIC POWER PLANT SITING: FEDERAL PERMITTING POLICY AND REPORT OF PROGRESS

Leroy M. Mohorich

Minerals Management Service
Menlo Park, CA 94025

ABSTRACT

The U.S. Minerals Management Service (MMS) approves power plant site and facility design, construction criteria, and compliance plans for projects on Federal geothermal leases. Rather than maintain in-house staff expertise, the MMS relies on the applicant and his contractors for many aspects of project review and inspection. Nearly all siting experience has been in California, where two commercial facilities are under construction. The Federal Government and the State of California have formalized procedures for cooperation for site and facility approval and postapproval compliance review.

INTRODUCTION

For about the past four years the Federal Government has been involved in the approval of the design and construction of commercial geothermal electric power plants. During this time the Minerals Management Service (MMS) and its predecessor*, along with the Bureau of Land Management (BLM), approved construction of the nation's first two commercial power plants sited on public lands. This paper discusses the Federal Government's permitting policy and reviews its power plant siting experiences to date.

SUMMARY OF LEGISLATION

The Department of the Interior prepared regulations governing the siting of electric power plants on federally leased lands under the authority of the Geothermal Steam Act of 1970. The following briefly summarizes relevant portions of these regulations. Conover (1979) provides a more detailed discussion.

*On January 19, 1982, Interior Secretary James Watt abolished the Conservation Division of the U.S. Geological Survey. The duties of the Interior Department regarding electric power plant siting, as described herein, were transferred to the Department's Minerals Management Service.

The regulations provide that, in addition to obtaining a valid Federal lease, the lessee must secure separate permits to occupy the land surface, to construct, and to operate the facility. These permits consist of a land-use license issued by the BLM and a two-part Geothermal Utilization Permit (GUP) issued by the MMS.

The regulations address three scenarios for resource utilization and electrical facility siting. These include: either (1) individual production well facilities of not more than 10 megawatts net electrical capacity or heat energy equivalent, or (2) research and demonstration (R&D) facilities of 20 megawatts or less net electrical capacity or heat energy equivalent, or (3) commercial facilities of greater than 20 megawatts net electrical capacity or heat energy equivalent. In the first two cases only GUP's are required of the applicant. In the third case both a BLM license and a GUP are required. All proposals must include a Plan of Utilization (POU), which provides a detailed overview of the proposed facility and its operation.

POLICY APPROACH TO FACILITY APPROVAL

While the MMS has gained expertise in various aspects of power plants, the Agency does not maintain a staff of engineers devoted solely to power plant design, siting, or construction. Neither does the MMS presently have the resources to provide comprehensive independent engineering review and project control during the design and construction phases. Therefore, the MMS siting policy is directed toward identifying applicable criteria and following through with a program to ensure that these criteria are applied during the design, construction, and operation of the facility. This program involves spot work checks by MMS staff for areas within its expertise and "certification checks" for areas outside its expertise.

A certification check means review and approval of a statement stamped and signed by an appropriate registered professional who thereby certifies, prior to construction, that the design criteria, analysis methods, and intended construction practices are adequate to meet the applicable laws, standards, ordinances, and codes necessary to ensure safe and reliable construction and operation of the facility. Furthermore, this certification check verifies that these laws and standards have been incorporated into the project's final plans and specifications. At the conclusion of the project, a similar statement must certify that the finished product is accurately represented by the as-built drawings and that it does incorporate the agreed-upon design criteria, analysis methods, and construction practices. The MMS relies heavily on the construction and material inspection activities performed by local agencies or the applicant, supplemented by occasional MMS audits.

OPERATIONAL APPROACH TO FACILITY APPROVAL

The steps involved in obtaining MMS approval for the site and facility under a Plan of Utilization (POU) are for the most part similar to those for any other Plan of Operation.

After an applicant obtains a lease and proves commercial potential, he must submit a POU to the MMS. MMS Geothermal Resources Operational (GRO) Order Number 5 explains the requirements for POU's.

The approval process begins with a review of the POU. Once the MMS receives a complete POU, it informs the BLM or other appropriate surface management agency (SMA) and other interested parties of the proposal and arranges any necessary site inspections. Comments on the POU are solicited from all interested parties. After the MMS receives and considers all comments, it prepares draft and final environmental reports. A letter is then prepared and signed by the MMS and the SMA, indicating joint approval of the POU. After both parties sign the joint approval letter and the applicant receives approval of any other appropriate documentation, the MMS will approve and sign the first part of the GUP, authorizing plant construction and prestart-up testing. If the applicant rates the proposed facility as "commercial", as discussed in the regulations, then he must also obtain a license from the BLM before the MMS can approve the GUP.

The Geothermal Utilization Permit

The GUP contains two parts: one for construction approval, the other for operation approval. Prior to approval of the construction part, the applicant must submit to the MMS a list of all laws, ordinances, standards, codes, and other criteria to which the plant site and specific plant structures will be designed, constructed, and operated. This list of laws and standards must be all inclusive, taking into account all pertinent Federal, State, regional, county and local regulations. This information must also indicate which agencies will require permits of the applicant. The list must include at a minimum the following resources and activities: air and water quality, water resources, health, socioeconomics, visual/aesthetics, soils, structural engineering, geotechnical and civil engineering, biological resources, noise, safety, and waste management.

The applicant must then use the above information to develop a Compliance Verification Program (CVP). This program outlines all applicable laws and standards to be followed during the project's design, construction, and operational phases and what measures will be employed for compliance verification and enforcement. The MMS must approve the CVP before site preparation work and construction can begin.

As part of the CVP, the applicant is expected to develop a Quality Assurance Plan. This plan, when approved, allows responsibility for field and construction inspections to be delegated by the MMS to the applicant or his contractor. This plan should address all aspects of construction monitoring, including material testing, manufacturer or fabricator certification, preparation of as-built drawings, deviations from plans, changed conditions, and so forth. The plan should also provide for the filing of periodic compliance reports with the MMS.

As partial compliance with the operational aspects of the overall program, the MMS requires the applicant to submit various types of design, construction, and operational documentation either for approval or for informational purposes. All drawings and related structural design calculations prepared for site preparation and facility construction must be submitted for approval. As a rule, this type of submittal includes construction contract bid packages, consisting of bid books, drawings, and plans prepared for prospective contract bidders. Necessary submittals include those for site preparation, civil, structural, and concrete work; the power plant building (turbine generator); cooling tower supply and erection, and H_2S control equipment supply and installation, among others.

Some drawings and plans need to be submitted for information only and do not require approval. These include so-called interface drawings, or drawings prepared by vendors of equipment for use in facility design. Examples include drawings needed for foundation designs for switchyard equipment or H_2S abatement/control equipment.

Although the construction part of the GUP may have been approved, additional data requirements must be met by the applicant before any site or facility work can commence. The data are submitted with a Sundry Notice and two alternatives for submittal are available to the applicant. In the first, the MMS will approve a Sundry Notice containing all necessary information for all phases of site and facility design and construction. Usually, however, information regarding the entire facility is not available at the outset of complex projects, and the applicant therefore must utilize the second alternative whereby he submits a series of Sundry Notices corresponding to various phases of site and facility design and construction. In this case each Sundry Notice addresses a particular phase, and work in that phase cannot commence prior to approval of that particular Sundry Notice. The project can be broken into as many phases as the applicant feels appropriate. Each surface disturbing activity or component of design or construction, however, must be addressed in a Sundry Notice.

An applicant may submit Sundry Notices with the POU or at any time until the activity covered by the Sundry Notice is to be performed, taking into account review time for regulatory Agencies.

The final part of the approval sequence involves prestart-up testing and documentation that the applicant accurately depicted the finished work in the final plans and specifications and incorporated the accepted design and construction criteria, analysis methods, and construction practices. The MMS evaluates test results and verifies construction compliance. At this point, the second part of the GUP is signed, and the applicant may commence commercial operation of the facility.

CASE HISTORIES OF POWER PLANT APPROVALS - CALIFORNIA

The Federal Government has to date approved the construction of three commercial-sized power plants. Two are located in Sonoma County, California, in The Geysers Known Geothermal Resources Area (KGRA), approximately 100 kilometers north of the City of San Francisco. Review and approval of construction are on-going. The USGS performed regulatory activities prior to January 19, 1982, as described below. The MMS is now the responsible Agency.

Case No. 1

The first generating station to gain Federal construction approval was the Northern California Power Agency's (NCPA) 110-Mw facility. The NCPA is a consortium of 12 public agencies, including 11 northern California cities and one rural electric cooperative. Construction of this facility, known as the NCPA Geothermal Project No. 2, was approximately 80-percent complete when a fire destroyed the completed cooling tower and various nearby plant components. Commercial operation has been set back about four months to a new scheduled commercial operation date of February 1, 1983 (Whalen, 1982).

In addition to review and authorization by the MMS and BLM, the U.S. Department of Energy (DOE) was involved because the NCPA applied for a federally guaranteed loan for a portion of the project financing. Because the California Energy Commission (CEC) develops and carries out State policy and responsibility for electric power generation and distribution, the State of California was also involved in project review and authorization. As a result, three Federal Agencies and one State Agency cooperate in their overlapping roles in the environmental protection, site/facility certification and compliance review of the NCPA No. 2 Project.

The parties involved recognized early in the project that a unique opportunity existed to combine State and Federal review efforts in order to prepare a single environmental document and provide information necessary for each Agency's decision requirements. To take advantage of this opportunity, the MMS and the other Agencies developed and entered into a formal Memorandum of Understanding (MOU) which constituted an agreement to participate jointly in the project review. As a part of the MOU, the MMS agreed to participate in the State's detailed certification proceedings with the understanding that each Agency must still issue appropriate individual permits for the project in accordance with individual procedures.

During negotiations between the MMS and the CEC, the parties agreed that final authority for compliance verification for plant design, construction, and operation resided with the Federal Government. However, the two Agencies developed and signed a formal Letter of Understanding (LOU) for postlicensing duties and responsibilities to insure fulfillment of State requirements for compliance verification. The LOU incorporated a system of primary and secondary CEC involvement. The CEC was primarily involved in: (1) review of final seismic design criteria, (2) structural design criteria for critical structures and components (as defined in the LOU), and (3) the evaluation of mitigation measures for any significant adverse geologic conditions (also defined in the LOU) which might be encountered during site preparation. Primary involvement means that the applicant submits all required drawings, reports, and analyses concurrently to the MMS and to the CEC. The MMS does not approve such drawings, reports, or analyses without expressly soliciting the advice and recommendations of the CEC. The MMS must provide a written explanation to the CEC of its reasons for not adopting any CEC recommendations on these items.

The CEC is secondarily involved for all noncritical structures and components and for mitigation measures as specified in the project's environmental analysis. Secondary involvement means the applicant submits all drawings, reports, and analyses to the MMS which forwards these to the CEC, which may or may not comment on these items.

In order to prepare the site and construct the facility, the NCPA elected to submit a series of Sundry Notices, with each Sundry Notice corresponding to a certain phase of its construction schedule. In general, each Sundry Notice covered all activities contained in a construction contract, although at times NCPA submitted additional Sundry Notices for specific activities within the contract. This method proved quite convenient and expedited regulatory review and approval.

Case No. 2

The second generating station to receive Federal construction approval was a 72-Mw facility to be operated by the Sacramento Municipal Utility District (SMUD). Known as the SMUDGEO No. 1 Project, the plant is approximately 65-percent complete and is scheduled to go on-line about December, 1983 (Knierim, 1982).

Essentially, SMUD repeated for their project all Agency review and approval procedures developed for the NCPA project. The parties again developed a MOU for preparation of an environmental document and the Federal Government again participated in the State's certification proceedings. Each Agency again agreed to issue its own separate permit/license after the certification proceedings concluded. Another LOU was developed for postlicensing duties and responsibilities, with all Agencies basically retaining the roles and responsibilities established for the NCPA project. The DOE, however, was not a party to this project.

As far as site preparation and plant design and construction are concerned, SMUD, like NCPA, decided to submit a series of Sundry Notices. While NCPA broke its project into many phases, SMUD initially intended to submit only three Sundry Notices, each one corresponding to one of three large construction contracts. The inclusion of so many design, construction, and analysis activities into three phases caused considerable potential for delay because of the volume of material submitted with each Sundry Notice and the MMS's limited resources. In this case, therefore, the MMS met with SMUD and requested a further breakdown of activities which would more readily provide for review within shorter timeframes. The parties agreed to a series of Sundry Notices giving partial approval of each of the three original Sundry Notices. In this way, various activities proceed while other later activities are under review, and the total of the various Sundry Notices constitutes a final, unconditional approval of the three original Sundry Notices.

CASE HISTORY OF POWER PLANT APPROVAL - OUTSIDE CALIFORNIA

At the present time, Federal experience with power plants on public lands outside the State of California consists of one project, the Milford No. 1 Unit proposed by Utah Power & Light (UP&L) Company. The 20-Mw facility will be located in the Roosevelt Hot Springs KGRA, in southwestern Utah's Beaver County, about 19 km northeast of the town of Milford. The facility is about 5-percent complete and commercial operation is presently scheduled for April, 1984 (Brown, 1982).

During preliminary discussions, UP&L expressed concern regarding the requirement for two approval signatures on the GUP. To alleviate their concern, the MMS developed an approval procedure whereby site preparation takes place under a conditionally-approved GUP calling for a follow-up series of Sundry Notices. While site preparation is in progress, UP&L prepares and submits civil and structural design information on an as-completed basis. The MMS authorizes no work other than site preparation while this material is under review. Upon approval of a complete civil/structural package, the MMS will sign the second part of the GUP, authorizing commercial operation of the facility. Attached to this signature approval, however, will be a list of conditions UP&L must meet before any actual commercial operations can begin. This procedure essentially is a combination of the phased and single approval approach.

Because Utah has no State agency performing functions comparable to those of the State of California, local agencies are taking a greater role in regulatory review for this project. Because the project has only recently commenced, operational procedures are still evolving, so a more detailed discussion about them is not possible at this time.

SUMMARY

Since mid-1978 a policy has evolved regarding the Federal Government's approval and permitting of geothermal power plants. During this period, the MMS helped develop and now participates in cooperative working agreements with other regulatory Agencies. These agreements provide an expeditious yet comprehensive regulatory framework by which resource utilization proceeds. As the future of geothermal development and utilization grows we anticipate continuing our efforts nationwide to streamline and otherwise improve the regulatory environment within which we accomplish our mission. The Federal Government's power plant experience may not now be extensive, but we are learning and have established a base upon which we will continue to help the Nation meet its energy goals.

REFERENCES

Brown, D. R., 1982, pers. comm., Utah Power & Light Co., 1359 West No. Temple, Salt Lake City, Utah, 84110.

Conover, R. D., 1979, Electrical power plant siting under the Geothermal Steam Act of 1970: Transactions, 1979 Geothermal Resources Council Annual Meeting, Reno, Nevada, Vol. 3, p. 129-131.

Knierim, R., 1982, pers. comm., Sacramento Municipal Utility District, 6201 S Street, Sacramento, CA., 94813.

Whalen, J. W., 1982, pers. comm., Northern California Power Agency, 8421 Auburn Blvd., Suite 160, Citrus Heights, CA 95610.

Geothermal Resources Council, TRANSACTIONS Vol. 6, October 1982

ENVIRONMENTAL NOISE NEED NOT HINDER GEOTHERMAL POWER DEVELOPMENT

Thomas R. Norris

Consultants in Engineering Acoustics
25 Drumm Street, Suite 202
San Francisco, CA 94111

ABSTRACT

Environmental noise issues have hindered some geothermal power developments located near residents by delaying necessary regulatory approvals. However, with full use of demonstrated noise control technology, noise can be reduced to levels acceptable to most quiet rural communities at a distance of about 1000 feet. Thus, it may be feasible to drill closer to residences than is often presumed.

INTRODUCTION

The purpose of this paper is to show that geothermal steam power can be developed rather close to noise sensitive land uses, provided that available noise control technology is fully utilized. This conclusion was developed as follows: First, noise levels were identified that would be acceptable to a quiet rural community using the "CNR" methods in Reference 1. These methods are based on the use of octave band noise spectra rather than more familiar A-weighted sound levels. This makes the calculations more complex, but the increased accuracy has gained this method acceptance in the electric power industry.

The second step was to estimate at several distances noise levels for geothermal noise sources, based on file data of best demonstrated noise control technology. That is, these methods have been used in geothermal or other industries.

The third step was to compare the noise level estimates to the criteria. Where the noise level estimates in each octave band are at or below the criteria for a given distance, then no adverse community reaction would be anticipated, and no valid complaints of noise expected.

NOISE CRITERIA

The CNR noise criteria method is described in detail in Reference 1. For purposes here, the method is the best known for predicting the likelihood of noise complaints due to industrial noise.(2) It takes into account background noise levels, previous industrial noise exposure, the time of day, duration, and character of the noise, as well as how good relations are between the geothermal developer and his neighbors. The bottom line of Table 1 shows the allowable noise source levels considered acceptable. The remainder of Table 1 shows allowable noise source levels for greater distances.

In calculating noise transmission for Table 1, we conservatively assumed worst case conditions--that is, what would be expected with an atmospheric inversion layer. The assumed temperature was 59°F and the humidity 70 percent.

We presumed that community relations are neither good nor bad. This assumption probably requires the developer to take advantage of the EIR process to show that the project plan minimizes noise impacts. Examples are locating drill sites and access roads as far from residences as possible and preferably behind ridges. Another example is to use as few drill sites as is feasible so that fewer people are noise-impacted. A third is to plan the drill site so that noise is projected away from residences. An often overlooked possibility, communicating factually about the project with the public, may help build project support. One savy developer spoke at a local service club meeting, apparently with significant benefit.

According to the CNR criteria, with poor community relations the residents would be approximately 5-10dB more sensitive to noise. Thus is shown the advisability of taking the above steps.

The background noise level assumed is equivalent to that in a very quiet rural community at night, typically in the mid 20 dBA range. For reference, this is similar to quieter areas near The Geysers, California.

Based on our geothermal experience, the criteria shown on Table 1 are conservative. That is, the community would accept 0-5dBA more noise than is assumed in this paper without complaints.

TABLE 1: OCTAVE BAND NOISE SOURCE LEVELS, MEASURED AT 100 FEET, THAT WOULD BE ACCEPTABLE TO MOST QUIET RURAL COMMUNITIES AT SPECIFIED DISTANCES AT NIGHT. (EQUIVALENT TO 38 dBA IN THE COMMUNITY)

DISTANCE FROM NOISE SOURCE TO COMMUNITY	FREQUENCY								
	31.5Hz	63	125	250	500	1000	2000	4000	8000
4000 FEET	96dB	86	79	72	70	69	71	86	107
2000 FEET	90	80	72	66	62	60	58	65	74
1000 FEET	84	74	66	59	55	52	50	51	54
500 FEET	78	68	62	52	49	45	52	41	41
100 FEET*	63	53	45	38	34	30	26	23	20

*FOR DAYTIME-ONLY NOISES OF SEVERAL DAY'S DURATION, ADD 10 dB, SUCH AS FOR WELL TESTING.

TABLE 2: OCTAVE BAND AND A-WEIGHTED GEOTHERMAL NOISE SOURCE LEVELS FOR STANDARD SILENCING, NORMALIZED TO 100 FOOT MEASUREMENT DISTANCE.

		FREQUENCY								
	dBA	31.5	63	125	250	500	1000	2000	4000	8000
1. 1-1/2 inch open valve, measured downstream, approx. 425 psig	126dBA	-	-	92	102	106	119	120	120	117
2. 6 inch open valve measured downstream, approx. 150 psig	131	-	103	111	110	118	125	125	125	120
3. Drilling in Mud	75-87	82-90	82-86	83-89	70-88	70-86	69-82	66-75	56-71	-
4. Maximum noise of pipe impacts in rig during roundtripping	87	-	-	-	74	84	87	85	71	-
5. Steady noise of two or four air compressors	83-86	88-91	88-91	87-90	79-81	70-73	70-73	67-70	65-68	63-68
6. Steady noise of steam venting into high efficiency drilling muffler w/ water injections, 210 klb/hr	81	89	88	80	73	70	67	66	61	50
7. Same, but for typical muffler at approx. 160 klb/hr, well testing	101	90	91	89	92	89	91	94	95	95
8. Wooden forced draft cooling towers, for 110-135 MW Power Plant at the Geysers, in direction of maximum noise	84	83	86	86	84	79	78	75	73	65

NOISE SOURCE ASSUMPTIONS

Two degrees of noise control are considered for each geothermal development activity. The first is "standard practice", such as is in general use. The second is with the best demonstrated or known, practical technology. Only the major noise sources are considered in this paper. Minor noise sources, such as 100HP generators, are not of concern here, because it is widely known that existing units can be quieted to the degree necessary.

In this report, we will not discuss construction activities or traffic noise. Suffice it to say that the noise impact from most of these sources can be reduced, possibly at some loss in efficiency, provided there is sufficient economic incentive.

EXPLORATORY AND DEVELOPMENTAL DRILLING

In drilling with "mud", the standard noise control is for all engines to have mufflers so that exhaust noise is not prominent. For the silenced operation, it is presumed that better mufflers are in use, that the engines are enclosed or behind barriers, that radiator cooling air openings do not provide excessive noise leaks, that a "winterization" kit is in use, and that "roundtripping" and unnecessary pipe impacts are avoided at night. Small noise barriers, or other means, may be in use to minimize hoist brake squeal. These measures reduce the drilling engine and hoist brake noise by about 15 dB. For the pipe impacts, by slower handling the noise can be reduced by at least 5 dB, except possibly during roundtripping. The standard and reduced noise source spectra are shown on Tables 2 and 3.

DRILLING WITH AIR AND IN STEAM

The dominant noise sources are large air compressors, compressed air releases, and steam venting. The air compressors are normally equipped with a modest sized muffler and no other exhaust noise controls. In this paper, we assumed that at least 20 dB of noise control can be obtained using properly designed engine enclosures and better mufflers.

Compressed air releases may occur when the drill pipe is disconnected. In this paper, we assumed that the drill pipe is always depressurized through an effective muffler. This renders the compressed air releases inaudible in the presence of other noise sources.

For the steam releases, the standard noise control is a cyclonic muffler-separator equipped with an effective water injection system. Also, the inlet piping is properly insulated. For the silenced steam releases, noise barriers, a quieter muffler inlet and better water injection are used. In this paper, we presume that, in combination, these reduce the noise by an additional 10 dB.

WELL TESTING

Some developers have demonstrated that they can flush out and flow test wells through the same muffler as is used in drilling. Sound levels would typically be 10 dB higher than those during drilling, for both the standard and specially silenced mufflers. However, testing normally can be scheduled as a daytime-only activity, so that the noise impact is no greater than that of the steam releases during drilling.

TABLE 3: OCTAVE BAND AND A-WEIGHTED GEOTHERMAL NOISE SOURCE LEVELS, SPECIALLY SILENCED WITH KNOWN TECHNOLOGY, NORMALIZED TO 100 FOOT MEASUREMENT DISTANCE.

		FREQUENCY								
	dBA	31.5	63	125	250	500	1000	2000	4000	8000
1. Drilling in Mud	60	67	67	68	55	55	54	51	41	-
2. Maximum noise of pipe impacts in rig during roundtripping	82	-	-	-	69	79	82	80	66	-
3. Steady noise of two or four air compressors	63-66	68-71	68-71	67-70	59-61	50-53	50-53	47-50	45-48	43-48
4. Steady noise of steam venting into high efficiency drilling muffler w/ water injections, 210 klb/hr	71	79	78	70	63	60	57	56	51	40
5. Wooden forced draft cooling towers, for 110-135 MW Power Plant at the Geysers, in direction of maximum noise	79	78	81	81	79	74	73	70	68	60

PRODUCTION NOISE

The dominant noises during steam production and power generation are control valve noise, noise from separator drains and driplegs when vented, and master valve replacements. Our experience is that with valve noise control techniques and careful pipeline location, valve noise is not environmentally significant.

Separator drains and other steam releases can be extremely noisy and are partly responsible for the noisy reputation earned by early geothermal developments. Table 2 shows noise levels from 1-1/2 inch and 6 inch valves being opened to the atmosphere. These noise levels are so high that they are unacceptable, by comparison of Tables 1 and 2, even at the largest distance shown. For the silenced case, all driplegs, separator drains and miscellaneous steam releases are silenced to insignificance--possibly using small rock mufflers or other mufflers.

For master valve replacements, when feasible, it is presumed that the well bore is plugged during valve replacement, so that any steam releases are of short duration and small, and resulting noise is slight.(3) Valves are replaced so seldomly that a community may temporarily tolerate even high noise levels, if need be.

POWER PLANT

For a carefully designed geothermal plant, the mechanical draft cooling towers are the only noise source for which there is no readily available noise control technology that is highly effective, yet economical. In this paper, we assume that cooling tower noise can be reduced by approximately 6 dB, mostly by using special fans and possibly adding additional cells. An additional 4-6 dB of noise reduction is possible in some cases, but is not assumed for this paper. Also, yet more noise control is possible, but at higher cost.

CONCLUSIONS

Presuming a clear line of sight between the geothermal development and residences, the following conclusions are valid. First, the standard practice noise levels (Table 2) are just above those acceptable at 4000 feet (Table 1) for drilling with mud, air, or in steam, and for the cooling tower. Thus, limited noise complaints would be expected, which is consistent with experience. Noise shielding by terrain, forests, equipment or a lip of the pad would be necessary to eliminate the complaints and may allow a shorter distance.

For the silenced drilling noise source levels shown on Table 3, the noise limits for 1000 feet are just exceeded for 125 and 1000 Hertz. Thus the minimum distance for no complaints is a bit above 1000 feet. However, by shielding some noise sources with equipment and by some use of noise barriers, a margin of safety could be achieved, even at 1000 feet. This would seem wise given Murphy's Law.

The cooling tower noise would be above the criteria, at 4000 feet, and would be the dominant power plant noise source. Thus, the cooling tower noise would be the major problem in locating the power plant close to the "very quiet" community. Fortunately, in actual practice, it is usually possible to reduce noise by using building and terrain shielding, noise source directivity, and possibly other methods. Also, the efficient atmospheric sound transmission assumed in this paper is not likely at all sites.

The noise reductions presumed would not be likely by casually muffling and enclosing engines and steam vents, and installing rock mufflers on all pads. Instead, all factors would have to be considered, and the final noise controls methodically designed and installed. Developers and regulators should also realize that the noise reductions would require continuing attention and supervision, and in some cases may have significant hidden costs. Furthermore, at a distance between about 1000 and 2000 feet, many minor nighttime noise sources may require attention, such as hammering, swearing and other activities. Thus, the degree of noise reduction assumed for 1000 feet in this paper should not be offered or required without carefully documented justification.

The "single complaint" noise criteria used in this paper is conservative. While in some cases, adopting this criteria as a project design goal may be prudent, meeting the goal at all times for all noise sources is not realistic or expected. This is why noise ordinances and use permits typically allow higher noise limits for residential, transportation and industrial noise sources - out of practical and economic necessity.

It is believed that the conclusions in this paper would also apply to electrical power from hot water dominated resources, as well as dry steam developments.

REFERENCES

(1) Stevens, K.N., Rosenblith, W. A., and Bolt, R. H., 1955, A Communities Reaction to Noise: Can It Be Forecast:, "Noise Control" magazine, Vol. 1 No. 1.

(2) Edison Electric Institute, 1978, Electric Power Plant Environmental Noise Guide, Volume I.

(3) Personal communication with Mr. Tebow of WKM.

Geothermal Resources Council, TRANSACTIONS Vol. 6, October 1982

RESULTS OF THE GEOTHERMAL RESERVOIR INSURANCE STUDY

Donald M. Routh

Alexander Grant & Company
One California Street
San Francisco, California 94111

ABSTRACT

On June 30, 1980, Public Law 96-294, referred to as the Energy Security Act, was enacted by the Congress of the United States. Subtitle B of Title VI (Geothermal Energy) of the Act required that a reservoir insurance program study be conducted. The objective of the study was to provide an analysis of the need for and feasibility of a geothermal reservoir insurance program. In conjunction with the analysis, the appropriate level of federal support, if any, was to be determined.

INTRODUCTION

The commercial scale developer and user of geothermal energy, either for generation of electricity or for direct-use, face a number of significant risks that can inhibit development. In the early stages of field development the size and character of the reservoir are highly uncertain, yet the costs to the developer and user in terms of initial wells, surface facilities and power plants are significant. Many of the risks that tend to inhibit development can be transferred from the developer and user to an insurer for a set price thereby reducing economic uncertainties.

One hypothesis is that a geothermal reservoir insurance program would be an incentive for increasing geothermal development and improving geothermal technology. The purpose of the Goethermal Reservoir Insurance Study was to analyze this potential incentive on its own merits -- not to attempt to determine the single best incentive that might be provided to the geothermal industry.

The study involved five major tasks: (1) determine perception of risk by major market sectors, (2) determine status of private sector insurance programs, (3) analyze reservoir risks, (4) analyze alternative government roles, and (5) provide recommendations.

PERCEPTION OF RISK BY MAJOR MARKET SECTORS

More than forty interviews were conducted with representatives of the developer, user and lender sectors of the geothermal industry. The objectives of the interviews were to: (a) identify the industry representative's perception of the primary risks associated with geothermal development and production, and (b) obtain their perception of the need for a federal geothermal insurance program and its potential impact on geothermal development. The perceived priority risks varied by size of firm, type of resource being developed, size of development and respondent's role in the project. The priority risks commonly identified were:

- Reservoir decline;
- Well failure or damage;
- Environmental, legal and institutional delays;
- Physical damage to plant;
- Financial impediments; and
- Inability of developers and utilities to secure satisfactory long-term sales agreements.

Although the individuals interviewed had differing opinions on the appropriateness and need for a federal geothermal insurance program, they generally agreed that the availability of insurance would speed geothermal development. Insurance would address the uncertainty surrounding this resource and as a result would overcome some of the reluctance to become involved in geothermal projects.

STATUS OF PRIVATE SECTOR INSURANCE PROGRAMS

More than sixty interviews with representatives of the various segments of the private insurance community were conducted to determine the current status of applicable insurance programs and to determine their perception of the insurability of the specific geothermal risks previously identified by the major market sectors. Although only one primary insurer has actively sought to provide reservoir performance coverage, many of the firms interviewed indicated a willingness to provide insurance coverage for certain of the geothermal risks including reservoir performance.

The key reasons identified for the current lack of broad participation in insuring geothermal developments were;

- Lack of historical performance data;
- Questionable reliability of available data;
- Potential for unusually large loss;
- Unacceptability of desired policy term; and
- Lack of communication between geothermal and insurance industries.

In considering the insurability of the priority risks identified by the major market sectors, certain specific risks were perceived as uninsurable (e.g., marketability). The remaining risks, although opinions differed widely as to insurability, comprise the set of geothermal risks considered for further analysis.

Although there is currently only limited participation in providing insurance protection for geothermal risks, it is not unusual for developing technologies to be served by only a few insurers. If some insurers gain positive experience in the geothermal industry other insurers have indicated they would be willing to become more involved, thereby increasing competition.

ANALYSIS OF RESERVOIR RISKS

The risks of geothermal development were analyzed to estimate the approximate level of insurance premiums necessary to cover potential losses. Prior to performing actuarial analyses resulting in premium estimates, specific risks were identified and their probability of occurrence and estimated cost consequences were determined. The comprehensive set of individual risks identified for analysis are contained in the following five major risk categories:

- Well Risks
- Reservoir Performance Risks
- Power Plant Risks
- Surface Facility Risks
- Acts of God

Probabilities of occurrence and cost consequences were estimated based on available data and subjective probability assessments of geothermal reservoir experts. On the basis of the probabilistic analysis, an expected loss and loss distribution for each risk was estimated in terms of:

- Direct Cost to Developer - direct costs to replace or add wells, surface piping, etc.

- Indirect Costs to Developer - loss of revenue from reduced steam sales.

- Direct Cost to User - repair costs from damage to plant or turbines, and unamortized value of plant resulting from total or partial abandonment.

- Indirect Cost to User - Excess cost of replacement power resulting from shut down or reduced capacity.

The expected losses and distributions were estimated for each of three different stages of development (field development, initial operation and full operation) and for each of seven geologic project types (e.g., vapor dominated). Insurance premiums were then estimated using appropriate actuarial methods as a function of the expected losses and loss distribution for each coverage category, along with a provision for administrative expenses.

ALTERNATIVE GOVERNMENT ROLES

After examining the range of possible programs, the following five alternative government roles were selected for detailed evaluation:

(1) Private market insurance program exclusive of any government involvement.

(2) Private market insurance program with government providing excess catastrophe reinsurance.

(3) Private market insurance program with government making available specific excess reinsurance.

(4) Private market insurance program with primary government insurance to cover those risks not insured by the private sector.

(5) Government primary insurance program contracted to a third party for underwriting and administration.

Detailed analyses were performed on each alternative including: (a) impact on private insurance sector, (b) financial impact on geothermal industry,

(c) estimated cost to government, and (d) interaction with other government programs.

RECOMMENDATION

The recommended geothermal reservoir insurance alternative was determined on the basis of its responsiveness to the perceived need for geothermal reservoir insurance and its effectiveness in stimulating development of geothermal resources. Although current means exist to reduce certain aspects of the financial uncertainty of loss to geothermal developers and users (e.g., Geothermal Loan Guaranty Program), there is room for complementing these programs. The Geothermal Reservoir Insurance Study has shown that insurance would provide a means of protecting against the financial uncertainties of geothermal development. The study has also shown that insurance would most likely be a cost effective means of dealing with geothermal project financial uncertainties.

The study determined that there is a viable role for the government to help accelerate the emergence of geothermal insurance supplied through the private sector. Given that:

- it is desirable to provide incentives for the development of geothermal energy as an alternative energy source,

- there are significant risks associated with geothermal development,

- insurance provides incentives for geothermal development by reducing the financial uncertainty of geothermal risks to the insured,

- the geothermal constituency believes that a properly structured insurance program would speed the development of geothermal resources, and

- the private insurance sector currently lacks broad participation in insuring geothermal developments,

there is a need for a temporary government role in a geothermal reservoir insurance program until such time as private insurers are actively providing adequate coverage on a broad basis. In addition, because (a) the significant risks associated with geothermal development can be insured, and (b) there is historical precedent for the government playing a role in insuring highly technical or emerging indistries, it is feasible for the government to have a role in a geothermal reservoir insurance program.

Because of the established need for and feasibility of a federally supported geothermal reservoir insurance program, the study recommends:

A private market insurance program for insurable risks underwritten by private insurers should be encouraged. The federal government should support this effort by making available limited excess reinsurance at a specified level decreasing over time. Additionally, through cost support, the price to insurers should be substantially less than what the private reinsurance market might provide.

ACKNOWLEDGEMENTS

This paper has been excerpted from the Geothermal Reservoir Insurance Study prepared for the United States Department of Energy by Coopers & Lybrand and their subcontractor GeothermEx, Inc. The author was Coopers & Lybrand's project manager responsible for the study and is now the National Manager of Alternative Energy Consulting for Alexander Grant & Company. More than forty professionals participated in the study. The author expresses particular thanks to Dr. Richard Fallon and James A. Hall III of Coopers & Lybrand and to Dr. Subir Sanyal of GeothermEx, Inc.

The reader is advised to refer to the Geothermal Reservoir Insurance Study for detailed documentation supporting and expanding on the statements made in this paper.

REFERENCE

"Geothermal Reservoir Insurance Study - Final Report," 1981, United States Department of Energy, Contract No. DE-AC03-81SF11501, 509 pp.

Geothermal Resources Council, TRANSACTIONS Vol. 6, October 1982

AIR QUALITY MODELING OF GEOTHERMAL POWER PLANTS IN COMPLEX TERRAIN

T. W. Tesche

Systems Applications, Inc.
San Rafael, California 94903

ABSTRACT

Commercial energy development at the Geysers KGRA, California--the world's largest producing geothermal field--is confronted with increasingly more stringent air quality impact assessment requirements for not only regulated pollutants (e.g., hydrogen sulfide and primary particulates) but also unregulated species including benzene, ammonia, arsenic, Radon^{222}, boron, and mercury. An integrated program of ambient aerometric monitoring, tracer diffusion experiments, cooling tower plume rise studies, and numerical air quality simulation modeling was conducted for this mountainous region to provide estimates of potential short term (i.e., $\sim$ 1 hour) impacts attributable to development of the Geysers steam resource. A series of atmospheric tracer diffusion experiments, focusing on aerodynamic downwash, nocturnal drainage, fumigation, and limited vertical mixing were conducted to provide a basis for air quality model performance evaluation. Upper air meteorological measurements and cooling tower plume rise studies were performed to develop data useful in evaluating windfield and plume rise submodels.

INTRODUCTION

Commercial energy development at the Geysers KGRA, California--the world's largest producing geothermal field--is confronted with increasingly more stringent air quality impact assessment requirements for not only regulated pollutants (e.g., hydrogen sulfide and primary particulates) but also unregulated species including benzene, ammonia, arsenic, Radon^{222}, boron, and mercury. Notwithstanding ongoing abatement research and pilot-scale testing, malodorous H_2S emissions continue to promote violations of the California 1-hour ambient standard. In support of permit applications by state and municipal agencies, an integrated program of ambient aerometric monitoring, complex terrain tracer diffusion experiments, cooling tower plume rise studies, and numerical air quality simulation modeling was conducted to provide estimates of short-term impacts attributable to further development of the Geysers steam resource. Details of the historical data analyses, field experiments, and numerical modeling are discussed by Tesche et al. (1980) and Knuth et al. (1979).

COMPLEX TERRAIN TRACER DIFFUSION EXPERIMENTS

A series of atmospheric tracer experiments were designed and carried out at the proposed California Department of Water Resources' South Geysers site and at the Sacramento Municipal Utility Districts' SMUDGEO#1 site. Objectives of the tracer tests were to provide direct source-receptor relationships for estimating future H_2S impacts via scale-up of cooling tower emissions, and to provide data sufficient for air quality model performance evaluation. Criteria for the performance of the tests were based upon a detailed statistical analysis of historical aerometric data collected at the Geysers from 1976-1980. Dual tracers, SF_6 and Freon, were released under adverse limited mixing, fumigation, downwash, and nocturnal drainage conditions. Tracer sampling, at the ground and aloft, continued for 12 hours at more than two dozen downwind locations. Aircraft sampling of existing cooling tower plumes near the release sites was also performed to establish the proper tracer release height and to provide data for plume rise model verification.

MODEL PERFORMANCE EVALUATION STUDIES

Due to the complex topography (1400 m terrain relief) and moderate source-receptor distances (i.e., 5-10 km), the SAI Hybrid Model, employing both semi-Lagrangian and Eulerian concepts, was developed and applied to the geothermal field. The model simulates the sequential downwind transport and turbulent diffusion of point source emissions puffs until they attain a size commensurate with a specified Eulerian grid; at this point the Eulerian concept is invoked to characterize further transport and dispersion. Rise of buoyancy-generating moist cooling tower plumes is simulated numerically. Transport fields are computed by a diagnostic wind field model which relies, in part, on the high resolution terrain data available from the National Cartographic Information Center. The basic modeling approach consists of two parts, simulation of the flow-field and simulation of plume rise and dispersion. This dual approach received broad acceptance at a recent EPA-sponsored Workshop on Atmospheric Dispersion Models in Complex Terrain (EPA, 1979). The hybrid model is composed of the following sub-models:

> An Eulerian, grid-based dispersion model [the SAI Airshed Model (Reynolds et al., 1979)]
> A three-dimensional, mass-consistent wind model [the SAI Complex Terrain Wind Model [Yocke and Liu (1978)]
> A numerical cooling tower plume rise model [Winiarski-Frick (1978)]
> A Langrangian, sub-grid-scale Gaussian puff model.

This modeling methodology was selected over the Gaussian approach because it provides a more realistic treatment of the relevant physical processes occurring in complex terrain.

Two numerical cooling tower plume rise models--Hanna (1976) and Winiarski and Frick (1978)--were evaluated using the results of 23 plume rise experiments performed at the Geysers. Examination of model predictions and observations revealed that the average errors associated with the Winiarski and Frick model vary between 22 and 27 percent. For the Hanna model, errors range between 30 and 36 percent. The average errors exhibit an overall tendency toward underestimation.

Policastro et al. (1977) reported that the Hanna and the Winiarski and Frick model's predictions of visibile plume height were within a factor of 2 of the observed values in at least 50 percent of the cases they studied. From results obtained in application to the Geysers cooling towers, we found that the Hanna model is within a factor of 2 of the observed values 48 percent of the time, and the Winiarski and Frick model is within a factor of 2 of the observed values 57 percent of the time. These results, consistent with the findings of Policastro et al., are encouraging, particularly in light of the fact that a far more detailed cooling tower data base was available to the Argonne group in their model verification work.

The Hybrid Model was evaluated using data sets from five tracer experiments and one historical Geysers H_2S eipsode. The tests were carried out under adverse limited mixing, drainage, downwash, and fumigation conditions. Model evaluation using the tracer data was intended to examine model performance in simulating single point source plumes. The historical H_2S episode was simulated to determine the model's capability of predicting the dispersion of emissions from 22 nearby geothermal power plants, leading to cumulative H_2S concentration levels.

Several measures of model performance were investigated with the Hybrid Model, including accuracy of the peak concentration prediction, bias, gross error, overall accuracy and precision, and temporal and spatial correlation. In general, agreement between the peak predicted and observed concentrations was within a factor of 1.5 to 2.5. Figure 1 exemplifies estimates of model bias and error (signed and absolute normalized residuals) as a function of concentration level for one of the limited mixing simulations. For the NS-1 experiment, the model tends to underestimate, on the average, by 50 percent. Overall accuracy and precision may be defined as the mean and standard deviation of the distribution of residuals (the difference between predicted and observed equivalent H_2S concentrations). In general, the

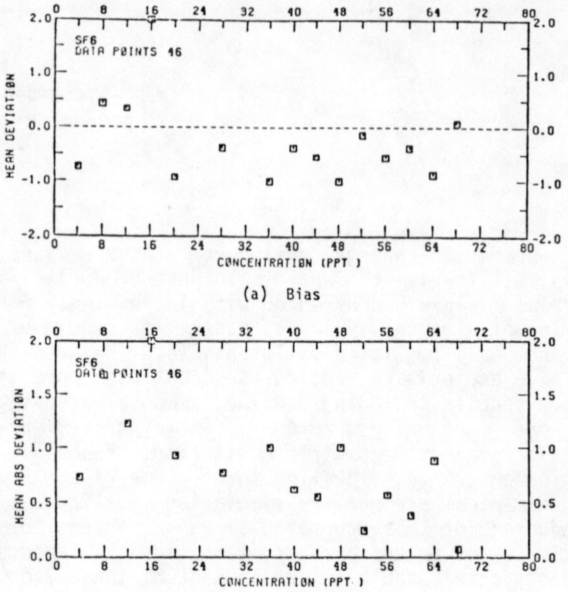

(a) Bias

(b) Gross Error

Figure 1. Bias and Error Estimates as a Function of SF_6 Tracer Concentration Level for the NS-1 Diffusion Experiment [Mean and absolute deviations (residuals) are normalized by the observed concentration level.]

accuracy of model predictions is approximately 1-13 ppt depending upon which test is simulated. (There exists a slight tendency toward overestimation.) Further, the precision is approximately ± 12 to 27 ppt. (Typically, the range in peak observed tracer concentrations is 75-125 ppt.) In the simulation of the historical 28 July 1976 H_2S episode, the average overall error (i.e., the average normalized difference between maximum predicted and observed H_2S concentrations at the seven SRI stations) was 37 percent.

CONCLUSIONS

Comparison of the peak prediction and observation at each monitor for the five tracer tests indicated agreement within a factor of 2 over 60 percent of the time. This is comparable with a 70 percent figure obtained in recent application of the model to the Harry Allen Generating Station in southern Nevada. The model's performance, discussed in the recent studies of Burton et al. (1980) and Tesche et al. (1980), has led to its recent acceptance for use in estimating future impacts of proposed new sources in the West by local, state, and regional air regulatory bodies.

REFERENCES

Burton, C. S., et al. (1980), "Harry Allen Station Complex Terrain Grid Model Performance; Further Comparison of Observed and Calculated Tracer Concentrations with Emphasis on Three-Hour Averages," SAI No. EI80-115, Systems Applications, Inc., San Rafael, California.

Environmental Protection Agency (1979), "Workshop on Atmospheric Dispersion Models in Complex Terrain," EPA-600/9-79-041, Research Triangle Park, North Carolina.

Hanna, S. R. (1976), "Predicted and Observed Cooling Tower Plume Rise and Visible Plume Length at the John E. Amos Power Plant," Atmos. Environ., Vol. 10, pp. 1043-1052.

Knuth, W. R., et al. (1979), "Geysers-Cobb Valley Air Quality Impact Study," MRI 79 FR-1670, Meteorology Research, Incorporated, Altadena, California.

Reynolds, S. D., et al. (1979), "Photochemical Modeling of Transportation Control Strategies --Vol. 1. Model Development, Performance, Evaluation, and Strategy Assessment," SAI No. EF79-37R, Systems Applications, Inc., San Rafael, California.

Tesche, T. W., et al. (1980), "Assessment of the Sacramento Municipal Utility District's Proposed Geothermal Power Plant--Volume III, Final Worst Case Air Quality Analysis," SAI No. EF80-180, Systems Applications, Inc., San Rafael, California.

Winiarski, L. D., and W. E. Frick (1978), "Methods of Improving Plume Models," Proc. of Cooling Tower Environment--1978.

Yocke, M. A., and M. K. Liu (1978), "Modeling Wind Distributions Over Complex Terrains," SAI No. EF78-78, Systems Applications, Inc., San Rafael, California.

EFFECTS OF H_2S ON CROP AND FOREST PLANTS

C. Ray Thompson, G. Kats, P. J. Dawson

Statewide Air Pollution Research Center
University of California
Riverside, California 92521

ABSTRACT

Continuous fumigation of alfalfa, grapes, lettuce, sugar beets, California buckeye, ponderosa pine, and Douglas fir with 3000 parts per billion (ppb) H_2S caused leaf lesions, defoliation, reduced growth, and death of sensitive species. Three hundred ppb caused lesser but similar effects. Sulfur accumulated in leaves depending upon dosage. Lower levels of H_2S, 30 ppb and sometimes 100 ppb, caused significant stimulation in growth of lettuce, sugar beets, and alfalfa. The stimulation occurred at certain times of year. Addition of 50 parts per million (ppm) of CO_2 to the fumigation atmospheres overcame growth reduction caused by 300 ppb H_2S in lettuce and sugar beets. Two varieties of grapes and one variety each of pears and walnuts were fumigated with the highest ambient levels of H_2S found near The Geysers, treble ambient H_2S and amounts of SO_2 equivalent to the treble levels of H_2S. No deleterious growth or fruiting symptoms were observed which could be ascribed to the treatments.

INTRODUCTION

Hydrogen sulfide is a malodorous gas produced by many industrial and biological processes and is often a major component of geothermal emissions. It is highly toxic to man. Extensive power development at The Geysers, Calif. has caused major complaints because of odor and ranchers fear that crops are being injured. Limited studies (1,2) with high level, short term fumigations have shown injury to some crop and weed species but recent studies at the University of California, Riverside (3,4,5) are the only long term, lower level studies designed to assess injury to crop and forest species from geothermal emissions such as those occurring at The Geysers.

EXPERIMENTAL

A. CONTINUOUS CONSTANT LEVEL FUMIGATIONS.

Procedure

To establish basic acute toxicity of H_2S to a number of crop and forest plants growing in The Geysers area, alfalfa (*Medicago sativa* L), Thompson seedless grapes (*Vitis vinifera*), lettuce (*Lactuca sativa*), sugar beets (*Beta vulgaris*), California buckeye (*Aesculus californica*), ponderosa pine (*Pinus ponderosa*), and Douglas fir (*Pseudotsuga menziesii*) were exposed to serial concentrations of H_2S.

Continuous, uniform level fumigations were conducted in four greenhouses. Hydrogen sulfide was injected into the carbon-filtered, incoming air stream to provide the required concentrations of fumigant.

Three fumigation procedures were employed. In the first series of experiments during 1975, alfalfa, Thompson seedless grapes, ponderosa pine, and California buckeye were exposed to concentrations of 0, 30, 300, and 3000 parts per billion (ppb) of H_2S. Lower pollution levels (0, 30, 100, and 300 ppb) were used in the second fumigation series during 1976 which involved alfalfa, lettuce, sugar beets, Douglas fir, and a second set of Thompson seedless grapes. The third fumigation attempted to simulate effluent gases from The Geysers which contain considerable levels of CO_2 in addition to the H_2S. This study used 0 (control), 30, 100 and 300 ppb H_2S, 300 ppb H_2S + 50 parts per million (ppm) CO_2 and 300 ppb SO_2. The SO_2 was compared because H_2S is oxidized to SO_2 in the atmosphere.

All species of plants were grown in pots. A soil mix consisting of peat moss, redwood shavings, and silt (1-1-1) was used. Mineral salts were added to provide initial fertilization. The plants were irrigated with one-half strength Hoagland's solution twice weekly. All data, where replicated, were analyzed statistically by an analysis of variance and a multiple-range test.

Results

Effects of Continuous H_2S Fumigation on Alfalfa. During 1975 when fumigation levels were 0, 30, 300, and 3000 ppb H_2S, green alfalfa showed white marginal leaf lesions on mature leaves within 5 days at the highest level of H_2S. To determine yield, the plants were cut at 28-35-day intervals. Growth was reduced during the first growing period at both 300 and 3000 ppb H_2S with Hayden (Table 1). The highest level reduced growth of Eldorado, and a trend was shown at 300 ppb. No effect on growth was caused by 30 ppb. During the subsequent growth period, most of the plants died at 3000 ppb, and growth was reduced in both varieties at 300 ppb. Samples of dried tissue from these two cuttings were analyzed for total sulfur and showed graduated accumulation corresponding to the levels of H_2S in the different atmospheres.

Table 1. Effects of Continuous H_2S Fumigation on Growth and Sulfur Accumulation in Alfalfa[a]

		Alfalfa variety			
		Eldorado		Hayden	
	H_2S, ppb	Av dry wt/ pot, g	Total S as SO_4, %	Av dry wt/ pot, g	Total S as SO_4, %
Cutting #1	0	52 y	1.03	52 x	0.92
13 Aug 1975	30	51 y	1.23	52 x	1.36
	300	42 y	2.45	32 y	2.44
	3000	16 z	4.85	11 z	5.20
Cutting #2	0	45 y	0.94	46 y	1.00
9 Sept 1975	30	46 y	1.10	43 y	1.29
	300	31 z	3.00	28 z	3.41

[a] Values followed by different letters are different at the 1% level.

During 1976, fumigation at 3000 ppb was discontinued, and a treatment at 100 ppb H_2S was added in an attempt to determine more precisely the amount of H_2S which caused reduced growth in this crop. Eight successive cuttings were harvested from May until October. The results showed that 300 ppb H_2S caused significantly reduced yields in all cuttings except August 3 and September 21, while 100 ppb had no statistical effect on yield. The lowest level, 30 ppb, significantly increased yields in the late summer, August 3, August 26, and September 21, of 136, 118, and 132%, respectively, of the alfalfa grown in carbon-filtered air. Thus, continuous fumigation with H_2S at 3000 and 300 ppb caused foliar injury and reduced growth. At 100 ppb neither of these effects occurred, and with 30 ppb a significant growth stimulation during summer was observed.

Effects on Grapes. The immediate effect of the 3000 ppb level on grapes was similar to that seen on alfalfa in 1975. White to yellow lesions, which later turned brown, appeared on leaves. Defoliation began to occur after about 4 weeks at the highest level and within several more days almost all leaves were lost. However, the plants were not killed but continued growth of canes and put out small chlorotic leaves until the exposure was terminated. The 300 ppb level caused lesser but similar, readily observable injury. No foliar effects were seen at the 30 ppb level.

Cane length was reduced at the highest H_2S level as compared to the other treatments, but less dieback occurred at 30 and 300 ppb H_2S, Table 2.

Table 2. Effects of Increasing Levels of H_2S on Cane Length, Dead Length, and Dry Weight of Thompson Seedless Grapes[a] 1975

H_2S, ppb	Total length, cm	Dead length, cm	% Dead	Total dry wt, g	Total sulfur as SO_4 in leaves, %
0	1252.9 y	191.3 z	15.5 z	145.0 x	0.78
30	1160.0 y	212.4 z	18.7 z	143.1 x	1.26
300	1090.1 y	572.1 x	53.3 x	72.5 y	3.33
3000	673.5 z	422.2 y	63.1 y	38.0 z	4.50

[a] Values followed by different letters are different at the 1% level.

Total dry weight of cane was reduced to one-half with 300 ppb as compared to the control, but the total length was the same statistically, showing that the H_2S caused a thin spindly growth of canes. With 3000 ppb the cane length was about one-half the control, but weight was one-fourth, showing the same effect. Composite leaf samples were analyzed for total sulfur and showed accumulation in grape leaves similar to those in alfalfa.

Significant increase of fresh leaf weight was induced with 30 and 100 ppb H_2S over the controls with the same numerical trend in dry weight (Table 3) in 1976. Fresh weight of canes

Table 3. Effects of Increasing Levels of H_2S on Leaf and Cane Weights of Grapes[a] 1976

	Leaves		Canes	
H_2S, ppb	Fr wt, g	Dry wt, g	Fr wt, g	Dry wt, g
0	265.6 a	78.6 a	267.2 a	123.8 a
30	375.6 b	97.0 a	251.8 a	115.8 ab
100	298.4 b	80.6 a	219.6 a	86.0 bc
300	260.4 a	57.0 b	152.0 b	62.8 c

[a] Values followed by different letters are different at the 5% level.

was reduced significantly by 300 ppb, and dry weight with both 300 and 100 ppb. Thus, H_2S caused severe foliar injury and defoliation but was not lethal. The stimulation in fresh weight of leaves with lower levels of H_2S was similar to the effect observed with fresh weight of alfalfa in summer.

Effects on Ponderosa Pine. Little effect of fumigation was observed for 4-6 weeks, but later at the 3000 ppb H_2S level a progressive tip burn was observed with defoliation after 10 weeks of exposure. No visible effect was caused by 30 ppb.

Effects on California Buckeye. The buckeye proved quite resistant to H_2S. At the 3000 ppb level, a bronzing of the leaves occurred after 4 weeks exposure and some defoliation occurred at the highest level after 8 weeks.

Effects on Sugar Beets. Beets were grown, harvested, and analyzed for fresh and dry weight of leaves, percent total sulfur, fresh weight of roots, and percent sugar. Results with individual beets (Table 4) showed that 30 ppb H_2S stimulated fresh weight of leaves and roots 41 and 51%, respectively, but 300 ppb inhibited these two responses significantly and also reduced sugar content in roots. Leaves had much more sulfur per unit weight than roots.

Statistical analyses for the 1976 crop of sugar beets (Table 4) showed the leaf growth was stimulated significantly by both 30 and 100 ppb H_2S. Root weight was significantly greater in 100 ppb H_2S than in the control atmosphere but was statistically the same in the 30 and 300 ppb H_2S.

Effects on Lettuce. Results with the first lettuce crop (Table 5) showed a large stimulation

Table 4. Yield of Leaves, Roots, and Sugar Content of Sugar Beets Exposed to Increasing Levels of H_2S [a]

Individual beets, leaves

H_2S, ppb	Fresh wt, g	Dry wt, g	% Sulfate
0	149.6 bc	19.8 bc	0.66
30	210.5 a	25.2 b	0.93
100	200.8 ab	23.3 b	1.33
300	127.6 c	15.2 c	1.88

Individual beets, roots

H_2S, ppb	Fresh wt, g	% Sugar	% Sulfate
0	291.3 bc	19.3 a	0.05
30	440.6 a	18.3 a	0.06
100	370.7 ab	18.2 a	0.08
300	219.6 c	15.8 b	0.12

Total, 3 beets/pot

H_2S, ppb	Leaves Fresh wt, g	Leaves Dry wt, g	Roots, fresh wt, g
0	1242.0 c	126.2 c	643.4 bc
30	2037.6 a	194.0 a	881.0 ab
100	1874.8 ab	176.8 ab	1034.2 a
300	1640.8 abc	137.0 bc	502.2 c

[a] Values followed by different letters are different at the 5% level.

Table 5. Effects of Increasing Levels of H_2S on Fresh Weight, Diameter, and Sulfur Content of Dark Green Boston Head Lettuce [a]

H_2S, ppb	Fresh wt, g	Diam, cm	Sulfur as SO_4
0	104.4 x	26.9 x	0.78
30	167.7 y	29.9 y	0.82
100	97.3 x	25.5 x	1.30
300	34.7 z	19.9 z	1.77

[a] Values having different letters are different at the 1% level. Growth period 3/13/76–5/10/76.

in growth with 30 ppb H_2S, but reduced growth at the highest level. Dried samples of the lettuce were analyzed for total sulfur and show that additional sulfur accumulated in leaves in direct relationship to the H_2S in the atmosphere. Limited organoleptic tests of flavor of lettuce from control and fumigated treatments showed no differences. A second trial during summer failed to show the growth stimulation with 30 ppb H_2S but a third trial in fall gave a stimulation but the amount was less.

Effects on Douglas Fir. Color of the Douglas fir foliage was not affected by 30 ppb H_2S. All needles remained bright green with no "tip burn." A slight burn was observed with 100 ppb H_2S but 300 ppb caused very extensive foliar injury. This species showed overt injury to the pollutant more clearly than all other plants tested. Growth measurements showed that 300 ppb H_2S caused reduction in growth and dry weight but at 100 ppb these effects were insignificant.

When 50 ppm CO_2 was added to 300 ppb H_2S in the third series of continuous fumigations, the CO_2 overcame some of the growth reduction caused by the H_2S in one crop of head lettuce but not in another. Similar results were shown with sugar beets. Comparisons of the growth reducing effects of 300 ppb of H_2S vs SO_2 on lettuce and sugar beets showed greater effects of H_2S than SO_2. However, a weather front with high humidity, i.e., 80%, in the greenhouse caused extensive foliar injury to sugar beets with SO_2 but none with H_2S.

B. SIMULATED FIELD LEVEL FUMIGATIONS.

To find out whether the actual amounts of H_2S that occur at The Geysers are causing injury to crop plants, potted vines of 'Cabernet Sauvignon' and 'White Riesling' grapes (*Vitis vinifera* L.) and potted 'Bartlett' pear (*Pyrus communis* L.) and 'Hartley' English walnut trees (*Juglans regia* L.) were grown in greenhouses for 2 seasons and fumigated with ambient levels of H_2S, treble ambient H_2S, and amounts of SO_2 equivalent to the treble levels of H_2S.

Procedure

One-year-old rooted cuttings of 'White Riesling' and 'Cabernet Sauvignon' grapevines, 1-year-old 'Hartley' English walnuts on black walnut rootstocks, and 1-year-old 'Bartlett' pear trees grafted on *Pyrus betulafolia* rootstocks were grown in pots.

Eight separate greenhouses equipped with air conditioning, gas dispensing lines, air sampling lines, wet-dry bulb thermocouples, and irrigation systems were divided into 4 groups of 2 each. Group 1 received carbon-filtered air; group 2, carbon-filtered air plus simulated H_2S concentrations which occur near The Geysers at "Anderson Ridge," the location which receives the highest measured levels of ambient H_2S as recorded by SRI International; group 3, carbon-filtered air plus 3 times the Anderson Ridge levels; and group 4, carbon-filtered air plus amounts of SO_2 equivalent to the H_2S in group 3.

The pollutant levels were programmed on an hourly basis into a dispensing system; i.e., the same levels which occurred each calendar day in 1977 and 1978 were given to the test plants in 1979 and 1980, respectively.

The simulation of these field dosages of H_2S and SO_2 were metered into the various houses for 1 hr at a time, giving a close approximation to the recorded field levels. The treble H_2S and equal levels of SO_2 showed a similar degree of agreement with the projected levels. Five grapevines of each of the 2 cultivars (10 total), 5 pear trees, and 5 walnut trees in 1979 (4 in 1980) were placed in each greenhouse.

The plants were grown for two seasons. For over 90% of the simulated ambient exposure time, the plants received 20 ppb or less of H_2S and only rarely exceeded 20 ppb. The treble H_2S and corresponding SO_2 fumigations were proportionally greater.

Growth parameters for grapes in 1979 included linear growth, all shoots 10 cm or greater; the weight of prunings removed prior to 1980, total dry weight of leaves, number of flowers, and percent soluble solids in berries as determined by hand refractometer. The same measurements were made in 1980 plus fresh weight of berries and total dry weight of the entire vine including the root ball.

Measurements on pear and walnut trees included total length of shoot growth per tree, number of shoots, dry weight of leaves, total growth as determined by dry weight prunings, and in 1980 the number of flower clusters (pears only) and dry weight of roots.

No visible symptoms of injury occurred on foliage or other plant parts which could be attributed to treatment. None of the parameters measured on the 3 fruit species showed any statistically significant effects caused by the H_2S or SO_2 fumigations over the controls.

These studies show that fumigation in greenhouses for 2 growing seasons of 2 cultivars of grapes, and 1 cultivar each of pears and walnuts, which are grown widely near The Geysers, are unaffected by the highest ambient levels of H_2S that occur in the area, and also treble these levels or amounts of SO_2 equal to the threefold levels.

REFERENCES

1. Benedict, H. M. and Breen, W. H., 1955, The use of weeds as a means of evaluat- vegetation damage caused by air pollution. Proc. 3rd Nat. Air Poll. Symp. Pasadena, Calif. p. 18-20.

2. McCallan, S. E., Hartzel, A. A., and Wilcoxson, F, 1936, Hydrogen sulphide injury to plants. Contrib. Boyce Thompson Inst. 8:189-197.

3. Thompson, C. R. and Kats, G., 1978, Effects of continuous H_2S fumigation on crop and forest plants. Environ. Sci. Technol. 12:550-553.

4. Thompson, C. R., Kats, G., and Lennox, R. W., 1979, Effects of fumigating crops with hydrogen sulfide or sulfur dioxide. Calif. Agr. 33 (3):9-10.

5. Thompson, C. R., Kats, G., and Dawson, P. J., 1982, Low-level effects of H_2S and SO_2 on grapevines, pear and walnut trees. Hort. Science. In Press.

Acknowledgment

Tables 1, 2, 3, 4 and 5 reprinted with permission of ENVIRONMENTAL SCIENCE AND TECHNOLOGY 12, 550-553, Copyright of the American Chemical Society 1978.

Geothermal Resources Council, TRANSACTIONS Vol. 6, October 1982

UPDATE ON FEDERAL TAXATION OF GEOTHERMAL ENERGY

Sharon C. Wagner and Charles Stepkin

O'Brien and Hallisey

One California Street, San Francisco, Ca. 94111

ABSTRACT

We are presenting a brief overview and update of federal taxation of geothermal energy. In our oral presentation at the Technical Session we will discuss how these tax laws interrelate and apply to a variety of taxpayers.

UPDATE ON FEDERAL TAXATION OF GEOTHERMAL ENERGY

Prior to the passage of the Energy Tax Act of 1978, 1/ the federal tax treatment of geothermal resources was based mainly on judicial decisions, not statutory authority. The 1978 Act eliminated most of the uncertainties of tax treatment of geothermal exploration and development. The Act's definition of geothermal deposits - "a geothermal reservoir consisting of natural heat which is stored in rocks or in an aqueous liquid or vapor (whether or not under pressure)" 2/ - is broad enough to include all known forms of geothermal energy including steam, hot water or brines, or dry hot rocks. The reservoir must be located within the United States or a possession. The Act changed three areas of taxation as applied to geothermal resources: intangible drilling costs, depletion, and tax credits.

I. INTANGIBLE DRILLING COSTS

A. Option to Deduct Intangible Drilling Costs

Section 263(c) of the IRC now allows a taxpayer the option to deduct as expenses intangible drilling costs (called "intangibles" or IDCs). 3/ The costs of drilling and completing a geothermal well are divided for tax purposes into two classes: intangible drilling costs and equipment costs. The equipment costs must be capitalized and "recovered" through depreciation or depletion. Intangible drilling costs may be treated in two ways. 4/ They may be deducted as expenses (in tax terminology they may be expensed) in the year in which they are incurred or they may be capitalized and deducted over a certain period of time as depreciation or depletion. 5/

The taxpayer must make his election to expense or to capitalize intangibles in his first taxable year in which he incurs such costs. 6/ Once the election is made, the taxpayer must treat such expenditures on all geothermal properties in the same manner for all future years. 7/ However, if the taxpayer elects to capitalize his intangibles, he is granted a second election or dry or unproductive wells. 8/ For example, if Taxpayer (T) 9/ has spent $50,000 on intangible costs, T may claim as a deduction on his income tax return $50,000. But if T decides to capitalize intangible drilling costs T will not take a $50,000 deduction for the tax year, but instead will deduct this amount over a given period of time as depreciation or depletion.

The costs incurred in drilling a nonproductive well may be deducted by the taxpayer as an ordinary loss provided a proper election is made. If a clear election is not made, such costs can be recovered only through depreciation and depletion.

Taxpayers who decide to expense intangibles, however, will be subject to the following: the minimum tax, a limitation on deductions to the amount "at risk" and recapture.

B. Preference Income-Minimum Tax

Some types of income are given preferential treatment by special provisions of the tax law. A minimum tax applies to a number of items that are considered to be of a tax preference nature. These types of income include income offset by depletion and intangible drilling costs. The tax is computed by totaling all of the items of tax preference, then reducing this amount by the greater of $10,000 or one-half the taxpayer's regular income tax after reduction by credits. A flat 15% rate is then applied against the balance.

If a taxpayer has "excess intangible drilling costs" that exceed net geothermal income, he will have preference income subject to the minimum tax. Intangible drilling costs which are deducted in any given tax year constitute a preference item to the extent that such costs are greater than the sum of (1) the amount allowable if the costs had been capitalized and straight-line recovery of the intangibles had been used and (2) the net income for the tax year from the geothermal property. This preference does not apply to nonproductive wells. Special rules apply to corporations in computing their minimum tax, in particular excess intangible drilling costs are not a preference item for a regular corporation.

The preference amount with respect to the depletion deduction is the excess of the depletion deduction over the adjusted basis of the property at the end of the year (determined without regard to the depletion deduction for the year).

In effect what this provision does is to lessen the benefit of percentage depletion and the option to expense intangible drilling costs. Few taxpayers now have geothermal income and if they chose to expense intangibles, they will have preference income.

C. Losses Limited to Amount at Risk

The 1976 Tax Reform Act limited the tax benefits available to persons engaging in oil and gas operations. These same limitations with some changes were extended to geothermal operations by the 1978 Energy Tax Act.

Before passage of the 1976 Act a taxpayer could take deductions up to the amount of his cost (or "basis") in a business or investment venture. But the basis of a taxpayer often included expenditures financed by nonrecourse loans for which the taxpayer had no personal liability (i.e., he had nothing "at risk" because of the way the loan was made to him or to an investment group). Such leveraged nonrecourse loans were often employed by investors to finance drilling and development costs of oil and gas activities. Since a taxpayer could elect to expense intangible drilling costs, he could take deductions far in excess of his own actual investment. This kind of investment was desirable for a high bracket taxpayer because the large deductions for intangibles could be used to offset income earned from other sources.

The taxpayer's total deduction in a given venture cannot exceed the total amount the taxpayer has <u>at risk</u>. A loss is the excess of the allowable deductions allocable to a particular activity over the income derived from the activity during the taxable year, and thus includes deductions taken for intangibles. In addition, previously allowed losses must be recaptured when the taxpayer's "at risk" amount is reduced below zero, but only to the extent of the excess of the losses previously allowed in a particular "at risk" activity over any amounts previously recaptured. Recaptured losses may be deductible in a later year if the amount "at risk" is later increased. The practical effect of these "at risk" provisions is to eliminate the use of nonrecourse financing to increase available deductions.

D. Recapture of Expensed Intangible Costs As Ordinary Income on Disposition of Geothermal Property

Probably the most far-reaching change of the 1976 Tax Reform Act affecting taxpayers is the requirement that upon the disposition of oil and gas property taxpayers are required to recapture all or some part of any expensed intangible costs as ordinary income if the property is disposed of at a <u>gain</u>. These recapture provisions now extend to intangible drilling costs expensed in connection with geothermal deposits. This recapture provision applies only to intangibles which the taxpayer elects <u>to expense</u> in the year in which they were incurred and does not apply to intangibles which were <u>capitalized</u>. The amount recaptured and taxed as ordinary income is the amount that the intangibles deducted exceed that which would have been allowed had the intangibles been capitalized and amortized on a straight-line basis from the time the property went into production, but not to exceed the amount of the gain.

II. PERCENTAGE DEPLETION

The IRC provides two methods of computing a depleton allowance: cost depletion and percentage depletion. Cost depletion provides for a deduction for the taxpayer's basis (cost) in the property. On the other hand, percentage depletion is a statutory concept that provides for a deduction of specified percentages of the gross income from the property. A taxpayer is required to compute depletion both ways and to claim the larger of the two amounts, on a property by property basis.

The percentage depletion allowance reduces the taxpayer's basis in a property but the total amount taken as a depletion allowance is not limited to the taxpayer's basis. Even though cost depletion will be zero (for example, T deducts $5,000 per year for five years for a total of $25,000 - the amount of his original investment), the taxpayer may continue to claim a percentage depletion based on gross income from the property. 10/

Section 403 of the 1978 Energy Tax Act granted percentage depletion on income from geothermal deposits. The rate through 1980 is 22%. It decreases by 2% yearly until 1983 and thereafter the rate is 15%. This percentage depletion allowance is much more favorable than the one allowed oil and gas. It is not limited in any way to a specified amount of production. It has no 65% of taxable income limitation nor is it restricted to independent producers. However, the percentage depletion deduction in any year cannot exceed 50% of the taxable income from the property and is subject to the minimum tax-preference income rules. 11/

Two problems have developed with respect to percentage depletion for direct-use applications of geothermal resources. First, neither Section 613(c) of the code nor its accompanying regulations define "gross income" for the geothermal property. Treas. Reg. Section 1.613-3(a) defines gross income for oil and gas property and Treas. Reg. Section 1.613-4 defines gross income for "other minerals". Secondly, there is some question about the availability of percentage depletion deduction for geothermal resources consumed by the producer of such resource.

III. TAX CREDITS

A. Residential Energy Credit

IRC Section 44(C) provides for a nonrefundable tax credit for certain expenditures incurred for equipment which uses geothermal energy (deemed a "renewable energy source") in a taxpayer's principal residence in the United States. The equipment must be new and must meet certain performance and quality standards; it must reasonably be expected to remain in production for five years. The credit for tax years beginning after 1979 is 40 percent of the first $10,000 invested with a maximum credit of $4000. The credit may be carried over to future years for equipment purchased after April 20, 1977 and before January 1, 1986.

B. Additional Investment Tax Credit for Alternative Energy Property

In addition to the existing investment tax credit of 10 percent for businesses (IRC Section 46), there is also available a nonrefundable 15 percent investment tax credit for geothermal equipment which qualifies as either "alternative energy property" or "specially defined energy property." "Alternative energy property" is defined as equipment "used to produce, distribute or use energy derived from a geothermal deposit...but only in the case of electricity generated by geothermal power up to (but not including) the electrical transmission stage." "Specially defined energy property" is a list of enumerated items including a heat exchanger, etc., "the principal purpose of which is reducing the amount of energy consumed in any existing industrial or commercial process and which is installed in connection with an existing industrial or commercial facility." Public utilities cannot benefit to the extent of "alternative energy property" but can use the credit for "specially defined energy property."

The energy credit applies up to 100 percent of the taxpayer's tax liability, but it may be taken only after the "regular" investment credit is taken. However, the unused portion of the energy credit may be carried back 3 years and forward 15 years.

The 1981 Economic Recovery Tax Act (1) changed the concepts of "useful life" and the "percentage of investment" in the application of the investment tax credits, and (2) made the "at risk" rules applicable to the "regular" investment tax credit and the energy investment tax credit. Under ERTA if the qualifying the property purchased by the taxpayer has a useful life of 3 years, the taxpayer is entitled to a credit on 60 percent of the cost of the property. If the useful life is 5 years or more, then the credit may be calculated on 100 percent of the cost of the property. The "at risk" rules for the regular investment tax credit and the rules for the energy investment tax credit vary somewhat. In addition, the energy credit will be recaptured if the property is disposed of, ceases to be energy property, or ceases to be "at risk."

1/ Pub. L. No. 95-618, 92 Stat. 3174.

2/ I.R.C. Section 613 (e)(3).

3/ Reg. Section 1.612-5 defines intangible drilling costs as any cost incurred which in itself has no salvage value and which is "incident to and necessary for the drilling of wells and the preparation of wells for the production of geothermal steam or hot water." Such expenditures expressly include labor, fuel, repairs, hauling, and supplies that are used "(1) in the drilling, shooting and cleaning of wells, (2) in such clearing of ground, road making, surveying, and geological works as are necessary in preparation for the drilling of wells, and

(3) in the construction of such derricks, tanks, pipelines, and other physical structures as are necessary for the drillin of wells and the preparation of wells for the production of geothermal steam or hot water."

4/ A taxpayer may make one kind of election for his geothermal deposits and a different one for his oil and gas wells. For example, he could decide to expense intangibles for both geothermal and oil and gas properties or he could capitalize oil and gas and expense geothermal intangibles.

5/ Reg. Section 1.612-5, supra note 3, states that intangibles, if capitalized, are to be separated and recovered as depreciation or depletion. Intangibles not represented by physical property (clearing ground, draining, road making, surveying geological work, excavating, grading, and the drilling, shooting, and cleaning of wells) are to be recovered through depletion. But intangible expenditures represented by physical properties (wages, fuel, repairs, hauling, supplies, etc.) are to be recovered through depreciation.

6/ A taxpayer must make a clear election either to expense or to capitalize. If he does not, the IRS will hold that he elected to capitalize intangibles. It is best that if a taxpayer desires to expense intangibles, he include with his income tax return an express statement of electin to expense in accordance with the option.

7/ Reg. Section 1.612-5(d).

8/ But this second election need not be exercised until the first year in which a dry hole is drilled.

9/ The owner of the operating rights in a property who has the responsibility to develop the property is granted the option of expensing intangibles. But each taxpayer, regardless of his relationship to another taxpayer, is entitled to a separate election. Thus each partner in a partnership is entitled to a separate election. Trusts as separate taxpayers are entitled to an election regardless of the kind of election made by the beneficiaries.

10/ A depletion allowance is available only to the owner of an economic interest in that property. An owner of an economic interest can be an owner of mineral interests, royalties, working interests, overriding royalties, net profits interests or certain kinds of production payments. Note that the depletion allowance is available to holders of nonworking interests and is, therefore, available to taxpayers ineligible to expense IDCs.

11/ The excess of the depletion deduction over the adjusted basis of the property at the end of the year (determined without regard to the depletion deduction for the year) is what would be preference income.

Bibliography

Internal Revenue Code Sections 57,263, 465, 611, 612, 613, 1254 and accompanying Regulations.

Energy Tax Act of 1978.

Windfall Profit Tax Act of 1980.

Economic Recovery Tax Act of 1981.

Geothermal Resources Council, TRANSACTIONS Vol. 6, October 1982

EVALUATION OF CONTROL NEEDS FOR HAZARDOUS AIR EMISSIONS (NON-H_2S) FROM GEOTHERMAL PROCESSES

Stephen C. Wilson

Monsanto Research Corporation
Dayton Laboratory
Dayton, Ohio 45418

ABSTRACT

Control technology needs for geothermal air emissions other than hydrogen sulfide have not been considered in federal geothermal research programs, nor to a great extent by private developers. This study (sponsored by the U.S. Environmental Protection Agency under EPA Contract 68-03-3025, Work Assignment SDM-04) was designed to identify the potential emission points, resulting concentrations in ambient air, and the impacts on the environment in the vicinity of geothermal operations.

This paper presents the results of the study in which ammonia, arsenic, benzene, boron, fluoride, hydrocarbons (other than benzene), mercury, radon, and sulfur oxide emission information was compiled for over 40 national and international geothermal development sites. The emission points within the geothermal processes, the compounds emitted, and the need for control technology are discussed.

INTRODUCTION

Geothermal energy is usually found in the form of heat contained in steam, hot water, or heated rock located beneath the earth's surface. The first two sources (steam and hot water) contain substances from the surrounding rocks within the geothermal reservoir. These substances include a variety of potentially toxic gases. To date, hydrogen sulfide (H_2S) has attracted a considerable amount of attention, however, other gases are emitted from geothermal developments. These potential emissions (ammonia, arsenic, benzene, boron, fluorine, hydrocarbons, mercury, radon, and sulfur dioxide/sulfur peroxide) are of concern to the Environmental Control Panel of the Interagency Geothermal Coordinating Council. In 1981, the Panel recognized the need for characterizing these emissions and recommended that control strategies be developed for these emissions, if needed.

Monsanto Research Corporation (MRC), under EPA Contract 68-03-3025, Work Assignment SDM-04, was assigned the task of estimating potential loadings, resulting concentrations in air, and the expected environmental impact of non-H_2S emissions resulting from the use of geothermal energy.

The objectives of this study were to:

1. Identify and quantify the potential non-H_2S air emissions from a representative variety of geothermal sites;
2. Describe health and environmental effects of identified contaminants and the associated exposure levels, and
3. Describe control concepts and approaches that are or may be applicable to non-H_2S emissions which need to be controlled.

PROJECT APPROACH

The three (3) objectives of the project were considered as spearate tasks and were performed consecutively. Table 1 gives the specific activities associated with each task.

Task A

Task A involved gathering and evaluating all available published and unpublished information on geothermal site developments, engineering data on the resource, and gaseous emissions characterizations. The data gathering included U.S. and world-wide geothermal developments. The data were obtained from twelve computerized data bases, Department of Energy (DOE) and Environmental Protection Agency (EPA) reports, and geothermal energy developers and experts within DOE, EPA, and the academic community. Information on each development site was collected regarding site name, location, geothermal well information, present activity, past activity, planned future development, environmental controls in place or antiticipated, process schematic, and gaseous emission points, characteristics and volumes. The 44 sites on which information was gathered in Task A are listed in Table 2.

Before initiating Task A, the EPA had directed MRC to pay particular attention to gaseous emissions of arsenic, boron, hydrocarbons, mercury, and radon. During the performance of Task A, ammonia, benzene, fluorine, sulfur dioxide, and sulfur peroxide were added, as the data indicated that these compounds were being emitted also.

TABLE 1. PROJECT ACTIVITIES BY TASK

TASK A: Identification and Quantification of of Air Emissions

- A.1 Compilation of Existing Site Location Data
  - A.1.1 Determination of U.S. site locations, process types and current and past activities
  - A.1.2 Determination of major foreign sites and activities
- A.2 Compilation of Existing Engineering Data by Process Type (including end use type)
- A.3 Compilation of Existing Emission Data by Process Type
  - A.3.1 Determination of emission points for each process and end use type
  - A.3.2 Determination of volumetric (mass) flow rates and physical parameters at each emission point
  - A.3.3 Determination of emission composition

TASK B: Description of Potential Health and Environmental Effects of Emissions

- B.1 Evaluation of Effects of Emissions from Zero Concentration to Expected Concentration
  - B.1.1 Description of worker effects
  - B.1.2 Description of general population effects
  - B.1.3 Description of plant life effects
  - B.1.4 Description of animal life effects
- B.2 Determination of Emission Fate Pathways
  - B.2.1 Description of emission decay and daughter molecules
  - B.2.2 Description of emission dispersion
  - B.2.3 Description of emission decomposition
  - B.2.4 Description of synergistic effects
- B.3 Determination of Candidate Emission Sources Needing Control

TASK C: Description of Emission Control Concepts

- C.1 Evaluation of Emissions Control Options for Primary and Secondary Applications
  - C.1.1 Identification of industries with similar emission characteristics
  - C.1.2 Determination of the most feasible candidate industries
  - C.1.3 Identification and evaluation of the applicability of analogous control schemes
- C.2 Integration of Control Unit Operations
  - C.2.1 Description of process
  - C.2.2 Description of chemical and physical reactions
  - C.2.3 Description of hazardous nature of treated gases
  - C.2.4 Description of hazardous residues generated by control
  - C.2.5 Estimation of capital and O&M costs
- C.3 Recommendations for Research on New Treatment Methods
- C.4 Recommendations for Further Development of Current Control Technology

Task B

In Task B, existing toxicological and environmental effects data and publications on geothermal energy toxic effects were evaluated for each of the above compounds to evaluate the potential impacts on human health and the environment. The pollutant concentrations determined in Task A were compared to concentration levels reported in the literature which had previously produced a toxic effect. In addition to toxic effects data, Threshold Limit Values (TLVs) and published emission standards were also used. The order of preference of health effects data to compare the emissions to was as follows:

- Published Standards (EPA)
- Proposed Standards (EPA)
- Other Standards (OSHA, etc)
- TLV Values
- Toxicity Data
- Industry Practices

TABLE 2. GEOTHERMAL SITES INVESTIGATED IN TASK A

Site	Location
U.S. LOCATIONS	
Electric Power Production	
Baca	Baca Ranch, New Mexico
Beowawe	Beowawe, Nevada
Brawley	Imperial County, CA
Fenton Hill	LaCueva, New Mexico
Geo Loop Exp Facility (GLEF)	Niland, CA
The Geysers	Sonoma & Lake Counties, CA
Heber	Heber, CA
Magma Max	East Mesa, CA
Mono-Long Valley	Casa Diablo, CA
Puna	Puna, Hawaii
Raft River	Burley, Idaho
Republic Geothermal	East Mesa, CA
Republic Geothermal	Westmorland, CA
Roosevelt Hot Springs	Milford, Utah
Salton Sea	Salton Sea, CA
Industrial/Residential Uses	
Boise	Boise, Idaho
Brady Hot Springs	Fernley, Nevada
Gulf Coast	Texas and Louisiana
Klamath Falls	Klamath Falls, Oregon
Madison Aquifer	Philip, South Dakota
North Bonneville	Bonneville, Washington
Susanville	Susanville, CA
FOREIGN LOCATIONS	
Electric Power Production	
Ahuachapan	Ahuachapan, El Salvador
Cerro Prieto	Cerro Prieto, Mexico
China	China
El Tatio	El Tatio, Chile
Hatchobaru	Hatchobaru, Japan
Krafla	Krafla, Iceland
Larderello	Larderello, Italy
Matsukawa	Matsukawa, Japan
Monte Amiata	Monte Amiata, Italy
Ohaki	North Island, New Zealand
Otake	Otaka, Japan
The Philippines	The Philippines
Svartsengi	Svartsengi, Iceland
Travale	Travale, Italy
Wairakei	Wairakei, New Zealand
Industrial/Residential Uses	
Carwynnen and Rosmanowas	Cornwall, England
Kodari	Nepal
Low Temperature Areas	Iceland
Mayangdi Khola	Nepal
Namafjall	Namafjall, Iceland
Rear Thermal Spring	Nepal
Surai Khola	Nepal

The rationale for this order is that currently enforceable emission standards will have the greatest impact on the selection of control technology.

At the end of Task B, the emission points which (based on their expected toxic effects) had a potential for negative environmental impact were to be defined and candidate emission sources needing control were to be listed.

Task C

Task C was originally designed to identify applicable control technology approaches and related costs for the control of non-H_2S emissions. However, the results of Tasks A and B indicated that non-H_2S emissions are either not of environmental significance or are controllable with currently available technology and that Task C was not necessary at this time (see conclusions).

RESULTS

Task A

The results of Task A on emission locations and their significance are as follows:

- Naturally occurring emissions from fumaroles and hot springs near geothermal developments, in some instances, dwarf emissions from the exploitation of geothermal energy. In those instances, control of non-H_2S emissions from geothermal developments may have no significant impact on improving environmental quality.

- Non-electrical uses of geothermal energy (projected to be up to 60,000 times greater users of geothermal energy than electrical users) are "closed loop" with fluid injection and little or no air emissions [1].

- The emissions from well drilling, testing, clean-out, and stacking (venting of the well during power plant shutdown) are expected to be less than 1 percent of the overall process emissions because: (1) they do not occur often, (2) they are relatively short when they do occur, and (3) the emission rates are reduced by industry practices (throttling back flow from the well if venting is necessary for over two hours, redirection of steam to other users if venting is necessary for more than 24 hours). Thus these emissions should not require controls [2].

- For liquid-dominated reservoirs using steam flashing for producing electricity, the major air emissions are non-condensible gases from the vent off of the direct contact steam condensers, cooling tower drift, and stripping of volatile compounds in the cooling tower (steam condensate is used as cooling tower make-up). If controls are necessary, these three sources should be considered. Emission control problems are similar for power generating plants using direct steam from vapor-dominated sources as those using steam from flashing of superheated water from liquid-dominated sources.

- The binary cycle type of power plant, operating with superheated water, and an organic working fluid, should not, in principle have any gaseous emissions to the atmosphere, as the geothermal fluid is in a closed heat exchange system, including injection back to the geothermal reservoir.

The following results were obtained for individual pollutant concentrations.

- Although CO_2 is the major constituent (over 90 percent) of the non-condensible gases, it should not be considered as a pollutant.

- The concentrations of the nine pollutants investigated varied widely between reservoir locations.

- Ammonia concentrations range between 0.001 ppm and 1,000 ppm, with a majority of concentrations at 100 ppm , ±20 ppm.

- Arsenic concentrations range between 0.002 and 0.2 ppm with a majority of concentrations at 0.02 ppm, ±0.01 ppm.

- Benzene concentrations ranged between 45 ppm and 370 ppm; however benzene was reported to be present in the geothermal fluid at two locations. The existence of benzene in other geothermal fluids and its transfer to ambient air is unknown at this time.

- Boron concentrations ranged between 2 ppm and 250 ppm with the majority of concentrations at 21 ppm, ±10 ppm.

- Fluorine concentrations ranged between 0 ppm and 0.2 ppm.

- Hydrocarbon concentrations ranged between 4 ppm and 1,000 ppm with the majority of concentrations at 190 ppm, ±10 ppm.

- Mercury concentrations range from 0.001 ppb to 2 ppb with the majority of the concentrations at 0.005 ppb, ±0.004 ppb.

- Radon concentrations range between 10 pCi/L and 30,000 pCi/L with the majority of concentrations at 480 pCi/L, ±200 pCi/L.

- Sulfur dioxide/sulfur peroxide concentrations are below 50 ppm in nearly all cases.

TASK B

The results of Task B can be summarized as follows:

- Ammonia - Although ammonia has been measured in high concentrations in the steam at some sites, the literature published to date does not indicate that ammonia is reaching the atmosphere in significant quantities. The local effects of ammonia emissions should be considered as a part of individual development site pre-operational environmental impact assessments to determine if high concentrations will be emitted and what their impact will be.

- Arsenic - Concentrations of arsenic in geothermal emissions are generally at least one order of magnitude below toxic threshold concentrations (0.2 mg/m^3 vs. 2.50 mg/m^3 for humans). Therefore, no human health problem is anticipated due to arsenic. However, the very low toxic threshold of honey bees (0.02-0.1 μg/m^3) may be of concern in some areas due to the importance of bees to the agricultural industry. The local effects of arsenic emissions on bees should be evaluated during the performance of pre-operational environmental impact assessments.

- Benzene - In geothermal fluids where benzene is present, it may pose an occupational exposure problem. However, dispersion studies have indicated that benzene will rapidly disperse to levels below the toxic threshold at developments with a flat geography. This may not be the case in mountainous terrain. Benzene should then be investigated further as a part of pre-operational environmental impact assessments if it is present in the geothermal fluid or steam.

- Boron - Although boron air emissions produced toxic effects on vegetation at geothermal developments, these effects are attributed solely to boron contained in cooling tower drift. Cooling tower drift control is a well developed technology, and if implemented, boron air emissions from geothermal developments are not expected to present any hazard to man or the environment.

- Fluorine - Emissions of fluorine are very low (0.2 ppm max.) and are not expected to present any problems to man or the environment with the possible exception of honey bees (toxic threshold of 4 ppm). Although concentrations are expected to be below the threshold, as with arsenic, fluorine effects on bees should be further investigated during the the performance of pre-operational environmental impact assessments.

- Hydrocarbons (other than benzene) - The generally non-toxic nature of C_1 through C_8 emitted hydrocarbons (20,000 ppm toxic threshold), their low level in geothermal emissions, and their rapid atmospheric degradation, indicate that emission controls for these compounds will not be necessary.

- Mercury - The very low concentrations of mercury measured to date at the majority of sites (0.005 ppb), and the lack of any evidence to the contrary, indicate that mercury emissions from geothermal developments are not of environmental significance (this level is 10,000 times lower than observed toxic thresholds).

- Radon - The levels of radon emitted from the majority of sites (480 pCi/L) are similar to natural emissions of radon. Radon is not expected to pose an environmental hazard from geothermal developments. However, the tendency for radiation to amass inside equipment may require some type of maintenance worker protection.

- Sulfur Dioxide/Sulfur Peroxide - Since all new U.S. geothermal developments are expected to employ some type of hydrogen sulfide (H_2S) control technology, sulfur dioxide and sulfur peroxide are not expected to present an environmental problem because they will be controlled by H_2S control systems, such as Stretford units, currently planned for the industry.

CONCLUSIONS OF THE STUDY

The results of this study indicate that most non-H_2S emissions are either not of environmental significance or are controllable with currently available technologies. Emission of ammonia, hydrocarbons (other than benzene), mercury, radon, sulfur dioxide, and sulfur peroxide are not viewed as significant problems, while arsenic and fluorine may be a problem where honey bees are an important element in the agricultural industry. Benzene, if present in the fluid may pose an occupational exposure problem. These should be investigated further during the performance of pre-operational environmental impact assessments. Emissions of boron, while significant, can be readily controlled by currently available technology such as cooling tower demisters.

ACKNOWLEDGEMENTS

The author would like to thank the MRC staff members who contributed to the success of this project, Bharat O. Desai, Sam C. Cheng, Carol Smith, Carl R. McMillin, William H. Hedley, and Tom E. Ctvrtnicek. The assistance provided by The Ben Holt Company in providing consulting services is appreciated. This work was sponsored by the U.S. Environmental Protection Agency, Industrial Environmental Research Laboratory, Cincinnati, Ohio. The EPA project officer was Robert P. Hartley.

REFERENCES

1. Geothermal Energy Prospects for the Next 50 Years, EPRI Special Report, ER-611-SR, February 1978.

2. Personal communication to S. C. Wilson by M. Scheve, U.S. Department of Energy, November 18, 1981.

LBL-14575

LAPA - A Graphical Analysis Technique for Prioritizing Geothermal Leasing Activities on Public Lands*

by
Winifred W.S. Yen, Energy and Environment Division
William H. Benson, Physics, Computer Science & Mathematics Division
Lawrence Berkeley Laboratory
University of California, Berkeley 94720

ABSTRACT

This paper describes the development of a composite indicator for prioritizing geothermal leasing activities on public lands. LAPA utilizes fuzzy set software developed by the GRAD/SEEDIS projects at the Lawrence Berkeley Laboratory. It is an automated procedure which allows the analyst to use subjective categories to prioritize geothermal areas according to their unleased energy potential, the level of private response, and the potential for pre-lease environmental delays.

THE NEED FOR DECISION-SUPPORT SYSTEMS

Computerized information systems are now routinely used to monitor and evaluate geothermal development in both government agencies and in industry. Effective use of these systems to support decision-making requires the development of software tools that can accommodate the analyst's need to manipulate large amounts of data, incorporate appropriate subjective criteria, and communicate results in an effective manner. The advantage of interactive software is that it allows a decision-maker to iteratively examine the latent information content of large data sets. Such tools can be easy to use; they benefit the analyst by allowing him to probe data without disrupting the "think continuous" process inherent in analysis.[1]

This paper describes an experimental interactive graphics procedure for prioritizing geothermal leasing on public lands. The authors would like to express their appreciation to the staff of the Minerals Management Service (formerly Conservation Division, U.S. Geological Survey), members of the Bureau of Land Management, and the U.S. Forest Service for their assistance. The results presented in this paper should be regarded as preliminary and illustrative. They are the sole responsibility of the authors and do not in anyway reflect the positions or opinions of these agencies.

USING SUBJECTIVE CATEGORIES TO SUMMARIZE DATA

One aspect of data analysis is the need to get quick answers to simple questions. Data can be characterized quickly and easily using subjective categories expressed in everyday language. Some examples are: "high" unemployment rate, "high" energy potential, "positive" response, "minimal" delay.

Fuzzy set theory and operations provide a model for transforming numerical data into subjective categories.[2,3] In traditional set theory, objects may either belong or not belong to the set in question. The essential idea in fuzzy set theory is that each element in a universe of discourse is associated with a degree of membership: a number between 0 and 1. The "more" an element belongs to a set, the higher the degree of membership.

Figure 1 shows how a data set fits into a characterization of "high" unemployment. Unemployment rates below 6% do not fit that characterization at all; rates from 6 to 10 fit increasingly better; rates above 10 fit best.

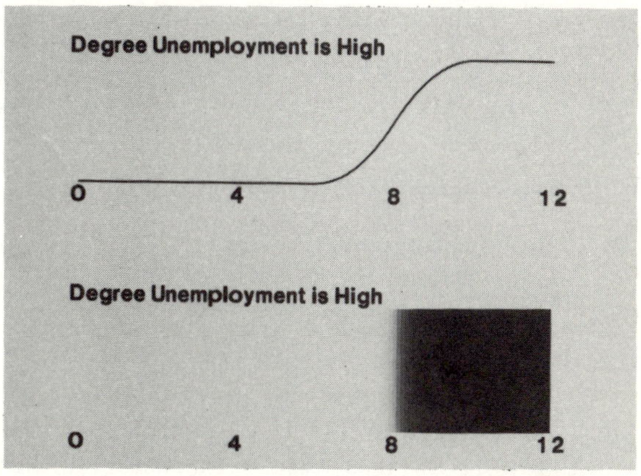

Figure 1. "High" Unemployment

The bottom of the figure shows how degrees of membership can be conveyed by a continuous black and white scale. Notice that one can only differentiate among rates in the area between 6 and 10. There are clear-cut cases of full membership above 10, and clear-cut cases of non-membership below 6. Thus, fuzzy set graphical analysis can be used as a technique that allows the analyst to focus his attention on areas where it is needed, and to ignore unnecessary detail.

* This work is supported by the Assistant Secretary for Renewable Technology, Office of Renewable Technology, Division of Geothermal and Hydropower Technologies of the U.S. Department of Energy under contract DE-AC03-76sF00098.

FUZZY SET OPERATIONS

Fuzzy sets can be combined in several ways that are equivalent to the linguistic expressions "or", "and", and "not".[4]

The <u>union</u> of two fuzzy sets A and B is denoted by A U B and can be interpreted as the <u>maximum</u> of "A or B or both".

The <u>intersection</u> of two fuzzy sets A and B is denoted by A B, whose membership function is the <u>minimum</u> of each corresponding membership value.

The <u>negation</u> of a fuzzy set A is a fuzzy set whose membership function is the <u>complement to 1</u> of A's membership function.

PRIORITIZATION OF GEOTHERMAL AREAS ON FEDERAL LANDS

The prevailing federal policy is to increase leasing of federal lands for geothermal exploration and development consistent with the preservation of environmental values.[4] In order to maximize the benefit (early contribution to power-on-line) of lease sales, it is useful to prioritize leasing of identified resource areas according to their potential for successful development. In the development of the Leasing Action Priority Area (LAPA) indicator, we have focused on the idea of the marginal lease, i.e., where should the surface management agency make available additional acres of federal lands?

Preliminary assessments of geothermal resources on federal lands have been carried out by the U.S. Geological Survey and in cooperation with state agencies.[5] Using information available from federal agencies,[6,7,8] a composite indicator consisting of three factors and incorporating six quantitative measures was developed to reflect leasing decision criteria.

An ideal geothermal area would have:

- o High unleased energy potential;
- o Historically positive response from private developers;
- o Minimal environmental delay due to incomplete environmental impact studies.

Evaluating geothermal areas against these criteria is not a well defined task. The numerical precision of indicators may be of limited value because available data may be imprecisely measured, or inherently imprecise, as in the use of estimates, projections, or proxy variables. In addition, there may be differences of opinion as to relative importance of the criteria and differences as to how faithfully the indicators represent the criteria. Un-representatable criteria, such as political considerations, (e.g. "This area has high unemployment and needs the influx of jobs which could be created by leasing this geothermal tract") need to be folded in by the decision maker.[4]

METHODOLOGY

The composite LAPA indicator is obtained using the GRAD/SEEDIS programs in five steps:

1. Select and compute numerical indicators to represent the criteria.
2. Describe the ideal case for each indicator.
3. Modify membership values by importance weights.
4. Combine indicators using fuzzy set operations, and display degree of fit to the ideal case using a continuous color scale.

These steps are performed with the assistance of an 'expert'; someone who is knowledgeable about the data, the criteria, relative importance, etc. An analyst might interactively modify the indicators or the descriptions of ideal cases.

1. <u>Select and Construct Indicators</u>

o How much <u>unleased energy</u> exists in the geothermal area?

The unleased factor (UF) is defined as

$$UF_i = \frac{\text{acreage not yet offered}_i \cdot \text{energy potential}_i}{\text{Total acreage offered for lease}_i}$$

Notice that this indicator (UF) only roughly represents the concept "unleased energy potential still available for private development". The imprecision in this quantity comes, for example, from the fact that the geothermal resource may not be evenly distributed, and the energy potential is only an estimate.

o What is the level of <u>private interest</u> in leasing the area?

The private response (PR) is obtained from four indicators: the number of bids per lease sale in the most recent year (BL); the number of noncompetitive applications (NC), if any, in the area; the average bonus per acre offered paid in the most recent year (BB); the ratio of acres relinquished or terminated (after a lease) to the total acreage leased (AR).

BL is a measure of how interesting the land offered for lease has been to the private sector recently. NC is used to capture the same dimension for lands which have not been offered for lease on a competitive basis. We constructed a private interest indicator (PI) by combining these two indicators. In other words, if either BL or NC or both are high, we consider that private interest for the area has been manifested.

BB is a measure of the marginal value of land in the area to the private investor -- lease sales in the most recent year are selected as the best approximation.

AR is a measure of how successful post-lease private activity has been in the area. For the purpose of constructing this indicator, we have assumed that relinquishment/termination follow unsuccessful drilling and exploration, but it should be considered that other reasons for relinquishment/termination may be that the tract was leased on a speculative basis or that the investor's acreage limitation has been reached.

PR is computed as the intersection (minimum) of PI(BL and NC), BB, and AR.

o What is the potential for pre-lease delay from environmental factors?

The environmental delay factor (EF) is measured by the amount of unassessed, unoffered acreage over unoffered acreage in the geothermal area

$$UU = \frac{\text{unassessed, unoffered acreage}}{\text{Unoffered acreage}}$$

UU is a measure of the potential delay incurred in the pre-lease process rather than a measure of the potential costs and benefits of environmental regulations.

Following the procedures for fuzzy set operations, a composite LAPA indicator is obtained as the intersection of the three main decision criteria as follows:

$$LAPA_i = \text{Min}[UF_i,\ \text{Max}(BL_i, NC_i),\ AR_i, BB_i, UU_i]$$

Sample data for individual indicators in 10 KGRAs are shown in Table 1.

Table 1. Sample Data for LAPA

UF Unleased Factor
BB Bonus Bids per acre
BLNC either Bids per Lease
Non-Competitive leases
AR Acreage Relinquished
UU Unleased, Unassessed

KGRA	UF	BB	BL	NC	AR	UU
CA Lake City	460	215	6	42	.75	0
CA Mono-Long V.	2018	116	10	6	.00	0
CA Salton Sea	2400			10		1
ID Castle Creek	59	8	23		.31	0
ID Crane Creek	195	17	1	19	.00	0
ID Raft River	26	2	1		.00	0
NV Gerlach	3	9	1		.23	0
NV Stillwater	352	18	2	0	.25	0
OR Alvord	45	64	13	23	.39	0
OR Newberry	700			151		0

2. Describe the Ideal Case for Each Indicator.

The numerical indicators constructed act as proxy variables for leasing criteria. At this point, analysis can be shifted from the numerical values themselves to broad categories appropriate for each indicator, such as "high" or "low". The primary motivation for a shift in focus is the opportunity for reducing the cognitive load on the analyst, that is, the amount of information needed to be recalled in "sizing up the situation" for the decision at hand.

For example, the analyst is searching for areas with high values for UF, and low values for UU. Attention can be directed to those areas in the sample where UF is high to some degree, and UU is low to some degree.

"High" in UF is defined relative to a range and a reference point, the median.

This definition of "high" was chosen to approximate users' intuitions--interviews with experts indicated that the threshold value of interest was about 300 MWe for electric power production. Further, the median is a robust measure of location. Exactly half the cases will fit the ideal case for each indicator, at least to some degree. In fact, experts preferred not to differentiate cases above the median for this indicator. Because of this, the more inclusive fuzzy set "not low" was used, so that all cases above the median fit perfectly, and all except the very lowest have some (non-zero) degree of fit.

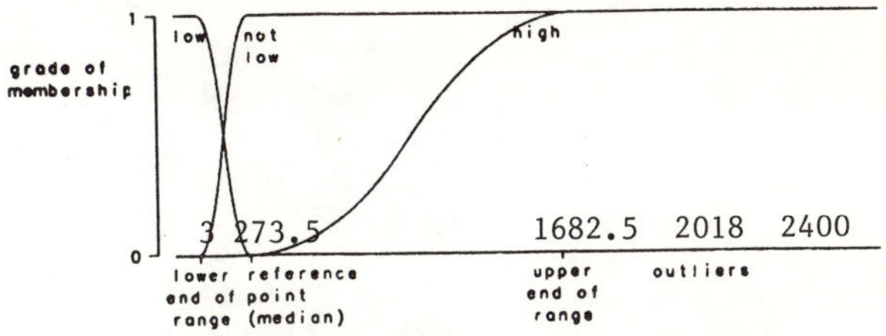

Figure 2. Membership Function Curves for UF

3. Importance Weights.

Importance weights are applied to each indicator by raising membership values to a constant power (the importance weight for that indicator). The less important the indicator, the closer to zero is the weight, and the nearer to one is the modified membership value.[9] Raising membership values according to importance helps overcome the effect of poor scores on less important indicators. We deferred the assignment of weights to experts involved in the leasing process but combine their judgements in a systematic way according to fuzzy set rules.

Combine the Indicators.

The indicators can be combined in two ways: using fuzzy logical operators, or integrating the degree of fit information visually.

Table 2 shows the membership values for unadjusted indicators.

Table 2.

Not Adjusted for Importance

CI	KGRA	UF	BB	BLNC	AR	UU
.00	CA Lake City	1.00	1.00	1.00	.00	1.00
1.00	CA Mono-Long V.	1.00	1.00	1.00	1.00	1.00
.00	CA Salton Sea	1.00		.51		.00
.07	ID Castle Creek	.07	.15	1.00	.96	1.00
.79	ID Crane Creek	.79	.99	1.00	1.00	1.00
.00	ID Raft River	.01	.00	.00	1.00	1.00
.00	NV Gerlach	.00	.33	.00	1.00	1.00
.11	NV Stillwater	1.00	.99	.11	1.00	1.00
.03	OR Alvord	.03	1.00	1.00	.84	1.00
1.00	OR Newberry	1.00		1.00		1.00

In this example, default rules for logical operators were used to produce the combined indicators for 10 KGRAs. These default rules model the case where there is no interaction among the indicators. That is, they are equally important and no consideration is given to tradeoffs, exceptions, or missing values.

Figure 3 is a display of the combined[11] indicators using a continuous grey scale. It is easy to see from the leftmost column that Mono, Crane Creek, and Newberry satisfy all the criteria. Three others, Castle Creek, Stillwater, and Alvord fit the ideal case, but not as well.

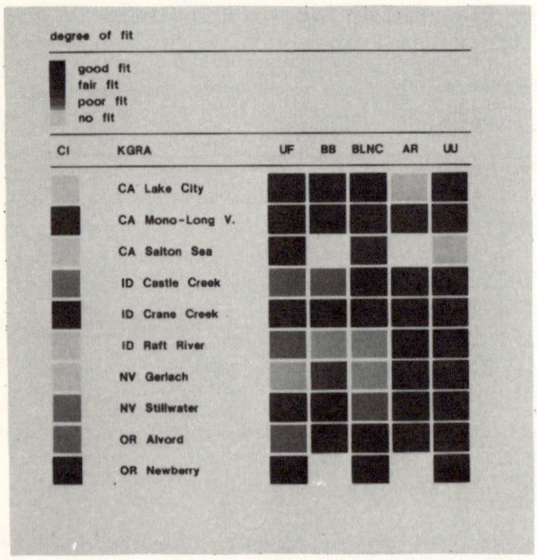

Figure 4. Composite LAPA Indicator

Degree of fit to the ideal case for each indicator is shown at the right. The value of this display is that the analyst may prefer to consider importance, tradeoffs, exceptions, and missing values in the context of a particular data set.

For example, comparing the first two rows (Lake City and Mono) brings up the question: can you make an exception of AR? Comparing each one vs. the last row (Newberry) calls for a judgment about missing values with BB and AR. How should missing values be interpreted? It is legitimate and often easier to evaluate choices when faced with real data, rather than hypothetical situations. The trade-off in using a descriptive approach for exploratory data analysis is that one can avoid the up-front investment involved in a prescriptive method, such as setting up a model of rational behavior involving many judgments about exceptions and tradeoffs in importance for every conceivable situation.

CONCLUSION

To summarize, LAPA can be used as a decision-support tool that allows the analyst to incorporate judgement in the evaluation of data relevant to lease actions. It provides an automated procedure for summarizing information in a manner that reduces the load on the analyst's short-term memory. Data is presented according to how well it matches a given set of leasing decision criteria, and irrelevant numerical precision is dispensed with. Both the clear-cut match and cases having different degrees of fit are easily summed up by color graphics that support almost spontaneous visual tradeoffs between objectives. The ideal of cognitive economy in information systems is served by giving the decision-maker access to better summary information with less effort.

REFERENCES

1. Filliben, J. (1981). "Dataplot--An Interactive High-Level Language for Graphics, Non-Linear Fitting, Data Analysis, and Mathematics," Computer Graphics, Volume 15, Number 3, August 1981.

2. Zadeh, L. (1973). "Outline of a New Approach to the Analysis of Complex Systems and Decision processes," IEEE Transactions: Man, Systems, Cybernetics, Volume SMC 3, No. 1, January 1973.

3. Yager, R. and D. Basson (1975). "Decision-Making with Fuzzy Sets." Decision Sciences, Volume 6, 1975.

4. Interagency Geothermal Coordinating Council, (1980). Fourth Annual Report: Geothermal Energy Research, Development and Demonstration Program, DOE-RA-0050, IGCC-5.

5. Muffler, L.J.P. (1979), edited. Assessment of Geothermal Resources of the United States - 1978. Geological Survey Circular 790, USGS, 1979.

6. Beeland, G.V., E. Schumann, and M. Wieland, (1980). Geothermal Leasing and Permitting Database: A Tool for Future Planning, Wapora, Inc., December 1980.

7. Burkhardt, H.E., C.A. Brook, and F.W. Smith (1980). Selected Administrative Land and Resource Data for KGRA's in Arizona, California, Idaho, Nevada, Oregon, and Washington. USGS Open File Report 80-1290, 1980.

8. Bureau of Land Management, Area Environmental Assessment Records and Reports, various dates.

9. Yager, R. (1980). "A Linguistic Variable for Importance of Fuzzy Sets," Journal of Cybernetics, Volume 10, pp. 249-260, 1980.

10. Yen, W. W.S., G. Carbonaro, and W.H. Benson (1982). "LAPA: A Composite Indicator for Prioritization of Geothermal Leasing Activities on Federal Lands in the United States," LBL-13038, February 1982.

11. Benson, W. (1981). "An Application of Fuzzy Set Theory to Data Display," LBL-11590, Revised. Preprint.

ALPHABETICAL INDEX OF AUTHORS

A

B

C

D

E

F

J

K

L

M

R

S

T